T0343181

The Ecology of Large Mammals in Central Yellowstone

Sixteen Years of Integrated Field Studies

Volume 3 in the Academic Press | TERRESTRIAL ECOLOGY SERIES

Editor-in-Chief

James R. Ehleringer, University of Utah, USA

Editorial Board

James MacMahon, Utah State University, USA
Monica G. Turner, University of Wisconsin, USA

Published Books in the Series

Todd E. Dawson and Rolf T. W. Siegwolf: *Stable Isotopes as Indicators of Ecological Change*, 2007

Keith A. Hobson and Leonard I. Wassenaar: *Tracking Animal Migration with Stable Isotopes*, 2008

Robert A. Garrott, P. J. White and Fred G. R. Watson: *Large Mammal Ecology in Central Yellowstone: A Synthesis of 16 Years of Integrated Field Studies*, 2008

The Ecology of Large Mammals in Central Yellowstone

Sixteen Years of Integrated Field Studies

Edited by

Robert A. Garrott

Fish and Wildlife Management Program
Department of Ecology
Montana State University
Bozeman, Montana

P. J. White

National Park Service
Yellowstone National Park
Mammoth, Wyoming

Fred G. R. Watson

Division of Science and Environmental Policy
California State University Monterey Bay
Seaside, California

ELSEVIER

AMSTERDAM • BOSTON • HEIDELBERG • LONDON
NEW YORK • OXFORD • PARIS • SAN DIEGO
SAN FRANCISCO • SINGAPORE • SYDNEY • TOKYO
Academic Press is an imprint of Elsevier

Academic Press is an imprint of Elsevier
525 B Street, Suite 1900, San Diego, CA 92101-4495, USA
30 Corporate Drive, Suite 400, Burlington, MA 01803, USA

First edition 2009

Copyright © 2009 Elsevier Inc. All rights reserved

No part of this publication may be reproduced, stored in a retrieval system
or transmitted in any form or by any means electronic, mechanical, photocopying,
recording or otherwise without the prior written permission of the publisher

Permissions may be sought directly from Elsevier's Science & Technology Rights
Department in Oxford, UK: phone (+44) (0) 1865 843830; fax (+44) (0) 1865 853333;
email: permissions@elsevier.com. Alternatively you can submit your request online by
visiting the Elsevier web site at http://elsevier.com/locate/permissions, and selecting
Obtaining permission to use Elsevier material

Notice
No responsibility is assumed by the publisher for any injury and/or damage to persons
or property as a matter of products liability, negligence or otherwise, or from any use
or operation of any methods, products, instructions or ideas contained in the material
herein. Because of rapid advances in the medical sciences, in particular, independent
verification of diagnoses and drug dosages should be made

British Library Cataloguing in Publication Data
A catalogue record for this book is available from the British Library

Library of Congress Cataloging-in-Publication Data
A catalog record for this book is available from the Library of Congress

ISBN–13: 978-0-12-374174-5

For information on all Academic Press publications
visit our website at elsevierdirect.com

Printed and bound by CPI Group (UK) Ltd, Croydon, CR0 4YY
Transferred to Digital Print 2012

Working together to grow
libraries in developing countries
www.elsevier.com | www.bookaid.org | www.sabre.org
ELSEVIER   BOOK AID International   Sabre Foundation

Cover Image: Bull elk along the Firehole River on a cold mid-winter day feeding
on aquatic plants and the band of snow-free vegetation just above the waterline
exposed due to the geothermally-heated water. Photo by Jeff Henry.

We dedicate this book to our children and all the professionals that have worked to conserve the resources of Yellowstone National Park since its inception.

Contents

Contributors

Numbers in parentheses indicate the pages on which the authors' contributions begin

SUSAN E. ALEXANDER (651) Division of Science and Environmental Policy, California State University Monterey Bay, Seaside, California 93955

THOR N. ANDERSON (53, 67, 85, 113, 651) Division of Science and Environmental Policy, California State University Monterey Bay, Seaside, California 93955

KEITH E. AUNE (255) Montana Department of Fish, Wildlife and Parks, Helena, Montana 59620

MATTHEW S. BECKER (283, 305, 339, 373, 451, 519) Fish and Wildlife Management Program, Department of Ecology, Montana State University, Bozeman, Montana 59717

JAMES G. BERARDINELLI (157, 477) Department of Animal and Range Sciences, Montana State University, Bozeman, Montana 59717

ERIC J. BERGMAN (305, 339) Fish and Wildlife Management Program, Department of Ecology, Montana State University, Bozeman, Montana 59717

DANIEL D. BJORNLIE (603) Fish and Wildlife Management Program, Department of Ecology, Montana State University, Bozeman, Montana, 58717

JOHN J. BORKOWSKI (157, 339, 477, 581, 603) Department of Mathematical Sciences, Montana State University, Bozeman, Montana 59717

JASON E. BRUGGEMAN (217, 603, 623) Fish and Wildlife Management Program, Department of Ecology, Montana State University, Bozeman, Montana 59717

MAURICE A. CHAFFEE (177) Geologic Division, U.S. Geological Survey, Denver Federal Center, Lakewood, Colorado 80225

STEVE CHERRY (137, 401, 451) Department of Mathematical Sciences, Montana State University, Bozeman, Montana 59717

SIMON S. CORNISH (67, 85, 113, 373, 651) Division of Science and Environmental Policy, California State University Monterey Bay, Seaside, California 93955

JULIE A. CUNNINGHAM (541) Montana Fish, Wildlife and Parks, Bozeman, Montana 59717

TROY DAVIS (581) National Park Service, Yellowstone National Park, Mammoth, Wyoming 82190

JON DETKA (67, 651) Division of Science and Environmental Policy, California State University Monterey Bay, Seaside, California 93955

JULIE A. FULLER (237, 255) Fish and Wildlife Management Program, Department of Ecology, Montana State University, Bozeman, Montana, 58717

ROBERT A. GARROTT (3, 53, 137, 157, 177, 191, 217, 237, 255, 305, 339, 373, 401, 423, 451, 477, 489, 519, 541, 581, 603, 623, 651, 671) Fish and Wildlife Management Program, Department of Ecology, Montana State University, Bozeman, Montana 59717

CHRIS GEREMIA (255) National Park Service, Yellowstone National Park, Mammoth, Wyoming 82190

CLAIRE N. GOWER (305, 339, 373, 401, 423, 451, 519) Fish and Wildlife Management Program, Department of Ecology, Montana State University, Bozeman, Montana 59717

KENNETH L. HAMLIN (477, 541) Montana Fish, Wildlife and Parks, Bozeman, Montana 59717

ROSEMARY JAFFE (305, 339) Fish and Wildlife Management Program, Department of Ecology, Montana State University, Bozeman, Montana 59717

MARC KRAMER (67) Earth and Planetary Sciences, University of California, Santa Cruz, Albany, California 94710

RYAN E. LOCKWOOD (53, 113) Division of Science and Environmental Policy, California State University Monterey Bay, Seaside, California 93955

TAD MASEK (67) Division of Science and Environmental Policy, California State University Monterey Bay, Seaside, California 93955

D. CRAIG McCLURE (581) National Park Service, Yellowstone National Park, Mammoth, Wyoming 82190

ERIC MEREDITH (137, 451) Department of Mathematical Sciences, Montana State University, Bozeman, Montana 59717

DAVID R. MERTENS (157) U.S. Department of Agriculture, Agriculture Research Service, U.S. Dairy Forage Research Center, University of Wisconsin, Madison, Wisconsin 53706-1108

MATHEW A. MESSER (137) Fish and Wildlife Management Program, Department of Ecology, Montana State University, Bozeman, Montana, 58717

STEVE W. MOORE (67) Division of Science and Environmental Policy, California State University Monterey Bay, Seaside, California 93955

WENDI B. NEWMAN (17, 37, 53, 85, 113, 651) Division of Science and Environmental Policy, California State University Monterey Bay, Seaside, California 93955

S. THOMAS OLLIFF (671) National Park Service, Yellowstone National Park, Mammoth, Wyoming 82190

JAMES K. OTTON (177) Geologic Division, U.S. Geological Survey, Denver Federal Center, Lakewood, Colorado 80225

ANDREW C. PILS (157) Fish and Wildlife Management Program, Department of Ecology, Montana State University, Bozeman, Montana, 58717

SALLY PLUMB (651) National Park Service, Yellowstone National Park, Mammoth, Wyoming 82190

DANIEL P. REINHART (581) National Park Service, Yellowstone National Park, Mammoth, Wyoming 82190

JAY J. ROTELLA (191, 489) Fish and Wildlife Management Program, Department of Ecology, Montana State University, Bozeman, Montana 59717

DOUGLAS W. SMITH (283) National Park Service, Yellowstone National Park, Mammoth, Wyoming 82190

DANIEL R. STAHLER (283) National Park Service, Yellowstone National Park, Mammoth, Wyoming 82190

THOMAS R. THEIN (85, 113) Division of Science and Environmental Policy, California State University Monterey Bay, Seaside, California 93955

JOHN TREANOR (255) National Park Service, Yellowstone National Park, Mammoth, Wyoming 82190

RICK W. WALLEN (255, 623) National Park Service, Yellowstone National Park, Mammoth, Wyoming 82190

FRED G. R. WATSON (17, 37, 53, 67, 85, 113, 137, 217, 373, 451, 603, 623, 651) Division of Science and Environmental Policy, California State University Monterey Bay, Seaside, California 93955

P. J. WHITE (3, 137, 157, 177, 191, 217, 237, 255, 305, 339, 373, 401, 423, 451, 477, 489, 519, 541, 581, 603, 623, 671) National Park Service, Yellowstone National Park, Mammoth, Wyoming 82190

NIGEL G. YOCCOZ (401) Department of Biology, University of Tromsø, Norway

Acknowledgements

The foundation for the success of this program was based on the dedication and commitment of the field teams that worked very long and arduous days nearly nonstop throughout each 5–6-month winter field season. Montana State University graduate students leading teams for multiple years in the central Yellowstone study area included Matthew Becker (Ph.D.), Eric Bergman (M.S.), Daniel Bjornlie (M.S.), Jason Bruggeman (Ph.D.), Claire Gower (Ph.D.), Matthew Ferrari (M.S.), Amanda Hardy (M.S.), Rosemary Jaffe (M.S.), Mathew Messer (M.S.), and Andrew Pils (M.S.). Field teams in the comparative study area in the Madison Valley west of the park were lead by Julie Fuller (M.S.), Jamin Grigg (M.S.), and Justin Gude (M.S.). Steve Hess (Ph.D.) conducted aircraft-based research to design rigorous and effective methodologies for estimating the bison population within the Park. We were fortunate to attract many outstanding young people near the start of their careers that worked side-by-side with the graduate students as field technicians and team leaders, including Michael Boyce, Mcrae Cobb, Matthew Coller, Thain Cook, Shana Dunkley, Derek Fagone, Jonathan Felis, Vince Green, Aaron Hasch, Mark Johnston, Elizabeth Joyce, Chris Kenyon, Clint Kolarich, Greg Pavellas, Kevin Pietrzak, Jennifer Pils, Jesse Rawson, Ellen Robertson, John Salerno, Terra Scheer, Tim Shafer, Ty Smucker, Derek Thompson, Renee Wulff, Veronica Yovovich, and Steve Yu. Laboratory, computer, and mathematical assistance were provided by Ron Adair, Larissa Jackie, Jarina Jechova, Lynn Marschke, Sangita Nayak, Sarah Olimb, Megan O'Reilly, Kamal Paudel, Scott Schmieding, and Shaun Tauck.

We could not have completed our studies in Yellowstone without the support of many National Park Service professionals and volunteers. Those not listed as authors on chapters included Bernie Adams, Amy Black, Tami Blackford, Brian Buchanan, Tom Cawley, Alana Dimmick, Mary Ann Donovan, Deb Elwood, Julie Garton, Deb Guernsey, Kerry Gunther, Hank Heasler, Christie Hendrix, Justin Hoff, Scott Jawors, Craig Johnson, Montana Lindstrom, John Mack, Wendy Maples, Melissa McAdam, Terry McEneaney, Mary McKinney, Steve Miller, Kerry Murphy, Joy Perius, Pat Perrotti, Glenn Plumb, Terri Reese, Roy Renkin, Jennifer Wipple, Ann Rodman, John Sacklin, Christine Smith, Jeremiah Smith, Denise Swanke, Brian Teets, John Varley, Janine Waller, Bob Weselmann, Becky Wyman, Travis Wyman, Mike Yochim, and Linda Young. NPS rangers that were always helpful in contributing to our field efforts and our safety included Lane Baker, Rick Bennett, Curt Dimmick, Les Inafuku, Tom Mazzarisi, Gary Nelson, Dave Page, Jan Page, Robert Siebert, Bonnie Schwartz, Tom Schwartz, Mary Taber, Tim Townsend, Matt Vandzura, and Dennis Young. We specifically thank Michael Keator for his help to develop a rigorous safety protocol, provide us with avalanche training, and volunteering his time to teach us backcountry skills. We thank NPS interpreter Katie Duffy, and her late husband Patrick, for their long-term friendship to the project and their perpetual support and encouragement. Many other members of the Old Faithful interpretive division also provided valuable assistance. A special thanks to NPS mechanics Craig Van de Polder and the late Randy Abbeglen, who volunteered their personal time and expertise to help us keep snowmobiles and trucks operating throughout the entire duration of the 16-year study. Without their help many days in the field would have been lost. We would also like to thank the Canyon, Madison, and Old Faithful maintenance personnel, particularly Victor Cavalier, Pat Connors, and Woody Wimberley for their logistical support.

California State University Monterey Bay staff and interns contributing to the studies other than those listed as chapter authors included Jason Maas-Baldwin, Joel Casagrande, Julie Casagrande,

Alberto Guzman, Don Kozlowski, Jordan Plotsky, Morgan Wilkinson, and Brian Wilson. The CSUMB team members are grateful for voluntary field assistance from Singli Agnew, Don Alexander, Ginger Alexander, Jennifer Anderson, Martel Anderson, Mark Angelo, Anthony Brantley, Rob Burton, Bree Candiloro, Cindy Detka, Carla Engalla, David Frank, Miguel Angel Gomez, Ben Haberthur, Nick Huerta, Frank Lafon, Karl Landorf, Flower Moye, Malcolm O'Keeffe, Anthony Propernick, John Slusser, Kristy Uschyk, Lisa Uttal, Betty Watson, Jessica Watson, and Monica Woo. NASA program staff supporting the CSUMB efforts included Joseph Coughlan, Rodney McKellip, Ed Sheffner, and Woody Turner.

Montana Fish, Wildlife, and Parks personnel that contributed to the studies included Kurt Alt, Neil Anderson, Steve Ard, Val Asher, Mark Atkinson, Bob Brannon, Mark Duffy, Kevin Frey, Jeff Herbert, Dave Hunter, Craig Jourdonnais, Fred King, Tom Lemke, Coleen O'Rourke, Mike Ross, Carolyn Sime, Jenifer Verschyul, and Harry Whitney. Montana State University faculty other than those listed as chapter authors who contributed to the studies in various ways included Robert Boik, Scott Creel, Mark Greenwood, William Inskeep, Lynn Irby, Steven Kalinowski, Billie Kearns, Timothy McDermott, Thomas McMahon, Harold Picton, David Roberts, David Willey, and Al Zale. Other MSU graduate students and staff who donated time and expertise to our studies or ably administered various aspects of the studies included David Christianson, Gina Himes-Boor, Joan Macdonald, Tom Olenicki, Douglas Ouren, Linda Phillips, Kelly Proffitt, Julia Sharp, David Staples, Emelyn Udarbe, Judy Van Andel, and John Winnie. University of Wisconsin-Madison faculty, staff, and graduate students that contributed to the studies included John Cary, Scott Craven, Dennis Heisey, Wally Jakubas, Robert Ruff, Don Rusch, and David Vagnoni. Collaborators associated with U.S. Geological Survey included Peter Gogan, Robert Klaver, and Ed Olexa. Steve Monfort and Kendall Mashburn of the Smithsonian Institute aided in the development of fecal steroid assays. John and Rachael Cook of the Forestry and Range Sciences Laboratory, La Grande, Oregon, collaborated on captive elk nutrition and reproductive physiology studies. We are grateful for the assistance of David Tarboton, author of the UEB model upon which the Langur snow model is based, and to USDA Natural Resources Conservation Service professionals Jerry Beard and Phil Farnes for assistance with climate data and snowpack coring equipment. For contributions to the video products described in Chapter 29 we owe a special debt of gratitude to the voluntary work of David Oebker, narrator, Richard Newman, composer of original music, and Bob Landis for generous donation of natural sound effects. Chapters in this book benefited from reviews and criticisms from peers, in particular comments from Layne Adams, Anne Carlson, Bruce Dale, Jed Murdoch, and Nigel Yoccoz.

Mike Bjorklund donated generously of his time to travel from Minnesota to Yellowstone throughout the studies to provide veterinary expertise when we were immobilizing and collaring animals. Thanks to the darting and handling protocols he helped us develop we performed 277 chemical immobilizations of animals and experienced only two mortalities over the course of our studies. Fixed-wing pilots Randy Arment, Doug Chapman, Steve Monger, Dave Stradley and Rodger Stradley; and helicopter pilots Gary Brennan, Bob Hawkins, Dan Williamson, and Mark Duffy provided efficient and safe aerial animal monitoring and capture. The comparative wolf-ungulate studies conducted outside of Yellowstone National Park that are highlighted in Chapter 25 could not have been completed without access to private lands and we are grateful for the cooperation of the owners and managers of CB, Carroll, Corral Creek, Elk Meadows, Elkhorn, High Valley, and Sun Ranches, as well as Rising Sun Mountain Estates and Sun West Estates. Special thanks to ranch managers Todd Graham, David Henderson, Rich and Shelby Hewitt, Mark and Peggy Jasmann, and Marina Smith.

Thanks to Jeff Henry for graciously allowing us to select many stunning photographs from his library for publication in this book, and to Jennifer Pils, who created the beautiful drawings depicting scenes from our field studies that introduce each section of the book. We are also appreciative of Bob Weselmann and all the other folks that volunteered their photographs from the field that have contributed substantially to the aesthetics of this book.

During those first rough and bitterly cold couple of weeks in December of 1991, when we were collaring our first elk and nothing seemed to go right, Les Eberhardt, L. Lee Eberhardt, and Ted Hammond stuck with us until the job got done and our studies were launched. Those weeks will always be remembered. A special thanks to L. Lee Eberhardt for his many years of collaboration and mentoring of both R.A. Garrott and P.J. White. We learned much about science, objectivity, and professionalism from Lee, and his over half century of contributions to wildlife ecology will always be an inspiration. R.A. Garrott would also like to acknowledge the patience and support of Diane Garrott for accepting the many long absences in the field, conversations during all the rough times when things were not going well, and for tolerating my evening retreats to the computer during the two years leading up to the publication of this book.

Major funding was provided by the National Science Foundation (DEB-9806266, DEB-0074444, DEB-0413570, DEB-0716188, two Research Experience for Undergraduates supplements, two Graduate Research Fellowships, and one GK-12 Fellowship), the NASA Office of Earth Science 'Research Education Applications Solutions Network' (NCC13-03009), the NASA Intelligent Data Understanding program (NCC2-1186), the National Park Service, and Montana Fish, Wildlife and Parks (Federal Aid in Wildlife Restoration Project W-120-R and Survey and Inventory Projects). Additional funding was provided by Department of Wildlife Ecology-University of Wisconsin Madison, Department of Ecology-Montana State University Bozeman, Bob and Annie Graham, Rodger and Cindy Lang, Max McGraw Wildlife Foundation, Rob and Bessie Welder Wildlife Foundation, Rocky Mountain Elk Foundation, Thermal Biology Institute-Montana State University Bozeman, Yellowstone Park Foundation, U.S. Fish and Wildlife Service, U. S. Geological Survey, and the Zoological Society of Milwaukee County. Additional funding for the National Park Service wolf studies, which contributed to the success of our research program, included the Yellowstone Park Foundation wolf fund, an anonymous donor, the Tapeats Foundation, the Perkins-Prothro Foundation, and NSF grant DEB-0613730. Patagonia, Master Foods, Canon Corporation, the California Wolf Center, numerous collar sponsors, Frank and Kay Yeager, and Yvon and Malinda Chouinard also contributed significantly. In addition, we thank Kathy Tonnessen and the Rocky Mountains Cooperative Ecosystem Studies Unit in Missoula, Montana, for facilitating cooperative funding agreements between collaborators during this project. Field equipment and supplies were donated by Atlas Snowshoe Company, Cascade Designs Inc. (snowshoes), Lloyd Laboratories (drugs), Polaris Industries (snowmobiles), Tubbs Snowshoe Company, and Yellowstone Alpine Guides (logistics).

Such an extensive list of people contributing their time, effort, and expertise speaks to the collaborative nature of this project, and we truly hope we have not overlooked anyone; any omissions are unintentional and sincerely regretted. The views and opinions in this book are those of the authors and should not be construed to represent any views, determinations, or policies of the National Park Service.

Section 1

Introduction

CHAPTER 1

Integrated Science in the Central Yellowstone Ecosystem

Robert A. Garrott[*] and P. J. White[†]

[*]Fish and Wildlife Management Program, Department of Ecology, Montana State University
[†]National Park Service, Yellowstone National Park

Yellowstone National Park is one of the most well-known natural areas in the world, visited annually by approximately three million people from around the globe to experience the natural wonders and famous wildlife populations. The park is relatively large at slightly over 8990 km² and represents the heart of a much larger 56,000 km² landscape known as the Greater Yellowstone Ecosystem (Keiter and Boyce 1991). Most of the land in this ecosystem is under public ownership and managed by federal natural resource agencies such as the National Park Service, Forest Service, Fish and Wildlife Service, and Bureau of Land Management. This diverse landscape represents the largest intact temperate ecosystem remaining on the planet. The region has high scenic values with its expanses of forested mountains and plateaus, high-elevation grassland valleys and meadows, and pristine rivers (Figure 1.1), and contains one of the world's highest concentrations of unaltered geothermal features (Fournier 1989). The Greater Yellowstone Ecosystem also contains the largest concentration of wild ungulates and large carnivores in the lower 48 states (Smith *et al.* 2004) and has a complete compliment of the organisms believed to have been present during European settlement. Thus, there is an abundance of large mammal species with strong ecological influences.

Protected natural areas are particularly attractive to ecologists for field investigations (National Research Council 1992, Parsons 2004, Pringle and Collins 2004), with highly diverse and productive research programs conducted in some of the world's most famous national parks and preserves (Sinclair and Norton-Griffiths 1979, Sinclair and Arcese 1995, Jędrzejewska and Jędrzejewski 1998, DuToit *et al.* 2003). The attributes of the Greater Yellowstone Ecosystem make it an area of high conservation value as well as an outstanding natural laboratory where ecological processes are still allowed to operate at large scales. Examples of large scale ecological processes include major episodic disturbances such as the wildfires that burned nearly half of Yellowstone's forests and grasslands during the summer and fall of 1988. The high densities of ungulates, combined with the harsh climate of the northern Rockies also periodically result in winter kills of ungulates that can involve thousands of animals (*e.g.*, Singer *et al.* 1989) and drive dramatic movements of large numbers of animals outside of the park. The combination of broad-scale ecological events, an abundance of high-profile animals, and a strong sense of public ownership in Yellowstone National Park and the surrounding lands, results in a tremendous amount of public attention and dialogue about park policy and management of its resources. Because the public is represented by a very wide array of constituencies with divergent

The Ecology of Large Mammals in Central Yellowstone
R. Garrott, P. J. White and F. Watson
ISSN 1936-7961, DOI: 10.1016/S1936-7961(08)00201-7

Copyright © 2009, Elsevier Inc.
All rights reserved.

FIGURE 1.1 Yellowstone National Park and the lands surrounding the park known as the Greater Yellowstone Ecosystem represent the largest relatively intact temperate ecosystem remaining on the planet. The park is world renowned for its sweeping scenic landscapes, diverse and abundant geothermal features, and wildlife populations (Photo by Jeff Henry).

philosophies and goals, Yellowstone is nearly always embroiled in controversy (Chase 1987, Schullery 1997, Wagner 2006, Figure 1.2). The seemingly unending debates surrounding natural resource management in the Yellowstone region create a need for objective scientific information about its organisms, populations, and biological processes which, in turn, provides opportunities to formulate and carry out ecological research. This information then contributes to policy and management decisions.

The vast majority of terrestrial ecological research conducted in Yellowstone National Park has been focused on a series of interconnected, broad grassland-sagebrush valleys known as the northern range (Houston 1982). This 1500 km² expanse, spanning the northern border of the park along the Lamar and Yellowstone River valleys, is defined by the winter range of one of the largest migratory populations of elk in North America and has been at the heart of debates for the past century over the appropriate management of ungulate densities and the impacts of elk herbivory on grassland, forest, and riparian communities (National Research Council 2002, Wagner 2006). This productive and interesting debate has resulted in a rich legacy of publications addressing many aspects of wildlife ecology, management, and public policy (*e.g.*, Cahalane 1943, Houston 1982, Frank and McNaughton 1992, Coughenour and Singer 1996, Yellowstone National Park 1997, Meagher and Houston 1998, National Research Council 2002, Taper and Gogan 2002, Barmore 2003). Since the reintroduction of wolves into the system over a decade ago the scientific and public controversy has shifted to evaluating the impacts of large predators on their ungulate prey (National Research Council 2002, Eberhardt *et al.* 2003, Smith *et al.* 2003, Vucetich *et al.* 2005, White and Garrott 2005) and potential predator-induced trophic cascades (Ripple *et al.* 2001, Beschta 2003, Wilmers *et al.* 2003, Fortin *et al.* 2005).

While the public policy issues and the ecological richness of the northern range certainly justify this intensive scientific attention, looking at a map of the entire park it is striking to note that the northern range represents a relatively small proportion of Yellowstone (Figure 1.3). When we began discussions with National Park Service administrators to establish a research project in Yellowstone we were drawn to the valleys, forested plateaus, and river systems south of the northern range in the heart of the park.

FIGURE 1.2 There is a strong sense of public ownership and interest in Yellowstone National Park with the diversity of opinions, philosophies, and values resulting in nearly continuous protests, debates, and controversies about natural resource management and policy (Photo by Jeff Henry).

Initially, our studies focused on the elk herd occupying the Madison headwaters area in the west-central part of the park, but as we began incorporating studies of the migratory bison population into our program we expanded the spatial scale of our activities to include the upper portions of the Yellowstone River drainage to the east. Similar to the northern range that is defined by the range of elk wintering in that area, an analogous "ecological unit" of comparable size, which we call the central range, can be circumscribed by the range of the large (2000–3500) migratory central bison herd that summers in the high-elevation Pelican and Hayden valleys in east-central Yellowstone, with winter ranges along the headwaters of the Madison River near the western edge of the park (Meagher 1973, 1989).

With the exception of studies focused on forest and sagebrush-grassland community dynamics, stimulated by the large wildfires of 1988 (Turner *et al.* 2003, Wallace 2004), and pioneering work on bison, elk, and grizzly bear biology and natural history (Meagher 1973; Craighead *et al.* 1973, 1995), terrestrial ecologists have paid comparatively little attention to this unique and diverse area of the park and the opportunity it presents for studying ecological processes in a large intact landscape that is not actively manipulated. The landscape of central Yellowstone is heterogeneous with considerable topographic relief resulting in a complex physiography of varying slopes, aspects, and elevations (Figure 1.4). Grassland habitats range from the narrow meadow complexes along the Madison River drainage to the broad, sweeping grassland/sedge valleys of the Pelican and Hayden. Forests are dominated by lodgepole pine, but form complex mosaics ranging from old growth to broad sweeps of early successional forests of varying attributes owing to the extensive wildfires of 1988. The central range is also a region of climatic extremes with marked annual variation in both warm and cold season

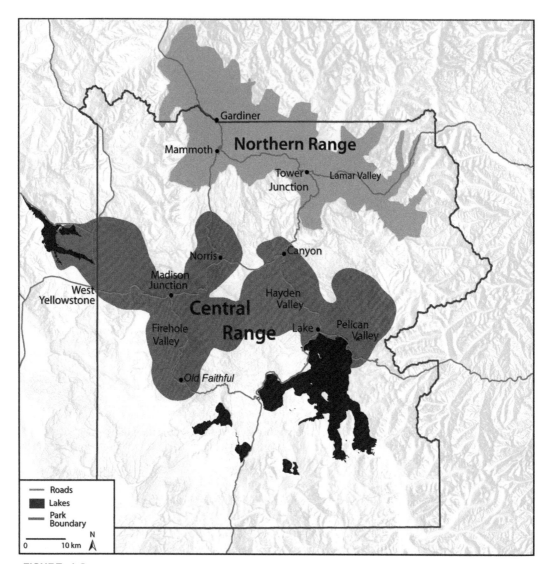

FIGURE 1.3 The northern range of Yellowstone National Park is generally defined by the wintering area of the northern elk herd and has been the focus of a nearly century-long debate about the management of large mammals. The central range is a similar ecological unit defined by the range of the migratory central bison herd and includes the range of the nonmigratory elk herd that occupies the headwater drainages of the Madison River along the western edge of the park.

climatic drivers. Winters are typically prolonged with much deeper snow pack than is experienced on the northern range. However, snow pack can be light in some years, and occasional Chinook rains in the middle of winter can result in the formation of dense ice layers in the snow pack that have severe consequences for wintering ungulates. Annual variation in summer climate is also considerable, with droughts sometimes extending for multiple years, while at other times summer rains can prolong the growing season into August. This area also contains the major geothermal basins in the park, and the heat associated with these areas provides considerable relief from deep snow packs for many wintering ungulates. The concentration of geothermal features, combined with the abundance of wildlife in the central Yellowstone region, attract the vast majority of people who visit the park each year. Thus the potential conflicts with the park's dual mandate of protection of natural resources while providing for human enjoyment by the public is accentuated in this area.

Our research program in central Yellowstone began modestly in 1991 and grew incrementally as we gained insights, formulated additional hypotheses, tested new data collection techniques, and accrued additional resources and collaborators. When we started this effort we had no grand vision for a long-term ecological research program. Instead, we struggled to find resources to support a single dissertation project in what we thought was an extremely interesting system that we personally wanted to work in to gain a better understanding of a small subset of ecological processes. The more we worked, the more questions we formulated, the more interesting the system became, and the more opportunities we saw for gaining ecological insight. So we exploited many options to maintain the studies and eventually found ourselves managing what could be considered a long-term ecological study. While the long-term nature of our studies was initially unplanned, we purposefully orchestrated the slow expansion of our collaborations and data collection to ensure a tight integration of our entire research program. Throughout this effort, we maintained a common vision of building an integrated and multidisciplinary research program dedicated to (1) producing objective science with the goal of advancing our knowledge of basic ecological processes within Yellowstone's central range, (2) supporting sound natural resource management, and (3) communicating our knowledge and discoveries to the public to enhance their experience, appreciation, and enjoyment of the park and the process of ecological research.

Attempting to understand ecological processes is a daunting task because understanding the complexities of ecosystem linkages requires expertise from a wide variety of scientific and technical disciplines. To meet this challenge we had the good fortune over the course of this research effort of being able to attract 66 scientists and professionals with complementary skills and expertise, as well as 15 highly motivated and talented graduate students to contribute to this effort. Although these numbers may seem to represent a plethora of intellectual power and expertise, our collective efforts tackled only a limited set of ecological questions and processes, because few ecological studies can encompass the myriad biotic and abiotic factors that operate on numerous temporal and spatial scales. Thus, we admittedly confined our studies to large mammal ecology. Essentially, this book is a synthesis of our attempts to understand the influence of climate and landscape attributes on the spatial and population dynamics of bison and elk, the two major large herbivores occupying central Yellowstone, and the predominant predator of these herbivores, the wolf. It quickly became apparent in our studies that environmental effects strongly influenced large mammal dynamics. Thus, we devoted the first section of this book to detailed descriptions of the central Yellowstone environment before addressing the large mammal interactions that played out on this stage. In addition, while a host of ecological questions can be investigated with these wildlife species, they are also often at the center of some of the most contentious management controversies in the park. Consequently, throughout the study we contributed objective scientific data to help resolve management conflicts in Yellowstone. In this book we present an example of our investigations into a major and ongoing management and policy controversy for the park, namely the effects of winter recreation on wildlife behavior and populations in central Yellowstone. We were in a position to contribute to this debate because we had been working in the system and had baseline data on the animal populations that were the primary focus of the controversy. Lastly, because virtually all natural resource controversies involve people and their values, and most people are not scientists, there is a growing awareness of the need for ecologists to effectively communicate their results, conclusions, and ideas to a broader audience beyond their specific disciplines. The unique resources and strong sense of public ownership in the park provided an exceptional opportunity for science education to elucidate what we perceive to be the role of science in national parks, and these topics comprise the final section of this book.

There is ever-increasing discussion within the ecological community of the need to develop long-term, integrated and interdisciplinary research programs (Likens 1989), but examples of such programs are relatively rare. Perhaps the most concerted effort to meet this ambitious objective is the

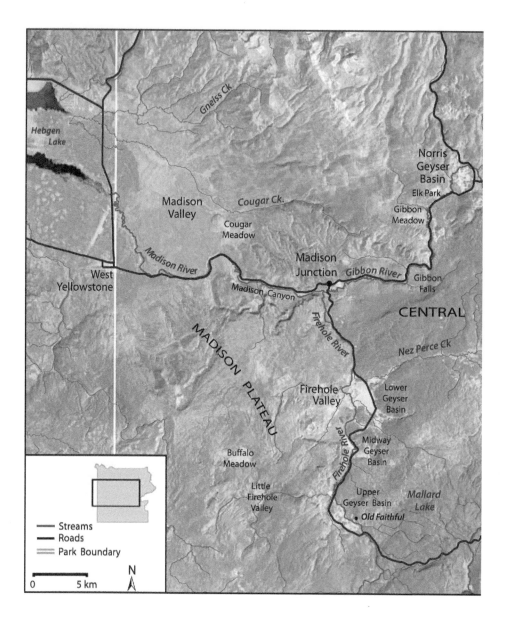

National Science Foundation's Long Term Ecological Research Program (National Science Foundation 2002), which has resulted in productive and exciting gains in ecological knowledge (Ross *et al.* 1996, Knapp *et al.* 1998, Priscu 1998, Bowman and Seastedt 2001, Chapin *et al.* 2006). However, sustaining long-term ecological studies of large mammal populations is challenging and expensive because they necessitate working at the relatively large spatial scales required by these animals to meet their biological demands (Hobbs 1996). In addition, ethical constraints severely limit potential manipulation of such organisms in ways and numbers adequate to evaluate biological processes. Combined, these restrictions seriously limit opportunities to conduct rigorous experimental studies that are the foundation of the scientific process. Ecologists nevertheless must confront these constraints because large herbivores are prominent components in many grassland, shrub-steppe, and tundra ecosystems and have important effects on nutrient cycling, energy transfer, fire regimes, soil erosion, hydrology, structuring plant communities, and maintenance of species diversity (McNaughton 1985, Frank and

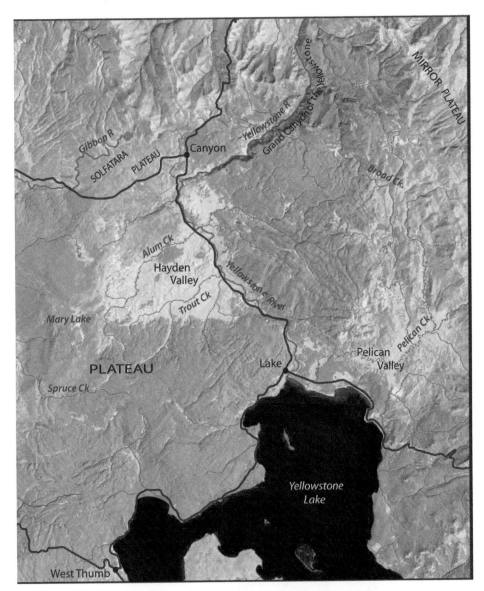

FIGURE 1.4 The central range of Yellowstone National Park.

McNaughton 1992, Hobbs 1996, Collins 1998, Danell *et al.* 2006). In turn, the predators of large herbivores have the potential to influence the spatial distribution and abundance of large herbivore populations (Jędrzejewski *et al.* 2002, Hebblewhite *et al.* 2005, Andersen *et al.* 2006), thereby directly and indirectly influencing ecosystem structure and function (Schmitz *et al.* 2000, Ripple and Beschta 2004, Fortin *et al.* 2005). In addition, landscape modifications, persecution, and overexploitation have reduced the viability or totally extirpated many large mammal populations and predator-prey systems, necessitating the formulation of sound conservation and management strategies. Therefore, knowledge of the mechanisms that limit and/or regulate large mammal populations and influence their distributions on landscapes is important not only to our theoretical understanding of population, community, and ecosystem patterns and processes, but also to our ability to effectively maintain assemblages of these organisms and the ecological processes that they facilitate.

Fundamental to the development of our program was the principle that the primary foundation for understanding wildlife communities and ecosystem processes is a strong field presence. We accomplished this by deploying a research team of three to four graduate students and technicians for five to six months each winter, as well as periodic, short-duration, field efforts through the year by teams focused on studies of landscape and climate attributes. These efforts translated into more than 9000 person-days of fieldwork over the past 16 years (Figure 1.5). Using consistent and rigorous sampling methodologies that were designed to address specific hypotheses about physical and ecological processes, we have amassed some of the most extensive landscape and ecological databases ever collected in Yellowstone. Field data to evaluate snow pack properties and processes include approximately 4300 snow cores, 150 snow pits, seasonal temperature profiles for seven localities, and >24,000 h of wind data collected from arrays of 20–45 custom-designed anemometers deployed at 104 sites. Vegetation, ground cover, and plant phenology studies involved >17,000 point measurements from approximately 360 plots. Population distribution, behavioral, and individual movement databases include >15,000 records for the Madison headwaters elk, >90,000 records for central herd bison, and >1900 wolf pack records, with an additional approximately 4175 km of wolf travel vectors obtained from snow tracking, and over 750 wolf-killed elk and bison. Ungulate and wolf demographic data include multiple annual population estimates, indices of annual pregnancy rates, reproduction, and recruitment, and estimates of survival and cause-specific mortality. Our field efforts also included the collection of biological samples including >6500 urine-saturated snow samples for assessing nutrition (White *et al.* 1997, Pils *et al.* 1999), >2000 fecal samples for stress and pregnancy assessments (Garrott *et al.* 1998, Cook *et al.* 2002, Creel *et al.* 2002), >150 blood samples for disease and pregnancy studies (Ferrari and Garrott 2002), >1000 plant samples for nutritional assays and biomass assessments (Jakubas *et al.* 1994, Bruggeman 2006), and >120 animal tissue samples for chemical analyses to address geochemical linkages in the system (Garrott *et al.* 2003). In addition to the data we collected ourselves, we had access to pertinent current and historic databases from National Park Service files

FIGURE 1.5 The foundation of our science program has been the commitment to intensive and integrated field studies with extremely dedicated research teams working nearly every day of a 5–6 months winter field season. These efforts have translated into more than 9000 person-days of fieldwork over the past 16 years and the accumulation of some of the largest and most diverse ecological databases collected in the park (Photo by Claire Gower).

and archives, as well as internal documents and expert knowledge that provided crucial insights into the strengths and weaknesses of this information so that these data could be appropriately integrated into our overall ecological studies.

Inter-annual variation in climate in our study system is quite high and one particularly powerful aspect of the databases and biological samples amassed is the fact that over the first seven years of our field work, prior to wolf reestablishment, the study system experienced nearly the entire range of climate conditions recorded in central Yellowstone since modern climate instrumentation sites were established in the region. We were equally fortunate during the subsequent nine years of study after wolves became established to have experienced a nearly comparable range of climate conditions. The intensity of our field efforts to collect a wide diversity of data types, combined with the relatively long duration of this study, the climate variation experienced, and the natural experiment imposed on our study system due to the reestablishment of wolves, enhanced our ability to effectively partition the effects of climate drivers from other biotic and abiotic mechanisms affecting the ecological processes we investigated, and to identify potential interactions between these effects.

Ecological research in protected areas can serve numerous conservation and management purposes for the particular system, but generalizing ecological insights can also be criticized as limited, perhaps even irrelevant, due to the uniqueness of many of these systems. While the central Yellowstone system is certainly unique in many facets, such uniqueness is also arguably its strength. Patterns exist in nature and investigations into a relatively simple, tractable, and highly heterogeneous large mammal system largely free from development and human harvest provided unique opportunities to better discern the ecological processes and mechanisms underlying these patterns without the complexity and confounding effects prevalent in most systems. Unique systems have always been the subject of fruitful ecological research, beginning with the foundational studies of evolution and natural selection in the Galapagos Islands and can provide useful applications and comparisons to more complex systems with pronounced human impacts. The intent in synthesizing our extensive and interrelated research efforts in a book was to provide an integrated interpretation of this central Yellowstone system, to evaluate current ecological theory with an extensive collection of empirical databases, and to contrast the dynamics we document in this study system with those described for similar large mammal communities, as well as for smaller taxa systems traditionally the focus of much ecological research. Potentially, this not only yields immediate applied management and conservation insights for large mammal systems, but also contributes to the growing need for contingent theory in ecology (Holt and Lawton 1994) by providing insights into community relationships and the significant direct and indirect effects of top predators. In addition, the book is intended to provide objective scientific data for some of the most contentious management issues in the park, as well as provide a strong outlet for public science education in general and the central Yellowstone ecosystem in particular. It is our sincere hope that the collective efforts of everyone involved in this 16-year effort will provide a worthy contribution to these ends.

REFERENCES

Andersen, R., J. D. C. Linnell, and E. J. Solberg. 2006. The future role of large carnivores in terrestrial interactions: The northern temperate view. Pages 413–448 *in* K. Danell, R. Bergström, P. Duncan, and J. Pastor (Eds.) *Large Herbivore Ecology, Ecosystem Dynamics and Conservation.* Cambridge University Press, New York, NY.

Barmore, W. J. 2003. *Ecology of Ungulates and their Winter Range in Northern Yellowstone National Park: Research and Synthesis 1962–1970.* National Park Service, Mammoth Hot Springs, WY.

Beschta, R. L. 2003. Cottonwoods, elk, and wolves in the Lamar Valley of Yellowstone National Park. *Ecological Applications.* **13**:1295-1309.

Bowman, W. D., and T. R. Seastedt. 2001. *Structure and Function of an Alpine Ecosystem: Niwot Ridge, Colorado.* Oxford University Press, New York, NY.

Bruggeman, J. E. 2006. Spatio-temporal dynamics of the central bison herd in Yellowstone National Park. Ph.D. dissertation. Montana State University, Bozeman, MT.

Cahalane, V. H. 1943. Elk management and herd regulation-Yellowstone National Park. *Transactions North American Wildlife Conference* **8**:95–101.

Chapin, F. S., M. W. Oswood, K. van Cleve, L. A. Viereck, and D. L. Verbyla. 2006. *Alaska's Changing Boreal Forest*. Oxford University Press, New York, NY.

Chase, A. 1987. *Playing God in Yellowstone*. Harcourt Brace Jovanovich, San Diego, CA.

Collins, S. L., A. K. Knapp, J. M. Briggs, J. M. Blair, and E. M. Steinauer. 1998. Modulation of diversity by grazing and mowing in native tallgrass prairie. *Science* **280**:745–747.

Cook, R. C., J. G. Cook, R. A. Garrott, and S. L. Monfort. 2002. Effects of diet and body condition on fecal progestagen excretion in elk. *Journal of Wildlife Diseases* **38**:558–565.

Coughenour, M. B., and F. J. Singer. 1996. Elk population processes in Yellowstone National Park under the policy of natural regulation. *Ecological Applications* **6**:573–593.

Craighead, J. J., F. C. Craighead, Jr, R. L. Ruff, and B. W. O'Gara. 1973. Home ranges and activity patterns of nonmigratory elk of the Madison drainage herd as determined by biotelemetry. *Wildlife Monographs* **33**:1–50.

Craighead, J. J., J. S. Sumner, and J. A. Mitchell. 1995. *The Grizzly Bears of Yellowstone: Ecology in the Yellowstone Ecosystem, 1959–1992*. Island Press, Washington, DC.

Creel, S. C., J. E. Fox, A. Hardy, J. Sands, R. A. Garrott, and R. O. Peterson. 2002. Snowmobile activity and glucocorticoid stress responses in wolves and elk. *Conservation Biology* **16**:1–7.

Danell, K., R. Bergström, P. Duncan, and J. Pastor. 2006. *Large Herbivore Ecology, Ecosystem Dynamics and Conservation*. Cambridge University Press, New York, NY.

DuToit, J. T., K. H. Rogers, and H. C. Biggs. 2003. *The Kruger Experience: Ecology and Management of Savanna Heterogeneity*. Island Press, Washington, DC.

Eberhardt, L. L., R.A. Garrott, D.W. Smith, P.J. White, and R.O. Peterson. 2003. Assessing the impact of wolves on ungulate prey. *Ecological Applications* **13**:776-783.

Ferrari, M. J., and R. A. Garrott. 2002. Bison and elk: Brucellosis seroprevalence on a shared winter range. *Journal of Wildlife Management* **66**:1246–1254.

Fortin, D., H. L. Beyer, M. S. Boyce, D. W. Smith, T. Duchesne, and J. Mao. 2005. Wolves influence elk movements: Behavior shapes a trophic cascade in Yellowstone National Park. *Ecology* **86**:1320–1330.

Fournier, R. O. 1989. Geochemistry and dynamics of the Yellowstone National Park hydrothermal system. *Annual Review of Earth and Planetary Sciences* **17**:13–53.

Frank, D., and S. J. McNaughton. 1992. The ecology of plants, large mammalian herbivores, and drought in Yellowstone National Park. *Ecology* **73**:2043–2058.

Garrott, R. A., L. L. Eberhardt, P. J. White, and J. Rotella. 2003. Climate-induced limitation of a large herbivore population. *Canadian Journal of Zoology* **81**:33–45.

Garrott, R. A., S. L. Monfort, P. J. White, K. L. Mashburn, and J. G. Cook. 1998. One-sample pregnancy diagnosis in elk using fecal steroid metabolites. *Journal of Wildlife Diseases* **34**:126–131.

Hebblewhite, M., E. H. Merrill, and T. L. McDonald. 2005. Spatial decomposition of predation risk using resource selection functions: An example in a wolf-elk predator–prey system. *Oikos* **111**:101–111.

Hobbs, N. T. 1996. Modification of ecosystems by ungulates. *Journal of Wildlife Management* **60**:695–713.

Holt, R. D., and J. H. Lawton. 1994. The ecological consequences of shared natural enemies. *Annual Review of Ecology and Systematics* **25**:495–520.

Houston, D. B. 1982. *The Northern Yellowstone Elk: Ecology and Management*. Macmillan, New York, NY.

Jakubas, W. J., R. A. Garrott, P. J. White, and D. R. Mertens. 1994. Elk consumption of burnt lodgepole pine as influenced by digestibility and secondary metabolites. *Journal of Wildlife Management* **58**:35–46.

Jędrzejewska, B., and W. Jędrzejewski. 1998. *Predation in Vertebrate Communities: The Białowieża Primeval Forest as a Case Study*. Springer, Berlin, Germany.

Jędrzejewski, W., K. Schmidt, J. Theuerkauf, B. Jędrzejewska, N. Selva, K. Zub, and L. Szymura. 2002. Kill rates and predation by wolves on ungulate populations in Bialowieza Primeval Forest (Poland). *Ecology* **83**:1341–1356.

Keiter, R. B., and M. S. Boyce (Eds.). 1991. *The Greater Yellowstone Ecosystem*. Yale University Press, New Haven, CT.

Knapp, A. K., J. M. Briggs, D. C. Hartnett, and S. L. Collins. 1998. *Grassland Dynamics: Long-Term Ecological Research in Tallgrass Prairie*. Oxford University Press, Washington, DC.

Likens, G. E. (Ed.). 1989. *Long-Term Studies in Ecology: Approaches and Alternatives*. Springer, New York, NY.

McNaughton, S. J. 1985. Ecology of a grazing system: The Serengeti. *Ecological Monographs* **55**:259–294.

Meagher, M. M. 1973. *The Bison of Yellowstone National Park*. National Park Service Scientific Monograph Series No. 1. U.S. Government Printing Office, Washington, DC.

Meagher, M. M. 1989. Range expansion by bison of Yellowstone National Park. *Journal of Mammalogy* **70**:670–675.

Meagher M. M., and D. B. Houston. 1998. *Yellowstone and the Biology of Time: Photographs Across a Century*. University of Oklahoma Press, Norman, OK.

National Research Council.1992. *Science and the National Parks*. National Academy Press, Washington, DC.

National Research Council. 2002. *Ecological Dynamics on Yellowstone's Northern Range.* National Academy Press, Washington, DC.

National Science Foundation. 2002. *Long-Term Ecological Research Program: Twenty-Year Review.* National Science Foundation, Arlington, VA.

Parsons, D. J. 2004. Supporting basic ecological research in U.S. national parks: Challenges and opportunities. *Ecological Applications* **14**:5–13.

Pils, A. C., R. A. Garrott, and J. Borkowski. 1999. Sampling and statistical analysis of snow-urine allantoin:Creatinine ratios. *Journal of Wildlife Management* **63**:1118–1131.

Pringle, C. M., and S. L. Collins. 2004. Needed: A unified infrastructure to support long-term scientific research on public lands. *Ecological Applications* **14**:18–21.

Priscu, J. C. 1998. *Ecosystem Dynamics in a Polar Desert: The McMurdo Dry Valleys, Antarctica.* American Geophysical Union, Washington, DC.

Ripple, W. J., and R. L. Beschta. 2004. Wolves and the ecology of fear: Can predation risk structure ecosystems? *Bioscience* **54**:755–766.

Ripple, W. J., E. J. Larsen, R. A. Renkin, and D. W. Smith. 2001. Trophic cascades among wolves, elk, and aspen on Yellowstone National Park's Northern Range. *Biological Conservation* **102**:227–234.

Ross, R. M., E. E. Hufmann, and L. B. Quetin. 1996. *Foundations for Ecological Research West of the Antarctic Peninsula. Antarctic Research Science.* Vol. 70. American Geophyscial Union, Washington, DC.

Schmitz, O. J., P. A. Hamabaeck, and A. P. Beckerman. 2000. Trophic cascades in terrestrial ecosystems: A review of the effects of carnivore removals on plants. *American Naturalist* **155**:141–153.

Schullery, P. 1997. *Searching for Yellowstone: Ecology and Wonder in The Last Wilderness.* Houghton Mifflin, New York, NY.

Sinclair, A. R. E., and P. Arcese (Eds.). 1995. *Serengeti II: Dynamics, Management, and Conservation of an Ecosystem.* University of Chicago Press, Chicago, IL.

Sinclair, A. R. E., and M. Norton-Griffiths (Eds.). 1979. *Serengeti: Dynamics of an Ecosystem.* University of Chicago Press, Chicago, IL.

Singer, F. J., W. Schreir, J. Oppenheim, and E. O. Garton. 1989. Drought, fires and large mammals. *Bioscience* **39**:716-722.

Smith, B., E. Cole, and D. Dobkin. 2004. *Imperfect Pasture: A Century of Change at the National Elk Refuge in Jackson Hole, Wyoming.* Grand Teton Natural History Association, Moose, Wyoming.

Smith, D. W., R. O. Peterson, and D. B. Houston. 2003. Yellowstone after wolves. *Bioscience* **53**:330–340.

Taper, M. L., and P. J. P. Gogan. 2002. The northern Yellowstone elk: density dependence and climatic conditions. *Journal of Wildlife Management* **66**:106-122.

Turner, M. G., W. H. Romme, and D. B. Tinker. 2003. Surprises and lessons from the 1988 Yellowstone fires. *Frontiers in Ecology and Environment* **1**:351–358.

Vucetich, J. A., D. W. Smith, and D. R. Stahler. 2005. Influence of harvest, climate, and wolf predation on Yellowstone elk, 1961–2004. *Oikos* **111**:259–270.

Wagner, F. H. 2006. *Yellowstone's Destabilized Ecosystem: Elk Effects, Science, and Policy Conflict.* Oxford University Press, New York, NY.

Wallace, L. L. 2004. *After the Fires: The Ecology of Change in Yellowstone National Park.* Yale University Press, New Haven, CT.

White, P. J., R. A. Garrott, and D. M. Heisey. 1997. An evaluation of snow-urine ratios as indices of ungulate nutritional status. *Canadian Journal of Zoology* **75**:1687–1694.

White, P. J., and R. A. Garrott. 2005. Northern Yellowstone elk after wolf restoration. *Wildlife Society Bulletin* **33**:942–955.

Wilmers, C. C., R. L. Crabtree, D. Smith, K. M. Murphy, and W. M. Getz. 2003. Trophic facilitation by introduced top predators: gray wolf subsidies to scavengers in Yellowstone National Park. *Journal of Animal Ecology* **72**:909-916.

Yellowstone National Park. 1997. *Yellowstone's Northern Range: Complexity and Change in a Wildland Ecosystem.* National Park Service, Mammoth Hot Springs, WY.

Section 2

Landscape and Climate

CHAPTER 2

The Central Yellowstone Landscape: Terrain, Geology, Climate, Vegetation

Wendi B. Newman and Fred G. R. Watson

Division of Science and Environmental Policy, California State University Monterey Bay

Contents

Theme

Ecological research can be defined as the study of interactions between organisms and their environment. While the interactions between individuals, species, and communities of organisms are typically dynamic and complex, the backdrop of landscape and climate variables upon which these interactions occur can also assume significant importance and equal complexity. Variation in a system's terrain, geology, soils, climate, snow pack, and vegetation at fine and broad spatial and temporal scales can create considerable landscape heterogeneity, potentially affecting the distribution and abundance of resources and influencing the spatial, demographic, and behavioral dynamics of organisms. Therefore, adequately characterizing the heterogeneity of a landscape is an important aspect of understanding the ecological processes that occur upon it. We described a highly heterogeneous landscape strongly influenced by severe winters, geological processes, and natural disturbances in the central portion of Yellowstone National Park. Using data on climate, geology, soils, hydrology, terrain, and vegetation, we created detailed Landsat maps of forest cover and vegetation type to provide a baseline upon which the composite dynamics of landscape, climate, and large mammal communities can be evaluated.

The Ecology of Large Mammals in Central Yellowstone
R. Garrott, P. J. White and F. Watson
ISSN 1936-7961, DOI: 10.1016/S1936-7961(08)00202-9

Copyright © 2009, Elsevier Inc.
All rights reserved.

I. OVERVIEW AND GEOGRAPHIC SETTING

The central Yellowstone study area is part of the Greater Yellowstone Ecosystem, which encompasses some 56,000 km² of the northern Rocky Mountains including Yellowstone and Grand Teton National Parks, and six national forests overlapping the borders between Wyoming, Idaho, and Montana (Patten 1991). The ecosystem extends approximately 450 km north to south and 250 km west to east, and primarily includes areas above 1500 m in elevation (Figure 2.1). For thousands of years, this area has experienced major changes as a result of climate shifts, geological processes, and other disturbances such as fire, severe winters, droughts, floods, blow downs, and outbreaks of diseases and insects that affect plants and animals (Meagher and Houston 1998). It is also part of a growing human community in which people are having an increasing impact on both sides of the park boundary (Wright Parmenter *et al.* 2003). Because the Greater Yellowstone Ecosystem spans the jurisdictions of several state and federal agencies, different portions of the area are managed according to different policies and land use practices. This especially affects migratory animals such as ungulates, many of which winter in lower-elevation areas on the public–private land interface.

The central Yellowstone study area comprises 3959 km², spanning the central and western portions of Yellowstone National Park, and extended slightly over its western boundary (Figures 1.3 and 1.4 in Chapter 1). The area has a relatively simple faunal complex, with only two abundant ungulate species (*i.e.*, bison (*Bison bison*) and elk (*Cervus elaphus*)) and two significant predators (*i.e.*, grizzly bears

FIGURE 2.1 Geographic context of the central Yellowstone study area within the Greater Yellowstone Ecosystem (GYE). The area shown is approximately one fifth of the GYE (Photo by Fred Watson).

(*Ursus arctos horribilis*) and wolves (*Canis lupus*); Chapter 16 by Becker *et al.*, this volume). The central Yellowstone elk population occupies the headwaters of the Madison River, including the Madison, Gibbon, and Firehole drainages. These elk are nonmigratory, staying within the borders of the park year-round, with the population remaining at approximately 600–800 animals from the mid 1960s to the start of these studies (Craighead *et al.* 1973, Garrott *et al.* 2003; Chapter 11 by Garrott *et al.*, this volume). The range of the migratory central bison herd is also within the study system, with summer range in the Hayden and Pelican Valleys just north of Yellowstone Lake, and a major winter range in the Madison headwaters area (Hess 2002; Chapter 12 by Bruggeman *et al.*, this volume). This central bison herd has varied in abundance from approximately 1000–4000 over the past three decades (Fuller *et al.* 2007; Chapter 13 by Fuller *et al.*, this volume). Grizzly bears are seasonally common from spring through autumn. Wolves were reintroduced to the Firehole drainage in 1995–1996 and became established during 1998. Four packs totaling approximately 45 wolves completely occupied the area by winter 2004–2005 (Chapter 15 by Smith *et al.*, this volume). Coyotes (*Canis latrans*) are common and black bears (*U. americanus*) and mountain lions (*Puma concolor*) also occur in the area, but only infrequently and at low densities (Garrott *et al.* 2003).

Much of the study area consists of the Central Plateau and Madison Plateau between about 2150–2700 m above sea level (Figure 2.2). The Central Plateau forms a drainage divide at about 2500 m, with the headwaters of the Madison River to its west and tributaries of the Yellowstone River to its east. The upper Madison headwaters are drained by the Firehole and Gibbon Rivers, which flow through a system of shallow basins in the plateaus above 2150 m. Below 2150 m, the Firehole and Gibbon rivers plunge steeply down to their confluence at 2070 m in Madison Canyon (Figure 2.3). The Madison River runs through this canyon for a short distance before emerging into more open terrain along the western border of the park near the town of West Yellowstone. The high-elevation Hayden and Pelican valleys lie to the east of the Central Plateau at about 2350–2400 m. The Yellowstone River flows north

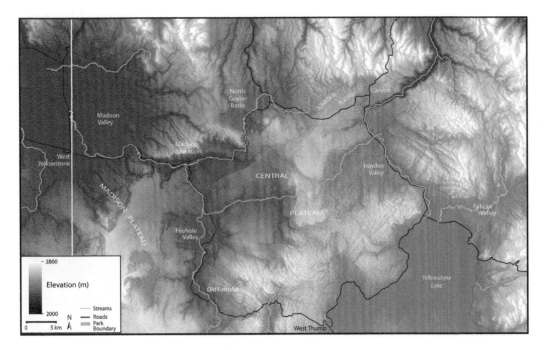

FIGURE 2.2 Shaded-relief digital elevation model of the central Yellowstone study area, depicting the Central Plateau and higher valleys above about 2150 m, drained via canyons leading west and north to lower terrain.

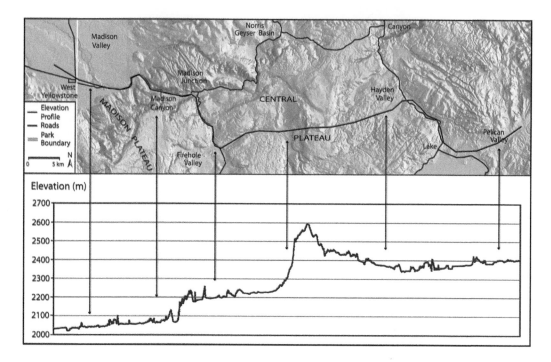

FIGURE 2.3 Eastward trending elevation transect through the study area. From the western border of the park in the Madison drainage (2050 m), one encounters a steep climb up to the Firehole drainage (2200 m) and Nez Perce Creek (2230 m), and then again steeply over Mary Mountain pass (2530 m), to the Hayden and Pelican Valleys (2350–2400 m).

between these valleys from Yellowstone Lake in the south to the Grand Canyon of the Yellowstone in the north, eventually reaching the northern portion of Yellowstone, which is beyond our study area. A transect connecting the major meadow complexes together reveals distinct jumps in elevation between the low-elevation Madison drainage at the western border of the park, mid-elevation Firehole and Gibbon drainages, and higher-elevation Hayden and Pelican valleys (Figure 2.3).

The landscape pattern is complex, resulting from interacting influences of climate, geology, soils, hydrology, terrain, vegetation, and human impacts (Figure 2.4). These interactions determine the boundaries of the wildlife ranges. Bison strongly favor meadow vegetation in winter, while elk are more generalists and routinely occupy both forests and meadows (Chapter 8 by Messer *et al.*, Chapter 21 by White *et al.*, and Chapter 28 by Bruggeman *et al.*, this volume). Meadows primarily occur on alluvial plains where relatively fertile soils were deposited between adjacent lava flows and at the bottom of Madison Canyon. The volcanic terrain that borders the meadows is forested with lodgepole pine, which thrives on the poor rhyolitic soils that derive directly from rhyolitic tuffs and lava flows. Winters are long and cold and deep snow packs form in the high country, especially to the west. Mitigating this effect is active geothermal influence, which reduces or even precludes the snow pack in many places.

II. GEOLOGY

The geologic events and processes that shaped Central Yellowstone area have a long and complex history (Keefer 1971). The oldest surface rocks in and near the study area are found in the Gallatin Range to the northwest. These include Precambrian metamorphic basement rocks formed 2700 million years ago (mya), Paleozoic (570–270 mya) marine sediments, and some Mesozoic (*ca.* 100 mya) sediments

FIGURE 2.4 Gibbon Meadow is typical of the study area. It is a flat alluvial plain with some geothermal influence on the snow pack, surrounded by the undulating and occasionally steep terrain formed by lava flows and tuff deposits, now forested with a mixture of burned and unburned lodgepole pine forest, and with a river and a road running through it (Photo by Fred Watson).

associated with the formation of the Rocky Mountains. A major volcanic period occurred during the Eocene epoch (56–34 mya), resulting in the breccias of the Absaroka Range to the east of the study area. The Gallatin and Absaroka Ranges now form natural ecological barriers on two sides of the study area.

The Yellowstone caldera erupted much more recently (0.6 mya). This was the most recent major eruption above the Yellowstone Hotspot, a region of the mantle with associated eruptions traceable in a line back to southeastern Oregon approximately 16 mya (Pierce and Morgan 1992, Christiansen 2001, Christiansen *et al.* 2007). The majority of the study area is now covered with volcanic material associated with the caldera (Figure 2.5). Tuffs from the caldera eruption itself are found just outside the caldera boundary to the north. Post-caldera (150–75 thousand years ago) lava flows now fill the depression left by the caldera, spilling over its boundary to the west. Most of the tuffs and lava flows have rhyolitic chemical composition; with basaltic and (intermediate) andesitic composition being limited to just a few small areas in the north of the study area. Rhyolite is low in available plant nutrients and is one reason for the presence of large areas of lodgepole forest stands on these flows (Despain 1990). Most of the forested portion of the study area is classified as having rhyolitic soils.

The Pinedale glaciation occurred after the most recent lava flows, from between 25 and 50 thousand years ago until just over ten thousand years ago (Smith and Siegel 2000). An ice cap covered most of what is now Yellowstone National Park during this period, and many key landscape features were created by the consequences of this glaciation. The Hayden Valley's hummocky terrain resulted from rocky glacial debris that accumulated in the ice cap and was partly covered by lake sediments when an ice dam formed

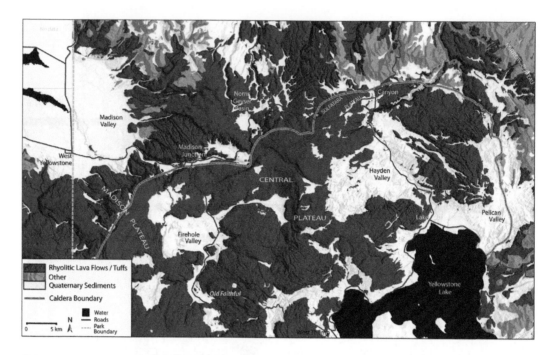

FIGURE 2.5 Simplified geologic map of central Yellowstone emphasizing the key geologic distinctions that influence vegetation. Meadow vegetation typically forms on the Quaternary sediments. Forest vegetation is typically found on the Pleistocene rhyolitic lava flows and tuffs associated with the Yellowstone caldera. Other forested and often mountainous units include Pleistocene basalts as well as the much older Tertiary volcanics mainly in the east and Paleozoic sediments and Precambrian metamorphics in the northwest (adapted from Christiansen 1999).

at the head of the Grand Canyon of the Yellowstone (Pierce 1979). When this dam burst, the resulting floodwaters enlarged the canyon. Similar floodwaters in the Madison Canyon formed the vast, relatively flat, glacial outwash plain to the west of the study area. Most of the modern undulating lava flow terrain of the Central Plateau is now covered with a thin veneer of glacial till (Shovic 1996).

The influence of the Yellowstone Hotspot continues to be felt. Two active resurgent domes are located within the study area, including Mallard Lake Dome to the east of Firehole Valley, and Sour Creek Dome to the east of Hayden Valley (Smith and Siegel 2000). Active geothermal areas are widely distributed within and beyond the caldera boundary. They are often characterized by white to grey soils known as sinter or geyserite (primarily siliceous due the abundance of silica-rich rhyolite).

The west-central portion of Yellowstone National Park represents one of the largest concentrations of active geothermal features in the world and contains thousands of geysers, fumaroles, hot springs, and mud pots. Geothermal influences reduce local snow pack and maintain adjacent rivers and creeks in a perennially ice-free state, thereby allowing photosynthesizing plant communities to grow through the winter (Despain 1990; Chapter 4 by Watson *et al.*, this volume). Thus, these geothermally influenced areas are of great ecological relevance to herbivores that are nutritionally constrained in their ability to forage and move in this harsh winter environment (Chapter 8 by Messer *et al.*, Chapter 18 by Gower *et al.*, and Chapter 21 by White *et al.*, this volume). These influences are concentrated around several major geyser basins in the Firehole and Gibbon drainages, but isolated patches are distributed through the study area. The pattern of geothermal areas is frequently linear, and aligned with the same lava-flow boundaries as the meadows, leading to long corridors of meadow habitat with reduced winter severity, such as along Nez Perce Creek, which facilitates a winter-accessible route for bison to move between the Hayden and Firehole valleys (Chapter 12 by Bruggeman *et al.*, this volume).

The resulting modern landscape is dominated by rhyolitic volcanic geology, including post-caldera lava flows and tuff deposits from the caldera eruption itself (Figure 2.5). The viscous nature of the lava flows is clearly visible in the shaded relief image of Figure 2.2, in particular in the high country spanned by the Madison and Central plateaus. Core wildlife areas such as the Firehole Valley and the Hayden Valley are clearly associated with quaternary sediments filling the voids between volcanically defined higher terrain.

III. CLIMATE

Yellowstone's climate is characterized by long, cold winters and short, cool summers (Figure 2.6). The mean annual temperature averaged 3 °C between water years 1999 and 2007 (data averaged between West Yellowstone and Madison Plateau snow telemetry sites (SNOTEL) in Montana at 2042 and

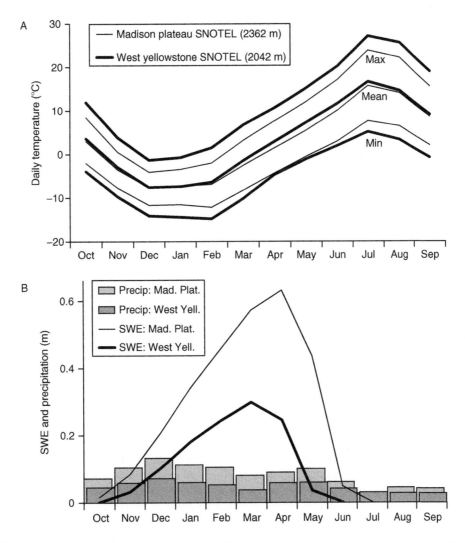

FIGURE 2.6 Seasonal climate variation in the central Yellowstone study area averaged over water years 1999–2007 using data from the West Yellowstone and Madison Plateau SNOTEL stations in Montana.

2362 m, respectively). Mean daily minimum temperatures averaged below freezing for eight months of the year (Figure 2.6A). Mean daily temperatures averaged above 10 °C only in June, July, and August (Figure 2.6A). A snow pack is typically present for 6–7 months of the year at West Yellowstone and for 8–9 months of the year at Madison Plateau (Figure 2.6B).

Mean annual precipitation varies both with elevation and distance downwind (northeastward) of the Madison and Pitchstone plateaus on the southwestern margins of the study area (Despain 1987; Chapter 3 by Watson and Newman, this volume). On these plateaus, mean annual precipitation is estimated to reach 1900 mm, whereas in the relatively high-elevation Hayden Valley in the east it decreases to about 550 mm. In the lowest elevations of the study area in the Madison drainage, mean annual precipitation decreases to 460 mm.

The snow pack accumulates from November, peaks in about April, and typically by June has completely ablated. Ablation occurs about a month earlier at the lowest elevations and a month later at the highest elevations (Figure 2.6; Chapter 6 by Watson *et al.*, this volume). The spatial distribution of peak snow pack accumulation is dictated at large scales by the precipitation patterns described above, and reduced at smaller scales by geothermal influence, wind, and forest cover (Chapters 4 and 5 by Watson *et al.*, this volume). Snow pack accumulation is typically measured in terms of snow-water equivalent (SWE) (*i.e.*, the amount of water in a column of snow, and values may be multiplied by three to obtain a crude estimate of snow pack depth; Chapter 6 by Watson *et al.*, this volume). Measurements of SWE thus reflect both the depth and the density of the snow pack. The West Yellowstone SNOTEL snow pack monitoring site is a typical low-elevation site just outside the park near the western edge of the study area (2042 m). Its peak annual SWE ranged between 15 and 40 cm in water years 1999 and 2007. The Madison Plateau site, which is also located just outside the western boundary of the park, is more characteristic of the highest elevations on the orographic barrier to the southwest. Its peak annual SWE ranged between 32 and 113 cm in water years 1968 and 2007 (and for comparison to West Yellowstone, between 39 and 86 cm in water years 1999–2007). Such accumulations have significant effects on ungulate movements, nutrition, and susceptibility to predation (Chapter 8 by Messer *et al.*, Chapter 9 by White *et al.*, Chapter 16 by Becker *et al.*, and Chapter 28 by Bruggeman *et al.*, this volume).

There is scattered evidence for long-term increases in temperature and, possibly, precipitation over the past century (Figure 2.7). The United States Historical Climatology Network (USHCN) includes three stations near the study area (Williams *et al.* 2007). Temperatures have increased in the USHCN record at Mammoth Hot Springs (1902 m) at a rate of approximately one to two degrees Celsius per century (maximum temperatures $R^2 = 0.27$, $F_{1,99} = 36$, $P < 0.001$, minimum temperatures $R^2 = 0.13$, $F_{1,99} = 15$, $P < 0.001$, and average temperatures $R^2 = 0.09$, $F_{1,99} = 10$, $P = 0.002$). Closer to the study area at West Yellowstone (2033 m), the record only shows increases in minimum temperatures (2 °C per century; $R^2 = 0.38$, $F_{1,99} = 61$, $P < 0.001$). At Lake Yellowstone (2368 m) in the east of the study area, the USHCN record reports a dramatic increase in precipitation of 0.27 m per century, but only after a correction step ("FILNET") involving gap-filling based on nearby stations. Meagher and Houston (1998) reported decreased precipitation in the (presumably raw) Mammoth data, but we did not detect this change in the corresponding raw USHCN record, perhaps because of differences in the period of record examined. Long-term trends in snow pack cannot be immediately inferred from these data, since increases in temperature and precipitation have opposite effects on snow pack. Wilmers and Getz (2005) observed long-term reductions in snow pack in the northern portion of Yellowstone, but this was at lower elevations than our central Yellowstone study area. The balance between temperature effects and precipitation effects on snow pack accumulation may differ between lower elevation and higher elevation sites. Climate-driven increases in conifer cover at lower elevations (<2000 m) fore-shadowed by Romme and Turner (1991) and observed by Powell and Hansen (2007) may not be significant in the central Yellowstone study area, given that conifer cover appears to be defined more by geologic boundaries (Figure 2.5) than climate, except perhaps in the Madison Valley in the far west of the study area.

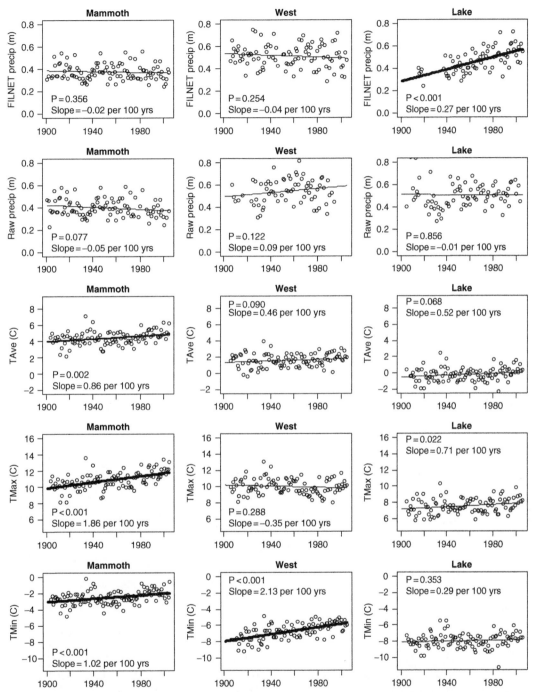

FIGURE 2.7 Scattered evidence for long-term, historical increases in temperature and precipitation at three locations near our central Yellowstone study area. We used data from United States Historical Climate Survey records corrected specifically for long-term climate analysis (Williams *et al.* 2007) at Mammoth Hot Springs (1902 m), West Yellowstone (2033 m) and Lake Yellowstone (2368 m). Simple linear regression was used to fit trend lines and calculate the statistical significance of the lines, as indicated by *P*-values and the thickness of the line (thick: $P < 0.001$; medium: $P < 0.01$; thin: $P \geq 0.01$). Both raw and corrected ("FILNET") precipitation records are shown because the trends in these data differed at the Lake Yellowstone station.

IV. VEGETATION

The vegetation of the study area is characterized by large tracts of forest within which isolated meadow complexes interconnect along drainage lines (Figure 2.8). Approximately 80% of the landscape is forest, and about 95% of the forest is dominated by lodgepole pine differentiated mainly on the basis of successional stage (Table 2.1). The non-forested area can be broadly divided into wet meadows, dry meadows and slopes, and geothermal areas.

A. Wet and Dry Meadows

Meadows occupy about 10–20% of the landscape, depending on the mapping method (Table 2.1). Wet meadows in the study area are characterized, in decreasing order of moisture availability by willows (*Salix* spp.), sedges (*Carex* spp.), tufted hairgrass (*Deschampsia cespitosa*), silver sage (*Artemesia cana*), and some Idaho fescue (*Festuca idahoensis*). Dry meadows are characterized more by Idaho fescue and big sagebrush (*Artemesia tridentata*). Wet meadows in the study area occur along stream-sides, seeps, and poorly drained areas (*e.g.*, Madison Canyon, Gibbon Meadow, Nez Perce Creek, stretches of the Firehole River, topographic lows in the Hayden Valley, and much of the Pelican Valley). Dry meadows occur on better-drained, slightly elevated sites such as the higher points in the Hayden Valley and most of Cougar Meadow. Associations between Idaho fescue and big sagebrush or silver sage constitute the

FIGURE 2.8 The central portion of Yellowstone National Park exhibits a mosaic of habitats, as illustrated by this photo of the Firehole Valley looking south from Purple Mountain. Rivers flow through alluvial meadows, surrounded by volcanic hills covered with a mixture of mature and regrowth lodgepole pine forest (Photo by Fred Watson).

TABLE 2.1	Vegetation-type statistics for the central Yellowstone study area and the smaller bison and elk ranges. Statistics were computed for Despain's post-1988 cover type map (Despain 1990) and the remotely sensed vegetation type described in this chapter

Vegetation type	Fraction of Central Yellowstone study area	Fraction of MGFHP bison range and surrounding areas	Fraction of MGF elk range and surrounding areas	Data source
Forest, any kind	85%	79%	90%	Despain (1990)
Forest, any kind	80%	76%	80%	This chapter
Forest, primarily lodgepole pine	78%	75%	88%	Despain (1990)
Forest, post-disturbance, any species	36%	39%	54%	Despain (1990)
Forest, post-disturbance, any species	19%	21%	25%	This chapter
Meadow, *i.e.*, non-forest, non-water	8%	12%	10%	Despain (1990)
Meadow	14%	20%	19%	This chapter
Meadow-sagebrush and other shrubs	3%	5%	1%	This chapter
Fraction of forest that is primarily lodgepole pine	92%	95%	97%	Despain (1990)
Fraction of forest that is post-disturbance	42%	49%	60%	Despain (1990)
Fraction of forest that is post-disturbance	24%	27%	31%	This chapter
Total area (km^2)	3948	1815	362	

Differences between the two data sources reflect differences in spatial resolution and the way in which mixed forest and meadow areas were accounted. The higher resolution remote sensing data are likely to be more accurate indicators of total meadow cover, but less accurate indicators of different types of forest. MGF: Madison, Gibbon, and Firehole drainages; HP: Hayden and Pelican valleys.

most common meadow habitats in the park. In the study area, they represent about 80% of the open country in the Hayden Valley and to a lesser extent the Pelican Valley, but only about 7% of meadows in the Madison, Gibbon, and Firehole drainages where streams and geothermal areas are in close proximity. Small conifer stands also occur in many meadow complexes.

B. Geothermal Area Vegetation

Geothermal heat reduces or eliminates the snow pack in many of the non-forested portions of the Madison, Gibbon, and Firehole drainages and in isolated parts of the Hayden and Pelican valleys (Chapter 4 by Watson *et al.*, this volume). Such areas become winter refuges for elk and bison (Chapter 8 by Messer *et al.* and Chapter 28 by Bruggeman *et al.*, this volume) and, as such, the forage they provide is of particular importance (Chapter 9 by White *et al.*, this volume). Soils, stream sediments, and surface waters in geothermally influenced drainages of the Madison headwaters area contain anomalously high levels of arsenic, fluoride, and silica derived from magmatic sources. In turn, plants growing on soils derived from geothermally altered rocks or irrigated by geothermal waters absorb high levels of these elements or become contaminated on their surface by soil particles (Garrott *et al.* 2002; Chapter 10 by Garrott *et al.*, this volume). The plants found in geothermal areas also tend to exhibit adaptations such as salt tolerance and low stature to tolerate and take advantage of warm soils saturated with geochemical salts. Despain (1990) lists plant species growing along an increasing heat gradient, including herbs such as golden-aster (*Chrysopsis villosa*) and sheep sorrel (*Rumex acetosella*), grasses such as Nuttall's alkali-grass (*Puccinellia airoides*) and cheatgrass (*Bromus tectorum*), mosses, and algae (see also Stout and Al-Niemi 2002).

C. Fire and Forests

Fires have been part of the study area for millennia, with major fires occurring every few hundred years and smaller fires occurring every few decades before the initiation of fire-control measures in the late 1800s (Romme 1982, Romme and Despain 1989, Meagher and Houston 1998, Turner *et al.* 2003). Natural fires were suppressed in Yellowstone through most of the twentieth century until the policy changed in 1972. During a particularly dry period in 1988, fire swept through Yellowstone and burned more than 3200 km². Over 50% of forested areas burned and many large stands were destroyed. About one-half of the forested area is now characterized by downed logs and dense lodgepole pine regrowth following those fires (Figure 2.9). Thus, the forest is a complex mosaic of burned and unburned forests

FIGURE 2.9 Much of the forest in the central portion of Yellowstone National Park is recent regrowth, primarily following the 1988 fires: (A) one year after 2003 fire near Mary Bay; (B, C, D) light, medium, and dense regrowth 16–19 years after 1988 fires (Photos by Fred Watson).

at different stages of succession (Romme and Despain 1989). Fire variability, particularly in large events like crown fires, plays an important role in landscape heterogeneity, leaving behind stands of mixed classes of trees due to survival and regeneration (Turner *et al.* 1997, 2003). Mixed classes of forest are better protected from future catastrophic fires as older classes of lodgepole pine are generally more susceptible to fire than younger age classes (Romme 1982, Romme and Knight 1982, Turner 1989).

Common tree species in the study area include lodgepole pine (*Pinus contorta*), whitebark pine (*Pinus albicaulis*), subalpine fir (*Abies lasiocarpa*), Engelmann spruce (*Picea engelmannii*), Douglas fir (*Pseudotsuga mensiesii*), and aspen (*Populus tremuloides*). The most common forested cover types mapped by Despain (1990) in our study area are, in descending order of abundance, lodgepole pine (78%; post-disturbance, middle successional, late successional, and climax), whitebark pine (3.9%; mainly post-disturbance and climax), Engelmann spruce with subalpine fir (1.6%; climax), Douglas fir (1.5%; mainly post-disturbance and climax), and aspen (0.2%). The lodgepole pine cover types are by far the most widespread. The post-disturbance lodgepole stage dates mainly from 1988 fires and represents approximately 34% of the study area and 53% of the Madison headwaters elk range (Chapter 8 by Messer *et al.*, this volume). This stage is dominated by lodgepole pine, with occasional aspens typically less than a meter high and amidst a sea of downed logs (Figure 2.9). The remaining successional stages of lodgepole pine (*i.e.*, early, middle, late, and climax) occupy 44% of the study area and 34% of the elk range in roughly equal proportions of different successional stages. Sub-dominant tree species typically include subalpine fir, Engelmann spruce, and whitebark pine. These latter species occur as dominants only in isolated stands scattered through the upper elevations of the study area (3.1% of study area). Stands of Douglas fir are more common at lower elevations near the western boundary of the park, both in post-disturbance and climax stages. Small groves of mature aspen also occur in this area at the low-elevation treeline and in talus.

The understory of open and regenerating forests often supports a substantial cover of graminoids, including Idaho fescue. This understory changes as forests mature toward canopy closure. Typical species in well-shaded forests include grouse whortleberry (*Vaccinium scoparium*), elk sedge (*Carex geyeri*), and pinegrass (*Calamagrostis rubescens*) (Marston and Anderson 1991).

V. MAPPING OF VEGETATION TYPE AND FOREST COVER

A. Mapping of Vegetation Type

A number of the analyses in this book examined the association between wildlife and vegetation type, and required characterization of vegetation type at random locations (referred to as "habitat" in Chapter 8 by Messer *et al.*, Chapter 21 by White *et al.*, and Chapter 28 by Bruggeman *et al.*, this volume). For this purpose, an objectively defined map of vegetation type was required. We created such a vegetation type map from Landsat satellite imagery, which has a large spatial extent (180 km) and fine resolution (28.5 m), for use in a variety of analyses presented in other chapters of this book. A Landsat scene acquired on September 23, 2002 was chosen for its lack of clouds and autumn seasonality, which is when maximum contrast is observed between meadow and forest types. We first used an automated, unsupervised classification scheme to categorize the image into 400 spectral classes, each with similar spectral properties (*K*-means algorithm, Peña *et al.* 1999, TNTMips, MicroImages, Lincoln, Nebraska, applied to Bands 1 through 5 and a terrain slope layer with a 9×9 pixel smoothing kernel applied). We then examined maps of each of the 400 classes (spectrally similar areas) and manually merged pairs of classes that could not be discerned from aerial imagery as being representative of different

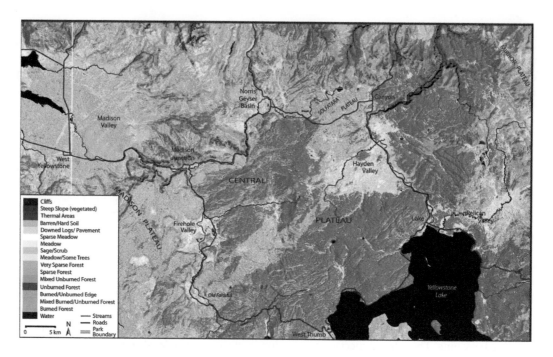

FIGURE 2.10 Vegetation types ($n = 17$) of the central portion of Yellowstone National Park that were mapped using Landsat remote sensing acquired on September 23, 2002.

vegetation classes. This resulted in a 17-class map of vegetation type (Figure 2.10). The *K*-means algorithm has been used to predict vegetation before in Yellowstone based on topo-climatic variables (Burrough *et al.* 2001) and MODIS imagery (Wessels *et al.* 2004), both of which have lower spatial resolution than Landsat imagery.

Seventeen classes is too many for logistic regression analyses that depend on mapped vegetation type, since these analyses require that each type represents a distinct predictor variable, and a goal in the parsimony of the analysis is to minimize the number of predictor variables (Chapter 8 by Messer *et al.*, Chapter 21 by White *et al.*, and Chapter 28 by Bruggeman *et al.*, this volume). Therefore, we further merged classes to yield a 5-class version of the map (Figure 2.11). The five classes were meadow, unburned forest, burned forest, geothermal, and other. Geothermal areas were initially under-estimated in this image relative to our geothermal map derived from thermal imaging (Chapter 4 by Watson *et al.*, this volume), so we overlaid a binary version of the geothermal map onto the original 5-class vegetation map to create our final 5-class vegetation map. This simpler level of classification reflected the habitat effects that we hypothesized could be reasonably discerned in our elk and bison habitat selection analyses (*e.g.*, Chapter 8 by Messer *et al.*, Chapter 21 by White *et al.*, and Chapter 28 by Bruggeman *et al.*, this volume).

B. Mapping of Forest Cover

The snow pack model used in this book (Chapter 6 by Watson *et al.*, this volume) considers forest cover to be a continuous variable, not a categorical one, as in the vegetation type map just described. Therefore, we also created a map that estimates percent forest cover from the same Landsat imagery.

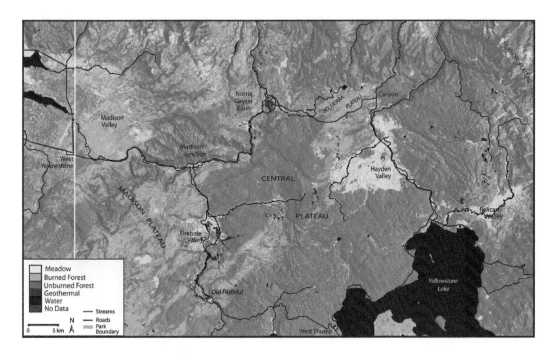

FIGURE 2.11 Vegetation types ($n = 5$) of the central portion of Yellowstone National Park that were mapped using Landsat remote sensing acquired on September 23, 2002.

We employed what is perhaps an original technique that we refer to as "spectral kerneling." The cover of each pixel, p_0, in the scene was estimated from its spectral properties as the weighted mean of the cover, p_i, of a number of reference sites, i, with the weighting determined by a Gaussian kernel $K(x)$ on the Euclidean spectral distance, $d_{0,i}$, between the pixel and each site:

$$p_0 = \frac{\sum_i^N p_i K\left(\frac{d_{0,i}}{h}\right)}{\sum_i^N K\left(\frac{d_{0,i}}{h}\right)}$$

$$d_{0,i} = \sqrt{\sum_{b=1}^5 \left(R_{b,0} - R_{b,i}\right)^2}$$

$$K(x) = \frac{1}{\sqrt{2\pi}} \exp\left(\frac{-x^2}{2}\right)$$

where N is the number of reference sites, $R_{b,i}$ is the reflectance in band b at site i measured in raw digital numbers (DN), and $h = 6$ DN is a bandwidth parameter.

A total of 79 reference sites were manually selected and ascribed an approximate percent forest cover based on aerial imagery. The sites were selected to represent a wide range of forest cover (from non-forested through to densely forested sites), and a wide range of background textures such as grass, shrubs, talus, alpine tundra, fallen logs, and pavement. The resulting map clearly identified the expected variation in forest cover from open country (mean = 10% cover), through post-disturbance forest (mean = 25% cover), to more mature forest (early successional through

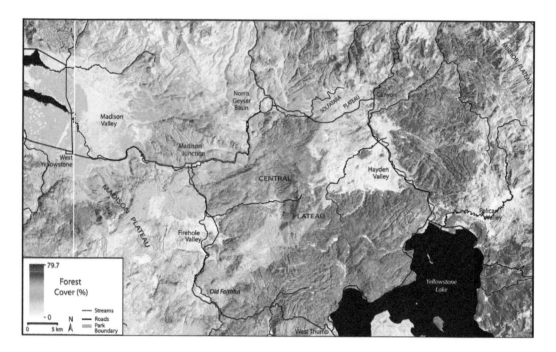

FIGURE 2.12 Percent forest cover of the central portion of Yellowstone National Park, mapped using spectral kerneling of Landsat imagery acquired on September 23, 2002.

climax; mean = 55% cover) (Figure 2.12). These statistics were computed within the pure-age classes mapped by Despain (1990).

We validated the forest cover map against ground-based measurements. We manually selected a list of 23 diverse validation sites from relatively homogeneous areas in the vegetation type map (Figure 2.10). We demarcated a 15×15 m area at each site and measured the presence or absence of canopy cover above breast height at (typically) 64 points arranged in a nested pattern (4×4×4). The points were located objectively using the scheme described in Chapter 7 by Thein *et al.*, this volume. Each point was measured using a Densitometer (Geographic Resource Solutions, Arcata, California) placed on a tripod-mounted arm (Figure 2.13). We also used data from 252 non-canopied 15×15 m sites sampled in closely spaced groups of (typically) five sites as part of our MODIS validation exercise (Chapter 7 by Thein *et al.*, this volume).

The forest cover validation results confirmed the general quantitative accuracy of the map (Figure 2.14). Remotely sensed estimates were highly correlated with ground-based estimates at sites with trees above breast height ($R^2 = 0.74$, $F_{1,22} = 60$, $P < 0.001$), but were biased toward slight over-estimation because trees below breast height were represented in the imagery but not the ground data. Also, spatial registration errors between field locations and corresponding remotely sensed pixels were a likely source of variation in the correlation.

VI. SUMMARY

1. The central Yellowstone study area comprises 3959 km², mostly within Yellowstone National Park (8991 km²), and entirely within the Greater Yellowstone Ecosystem (56,000 km²). Elevations range mainly between 2000 and 2700 m, including the headwaters of the Madison River, Hayden Valley,

FIGURE 2.13 Field measurement of forest cover in the central portion of Yellowstone National Park during 2003 and 2004. Point measurements of canopy cover were made using a densitometer that was objectively located in a nested pattern of 64 points using a tripod-mounted guide arm and laser range-finding equipment (Photo by Fred Watson).

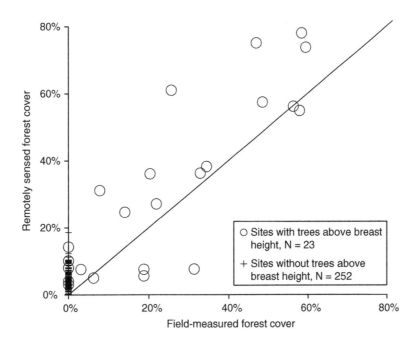

FIGURE 2.14 Validation of remotely sensed forest cover map against ground-based measurements for the central portion of Yellowstone National Park. The presence of trees shorter than breast height causes the remotely sensed values to be slightly higher than the ground-based values measured above breast height.

Pelican Valley, Central Plateau, and Madison Plateau. This area includes the range of the central elk herd, known as the Madison headwaters herd, and the central bison herd.

2. The area includes much of the volcanic terrain associated with the Yellowstone caldera, including rhyolitic lava flows and tuff deposits with poor soils upon which lodgepole pine forests mainly grow. The topographic voids between volcanic features are filled with quaternary alluvial and glacial deposits supporting primarily meadow vegetation.

3. The climate is characterized by long, cold winters and short, cool summers. Night-time freezing conditions occur for eight months of the year and snow pack is present from about November to May. Average snow pack depths are about one meter, but can reach several meters at higher elevations during some winters. The snow pack is greatly reduced in numerous active geothermal areas.

4. About 80% of the landscape is forested, mostly with lodgepole pine as the dominant species. About one-half of the forested area is recent regrowth from widespread, intense fires in 1988. Meadows occupy about 10–20% of the landscape and occur along moisture gradients from wet meadows dominated by sedges and willows to drier meadows dominated by grasses and sagebrush. Geothermal meadows support a somewhat distinct flora, influenced by gradients in heat.

5. We developed remotely sensed maps of vegetation type and forest cover using Landsat imagery at 28.5 m spatial resolution. The forest cover map was validated against field data from 23 forested and 252 non-forested sites, each sampled using 64 nested point-measurements (typically). Remotely sensed forest cover was highly correlated with field estimates. The maps of vegetation type and forest cover were used as inputs to many of the wildlife habitat selection analyses presented later in this book, both directly and indirectly via their role in our snow pack model (Chapter 6 by Watson *et al.*, this volume).

VII. REFERENCES

Burrough, P. A., J. P. Wilson, P. F. M. van Gaans, and A. J. Hansen. 2001. Fuzzy *k*-means classification of topo-climatic data as an aid to forest mapping in the Greater Yellowstone Area. *Landscape Ecology* **16**:523–546.

Christiansen, R. L. 1999. *Digital Geologic Map for Yellowstone National Park, Idaho, Montana, and Wyoming and Vicinity*. U.S. Geological Survey, Denver, CO. U.S. Geological Survey Open-File Report 99–0174.

Christiansen, R. L. 2001. The quaternary and pliocene Yellowstone Plateau volcanic field of Wyoming, Idaho, and Montana. U.S. Geological Survey, Reston, VA. U.S. Geological Survey Professional Paper 729-G.

Christiansen, R. L., J. B. Lowenstern, R. B. Smith, H. Heasler, L. A. Morgan, M. Nathenson, L. G. Mastin, L. J. P. Muffler, and J. E. Robinson. 2007. Preliminary assessment of volcanic and hydrothermal hazards in Yellowstone National Park and vicinity. U.S. Geological Survey, Reston, VA. U.S. Geological Survey Open-File Report 2007–1071.

Craighead, J. J., F. C. Craighead Jr., R. L. Ruff, and B. W. O'Gara. Home ranges and activity patterns of nonmigratory elk of the Madison drainage herd as determined by biotelemetry. *Wildlife Monographs* **33**.

Despain, D. G. 1987. The two climates of Yellowstone National Park. *Proceedings of the Montana Academy of Science* **47**:11–20.

Despain, D. G. 1990. *Yellowstone Vegetation: Consequences of Environment and History in a Natural Setting*. Roberts Rhinehart, Boulder, CO.

Elmore, A. J., J. F. Mustard, S. J. Manning, and D. B. Lobell. 2000. Quantifying vegetation change in semiarid environments: Precision and accuracy of spectral mixture analysis and the normalized difference vegetation index. *Remote Sensing of Environment* **73**:87–102.

Fuller, J. A., R. A. Garrott, and P. J. White. 2007. Emigration and density dependence in Yellowstone bison. *Journal of Wildlife Management* **71**:1924–1933.

Garrott, R. A., L. L. Eberhardt, J. K. Otton, P. J. White, and M. A. Chaffee. 2002. A geochemical trophic cascade in Yellowstone's geothermal environments. *Ecosystems* **5**:659–666.

Garrott, R. A., L. L. Eberhardt, P. J. White, and J. Rotella. 2003. Climate-induced variation in vital rates of an unharvested large-herbivore population. *Canadian Journal of Zoology* **81**:33–45.

Gilabert, M. A., J. González-Piqueras, F. J. García-Haro, and J. Meliá. 2002. A generalized soil-adjusted vegetation index. *Remote Sensing of Environment* **82**:303–310.

Hess, S. C. 2002. Aerial survey methodology for bison population estimation in Yellowstone National Park. Dissertation. Montana State University, Bozeman, MT.

Keefer, W. R. 1971. *The Geologic Story of Yellowstone National Park*. U.S. Geological Survey, Reston, VA. U.S. Geological Survey Bulletin 1347.

Marston, R. A., and J. E. Anderson. 1991. Watersheds and vegetation of the Greater Yellowstone ecosystem. *Conservation Biology* **5**:338–346.

Meagher, M., and D. B. Houston. 1998. *Yellowstone and the Biology of Time: Photographs Across a Century*. University of Oklahoma Press, Norman, OK.

Patten, D. T. 1991. Defining the greater Yellowstone ecosystem. Pages 19–26 *in* R. B. Keiter and M. S. Boyce (Eds.) *The Greater Yellowstone Ecosystem: Redefining America's Wilderness Heritage*. Yale University Press, New Haven, CT.

Peña, J. M., J. A. Lozano, and P. Larrañaga. 1999. An empirical comparison of four initialization methods for the *K*-means algorithm. *Pattern Recognition Letters* **20**:1027–1040.

Pierce, K. L. 1979. *History and Dynamics of Glaciation in the Northern Yellowstone National Park Area*. U.S. Geological Survey, Reston, VA. U.S. Geological Survey Professional Paper 729-F.

Pierce, K. L., and L. A. Morgan. 1992. The track of the Yellowstone hot spot: volcanism, faulting, and uplift. Pages 1–53 *in* P. L. Link, M. A. Kuntz, and L. B. Platt (Eds.) *Regional Geology of Eastern Idaho and Western Wyoming*. Geological Society of America, Boulder, CO. Geological Society of America Memoir 179.

Powell, S. L., and A. J. Hansen. 2007. Conifer cover increase in the Greater Yellowstone Ecosystem: Frequency, rates, and spatial variation. *Ecosystems* **10**:204–216.

Romme, W. H. 1982. Fire and landscape diversity in subalpine forests of Yellowstone National Park. *Ecological Monographs* **52**:199–221.

Romme, W. H., and D. G. Despain. 1989. Historical perspective on the Yellowstone fires of 1988. *BioScience* **39**:695–699.

Romme, W. H., and D. H. Knight. 1982. Landscape diversity: the concept applied to Yellowstone Park. *BioScience* **32**:664–670.

Romme, W. H., and M. G. Turner. 1991. Implications of global climatic change for biogeographic patterns in the greater Yellowstone ecosystem. *Conservation Biology* **5**:373–386.

Shovic, H. F. 1996. *Landforms and Associated Surficial Materials of Yellowstone National Park*. National Park Service, Yellowstone National Park, Mammoth, WY. <http://www.nps.gov/gis/metadata/yell/yell_landform.html.>.

Smith, R. B., and L. J. Siegel. 2000. *Windows into the Earth: The Geologic Story of Yellowstone and Grand Teton National Park*. , New York, NY: Oxford University Press, New York, NY.

Stout, R. G., and T. Al-Niemi. 2002. Heat-tolerant flowering plants of active geothermal areas in Yellowstone National Park. *Annals of Botany* **90**:259–267.

Turner, M. G. 1989. Landscape ecology: The effect of pattern on process. *Annual Review of Ecology and Systematics* **20**:171–197.

Turner, M. G., W. H. Romme, R. H. Gardner, and W. W. Hargrove. 1997. Effects of fire size and pattern on early succession in Yellowstone National Park. *Ecological Monographs* **67**:411–433.

Turner, M. G., W. H. Romme, and D. B. Tinker. 2003. Surprises and lessons from the 1988 Yellowstone fires. *Frontiers in Ecology and the Environment* **1**:351–358.

Wessels, K. J., R. S. De Fries, J. Dempewolf, L. O. Anderson, A. J. Hansen, S. L. Powell, and E. F. Moran. 2004. Mapping regional land cover with MODIS data for biological conservation: Examples from the Greater Yellowstone Ecosystem, USA and Pará State, Brazil. *Remote Sensing of Environment* **92**:67–83.

Williams, C. N., Jr., M. J. Menne, R. S. Vose, and D. R. Easterling. 2007. United States Historical Climatology Network monthly temperature and precipitation data. Carbon Dioxide Information Analysis Center, Oak Ridge National Laboratory, U.S. Department of Energy, Oak Ridge, TN.

Wilmers, C. C., and W. M. Getz. 2005. Gray wolves as climate change buffers in Yellowstone. *Public Library of Science Biology* **3**:571–576.

Wright Parmenter, A., A. Hansen, R. E. Kennedy, W. Cohen, U. Langner, R. Lawrence, B. Maxwell, A. Gallant, and R. Aspinall. 2003. Land use and land cover change in the Greater Yellowstone Ecosystem: 1975–1995. *Ecological Applications* **13**:687–703.

CHAPTER 3

Mapping Mean Annual Precipitation Using Trivariate Kriging

Fred G. R. Watson and Wendi B. Newman

Division of Science and Environmental Policy, California State University Monterey Bay

Contents

Theme

Mean annual precipitation is arguably one of the most significant "bottom-up" influences in temperate large mammal systems, as precipitation typically governs plant phenology and productivity in spring and summer while limiting accessibility and distribution of forage in winter. Mean annual precipitation exhibits marked spatial patterns in mountainous regions such as Yellowstone National Park, resulting in strong spatial gradients in snow pack (Chapter 6 by Watson *et al.*, this volume) and meadow phenology (Chapter 7 by Thein *et al.*, this volume). These spatial patterns have strong influences on large herbivore population dynamics, distribution and movement, and may play an important role in the interactions between predators and prey. Accurately capturing regional variation in precipitation patterns therefore is fundamental to understanding wildlife dynamics and ecosystem processes in general. We produced a map of mean annual precipitation using objective methods involving the development of a variant of kriging, a statistical technique for spatial interpolation between values at known locations. This appears to be a new approach to precipitation mapping, that we termed trivariate zonal ordinary kriging. Rather than treating elevation as a secondary variable, as in cokriging, or a deterministic influence, as in universal kriging, trivariate zonal ordinary kriging considers elevation as a component of the spatial separation between any two locations.

The Ecology of Large Mammals in Central Yellowstone
R. Garrott, P. J. White and F. Watson
ISSN 1936-7961, DOI: 10.1016/S1936-7961(08)00203-0

Copyright © 2009, Elsevier Inc.
All rights reserved.

I. INTRODUCTION

Climatic conditions such as precipitation and temperature can have important consequences for large herbivore population dynamics due to the significant role these variables play in the production and quality of available food (Owen-Smith and Gout 2003). Because large herbivores are often limited by availability of food, climatic variations can determine the carrying capacity for the animal's range (Coe *et al.* 1976). In addition to the demographic parameters that are often linked to density-independent weather effects (Sæther 1997; Gaillard *et al.* 1998, 2000), the distribution and movements of large herbivore populations are also influenced by precipitation patterns. For example, in grazing ecosystems such as the Serengeti, the spatial dynamic processes of large herbivores such as migration and aggregation can be associated with temporal and spatial changes in forage as a result of rainfall patterns (Fryxell 1995, McNaughton and Banyikwa 1995). In high latitude environments seasonal shifts in distribution similarly occur due to temporal and seasonal changes in forage, but also due to the accumulation of deep snow (see Chapter 8 by Messer *et al.*, Chapters 12 and 28 by Bruggeman *et al.*, this volume). Snow pack can also contribute considerably to an individual's vulnerability to predation by impeding escape and/or weakening prey animals (Peterson 1977; Nelson and Mech 1986; Huggard 1993; Bergman *et al.* 2006; Chapters 16 and 17 by Becker *et al.*, Chapter 18 by Gower *et al.*, and Chapter 21 by White *et al.*, this volume). In Yellowstone, precipitation can have both a positive influence on herbivores through its role in forage production, and a negative influence through its role in snow pack accumulation. Gradients of increasing precipitation lead to gradients of decreasing aridity and increasing snow pack, and most large herbivores move back and forth seasonally in response to this. Robust, quantitative, spatial analyses of these effects require an objective map of mean annual precipitation.

Mean annual precipitation varies widely throughout Yellowstone National Park (Figure 3.1). The northern boundary near Gardiner, Montana is an arid steppe while most of the interior supports subalpine forest interspersed with meadow complexes (Whittaker 1975; Peel *et al.* 2007; Chapter 2 by

FIGURE 3.1 Precipitation is a fundamental abiotic driver of ecosystem processes and precipitation patterns are not uniform across the park. Yellowstone's mountain ranges and high elevation plateaus receive the most precipitation. Here a summer thunderstorm drops rain on Yellowstone Lake and the surrounding forests (Photo by Jeff Henry).

Newman *et al.*, this volume). To our knowledge, no objective precipitation map has been published specifically for the Yellowstone region. Farnes *et al.* (1999) subjectively fitted isohyetal contour lines to mean annual precipitation observations from Snow Telemetry (SNOTEL) and climate (CLIM) stations in the Yellowstone region. These contour lines were later digitized and interpolated by the National Park Service, resulting in the precipitation map that is available from the official National Park Service Data Store (<http://science.nature.nps.gov/nrdata/quickoutput.cfm?statecode=wy&Parkcode=yell>). No accuracy information accompanies these data. Alternatively, the PRISM group produces detailed climate maps for the entire United States using objective methods (Daly *et al.* 2002). Their latest precipitation map has 800 m horizontal resolution (<http://www.prism.oregonstate.edu/>). However, we found that this map lacks sufficient detail to represent important key fine-scale patterns in central Yellowstone. We reached this conclusion after using the map to drive the Langur snow pack model, comparing the resulting snow water equivalent predictions to a random coring data set, and finding regionalized error bias in areas of highly incised terrain (Chapter 6 by Watson *et al.*, this volume). Guan *et al.* (2005) noted similar resolution limitations in using PRISM estimates at sub-regional scales, as well as lower accuracy than kriging variants. No region-specific accuracy information accompanies the PRISM data.

In the context of this book, a suitable method for mapping precipitation should be objective, documented, as accurate as possible, and accompanied by some quantification of the spatial distribution of estimation error. The method should also be executable automatically and repeatedly to allow sensitivity analysis and calibration of the effects of precipitation mapping uncertainty on snow pack model function (see "phantom gauge" used in Chapter 6 by Watson *et al.*, this volume). The goal of the work described in this chapter was to provide an objectively created map of mean annual precipitation in Yellowstone National Park for use in our snow pack model (Chapter 6 by Watson *et al.*, this volume) and as a general descriptor of one of the key components of landscape heterogeneity in the park. The chapter develops and documents a new method for estimating mean annual precipitation in mountainous terrain. The map was used, via the snow pack model, as a key explanatory variable in a number of the wildlife analyses presented throughout this book.

II. AVAILABLE CLIMATE STATION DATA

Precipitation data are available from six different data sources employing a variety of instrumentation (Figures 3.2 and 3.3, Appendix 3A.1). Data sources with temporally standardized periods (*i.e.*, 1971–2000 normals) were given priority in cases where multiple data sources were available for a single site, noting that temporally inconsistent data sets can induce regional biases of 5–10% (Hulme and New 1997). The priority we adopted was as follows (from highest to lowest): (1) published National Climate Data Center (NCDC) CLIM precipitation normals during 1971–2000, (2) published Natural Resources Conservation Service (NRCS) SNOTEL precipitation normals during 1971–2000, (3) NCDC CLIM precipitation mean of record through September 30, 2005, (4) published NRCS Snow Course snow water equivalent normals during 1971–2000 for March 1 of each year for Snow Course stations where no SNOTEL data were available, re-scaled to become precipitation normals by linear regression of SNOTEL precipitation normals against Snow Course March 1 snow water equivalent for stations where both SNOTEL and Snow Course data were available ($n = 9$, $R^2 = 0.95$), (5) NRCS Snow Course, Aerial Marker, CLIM, and SNOTEL mean annual precipitation estimated by Farnes *et al.* (1999) for a range of stations for which 1971–2000 normals were not available, and (6) a mean annual precipitation estimate for the Madison Ranger Station derived from daily ranger observations of snow pack depth, by re-scaling the NCDC CLIM precipitation normal during 1971–2000 for West Yellowstone by the ratio of mean February 1 snow depths at the Madison and West Yellowstone ranger stations during 1992–2005.

FIGURE 3.2 A wind-shielded storage precipitation gauge at the Canyon SNOTEL station in Yellowstone National Park (Photo by Fred Watson).

Precipitation is well-known to be log-normally distributed (Hevesi *et al.* 1992, Phillips *et al.* 1992, Diodato 2005) and we confirmed this for central Yellowstone using a Kolmogorov–Smirnov quantile–quantile plot. Since kriging methods generally assume normally distributed variables, the mean annual precipitation data were log-transformed before further analysis.

III. KRIGING

Kriging is a statistical procedure for estimating the values of some spatial field, $p(\mathbf{x})$ by interpolation between a sample of n locations $\mathbf{X} = \mathbf{x}_1,\ldots,\mathbf{x}_n$ with known values. Here, we apply kriging to the problem of estimating the spatial distribution of mean annual precipitation by interpolation between mean annual precipitation sampled at climate stations. Initially, we describe the ordinary kriging system, which appears in many texts (*e.g.*, Isaacs and Srivastava 1989, Goovaerts 1997, Kitanidis 1997, Deutsch and Journel 1998, Chiles and Delfiner 1999, Webster and Oliver 2001), to establish a frame of reference for the description of trivariate kriging of mean annual precipitation.

The initial step in kriging is to model the semi-variogram, which expresses how the property of interest varies between two locations as a function of the difference in their locations:

$$\gamma_{i,j} = g(\mathbf{\Delta}), \quad \mathbf{\Delta} = \mathbf{x}_i - \mathbf{x}_j$$

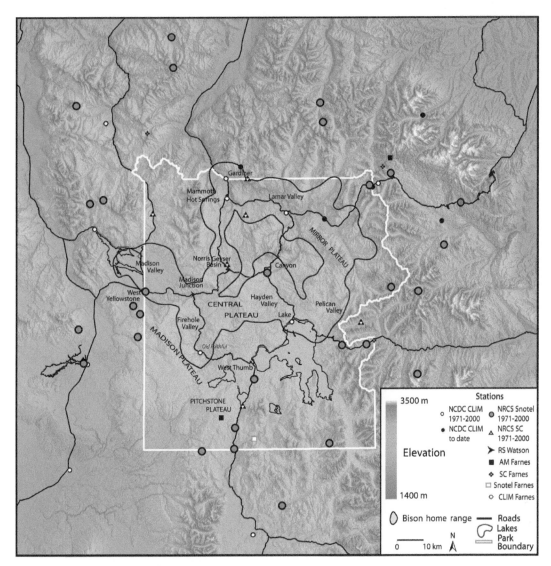

FIGURE 3.3 A network of climate stations in the Greater Yellowstone Area provided the basic data used to develop a map of mean annual precipitation for Yellowstone National Park.

where $\mathbf{x}_i$ and $\mathbf{x}_j$ are two locations, Δ is the difference in their locations, $g(.)$ is the semi-variogram model, and $\gamma_{i,j}$ is the estimated semi-variance between the two locations. Commonly, the semi-variogram model is simply a function of the two-dimensional (2D) Euclidean distance, h_{xy}, between the locations, in which case the above expressions become:

$$\gamma_{i,j} = g\left(h_{xy}\right), \quad h_{xy} = \left|\mathbf{x}_i - \mathbf{x}_j\right|$$

For stationary fields, variogram models are characterized by three parameters: (1) nugget variance, (2) sill variance, and (3) range. Modeled semi-variance starts at the nugget value (c_0) at distance $h = 0$, and then increases to a sill value (c), which is reached as h passes the range distance (a). An appropriate variogram model for precipitation is a cubic model:

$$\gamma_{i,j} = c_0 + (c - c_0)\mathrm{Cubic}\left(\frac{h}{a}\right)$$

with

$$\mathrm{Cubic}(h') = \left\{ \begin{array}{ll} 7h'^2 - 8.75h'^3 + 3.5h'^5 - 0.75h'^7 & h' < 1 \\ 1 & h' \geq 1 \end{array} \right\}$$

We argue that the nugget should be zero for mean annual precipitation, since mean annual precipitation must surely be a spatially continuous field. The cubic function rises gradually from the nugget, which is also consistent with smooth fields. A cubic variogram fitted to the log-transformed mean annual precipitation data is shown in Figure 3.4.

Once the semi-variogram is modeled, the kriging system yields an estimate of the property of interest, p_0^*, at any location, $\mathbf{x}_0$, as:

$$p_0^* = \boldsymbol{\lambda}\mathbf{p}$$

where $\boldsymbol{\lambda}$ is an $(n+1)\times 1$ matrix of kriging coefficients and a Lagrange multiplier for each gage, and $\mathbf{p}$ is an $(n+1)\times 1$ matrix of n sampled $p(\mathbf{x})$ values at locations $\mathbf{X}$, and a final zero value (Webster and Oliver 2001). In turn, the kriging coefficients are computed from:

$$\boldsymbol{\lambda} = \mathbf{A}^{-1}\mathbf{b}$$

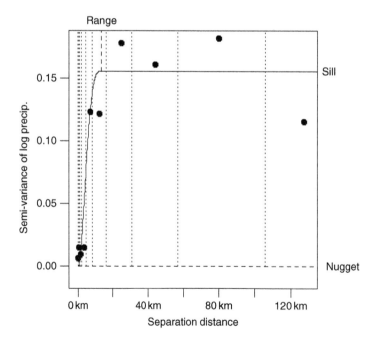

FIGURE 3.4 Cubic variogram model fitted to experimental variogram of log mean annual precipitation. Experimental variogram represents means calculated within intervals are delineated by dotted lines. The model fit minimized mean absolute error between experimental and modeled variogram.

with

$$
\mathbf{A} = \begin{bmatrix}
\gamma_{1,1} & \gamma_{1,2} & \cdots & \gamma_{1,n} & 1 \\
\gamma_{2,1} & \gamma_{2,2} & & \gamma_{2,n} & 1 \\
\vdots & & \ddots & \vdots & \vdots \\
\gamma_{n,1} & \gamma_{n,2} & \cdots & \gamma_{n,n} & 1 \\
1 & 1 & \cdots & 1 & 0
\end{bmatrix} \text{ and } \mathbf{b} = \begin{bmatrix}
\gamma_{1,0} \\
\gamma_{2,0} \\
\\
\gamma_{n,0} \\
1
\end{bmatrix}
$$

where $\mathbf{A}^{-1}$ is the inverse of the $(n+1) \times (n+1)$ matrix including modeled semi-variances between all n locations $\mathbf{X}$ where $p(x)$ was sampled, and $\mathbf{b}$ is an $(n+1) \times 1$ matrix including modeled semi-variances between the sampled locations $\mathbf{X}$ and the location to be estimated, $\mathbf{x}_0$. Thus, once one defines the function $g(\cdot)$, a map of estimates p_0^* directly follows from one matrix inversion and two matrix products for each location to be estimated. The kriging error at location $\mathbf{x}_0$ is given by:

$$
\sigma^*(\mathbf{x}_0) = \sqrt{\mathbf{b}^{\mathrm{T}} \boldsymbol{\lambda}}
$$

This provides a simple measure of the error associated with the kriging estimator, based only on the sample locations (it is simple in the sense that a better measure would also depend on the sample values; Chiles and Delfiner 1999).

IV. KRIGING VARIANTS AS APPLIED TO ESTIMATION OF MEAN ANNUAL PRECIPITATION

A number of variations on the simple kriging system may be considered. Ordinary kriging in 2D yields estimated spatial fields that exhibit a spatially constant value except within range of sampled locations, where the estimated values smoothly interpolate to reach the known values. 2D ordinary kriging has been applied to precipitation estimation by Karnieli (1990), Phillips *et al.* (1992), Boer *et al.* (2001), Diodato (2005), and others. In sparse data sets such as the precipitation data sets investigated in our study, the variogram range is short relative to the typical separation distance between sampled locations. Many locations to be estimated are beyond the horizontal range of sampled locations. Thus, these fields have the appearance of a flat surface with isolated bumps and troughs. The flat background is unsatisfactory because we generally expect precipitation to vary with elevation, even when we are out of horizontal range of any climate stations. Thus, we sought some way of incorporating elevation dependence into the kriging system.

A number of comparative studies have determined that incorporation of elevation information improves kriging estimates of precipitation (Goovaerts 1999a, Boer *et al.* 2001, Diodato 2005, Diodato and Ceccarelli 2005). Cokriging methods incorporate dependence on secondary variables like elevation by characterizing the spatial correlation between primary and secondary variables, as well as the spatial variance in each variable (Goovaerts 1999b). Ordinary cokriging is the standard approach, and is generally applied in cases when the primary variable is relatively under-sampled and the secondary variable is more densely sampled (Webster and Oliver 2001). It has been applied to precipitation estimation with elevation as the secondary variable by Hevesi *et al.* (1992), Phillips *et al.* (1992), Boer *et al.* (2001), Diodato (2005), Diodato and Ceccarelli (2005), and others. However, Goovaerts (1997) notes that when the secondary variable is much more densely sampled, most of the elevation information is redundant and its' influence tends to screen out that of the primary variable. Collocated ordinary cokriging is a variant that addresses this problem by only using secondary variable information from the locations where the primary variable was sampled, and the location to be estimated

(Goovaerts 1997). Goovaerts (1999a,c) applied collocated ordinary cokriging to estimation of rainfall erosivity and mean rainfall, resulting in superior performance over ordinary kriging and universal kriging (below). We applied this method, with a resulting map of mean annual precipitation that resembled an elevation map. Retrospectively, this is not surprising since the method involves a constraint that the correlation between the primary and secondary variables is linear and spatially uniform (Goovaerts 1997). In reality, the rate at which elevation modulates precipitation patterns after accounting for strictly horizontal patterns is probably not constant (Phillips *et al.* 1992) and we would intend that our method should allow for this.

Universal kriging, which is also known as "kriging with external drift," or "kriging with a trend," and is similar to "regression kriging" and "kriging with locally varying means," adds together the stochastic variance component of (usually 2D) ordinary kriging, with an additional deterministic component, usually some polynomial on horizontal location or possibly elevation (Webster and Oliver 2001). The approach has been applied to precipitation estimation by Phillips *et al.* (1992), Goovaerts (1999a), Boer *et al.* (2001), Kyriakidis *et al.* (2001), Guan *et al.* (2005), Diodato (2005), and Haberlandt (2007). A particularly elaborate variant was implemented by Guan *et al.* (2005), involving deterministic components created from monthly multi-scale regressions of precipitation against elevation and terrain aspect. We did not explore the universal kriging approach in our study area, but expected that it would provide similar results to collocated ordinary cokriging since the effect of elevation is only incorporated through simple linear or possibly quadratic relationships that are invariant horizontally. Universal kriging in essence would model precipitation patterns as a simple horizontally fixed precipitation-elevation correlation, with spatially auto-correlated noise.

V. TRIVARIATE ZONAL ORDINARY KRIGING

Ultimately, elevation is simply another spatial dimension. Thus, rather than treating elevation as a secondary variable that happens to correlate both with precipitation and horizontal location, we explored an approach that treats it as part of the definition of location. Specifically, this approach incorporates both vertical separation distance and horizontal separation distance into the basis of variograms. The simplest method under this approach is a trivariate implementation of ordinary kriging. Separation distance is computed as a 3D Euclidean distance, instead of a 2D one. To allow for vertical separation to influence variance at a different rate to horizontal separation (known as geometric anisotropy), a vertical scaling factor is applied to vertical locations before computation of Euclidean distance. This approach has been applied to precipitation estimation by Boer *et al.* (2001). The method yielded intuitively satisfactory results over our study area, but involved contradictory constraints with respect to estimation of mean annual precipitation fields. We argue that the mean annual precipitation variogram on horizontal distance should have zero nugget, since mean annual precipitation is surely a horizontally continuous spatial variable. However, the mean annual precipitation variogram on vertical distance should have a non-zero nugget since two horizontally separate locations of equal elevation would be expected to have different mean annual precipitation. This results in a vertical variogram that has a different shape to the horizontal variogram, a condition known as zonal anisotropy.

We represented zonal anisotropy by modeling the variogram as the sum of two separate variograms for the horizontal, h_{xy}, and vertical, h_z, components of the 3D separation vector, $\mathbf{h}$, respectively:

$$\gamma(\mathbf{h}) = \gamma_{xy}(h_{xy}) + \gamma_z(h_z)$$

with

$$h_{xy} = \sqrt{h_x^2 + h_y^2}$$

Specifically, we combined two cubic variograms, with different nuggets, sills, and ranges.

$$\gamma(\mathbf{h}) = n_{xy} + \left(s_{xy} - n_{xy}\right) Cubic\left(\frac{h_{xy}}{r_{xy}}\right) + n_z + \left(s_z - n_z\right) Cubic\left(\frac{h_z}{r_z}\right)$$

Note that strict partitioning of horizontal and vertical coordinates, as is done here, can lead to non-invertible matrices in the kriging system (Myers and Journel 1990, Chilès and Delfiner 1999), but ours fortunately did not. A more robust approach would be a zonal model that summed 3D isotropic and vertical zonal components as opposed to horizontal zonal and vertical zonal components. Under our zonal variogram model, all variance is viewed as the sum of variance components resulting from horizontal separation and vertical separation, respectively. A zonal variogram fitted to the log-transformed mean annual precipitation data is shown in Figure 3.5 (compare with the standard 1Dvariogram shown in Figure 3.4). When combined with ordinary kriging, the resulting method could be termed "trivariate zonal ordinary kriging." We are not aware that trivariate zonal ordinary kriging has previously been applied to precipitation estimation, though it may have been applied to mineral mapping where 3-D zonal anisotropy would be expected due to structures such as bedding planes. We implemented it within the Tarsier Environmental Modeling Framework (Watson and Rahman 2003).

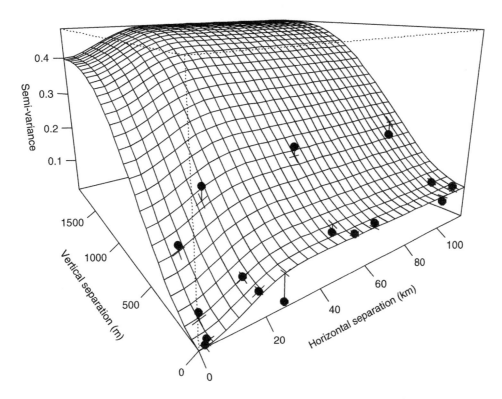

FIGURE 3.5 Zonal semi-variogram of log mean annual precipitation on horizontal and vertical separation distance. The solid circles represent experimental variogram means within intervals, while the gridded surface depicts the modeled variogram.

VI. FITTING THE ZONAL VARIOGRAM

To fit the zonal variogram model, one must first develop the experimental variogram in two dimensions simultaneously (horizontal and vertical). The experimental variogram function is the sequence of mean semi-variances within specified intervals of separation. To reduce variogram estimation error at high separations (Kitanidis 1997), we specified interval sizes to increase exponentially according to

$$H_i = h_{\min} + (h_{\max} - h_{\min}) \left(1 - \frac{\exp\left(\theta\left(\frac{i}{m} - 1\right)\right) - 1}{\exp(-\theta) - 1} \right)$$

where H_i defines the boundaries $i \in \{0,1,\ldots,m\}$ of m intervals, θ is a parameter expressing the degree of bias toward having more intervals at lower separation distances, and $h_{\min}$ and $h_{\max}$ are the minimum and maximum sampled separation distances respectively. Both m and θ have distinct values for horizontal and vertical separation distances: m_{xy}, m_z, θ_{xy}, θ_z. We chose a small number of intervals in each dimension, $m_{xy} = m_z = 4$, following the "three to six" heuristic provided by Kitanidis (1997:32). We selected $\theta_{xy} = 3.7$ and $\theta_z = 7.2$ by trial and error as values that ensured no missing values in the experimental variogram (*i.e.*, that at least one pair of sampled horizontal and vertical separation distances fell within each of the $4 \times 4 = 16$ intervals). Subsequent fitting of variogram parameters was relatively insensitive to these choices governing the domain of the experimental variogram.

Once the intervals were specified and the experimental variogram was computed, we fit the model variogram by using pattern search optimization (Hookes and Jeeves 1961) to minimize the mean absolute error between modeled and experimental variogram. We then rounded the resulting parameter values so as not to give a false sense of precision. The resulting variogram model is illustrated in Figure 3.5 with the following parameter values: (1) $n_{xy} = 0$ which is rational since mean annual precipitation varies smoothly, (2) $n_z = 0.01$ which is rational, since distant sites can have the same elevation and different mean annual precipitation, (3) $s_{xy} = 0.08$ which is well-defined in the optimization results, (4) $r_{xy} = 45$ km which was relatively insensitive in the range 15–70 km and narrowed down to 45 km using ordinary cross-validation (see below), (5) $s_z = 0.4$ (see below), and (6) $r_z = 2000$. The vertical sill and range were co-dependent in the optimization because mean annual precipitation did not exhibit a well-defined sill with respect to vertical separation. The experimental variogram was approximately linear on h_z. Thus, any cubic variogram with approximately the same ratio s_z/r_z would be equivalent, provided that both the sill and range were beyond the range of the data. The underlying physical phenomenon is that precipitation was non-stationary over elevation within the study area. However, a model with a sill is still justified because clearly there is an upper and lower limit to precipitation in nature. The value $s_z = 0.4$ was chosen to be just beyond the range of the data. Given this, r_z optimized to 2000 m.

VII. OTHER DEVELOPMENTS

Several other ideas were tested during the development of the trivariate zonal ordinary kriging system described above. We explored optimizing the variogram model by optimizing the ordinary cross-validation between kriged and observed mean annual precipitation (as opposed to between modeled and experimental semi-variance). We explored optimization of the range and ratio of nugget to sill since the kriging estimate is actually independent of scaling the nugget and sill together. We found that ordinary cross-validation was sensitive enough to r_{xy} that a value of $r_{xy} = 45$ km could be concluded, where earlier it was relatively insensitive in the range 15 km to 70 km. Also, we explored the idea that mean annual precipitation depends not so much on the local elevation, as on the local elevation offset either forward or backward along some prevailing wind direction as a result of orographic wind flow effects. An air stream tends to move upward before encountering a mountain range, which could account for an offset in one direction. Conversely, precipitation may be blown some distance downwind from the orographic influence that caused it, accounting for an offset in the other direction. We included such an offset as a vector parameter in ordinary cross-validation optimizations. Local optima occurred on opposite sides of the zero vector, but these could not be independently reproduced in subsequent optimizations against random snow pack coring data (Chapter 6 by Watson *et al.*, this volume). Thus, the effort was inconclusive and, potentially, merely a result of orographic bias in the location of the mean annual precipitation stations.

In addition, we explored the idea that mean annual precipitation depends on mean elevation within some horizontal range, since localized terrain features may not have the same effect on precipitation as larger terrain features. This "window averaged" elevation dependency has been previously exploited in precipitation mapping (Kyriakidis *et al.* 2001, Guan *et al.* 2005). We explored this by optimizing a parameter for the length scale for the mean elevation calculation. The ordinary cross-validation improved, but not enough to warrant the additional complexity. Finally, we explored tri-cubic thin-plate smoothing splines (Hutchinson 1998) as a completely different approach. The results suffered from under-shoot and over-shoot artifacts on steep mountain ranges.

VIII. FINAL MAPS

The estimated distribution of mean annual precipitation throughout Yellowstone National Park made using the trivariate zonal ordinary kriging method is shown in Figure 3.6 (back-transformed from initially log-transformed data). Estimated precipitation varies from 260 to 1830 mm across the area shown. The figure also shows a 99% kernel home range on all bison groups surveyed aerially between 1998 and 2007, based on a 1850 m bandwidth. The bison range is clearly correlated with low precipitation and demarcated by areas of high precipitation. Apart from being generally mountainous, these areas exhibit a relatively persistent seasonal snow pack and, thus, shorter growing seasons. We performed a brief analysis with bison aerial survey data from 1998 to 2007, which showed that 97% of all bison locations occurred in locations with mean annual precipitation lower than 800 mm (bison data from C. Geremia and R. Wallen, National Park Service).

The estimated kriging error in log-transformed mean annual precipitation is shown in Figure 3.7, which reveals portions of the study area where there is the greatest uncertainty in estimated mean annual precipitation. Specifically, precipitation on the Mirror Plateau and Central Plateau (Mary Mountain) is the least accurately estimated. The Mary Mountain corridor between the Firehole Valley and the Hayden Valley is of particular relevance to our study of the large mammals of central Yellowstone. This is the main corridor used by bison to move between summer and winter ranges (Chapters 12 and 28 by Bruggeman *et al.*, this volume). Limitations in precipitation mapping accuracy in these areas translate to limitations in modeled snow pack and the precision of wildlife analyses pertinent to these areas. Another area of note is at Madison junction, which lies at the center of our study area in the Madison headwaters portion of Yellowstone. Mapped precipitation there was distinctly lower than the surrounding area due to the influence of the mean annual precipitation estimate from Madison ranger station. This estimate may not be as reliable as the 1971–2000 climate normals available for many of the other stations because it is based on snow-depth observations that were possibly affected by tree cover.

IX. SUMMARY

1.　Precipitation has an important influence on wildlife dynamics through its role in determining snow pack accumulation and forage quality and productivity.

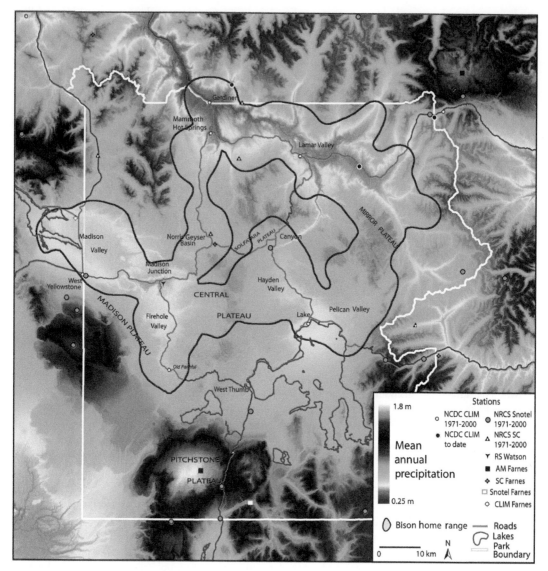

FIGURE 3.6   Mean annual precipitation estimated within Yellowstone National Park using the trivariate zonal ordinary kriging method and data from regional climate stations.

2. We developed a variant of methods for the objective mapping of mean annual precipitation. The variant is termed trivariate zonal ordinary kriging and has apparently not been previously applied to precipitation estimation. The method interpolates precipitation in both horizontal and vertical space, but unlike cokriging and universal kriging, it makes no assumptions about the nature of the precipitation-elevation correlation.

3. We used trivariate zonal ordinary kriging to produce the first objectively created digital map of mean annual precipitation as a continuously varying property specifically for Yellowstone

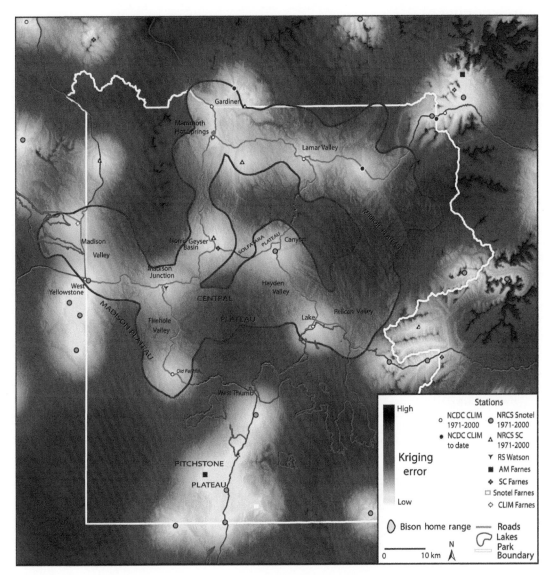

FIGURE 3.7 Kriging error in estimated log-transformed mean annual precipitation (*i.e.*, the kriging error is the square root of the kriging variance).

National Park. For input, the method used a digital elevation model and mean annual precipitation estimates for 64 observing stations in the Yellowstone area.

4. Estimated precipitation varied throughout the park from 260 mm along the arid low-elevation sections of the northern boundary, to 1830 mm on Pitchstone Plateau, one of the first high mountain masses encountered by storm systems reaching the park from the southwest.

5. The map was used as a key input to our snow pack model (Chapter 6 by Watson *et al.*, this volume), which was used to provide snow pack covariates to numerous wildlife analyses in this book.

X. REFERENCES

Bergman, E. J., R. A. Garrott, S. Creel, J. J. Borkowski, and R. M. Jaffe. 2006. Assessment of prey vulnerability through analysis of wolf movements and kill sites. *Ecological Applications* **16**:273–284.

Boer, E. P. J., K. M. de Beurs, and A. D. Hartkamp. 2001. Kriging and thin plate splines for mapping climate variables. *Journal of Applied Earth Observation and Geoinformation* **3**:146–154.

Chilès, J.-P., and P. Delfiner. 1999. *Geostatistics: Modeling Spatial Uncertainty.* Wiley, New York, NY.

Coe, M. J., D. H. Cumming, and J. Phillipson. 1976. Biomass and production of large African herbivores in relation to rainfall and primary production. *Oecologia* **22**:341–354.

Daly, C., W. P. Gibson, G. H. Taylor, G. L. Johnson, and P. Pasteris. 2002. A knowledge-based approach to the statistical mapping of climate. *Climate Research* **22**:99–113.

Deutsch, C. V., and A. G. Journel. 1998. *GSLIB: Geostatistical Software Library and User's Guide.* 2nd edn Oxford University Press, New York, NY.

Diodato, N. 2005. The influence of topographic co-variables on the spatial variability of precipitation over small regions of complex terrain. *International Journal of Climatology* **25**:351–363.

Diodato, N., and M. Ceccarelli. 2005. Interpolation processes using multivariate geostatistics for mapping of climatological precipitation mean in the Sannio Mountains (southern Italy). *Earth Surface Processes and Landforms* **30**:259–268.

Farnes, P. E., C. Heydon, and K. Hansen. 1999. *Snowpack Distribution Across Yellowstone National Park.* Earth Sciences Department, Montana State University, Bozeman, MT.

Fryxell, J. M. 1995. Aggregation and migration by grazing ungulates in relation to resources and predators. Pages 257–273 *in* A. E. Sinclair and P. Arcese (Eds.) *Serengeti II Dynamics, Management, and Conservation of an Ecosystem.* University of Chicago Press, Chicago, IL.

Gaillard, J.-M., M. Festa-Bianchet, and N. G. Yoccoz. 1998. Population dynamics of large herbivores: Variable recruitment with constant adult survival. *Trends in Ecology and Evolution* **13**:58–63.

Gaillard, J.-M., M. Festa-Bianchet, N. G. Yoccoz, A. Loison, and C. Toïgo. 2000. Temporal variation in fitness components of population dynamics in large herbivores. *Annual Review of Ecology and Systematics* **31**:367–393.

Goovaerts, P. 1997. *Geostatistics for Natural Resources Evaluation.* Oxford University Press, Oxford, United Kingdom.

Goovaerts, P. 1999a. Geostatistical approaches for incorporating elevation into the spatial interpolation of rainfall. *Journal of Hydrology* **228**:113–129.

Goovaerts, P. 1999b. Geostatistics in soil science: State-of-the-art and perspectives. *Geoderma* **89**:1–45.

Goovaerts, P. 1999c. Using elevation to aid the geostatistical mapping of rainfall erosivity. *Catena* **34**:227–242.

Guan, H., J. L. Wilson, and O. Makhnin. 2005. Geostatistical mapping of mountain precipitation incorporating autosearched effects of terrain and climatic characteristics. *Journal of Hydrometeorology* **6**:1018–1031.

Haberlandt, U. 2007. Geostatistical interpolation of hourly precipitation from rain gauges and radar for a large-scale extreme rainfall event. *Journal of Hydrology* **332**:144–157.

Hevesi, J. A., J. D. Istok, and A. L. Flit. 1992. Precipitation estimation in mountainous terrain using multivariate geostatistics. Part I: Structural analysis. *Journal of Applied Meteorology* **31**:661–676.

Hookes, R., and T. Jeeves. 1961. "Direct Search" solution of numerical and statistical problems. *Journal of the Association for Computing Machinery* **8**:212–229.

Huggard, D. J. 1993. Effect of snow depth on predation and scavenging by gray wolves. *Journal of Wildlife Management* **57**:382–388.

Hulme, M., and M. New. 1997. Dependence of large-scale precipitation climatologies on temporal and spatial sampling. *Journal of Climate* **10**:1099–1113.

Hutchinson, M. I. 1998. Interpolation of rainfall data with thin plate smoothing splines: Analysis of topographic dependence. *Journal of Geographic Information and Decision Analysis* **2**:168–185.

Isaacs, E. H., and R. M. Srivastava. 1989. *Applied Geostatistics.* Oxford University Press, New York, NY.

Karnieli, A. 1990. Application of kriging technique to areal precipitation mapping in Arizona. *GeoJournal* **22**:391–398.

Kitanidis, P. K. 1997. *Introduction to Geostatistics: Applications to Hydrogeology.* Cambridge University Press, Cambridge, United Kingdom.

Kyriakidis, P. C., J. Kim, and N. L. Miller. 2001. Geostatistical mapping of precipitation from rain gauge data using atmospheric and terrain characteristics. *Journal of Applied Meteorology* **40**:1855–1877.

McNaughton, S. J., and F. F. Banyikwa. 1995. Plant communities and herbivory. Pages 49–70 *in* A. E. Sinclair and P. Arcese (Eds.) *Serengeti II Dynamics, Management, and Conservation of an Ecosystem.* University of Chicago Press, Chicago, IL.

Myers, D. E., and A. Journel. 1990. Variograms with zonal anisotropies and noninvertible kriging systems. *Mathematical Geology* **22**:779–785.

Nelson, M. E., and D. L. Mech. 1986. Relationship between snow depth and gray wolf predation on white-tailed deer. *Journal of Wildlife Management* **50**:471–474.

Owen-Smith, N., and J. Ogutu. 2003. Rainfall influences on ungulate population dynamics. Pages 310–331 *in* J. T. Du Toit, K. H. Rogers, and H. C. Biggs (Eds.) *The Kruger Experience.* Island Press, Washington, DC.

Peel, M. C., B. L. Finlayson, and T. A. McMahon. 2007. Updated world map of the Köppen–Geiger climate classification. *Hydrology and Earth System Sciences* **11**:1633–1644.

Peterson, R. O. 1977. *Wolf Ecology and Prey Relationships on Isle Royale.* U.S. National Park Service Scientific Monograph Series 11, U.S. Govt. Printing Office, Washington, DC.

Phillips, D. L., J. Dolph, and D. Marks. 1992. A comparison of geostatistical procedures for spatial analysis of precipitation in mountainous terrain. *Agricultural and Forest Meteorology* **58**:119–141.

Sæther, B. E. 1997. Environmental stochasticity and population dynamics of large herbivores: A search for mechanisms. *Trends in Ecology and Evolution* **12**:143–149.

Watson, F. G. R., and J. M. Rahman. 2003. Tarsier: A practical software framework for model development, testing and deployment. *Environmental Modelling and Software* **19**:245–260.

Webster, R., and M. Oliver. 2001. *Geostatistics for Environmental Scientists.* Wiley, West Sussex, United Kingdom.

Whittaker, R. H. 1975. *Communities and Ecosystems.* Macmillan, London, United Kingdom.

APPENDIX

APPENDIX 3A.1	Estimated mean annual precipitation for 64 stations in the Yellowstone region

Station name	Station type	Mean annual precipitation (m)	Data source	Easting (m)	Northing (m)
Ashton	CLIM	0.508	NCDC 1971–2000	463,961	4,879,160
Aster Creek	SC	1.121	SC 1971–2000	529,700	4,902,800
Base Camp	SNOTEL	0.843	NRCS 1971–2000	544,397	4,865,789
Beartooth Lake	SNOTEL	0.894	NRCS 1971–2000	613,000	4,977,400
Beaver Creek	SNOTEL	0.942	NRCS 1971–2000	471,713	4,977,186
Big Sky 3S	CLIM	0.52	NCDC 1971–2000	477,753	5,006,841
Black Bear	SNOTEL	1.516	NRCS 1971–2000	489,823	4,928,124
Blackwater	SNOTEL	0.98	NRCS 1971–2000	596,132	4,914,201
Box Canyon	SNOTEL	0.653	NRCS 1971–2000	558,823	5,014,329
Canyon	SNOTEL	0.704	NRCS 1971–2000	538,800	4,951,700
Carrot Basin	SNOTEL	1.255	NRCS 1971–2000	476,810	4,978,545
Cooke City 2W	CLIM	0.647	NCDC 1971–2000	581,000	4,984,600
Cooke Station	SC	1.016	Farnes	586,400	4,986,900
Coulter Creek	SNOTEL	1.039	Farnes	534,000	4,890,500
Crandall Creek	CLIM	0.381	NCDC to date	605,303	4,970,638
Crevice Mountain	SC	0.638	SC 1971–2000	531,500	4,986,500
East Entrance	SC	0.584	Farnes	579,400	4,926,700
Evening Star	SNOTEL	1.125	NRCS 1971–2000	596,400	4,944,800
Fisher Creek	SC	1.499	Farnes	582,700	4,990,700
Gardiner	CLIM	0.25	NCDC 1971–2000	523,300	4,986,400
Grassy Lake	SNOTEL	1.369	NRCS 1971–2000	514,300	4,886,000
Hebgen Dam	CLIM	0.744	NCDC 1971–2000	473,500	4,967,800
Island Park	CLIM	0.702	NCDC 1971–2000	470,808	4,918,002
Island Park	SNOTEL	0.79	NRCS 1971–2000	469,341	4,918,412
Jardine	CLIM	0.445	NCDC to date	528,900	4,990,700
Lake Camp	SC	0.629	SC 1971–2000	574,400	4,933,500

(*continued*)

APPENDIX 3A.1 (*continued*)

Station name	Station type	Mean annual precipitation (m)	Data source	Easting (m)	Northing (m)
Lake Yellowstone	CLIM	0.518	NCDC 1971–2000	548,100	4,933,600
Lamar RS	CLIM	0.343	NCDC to date	560,400	4,971,400
Lewis Lake Divide	SNOTEL	1.356	NRCS 1971–2000	526,800	4,894,500
Lick Creek	SNOTEL	0.848	NRCS 1971–2000	502,637	5,038,735
Lone Mountain	SNOTEL	0.991	NRCS 1971–2000	466,512	5,013,273
Lupine Creek	SC	0.623	SC 1971–2000	530,800	4,973,100
Madison Junction	RS	0.442	Watson	511,182	4,943,713
Madison Plateau	SNOTEL	1.069	NRCS 1971–2000	490,700	4,936,500
Monument Peak	SNOTEL	0.968	NRCS 1971–2000	559,900	5,007,200
Moran 5 WNW	CLIM	0.639	NCDC 1971–2000	533,491	4,855,081
Mystic Lake	CLIM	0.633	NCDC to date	598,118	5,009,631
Norris Basin	SC	0.655	SC 1971–2000	523,748	4,955,006
Norris Basin (Old)	SC	0.813	Farnes	524,100	4,953,000
Northeast Entrance	SNOTEL	0.66	NRCS 1971–2000	577,700	4,983,900
Old Faithful	CLIM	0.62	NCDC 1971–2000	513,600	4,922,300
Parker Peak	SNOTEL	0.787	NRCS 1971–2000	586,000	4,946,300
Pitchstone Plateau	AM	1.791	Farnes	521,600	4,898,200
Rock Creek Meadows	SC	0.803	Farnes	493,500	5,003,100
Shower Falls	SNOTEL	1.29	NRCS 1971–2000	503,300	5,027,300
Snake River	CLIM	0.801	NCDC 1971–2000	526,709	4,886,860
Snake River Station	SNOTEL	0.897	NRCS 1971–2000	526,465	4,886,766
Star Lake E	AM	1.613	Farnes	585,600	4,993,800
Sylvan Lake	SNOTEL	0.947	NRCS 1971–2000	567,100	4,925,000
Sylvan Road	SNOTEL	0.77	NRCS 1971–2000	576,500	4,925,300
Thumb Divide	SNOTEL	0.775	NRCS 1971–2000	534,000	4,912,500
Tower Falls	CLIM	0.415	NCDC 1971–2000	545,900	4,973,700
Twenty-one Mile	SC	0.796	SC 1971–2000	495,600	4,973,600
Two Ocean Plateau	SNOTEL	1.146	NRCS 1971–2000	562,300	4,888,800
West Yellowstone	CLIM	0.552	NCDC 1971–2000	492,000	4,945,100
West Yellowstone	SNOTEL	0.655	NRCS 1971–2000	492,800	4,944,800
West Yellowstone9N / 9NNW	CLIM	0.532	Farnes	489,500	4,959,100
Whiskey Creek	SNOTEL	0.894	NRCS 1971–2000	488,100	4,939,500
White Elephant	SNOTEL	1.204	NRCS 1971–2000	467,356	4,930,903
White Mill	SNOTEL	1.153	NRCS 1971–2000	585,800	4,988,300
Wolverine	SNOTEL	0.653	NRCS 1971–2000	606,211	4,961,869
Yellowstone NatlPark NE Ent	CLIM	0.663	NCDC to date	578,818	4,983,220
Yellowstone Park	CLIM	0.372	NCDC 1971–2000	523,659	4,979,074
Younts Peak	SNOTEL	0.795	NRCS 1971–2000	594,900	4,864,800

CHAPTER 4

Effects of Yellowstone's Unique Geothermal Landscape on Snow Pack

Fred G. R. Watson,* Wendi B. Newman,* Thor N. Anderson,* Ryan E. Lockwood,* and Robert A. Garrott[†]

*Division of Science and Environmental Policy, California State University Monterey Bay
[†]Fish and Wildlife Management Program, Department of Ecology, Montana State University

Contents

Theme

The unique concentration of geothermal areas in Yellowstone National Park moderates winter severity for large herbivores by providing areas of reduced snow accumulation and access to photosynthesizing plants year round. Consequently these areas concentrate large herbivores and influence associated ecological processes in winter. While the well-known, dramatic hydrothermal features have received the most attention, the more extensive and less obvious areas of geothermal influence, where warming sufficiently reduces or eliminates snow pack accumulation without precluding forage production, can be expected to exert the most significant influence on large herbivore dynamics. We extend our earlier foundations in remote sensing of a thermal emittance anomaly (TEA; W/m^2) in Yellowstone to mapping of the geothermal heat flux (GHF; W/m^2). This facilitates mapping of the extent of geothermal snow pack reduction using a spatial snow pack simulation model that is sensitive to GHF inputs. The result is a quantification of the degree to which geothermal activity enhances winter habitat for large herbivores by reducing snow pack.

The Ecology of Large Mammals in Central Yellowstone
R. Garrott, P. J. White and F. Watson
ISSN 1936-7961, DOI: 10.1016/S1936-7961(08)00204-2

Copyright © 2009, Elsevier Inc.
All rights reserved.

I. INTRODUCTION

The existence of the elk and bison herds in the central portion of Yellowstone National Park is closely tied to geothermal heat flux (GHF). Heat flowing upward to the surface from the interior of the Earth prevents snow from accumulating in many places in winter (Figure 4.1). This absence of snow allows herbivores to access photosynthesizing forage plants year-round (Figure 4.2) and, in turn, influences their spatial dynamics and distribution across the landscape. The elk herd inhabiting the Madison headwaters area is non-migratory, remaining year-round in a range that encompasses many of the largest geothermal areas (Craighead *et al.* 1973; Chapter 8 by Messer *et al.* and Chapter 11 by Garrott *et al.*, this volume). Bison progressively accumulate in the major thermal valleys as winter progresses, selecting for the geothermal habitats in particular (Meagher 1973, Bruggeman *et al.* 2006, 2007; Chapters 12 and 28 by Bruggeman *et al.*, this volume).

Though Yellowstone is famous for its geothermal features and wildlife, the interactions between these resources have received relatively little attention. Garrott *et al.* (2002) traced geothermal influences through soils and plants to large herbivores, with chemical compounds from geothermal effluent resulting in early onset of senescence, reduced lifespan, and a modified age structure due to accelerated and abnormal tooth wear (Chapter 10 by Garrott *et al.*, this volume). Additional interactions between geothermal features and large mammal dynamics are related to heat and snow pack. In particular it was speculated that snow-free thermal areas surrounded by deep snow would contribute to the vulnerability of large herbivores following wolf recolonization (Figure 4.3). Wolves could potentially be more successful in capturing and killing prey in geothermal areas by exploiting hard boundaries between habitats, such as the snow-free and snow-covered habitat at the edges of geothermal areas. These boundaries would essentially act as escape barriers for fleeing prey—an idea that was subsequently supported by Bergman *et al.* (2006). Thus while geothermal areas represented preferred foraging areas for wintering ungulates they were areas of high predation risk, and evaluating the subsequent dynamics of predators and prey in such a system necessitated quantitative descriptions of the intensity and distribution of geothermal features on the landscape.

Quantitatively assessing geothermal impacts on large mammal dynamics requires detailed descriptions of the spatial pattern of geothermal areas and geothermally influenced snow pack. In particular, geothermal maps should objectively quantify the intensity of geothermal influence on a continuous

FIGURE 4.1 Geothermal areas in central Yellowstone are frequently devoid of snow in winter. This influence extends far into the broader landscape, beyond the obvious geysers and hot springs in named thermal basins (Photo by Fred Watson).

FIGURE 4.2 Geothermal areas provide large herbivores with access to photosynthesizing plants year round. Geysers are often surrounded by algal mats as shown in the runoff channel from Daisy Geyser in the Upper Geyser Basin (Photo by Jeff Henry).

scale (*i.e.*, from no geothermal activity to very intense activity) through a species' entire range, as geothermal influence frequently extends over large areas with only a slight intensity that is visually subtle, but important to wildlife.

As a spatial property, geothermal intensity is measured as upward GHF in Watts per horizontal square meter (W/m^2). GHF averages $0.075 W/m^2$ globally and between $0.02 W/m^2$ and $0.1 W/m^2$ in the western United States (Pollack *et al.* 1991, Blackwell and Richards 2004). Within a 50-km radius of central Yellowstone, GHF averages above $0.15 W/m^2$ (Blackwell and Richards 2004) due to the Yellowstone hotspot, a plume in the Earth's mantle that is presently directly beneath Yellowstone National Park (Smith and Braile 1994, Smith and Siegel 2000). Sixteen million years ago, this plume was beneath southeastern Oregon, but the North American plate has moved approximately 1000 km over the hotspot since then.

The most dramatic consequences of the hotspot in Yellowstone were major caldera eruptions 2.0, 1.3, and 0.6 million years ago (Christiansen 2001). These eruptions were preceded and followed by smaller eruptions as recently as 150 and 70 thousand years ago that filled the Yellowstone caldera and covered most of the central Yellowstone landscape with lava flows (Christiansen 2001). The area remains volcanically active, with modern measurements of crustal deformation, faulting, earthquakes, lack of seismic focal depths below about 3–4 km, and high heat flow (Fournier 1989). Much of the heat flow is advected away from surface sources in rivers (Fournier 1989), while the remainder is dissipated into the atmosphere or snow pack through radiation and convection.

The spatial pattern of heat flow at the surface is a complex outcome of the interaction between the pattern of geologic history and modern hydrothermal processes. Most geothermal areas lie inside the

FIGURE 4.3 Heat associated with Yellowstone's famous geothermal features has a dramatic affect on snow pack accumulation. This elevated boardwalk in the Norris Geyser Basin isolated falling snow from the heat emitted from the ground, allowing snow to accumlate and demonstrating both the magnitude of snow pack that can accumulate on the winter ranges in central Yellowstone as well as the important influence of geothermal heat to create areas of reduced snow pack that attract both elk and bison (Photo by Robert Garrott).

caldera boundary and occur in linear patterns along valleys comprised of alluvial sediments that fill the gaps between major lava flow margins (Christiansen 2001). However, some geothermal features occur on hilltops or in locations far removed from other geothermal areas. None are apparent at the highest elevations, where hydrostatic pressure precludes their existence (Fournier 1989). Some geothermal features have appeared in recent decades, while others are apparently dormant.

Until recently, the only maps of geothermal influence were either maps of point features or polygon features. Point features include the myriad geysers, hot springs, fumaroles, and mud pots scattered throughout the park. They are mapped on U.S. Geological Survey topographic maps and, also, in the Thermal Inventory data set compiled by the National Park Service in the past few years (A. Rodman, National Park Service, personal communication). Polygon features describe the boundaries of geothermally influenced areas, usually with only coarse categorical indication of the type of influence within each boundary (*e.g.*, acid or alkaline). An example is Shovic's (1996) landforms map based on air-photo interpretation. This map is extensive enough to include remote, little-known, geothermal areas in the backcountry, but it is categorical and does not measure intensity.

In previous work, we provided the first continuous map of the intensity of geothermal influence in Yellowstone National Park (Watson *et al.* 2007). We used the thermal band of the Landsat Enhanced Thematic Mapper (ETM) to estimate thermal emittance (W/m^2) at the Earth's surface. We corrected this map for non-geothermal effects such as solar radiation and elevation, yielding a final map that estimated just the terrestrially originated component of surface emittance, which we termed the terrestrial emittance anomaly (TEA, W/m²). The map combined methods from the high-intensity volcanic thermal imaging literature (Oppenheimer 1997, Flynn *et al.* 2001, Urai 2002, Lombardo *et al.* 2004, Donegan and Flynn 2004, Pieri and Abrams 2004, Kaneko and Wooster 2005) and the primarily visual hydrothermal imaging literature (Mazzarini *et al.* 2001, Pickles *et al.* 2001, Hellman and Ramsey 2004, Patrick *et al.* 2004, Vaughan *et al.* 2005).

In this chapter, we expand our previous work to derive maps of GHF from the original maps of TEA. We used a spatial snow pack model (Langur, Chapter 6 by Watson *et al.*, this volume) designed to be sensitive to variations in GHF. Specifically, we regressed GHF estimated by inversion of the snow pack model against remotely sensed TEA along 12 field-surveyed, snow-free perimeters of geothermal areas. We quantified the large-scale spatial distribution of geothermal influence on the snow pack by inputting the GHF map into the snow pack simulation model. We extracted simple cumulative frequency statistics to estimate the proportion of Madison headwaters elk range (Chapter 8 by Messer *et al.*, this volume, Figure 8.3) that was distinctly geothermal, and the proportion for which the snow pack was significantly reduced or ablated.

II. SPATIAL DISTRIBUTION OF GEOTHERMAL INTENSITY

Using Landsat ETM+ imagery from March 25, 2000 our previous work estimated a TEA as that portion of remotely sensed terrestrial emittance (M_{terr}, W/m²) that was not explained by the effect of elevation above sea level (*E*, m) or remotely sensed absorbed solar radiation (*A*, W/m²; Watson *et al.* 2007):

$$TEA = M_{terr} - \left(\beta'_0 + \beta'_E E + \beta'_A A \right)$$

with parameters β'_0, β'_E, and $\beta'_A A$ fitted by multiple linear regression within a spatial domain limited to snow-free areas as determined using a remotely sensed Normalized Difference Snow Index (NDSI, Hall *et al.* 1995). We argued that TEA estimates a lower bound for GHF (W/m²) with the relation:

$$TEA = \frac{1}{k} \left(GHF + \varepsilon_{src} - \varepsilon_{sink} \right)$$

where k is a constant of proportionality ($k \geq 1$), ε_{sink} is an error term representing variance about assumptions that terrestrial emittance is correlated with latent, sensible, and advective heat fluxes, and ε_{src} is an error term representing any bias in the relationship between *E* and *A* and heat sources including absorbed solar radiation, net downward atmospheric emittance, and change in subsurface heat storage.

We compared TEA to independent estimates of GHF provided by inversion of the Langur snow pack model (Watson *et al.* 2006). The accumulation of snow (measured as snow water equivalent, SWE) is continuously dependent on variation in GHF (Figure 4.4) and this dependency yields a means of estimating GHF from SWE measurements. We surveyed 18.5 km of snow-free perimeters around six geothermal areas on one to three different dates using Global Positioning Satellite equipment (Figure 4.5). The mean TEA for each of these perimeters was computed. The GHF at these perimeters was estimated using the snow pack model as the rate of heat flux into the soil from below that would be required to just melt the snow pack at the perimeter location on the survey date. The comparison revealed a correlation between TEA and GHF, and supported the argument that *k* should be greater than one because TEA represents only the radiative dispersion of geothermal heat, and does not measure convective or advective dispersion of that heat.

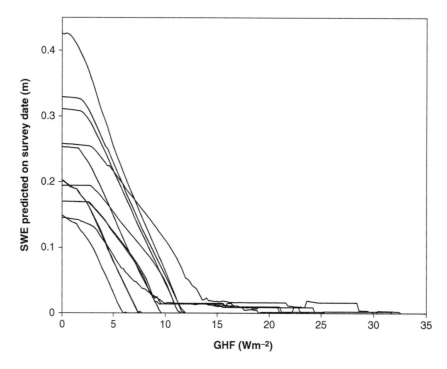

FIGURE 4.4 Modeled response of snow water equivalent (SWE) to variations in geothermal heat flux (GHF) intensity.

FIGURE 4.5 Example of field-mapped snow-free perimeters around a geothermal area on Solfatara Plateau during three different dates, compared with a visualization of remotely sensed GHF (yellow and red), snow pack distribution (white and gray), and forest cover (trees).

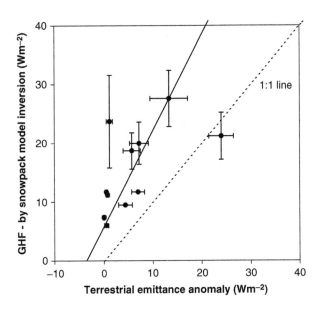

FIGURE 4.6 Relationship between mapped thermal emittance anomaly (TEA; W/m²) and modeled GHF (W/m²) at snow-free geothermal perimeters that was used to estimate the GHF map from the TEA map. Horizontal whiskers indicate the range of variation of TEA along each perimeter. Vertical whiskers represent the range of uncertainty in the snow pack model inversion estimate of GHF at each perimeter.

We extended this analysis to estimate GHF from TEA using the above correlation. Initially, we repeated the above comparison using updated GHF estimates based on an updated snow pack model (Figure 4.6). A simple linear regression was fitted:

$$\text{TEA} = \frac{1}{k^*}(\text{GHF} + \varepsilon)$$

with $k^* = 1.651 < k$ and $\varepsilon = 5.825$. The inequality between k^* and k denotes that k^* does not incorporate advective flux (i.e., liquid heated water draining away from geothermal areas).

This inverts to yield a means of estimating the spatial distribution of GHF from TEA (Figure 4.7A):

$$\text{GHF} = k^* \, \text{TEA} - \varepsilon$$

False positives were abundant at low elevations, so we limited the estimates to areas above 2250 m in the watershed of the Yellowstone River and above 1940 m elsewhere. This includes all major thermal areas in central, southern, western, and eastern Yellowstone National Park but excludes some thermal areas in the northern portion of the park (*e.g.*, Mammoth Hot Springs, Devil's Kitchen). We also excluded areas above 2700 m since hydrothermal activity is theoretically limited at high elevations (Fournier, 1989) and false positives were common on cliffs in these high-mountain areas. We expanded the spatial domain beyond the NDSI snow-free areas, so as to include more representation of the slight and marginal thermal influences that were excluded by the strict snow-free criterion.

The GHF map was a key spatial driver of the snow pack model, which in turn was used to simulate snow pack covariates for many chapters in this book (Chapter 8 by Messer *et al.*, Chapter 9 by White *et al.*, Chapters 10 and 11 by Garrott *et al.*, Chapter 13 by Fuller *et al.*, Chapters 16 and 17 by Becker *et al.*, Chapters 18, 19, and 20 by Gower *et al.*, Chapters 21 and 22 by White *et al.*, Chapter 23 by Garrott *et al.*, Chapters 27 and 28 by Bruggeman *et al.*, and Chapter 29 by Alexander *et al.*, this volume). These simulations used a slightly earlier version of the GHF map than the one described here

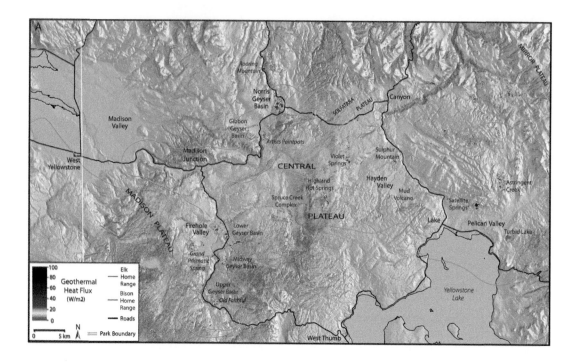

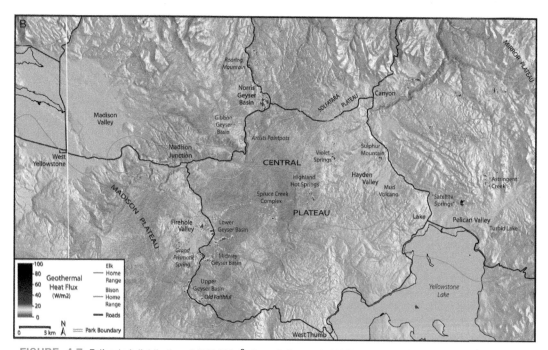

FIGURE 4.7 Estimated distribution of GHF (W/m^2) through the central Yellowstone study area: A) objective version based on Watson *et al.* (2007) with subsequent GHF versus TEA (W/m^2) relationship applied; and B) subjective version used in this book that was developed prior to Watson *et al.* (2007).

(Figure 4.7B), but its equivalence is demonstrated in the following section. This earlier version of the GHF map was produced from the following relation:

$$\text{GHF}_{\text{prev}} = M_{\text{terr}} - 12\cos I + 0.34 B_5 + \exp\left(\frac{E - 2200}{425}\right) - 295.5$$

where I is the angle of incidence of the sun on terrain slopes at the time of Landsat scene acquisition, and B_5 is the raw Landsat Band 5 digital number. The two methods are similar since absorbed solar radiation is determined partly by incidence-cosine and is computed in terms of non-thermal Landsat bands such as Band 5. The earlier version involved subjective specification of coefficients by manually fitting lines to two-dimensional histograms of residuals. This subjectivity was eliminated in the subsequent version by Watson *et al.* (2007), who specified coefficients using multiple linear regression. However, the objective version appears to suffer more from false positives. Other than in cliff terrain, false positives only tend to occur in the subjective version below about 7.5 W/m², whereas this detection criterion must be increased to about 15 W/m² to eliminate most of the false positives from the objective version.

III. STATISTICAL DISTRIBUTION OF GHF

It is helpful to understand the extent to which Yellowstone's wildlife ranges are geothermally affected. The existence of the non-migratory Madison headwaters elk herd is attributed to the presence of geothermal refugia in winter (Craighead *et al.* 1973). We constructed frequency distributions of GHF within the Madison headwaters elk range (Figure 4.8). The analysis provides quantitative support for the general equivalence of the subjective and objective maps (Figures 4.7A and B) because their estimates differ mainly above 60 W/m², which is inconsequential because snow packs do not form in

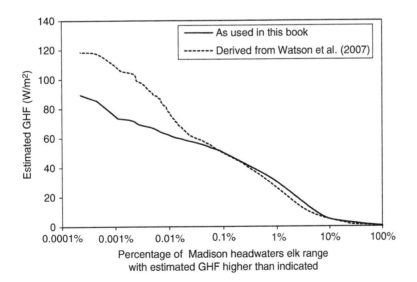

FIGURE 4.8 Cumulative frequency distributions of estimated GHF in the Madison headwaters elk range. The equivalence of two slightly different methods is examined. The method used in the book results in fewer false positives, but involves some subjectivity. The method derived from Watson *et al.* (2007) was completely objective and, thus, provides a useful verification that the subjective method does not suffer from undue bias.

such intensely geothermal areas (see Figure 4.4). One percent of the Madison headwaters elk range was estimated to have GHF higher than 30 W/m², which is the approximate threshold above which a snow pack does not form (Figure 4.4). Seven percent of the range was estimated to have a heat flux higher than 7.5 W/m², which is the range in which snow pack is reduced and false positives were minimal (*i.e.*, restricted to steep cliffs and talus slopes).

IV. EFFECT OF GHF ON SNOW PACK

We estimated the extent to which the snow pack is reduced or ablated by geothermal influences using the Langur snow pack simulation model (Chapter 6 by Watson *et al.*, this volume). We compared simulations of spatial patterns in SWE both with and without geothermal influences. The spatial distribution of SWE through the Madison headwaters elk range was estimated for March 1, 2006, just prior to peak snow pack accumulation. During simulation, the snow pack model maintained state variables representing the heat and water content of the soil and two snow pack layers. To simulate the geothermal influence on the water and energy balance of the snow pack-soil system, heat was continually added to the soil at a rate determined by the GHF map. Heat was transferred between soil and snow according to temperature gradients and the thermal conductance of the snow and ice mixture (Watson *et al.* 2007). Snow melt and sublimation were simulated in a manner that was sensitive to snow pack temperature. To simulate the snow pack without geothermal influence, all modeled processes were identical, except that the GHF was set to zero.

The results of this analysis of spatial sensitivity to GHF are shown in Figure 4.9, which plots both the frequency distribution and the cumulative frequency distribution of simulated SWE in the Madison headwaters elk range. Geothermal influence increased the snow-free area from 0.4% of the range to 5.9% of the range. Also, geothermal influences increased the area with SWE less than 15 cm from 14% to 23% of the range.

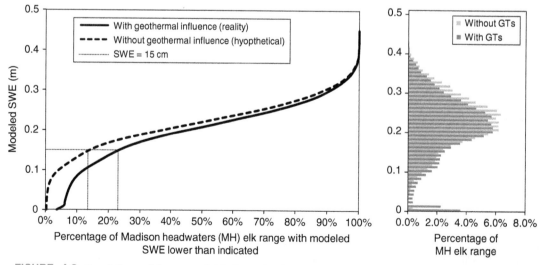

FIGURE 4.9 Cumulative and frequency distributions of estimated SWE within the Madison headwaters elk range under actual conditions on March 1, 2006 and under a hypothetical scenario with no geothermal influence.

V. DISCUSSION

The map developed in this chapter is the first calibrated, quantitative, digital layer representing the intensity of geothermal areas through Yellowstone National Park, and the first that represents GHF (W/m^2). Heasler *et al.* (2004) noted some early remote sensing investigations in the park using Landsat and other platforms. More recently, Jaworowski *et al.* (2006) used airborne spectral remote sensing to map surface temperature of a 104 km^2 area including the Norris Geyser Basin. Hellman and Ramsey (2004) used satellite and airborne sensors to map a suite of largely qualitative properties of hydrothermal basins in the Firehole Valley (<100 km^2). Our map estimates GHF over 6807 km^2, an area including the entire ranges of the central elk and bison herds and all the major geothermal areas in the park areas except Mammoth Hot Springs and Devil's Kitchen. Our calibration data included 18.5 km of ground survey routes around a total of 12 perimeters of six geothermal areas discretized into 6180 individual survey points.

The GHF map clearly discriminated the major geothermal areas of central Yellowstone and uniquely revealed the spatial pattern of differences in radiated geothermal heat both within and between geothermal areas. Each geothermal area contained spatially concentrated centers of activity from which the most heat was radiated on the date of scene acquisition (*i.e.*, March 25, 2000), usually more than 30 W/m^2 estimated GHF (Figure 4.7). GHF then declined along horizontal gradients outward from the centers, finally reaching 0 W/m^2 up to a kilometer from each center. Below about 7.5 W/m^2, the classification of a pixel as "geothermal" became ambiguous when viewed in isolation, but when these low-intensity pixels formed concentric spatial patterns around the major centers, the ambiguity was reduced and the great extent of subtle geothermal influence was more readily apparent.

FIGURE 4.10 The hottest hectare in Yellowstone National Park on March 25, 2000 was a feature we informally named "Satellite Springs" in the Sulfur Hills north of Pelican Valley. This 3-dimensional rendering is based on the GHF map described in this chapter (yellow and red coloring in foreground), with background terrain described from true color Landsat imagery and Landsat-based estimates of forest cover (Chapter 2).

Interestingly, the highest GHF was estimated not in the well-known intensely geothermal areas of the Firehole Valley or Norris Basin, but from an unnamed feature in the Sulfur Hills north of Pelican Valley that we informally named "Satellite Springs" (Figures 4.7 and 4.10). Here, estimated GHF reached 161 W/m^2 at one 28.5×28.5 m location and averaged 146 W/m^2 over a 114×114 m area.

From the wildlife perspective, it is the low-intensity geothermal areas that are perhaps most important. In central Yellowstone, geothermal influence converts the winter landscape from being more or less completely snow-bound (Figure 4.11B) to being a somewhat interconnected mosaic of snow-free or snow-reduced refugia (Figure 4.11A). In an average year at the time of peak snow pack, we estimated that 21 km^2 of the elk range was snow free, compared to 1 km^2 in the hypothetical case where geothermal influence was absent. A further 13 km^2 was estimated to be reduced below a SWE threshold of 15 cm as a result of geothermal influence. Thus, a total of approximately 34 km^2 of additional habitat was maintained through geothermal influence. About one-third of this habitat was in areas that were not obviously geothermal, but were distinguishable as geothermally influenced only by reduced snow pack and concentric spatial association with more intensely geothermal areas. Insights gained from quantification of these geothermal influences provided valuable insights into numerous investigations of large mammal dynamics such as ungulate distribution (Chapters 8 by Messer *et al.*, Chapter 12 by Bruggeman *et al.*, and Chapter 21 by White *et al.*, this volume), resource selection (Chapter 8 by Messer *et al.*, Chapter 9 by White *et al.*, Chapter 10 by Garrott *et al.*, and Chapters 21 and 22 by

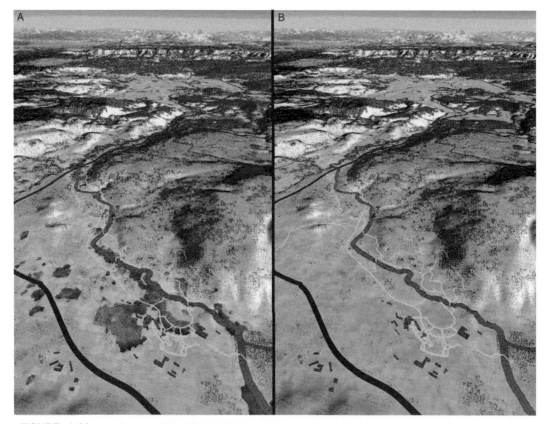

FIGURE 4.11 Visualization of the effect of GHF to create snow-free and reduced snow pack during winter in the Firehole Valley (looking north from Old Faithful). The left panel (A) shows modeled snow pack under actual conditions, while the right panel (B) shows modeled snow pack in the hypothetical absence of geothermal influence. White denotes snow, yellow denotes shallow snow, and pink denotes deep snow.

White *et al.*, this volume), and spatial dynamics (Chapter 18 by Gower *et al.* and Chapter 28 by Bruggeman *et al.*, this volume) in the presence and absence of predators.

The subtleties of geothermal influence are only observable as a result of mapping GHF as a continuous variable, as opposed to a categorical property as was the case with previous maps such as that of Shovic (1996). Our snow pack model sensitivity analyses showed that snow pack can accumulate on geothermal areas, but to a degree that is influenced both by the intensity of GHF, the amount of snow fall, and the air temperature. Thus, the snow-free and snow-reduced areas available to wildlife change within and between seasons, and only the more intense geothermal areas remain snow free during periods of heavy snow fall and cold air temperatures.

VI. SUMMARY

1. We produced the first calibrated, digital, map layer of the intensity of geothermal influence through Yellowstone National Park using Landsat ETM+ remote sensing of thermal emittance, after removing non-geothermal effects correlated with elevation and absorbed solar radiation.

2. We calibrated our map using field surveys of 18.5 km of perimeters around six geothermal areas on one to three dates each, and estimated GHF using snow pack model inversion.

3. Approximately seven percent of the Madison headwaters elk range was estimated to be distinctly geothermally affected (GHF > 7.5 W/m^2). This geothermal influence increases the snow-available (SWE < 15 cm) fraction of the range by 64% of the hypothetical non-geothermal amount (*i.e.*, from 14% to 23% of the range).

4. These results both quantify and underscore the postulate that geothermal activity is one of the most important influences on wildlife distribution and dynamics in central Yellowstone.

5. The geothermal map provided a key input to our snow pack simulation model (Chapter 6 by Watson *et al.*, this volume) which, in turn, was used to provide input to numerous wildlife analyses presented throughout this book.

VII. REFERENCES

Bergman, E. J., R. A. Garrott, S. Creel, J. J. Borkowski, F. G. R. Watson, and R. M. Jaffe. 2006. Assessment of prey vulnerability through analysis of wolf movements and kill sites. *Ecological Applications* **16**:273–284.

Blackwell, D. D., and M. Richards. 2004. Calibration of the AAPG geothermal survey of North America BHT data base. American Association of Petroleum Geologists, Annual Meeting, Dallas, TX. Paper 87616.

Bruggeman, J. E., R. A. Garrott, D. D. Bjornlie, P. J. White, F. G. R. Watson, and J. J. Borkowski. 2006. Temporal variability in winter travel patterns of Yellowstone bison: The effects of road grooming. *Ecological Applications* **16**:1539–1554.

Bruggeman, J. E., R. A. Garrott, P. J. White, F. G. R. Watson, and R. W. Wallen. 2007. Covariates affecting spatial variability in bison travel behavior in Yellowstone National Park. *Ecological Applications* **17**:1411–1423.

Christiansen, R. L. 2001. The Quaternary and Pliocene Yellowstone plateau volcanic field of Wyoming, Idaho, and Montana. U.S. Geological Survey Professional Paper 729-G.

Craighead, J. J., F. C. Craighead. Jr., R. L. Ruff, and B. W. O'Gara. 1973. Home ranges and activity patterns of nonmigratory elk of the Madison drainage as determined by biotelemetry. Wildlife Monographs 33.

Donegan, S. J., and L. P. Flynn. 2004. Comparison of the response of the Landsat 7 enhanced thematic mapper plus and the Earth observing-1 advanced land imager over active lava flows. *Journal of Volcanology and Geothermal Research* **135**:105–126.

Flynn, L. P., A. J. L. Harris, and R. Wright. 2001. Improved identification of volcanic features using Landsat 7 ETM+. *Remote Sensing of Environment* **78**:180–193.

Fournier, R. O. 1989. Geochemistry and dynamics of the Yellowstone National Park hydrothermal system. *Annual Review of Earth and Planetary Sciences* **17**:13–53.

Garrott, R. A., L. L. Eberhardt, J. K. Otton, P. J. White, and M. A. Chaffee. 2002. A geochemical trophic cascade in Yellowstone's geothermal environments. *Ecosystems* **5:**659–666.

Hall, D. K., G. A. Riggs, and V. V. Salomonson. 1995. Development of methods for mapping global snow cover using moderate resolution imaging spectroradiometer data. *Remote Sensing of Environment* **54:**127–140.

Heasler, H., C. Jaworowski, and D. Susong. 2004. *A Geothermal Monitoring Plan for Yellowstone National Park.* National Park Service, Mammoth, Wyoming http://www.esri.umt.edu/Research/CESU/NEWCESU/Assets/Individual%20Project%20Reports/NPS%20Projects/Wyoming/2003/Heasler_YNPGeothermal06Jan.pdf

Hellman, M. J., and M. S. Ramsey. 2004. Analysis of hot springs and associated deposits in Yellowstone National Park using ASTER and AVIRIS remote sensing. *Journal of Volcanology and Geothermal Research* **135:**195–219.

Jaworowski, C., H. P. Heasler, C. C. Hardy, and L. P. Queen. 2006. Control of hydrothermal fluids by natural fractures at Norris Geyser Basin. *Yellowstone Science* **14:**13–23.

Kaneko, T., and M. J. Wooster. 2005. Satellite thermal analysis of the 1986 Izu-Oshima lava flows. *Journal of Volcanology and Geothermal Research* **148:**355–371.

Lombardo, V., M. F. Buongiornoa, D. Pierib, and L. Meruccia. 2004. Differences in Landsat TM derived lava flow thermal structures during summit and flank eruption at Mount Etna. *Journal of Volcanology and Geothermal Research* **134:**15–34.

Mazzarini, F., M. T. Pareschi, A. Sbrana, M. Favalli, and P. Fulignati. 2001. Surface hydrothermal alteration mapping at Vulcano Island using MIVIS data. *International Journal of Remote Sensing* **22:**2045–2070.

Meagher, M. M. 1973. *The Bison of Yellowstone National Park.* National Park Service Scientific Monograph Series Number 1, National Park Service, Washington, DC.

Oppenheimer, C. 1997. Remote sensing of the colour and temperature of volcanic lakes. *International Journal of Remote Sensing* **18:**5–37.

Patrick, M., K. Dean, and J. Dehn. 2004. Active mud volcanism observed with Landsat 7 ETM+. *Journal of Volcanology and Geothermal Research* **131:**307–320.

Pickles, W. L., P. W. Kasameyer, B. A. Martini, D. C. Potts, and E. A. Silver. 2001. *Geobotanical Remote Sensing for Geothermal Exploration.* Proceedings of the Geothermal Resources Council, 2001 Annual Meeting, San Diego, CA.

Pieri, D., and M. Abrams. 2004. ASTER watches the world's volcanoes: A new paradigm for volcanological observations from orbit. *Journal of Volcanology and Geothermal Research* **135:**13–28.

Pollack, H. N., S. J. Hurter, and J. R. Johnson. 1991. Heat flow from the Earth's interior: Analysis of the global data set. *Reviews of Geophysics* **31:**267–280.

Shovic, H. F. 1996. *Landforms and Associated Surficial Materials of Yellowstone National Park.* National Park Service, Yellowstone National Park, Mammoth, WY. <http://www.nps.gov/gis/metadata/yell/yell_landform.html.>.

Smith, R. B., and L. W. Braile. 1994. The Yellowstone hotspot. *Journal of Volcanology and Geothermal Research* **61:**121–187.

Smith, R. B., and L. J. Siegel. 2000. *Windows into Earth: The Geological Story of Yellowstone and Grand Teton National Parks.* Oxford University Press, Oxford, United Kingdom.

Urai, M. 2002. Heat discharge estimation using satellite remote sensing data on the Iwodake volcano in Satsuma-Iwojima, Japan. *Earth Planets Space* **54:**211–216.

Vaughan, R. G., S. J. Hook, W. M. Calvin, and J. V. Taranik. 2005. Surface mineral mapping at Steamboat Springs, Nevada, USA, with multi-wavelength thermal infrared images. *Remote Sensing of Environment* **99:**140–158.

Watson, F. G. R., R. E. Lockwood, W. B. Newman, T. N. Anderson, and R. A. Garrott. 2007. Development and comparison of Landsat radiometric and snowpack model inversion techniques for estimating geothermal heat flux. *Remote Sensing of Environment* **112:**471-481.

Watson, F. G. R., W. Newman, J. C. Coughlan, and R. A. Garrott. 2006. Testing a distributed snowpack simulation model against spatial observations. *Journal of Hydrology* **328:**453–466.

CHAPTER 5

Effects of Wind, Terrain, and Vegetation on Snow Pack

Fred G. R. Watson,* Thor. N. Anderson,* Marc Kramer,[†] Jon Detka,*
Tad Masek,* Simon S. Cornish,* and Steve W. Moore*

*Division of Science and Environmental Policy, California State University Monterey Bay
[†]Earth and Planetary Sciences, University of California Santa Cruz

Contents

Theme

Snow pack and topography are widely recognized as factors influencing the abundance and distribution of resources for large herbivores in mid to high latitude environments. The influence of wind, however, and its interactions with snow and landscape variables, is poorly understood and difficult to describe but can also assume considerable importance. Wind alters the distribution of snow, thereby influencing the availability of resources and creating spatial patterns in foraging opportunities for large herbivores. In landscapes that exhibit undulating topography with alternating sequences of exposed and sheltered terrain, wind action can strongly

The Ecology of Large Mammals in Central Yellowstone
R. Garrott, P. J. White and F. Watson
ISSN 1936-7961, DOI: 10.1016/S1936-7961(08)00205-4

Copyright © 2009, Elsevier Inc.
All rights reserved.

influence snow pack distribution, with exposed terrain typically characterized by absent or shallow snow pack relative to the deeper snow pack of more sheltered terrain, and all areas strongly dependent on what lies upwind. Consequently, the mean snow depth over a large area is perhaps less important to habitat selection by large herbivores than the heterogeneity in snow pack caused by wind effects. Quantitative analyses of wildlife responses to wind effects on the snow pack first require quantification of the wind effects themselves. In central Yellowstone National Park wind strongly influences snow pack heterogeneity across the landscape, with large ungulates such as bison foraging heavily on exposed areas and establishing connecting trails through the deeper snow between them. We developed a model that allowed us to map spatial patterns in wind and wind effects on snow throughout central Yellowstone. The method combined separate terms for forest and terrain influences on wind. We incorporated this wind-effects model into our snow pack model (Chapter 6 by Watson *et al.*, this volume), which was then used as an input to several ecological analyses later in this book.

I. INTRODUCTION

Wind is a key influence on snow, which, in turn, is a key influence on the wildlife of central Yellowstone (Chapter 8 by Messer *et al.*, Chapter 12 by Bruggeman *et al.*, Chapter 16 by Becker *et al.*, and Chapter 28 by Bruggeman *et al.*, this volume). Wind redistributes fallen snow from exposed to sheltered places, enhances sublimation of snow (Fassnacht 2004), and tends to increase snow hardness and density by mechanically altering crystal shapes (Kind 1981). The snow pack restricts ungulate access to forage, and alters the locomotive energy requirements of both ungulates and their predators. The total amount of snow is often quantified in terms of snow water equivalent (SWE), which represents the combined influence of snow depth (in meters) and snow pack density (kg/m^3). SWE has been shown to influence ungulate vital rates (Garrott *et al.* 2003; Chapter 11 by Garrott *et al.*, this volume), nutrition (Chapter 9 by White *et al.*, this volume), migrations (Chapter 12 by Bruggeman *et al.*, this volume), and habitat selection (Chapter 8 by Messer *et al.* and Chapter 21 by White *et al.*, this volume) as well as vulnerability to predation (Bergman *et al.* 2006; Chapter 16 by Becker *et al.*, this volume) in central Yellowstone (Figure 5.1; Bruggeman *et al.* 2006, 2007). Our previous work using simulation models to

FIGURE 5.1 Wind reduces snow in exposed areas making forage available to ungulates in winter (Photo by Jeff Henry).

map SWE throughout central Yellowstone was limited by an inability to quantify spatial wind effects (Watson *et al.* 2006b).

The spatial patterns of wind effects on snow are commonly observed, yet poorly understood in quantitative terms that are applicable over large landscapes. During winter, readily visible cornices build up on the calm, lee side of steep ridges. In high wind areas, the shallowness of the snow is revealed by vegetation that is often seen protruding through the snow surface, and in the highest wind areas, a snow pack is absent. The patterns present during winter become most obvious in spring, when snow bound landscapes transform into a patchwork of snow-covered and snow-free areas (Figure 5.2). The visible alignment of these snow pack patterns is often related to the direction of the prevailing wind and the airflow patterns resulting from wind obstruction by terrain and vegetation.

Studies that have quantified and modeled the effects of terrain and vegetation on wind, and of wind on snow, have generally been limited to small spatial scales (<10 km^2) by theoretical and computational constraints (*e.g.*, Liston and Sturm 1998, Pomeroy and Li 2000, Winstral and Marks 2002, Erickson *et al.* 2005, Hiemstra *et al.* 2006, Lehning and Fierz 2008). Essery *et al.* (1999) achieved physically based modeling of airflow and snow dynamics over a 168 km^2 area by sub-dividing the area into 42 overlapping, 26-km^2 sub-areas. More empirical approaches have reached larger scales, such as the index-based approach of Purves *et al.* (1999), which was applied to a 300-km^2 mountainous area. The focal central Yellowstone study area for this book was 3959 km^2, and the modeling domain of our larger study area exceeded 14,000 km^2 (Chapter 12 by Bruggeman *et al.*, this volume). Our snow pack modeling resolution was typically 28.5 or 57 m, thus the scale and resolution was beyond the range of previous studies.

Our goal for this chapter was to develop a simple approach to mapping mean relative wind speed variation and wind-induced snow pack variation across the central Yellowstone study area. This approach was then extended to mapping wind and its effects on snow pack through the entire park, which, in turn, provided input to studies addressing the influence of snow pack on important wildlife populations.

FIGURE 5.2 Wind increases heterogeneity in snow pack depth in undulating terrain (*e.g.*, Sulphur Mountain, Hayden Valley; Photo by Fred Watson).

II. MODEL DESCRIPTION

We asserted that the two major spatial influences on wind in the study area are forest cover and terrain. We considered the forest influence first.

A. Forest

Based on our previous qualitative field observations, we developed the following postulate:

> The effect of forests to decrease wind speed at a point relative to a wide open area is proportional both to the density and proximity of forest cover.

A Gaussian smoothing model describes this effect:

$$F_{\mathbf{x}_0}(h_F) = \int\int C_{\mathbf{x}} K(|\mathbf{x} - \mathbf{x}_0| h_F^{-1}) dx\ dy, \quad \mathbf{x} = (x, y)$$

where $F_{\mathbf{x}_0}$ is an estimator of the forest effect (dimensionless), $C_{\mathbf{x}}$ is the percent forest cover at location $\mathbf{x}$, $|\mathbf{x} - \mathbf{x}_0|$ is the distance between $\mathbf{x}$ and $\mathbf{x}_0$, h_F is a scale parameter (in meters), and the function $K(x)$ is a standard bivariate Gaussian kernel defined as:

$$K(d) = \frac{1}{2\pi}\ \exp\left(-\frac{d^2}{2}\right)$$

A discrete approximation to this model is:

$$F_{\mathbf{i}_0}(h_F) = \frac{\sum_{\mathbf{i} \in D} C_{\mathbf{i}} K(|\mathbf{i} - \mathbf{i}_0| h_F^{-1})}{\sum_{\mathbf{i} \in D} K(|\mathbf{i} - \mathbf{i}_0| h_F^{-1})}$$

where D is the set of all raster cells $\mathbf{i}$ in a neighborhood of the cell of interest, $\mathbf{i}_0$, defined to be large enough such that the kernel was negligibly close to zero at its boundaries. Thus, our first hypothesis was that mean wind speed over some period of time, $W_{\mathbf{i}}$, at a pixel $\mathbf{i}$ is reduced from some reference value $W_{0,F}$ in proportion to the estimated forest effect, $F_{\mathbf{i}}$, in the absence of other effects (e.g., terrain):

$$H_F : W_i = \max\left(0, W_{0,F}\left(1 - \beta_F F_{\mathbf{i}}(h_F)\right) + \varepsilon_F\right)$$

with positive parameters $W_{0,F}$ (m/s), β_F (dimensionless), and h_F (m), and uncorrelated error ε_F.

B. Terrain

Based on our previous qualitative field observations, we developed the following postulate:

> The effect of terrain to increase wind speed at a point relative to a wide flat area is proportional to the protrusion of the point above the mean elevation of the surrounding area, at some scale.

A model that describes this effect is the difference between the elevation of a point and the Gaussian smooth of the surrounding elevation:

$$T_{\mathbf{x}_0}(h_T) = E_{\mathbf{x}_0} - \iint E_{\mathbf{x}} K(|\mathbf{x} - \mathbf{x}_0| h_T^{-1}) dx\ dy, \quad \mathbf{x} = (x, y)$$

where $T_{\mathbf{x}_0}$ is an estimator of the terrain effect (in meters) at two-dimensional location $\mathbf{x}_0$, $E_{\mathbf{x}}$ is the elevation of location $\mathbf{x}$, h_T is a scale parameter (in meters). The discrete approximation is:

$$T_{\mathbf{i}_0}(h_T) = E_{\mathbf{i}_0} - \frac{\sum_{\mathbf{i} \in D} E_{\mathbf{i}} K(|\mathbf{i} - \mathbf{i}_0| h_T^{-1})}{\sum_{\mathbf{i} \in D} K(|\mathbf{i} - \mathbf{i}_0| h_T^{-1})}$$

where D is the set of all raster cells $\mathbf{i}$ in a neighborhood of the cell of interest, $\mathbf{i}_0$, defined to be large enough such that the kernel was negligibly close to zero at its boundaries. Thus, our second hypothesis was that mean wind speed over some period of time is increased from some reference value, $W_{0,T}$, in proportional to the estimated terrain effect, T, in the absence of other effects (e.g., forest):

$$H_T : W_i = \max\left(0, W_{0,T}\left(1 + \beta_T T_{\mathbf{i}}(h_T)\right) + \varepsilon_T\right)$$

with positive parameters $W_{0,T}$ (m/s), β_T (1/m), and h_T (m), and uncorrelated error ε_T. Wind direction was not included in these hypotheses, though it was recognized as an influence that should be considered at some point in the future.

C. Combining Forest and Terrain Effects

We speculated that the terrain and forest effects are multiplicative and, as a result, a combined model of long-term mean winter wind speed in the study area is simply the product of our two empirical models:

$$W_{\mathbf{i},combined} = \max\left(0, W_{0,T}\left(1 + \beta_T T_{\mathbf{i}}(h_T)\right) + \varepsilon_T\right) \times \max\left(0, W_{0,F}\left(1 - \beta_F F_{\mathbf{i}}(h_F)\right) + \varepsilon_F\right)$$

D. Snow Pack

Based on our previous qualitative field observations, we developed the following postulate:

> In general, wind has a negative effect on SWE at any point due both to redistribution of blowing snow during and immediately after snowfall, and by sublimation thereafter.

A full exploration of cause and effect underlying this postulate is beyond the present scope. Of interest here was whether or not a correlation was observable and empirically estimable between SWE and wind. Thus, our third hypothesis was that SWE, $S_{\mathbf{i}}$, at a point relative to a reference regional mean, S_0, is inversely related to wind speed:

$$H_{S1} : \frac{S_{\mathbf{i}}}{S_0} = \max(0, \alpha_{S1} - \beta_{S1} W_{\mathbf{i}} + \varepsilon_{S1})$$

with parameters α_{S1} (dimensionless), β_{S1} (s/m), and uncorrelated error ε_{S1}.

Wind-driven transport of fresh snow is generally understood to occur above an incipient motion threshold of approximately 4 m/s (Kind 1981; measured at 0.5 m above ground level). Thus, we also considered a hypothesis that accounts for threshold effects according to a continuous function that decreases at a rate, β_{S2} (s^2/m^2), parabolically from unity above a threshold wind speed parameter, W_B (m/s):

$$H_{S2} : \frac{S_i}{S_0} = \max\left(0, 1 - \beta_{S2}\left(\max(0, W_i - W_B)\right)^2 + \varepsilon_{S2}\right)$$

III. METHODS

We examined our hypotheses using linear regression of observed wind speed against terrain effects, forest effects, and snow pack. Wind speed was measured using anemometers, while snow pack was measured using corers. Field sampling was stratified by forest effects and terrain effects to ensure a uniform distribution of the independent variables in each regression. We also stratified sampling by geographic location to minimize bias to a localized portion of the study area. Observations of mean wind speed and snow pack were made at multiple sites, typically over multiple days, as described below. Wind speed observations were averaged over the duration of each sampling design (*i.e.*, 19 hours to 18 days as determined by anemometer logging capabilities that improved during the project). Temporal variation in spatial effects was not examined, but assumed to be zero for the purpose of combining results from different spatial sampling designs.

A. Sampling Design for Forest Effects

Forest effects were sampled using a stratified subjective design (Figure 5.3). We first produced a map of the hypothesized forest effect, $F_i(h_F)$, based on the forest cover map described in Chapter 2 by Newman and Watson, this volume, under an initial assumption of $h_F = 100$ m. This assumption was based on our qualitative observations that as one moves from a dense forest out into an open meadow, the increase in wind speed levels off once the forest is approximately 100–200 m distant. We then visually examined this map and selected two sub-areas that exhibited a wide range in $F_i(h_F)$, and minimal influence of non-forest effects such as terrain. The first sub-area covered a fragmented boundary between the non-forested meadows of the Hayden Valley and the adjacent forested terrain to the south (Figure 5.3A). The second sub-area covered a similarly fragmented boundary between Gibbon Meadow and the surrounding forests (Figure 5.3B). Sampling points were subjectively located within these areas to represent a range of situations, such as a range of different sized clearings in forest, and a range of different sized forest patches in a meadow. For logistical reasons, the two sub-areas were sampled sequentially. Twenty-nine anemometers were deployed in the Hayden sub-area for 19 h from March 15–16, 2003, and 24 anemometers were deployed in the Gibbon sub-area for three days from March 21–24, 2003. To allow data from these nonoverlapping sampling periods to be analyzed as one data set, we used data from a nearby long-term weather station (described below) to re-scale all observed wind speeds, W_i, as estimates of the long-term mean, W'_i:

$$W'_i = W_i \frac{W_{ref,LT}}{W_{ref,SP}}$$

where $W_{ref,SP}$ and $W_{ref,LT}$ denote mean wind speed at the long-term reference site, respectively, during the sampling period and over the long-term during the winter months (November–March). This re-scaling assumed temporal invariance of spatial patterns. Additionally, two outliers were removed from the Hayden data after being noted during field deployment as sites where small-scale terrain features

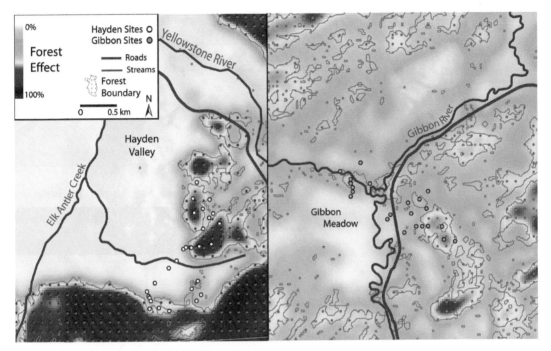

FIGURE 5.3 Sampling sites for forest effects on wind (circles) were located to sample a range of conditions from open exposed meadows to sheltered calm forest, as indicated by modeled forest effect.

funneled the wind to a degree that contradicted our intent to control for non-forest effects in this particular sampling design.

B. Sampling Design for Terrain Effects

We implemented a stratified random sampling design to sample terrain effects within the Hayden Valley (Figure 5.4). The Hayden Valley is an undulating, mostly unforested area about 8 km wide. Stratification was implemented in two stages. The primary stratification variable was a map of predicted wind speed for the Hayden Valley estimated using the Overflow computational fluid dynamics model (Jespersen *et al.* 1997) for 0.5 m above ground level based on a westerly regional incoming wind speed at 29 km/h. We stratified our sampling to distribute an even number of samples within each of six wind speed classes that we derived from this map (cutoff points: 16.0, 22.5, 29.0, 35.4, and 41.8 km/h). The secondary stratification variable was a map of geographically separate regions of the Hayden Valley, which we used to distribute sampling through the valley while ensuring some clustering of sampling sites to facilitate logistics (Figure 5.5). This map identified five circular sub-areas within the Hayden Valley where a variety of unforested terrain (*e.g.*, hilltops, gullies, flats) occurred within a relatively small radius (between 400 and 1250 m) of a central logistical staging point. We randomly located sampling points at least 100 m apart within each of the strata defined by these two variables. This resulted in 51 sites sampled, with between 5–8 sites per wind class and 4–21 sites per sub-area. The design was executed with the successful deployment of 51 anemometers for 18 days between 26 March and 13 April, 2005.

We normalized the observed wind speeds to long-term equivalents as described above for the sampling of vegetation effects. We also corrected for forest effects. Though the sampling sites for

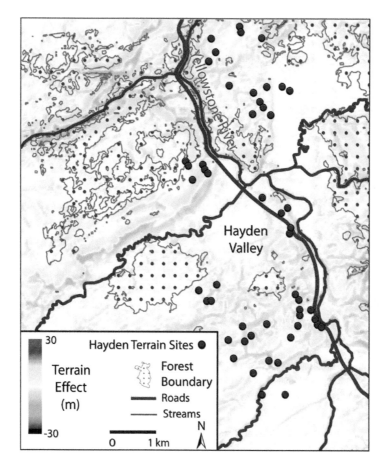

FIGURE 5.4 Sampling sites for terrain effects on wind (circles), were located to sample a range of conditions from exposed high points to sheltered low points, as indicated by modeled terrain effect.

terrain effects were selected *a priori* to be in non-forested terrain, clumps of trees occurred near some of the sites. We corrected the wind speed data to yield long-term non-forested (NF) equivalents, $W'_{i,NF}$:

$$W'_{i,NF} = \frac{W'_i}{1 - \beta_F F_i(h_F)}$$

where $W_{0,F}$, β_F, and h_F are the vegetation effect parameters defined earlier, and fitted as described below.

C. Anemometers

We designed and built 60 compact computerized logging anemometers to meet the needs of the study (Figure 5.6). The anemometers were designed to log a histogram of wind speed every 15 min (based on recordings made every five seconds), to be sensitive and accurate in the range 1–30 m/s, portable enough for a pair of field staff to be able to transport 20 of them on foot over snow, and resilient to

FIGURE 5.5 Hauling snow and wind sampling equipment to a remote staging location (Pelican Valley) (Photo by Fred Watson).

FIGURE 5.6 We designed and built 60 logging anemometers that could be deployed at remote locations to log wind speed for the following few weeks in inclement weather (Photo by Fred Watson).

moisture, freeze–thaw cycles, and inspection by wildlife. The sensing elements were 3-cup assemblies supplied by Davis Instruments (Hayward, California), mounted on bearings that we customized from lightweight DC motors, secured within a custom PVC housing, suspended inverted from steel tripods 0.5 m above the snow surface (Figure 5.6).

Some of the anemometers were withheld from operational use and used for testing accuracy and resistance to stress. Some were stress tested by mounting them to a vehicle and driving 5000 km at highway speeds. Others were tested during trial field deployments in conditions ranging from blizzard to spring freeze–thaw, rain-sun cycles. A final set of anemometers were used to derive a calibration curve for the remaining units by mounting them in a clean air stream on a pole extending 4 m forward from a vehicle and driving an accurately measured 645 m section of road several times in both directions on still mornings. As a result of these steps, the accuracy of the anemometers was measured to be ±10% in the range 1.5–21 m/s.

D. Temporal Normalization

We used a long-term wind recording station to normalize certain wind speed observations to compensate for the influence of regional temporal wind speed variation on our results, which were obtained during nonoverlapping, relatively short sampling periods. Long-term wind monitoring stations were scarce in the region. The closest station to the study area was an aviation weather reporting station operated by the National Oceanic and Atmospheric Administration (Station Identification KP60) near Lake Village, 7 km south of the southern edge of the study area. The station was located in a maintenance area surrounded by dense forest 12 m high. The anemometer was atop a mast approximately 10 m high. We compiled a wind rose for the station (*i.e.*, a radially plotted frequency histogram of wind direction and speed), which clearly revealed prevailing strong winds (>4 m/s) from the west, and some lighter winds (<3 m/s) from other directions that we judged to be due to local terrain. Specific normalization steps are described where they were used.

E. Snow Pack Coring

At each of the 104 anemometer deployment sites, nine snow pack cores were taken in a nested pattern of 10- and 1-m equilateral triangles centered on the anemometer. The snow at the anemometer itself was never sampled, because the coring activities would interfere with the observed wind until new snowfall or blowing snow smoothed out the disturbed area (Figure 5.7). The nested pattern maximized the accuracy of estimates of the mean snow pack within 10 m of the anemometer itself, relative to sampling effort (Watson *et al.* 2006a). The anemometer site was always approached from the prevailing down-wind direction and the anemometer was deployed using a telescopic boom, such that the anemometer itself and the upwind 120° perspective remained undisturbed by ski or snowshoe tracks.

Each snow pack core was taken using a standard Federal-type corer and weighed on a spring balance to yield a SWE measurement (in meters). SWE measurements for the nine cores at each site were averaged to yield a single estimate, S_i, for the site. To allow comparability between the three field deployment campaigns, the SWE estimates were normalized to a unit mean "relative SWE" (RSWE, dimensionless) by dividing each estimate by the mean of all estimates for that campaign, S_0.

F. Statistical Methods

None of the models were generalized linear models because they each included covariates whose definition, in turn, involved parameters (h_F, h_T, and W_B). Therefore, we used nonlinear optimization methods to estimate model parameters. Kernel computations involved significant spatial data processing that had to be accomplished outside standard statistical packages. For model H_F, we varied the kernel smoothing parameter h_F from 10 to 10,000 m in 10% increments, re-computed the forest effect at each sampling site, and used simple linear regression to estimate $W_{0,F}$ and β_F. We selected the value of h_F that yielded the maximum coefficient of determination (R^2) between observed wind speed (W'_i, long-term normalized) and $F_i(h_F)$. The same procedure was used in model H_T to estimate h_T, $W_{0,T}$, and β_T fitted to the normalized, long-term, non-forested, observed wind speed, $W'_{i,NF}$. For the RSWE models, H_{S1} and H_{S2} we estimated parameters α_{S1}, β_{S1}, β_{S2}, and W_B using the nonlinear modeling function *nlm()* in the *R* statistical package (R Core Development Team 2006). We compared the models

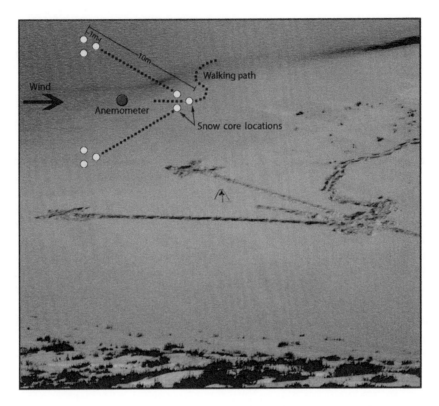

FIGURE 5.7 Anemometers were deployed so that the snow was undisturbed upwind of the instrument. The snow pack was cored at three locations in each of three areas surrounding the instrument, but not at the instrument itself. In the deployment shown, sastrugi indicated the prevailing wind direction was from the upper left. The tracks in the photo are those made by a sampling team wearing snowshoes.

to each other and a null model (*i.e.*, RSWE = *k*, where *k* is a parameter) using AIC methods (Burnham and Anderson 2002), drawing on an assumption of normally distributed residuals that we validated using Q–Q normal plots. We fitted the RSWE models using observed wind speed normalized to yield an estimate of the mean wind speed of the snow pack season to date, W_i'', since it is during this period that the observed snow pack has been subjected to wind effects:

$$W_i'' = W_i \frac{W_{\mathrm{ref,STD}}}{W_{\mathrm{ref,SP}}}$$

where $W_{\mathrm{ref,STD}}$ denotes the mean wind speed at the long-term reference site during the season to date (starting from November). We combined all sampling data (from forests effects sampling and terrain effects sampling) to fit the RSWE models.

IV. RESULTS

Our results are based on over 24,000 h of wind data collected using custom-designed anemometers deployed at 104 sites in the Gibbon and Hayden Valleys. The wind measurements were compared with snow pack measurements derived from 936 snow cores. Technological resources that we used included custom-developed Landsat remote sensing measurements of forest cover, U.S. Geological Survey terrain data, custom-coded regression and mapping software for optimization of kernel smoothing parameters and, to aid in the development of the stratified sampling design, 7000 h of CPU time on a NASA super-computer running the Overflow model.

A. Wind Speed Correlated with Modeled Terrain and Forest Effects

The correlation between long-term, normalized, observed wind speed and modeled terrain and forest effects was clearly sensitive to the scaling parameters (Figure 5.8), indicating that these effects are scale-dependent. The optimized scaling parameters were similar in value at 108 m for forest effects and 119 m for terrain effects. A linear relationship was evident in the optimal correlations, and thus hypotheses H_F and H_T were strongly supported by the data ($P < 0.001$; Figure 5.9). Conventional regression assumptions were upheld. Stratification assured uniformity in the independent variable, and there was no suggestion of skewed or heteroskedastic residuals. The fitted parameter values were: $W_{0,F} = 1.328$ m/s, $\beta_F = 1.624$, $h_F = 108$ m, $W_{0,T} = 5.243$ m/s, $\beta_T = 0.06211$/m, $h_T = 119$ m. The combined model of long-term mean winter wind speed given both forest and terrain influences was:

$$W_{i,\text{combined}} = \max\left(0, 5.243\left(1 + 0.06211\,T_i(119)\right)\right) \times \max\left(0, 1.328\left(1 - 1.624 F_i(108)\right)\right)$$

The long-term wind speed predicted by this model is mapped in Figure 5.10.

B. Snow Pack Correlated with Wind Speed

A clear nonlinear correlation was evident between RSWE and observed wind speed expressed as a season-to-date estimate (W_i''; Figure 5.11). The threshold–parabolic model (H_{S2}) was supported by the data to a far greater degree than the linear (H_{S1}) or constant models (AIC weights for the three models were 1.00, 0.00, and 0.00, respectively). The fitted parameter values were $\beta_{S2} = 0.03198$ s^2/m^2 and $W_B = 2.889$ m/s. Reasonable correlation was also observed between modeled wind speed and RSWE (Figure 5.12).

V. DISCUSSION

We achieved our goal of developing a simple method for mapping spatial patterns of wind speed induced by forest and terrain effects in heterogeneous landscapes, and correlating the variation in wind with variation in snow pack properties. Our methods were far simpler than those requiring a physically based model of air flow (Pomeroy and Li 2000). The wind speed correlations with forest and terrain were highly significant, and snow pack correlation with wind was strongly supported over a null model. We established the validity of these results with multiple, intensive, field sampling designs extending over many square kilometers of heterogeneous landscape. While detailed spatial sampling of snow pack properties has been used to support studies of wind effects on snow (Winstral *et al.* 2002), we are not aware of such detailed spatial measurements of wind speed itself in large landscapes. The model was simple enough that we could easily apply it to our largest 14,000 km^2 study area at 28.5 or 57 m resolution using a personal computer and only about 30 lines of code. This is a larger scale and simpler method than previous efforts of its kind (Purves *et al.* 1999, Hiemstra *et al.* 2006).

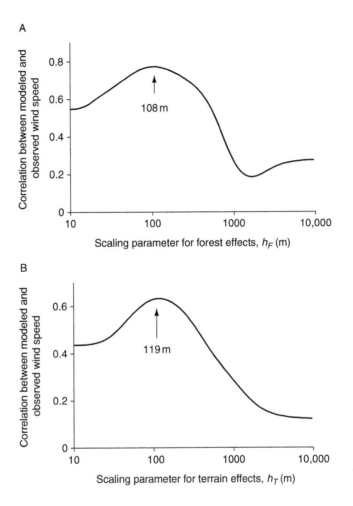

FIGURE 5.8 Calibration of scaling parameters for (A) terrain effect, and (B) forest effect. A clear scale-dependency in these effects is apparent, and the scale of the effect is similar for forest and terrain effects.

The optimal scaling parameters were similar in value (*i.e.*, 119 m for terrain effects and 108 m for forest effects), suggesting terrain and forests interfere with airflow in similar ways. Our analyses of forest effects and terrain effects were completely independent of each other. Observations were made in different years, in different locations. The spatial data on terrain patterns and forest patterns were derived by independent means, and the patterns themselves bore no obvious resemblance. The parameters were optimized within the range 10 m to 10,000 m, but both converged to just over 100 m. This suggests a characteristic length scale of about 100 m for spatial variance in wind speed, within a range up to a few kilometers (the extent of our sampling designs) regardless of the nature of the obstruction (forest or terrain). At longer scales, such as between entire mountains and valleys, we would expect additional variation and, perhaps, additional characteristic length scales in the 10–100 km range. Thus, it would be worthwhile to conduct a similar field effort over the entire park with much wider spacing between anemometers.

The resulting map of spatial patterns in mean, long-term, wind speed (Figure 5.10) captures the patterns we postulated. Large parts of the study area are densely forested, with almost no wind near the snow surface. However, the 1988 fires created large swaths of regenerating forest (Chapter 2 by

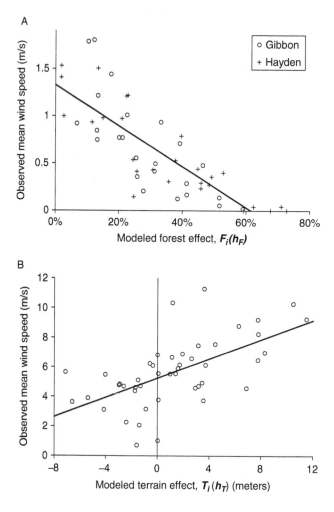

FIGURE 5.9 Correlation between observed mean wind speed and modeled terrain and forest effects. Wind speed values in (A) are expressed as long-term estimates (W_i') and, also, corrected in (B) to non-forest equivalents ($W_{i,NF}'$).

Newman and Watson, this volume) that are now more exposed to the wind, but generally not so exposed that significant ablation would occur (*i.e.*, <5 m/s). In contrast, some of the meadow complexes have many areas with wind speeds easily high enough to reduce the snow pack (*i.e.*, >7 m/s). Most notable are the central and western Hayden Valley and Cougar Meadow, 10 km WNW of Madison Junction. The Firehole and Norris meadow complexes are flatter, with fewer exposed areas. The most exposed terrain is on the rims of the canyons near Madison Junction and Canyon, and the steeper lava flow margins (*e.g.*, west of the lower Firehole Valley). These areas are often forested, but are predicted to be very windy nevertheless, and are often close enough to the meadow complexes to be potentially important foraging patches and movement routes for bison, elk, and wolves.

Our results indicate correlation, but do not indicate the cause of snow pack ablation observed at high wind speeds. They are perhaps most obviously consistent with an incipient motion hypothesis, whereby snow crystals lie immobile until a threshold wind speed is reached, whence they saltate away downwind and result in net scour in high wind speed areas (Kind 1981). A concurrent hypothesis is that the observed effect is at least partly due to a dependence of sublimation on wind speed (Fassnacht 2004).

FIGURE 5.10 Map of modeled spatial variation in long-term mean wind speed. Dark areas of low wind speed correspond to forests. White areas typically represent meadow complexes. Red areas represent exposed hill tops and canyon rims. Inset map shows three-dimensional perspective looking west from the eastern edge of the Hayden Valley.

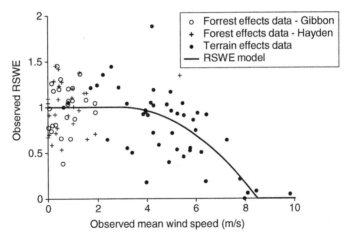

FIGURE 5.11 Correlation between observed relative snow water equivalent (RSWE) and observed mean wind speed (expressed as season-to-date estimate, W_i''). Snow is only predictably reduced above a certain wind speed threshold.

The ablation effect we observed is entirely due to the sampling of terrain effects, as opposed to the sampling of forest effects. The design for sampling forest effects was deployed in flat areas and resulted in wind speeds that were typically somewhat lower than the threshold wind speed. Thus, we observed no clear correlation between SWE and wind within our forest-effects data. This suggests that a forest effect on ablation would only manifest in forested areas that were considerably more terrain-exposed

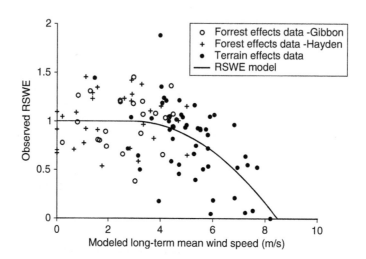

FIGURE 5.12 Correlation between observed RSWE and modeled long-term mean wind speed ($W_{i,combined}$). This correlation illustrates the accuracy with which we can simultaneously model spatial patterns in wind and their effect on snow pack.

than the areas we sampled. In this initial study, we deliberately avoided terrain-exposed forests to eliminate terrain effects while we were sampling forest effects.

Snow deposition effects were rare in our data. In the field, we observed physical evidence of wind-driven snow pack deposition at one or two of our 104 sampling sites, in the form of cornices. A depositional effect would manifest in our data as large RSWE at low wind speeds. However, the general pattern in the data was constant RSWE at low wind speeds, albeit with a large amount of unexplained variance. Depositional effects are either so localized that they are effectively rare outliers in the overall landscape, or are correlated with something other than the nondirectional wind speeds that we measured. Both explanations are likely because our unbiased sampling of terrain effects yielded many more field observations of ablation effects than depositional effects. Also, cornices are very clearly dependent on wind direction, which we did not attempt to model.

Overall, these results constitute a significant step toward improving our ability to quantify wildlife habitat relationships in Yellowstone. Bison and elk seek out low snow areas to access forage with minimal energetic costs (Chapter 8 by Messer *et al.*, Chapter 21 by White *et al.*, and Chapter 28 by Bruggeman *et al.*, this volume). Some of the most obvious low snow areas in the winter landscape are due to readily predictable wind effects, yet our previous snow pack modeling was not able to account for this. We now have a means of capturing the major increases in wind speed that occur primarily due to terrain exposure, but moderated by forest cover. Also, we can predict which specific 28.5 m pixels in a 14,000 km^2 landscape are likely to have reduced snow cover due to increased wind speed. Thus, if wildlife spatial data reveal animals are using those specific locations, then we have a quantitative explanation. While our data indicate that there are still places where we may incorrectly predict wind scouring or vice versa, such instances are much less likely under the models we developed. We can also predict large-scale reductions in snow cover in meadow complexes, such as the Hayden Valley, with a high proportion of exposed undulating terrain as compared to areas such as the Firehole Valley. The existence of such large-scale effects has not been documented at Yellowstone, but is consistent with blowing snow models in other areas (Bowling *et al.* 2004). We incorporated the map of long-term, mean, wind speed (Figure 5.10) and the ablation model into our snow pack model (Chapter 6 by Watson *et al.*, this volume), which, in turn, provided key inputs to many wildlife analyses through the remainder of the book.

VI. SUMMARY

1. Wind is a driver of ecological processes in central Yellowstone because it influences the accumulation of snow pack. Strong winds reduce snow pack in many locations, providing opportunities for ungulate foraging in winter. Spatial modeling of wind patterns and effects provides an important explanatory variable in wildlife analyses that seek to explain population and movement dynamics and habitat preferences.

2. We measured wind speed at 104 locations using continuously logging anemometers deployed for up to 18 consecutive days for a total of over 24,000 h of wind speed data. We measured snow pack properties using 936 cores at the same locations as the anemometers.

3. A simple model of terrain exposure effect with a characteristic horizontal length scale of 119 m explained a highly significant ($P < 0.001$) amount of the variation in wind speed in unforested portions of the Hayden Valley in central Yellowstone. At low terrain exposure, the terrain was never sheltered enough to completely preclude measurable air flow. At high terrain exposure, the increase in wind speed either completely ablated or precluded the formation of the snow pack.

4. A simple model of remotely sensed forest sheltering effect with a characteristic horizontal length scale of 108 m explained a highly significant ($P < 0.001$) amount of the variation in wind speed in low relief portions of Hayden Valley and Gibbon Meadows in central Yellowstone. Forest sheltering was sufficient to completely preclude measurable air flow at two sites.

5. Variations in wind speed resulted in a progressive influence on snow pack from nil to a threshold wind speed of approximately 2.9 m/s, after which ablation began to occur and was complete at the highest wind speeds (≥ 8.5 m/s). This threshold model was fully supported ($AIC_w = 1$) over non-threshold and null models.

6. Combining the models provides a computationally simple approach to estimating spatial patterns of wind speed and wind-ablated snow pack over large heterogeneous landscapes (at least 14,000 km^2).

7. Our models quantify one of the most important sources of snow pack heterogeneity. They were incorporated into our snow pack simulation model (Chapter 6 by Watson *et al.*, this volume), which was used to quantify snow as a spatial explanatory variable in many of the ecological analyses presented in this book.

VII. REFERENCES

Bergman, E. J., R. A. Garrott, S. Creel, J. J. Borkowski, F. G. R. Watson, and R. M. Jaffe. 2006. Assessment of prey vulnerability through analysis of wolf movements and kill sites. *Ecological Applications* 16:273–284.

Bowling, L. C., J. W. Pomeroy, and D. P. Lettenmaier. 2004. Parameterization of blowing-snow sublimation in a macroscale hydrology model. *Journal of Hydrometeorology* 5:745–762.

Bruggeman, J. E., R. A. Garrott, D. D. Bjornlie, P. J. White, F. G. R. Watson, and J. Borkowski. 2006. Temporal variability in winter travel patterns of Yellowstone bison: The effects of road grooming. *Ecological Applications* 16:1539–1554.

Bruggeman, J. E., R. A. Garrott, P. J. White, F. G. R. Watson, and R. Wallen. 2007. Spatial variability in Yellowstone bison use of a road and travel network. *Ecological Applications* 17:1411–1423.

Burnham, K. P., and D. R. Anderson. 2002. *Model Selection and Multi-Model Inference.* Springer, New York, NY.

Erickson, T. A., M. W. Williams, and A. Winstral. 2005. Persistence of topographic controls on the spatial distribution of snow in rugged mountain terrain, Colorado, United States. *Water Resources Research* 41:W04014.

Essery, R., L. Li, and J. Pomeroy. 1999. A distributed model of blowing snow over complex terrain. *Hydrological Processes* 13:2423–2438.

Fassnacht, S. R. 2004. Estimating alter-shielded gauge snowfall undercatch, snowpack sublimation, and blowing snow transport at six sites in the coterminous USA. *Hydrological Processes* 18:3481–3492.

Garrott, R. A., L. L. Eberhardt, P. J. White, and J. Rotella. 2003. Climate-induced variation in vital rates of an unharvested large herbivore population. *Canadian Journal of Zoology* **81**:33–45.

Hiemstra, C. A., G. E. Liston, and W. A. Reiners. 2006. Observing, modelling, and validating snow redistribution by wind in a Wyoming upper treeline landscape. *Ecological Modeling* **197**:35–51.

Jespersen, D. C., T. H. Pulliam, and P. G. Buning. 1997. *Recent enhancements to OVERFLOW*, AIAA-97-0644, AIAA Aerospace Sciences meeting, January 1997, Reno, NV.

Kind, R. J. 1981. Snow drifting. Pages 338–359 *in* D. M. Gray and D. H. Male (Eds.) *Handbook of Snow: Principles, Processes, Management and Use.* Pergamon, Ontario, Canada.

Lehning, M., and C. Fierz. 2008. Assessment of snow transport in avalanche terrain. *Cold Regions Science and Technology* **51**:241–242. doi:10.1016/j.coldregions.2007.05.012.

Liston, G. E., and M. Sturm. 1998. A snow-transport model for complex terrain. *Journal of Glaciology* **44**:498–516.

Pomeroy, J. W., and L. Li. 2000. Prairie and arctic areal snow cover mass balance using a blowing snow model. *Journal of Geophysical Research* **105**:26619–26634.

Purves, R. S., W. A. Mackaness, and D. E. Sugden. 1999. An approach to modelling the impact of snow drift on glaciation in the Cairngorm Mountains, Scotland. *Journal of Quaternary Science* **14**:313–321.

R Development Core Team. 2006. R: A language and environment for statistical computing. R Foundation for Statistical Computing, Vienna, Austria<http://www.R-project.org>.

Watson, F. G. R., T. N. Anderson, W. B. Newman, S. E. Alexander, and R. A. Garrott. 2006a. Optimal sampling schemes for estimating mean snow water equivalents in stratified heterogeneous landscapes. *Journal of Hydrology* **328**:432–452.

Watson, F. G. R., W. B. Newman, J. C. Coughlan, and R. A. Garrott. 2006b. Testing a distributed snowpack simulation model against diverse observations. *Journal of Hydrology* **328**:453–466.

Winstral, A., K. Elder, and R. E. Davis. 2002. Spatial snow modeling of wind-redistributed snow using terrain-based parameters. *Journal of Hydrometeorology* **3**:524–538.

Winstral, A., and D. Marks. 2002. Simulating wind fields and snow redistribution using terrain-based parameters to model snow accumulation and melt over a semi-arid mountain catchment. *Hydrological Processes* **16**:3585–3603.

CHAPTER 6

Modeling Spatial Snow Pack Dynamics

Fred G. R. Watson, Thor N. Anderson, Wendi B. Newman, Simon S. Cornish, and Thomas R. Thein

Division of Science and Environmental Policy, California State University Monterey Bay

Contents

Theme

Ecological processes are strongly influenced by climate drivers as well as by the landscape in which they are embedded, with the interactions between these influences resulting in complex phenomena. It is imperative to have a strong understanding of these attributes of an ecosystem as a foundation for ecological investigations. In mid- to high-latitude environments snow pack may be present for five to eight months each year and represents an abiotic attribute of fundamental importance to large mammal ecology. Capturing the dynamic heterogeneity of snow pack at temporal and spatial scales relevant to large mammal communities is extremely challenging. Thus, most studies of large mammal ecology have either ignored snow pack dynamics or used surrogates for snow pack such as elevation and aspect. However, snow pack attributes are determined by many influences. The large-scale orographic influence of mountain ranges and valleys determines the overall distribution of precipitation and, hence, snow water equivalent. But many influences such as solar radiation, dense forest cover, and exposure to high winds act at highly localized scales to reduce snow pack accumulation. The Madison headwaters system of central Yellowstone National Park provides a good template for evaluating these interactions, as it is a highly heterogeneous landscape experiencing heavy snow pack (Figure 6.1) and is strongly influenced by geothermals and high winds. We described the integration of climate and landscape databases, as well as rigorous field sampling of snow pack, to develop and validate a spatially distributed, physically-based snow pack simulation model. We used this model to provide a variety of snow pack metrics for 16 of the Chapters in this book (Chapter 8–11, 13, 16–17, 18–23, and 27–29).

The Ecology of Large Mammals in Central Yellowstone
R. Garrott, P. J. White and F. Watson
ISSN 1936-7961, DOI: 10.1016/S1936-7961(08)00206-6

Copyright © 2009, Elsevier Inc.
All rights reserved.

I. INTRODUCTION

In high-latitude systems, snow pack is arguably the primary abiotic factor influencing large mammal systems. Present for five to eight months of the year, snow pack severely reduces the availability of forage for large herbivores while significantly increasing the energetic costs of foraging and traveling (Parker *et al.* 1984, Fuller 1991, Chapter 8 by Messer *et al.*, Chapter 9 by White *et al.*, Chapter 27 by Bruggeman *et al.*, this volume). Such costs drive shifts in seasonal distribution of ungulate prey populations, (*e.g.* concentrating prey in response to diminishing per capita forage availability or prompting migration, Chapter 8 by Messer *et al.*, Chapters 12 and 28 by Bruggeman *et al.*, this volume), and often result in high winter mortalities due to starvation (Sæther 1997, Jędrzejewski *et al.* 1992, Garrott *et al.* 2003, Chapter 9 by White *et al.*, Chapter 11 by Garrott *et al.*, this volume). In addition, snow pack can assume considerable importance in predator-prey interactions by impeding escape and/or weakening prey animals, both of which can substantially increase an individual's vulnerability to predation (Peterson 1977, Nelson and Mech 1986, Huggard 1993, Bergman *et al.* 2006, Chapters 16 and 17 by Becker *et al.*, Chapter 18 by Gower *et al.*, Chapter 21 by White *et al.*, this volume). Thus, while primary productivity is essential to support herbivores, the direct influence of plants on large herbivore dynamics and herbivore-predator dynamics can also be strongly mediated by the spatial and temporal variability of snow pack in large mammal systems (Chapter 16 by Becker *et al.*, Chapter 18 by Gower *et al.*, Chapter 21 by White *et al.*, Chapter 24 by Garrott *et al.*, this volume).

There are various approaches to estimating snow pack characteristics that differ in spatial scale, reliability, and accuracy. These include snow pack telemetry stations, direct remote sensing, and spatial simulation using landscape characteristics. Snow pack is typically measured using snow water equivalents (*i.e.*, the amount of water in a column of snow; snow water equivalent (SWE)). This is seen as a more accurate integrator of snow pack characteristics than snow pack depth because it combines both snow pack depth and density. The U.S. Department of Agriculture operates the Snow Telemetry (SNOTEL) system, which is a network of remote, automatic, monitoring stations that yield online daily measurements of SWE, precipitation, temperature and, more recently, snow depth. There are nine SNOTEL stations in Yellowstone National Park, including two in the central Yellowstone area. These stations

FIGURE 6.1 Snow pack accumulation in the Yellowstone region can be extremely deep in heavy winters. Here, snow is being cleared from the roof of a cabin (Photo by Jeff Henry).

provide the best information on temporal variations in the geographic vicinity of the SNOTEL sites themselves, but their data are not indicative of spatial patterns of snow across a landscape. It is even debatable whether SNOTEL data should be used to estimate spatial averages for specific wildlife ranges in Yellowstone (*e.g.*, Gates *et al.* 2005), without taking account of the distinct differences between ungulate winter ranges with respect to key influences on the snow pack, such as geothermal intensity (Chapter 4 by Watson *et al.*, this volume) and exposure to wind (Chapter 5 by Watson *et al.*, this volume). Another key limitation of SNOTEL data is that they only date back to the 1970s and 1980s, whereas precipitation and temperature data in Yellowstone date back to the nineteenth century. Some of the major trends in wildlife dynamics at Yellowstone have been observed since that time, or in more detailed data since the 1960s.

Another commonly used method of describing snow pack is the use of direct, remote sensing of snow-covered area, although it is primarily a binary measure (snow cover vs no snow cover) of limited utility to wildlife analyses when compared with snow pack water equivalent (Simic *et al.* 2004). Imagery from the Moderate Resolution Imaging Spectroradiometer (MODIS) can be obtained daily for the study area at 250–500 m spatial scales (Chapter 7 by Thein *et al.*, this volume). This scale is perhaps too coarse for fine-scale habitat selection studies (Chapter 8 by Messer *et al.*, Chapter 20 by Gower *et al.*, Chapter 21 by White *et al.*, Chapter 28 by Bruggeman *et al.*, this volume), but may be adequate for generating areal averages. However, cloud cover is very common in winter (Chapter 7 by Thein *et al.*, this volume), so the imagery is unreliable as a source of temporally consistent information without some means of temporal interpolation. Landsat imagery has a finer scale (*circa* 28.5 m), but is acquired only every two weeks and cloud-free winter images are somewhat rare (Watson *et al.* 2007).

As an alternative to approaches that directly measure dynamic snow pack characteristics, a variety of empirical statistical models have been developed to map the spatial distribution of snow pack properties in terms of more-easily measurable influences such as terrain, vegetation, precipitation and, in some cases, wind exposure. To date, these models have lacked some of the characteristics required for evaluating large mammal dynamics, such as: a means of scaling up to large study areas (Erickson *et al.* 2005), a means of incorporating temporal dynamics (Elder *et al.* 1998), deterministic results (Marchand and Killingtveit 2005), or an objective, validated basis (Farnes *et al.* 1999). The Farnes *et al.* (1999) approach has made a substantial contribution to numerous ecological analyses in Yellowstone National Park (a recent example being Fortin *et al.* 2005). It combines SNOTEL data with a manually fitted contour map of mean annual precipitation, and a set of empirical equations for adjusting SWE based on terrain slope and aspect and forest cover. For our purposes, this approach is limited by its subjectivity and lack of basis for integration of multiple influences on snow pack, such as wind and geothermal heat, in addition to terrain and forest cover.

Spatially distributed simulation modeling has the advantage of being applicable at fine scales (*e.g.* 28.5 m, daily), with no gaps in spatial or temporal coverage and, depending on the model, for a range of dates extending back to the earliest basic climate data. This flexibility is balanced by reduced accuracy when compared to SNOTEL measurements and perhaps also when compared to remote sensing estimates. Thus, validation against spatially distributed field observations is a critical component of spatial snow pack modeling (Watson *et al.* 2006a,b).

Snow pack simulation models have generally been developed and used for purposes other than applications to ecological investigations. They are typically used in hydrological and climatological contexts, often at fairly coarse spatial resolutions designed to yield area-averaged estimates of snow pack properties and runoff over very large areas (Carroll *et al.* 1999, Melloh 1999, Etchevers *et al.* 2004). Snow pack models developed for fine-scale ecological purposes to date are limited in spatial extent (Hiemstra *et al.* 2006). In the most stringent applications, wildlife ecological analyses require a unique combination of high spatial and temporal resolution and extent to capture the complexity inherent in biological systems. For example, some of the analyses in this book (Chapter 12 by Bruggeman *et al.* and Chapter 14 by Geremia *et al.*, this volume) required daily estimates of SWE at 28.5 or 57 m resolution as far back as 1969, extending in areal coverage up to as much as 14,000 km^2 (the size of the rectangular bounding box for the Yellowstone bison ranges). Simulation models, if appropriately designed, are an ideal way to yield such estimates because they can operate prior to the SNOTEL record, they can incorporate multiple spatial influences at fine scales and large extents, they are objective and deterministic, and their outputs are customizable and free of gaps.

The purpose of the work described in this Chapter was to provide state-of-the-art estimates of spatial snow pack properties for analyses presented in many of the Chapters of this book to enhance our understanding of the effects of snow pack on large mammal abundance, distribution, and interactions. The central Yellowstone system is characterized by long, relatively severe winters, and the influence of snow pack distribution, accumulation, and duration is one of the primary factors driving the spatial and population dynamics of elk and bison, as well as their interactions with the system's top predator, wolves. While insights into these dynamics can be obtained through use of traditional methods for describing snow pack, evaluations often required a variety of scales and estimates that made these methods less applicable. For example, for many of these analyses, estimates of SWE and local heterogeneity of SWE were required at about 500,000 specific locations and dates within the Yellowstone study area at various times in the last 16 years (Chapter 8 by Messer *et al.*, Chapter 20 by Gower *et al.*, Chapter 21 by White *et al.*, Chapter 28 by Bruggeman *et al.*, this volume). For the remaining analyses, estimates of mean SWE over entire wildlife ranges were required either daily or annually for up to 37 years since 1969–1970 (Chapter 9 by White *et al.*, Chapter 11 by Garrott *et al.*, Chapter 12 by Bruggeman *et al.*, Chapter 14 by Geremia *et al.*, Chapters 16 and 17 by Becker *et al.*, Chapter 19 by Gower *et al.*, Chapter 22 by White *et al.*, Chapter 23 by Garrott *et al.*, Chapter 27 by Bruggeman *et al.*, this volume). Thus in this Chapter, we describe the snow pack simulation model used to support these wildlife analyses and assess its accuracy. In turn, this assessment facilitates the assessment of the subsequent wildlife analyses that use snow model output. The spatial domain of the snow pack model development included the western-central portion of Yellowstone National Park, but also extended to the remainder of the park and adjacent lands. The western-central portion of Yellowstone is localized within strong environmental gradients that are best observed and modeled by considering input, calibration, and testing data from a wider geographic area.

II. MODEL HISTORY AND ENHANCEMENTS

The Langur snow pack model was first described by Watson *et al.* (2006b). It was based on the UEB model (Tarboton 1994; Tarboton *et al.* 2000), which centers on the simultaneous simulation of the evolution of snow pack mass and energy balances during accumulation and ablation of seasonal snow packs. UEB is run on a sub-daily time step, whereas Langur runs on a daily time step to improve run time and place less demand on microclimate estimation at sites far away from climate stations. UEB is typically run in small experimental plots or experimental watersheds that are orders of magnitude smaller than our central Yellowstone study area. The Langur programming code was written in C++ with detailed reference to the UEB code written in FORTRAN. Langur is implemented as a module within the Tarsier Environmental Modeling Framework (Watson and Rahman 2003). To be applicable to large mountainous landscapes, the model relies on large-scale, microclimate, estimation algorithms developed by Bristow and Campbell (1984), Running *et al.* (1987), Band *et al.* (1993), Vertessy *et al.* (1996), Watson *et al.* (1999), and Dingman (2002). Watson *et al.* (2006b) compared this original version of the model to point-scale SNOTEL data and a spatial coring data set designed to quantify simple vegetation and terrain responses. As a result of this exercise, a number of improvements were recommended and partly addressed in a second version of the model (Watson *et al.* 2007). This Chapter addresses the remaining improvements.

Watson *et al.* (2007) studied geothermal heat flux by inversion of the snow pack model. This work incorporated a re-structuring of the model from a one-layer to a three-layer formulation of mass and energy balance (*i.e.*, two snow layers and one soil layer). This structural change was augmented by appropriate snow pack compaction and thermal conductance algorithms from Shapiro *et al.* (1997),

Koivusalo *et al.* (2001), and Gustafsson *et al.* (2004) to ensure realistic heat flow between the layers, and increasing heat flow as the snow pack aged and densified. This second version of the model was compared to a range of SNOTEL data, randomly located coring data yielding estimates of SWE and snow pack density, and snow pit temperature data. Broad agreement with the data was indicated, but with considerable scatter about the 1:1 lines. A key result of the exercise was the successful determination that the model simulated expected snow pack responses to subtle geothermal influences.

Three key enhancements were made following the version of Watson *et al.* (2007), leading to the third version of the model described in this Chapter. This version now incorporates all recommendations by Watson *et al.* (2006b). Two of the enhancements relate to the incorporation of spatially variable wind effects. The first is scouring and sublimation during and shortly after snowfall. The effect of wind on net snow accumulation was measured, modeled (Chapter 5 by Watson *et al.*, this volume), and incorporated into the snow model as part of the snowfall accumulation process. In previous versions, all snow falling on the snow pack surface became part of the snow pack. In the present version, at locations where mean wind speed is above a specified threshold, an increasing fraction of the snowfall does not get incorporated into the snow pack, but rather is assumed to have sublimated back into the atmosphere (potentially as part of a blowing snow process).

The second area where spatially variable wind effects are incorporated is in the modeling of convective fluxes of sensible heat and latent heat. These fluxes are important determinants of snow pack heat balance and, thus, the date of onset of substantial melt and sublimation (Koivusalo and Heikinheimo 1999, Gustafsson *et al.* 2001). Convective fluxes are modeled using standard boundary layer theory as being highly sensitive to wind speed. The resulting effect is spatial variation in melt rates in areas where the wind is spatially heterogeneous.

The third key enhancement is the incorporation of a more physically based snowfall canopy interception, unloading, and sublimation model. This is a daily version of the weekly model developed by Hedstrom and Pomeroy (1998). Snowfall is intercepted by the canopy up to a storage capacity that is influenced by forest cover and wind speed. Thereafter, it unloads progressively according to an exponential decay model and, also, may be sublimated back to the atmosphere.

The model application described in this Chapter also differs from previous versions with respect to the methods used to set values for model parameters. The details are described later, but a brief summary follows. In earlier versions, model parameters were either set from the literature, fitted to field data independently from the Langur model itself, or calibrated within Langur at the point scale using SNOTEL data. The model was then run spatially, and assessed by comparison to independent spatial data sets containing vegetation effects and terrain effects (Watson *et al.* 2006b), or to snow pack density and temperature information (Watson *et al.* 2007). For the present application, we calibrate the model against the data that were formerly used for independent assessment. Specifically, we calibrated the four key spatial parameters to the vegetation and terrain effects data from Watson *et al.* (2006a), as well as the wind effects data (Chapter 5 by Watson *et al.*, this volume). We then introduce a new "random coring" data set, part of which we use for fine-tuning calibration of parameter values and the remainder of which we use for completely independent testing. Thus, the development of the model follows the phased approach advocated by Watson *et al.* (2006b) of progressively improving the model and its parameters using existing data, but always collecting additional data and withholding some it from the calibration process for use in independent and objective accuracy assessment.

A. Model Structure and Inputs

The Langur model represents the landscape as a grid of cells. Each cell is characterized by several spatial variables: (1) Elevation (m), (2) Slope (°), (3) Aspect (°), (4) Mean annual precipitation (m), (5) Forest cover (%), (6) Long-term wind speed (m/s), and (7) Geothermal heat flux (W/m^2). The spatial scale is variable. The model has been run at cell widths ranging from 28.5 to 2 km. At a cell width of 28.5 m, the model can simulate heterogeneous snow pack dynamics over an area of approximately 120 by 120 km, using about 17 million cells on a computer with two gigabytes of random access memory. Within each cell, the model represents two snow pack layers (upper and lower), a soil layer, and a vegetation layer. The model keeps an account of the following state variables, snow pack water equivalent (m), snow pack depth (m), snow pack heat content (J), soil moisture content (m), soil heat content (J), snow pack albedo (%), and canopy-intercepted snow (m).

The model simulates the evolution of the above state variables daily. To drive this simulation through time, it first estimates three temporal inputs at each cell: (1) Daily precipitation (m/day), (2) Daily minimum temperature (°C), and (3) Daily maximum temperature (°C). These inputs are based on daily data from either SNOTEL stations or National Climate Data Center climate (CLIM) stations. These data are then scaled to be applicable to each cell. The precipitation data are scaled by the mean annual precipitation ratio between the cell and the climate station. The temperature data are scaled according to seasonally varying empirical temperature-elevation lapse rates. Estimates derived from multiple climate stations are combined using Gaussian distance weighting. Snow water equivalent data are not used as an input for driving the model. Instead, these data are used for calibrating the parameters of the model. This is because the scaling between climate observation stations and arbitrary points in the landscape is not straightforward with SWE data, unlike the case with precipitation and temperature data. It also allows the model to be run for regions and time periods where SWE data are not available.

The spatial pattern of landscape characteristics was input into the model using a set of raster maps of various characteristics. Elevation, slope, and aspect maps were derived from standard U.S. Geological Survey topographic quadrangle data. A map of mean annual precipitation (Chapter 3 by Watson *et al.*, this volume) was derived from trivariate kriging of precipitation values from a network of long-term climate observation stations. A forest cover map (Chapter 2 by Newman *et al.*, this volume) was based on spectral kerneling of a Landsat ETM+ scene in autumn 2002. A long-term wind speed map (Chapter 5 by Watson *et al.*, this volume) was based on kernel-smoothed terrain protrusions and forest cover, calibrated against an extensive field data set. A geothermal heat flux map (Chapter 4 by Watson *et al.*, this volume) was based on residual Landsat ETM+ thermal-band heat flux in March 2000 after accounting for estimated non-geothermal effects, and validated against field data of geothermal perimeters.

III. AN "INCREMENTAL RANDOM SAMPLING" DATA SET

A key goal of our snow pack modeling exercise was to produce spatial snow pack estimates in support of wildlife analyses with some objective, quantitative assessment of their accuracy. We conducted an extensive random snow pack coring exercise to provide testing data in support this goal. The requirements of the exercise were that snow pack measurements would be made at a large number of sites distributed in a completely objective and unbiased manner throughout the major wildlife ranges of interest. Some constraints were that, at the outset, we did know how many sites could be measured given the resources available, because we did not know how much effort would be required on average to visit any given site located randomly within the wildlife ranges. We also did not have, at the outset, an objective map of the boundary of the relevant wildlife ranges.

Consequently, we designed a novel scheme that we termed "incremental random sampling." We began by enumerating 10,000 random locations within Yellowstone National Park, a list of sites deliberately much higher than we expected to be able to visit. We numbered these sites in a fixed order, such that any sub-sample of size *N* of locations 1 through *N* would itself constitute a random sample. Moreover, any further sub-sample of just those locations within a defined polygon would constitute a random sample of locations within that polygon. Thus, by visiting sites within a defined polygon in numeric order, the resulting data set at any time represented a random sample. Sampling was then continued until resources were exhausted, without compromising the completeness or introducing bias into the resulting data set. This cannot similarly be done with a systematic sampling approach like sampling on a spatial grid, because under such a scheme, the total number of locations would need to be dictated in advance.

We employed incremental random sampling with some logistical modifications. We divided the sampling into discrete time periods during which we expected to be able to visit a "packet" of $n = 15–30$ sites. This time period was one week for large field crews (typically seven people) and one month for smaller crews (typically two people). At the start of each period, we drew the next packet of n sites on a map, and drafted a plan to visit all of these sites with the least travel effort, taking advantage where possible of the proximity of groups of sites to each other. Applying this proximity grouping in time would ordinarily lead to unwanted temporal bias (or more specifically, spatio-temporal correlation), but the packet constraint worked to prevent this.

In addition, we modified the sampling boundary slightly each year, for two reasons. First, our logistical competence improved over time as we became more adept at safely traveling further and further from the road network into areas more broadly representative of the wildlife ranges on the whole, and less biased by the fact that roads are often in open valleys (Figure 6.2). Second, our definition of the wildlife ranges improved with the development of objective methods for determining them. The final boundary (Figure 6.3) represents an objective estimate of the central and northern bison winter ranges (2765 km², which also encompass most of the elk winter ranges) that was made using an algorithm (selective computational diffusion) we developed during the course of these studies. These changes to the sampling boundary did not invalidate the objective randomness of our spatial sampling design, but they did introduce a small amount of temporal bias toward more distant locations being sampled later in the overall campaign.

The sampling domain was constrained within this boundary to exclude: (1) Cells mapped as being in lakes or rivers, (2) Cells mapped as being steeper than 25° (to avoid avalanche danger), (3) Locations in any active Bear Management Area for which we did not have special access permission, and (4) Cells mapped as being closer than 100 m to, or further than 4 km from, a groomed or plowed road, with the exception of cells mapped as being within two remote corridors across the Mary Mountain Trail and extending from the Pelican Valley across the Mirror Plateau to the Lamar Valley (Figure 6.3).

FIGURE 6.2 Overnight backcountry travel was required to access the more remote snow pack coring sites in Yellowstone National Park (Photo by Fred Watson).

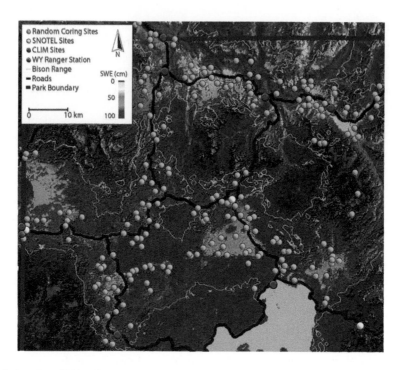

FIGURE 6.3 Location of 211 coring sites (representing 2532 coring measurements) distributed at random throughout the Yellowstone bison range. Also shown are the closest snow pack telemetry (SNOTEL) and climate (CLIM) stations, as well as the West Yellowstone Ranger Station. The background image shows modeled snow water equivalent for March 16, 1995, overlaid with forest cover.

All locations were located in the field to the nearest few meters using differential Global Positioning Systems. Locations were sampled regardless of the vegetation, terrain, or snow pack disturbance by wildlife that was encountered, except if a safety hazard was determined to exist upon arrival. This only occurred on two occasions due to perceived avalanche hazard. Another possibility would have been extreme geothermal activity. Wildlife hazards such as grizzly bears or dangerous bison behavior were dealt with by returning on a different day.

At each location, 12 snow pack cores were sampled in four 12 meter-spaced groups of three one-meter equilateral triangles, after Watson *et al.* (2006a; Figure 6.4). This nested pattern was computed using the methods of Watson *et al.* (2006a) as yielding the most accurate estimate of the mean SWE in a 28.5 m cell with one-hour of sampling effort. Additional data recorded included snow pit temperature profiles, tree stem density, tree height and vigor, terrain slope, terrain aspect, presence of downed logs and rocks, apparent geothermal influence (*e.g.*, sinter or thermophilic algae), and wildlife impacts. We also noted any unusual characteristics of the site or the coring measurements that would be expected to have an impact on the model-data comparison in ways that would not be expected to be represented by the model, or by the spatial maps supplied as input to the model. We refer to these cases below as "anomalies." The most common anomalies were cornices (*i.e.*, highly localized accumulations of deep wind-blown snow), bogs (*i.e.*, muddy slurries at the soil surface into which the snow pack corer readily plunged), and frozen open water that was too hard for the corer to penetrate. This effort yielded 2532 coring measurements from 211 random locations within the Yellowstone bison range (Figure 6.3), a sampling domain of 1346 km² within a range of 2765 km². The data were collected between January 2004 and May 2007.

FIGURE 6.4 Randomly located snow pack coring sites fell in a wide variety of terrain. We took 12 cores at each site. Each core was weighed to measure snow pack water equivalent (Photo by Fred Watson).

IV. MODEL PARAMETERIZATION

Model parameterization involved setting values for each of the 36 parameters used by the model. Some of these values were derived from the literature, while others were calibrated to match specific field observations. A specific parameterization sequence was followed (Appendix 6A.1) and each stage in the sequence is outlined in the following paragraphs:

Stage 1—Longwave radiation at each cell was estimated as a linear function of daily minimum temperature, with two parameters, a slope and an intercept. Parameter values (Appendix 6A.1) were based on linear regression of radiation data from a SURFRAD site at Fort Peck, Montana and are unchanged since Watson *et al.* (2006b), based on Watson (1999).

Stage 2—Atmospheric transmission of radiation at each cell was estimated as a function of daily temperature range with three parameters whose values (Appendix 6A.1) were unchanged since Watson *et al.* (1999) and Watson (1999), and were originally derived by fitting the transmission function to some Australian radiation data.

Stage 3—Daily minimum and maximum temperatures above each cell were estimated from the corresponding values at climate base stations, adjusted for the elevation difference between the cell and the base station according to temperature lapse rates. Separate lapse rates were used for maximum and minimum temperatures, and each lapse rate varied seasonally as a sine wave with parameters for the mean, amplitude, and phase (Watson 1999; Watson *et al.* 1999). The six parameter values (Appendix 6A.1) were unchanged since Watson *et al.* (2006b), where they were calibrated to match temperature-elevation differences among eight SNOTEL stations in the Yellowstone area. An exception was the simulations for Chapters 12 and 14, where because of low-elevation valley-bottom bias in the climate base stations (see Stage 4, below), lower lapse rates were used to avoid erroneously low temperature estimates at the highest elevations (Appendix 6A.1). This is consistent with previous findings that cold-air drainage leads to anomalously low minimum temperatures at valley-bottom sites (Laughlin 1982, Despain 1990, Rolland 2002, Chung *et al.* 2006).

Stage 4—Daily precipitation at each cell was estimated from the corresponding values at base stations, adjusted for the ratio between mapped mean annual precipitation (Chapter 3 by Watson *et al.*, this volume) at the cell and the base station. During daily simulation, the model produces a minimum and maximum temperature and precipitation estimate for each cell derived from each of a number of daily climate base stations. These are then combined into a single daily minimum and maximum temperature and precipitation for the cell by Gaussian horizontal distance weighting of the values derived from each base station. A parameter expresses the bandwidth for the Gaussian weighting, and its value was taken as the horizontal variogram range from the mean annual precipitation mapping exercise (Chapter 3 by Watson *et al.*, this volume).

Daily climate base station data were taken from up to 12 stations each day. Most of the simulations used in this book were driven by daily data from 11 automated SNOTEL stations in the Yellowstone region (Chapter 3 by Watson *et al.*, this volume), and one manual weather station at the West Yellowstone National Park Service ranger station. Some simulations started as early as 1969, before the earliest SNOTEL stations, in which case the entire simulation was driven by data from nine Yellowstone area climate stations (CLIM, Chapter 3 by Watson *et al.*, this volume). The SNOTEL stations are preferable, because they have associated snow pack data for model calibrating and testing and they are more evenly distributed throughout the range of elevations exhibited in the study area. National Climate Data Center CLIM sites in the region are biased to elevations lower than most of the study area, leading to less accurate temperature estimates at high elevations than under SNOTEL-driven simulations. Chapters using these longer-term, CLIM-driven simulations were Chapters 12 and 14.

Stage 5—In general, wind speed is a spatial input to Langur (see above). But during non-spatial runs such as calibration to SNOTEL snow pack data (see below), a single value is required. For model runs at SNOTEL stations, we set the wind speed at 50 cm above ground level to 0.17 m/s. This was the mean of observed wind speed in all sites sampled (Chapter 5 by Watson *et al.*, this volume) that were deemed to have landscape characteristics that matched the SNOTEL sites nearest to the study area (*i.e.*, small clearings in relatively flat, dense forest).

Stage 6—A number of parameter values were based directly on values in the literature (Appendix 6A.1). The radiative emissivity of snow and soil were unchanged since Watson *et al.* (2006b), taken from Oke (1987). The reflection coefficient for soil was also unchanged since Watson *et al.* (2006b), and was taken from Monteith and Unsworth (1990). Soil thermal conductivity varies between about 0.15 and 4.0 W/m/K (Jansson and Karlberg 2001, Hukseflux Thermal Sensors 2007). We set it to 2.0 W/m/K and found that our SNOTEL calibrations (below) were insensitive to this value in the range 0.15–4.0 W/m/K. The volume of liquid snow pack water retained against gravitational flow (relative to snow pack volume) is a parameter established by Tarboton *et al.* (1995; value = 0.05), which we found to be insensitive in daily simulations and optimal at any value lower than 0.02 (Watson *et al.* 2006b). A special parameter was used to indicate the thermal conductivity of ice to allow higher heat flow into snow packs with a warm base due to geothermal influence. The value was unchanged since Watson *et al.* (2007) and was taken from Engineering Toolbox (2007). The coefficient for the exponential model of canopy radiation extinction was unchanged since Watson *et al.* (2006b) and taken from Sampson and Smith (1993). For modeling canopy snowfall interception velocity, the vertical snowfall velocity and the canopy snow interception coefficient were taken from Hedstrom and Pomeroy's (1998) values for jack pine forest (jack pine is a 2-needle pine with canopy morphology not unlike the lodgepole forest of the study area). We also adopted Hedstrom and Pomeroy's unloading coefficient value of 0.678 (per half-week), but since theirs was a weekly model, we re-scaled it to a value suitable for use in our daily model as $-\log(0.678)/(7/2) = 0.11$ (per day), where $7/2 = 3.5$ represents the average number of days since snowfall at the time of Hedstrom and Pomeroy's weekly measurements.

Stage 7—Eight parameters required calibration against snow pack observations at SNOTEL stations (Appendix 6A.1). The definition of some of these parameters was at least partly model-dependent and, thus, literature values alone were not a sufficient basis for the parameter values. For other parameters, literature values were not precise enough to eliminate model sensitivity within the range of values that could be reasonably inferred from the literature. Fortunately, a large amount of snow pack data was available from multiple SNOTEL stations to use as a basis for calibration. Here, we used eight years of daily SWE data and one year of daily snow pack depth data from 11 SNOTEL stations. We computed snow pack density as a simple fraction by dividing SWE by depth. The calibration procedure used an automated optimization algorithm (Pattern Search; Hookes and Jeeves 1961) to find parameter values that maximized the model's accuracy in estimating the observed daily SWE and density data. As input to the optimization algorithm, we defined an objective function quantifying model accuracy as the weighted sum of the mean absolute error in estimating SWE and the mean absolute error in estimating density, with four times more weight given to the mean absolute error of SWE. Thus, an error of 1 cm in SWE would have four times more influence on the quantification of accuracy than a 1% error in density. This emphasized accuracy in SWE, but not without some degree of accuracy in density. Under the Pattern Search algorithm, a set of random initial parameters was chosen, and then each set was iteratively improved until it converged to a local optimum (*i.e.*, small parameter changes led to minimal further improvement). This was repeated until a global optimum was indicated by the coincidence of most of the local optima on a single point. This required 2342 eight-year runs, each one involving 11 SNOTEL stations. The final parameter set was selected from the best of all 2342 runs.

The resulting comparisons between modeled and observed SWE and density are illustrated for three of the 11 SNOTEL stations (Figures 6.5 and 6.6). The three stations shown are the closest to the Madison headwaters study area: Madison Plateau, Canyon, and West Yellowstone. They vary according to gradients in elevation (2362 m, 2420 m, 2042 m), mean annual precipitation (1.07 m, 0.70 m, 0.55 m), and the date on which snow melt is typically completed (late, intermediate, early). The model clearly reproduced most of the inter-annual variability in peak SWE at these stations (Figure 6.5). It also reproduced the overall shape of the seasonal time series of accumulation and melt, both in high-snow and low-snow years. At this long-term scale, there was no obvious bias to over-estimation or under-estimation of the peak SWE or melt date. Nor was there any obvious bias in prediction accuracy at high or low elevation sites. Figure 6.6 looks more closely at a specific year, and includes comparison with snow pack density data. In the year shown (2005–2006), the accumulation was well modeled. The melt was well modeled at Madison Plateau, but the date of complete ablation was estimated nearly a week too late at Canyon, and a few days too early at West Yellowstone. This was typical of the magnitude of error in other years, and at other SNOTEL stations (not shown). Density was well estimated during the periods when there is non-negligible snow pack. The overall magnitude, and gradual seasonal increase in density was accurately represented by the model, as was the existence and direction of short periods of more rapid change in density. There was some evidence that the density estimates are less accurate at the lower elevation station with a shallower snow pack (West Yellowstone).

Stage 8—Three parameters controlled the influence of spatial wind speed patterns on snow pack accumulation and melting. These parameters had little influence in simulations at SNOTEL sites, since most SNOTEL sites in the study area were sheltered, with very little wind (Figure 6.7). But once the model was run in the open meadows and exposed, undulating terrain typical of most of the wildlife ranges, these parameters became highly sensitive.

Two of the wind-related parameters governed the wind-driven scouring of snow pack accumulation and the third one linearly scaled sensible and latent heat flux from the snow/soil surface. The scouring parameters were initially defined and estimated (Chapter 5 by Watson *et al.*, this volume) as $W_B = 2.9$ m/s for the wind speed threshold and $\beta_{S2} = 0.032$ s^2/m^2 for the parabolic function for scouring beyond this threshold. Wind-driven snow pack accumulation at open sites was highly sensitive to these parameters. Therefore, we decided to re-calibrate the two scouring parameters within the snow pack model. This did not necessarily contradict the original values in Chapter 5, since those values were estimated from observations taken over a relatively short period.

Wind-driven sensible and latent heat flux were initially modeled according to Koivusalo *et al.*'s (2001) description of the UEB model based on boundary layer theory. This led to unrealistically cold snow packs and delayed melt-out in exposed areas, problems that were consistent with a history of somewhat arbitrary adjustments in the modeling of these convective fluxes (Koivusalo *et al.* 2001,

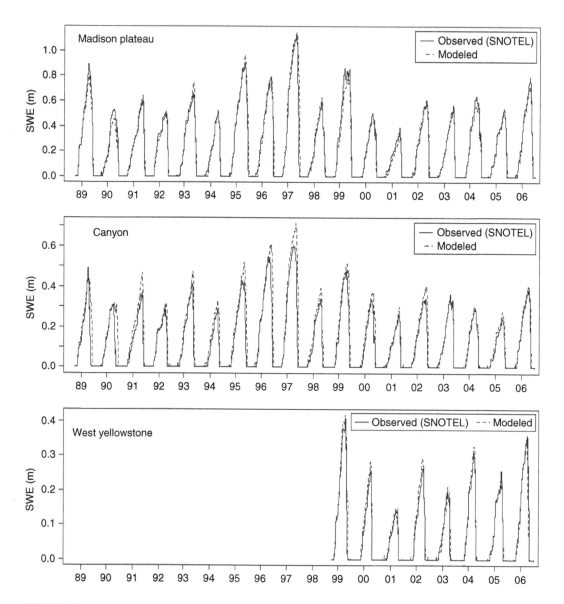

FIGURE 6.5 Modeled versus observed snow water equivalent at three of the 11 SNOTEL sites used for calibration of non-spatial parameters, for the entire period of SNOTEL record.

citing Jordan 1992, Tarboton and Luce 1996). We addressed this by introducing a calibration parameter that linearly scaled all sensible and latent heat flux.

We simultaneously optimized the three wind-related parameters to minimize mean absolute error in estimating SWE at 47 sites in the Hayden Valley sampled in Chapter 5 (we chose the 47 that were least affected by nearby vegetation, since calibration of vegetation-related parameters had not yet been done). As with the SNOTEL optimizations, a Pattern Search algorithm was used, with 100 random starting points resulting in 4500 model runs, each involving 47 modeled locations. The points are approximately grouped about the 1:1 line, indicating that the model accurately simulated wind effects by capturing some of the variation in SWE within this data set due to terrain-related effects (Figure 6.8). Other variables were held approximately constant, including time, vegetation,

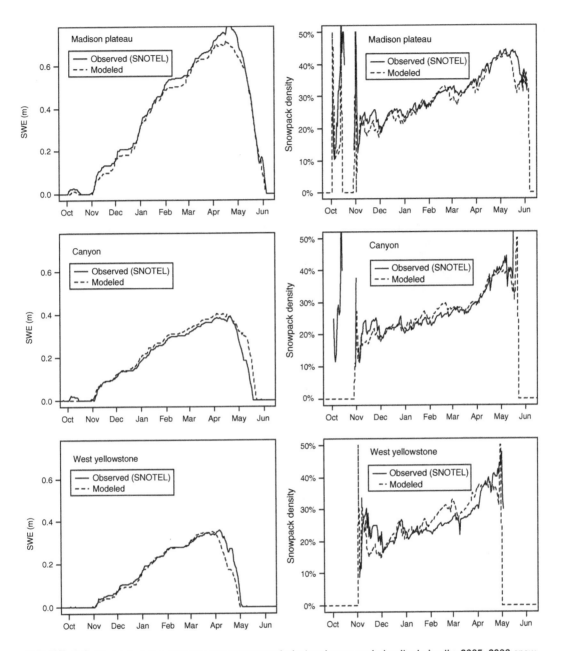

FIGURE 6.6 Modeled versus observed snow water equivalent and snow pack density during the 2005–2006 snow season.

precipitation, elevation, and geothermal influence. Estimated wind speed varied from one end of the 1:1 line to the other, suggesting that the effect that was evident was due to terrain influence on wind speed. Considerable scatter about these broad agreements was also evident, indicating room for future improvement of this component of the model.

Stage 9—Sublimation of canopy-intercepted snow was modeled simply as a fraction of the sublimation from the surface of the snow pack, where the fraction is a calibrated parameter. Thus, it was influenced by daily variations in temperature, radiation, and wind, but was allowed to be much higher

FIGURE 6.7 Snow pack telemetry (SNOTEL) stations operated by the U.S. Department of Agriculture like this one at Thumb Divide measure snow pack water equivalent as well as temperature and precipitation. Since late 2005, many of these stations also measure snow pack depth (Photo by Fred Watson).

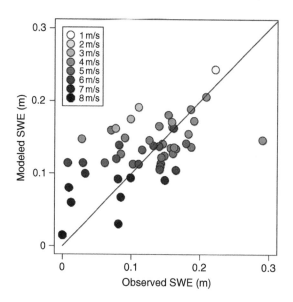

FIGURE 6.8 Correspondence between modeled and measured snow water equivalent (SWE) under a range of wind exposures in unforested terrain (Hayden coring data set; Chapter 5 by Watson *et al.*, this volume). The degree to which both modeled and observed SWE variation is influenced by wind variation is indicated by the shading of points according to estimated long-term wind speed.

than snow pack sublimation rates in general, due to the greater atmospheric exposure of the canopy. A more physically realistic approach would be to incorporate Pomeroy *et al.*'s (1998) model, but this involves considerable extra detail and still involves loose, sensitive parameters like "canopy exposure."

For optimization of the canopy sublimation parameter, we used the set of SWE observations made by Watson *et al.* (2006a) under a sampling design that was specifically intended to measure forest canopy effects on snow pack accumulation in a variety of terrain. Instead of using the mean absolute error between modeled and observed SWE as an objective function, we used the mean absolute error between modeled and observed differences in SWE between pairs of sampling strata. Each pair comprised a control and a treatment stratum, such as a meadow stratum compared with a forested stratum, with equal elevation and aspect. This approach allowed the calibration of vegetation effects to be more independent of inaccuracies in the model due to factors unrelated to vegetation. Watson *et al.* (2006a) estimated the mean SWE for each stratum from a nested pattern of usually 27 cores located in relatively homogeneous terrain. From Watson *et al.*'s (2006a) 27 strata surveyed in 2002, we extracted 21 control-treatment pairs (some control strata were used in more than one pair). The optimal value of the canopy sublimation parameter for reproducing these effects in the model was 190, which simulated that canopy sublimation rates were 190 times higher than snow pack sublimation rates per unit PLAI, scaled by how close the canopy snow pack was to the mechanical limit of interception capacity. Figure 6.9A and B show the degree to which modeled vegetation effects corresponded to measured vegetation effects after the optimization procedure. Watson *et al.* (2006b) estimated that 49% of the scatter in a plot such as Figure 6.9A was attributable to model functional error; with the remaining scatter being potentially indicative of parameter mapping error and pseudo-random variation (*i.e.,* variation that is uncorrelated with deterministic effects) in SWE at length scales larger than a single cell. So while there is considerable scatter about the 1:1 lines in Figure 6.9A and B, the forest cover effect itself is not particularly large relative to what might be termed stochastic snow pack variation. We conclude that the model-data comparison offers some confirmation that the model reproduced forest cover effects, and that there appears to be considerable room for improvement in the modeling of these effects. However, there may also be a practical limit to these improvements imposed by stochastic effects (for example, due to essentially random boundary layer turbulence during snowfall).

Stage 10—At this point in the calibration sequence we ran the model at the random coring locations from 2003–2004 through 2005–2006 (excluding the 2006–2007 locations). In most regions of the study area the model error appeared to be spatially independent. Over-estimates did not tend to be spatially clumped with other over-estimates or vice versa. But in one area, the error was strongly regionalized. The model consistently under-estimated SWE at all coring sites near the Mirror Plateau, suggesting that precipitation was poorly estimated in this remote part of the study area. This is consistent with the precipitation mapping error surface produced in Chapter 3 (Figure 3.7). We found that any attempt to calibrate the model to improve SWE estimates at the random locations was confounded by these outliers. Because the snow was so deep on the Mirror Plateau, the mean absolute error was strongly influenced by the under-estimation at just these few sites. The calibration procedure would then lead to over-estimation of all remaining sites to minimize under-estimation at the Mirror Plateau sites. Essentially, the snow pack model was being penalized, so to speak, for errors in the precipitation map.

To compensate for this, we introduced a "phantom precipitation gauge" on Mirror Plateau. This gauge did not physically exist, but we re-computed the map of mean annual precipitation as if it did exist. The gauge was defined by three parameters: its easting (m), northing (m), and mean annual precipitation (m). We then calibrated these parameters using Pattern Search to minimize the mean absolute error in the comparison between modeled and observed SWE at the random coring sites (2003–2004 through 2005–2006; 15 random starting points, 906 model runs, each on 122 locations). The phantom gauge alleviated an unrealistic constraint on the model and, thus, facilitated a more unbiased final calibration of spatial parameters.

Stage 11—To obtain improved model estimates, we re-calibrated the four spatial parameters from Stages 8 and 9 against the 2003–2004 through 2005–2006 random coring data (nine random starting points, 741 model runs, each on 122 locations). We used the phantom gauge during this re-calibration, but we then inactivated it for final testing below. For the re-calibration, we also excluded four

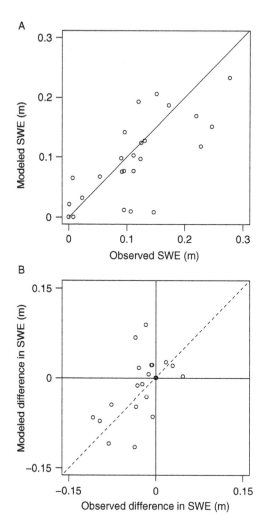

FIGURE 6.9 Correspondence between modeled and measured effects on snow water equivalent of forest cover (meadow, burned forest, or unburned forest) under a range of solar exposures (shady, flat, and sunny; measurements taken from Watson *et al.* 2006a).

field-identified anomalies (defined earlier). These last two stages (10 and 11) partially invalidated the independence of the 2003–2004 through 2005–2006 random coring data for testing purposes. However, the 2006–2007 random coring data remained as a completely independent test data set.

V. MODEL TESTING

The spatial pattern of snow pack model accuracy throughout central and northern Yellowstone National Park is shown in Figure 6.10. Not surprisingly, the least error occurred in the low-precipitation areas in the west and north. The largest errors occurred in higher elevation areas to the south and east, where there are few SNOTEL or CLIM sites from which to drive either daily climate driver data or mean precipitation estimates. There was some tendency for systematic under-estimation in the northeastern portion of the park, and systematic over-estimation in the Hayden Valley. These

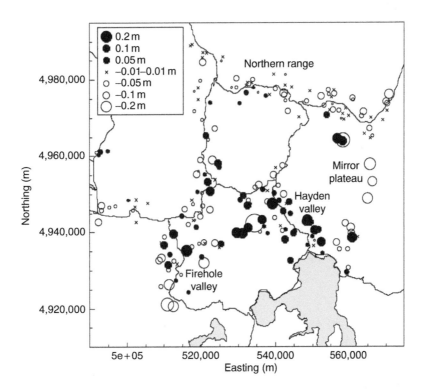

FIGURE 6.10 Spatial pattern of model accuracy in estimating snow water equivalent compared to randomly located coring data during 2003–2004 through 2006–2007 at 211 sites that included 2532 coring measurements.

errors were not strongly biased toward either regionalized under-estimation of over-estimation. Indeed the regionalized under-estimation that was evident on the Mirror Plateau in the 2003–2004 through 2005–2006 data was countered by over-estimation in the 2006–2007 data.

The same data are presented as a scatter plot in Figure 6.11, along with snow pack density and temperature comparisons. The SWE estimates appear to be unbiased through most of the observed SWE range. The largest SWE observations are under-estimated by the model. Four of these were flagged as anomalies in the field, and four more were on the Mirror Plateau where mean annual precipitation is suspected to be under-estimated. No obvious differences in model agreement are apparent between the 2003–2004 through 2005–2006 data, which were used in final model calibration, and the 2006–2007 data, to which the model was completely independent. In general, there was considerable scatter about the 1:1 line. This is most likely due to a combination of model error in representing spatial effects, field sampling error, parameter mapping error, and actual fine scale stochastic variation in SWE unrelated to any predictable influences (Watson *et al.* 2006b).

The Langur model was developed to elucidate wildlife-snow pack dependencies at scales as fine as 28.5 m. Mapping error and stochastic variation are fine-scale effects that would only be strongly apparent in spatial model testing at this fine-scale. Thus, while Figure 6.11 represents an unbiased, objective assessment of model accuracy at this scale, it would not be appropriate to compare this accuracy assessment to accuracy assessments made at larger scales, potentially in relation to other snow pack modeling approaches. Mapping error and stochastic variation would tend toward zero upon scaling up to larger cell sizes, so coarser-scale models would appear to be more accurate. Scale-dependency of model error would be a critical influence on any inter-comparison of models.

The comparisons to snow pack density and snow pack temperature data in Figure 6.11 are designed to provide some insight as to whether the internal functioning of the model is correct. The density

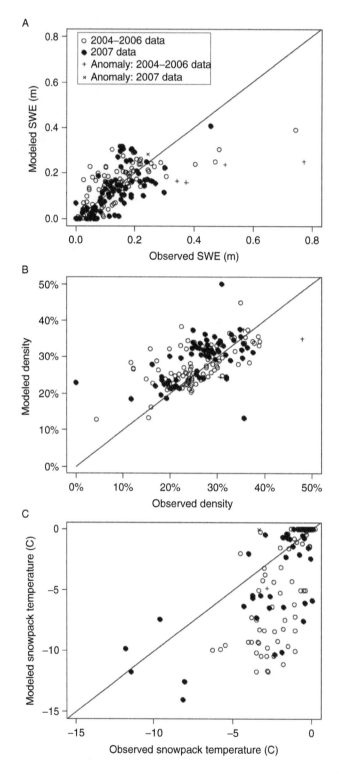

FIGURE 6.11 Model accuracy in estimating snow water equivalent, snow pack density, and snow pack temperature compared to randomly located coring data. Model parameterization was completely independent of the 2006–2007 data and partly dependent on the 2003–2004 through 2005–2006 data.

estimates are relatively unbiased with no widespread under-estimation, over-estimation, or lack of correlation with observed variation in density. The warmest and coldest snow pack temperatures were accurately predicted by the model, but temperatures in the mid-range between $-5\,°C$ and $-2\,°C$ were consistently under-predicted relative to the snow pit measurements (Figure 6.12). This over-cooling of the snow pack was also noted in the first version of the model (Watson *et al.* 2006b). It does not appear to have been corrected by expansion of the model to a 3-layer instead of a 1-layer formulation. Fortunately, as Watson *et al.* (2006b) observed, the over-cooling is later balanced by over-warming leading to a correct prediction of the timing of snow pack ripening and the onset of melt. This is supported by the timing indicated in Figure 6.5 and, also, the fact that Figure 6.11C indicates no situations where the observed snow pack was ripe ($0\,°C$) and the modeled snow pack was not so (*i.e.*, colder than about $-1\,°C$), or vice versa. Such situations would have been captured by this comparison, since the random coring data set, by design, has good coverage of the melt season (Figure 6.13) which ranges from approximately mid-March to late-May depending on the season and the elevation.

Since most of the error revealed in the comparisons presented in Figures 6.5–6.11 appeared to be related to spatial effects, it is informative to examine whether the magnitude of these errors depends systematically on the magnitude of mapped spatial effects. These dependencies are examined in Figure 6.14 for four of the most important spatial influences on the model. Not surprisingly, model error is larger where there is higher mean annual precipitation (Figure 6.14A) because more precipitation leads to more potential for error. However, there is little bias in this dependency. The confidence interval for the mean error almost always includes zero, except near the driest sites, where slight under-estimation of SWE is prevalent. Conversely, there is clear systematic bias in the model error with respect to forest cover (Figure 6.14B). The model tends to under-estimate SWE in areas with low to intermediate forest cover and over-estimate SWE in areas with very dense forest cover. This finding suggests that the canopy interception process should be strengthened, either by increasing the interception capacity parameter, decreasing the unloading rate parameter, increasing the canopy sublimation parameter, or making some alteration to wind-driven processes (since modeled wind and estimated forest cover are correlated by definition). This need for further refinement is consistent with the conclusions from Stage 9 of the parameterization process. Model error was relatively unbiased

FIGURE 6.12 Snow pits were dug to measure mean snow pack temperature (Photo by Anthony Brantley).

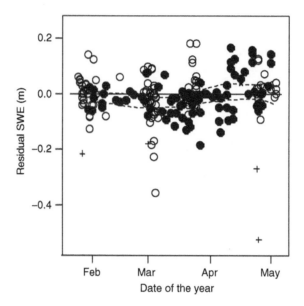

FIGURE 6.13 Relationship between model accuracy and time of year. Sampling during 2003–2004 through 2006–2007 occurred during snow pack accumulation, through initiation of melt, to complete ablation. See Figure 6.14 for explanation of symbols and lines.

with respect to geothermal heat flux (Figure 6.14C), which suggests that there were no erroneous responses or lack of response to mapped geothermal influence. The model exhibited a biased response to estimated spatial wind speed patterns (Figure 6.14D). SWE was over-estimated in areas where there was estimated to be no wind. Such areas are essentially dense forests, so this bias reflects a limitation in the modeling of either forest processes or wind processes. Under-estimation of SWE in higher-wind areas may simply be a compensation for the over-estimation in zero-wind areas. Thus, correcting the problem with zero-wind dense forests may, after re-calibration of the spatial wind parameters, lead to more widespread improvements.

Finally, we examined the response of SWE to the ensemble spatial pattern of landscape characteristics through comparison between aerial photographs and three-dimensional computer renderings of the simulated snow-covered landscape (Figure 6.15). Despite significant variation about the 1:1 line at the scale of individual cells (Figure 6.11A), the model is capable of correctly simulating the overall spatial pattern of snow pack distribution that results near the end of the season. Subtle effects of wind scouring and solar radiation on protruding and south-facing slopes are clearly reproduced by the model (Figure 6.15A), even if the timing in this example is off by one week. In Figure 6.15B, snow pack ablation due to subtle interactions between slight geothermal effects and forest cover are accurately reproduced, even where there is no clear visible evidence of geothermal influence. Figure 6.15C illustrates accurate response to larger scale influences such as elevation and exposure to solar radiation. It is these patterns of snow-free and snow-covered areas that are hypothesized to be exploited by elk and bison (Bruggeman *et al.* 2007, Chapter 8 by Messer *et al.*, Chapters 12 and 28 by Bruggeman *et al.*, and Chapter 21 by White *et al.*, this volume).

VI. SUMMARY

1. Spatial and temporal variations in SWE are important determinants of habitat selection, movement corridors, and vital rates in elk and bison. Objective, quantitative explanation of these relationships depends on objective, validated estimates of SWE.

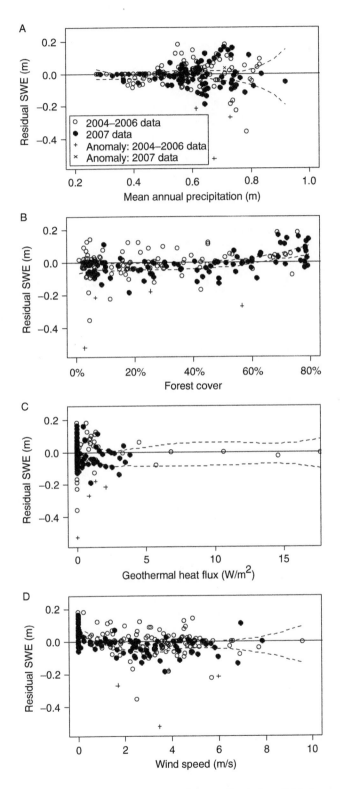

FIGURE 6.14 Relationship between model accuracy and four estimated spatial influences: (A) mean annual precipitation, (B) forest cover, (C) geothermal heat flux, and (D) long-term wind speed. Dashed lines represent 95% confidence intervals for the mean residual, estimated from the standard error of a local regression (fitted using Package locfit in R) multiplied by two-sided 95% standard normal quantiles (assuming local residuals are normally distributed).

FIGURE 6.15 Visual comparison between photographs of scenes with patchy snow cover (March 25, 2004), and computer renderings of the corresponding simulation of snow cover: (A) Cougar Meadow (simulation one week after photograph), (B) Spruce Creek Meadow, (C) Madison Junction (Photos by Fred Watson).

2. We developed and validated a physically based, spatially distributed, snow pack model that can estimate daily variation in SWE at 28.5 m resolution over landscapes extending to 14,000 km² for any period. For temporal inputs, the model requires only daily precipitation and temperature data, and does not require SWE data. For spatial inputs, the model uses maps of precipitation, forest cover, wind exposure, and geothermal intensity. The latter three of these were derived using remote sensing techniques specifically developed to support the snow pack model.

3. Non-spatial model parameters values were fixed using a combination of values sourced from the literature and calibration against daily data from 11 SNOTEL stations (eight years of daily SWE

data, and one year of daily density data). The model accurately reproduced temporal variation in SWE and snow density at SNOTEL sites.

4. Four spatial parameters were calibrated against our field coring data sets collected previously. These field data included 27 vegetation-effects strata (819 cores) and 104 wind-effects sites (936 cores). The wind effects were approximately reproduced by the model, but the reproduction of vegetation effects was weak and possibly confounded by a poor ratio of strength of effect to random or unaccounted influences.

5. For independent model validation, we conducted a new coring campaign at 211 sites (2532 cores and 151 snow pits) located at random within a 1346 km² sampling domain representing much of the 2765 km² Yellowstone bison range. One hundred and twenty-two of the random sites were used for final fine tuning of model parameters, as a well as validation, and the remaining 89 were retained for completely independent spatial model validation.

6. Modeled SWE approximately matched observed SWE, but with considerable scatter about the 1:1 line, and a tendency for under-estimation of the extreme lowest and highest observed SWE values. Bias in modeled SWE was unrelated to estimated precipitation and estimated geothermal intensity, but slight tendencies for systematic under-estimation and over-estimation were correlated with forest cover and wind exposure. These observations of bias indicate areas for future model improvement. The model accurately reproduced the spatial pattern of snow-covered area revealed in late-winter aerial photography.

7. A determination of whether the model is sufficiently accurate rests partly with the ecological analyses to which the model contributed (including 15 Chapters in this book). In applications to date, modeled SWE has frequently been selected by objective ecological model comparison exercises as an explanatory variable in some form (*i.e.*, either SWE itself, areal mean SWE, or local heterogeneity in SWE). This suggests that the model is sufficiently accurate for its intended use, not discounting the value of further improvements and notwithstanding the possibility of spurious correlation with unaccounted causal variables.

VII. REFERENCES

Band, L. E., P. Patterson, R. Nemani, and S. Running. 1993. Forest ecosystem processes at the watershed scale: Incorporating hillslope hydrology. *Agricultural and Forest Meteorology* **63**:93–126.

Bergman, E. J., R. A. Garrott, S. Creel, J. J. Borkowski, F. G. R. Watson, and R. M. Jaffe. 2006. Assessment of prey vulnerability through analysis of wolf movements and kill sites. *Ecological Applications* **16**:273–284.

Bristow, K. L., and G. S. Campbell. 1984. On the relationship between incoming solar radiation and daily maximum and minimum temperature. *Agricultural and Forest Meteorology* **31**:159–166.

Bruggeman, J. E., R. A. Garrott, P. J. White, F. G. R. Watson, and R. Wallen. 2007. Spatial variability in Yellowstone bison use of a road and travel network. *Ecological Applications* **17**:1411–1423.

Carroll, S. S., T. R. Carroll, and R. W. Poston. 1999. Spatial modeling and prediction of snow-water equivalent using ground-based, airborne, and satellite snow data. *Journal of Geophysical Research* **104**:19623–19629.

Chung, U., H. H. Seo, K. H. Hwang, B. S. Hwang, J. Choi, J. T. Lee, and J. I. Yun. 2006. Minimum temperature mapping over complex terrain by estimating cold air accumulation potential. *Agricultural and Forest Meteorology* **137**:15–24.

Despain, D. G. 1990. *Yellowstone Vegetation: Consequences of Environment and History in a Natural Setting.* Roberts Rhinehart, Boulder, CO.

Dingman, S. L. 2002. *Physical Hydrology.* Prentice Hall, Upper Saddle River, NJ.

Elder, K., W. Rosenthal, and R. E. Davis. 1998. Estimating the spatial distribution of snow water equivalence in a montane watershed. *Hydrological Processes* **12**:1793–1808.

Engineering Toolbox. 2007. Ice—Thermal Properties: Thermal and thermodynamic properties of ice—Density, thermal conductivity and specific heat at temperatures from zero to −100 °C. <http://www.engineeringtoolbox.com/ice-thermal-properties-d_576.html>Accessed October 21, 2007.

Erickson, T. A., M. W. Williams, and A. Winstral. 2005. Persistence of topographic controls on the spatial distribution of snow in rugged mountain terrain, Colorado, United States. *Water Resources Research* **41**:W04014.

Etchevers, P., E. Martin, R. Brown, C. Fierz, Y. Lejeune, E. Bazile, A. Boone, Y.-J. Dai, R. Essery, A. Fernandez, Y. Gusev, R. Jordan, *et al.* 2004. Validation of the energy budget of an alpine snowpack simulated by several models (SnowMIP project). *Annals of Glaciology* **38**:150–158.

Farnes, P., C. Heydon, and K. Hansen. 1999. *Snowpack Distribution Across Yellowstone National Park.* Montana State University, Department of Earth Sciences, Bozeman, MT.

Fortin, D., H. L. Beyer, M. S. Boyce, D. W. Smith, T. Duchesne, and J. S. Mao. 2005. Wolves influence elk movements: Behavior shapes a trophic cascade in Yellowstone National Park. *Ecology* **86**:1320–1330.

Fuller, T. K. 1991. Effect of snow depth on wolf activity and prey selection in north central Minnesota. *Canadian Journal of Zoology* **69**:283–287.

Garrott, R. A., L. L. Eberhardt, P. J. White, and J. Rotella. 2003. Climate-induced variation in vital rates of an unharvested large herbivore population. *Canadian Journal of Zoology* **81**:33–45.

Gates, C. C., B. Stelfox, T. Muhly, T. Chowns, and R. J. Hudson. 2005. *The Ecology of Bison Movements and Distribution in and Beyond Yellowstone National Park.* University of Calgary, Alberta, Canada.

Gustafsson, D., M. Stähli, and P. E. Jansson. 2001. The surface energy balance of a snow cover: Comparing measurements with two different simulation models. *Theoretical and Applied Climatology* **70**:81–96.

Gustafsson, D., P. A. Waldner, and M. Stähli. 2004. Factors governing the formation and persistence of layers in a subalpine snowpack. *Hydrological Processes* **18**:1165–1183.

Hedstrom, N. R., and J. W. Pomeroy. 1998. Measurements and modelling of snow interception in the boreal forest. *Hydrological Processes* **12**:1611–1625.

Hiemstra, C. A., G. E. Liston, and W. A. Reiners. 2006. Observing, modelling, and validating snow redistribution by wind in a Wyoming upper treeline landscape. *Ecological Modeling* **197**:35–51.

Hookes, R., and T. Jeeves. 1961. "Direct Search" solution of numerical and statistical problems. *Journal of Association for Computing Machinery* **8**:212–229.

Huggard, D.J. 1993. Effect of snow depth on predation and scavenging by gray wolves. Journal of Wildlife Management **57**:382–388.

Hukseflux Thermal Sensors. 2007. Thermal Conductivity Science. <http://www.hukseflux.com>Accessed October 21, 2007.

Jansson, P. E., and L. Karlberg. 2001. *CoupModel, Coupled Heat and Mass Transfer Model for Soil-Plant-Atmosphere Systems.* Royal Institute of Technology, Department of Civil and Environmental Engineering, , Stockholm, Sweden.

Jędrzejewski, W., B. Jędrzejewska, H. Okarma, and A.L. Ruprecht. 1992. Wolf predation and snow cover as mortality factors in the ungulate community of the Bialoweiza National Park, Poland. Oecologia **90**:27-36.

Jordan, R. 1992. *Estimating Turbulent Transfer Functions for Use in Energy Balance Modeling.* U.S. Army Cold Regions Research and Engineering Laboratory, Hanover, NH.

Koivusalo, H., and M. Heikinheimo. 1999. Surface energy exchange over a boreal snowpack: Comparison of two snow energy balance models. *Hydrological Processes* **13**:2395–2408.

Koivusalo, H., M. Heikinheimo, and T. Karvonen. 2001. Test of a simple two-layer parameterisation to simulate the energy balance and temperature of a snow pack. *Theory of Applied Climatology* **20**:65–79.

Laughlin, G. P. 1982. Minimum temperature and lapse rate in complex terrain: Influencing factors and prediction. *Archives for Meteorology, Geophysics, and Bioclimatology Series B* **30**:141–152.

Marchand, W.-D., and A. Killingtveit. 2005. Statistical probability distribution of snow depth at the model sub-grid cell spatial scale. *Hydrological Processes* **19**:355–369.

Melloh, R. A. 1999. *A Synopsis and Comparison of Selected Snowmelt Algorithms.* U.S. Army Corps of Engineers, Cold Regions Research and Engineering Laboratory, Hanover, NH.

Monteith, J. L., and M. H. Unsworth. 1990. *Principles of Environmental Physics.* Chapman and Hall, New York, NY.

Nelson, M. E., and D. L. Mech. 1986. Relationship between snow depth and gray wolf predation on white-tailed deer. *Journal of Wildlife Management* **50**:471–474.

Oke, T. R. 1987. *Boundary Layer Climates.* Routledge, London, United Kingdom.

Parker, K. L., C. T. Robbins, and T. A. Hanley. 1984. Energy expenditures for locomotion by mule deer and elk. *Journal of Wildlife Management* **48**:474–488.

Peterson, R. O. 1977. *Wolf Ecology and Prey Relationships on Isle Royale.* National Park Service Scientific Monograph Series No. 11. U.S. Government Printing Office, Washington, DC.

Pomeroy, J. W., J. Parviainen, N. Hedstrom, and D. M. Gray. 1998. Coupled modelling of forest snow interception and sublimation. *Hydrological Processes* **12**:2317–2337.

Rolland, C. 2002. Spatial and seasonal variations of air temperature lapse rates in alpine regions. *Journal of Climate* **16**:1032–1046.

Running, S. W., R. Nemani, and R. D. Hungerford. 1987. Extrapolation of synoptic meteorological data in mountainous terrain and its use for simulating forest evapotranspiration and photosynthesis. *Canadian Journal of Forest Research* **17**:472–483.

Sæther, B. E. 1997. Enviromental stochasticity and population dynamics of large herbivores: A search for mechanisms. *Trends in Ecology and Evolution* **12**:143–149.

Sampson, D. A., and F. W. Smith. 1993. Influence of canopy architecture on light penetration in lodgepole pine (*Pinus contorta* var. *latifolia*) forests. *Agricultural and Forest Meteorology* **64**:63–79.

Shapiro, L. H., J. B. Johnson, M. Sturm, and G. L. Blaisdel. 1997. *Snow Mechanics: Review of the State of Knowledge and Applications.* U.S. Army Corps of Engineers, Cold Regions Research and Engineering Laboratory, Hanover, NH.

Simic, A., R. Fernandes, R. Brown, P. Romanov, and W. Park. 2004. Validation of VEGETATION, MODIS, and GOES+ SSM/I snow-cover products over Canada based on surface snow depth observations. *Hydrological Processes* **18**:1089–1104.

Tarboton, D. G. 1994. *Measurement and Modeling of Snow Energy Balance and Sublimation from Snow.* International Snow Science Workshop Proceedings, Snowbird, UT.

Tarboton, D. G., G. Blöschl, K. Cooley, R. Kirnbauer, and C. Luce. 2000. Spatial snow cover processes at Kühtai and Reynolds Creek. Pages 158–186 *in* R. Grayson and G. Blöschl (Eds.) *Spatial Patterns in Catchment Hydrology: Observations and Modelling.* Cambridge University Press, Cambridge, United Kingdom.

Tarboton, D. G., T. G. Chowdhury, and T. H. Jackson. 1995. A spatially distributed energy balance snowmelt model. Pages 141–155 *in* K. A. Tonnessen, M. W. Williams, and M. Tranter (Eds.) *IAHS Proceedings of Symposium on Biogeochemistry of Seasonally Snow Covered Catchments.* IAHS, Boulder, CO.

Tarboton, D. G., and C. H. Luce. 1996. *Utah Energy Balance Snow Accumulation and Melt Model, Computer Model Technical Description and Users Guide.* Utah Water Research Laboratory and USDA Forest Service Intermountain Research Station, Odgen, UT.

Vertessy, R. A., T. J. Hatton, R. J. Benyon, and W. R. Dawes. 1996. Long term growth and water balance predictions for a mountain ash (*Eucalyptus regnans*) forest catchment subject to clearfelling and regeneration. *Tree Physiology* **16**:221–232.

Watson, F. G. R. 1999. Large scale, long term, physically based modelling of the effects of land cover change on forest water yield. Dissertation. Department of Civil and Environmental Engineering, University of Melbourne, Australia.

Watson, F. G. R., T. Anderson, W. Newman, S. Alexander, and R. A. Garrott. 2006a. Optimal sampling schemes for estimating mean snow water equivalents in stratified heterogeneous landscapes. *Journal of Hydrology* **328**:432–452.

Watson, F. G. R., R. E. Lockwood, W. B. Newman, T. N. Anderson, and R. A. Garrott. 2007. Development and comparison of Landsat radiometric and snowpack model inversion techniques for estimating geothermal heat flux. *Remote Sensing of Environment* DOI:10.1016/j.rse.2007.05.010.

Watson, F. G. R., W. Newman, J. C. Coughlan, and R. A. Garrott. 2006b. Testing a distributed snowpack simulation model against diverse observations. *Journal of Hydrology* **328**:453–466.

Watson, F. G. R., and J. M. Rahman. 2003. Tarsier: A practical software framework for model development, testing and deployment. *Environmental Modeling and Software* **19**:245–260.

Watson, F. G. R., R. A. Vertessy, and R. B. Grayson. 1999. Large scale modelling of forest hydrological processes and their long term effect on water yield. *Hydrological Processes* **13**:689–700.

APPENDIX

APPENDIX 6A.1	Summary of parameter values used as inputs to the snow pack model, and the parameterization stages during which each value was set				
Stage	**Sub-model**	**Parameter name**	**Value**	**Units**	**Source**
1	Climate–radiation	Atmospheric longwave intercept	275.8	W/m^2	Linear regression against daily minimum temperature at SURFRAD site, Fort Peck, MT (based on Watson 1999)
1	Climate–radiation	Atmospheric longwave slope	4.121	W/m^2/°C	Linear regression against daily minimum temperature at SURFRAD site, Fort Peck, MT (based on Watson 1999)

(*continued*)

APPENDIX 6A.1   (*continued*)

Stage	Sub-model	Parameter name	Value	Units	Source
2	Climate–radiation	Bristow-Campbell atmos. trans. factor	0.766	–	Maximum transmission observed by Watson (1999) using Australian data
2	Climate–radiation	Bristow-Campbell atmos. trans. factor	0.0327	–	Linear regression against daily temperature range by Watson (1999) using Australian data
2	Climate–radiation	Bristow-Campbell atmos. trans. factor	1.46	–	Linear regression against daily temperature range by Watson (1999) using Australian data
3	Climate–temperature	Lapse rate for maximum temperatures: mean	0.015 (0.007*)	°C/m	Sinusoidal fit to 8 SNOTEL stations (Watson *et al.* 2006b). *CLIM-driven value used for long-term runs (see text)
3	Climate–temperature	Lapse rate for minimum temperatures: mean	0.006 (0*)	°C/m	Sinusoidal fit to 8 SNOTEL stations (Watson *et al.* 2006b). *CLIM-driven value used for long-term runs (see text)
3	Climate–temperature	Lapse rate for maximum temperatures: amplitude	−0.0016	°C/m	Sinusoidal fit to 8 SNOTEL stations (Watson *et al.* 2006b)
3	Climate–temperature	Lapse rate for minimum temperatures: amplitude	−0.00274	°C/m	Sinusoidal fit to 8 SNOTEL stations (Watson *et al.* 2006b)
3	Climate–temperature	Lapse rate for maximum temperatures: phase	326	yearday	Sinusoidal fit to 8 SNOTEL stations (Watson *et al.* 2006b)
3	Climate–temperature	Lapse rate for minimum temperatures: phase	117	yearday	Sinusoidal fit to 8 SNOTEL stations (Watson *et al.* 2006b)
4	Climate–temperature & precip.	Gaussian distance-weighting bandwidth for daily climate data	45	Km	Assigned same value as horizontal MAP variogram range parameter (Chapter 3 by Watson *et al.*, this volume)
5	Climate–wind	Wind speed at 50 cm above ground level in SNOTEL sites	0.17	m/s	Mean of observed wind speed in SNOTEL-like sites (Chapter 5 by Watson *et al.*, this volume)

(*continued*)

APPENDIX 6A.1 (*continued*)

Stage	Sub-model	Parameter name	Value	Units	Source
6	Snowpack–energy balance	Emissivity of snow	0.97	–	Oke (1987, p. 12)
6	Soil–energy balance	Emissivity of soil	0.95	–	Oke (1987, p. 12)
6	Soil–energy balance	Reflection coefficient: soil	0.1	–	Monteith and Unsworth (1990, p. 84)
6	Soil–energy balance	Soil thermal conductivity	2	W/m/K	Jansson and Karlberg (2001); Hukseflux Thermal Sensors (online)
6	Snowpack–snowmelt	Vol. of liquid snowpack water retained against grav. flow	0.01	–	Initially based on Tarboton *et al.* (1995), then calibrated and optimal insensitive below 0.02 (Watson *et al.* 2006b)
6	Snowpack–energy balance	Ice thermal conductivity above warm soil	2.22	W/m/K	Engineering Toolbox. http://www.engineeringtoolbox.com/ice-thermal-properties-d_576.html
6	Vegetation–radiation	Canopy radiation extinction coefficient	0.43	–	Sampson and Smith (1993)
6	Vegetation–snowfall	Velocity of snow falling through canopy	0.8	m/s	Hedstrom and Pomeroy (1998)
6	Vegetation–snowfall	Canopy snow unloading coefficient	0.11	per day	Hedstrom and Pomeroy (1998), re-scaled from half-weekly to daily equivalent
6	Vegetation–snowfall	Specific snow interception coefficient	0.0066	m/PLAI	Hedstrom and Pomeroy (1998) (noting that 6.6 kg/m^2 is equivalent to 0.0066 m/PLAI)
7	Climate–precipitation	Precipitation multiplier	1.034	–	Calibrated. See also: sensitivity analysis in Watson *et al.* (2006b)
7	Climate–precipitation	Precip. phase temperature	0.412321	°C	Calibrated. See also: sensitivity analysis in Watson *et al.* (2006b)
7	Snowpack–accumulation	Maximum water equivalent of upper snow layer	0.0122	m	Calibrated
7	Snowpack–snowmelt	Snowpack saturated hydraulic conductivity	233	m/day	Calibrated. See also: sensitivity analysis in Watson *et al.* (2006b)
7	Snowpack–snowmelt	Fraction of upper-layer snowmelt that routes directly to soil	0.637	–	Calibrated

(*continued*)

APPENDIX 6A.1   (*continued*)

Stage	Sub-model	Parameter name	Value	Units	Source
7	Snowpack–energy balance	Snow reflection coefficient for new snow	0.762	–	Calibrated
7	Snowpack–energy balance	Snow reflection coefficient decay	0.98	per day	Insensitive when co-calibrated with the new snow reflection coefficient. Fixed at 0.98.
7	Snowpack–compaction	Snowpack viscosity: reference point	6.8. E + 06	kg/m/s	Calibrated within range consistent with Shapiro *et al.* (1997, Figure B4)
7	Soil–energy balance	Soil thermal depth	0.058	m	Calibrated
8 & 11	Snowpack–wind scouring	Snowpack scouring wind speed threshold	4.4	m/s	Calibrated in Stage 8 (0.53) and Stage 11 (4.4)
8 & 11	Snowpack–wind scouring	Snowpack scouring degree	0.039	s^2/m^2	Calibrated in Stage 8 (0.01) and then Stage 11 (0.039)
8 & 11	Snowpack & soil –turbulent fluxes	Scaling of sensible and latent heat flux	0.17	–	Calibrated in Stage 8 (0.23) and then in Stage 11 (0.17)
9 & 11	Vegetation–snowfall	Canopy sublimation factor	82		Calibrated in Stage 9 (190) and then in Stage 11 (82)
10	Climate–precipitation	Phatom precipitation gage on Mirror Plateau: Easting	561173	m	Calibrated against random coring data 2004–2006
10	Climate–precipitation	Phatom precipitation gage on Mirror Plateau: Northing	4968096	m	Calibrated against random coring data 2004–2006
10	Climate–precipitation	Phatom precipitation gage on Mirror Plateau: MAP	1.32	m	Calibrated against random coring data 2004–2006

CHAPTER 7

Vegetation Dynamics of Yellowstone's Grazing System

Thomas R. Thein,* Fred G. R. Watson,* Simon S. Cornish,* Thor N. Anderson,* Wendi B. Newman,* and Ryan E. Lockwood*

*Division of Science and Environmental Policy, California State University Monterey Bay

Contents

Theme

Primary production in Yellowstone National Park is constrained by long winters to a short, intense growing season characteristic of high-altitude, temperate environments. In systems like this, vegetation phenology plays a governing role in terrestrial herbivore ecology by determining temporal fluctuations in the quantity and quality of forage. Phenology and productivity vary considerably at the patch level, and this spatio-temporal heterogeneity in the forage base provides much of the impetus for the distributions and movement patterns of the dominant herbivores that structure this system (Frank *et al.* 2002, Gates *et al.* 2005). Phenology and productivity are modulated by climate variability and disturbances, thereby providing the trophic link between these exogenous influences and animal populations. The growing collection of fine-scale, spatial datasets describing animal locations and landscape attributes, coupled with increasingly sophisticated modern scientific analyses, allow a

The Ecology of Large Mammals in Central Yellowstone
R. Garrott, P. J. White and F. Watson
ISSN 1936-7961, DOI: 10.1016/S1936-7961(08)00207-8

Copyright © 2009, Elsevier Inc.
All rights reserved.

fuller understanding of wildlife-vegetation interactions at the patch scale. Similarly, population studies also benefit from a better understanding of the variations in phenological patterns and overall productivity at regional scales. Using the normalized difference vegetation index (NDVI) derived from daily satellite imagery spanning July 1, 2000 to December 31, 2006, we generated phenological profiles that characterize the dynamic patterns of vegetation green-up and senescence at the large-patch scale (*ca.* 250 m). We used these seasonal profiles to describe the vegetation dynamics of Yellowstone's grasslands and shrub-grasslands, paying particular attention to patterns of spring green-up.

I. INTRODUCTION

The meadow complexes and open ranges of Yellowstone National Park have been described as a "grazing ecosystem" (Frank *et al.* 1998). A prominent characteristic of these ecosystems is an extremely high rate of herbivory that funnels a large proportion of the system's energy toward large mammals. Migratory ungulates tend to dominate the trophic interactions and, to a large extent, determine ecosystem structure and functioning (Sinclair 1979, Frank *et al.* 1998, Pringle *et al.* 2007). The graminoids upon which they feed, though plentiful, exhibit formidable barriers to nutrient extraction, requiring highly evolved digestive systems and high intake to meet metabolic demands (Hickman *et al.* 2004). Consequently, small differences in forage conditions can have profound impacts on the fitness of herbivores (White 1983). The success of ungulates in these systems is largely due to their dynamic spatial use of the landscape, a strategy that allows them to exploit profitable foraging opportunities in heterogeneous environments (Sinclair 1979, Frank *et al.* 1998).

Grasslands are characteristically seasonal, cycling through periods of growth, reproductive production, senescence, and dormancy (Figure 7.1). Variations in the physical environment, as well as biotic

FIGURE 7.1 The seasonal cycle of a cold-climate grassland in Seven-Mile meadow in the Madison Canyon of Yellowstone National Park (Photos by Fred Watson).

interactions, introduce spatio-temporal heterogeneity to vegetation landscapes, one component of which is the variations in vegetation phenology (Frank *et al.* 1998, Pettorelli *et al.* 2005a). Green-up in Yellowstone has been described as a "green wave" that typically begins in March at low elevations and proceeds to high elevations following the receding snow line (Frank and McNaughton 1992, Merrill *et al.* 1993, Frank *et al.* 1998). The growth period generally lasts for one or two months, but may last longer in mesic environments (Frank and McNaughton 1992). Dead biomass begins to accumulate early in grasses (Archibold 1995), and green biomass concentrations decrease through the growing season (Frank and McNaughton 1992). In many patches the majority of the vegetation has senesced by autumn. However, in most grassland patches, which tend to occur on andesitic soils, soil moisture is sufficient to maintain some green vegetation as long as temperatures permit (Despain 1990). Vegetation communities in the vicinity of geothermal features are notable in that they can be productive year-round (Despain 1990).

Because the biomass, structure, and chemistry of vegetation can change dramatically through the seasons, phenology is perhaps the most important determinant of forage conditions at the patch scale. Through the growing season, protein and favorable mineral concentrations in forage species generally decrease, while indigestible fiber concentrations increase (Figure 7.2, Albon and Langvatn 1992, Frank *et al.* 1998, George and Bell 2001, Manske *et al.* 2003). Leaf to stem ratios also decrease as grasses develop, leading to a higher proportion of largely indigestible plant parts (Arzani *et al.* 2004). Thus, forage quality is highest at early phenological stages, both because the vegetation is comparatively nutrient rich and because it requires less time to ruminate (White 1983, Mysterud *et al.* 2001). Furthermore, the higher green biomass concentrations and the high quality of early season forage theoretically translate into higher foraging efficiency in terms of yield per bite (Frank and McNaughton 1992). While quality decreases through the growing season, standing crop climaxes later, often around the time of seed production.

Grassland productivity in Yellowstone also varies widely between patches largely according to edaphic, climatic, and topographic gradients. Frank and McNaughton (1992) reported annual above-ground net primary productivity estimates ranging from 16 g/m^2 at a high-elevation ridge top to 589 g/m^2 at a low-elevation sedge meadow over the same year. Also, Olenicki and Irby (2005) reported production estimates ranging from 81 to 331 g/m^2 between xeric and mesic plant

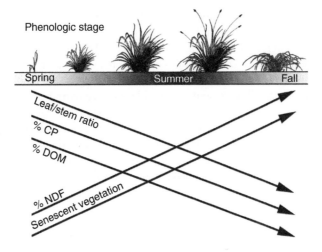

FIGURE 7.2 The seasonal development of grass species is accompanied by a steady decrease in forage quality. The percent digestible organic matter (% DOM), crude protein concentrations (% CP), and leaf to stem ratios decrease as the plants incorporate non-digestible fibers (NDF) such as lignin and green leaves are converted to dead material. Standing crop tends to peak around the time of seed production. Though the quality curves differ slightly between species, these general trends are virtually universal among graminoids.

associations in the Hayden Valley. Grazing feedback mechanisms are also important to grassland dynamics, promoting vegetation growth (Frank *et al.* 2002) and retarding phenological development (McNaughton 1979).

Spatial variations in phenology and productivity are a fundamental part of grazing ecosystem functioning, shaping the spatial patterns of use by herbivores and the forage contribution of patches. Herbivores respond spatially to changes in vegetation landscapes by timing their movements to take advantage of highly nutritious phenological stages and/or swards of plentiful forage (Frank *et al.* 1998, Van der Wal *et al.* 2000). Individuals that migrate along phenological gradients, selecting for patches in early phenological stages, prolong access to high quality forage (Frank and McNaughton 1992, Mysterud *et al.* 2001, Pettorelli *et al.* 2005a). It is believed that this strategy is adopted by herds of elk (*Cervus elaphus*) and bison (*Bison bison*) in Yellowstone as they track the wave of growing vegetation in the spring and early summer from low-elevation wintering grounds to their higher elevation summer ranges (Frank and McNaughton 1992, Gates *et al.* 2005). Herbivorous consumption in patches is positively related to above-ground production (Frank and McNaughton 1992), and Olenicki and Irby (2005) noted that, even though bison spent the majority of time foraging in less productive xeric habitats, sedges from mesic habitats made up a high proportion of their diet, indicating that forage could be harvested more efficiently per unit time in more productive patches (see also Bergman *et al.* 2001).

Climate variability and other disturbance factors such as fire alter productivity and phenological patterns (Knapp and Seastedt 1986, Pettorelli *et al.* 2005a). Shifts in seasonality, the degree of phenologic heterogeneity, rates of green-up, and above-ground productivity are all regarded as important properties of the vegetation landscape. Not surprisingly, these variations in the forage base have been linked to fluctuations in demographic and phenotypic aspects of herbivore populations (Albon and Langvatn 1992, Post and Stenseth 1999, Pettorelli *et al.* 2005a, 2006, 2007, Wittemyer *et al.* 2007).

While the functional importance of Yellowstone's vegetation dynamics is well understood, and generalities of seasonal patterns have been established via plot sampling (i.e., Frank and McNaughton 1992) and field observations (i.e., Despain 1990), a comprehensive quantitative description of the seasonal patterns in the landscape does not exist. To provide this information, we turned to a more synoptic vantage of Yellowstone's rangelands obtained through remote sensing techniques. We sought to characterize the temporal and spatial variation in green vegetation biomass within the grasslands and shrub-grasslands of Yellowstone using time-series of the normalized difference vegetation index (NDVI) generated from *ca.* 250-m resolution, daily satellite imagery.

The objectives of the work described in this chapter were to: (1) identify a relationship between the NDVI and the "greenness" of forage patches in Yellowstone; and (2) use serial NDVI imagery to extract seasonal parameters that characterized the patterns of phenology and productivity in the park's rangelands. The *ca.* 250-m resolution of the data describes the vegetation landscape at what might be called the local or "large-patch" scale and is suitable for behavioral and distribution studies of herbivores. We also present information that could aid researchers in the selection and interpretation of NDVI-based metrics for use as covariates in herbivore physiology and demographic studies. Our analysis is based on quantitative, remotely sensed descriptions of landscape dynamics, from which we make qualitative interpretations. Our goal was to compliment the description of static and dynamic landscape properties described in earlier chapters with a description of the grassland vegetation dynamics.

II. METHODS

A. Remotely Sensed Data

NDVI is a remotely sensed measure of vegetation that varies linearly with the fraction of photosynthetically active radiation and strongly correlates with green biomass (Huete *et al.* 2002, Pettorelli *et al.* 2005b). It is calculated from near infrared (NIR) and visible red (RED) reflectance values as

NDVI = (NIR − RED)/(NIR + RED). NDVI values range between − 1.0 and 1.0, where negative values typically indicate clouds, snow, or open water. Values near zero indicate bare ground and/or a mixture of water or snow effects and vegetation. In practice, NDVI values saturate near 0.9 under dense vegetation conditions (Huete *et al.* 2002).

The Moderate Resolution Imaging Spectroradiometers (MODIS) deployed on the National Aeronautics and Space Administration (NASA) satellites, Terra and Aqua, collect reflectance data for 36 spectral bands at resolutions up to 250 m. They achieve nearly global coverage every 24 h (Huete *et al.* 2002). The Terra satellite was launched first and began collecting data on February 24, 2000. MODIS data products are accompanied by quality assurance information that we used to filter the data. We produced serial rasters of NDVI from the 250 m resolution daily surface reflectance product of NASA's Terra MODIS data stream (MOD09GQK, <http://modis.gsfc.nasa.gov>) using only pixels classified by the MOD09 algorithm as "ideal quality." Existing MODIS vegetation products are available, but these use compositing techniques that reduce temporal and/or spatial resolution. Producing NDVI directly from the MOD09GQK data leverages the highest spatio-temporal resolution of the MODIS instrument, with significantly greater processing and memory overhead being the principal drawback. Our data set spans from July 1, 2000 to December 31, 2006 and covers 50,944 km^2 encompassing all of Yellowstone National Park and much of the surrounding region. Quality filtering removed most of the data points leaving gaps in time-series frequently exceeding 20 days. No resampling was done and the data remained in its native sinusoidal projection for analysis (results were nearest-neighbor resampled to 28.5 m UTM for display purposes only).

B. Smoothing and Gap-Filling

Time series of NDVI are commonly used to model seasonal vegetation changes (Justice *et al.* 1985, Markon *et al.* 1995, White *et al.* 1997, Pettorelli *et al.* 2005b). However, due to atmospheric phenomena, sun/view angle geometries, and sensor calibration problems, raw satellite based vegetation indices are very noisy, and cloud cover in particular often contaminates pixels (Huete *et al.* 2002, Jönsson and Eklundh 2004). Quality filtering removes most, but not all, of these errors. Therefore, a smoothing algorithm is widely considered a necessary step of NDVI time-series processing (Pettorelli *et al.* 2005b).

We processed each of the six complete years (2001–2006) of daily NDVI rasters individually using the TIMESAT program developed by Jönsson and Eklundh (2002, 2004). The program's key features include least squares fits of local or piecewise functions to the upper envelope of the data (noise in NDVI data is negatively biased; Jönsson and Eklundh 2002, 2004), spike detection, and phenological parameter extraction. We used the Adaptive Savitzky-Golay filter with a 30-day window, an-envelope adaptation strength of two, and three fitting iterations. The Adaptive Savitzky-Golay filter yields smoothed, continuous curves of NDVI at each pixel location, while preserving much of the temporal information of the original dataset (Figure 7.3). These curves are, in effect, phenological profiles depicting changes in green vegetation biomass. For each year, TIMESAT requires six months of the preceding and following years to correctly identify seasonal parameters. To process the 2006 season, we used the first six months of 2005 as a surrogate for the first six months of 2007. Hereafter within this chapter, NDVI refers to the smoothed, gap-filled data.

C. Field Sampling

To calibrate NDVI measurements to physical forage properties on the ground, we measured the percent green vegetation cover in summer and autumn of 26 meadow patches spanning a wide range of meadow characteristics. Each patch was 228 × 228 m in area, which is approximately the size of a

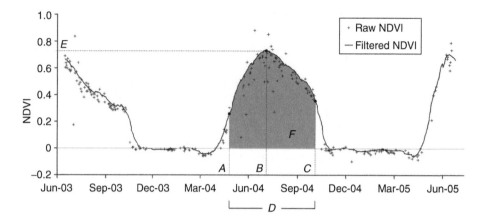

FIGURE 7.3 Two years of the raw normalized difference vegetation index (NDVI) values derived from "ideal quality" MOD09 data and the smoothed, gap-filled NDVI time-series produced by TIMESAT's Adaptive Savitzky-Golay filter. The six months to either side of the target year, in this case 2004, are necessary to allow the algorithm to correctly identify seasonal parameters, including Onset (A), Peak (B), Offset (C), LOS (D), Max (E), and INDVI (F). Definitions are provided in Table 7.1.

single MODIS pixel. The sampling design was first stratified by geographic location to include at least two patches from each of nine major meadow complexes within the Yellowstone bison range. We then stratified by NDVI of the previous year to sample patches exhibiting a range of NDVI values within each meadow complex. Finally, we stratified by season, sampling the 26 patches in July and again in October. We restricted the sampling domain to treeless areas, preferably with homogeneity between neighboring pixels to minimize error due to miss-registration of the imagery. At each pixel, percent green cover was estimated from point-intercept data collected using a novel laser-sampling device deployed in a nested sampling design similar to the designs developed by Watson *et al.* (2006). The surface cover of each pixel was characterized using 320 sampling points in a thrice-nested symmetric $5 \times 4 \times 4 \times 4$ design. The use of lasers facilitated rapid, unbiased sampling. Four low-frequency lasers were mounted to a horizontal arm on a tripod that could be swung into the four positions, marking out a total of $4 \times 4 = 16$ points on the ground at which observers manually recorded the surface cover (Figure 7.4). The tripod was then moved to $5 \times 4 = 20$ locations per pixel, using laser range finding and Global Positioning System equipment to eliminate observer bias in instrument location. The surface cover of each pixel was classified using a coded 65-class system that each observer memorized. For the present analyses, this was reduced to two classes—green vegetation or otherwise—which were averaged over all 320 points to yield an accurate estimate of percent green vegetation cover (some of the first-sampled pixels had fewer than 320 points due to slight adjustments in the nested pattern). The percent green cover data were compared to the processed NDVI to validate the relationship between NDVI and forage conditions in Yellowstone's rangelands.

D. Extracting Phenology and Productivity Information from NDVI Time Series

We characterized the seasonal vegetation patterns park-wide at each pixel location using six phenological and productivity metrics thought to be relevant to herbivore ecology (Table 7.1, Figure 7.3). These descriptive parameters were automatically computed by TIMESAT as the time-series were processed. Estimates for the start of the season (Onset) and the end of the season (Offset) are critical not only because they are key descriptors of seasonal dynamics themselves, but also because several other

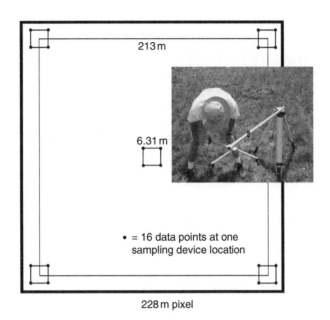

213 m

6.31 m

• = 16 data points at one
sampling device location

228 m pixel

FIGURE 7.4 Percent green vegetation cover was estimated from point-intercept data collected using a laser-sampling device deployed in a nested sampling design. The devise was positioned at each corner of the nested squares where the four lasers were rotated around the base to collect data at 16 points per corner (Photo by Jon Detka).

TABLE 7.1	Normalized difference vegetation index (NDVI)-derived seasonal parameters used to characterize vegetation dynamics in Yellowstone National Park, Montana, and Wyoming, USA

Parameters	Description	Ecological interpretation
Onset[a]	Time when spring values of the smoothed NDVI curve have reached 50% of the amplitude above the left minimum value	Approximates the start of the season: time when green forage becomes available; time of highest quality forage (Frank and McNaughton 1992, Mysterud *et al.* 2001)
Offset[a]	Time when fall values of the smoothed NDVI curve have declined to 50% above the right minimum value	Approximates the end of the season: time when seasonally active vegetation has effectively senesced or has been covered in snow; green forage becomes scarce
LOS[a]	Difference between Offset and Onset	Period that green forage is readily available
Peak	Time for the maximum smoothed NDVI value	Time of maximum green biomass (time of maximum forage biomass)
Max	Maximum smoothed NDVI value	Metric of green biomass (maximum forage biomass)
INDVI[a]	Sum of NDVI values from Onset to Offset	Metric for the annual net primary productivity (Goward *et al.* 1985, Jönsson and Eklundh 2004, Reed *et al.* 1994); above-ground net primary productivity co-varies with herbivorous offtake (Frank and McNaughton 1992)

[a] Derived using an arbitrary threshold of the seasonal amplitude (see discussion in main text).

phenological and productivity metrics are derived using these estimates (*e.g.*, length of season (LOS)). Onset and Offset are difficult to define in terms of phenological stage or other metrics commonly used in ground-based assessments (White *et al.* 1997, Jönsson and Eklundh 2004). For our purposes, Onset would ideally coincide with the start of spring green-up providing a useful metric of both a time when early stage forage first becomes readily available and the period of highest quality forage. Similarly, Offset would represent the time when the vast majority of seasonally active vegetation has senesced or become covered in snow. Relating these phenological dynamics to specific features of the NDVI profile is not straightforward. However, consistently derived Onset and Offset metrics that are reasonably close and relative to these respective targets represent satisfactory initial descriptors of their spatio-temporal dynamics, particularly for comparative purposes. For example, Reed *et al.* (1994) used the departure of a smoothed curve from a delayed moving average to identify Onset and reported that the method yielded early Onset dates for the northern Great Plains, thereby reflecting snowmelt dynamics rather than the start of green-up. However, the estimates were still useful for characterizing phenological patterns because green-up occurs as soon as the ground is snow free.

Numerous methods for identifying the start and end of a season from NDVI time-series have been developed for a variety of different purposes. These methods include absolute NDVI thresholds (Lloyd 1990, Fisher 1994, Markon *et al.* 1995), land-cover specific thresholds (Chen *et al.* 2000), the afore-mentioned delayed moving average method (Reed *et al.* 1994), the seasonal midpoint NDVI method (White *et al.* 1997), and a percent of the seasonal amplitude as with TIMESAT (Jönsson and Eklundh 2004). We selected a threshold of 50% of the seasonal amplitude, a value functionally equivalent to the seasonal midpoint NDVI method, which performed favorably against modeled spring indices and bud-break data for a deciduous forest (Schwartz *et al.* 2002). Other key descriptors of seasonal events include the time of maximum NDVI (Peak) and the LOS, calculated as the difference between Offset and Onset. It is important to note that Peak is not the same as the "time for the mid of the season" derived by TIMESAT and was achieved by reprocessing the time-series using either Onset or Offset with a threshold of 100%. Metrics of seasonal productivity include the maximum NDVI value (maximum green biomass), and NDVI values integrated over time (INDVI; a proxy for primary productivity, Goward *et al.* 1985, Reed *et al.* 1994, Jönsson and Eklundh 2004).

E. Phenological Patterns and Spatial Variations in Vegetation Performance

We constrained our analyses to grassland and sage-grassland habitats (hereafter referred to as range-land). Using the 5-class vegetation cover map described in Chapter 2 by Newman *et al.*, this volume (resampled to 250-m sinusoidal projection), we masked out all pixels with less than 50% meadow cover. The seasonal parameters were mapped and summarized by the major meadow complexes of the central region in Yellowstone (Figure 7.5). We defined meadow complex boundaries by combining aerial survey polygons that encompassed the known range of the Yellowstone bison herds (Hess 2002) until they contained regions related by spatial and elevational extent. Two a-priori modifications were made to separate the Upper Geyser Basin from the Lower Geyser Basin, and isolate the meadows in Madison Canyon and around Madison Junction from rangeland pixels further up the Gibbon River. We also summarized the seasonal parameters by region, again defining boundaries by combining aerial survey polygons for bison (Figure 7.5). Three regions were considered: (1) the entire bison survey area, (2) the northern range used by Yellowstone bison, and (3) the central range used by Yellowstone bison. Because elevation is known to be an important, albeit largely indirect, influence on phenology and productivity (Frank and McNaughton 1992, Hansen *et al.* 2000), we also plotted the seasonal para-meters for the rangeland pixels within the bison aerial survey polygons against a digital elevation map (resampled to 250-m sinusoidal from a 28.5 m U.S. Geological Survey digital elevation map).

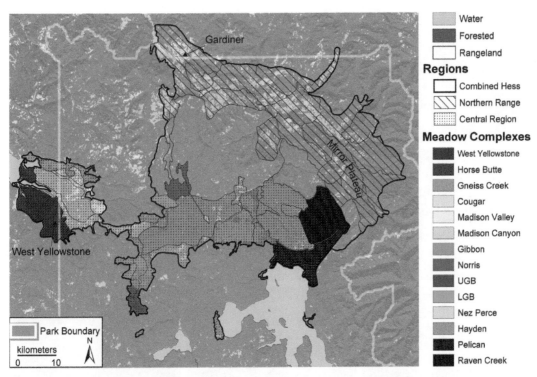

FIGURE 7.5 Boundaries of meadow complexes and regions in and near Yellowstone National Park. Aerial bison survey polygons (thin lines) developed by Hess (2002) were merged to define meadow complexes and regions for which seasonal parameters were summarized.

III. RESULTS

A. Smoothed, Gap-Filled NDVI

We reviewed the six years of daily NDVI imagery produced by the TIMESAT processing. The monthly sequence of NDVI images provides a quantitative, landscape-scale perspective of the vegetation dynamics of Yellowstone (Figure 7.6). In mid-winter (days 330–060), most rangeland patches had slightly negative NDVI values due to the presence of snow cover. Several areas did maintain slightly positive values, including concentrated patches around thermal basins and the area around Gardiner, Montana which receives very little precipitation, snow or otherwise (Chapter 3 by Watson *et al.*, this volume). In early spring (day 090), large expanses of the northern range and western portions of the central range exhibited rising NDVI values. Initial increases in NDVI were usually sharp and probably reflected the combined effects of new vegetation growth and snow melt, which by itself will cause a rise in the NDVI signal (Reed *et al.* 1994, Moulin *et al.* 1997). By late spring or early summer, most of the park's rangelands displayed rising NDVI values, though a few scattered low elevation patches had already peaked by this time (day 150). Yellowstone was at its greenest in mid-summer (approximately day 180), after which NDVI in most patches slowly decreased (days 210–240). Rapid park-wide decreases in NDVI marked the beginning of winter and the end of the growing season (days 270–330).

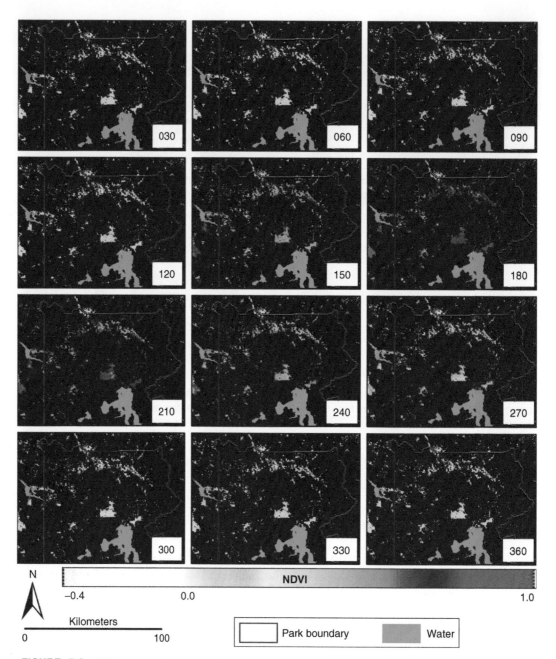

FIGURE 7.6 NDVI image-series produced from the 2006 smoothed, gap-filled NDVI, labeled by Julian day. The seasonality of grasslands in Yellowstone National Park is clearly evident in the NDVI sequence, shown here every 30 days. Compared to the other years we processed (2001–2005), this was an intermediate year in terms of phenology and NDVI magnitude.

B. NDVI as a Predictor of Percent Green Vegetation Cover

A total of 15,552 laser points were obtained from the July and October field campaigns in 2004. A few pixels were incompletely sampled to avoid thunderstorms and disturbing bison. Field-based green cover estimates at the 26 sites ranged from 1.3% in October at a site in the Lower Geyser Basin to 89.4% in July at a site in the Pelican Valley (mean = 37.3%, sd = 26.7%). The species composition ranged widely among sites; dominant vegetation classes varied between grasses, sedges, herbaceous dicotyledons, and sagebrush. A strong positive correlation was observed between percent green cover and the 2004 NDVI time-series data ($R^2 = 0.85$, $F_{1,50} = 287$, $P < 0.001$; Figure 7.7).

C. Patterns of Phenology and Productivity

The mapping of seasonal parameters to visualize spatial patterns of phenology and productivity (Figure 7.8) resulted in 36 rasters (6 parameters × 6 years). Rangeland pixels ($n = 7958$) contained within the bison aerial survey polygons were sampled from each raster, and summary statistics for the selected meadow complexes and regions were compiled (Appendix 7A.1). The green wave in the northern range described by Frank and McNaughton (1992) is clearly visible in the NDVI image-series and, as expected, closely follows the elevational gradient (Figures 7.6, 7.8A, and 7.9A). Onset in the lowest rangeland areas typically occurred approximately 100 days earlier than in the highest areas. Onset dates were earliest around the northern boundary in the vicinity of Gardiner, Montana. Then the wave swept upslope along the valleys to the south and east. The high meadows on the Mirror Plateau exhibited the latest Onset dates.

The patterns of onset in central Yellowstone were somewhat more complex (Figures 7.8A and 7.10) because the first patches to green-up were generally in the mid-elevation meadows (i.e., Lower Geyser Basin, Nez Perce Creek, Madison Canyon, and Norris) where geothermal features facilitate snowmelt (Chapters 4 and 6 by Watson *et al.*, this volume). This was followed shortly thereafter by the lower-elevation, Madison Valley region to the west. Next, the higher-elevation Hayden and Pelican Valleys

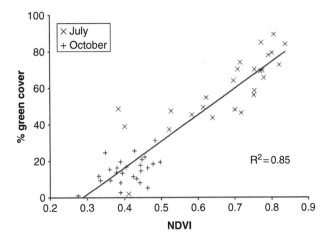

FIGURE 7.7 Comparison between remotely sensed and field-based estimates of green vegetation cover in Yellowstone National Park. Remotely sensed estimates are smoothed, gap-filled NDVI values. Field-based estimates are derived from point-intercept sampling of 26 meadow pixels (228 × 228 m each) in July and October 2004.

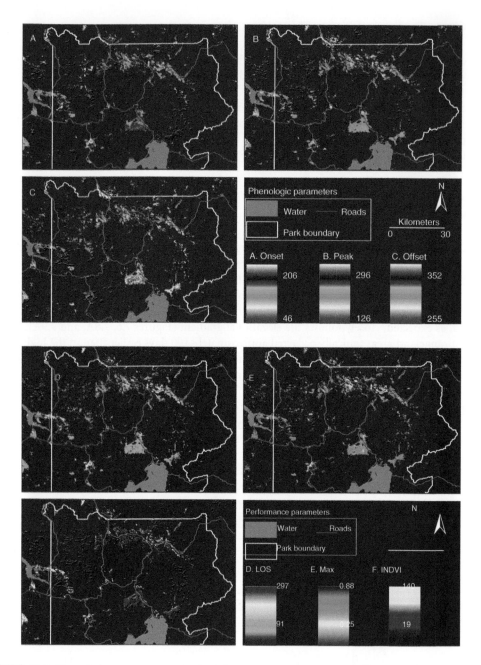

FIGURE 7.8 Mapped six-year averages of extracted seasonal parameters from normalized difference vegetation index values during 2001–2006 for rangeland pixels in Yellowstone National Park: (A) Onset: the time (Julian Day) when the NDVI signal rises through the seasonal midpoint, representing the time for the start of the growing season; (B) Peak: the time (Julian Day) when the NDVI signal reaches its seasonal maximum, representing the time of peak green biomass; (C) Offset: the time (Julian Day) when the NDVI signal descends through the seasonal midpoint, representing the end of the growing season; (D) LOS: the difference between Onset and Offset, representing the length of the growing season; (E) Max: the largest smoothed NDVI value of the season, representing the highest green biomass accumulation during a season; and (F) INDVI: the sum of daily NDVI values between Onset and Offset, representing annual productivity.

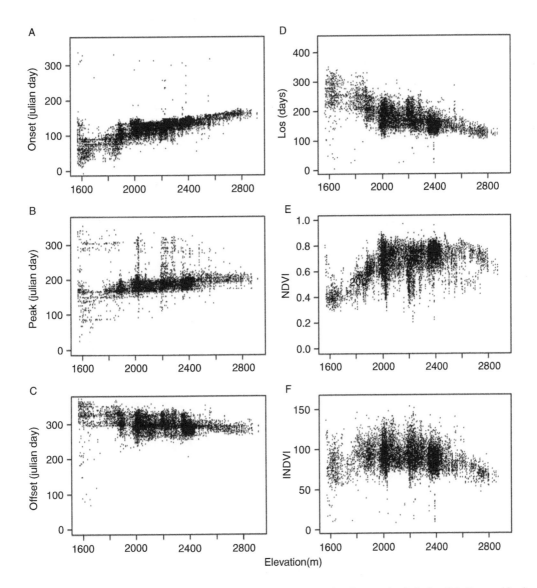

FIGURE 7.9 Comparison of the six phenologic parameters with elevation for rangeland pixels within the combined bison aerial survey polygons (Figure 7.5) during 2001–2006 in Yellowstone National Park: (A) Onset; (B) Peak; (C) Offset; (D) LOS; (E) Max; and (F) INDVI.

began to green in the east. Further east and higher still, the meadows along the upper reaches of Pelican and Raven creeks were consistently the last areas in the central range to green-up. An exception to this pattern occurred in 2003, with parts of the Madison Valley displaying the earliest onset dates in the central region.

The pattern of green-up was not simply a stepwise progression from one meadow to another. There was considerable local variation in Onset values leading to significant overlap between meadow complexes, such that after the growing season began in earnest, patches just starting to green-up could be found in a number of complexes spanning a range of elevations (Figures 7.8A, 7.9A, and 7.10). Substantial inter-annual variations in Onset dates were detected as well, with some years displaying Onset dates over a longer period. For example the range of Onset values within the central Yellowstone meadow complexes spanned 86 days in 2003, but only 71 days in 2006.

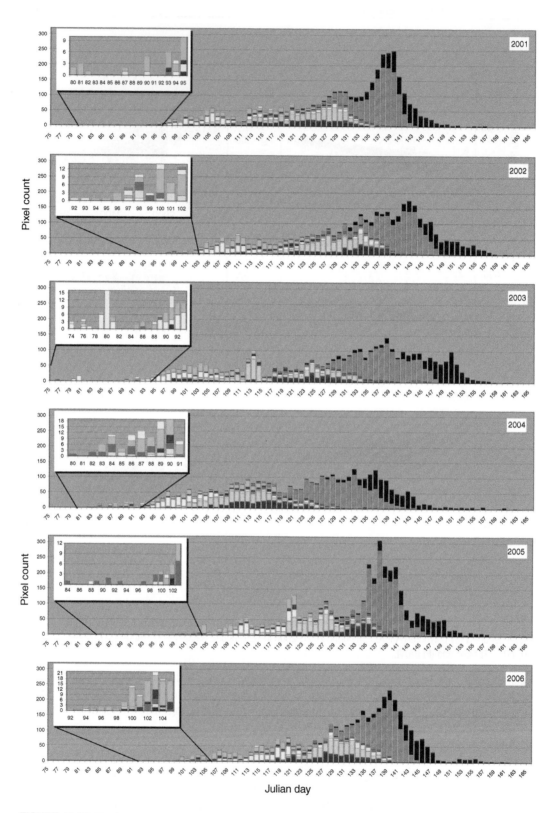

FIGURE 7.10 The frequency distribution of Onset date for different meadow complexes in the central portion of Yellowstone National Park (See Figure 7.5 for color codes), revealing a "wave" of green-up progressing longitudinally along sequences of neighboring meadows. The inset graphs show more detail in the earliest periods.

The date of peak NDVI was also correlated with elevation, with lower elevation rangeland pixels peaking approximately 30 days before higher elevation pixels (Figure 7.9B). Peak dates tended to occur approximately 70 days after Onset at low elevations and 40 days after Onset at high elevations. These results are also consistent with the periods of measurable primary production reported by Frank and McNaughton (1992). However, significant deviations from these trends occurred through the study area, which we attributed to a number of factors. Late peak dates were estimated sporadically throughout the study area. These may be indicative of mesic environments in some cases, but it is also possible that these pixels were actually forested and misclassified as rangeland due to image registration errors. Late peaks are more typical of forest pixels. False peaks from smoothing overshoots also occurred, though only rarely. This happened when large data gaps in the raw NDVI time-series followed the sharp spring rise in NDVI or preceded the end of season drop. Some pixels in the vicinity of geothermal regions displayed plateau-like time-series with no clear peak. Therefore, the peak dates identified in these pixels may not correspond to a distinct peak in vegetation biomass, but rather a high spot in the time-series arising from reflectance noise or smoothing overshoots. In the arid region around Gardiner, Montana, peak dates were highly variable, both spatially and inter-annually. It may be that the TIMESAT algorithm as applied does not reliably characterize the patterns of vegetation phenology in this area, perhaps because the vegetation signal is weak.

The earliest Offset dates occurred across a wide range of elevation and geographic extent (Figure 7.8C), but Offset occurred late at the lowest elevations and around geothermal features. An elevational trend was also evident, with Offset typically occurring at higher elevations about 40 days before lower areas (Figure 7.9C). The opposing elevational trends in Onset and Offset result in reduced LOS with increasing altitude. LOS in the lower areas was approximately double that of the highest elevations (Figure 7.9D). In five of the six years studied, Onset displayed greater heterogeneity than Offset and, thus, contributed more to spatial variation in LOS (Appendix 7A.1). However, Offset was more variable between years than Onset, and thus had greater influence on the inter-annual variation of LOS values (mean standard deviation within the combined aerial survey polygons for Offset and Onset were 13.1 and 8.8, respectively).

Maximum NDVI values were highly variable across the range of elevation, but were relatively low around Gardiner, Montana where a lack of precipitation limits production, near intense geothermal features, and in a few areas in Madison Valley (NDVI values 0.3–0.4). Maximum values reached levels reported for NDVI saturation (NDVI > 0.8) in many parts of the park, indicating very dense green vegetation cover at peak greenness. In particular, the Pelican Valley achieved high NDVI values, despite a relatively short growing season. Integrated NDVI values over time had a unimodal relationship with elevation (Figure 7.9F). At the low elevations around Gardiner, Montana the sum of NDVI values from Onset to Offset was limited by persistently low NDVI values, while at higher elevations it was limited by the short season. Despite the on-average shorter season of the central region, integrated NDVI values were higher than on the northern range. The Nez Perce corridor, Madison Canyon, and Gibbon Meadows displayed high integrated NDVI values (Appendix 7A.1). These three areas are all moderately geothermal (Chapter 4 by Watson *et al.*, this volume) and very moist, thereby benefiting from both adequate resources and a long growing season.

IV. DISCUSSION

Our understanding of Yellowstone's rangeland vegetation dynamics has primarily arisen from field-based studies that fall into two categories: (1) quantitative plot sampling, which is best exemplified by the studies of Frank and McNaughton (1992) and Frank *et al.* (1998) and (2) qualitative descriptions, such as the works of Despain (1990). While plot sampling yields highly detailed measurements of system processes, their scope is limited in spatial scale. Qualitative descriptions are broader in spatial scope, but are lacking in quantitative rigor. Both of these approaches have yielded important insight into the structure and functioning of grazing ecosystems, but their applicability to studies of ungulate movement patterns and population dynamics are limited.

The data produced by the present study provides quantitative results with complete coverage of Yellowstone National Park at daily, *circa* 250 m resolution—a relatively fine-scale given the spatial extent and temporal frequency. The smoothed, gap-filled NDVI explained a large proportion of the variation in the percent green herbaceous cover of rangeland patches, and many of the seasonal vegetation patterns described in the literature are visible in the NDVI image series and mapped seasonal parameters. These findings provide confidence that the methods we employed are suitable for describing and monitoring vegetation dynamics in Yellowstone rangelands. Yet, there are a few points of concern. The Adaptive Savitzky-Golay filter performs favorably where prolonged gaps in the raw NDVI time-series are rare and the seasonal vegetation signal is strong, as is the case most of the time over most of the park. However, certain conditions can negatively affect the parameterization of seasonal metrics. When data density is high, most of the noise is smoothed out of the time-series, but extended periods of cloud cover and missing data from sensor problems can lead to erroneous spikes and apparent shifts in seasonality.

The estimation of the start and end of the growing season from a threshold based on a percentage of the seasonal amplitude also has its drawbacks. This method is robust to evergreen components and other constant background contaminants that are problematic when using an absolute NDVI threshold (Reed *et al.* 1994, White *et al.* 1997), but it is more sensitive to NDVI fluctuations caused by snow conditions and variations in soil moisture, which also influence NDVI values (Adegoke and Carleton 2002, Huete *et al.* 2002, Delbart *et al.* 2006). These dynamic, non-vegetation influences on NDVI time-series are minimal when the seasonal vegetation signal is strong, but they are magnified in areas of low productivity or inter-seasonal stability. So, seasonal parameters may not describe actual biophysical vegetation conditions in some parts of the park such as the arid conditions at low elevations in the northern range near Gardiner, Montana, and in the vicinity of some of the park's geothermal features in central Yellowstone.

Integrated NDVI is commonly used as a proxy for net primary productivity (Goward *et al.* 1985, Reed *et al.* 1994), but as noted by Olenicki and Irby (2005), any attempt to remotely sense net primary productivity in grazed grasslands is a tenuous proposition. Studies have shown that grazing substantially increases net primary productivity (Frank and McNaughton 1993, Frank *et al.* 2002), yet this will lower green biomass and reduce NDVI values. Thus, it is likely that the use of previously reported relationships between integrated NDVI and net primary productivity would tend to underestimate net primary productivity in Yellowstone's rangelands.

Despite these caveats, there is clearly a wealth of information that can be obtained from NDVI time-series analyses, and we believe the application of the smoothed NDVI data and seasonal parameters to studies of the park's ungulates is warranted. Of key relevance to large mammal ecology is the ability to map green wave patterns and assess inter-annual variations in forage conditions. It is frequently suggested that the seasonal migrations undertaken by many of the region's ungulates are in response to the green wave of spring green-up (Russell 1932, Boyce 1991, Frank *et al.* 1998, Gates *et al.* 2005, White *et al.* 2007). As previously mentioned, the timing of spring green-up in Yellowstone is essentially a function of snowmelt. Orographic influences on temperature and precipitation result in greater snow accumulation and delayed melt-out with increasing elevation (Chapter 6 by Watson *et al.*, this volume) that, in turn, produce a strong elevational component in the green wave phenomenon that is thought to draw migrants from the low-elevation winter ranges back to the high-elevation summer ranges. This elevational wave can be seen in the NDVI imagery, as can the influence of geothermal features, which advance green-up and lengthen the growing season in proximal patches.

While we focused primarily on spatial patterns, inter-annual variations in the seasonal parameters reflect inter-annual variations in phenology and productivity. In this respect, NDVI time-series analysis has a long history of linking climate fluctuations and other perturbations with changes in vegetation dynamics (Davenport and Nicholson 1993, Reed *et al.* 1994, Anyamba and Eastman 1996, Moulin *et al.* 1997, Masek *et al.* 2000, Ichii *et al.* 2002, Pettorelli *et al.* 2005a). More recently, inter-annual variations in NDVI-based parameters have proven useful as covariates for studies of animal populations, and much of this research has focused on large herbivores (Pettorelli *et al.* 2005b). Indeed, as an index of forage conditions for herbivores, NDVI-based models have been shown to outperform models that rely on precipitation (Oesterheld *et al.* 1998, Rasmussen *et al.* 2006). Useful NDVI-based covariates in herbivore studies have included, but are not limited to, the NDVI on a consistent spring date (Pettorelli *et al.* 2005a, 2005c, 2006), maximum NDVI (Wittemyer *et al.* 2007), INDVI (Oesterheld *et al.* 1998), and the change in NDVI over a 15-day period (Pettorelli *et al.* 2007). Metrics should be selected with care because each metric measures different dynamics in the forage base and is subject to different interpretations.

Because this study was concurrent with most of the animal-based analyses presented in other chapters of this book and much of the data used in those analyses predates MODIS, the products of this research could not be incorporated into those studies. However, NDVI data collected with the National Oceanic and Atmospheric Administration-Advanced Very High Resolution Radiometer (NOAA-AVHRR) was processed in a similar manor by a colleague, Robert Klaver (U.S. Geological Survey, Center for Earth Resources Observation and Science, Souix Falls, South Dakota), and incorporated into many of the following chapters. MODIS data offers advancements in both resolution (250 m vs. 1 km) and atmospheric correction over AVHRR data, but the relatively recent deployment of MODIS precludes pairing this data with long-term studies that stretch back before the turn of the century. It is our hope that integrating of the methods employed here into future work will allow more rigorous testing of hypotheses relating to the spatial and performance responses of the park's ungulates to forage dynamics, and reveal trophic linkages between climate variability and animal populations in the future.

V. SUMMARY

1. Spatial and temporal variations in forage are an integral part of herbivore ecology in grassland ecosystems. Both vegetation phenology and productivity strongly influence the profitability of grassland patches.

2. Analyses of the effect of these variations on wildlife dynamics require quantitative descriptions of these patterns, either as temporal sequences of productivity maps or individual maps of phenological parameters such as the timing of onset of significant rangeland production.

3. Using remote sensing, we developed daily estimates of green vegetation cover from 2000–2006 at *ca.* 250-m scale extending throughout the bison range in Yellowstone National Park. Estimates were derived by applying a temporal gap-filling and smoothing algorithm to NDVI data derived from the MODIS instrument aboard NASA's Terra satellite.

4. We validated the remotely sensed estimates by comparing them to field estimates of percent green cover from 26 5-hectare meadow patches sampled in July and October, based on up to 320 point measurements per patch in a nested sampling pattern (15,552 points in total). The remotely sensed estimates strongly correlated with the field-based estimates.

5. Seasonal patterns of phenology and productivity revealed by the data corresponded to patterns described in the literature. The maps clearly revealed the annual cycle of green-up and senescence, and how this varies spatially throughout the park. A simple analysis confirmed that spatial patterns in phenology were correlated with elevation. Geothermal influences on vegetation dynamics were also detected.

6. The quantified spatial patterns in phenology aligned qualitatively with seasonal wildlife migratory routes. Thus, it appears likely that future analyses will be able to incorporate these patterns into quantitative comparisons of hypotheses about the drivers of migration.
7. The ability to assess inter-annual variations in the forage base also opens avenues for future research to couple vegetation-herbivore interactions with climate fluctuations.

VI. REFERENCES

Adegoke, J. O., and A. M. Carleton. 2002. Relations between soil moisture and satellite vegetation indices in the U.S. corn belt. *Journal of Hydrometeorology* **3**:395–405.

Albon, S. D., and R. Langvatn. 1992. Plant phenology and the benefits of migration in a temperate ungulate. *Oikos* **65**:502–513.

Anyamba, A., and J. R. Eastman. 1996. Interannual variability of NDVI over Africa and its relation to El Niño/Southern Oscillation. *International Journal of Remote Sensing* **17**:2533–2548.

Archibold, O. W. 1995. *Ecology of World Vegetation.* Chapman & Hall, London, United Kingdom.

Arzani, H., M. Zohdi, E. Fish, G. H. Zahedi Amiri, A. Nikkhah, and D. Wester. 2004. Phenological effects on forage quality of five grass species. *Journal of Range Management* **57**:624–629.

Bergman, C. M., J. M. Fryxell, C. Cormack Gates, and D. Fortin. 2001. Ungulate foraging strategies: Energy maximizing or time minimizing? *Journal of Animal Ecology* **70**:289–300.

Boyce, M. S. 1991. Migratory behavior and management of elk (*Cervus elaphus*). *Applied Animal Behaviour Science* **29**:239–250.

Chen, X., Z. Tan, M. D. Schwartz, and C. Xu. 2000. Determining the growing season of land vegetation on the basis of plant phenology and satellite data in Northern China. *International Journal of Biometeorology* **44**:97–101.

Davenport, M. L., and S. E. Nicholson. 1993. On the relation between rainfall and the normalized difference vegetation index for diverse vegetation types in East Africa. *International Journal of Remote Sensing* **14**:2369–2389.

Delbart, N., T. L. Toan, L. Kergoat, and V. Fedotova. 2006. Remote sensing of spring phenology in boreal regions: A free of snow-effect method using NOAA-AVHRR and SPOT-VGT data (1982–2004). *Remote Sensing of Environment* **101**:52–62.

Despain, D. G. 1990. *Yellowstone Vegetation: Consequences of Environment and History in a Natural Setting.* Roberts Rhinehart, Boulder, CO.

Fisher, A. 1994. A model for the seasonal variations of vegetation indices in coarse resolution data and its inversion to extract crop parameters. *Remote Sensing of Environment* **48**:220–230.

Frank, D. A., and S. J. McNaughton. 1992. The ecology of plants, large mammalian herbivores, and drought in Yellowstone National Park. *Ecology* **73**:2043–2058.

Frank, D. A., and S. J. McNaughton. 1993. Evidence for the promotion of aboveground grassland production by native large herbivores in Yellowstone National Park. *Oecologia* **96**:157–161.

Frank, D. A., M. M. Kuns, and D. R. Guido. 2002. Consumer control of grassland plant production. *Ecology* **83**:602–606.

Frank, D. A., S. J. McNaughton, and B. F. Tracy. 1998. The ecology of the earth's grazing ecosystems. *BioScience* **48**:513–521.

Gates, C. C., B. Stelfox, T. Mulhly, T. Chowns, and R. J. Hudson. 2005. *The Ecology of Bison Movements and Distribution in and Beyond Yellowstone National Park.* University of Calgary, Alberta, Canada.

George, M. R., and M. E. Bell. 2001. *Using stage of maturity to predict the quality of annual range forage.* University of California, Division of Agriculture and Natural Resources, Davis, CA. Rangeland Management Series 8019.

Goward, S. N., C. J. Tucker, and D. G. Dye. 1985. North American vegetation patterns observed with the NOAA-7 advanced very high resolution radiometer. *Plant Ecology* **64**:3–14.

Hansen, A. J., J. J. Rotella, M. P. V. Kraska, and D. Brown. 2000. Spatial patterns of primary productivity in the Greater Yellowstone Ecosystem. *Landscape Ecology* **15**:505–522.

Hess, S. C. 2002. Aerial survey methodology for bison population estimation in Yellowstone National Park. Dissertation. Montana State University, Bozeman, MT.

Hickman, C. P., L. S. Roberts, A. Larson, and H. l'Anson. 2004. *Integrated Principles of Zoology.* McGraw-Hill, New York, NY.

Huete, A., K. Didan, T. Miura, E. P. Rodriguez, X. Gao, and L. G. Ferreira. 2002. Overview of the radiometric and biophysical performance of the MODIS vegetation indices. *Remote Sensing of Environment* **83**:195–213.

Ichii, K., A. Kawabata, and Y. Yamaguchi. 2002. Global correlation analysis for NDVI and climatic variables and NDVI trends: 1982–1990. *International Journal of Remote Sensing* **23**:3873–3878.

Jönsson, P., and L. Eklundh. 2002. Seasonality extraction by function fitting to time-series of satellite sensor data. *IEEE Transactions on Geoscience and Remote Sensing* **40**:1824–1832.

Jönsson, P., and L. Eklundh. 2004. TIMESAT—A program for analyzing time-series of satellite sensor data. *Computers and Geosciences* **30**:833–845.

Justice, C. O., J. R. G. Townshend, B. N. Holben, and C. J. Tucker. 1985. Analysis of the phenology of global vegetation using meteorological satellite data. *International Journal of Remote Sensing* **6**:1271–1318.

Knapp, A. K., and T. R. Seastedt. 1986. Detritus accumulation limits productivity of tallgrass prairie. *BioScience* **36**:662–668.

Lloyd, D. 1990. A phenological classification of terrestrial vegetation cover using shortwave vegetation index imagery. *International Journal of Remote Sensing* **11**:2269–2279.

Manske, L. L., A. M. Kraus, S. A. Schneider, and L. J. Vance. 2003. *Biologically Effective Management of Grazinglands*. North Dakota State University, Dickinson Research Extension Center, Dickinson, ND.

Markon, C. J., M. D. Fleming, and E. F. Binnian. 1995. Characteristics of vegetation phenology over the Alaskan landscape using AVHRR time-series data. *Polar Record* **31**:179–190.

Masek, J. G., F. E. Lindsay, and S. N. Goward. 2000. Dynamics of urban growth in the Washington DC metropolitan area, 1973–1996, from Landsat observations. *International Journal of Remote Sensing* **21**:3473–3486.

McNaughton, S. J. 1979. Grassland-herbivore dynamics. Pages 46–81 *in* A. R. E. Sinclair and M. Norton-Griffiths (Eds.) *Serengeti: Dynamics of an Ecosystem*. University of Chicago Press, Chicago, IL.

Merrill, E. H., M. K. Bramble-Brodahl, R. W. Marrs, and M. S. Boyce. 1993. Estimation of green herbaceous phytomass from Landsat MSS data in Yellowtone National Park. *Journal of Range Management* **46**:151–157.

Moulin, S., L. Kergoat, N. Viovy, and G. Dedieu. 1997. Global-scale assessment of vegetation phenology using NOAA/AVHRR satellite measurements. *Journal of Climate* **10**:1154–1170.

Mysterud, A., R. Langvatin, N. G. Yoccoz, and N. C. Stenseth. 2001. Plant phenology, migration and geographical variation in body weight of a large herbivore: The effect of a variable topography. *Journal of Animal Ecology* **70**:915–923.

Oesterheld, M., C. M. DiBella, and H. Kerdiles. 1998. Relation between NOAA-AVHRR satellite data and stocking rate of rangelands. *Ecological Applications* **8**:207–212.

Olenicki, T. J., and L. R. Irby. 2005. *Determining Forage Availability and Use Patterns for Bison in the Hayden Valley of Yellowstone National Park*. Montana State University, Bozeman, MT.

Pettorelli, N., J. Gaillard, A. Mysterud, P. Duncan, N. C. Stenseth, D. Delorme, G. V. Laere, C. Toigo, and F. Klein. 2006. Using a proxy of plant productivity (NDVI) to find key periods for animal performance: The case for roe deer. *Oikos* **112**:565–572.

Pettorelli, N., A. Mysterud, N. G. Yoccoz, R. Langvatn, and N. C. Stenseth. 2005a. Importance of climatological downscaling and plant phenology for red deer in heterogeneous landscapes. *Proceedings Royal Society of London B* **272**:2357–2364.

Pettorelli, N., F. Pelletier, A. von Hardenberg, M. Festa-Bianchet, and S. D. Cote. 2007. Early onset of vegetation growth vs. rapid green-up: Impacts on juvenile mountain ungulates. *Ecology* **88**:381–390.

Pettorelli, N., J. O. Vik, A. Mysterud, J. Gaillard, C. J. Tucker, and N. C. Stenseth. 2005b. Using the satellite-derived NDVI to assess ecological responses to environmental change. *Trends in Ecology & Evolution* **20**:503–510.

Pettorelli, N., R. B. Weladji, Ø. Holand, A. Mysterud, H. Breie, and N. C. Stenseth. 2005c. The relative role of winter and spring conditions: Linking climate and landscape-scale plant phenology to alpine reindeer body mass. *Biology Letters* **1**:24–26.

Post, E., and N. C. Stenseth. 1999. Climatic variability, plant phenology, and northern ungulates. *Ecology* **80**:1322–1339.

Pringle, R. M., T. P. Young, D. I. Rubenstein, and D. J. McCauley. 2007. Herbivore-initiated interaction cascades and their modulation by productivity in an African savanna. *Proceedings National Academy of Sciences* **104**:193–197.

Rasmussen, H. B., G. Wittemyer, and I. Douglas-Hamilton. 2006. Predicting time-specific changes in demographic processes using remote-sensing data. *Journal of Applied Ecology* **43**:366–376.

Reed, B. C., J. F. Brown, D. VanderZee, T. R. Loveland, J. W. Merchant, and D. O. Ohlen. 1994. Measuring phenological variability from satellite imagery. *Journal of Vegetation Science* **5**:703–714.

Russell, C. P. 1932. Seasonal migration of mule deer. *Ecological Monographs* **2**:1–46.

Schwartz, M. D., B. C. Reed, and M. A. White. 2002. Assessing satellite-derived start-of-season measures in the conterminous USA. *International Journal of Climatology* **22**:1793–1805.

Sinclair, A. R. E. 1979. Dynamics of the Serengeti ecosystem. Pages 16–26 *in* A. R. E Sinclair and M. Norton-Griffiths (Eds.) *Serengeti: Dynamics of an Ecosystem*. University of Chicago Press, Chicago, IL.

Van der Wal, R., N. Madan, S. van Lieshout, C. Dormann, R. Langvatn, and S. D. Albon. 2000. Trading forage quality for quantity? Plant phenology and patch choice by Svalbard reindeer. *Oecologia* **123**:108–115.

Watson, F. G. R., T. N. Anderson, W. B. Newman, S. E. Alexander, and R. A. Garrott. 2006. Optimal sampling schemes for estimating mean snow water equivalents in stratified heterogeneous landscapes. *Journal of Hydrology* **328**:432–452.

White, R. G. 1983. Foraging patterns and their multiplier effects on productivity of northern ungulates. *Oikos* **40**:377–384.

White, P. J., T. L. Davis, K. K. Bamowe-Meyer, R. L. Crabtree, and R. A. Garrott. 2007. Partial migration and philopatry of Yellowstone pronghorn. *Biological Conservation* **35**:502–510.

White, M. A., P. E. Thornton, and S. W. Running. 1997. A continental phenology model for monitoring vegetation responses to interannual climatic variability. *Global Biogeochemical Cycles* **11**:217–234.

Wittemyer, G., H. B. Rasmussen, and I. D. Douglas-Hamilton. 2007. Breeding phenology in relation to NDVI variability in free-ranging African elephant. *Ecography* **30**:42–50.

APPENDIX 7A.1 Summary of seasonal phenology parameters by region and meadow complex in and near Yellowstone National Park during 2001–2006

Region		Onset (Julian Day)						Peak (Julian Day)						Offset (Julian Day)					
		2001	2002	2003	2004	2005	2006	2001	2002	2003	2004	2005	2006	2001	2002	2003	2004	2005	2006
Combined Hess	mean	124	130	121	115	123	124	187	188	176	198	189	185	303	295	293	306	289	299
7958 pixels	sd	20	19	22	24	24	24	29	20	21	30	21	29	23	17	18	22	18	22
Northern Range	mean	118	127	113	107	113	114	181	184	170	194	182	177	302	295	291	314	288	303
3667 pixels	sd	24	24	25	28	30	29	24	17	19	35	21	29	20	21	20	24	22	25
Central Region	mean	128	132	128	122	132	132	192	191	181	201	194	191	303	294	293	300	290	296
4047 pixels	sd	13	13	17	16	12	12	31	21	21	26	19	27	24	14	17	16	14	19
Meadow complex																			
West Yellowstone	mean	120	128	117	115	131	126	199	195	179	200	200	203	302	302	291	300	295	300
274 pixels	sd	10	8	11	7	7	7	52	36	35	33	21	45	30	8	23	11	12	22
Horse Butte	mean	125	134	128	118	129	130	180	187	176	197	189	183	307	294	294	307	287	300
113 pixels	sd	11	7	8	10	6	7	19	9	13	14	11	11	19	13	14	15	12	11
Gneiss Creek	mean	123	127	118	110	125	129	179	181	171	192	186	180	314	294	299	316	293	307
459 pixels	sd	7	5	8	6	5	4	14	6	8	19	7	11	19	15	18	14	9	13
Cougar	mean	124	130	119	116	127	132	175	180	170	196	189	179	314	291	285	298	277	296
131 pixels	sd	6	7	9	9	6	5	6	6	3	9	5	3	15	18	17	15	14	18
Madison Valley	mean	109	113	98	102	117	118	198	182	164	210	182	197	334	311	314	315	305	318
275 pixels	sd	10	8	9	13	6	6	53	26	27	43	19	43	7	6	5	18	6	5
Madison Canyon	mean	110	107	102	100	111	112	216	228	182	252	213	223	331	309	302	335	305	318
30 pixels	sd	10	10	13	14	8	7	52	43	45	43	29	42	3	8	36	6	4	4
Gibbon	mean	127	130	125	116	132	129	193	205	189	212	204	197	320	302	302	298	293	311
36 pixels	sd	4	6	5	8	6	5	30	21	21	22	15	17	12	6	5	11	8	7
Norris	mean	124	123	127	112	123	123	239	221	206	240	225	233	325	313	307	319	307	317
108 pixels	sd	14	12	14	25	17	11	58	44	34	46	30	49	10	13	6	20	18	10
UGB	mean	113	120	116	111	124	116	235	216	188	222	228	232	334	303	307	305	307	316
75 pixels	sd	10	6	7	10	7	9	63	36	29	28	36	39	5	2	3	10	4	4
LGB	mean	110	113	103	104	117	112	201	194	179	208	196	203	324	302	303	315	300	313
336 pixels	sd	13	7	8	11	9	8	40	25	32	40	33	37	12	5	6	14	6	5
Nez Perce	mean	113	109	104	95	113	112	205	190	189	213	196	208	329	306	308	333	307	315
36 pixels	sd	8	8	10	12	8	6	40	16	31	43	22	33	4	8	4	11	6	4
Hayden	mean	136	139	137	129	138	137	185	186	180	194	191	182	287	285	285	289	281	283
1389 pixels	sd	4	6	5	8	3	4	9	8	9	10	8	7	21	10	13	5	10	14
Pelican	mean	140	145	146	138	145	143	194	196	189	203	198	187	298	289	286	289	284	285
516 pixels	sd	9	8	6	5	7	13	19	8	8	10	11	13	16	12	17	7	11	14
Raven Creek	mean	142	150	148	143	144	147	192	202	197	210	205	198	288	295	301	292	294	289
166 pixels	sd	8	7	7	10	9	8	11	11	14	9	9	8	8	5	5	5	8	7

Region		LOS (Days)						Max						INDVI					
		2001	2002	2003	2004	2005	2006	2001	2002	2003	2004	2005	2006	2001	2002	2003	2004	2005	2006
Combined Hess	mean	179	164	171	191	165	175	0.675	0.662	0.649	0.669	0.703	0.685	89.51	82.05	84.41	100.13	89.36	90.97
7958 pixels	sd	37	30	32	41	35	39	0.116	0.114	0.102	0.112	0.116	0.110	13.83	12.29	12.92	14.85	11.34	13.33
Northern Range	mean	184	168	178	207	175	188	0.659	0.650	0.634	0.635	0.681	0.652	88.02	81.17	83.93	100.80	88.72	90.26
3667 pixels	sd	39	36	34	46	43	46	0.124	0.111	0.097	0.112	0.122	0.108	12.31	12.16	11.56	15.01	11.15	12.05
Central Region	mean	175	161	166	178	158	164	0.691	0.675	0.665	0.701	0.724	0.717	91.23	83.14	85.05	100.26	90.24	91.75
4047 pixels	sd	34	24	29	29	22	28	0.106	0.117	0.104	0.102	0.107	0.103	14.62	12.18	13.83	14.21	11.12	14.09
Meadow complex																			
West Yellowstone	mean	182	174	174	186	164	174	0.686	0.668	0.681	0.687	0.715	0.710	95.28	90.59	91.73	102.54	94.41	97.46
274 pixels	sd	34	15	29	15	17	27	0.134	0.138	0.130	0.127	0.140	0.136	15.83	12.68	15.77	14.43	13.58	17.24
Horse Butte	mean	182	160	166	189	158	170	0.707	0.741	0.708	0.732	0.772	0.745	96.96	90.51	90.64	111.11	95.66	97.97
113 pixels	sd	25	17	17	20	17	16	0.086	0.092	0.075	0.076	0.088	0.096	9.10	8.87	9.12	8.93	7.71	8.48
Gneiss Creek	mean	191	166	181	206	168	178	0.714	0.725	0.698	0.729	0.750	0.748	98.74	89.90	94.51	114.61	97.18	99.68
459 pixels	sd	23	17	22	17	12	15	0.079	0.082	0.078	0.070	0.064	0.071	9.63	9.00	11.55	9.34	8.17	9.93
Cougar	mean	191	161	166	182	151	164	0.729	0.734	0.743	0.762	0.804	0.769	102.22	88.92	93.57	112.13	94.68	96.50
131 pixels	sd	19	24	23	22	19	21	0.041	0.053	0.041	0.043	0.044	0.037	9.86	11.88	12.14	11.28	10.35	14.50
Madison Valley	mean	226	199	216	213	188	200	0.518	0.503	0.529	0.552	0.573	0.567	91.21	80.97	91.57	96.53	88.07	93.91
275 pixels	sd	15	12	12	22	10	9	0.094	0.098	0.085	0.094	0.095	0.093	12.66	11.15	11.40	15.71	10.82	13.21
Madison Canyon	mean	221	202	200	235	194	206	0.601	0.572	0.571	0.653	0.650	0.654	108.28	97.34	96.08	125.48	108.37	115.61
30 pixels	sd	11	15	37	18	10	10	0.074	0.070	0.066	0.067	0.070	0.082	12.10	10.50	21.02	15.49	9.31	12.18
Gibbon	mean	192	172	177	182	161	183	0.725	0.690	0.692	0.731	0.762	0.745	110.94	96.39	100.36	110.36	99.30	111.51
36 pixels	sd	13	10	9	17	11	9	0.070	0.066	0.076	0.062	0.070	0.061	8.42	5.95	6.30	5.78	4.38	7.44
Norris	mean	200	190	180	207	183	194	0.638	0.620	0.625	0.650	0.666	0.685	102.22	94.75	95.34	111.73	99.56	109.77
108 pixels	sd	20	22	18	38	31	17	0.110	0.123	0.109	0.107	0.118	0.117	13.84	12.06	11.57	17.98	12.15	14.11
UGB	mean	220	183	191	194	183	201	0.534	0.478	0.481	0.542	0.552	0.577	93.34	74.33	76.76	88.21	85.66	96.67
75 pixels	sd	12	7	7	19	9	11	0.074	0.080	0.076	0.088	0.084	0.084	11.31	10.72	10.60	12.51	12.23	12.18
LGB	mean	214	189	199	211	183	201	0.590	0.512	0.533	0.587	0.580	0.584	98.06	79.05	86.08	100.92	88.01	95.74
336 pixels	sd	18	9	10	21	12	10	0.103	0.095	0.096	0.109	0.098	0.099	16.76	13.94	15.74	18.76	14.08	15.49
Nez Perce	mean	216	197	204	238	194	203	0.684	0.663	0.660	0.709	0.708	0.715	120.48	106.82	112.25	137.83	115.95	121.41
36 pixels	sd	11	12	12	16	11	9	0.067	0.080	0.069	0.067	0.084	0.064	11.23	8.64	7.36	11.46	8.50	6.72
Hayden	mean	151	146	148	160	144	146	0.732	0.703	0.681	0.731	0.758	0.746	83.17	77.27	77.08	94.70	86.16	84.39
1389 pixels	sd	23	13	16	10	12	16	0.063	0.053	0.053	0.053	0.054	0.055	11.66	8.38	9.19	7.78	7.17	9.40
Pelican	mean	157	144	140	152	139	142	0.749	0.771	0.750	0.776	0.797	0.785	92.84	85.40	82.55	97.79	88.66	88.07
516 pixels	sd	19	16	19	10	15	19	0.059	0.066	0.069	0.055	0.054	0.055	12.14	11.39	13.08	6.63	9.85	11.95
Raven Creek	mean	147	146	154	149	150	142	0.691	0.698	0.682	0.716	0.734	0.732	82.21	82.80	85.67	89.59	89.69	85.84
166 pixels	sd	14	10	11	13	14	14	0.083	0.077	0.081	0.078	0.076	0.076	10.34	10.08	9.14	11.28	10.35	11.37

Section 3

Ungulate Spatial and Population Dynamics Prior to Wolves

CHAPTER 8

Elk Winter Resource Selection in a Severe Snow Pack Environment

Mathew A. Messer,* Robert A. Garrott,* Steve Cherry,[†] P. J. White,[‡]
Fred G. R. Watson,[§] and Eric Meredith[†]

*Fish and Wildlife Management Program, Department of Ecology, Montana State University
[†]Department of Mathematical Sciences, Montana State University
[‡]National Park Service, Yellowstone National Park
[§]Division of Science and Environmental Policy, California State University Monterey Bay

Contents

Theme

Animals utilize heterogeneous landscapes to obtain the resources needed to meet their physiological requirements, reproduce, and survive. Their movements and choices within these landscapes determine distributional
patterns and population dynamics. In most temperate and high latitude environments, including Yellowstone
National Park, snow is a dominant characteristic of the landscape for much of the year. During this time, large
herbivores such as elk (*Cervus elaphus*) experience a chronic energy deficit due to poor quality and restricted
availability of forage. Snow dramatically increases energetic demands as it is displaced to access underlying
forage and when moving among habitat patches. Though such constraints clearly influence patterns of habitat
selection and movement, understanding such responses to snow is challenging. Unlike relatively static landscape attributes such as elevation or plant community type, snow is a dynamic attribute that is highly variable
over a broad range of spatial and temporal scales. We combined our knowledge of snow pack dynamics in
central Yellowstone with seven years of location data from radio-collared elk in the Madison headwaters area to
better understand the influence of snow pack on large herbivore spatial dynamics. The severe winter climate of
this area, combined with the high concentration of geothermal features, results in an extremely heterogeneous

The Ecology of Large Mammals in Central Yellowstone
R. Garrott, P. J. White and F. Watson
ISSN 1936-7961, DOI: 10.1016/S1936-7961(08)00208-X

Copyright © 2009, Elsevier Inc.
All rights reserved.

landscape. Thus, animals were presented with a wide range of habitat choices in a severe environmental setting that, in turn, provided a unique opportunity to gain insights into the influence of snow on large herbivore ecology.

I. INTRODUCTION

Climatic variation causes dynamic environmental conditions at a variety of spatial and temporal scales that can influence plant and animal species (Fretwell 1972, Ackerly 2003). Understanding the influence of climate on animal behaviors requires carefully interpreting the biological scale of the behavioral response in question and examining that response at applicable environmental scales (Levin 1992, Karl *et al.* 2000). An animal's response is often assessed using locations as the biological scale, as animals distribute themselves across the landscape based on physiological requirements needed to survive. Determining the effects of the landscape components influencing those locations can be achieved by comparing attributes at animal locations to the attributes of those areas available across their normal range using geographic information system (GIS) data layers (Manly *et al.* 2002). However, these data layers are usually snapshots of the environment at a single point in time, making assessments difficult in environments that experience pronounced temporal variation.

Winter climate in temperate and high latitudes creates a dynamic landscape that can quickly change energetic requirements of animals and thereby influence their distributions (Fretwell 1972). Wind, temperature, and snow pack impact the energetic costs associated with movement, thermoregulation, and food acquisition (Parker *et al.* 1984, Root 1988). While some populations minimize these costs through migratory behavior or hibernation, others change resource use within their local environment (Berthold 2001, Humphries *et al.* 2003). The energetic cost of acquiring resources during winter can ultimately influence vital rates and, hence, trends in animal abundance (Sæther 1997, Loison and Langvatn 1998, Post and Stenseth 1999, Gaillard *et al.* 2000). These demographic responses to climate have been assessed at broad spatial and temporal scales, yet questions remain regarding the responses that are occurring at finer scales. Such questions require increasingly detailed biological and landscape data that are becoming available through innovations in computer models, technology, and remote sensing.

Assessing the influence of snow pack on animal distributions is challenging, since numerous components contribute to make snow accumulation and ablation highly variable across space and time (Coughlan and Running 1997, Marsh 1999). Variation in snow pack attributes results from precipitation interacting with abiotic components of the environment, such as elevation, aspect, temperature, wind, and solar radiation, as well as biotic components, such as vegetation structure (Coughlan and Running 1997, Greene *et al.* 1999, Marsh 1999; Chapters 5 and 6 by Watson *et al.*, this volume). Many studies that have examined how seasonal climatic changes influence animal distributions did not include snow covariates directly in their analyses, depending instead on other landscape attributes such as elevation as generalized surrogates for the influences of snow pack (Pearson *et al.* 1995, Terry *et al.* 2000, Johnson *et al.* 2002, Grignolio *et al.* 2004, Hebblewhite *et al.* 2005, Lele and Keim 2006). Other studies have incorporated snow pack covariates in their analyses, but the covariates were generalized across space and time due to limited physical measurements of attributes and/or precipitation patterns or unvalidated equations were used to interpolate snow measurements at a small number of locations across the landscape (Sweeney and Sweeney 1984, Schmidt 1993, Turner *et al.* 1994, Schaefer and Messier 1995, Arthur *et al.* 1996, Nellemann 1998, Doerr *et al.* 2005, Dussault *et al.* 2005, Fortin *et al.* 2005). The successful development and validation of a hydrological snow pack model for Yellowstone National Park (Watson *et al.* 2006a,b; Chapter 6 by Watson *et al.*, this volume) presents an opportunity to expand on these early efforts by generating daily estimates of snow pack attributes at fine spatial and temporal scales.

Elk in the Madison headwaters area of Yellowstone inhabit a mid-latitude montane ecosystem where climate limits the population. Snow is a major constraint to movements by these elk because depths routinely exceed 60 cm and the costs of locomotion increase curvilinearly as snow mass increases

(Robbins 1993). Snow pack also limits food availability, with starvation being the primary cause of death of adult female elk in this area prior to wolf restoration (Chapter 11 by Garrott *et al.*, this volume). In addition, snow had a pronounced affect on recruitment, with the most severe conditions resulting in the virtual elimination of the juvenile cohort (Chapter 11 by Garrott *et al.*, this volume). There were no significant predators of elk in this area during the winters of this study (1991–1998), with the exception of sporadic wolf (*Canis lupus*) presence in 1997–1998 (Chapter 15 by Smith *et al.*, this volume) and minimal grizzly bear (*Ursus arctos*) predation on adults in spring (Garrott *et al.* 2003; Chapter 11 by Garrott *et al.*, this volume). Further, hunting in the park is prohibited and, therefore, did not influence habitat selection by this nonmigratory herd.

We used fine-scale estimates of snow pack attributes to examine the response of elk distributions to changing snow conditions in Yellowstone (Figure 8.1). This examination compared snow pack attributes at known elk locations and locations selected at random from an area considered available to the population. These comparisons were made across several different categories of landscape-scale snow conditions. We also examined a suite of landscape components, including snow, to assess their explanatory power in differentiating between elk and random locations by evaluating suites of a priori models representing our biological hypotheses.

II. METHODS

A. Data Collection

Adult female elk were captured, instrumented, and monitored to collect a variety of physiological (Chapters 9 and 22 by White *et al.* and Chapter 10 by Garrott *et al.*, this volume), behavioral (Chapters 18, 19, and 20 by Gower *et al.*, this volume), demographic (Chapters 11 and 23 by Garrott *et al.*, this volume), and distributional data (Chapter 21 by White *et al.*, this volume). Elk were arbitrarily selected

FIGURE 8.1 A lone cow elk feeds in a small thermal pocket in the Gibbon Canyon. The heat associated with the many geothermal features present throughout the Gibbon and Firehole drainages dramatically reduced or eliminated snow pack, providing refuges for elk from the routinely deep snow pack. While a significant proportion of the elk population routinely occupied geothermal areas, elk were distributed throughout the study area during winter (Photo by Claire Gower).

for capture, while attempting to maintain an even distribution of individuals throughout the study area. Animals were captured using ground-based delivery of immobilization drugs via dart rifle and fitted with collar-mounted radio-transmitters. The number of animals monitored annually varied from 24 to 32 individuals, dependent upon mortality and research permit allowances. Animal locations were sampled from approximately 15 November until 30 April over seven winter seasons (1991–1992 to 1997–1998), following a stratified random sampling design. We divided the study area into several geographically defined strata defined by drainages and randomly selected a stratum to be sampled. Instrumented elk inhabiting the selected stratum were located in a randomly determined order and all remaining strata were subsequently sampled before re-sampling. Locations were obtained diurnally using hand-held telemetry equipment and homing procedures (White and Garrott 1990). Upon sighting, the location of the animal was recorded in Universal Transverse Mercator coordinates using United States Geological Survey (USGS) quadrangle maps.

The study area boundary was delineated by selecting that portion of the landscape considered available to the study population. The spatial extent of available landscape was defined using a kernel density estimator implemented within the Tarsier environmental modeling framework (Watson and Rahman 2003). A kernel bandwidth of 885 meters was chosen. This was the smallest value that resulted in a single home range polygon (99.5 percentile), as opposed to the fragmented home ranges that would result from smaller bandwidths. Using this boundary, we selected random locations to compare to instrumented elk locations as our binary response variable. Random locations were selected from within this boundary at a 20:1 ratio to instrumented elk locations and were assigned the same date as their corresponding elk location to ensure an equivalent temporal distribution (Messer 2003).

A USGS digital elevation model was used to acquire estimates of slope, aspect and elevation. Aspect presents difficulties in modeling because it is a continuous circular measure with 1° similar to 359°. While it can be treated as a categorical variable by designating parameters defined by the cardinal and sub-cardinal directions (Cooper and Millspaugh 1999), we chose to consider the biological implications of aspect related to solar exposure. Aspect and slope influence the solar exposure of a site, which can impact thermoregulation of organisms, vegetative productivity, and snow dynamics (Cook *et al.* 1998, Greene *et al.* 1999). To represent solar exposure we used a solar radiation index (SRI), calculated using latitude, slope and aspect (Iqbal 1983, Keating *et al.* 2007). Habitat categories were designated using vegetation cover type and geothermal data layers developed using Landsat 7 Enhanced Thematic Mapper satellite imagery acquired during March 2000 and September 2002 (Chapter 2 by Newman and Watson, this volume). Cover type classifications included meadow, burned forest, unburned forest and geothermal. In addition, geothermal pixels were assigned a ground heat flux (GHF) value ranging from 0 to 89 W/m^2, based upon estimates of GHF (Chapter 4 by Watson *et al.*, this volume).

Using a single static data layer is inadequate for representing snow, as it is highly variable over short time periods. Thus, we used the Langur snow pack model (Chapter 6 by Watson *et al.*, this volume) to estimate spatial and temporal heterogeneity of the snow pack each winter. Langur generated daily estimates of SWE throughout the study area using temporally dynamic precipitation and temperature data, and spatially explicit slope, aspect, elevation, GHF, vegetation cover type and mean annual precipitation data. Model parameter values controlling the temporal behavior of the model were calibrated using snow water equivalent and snow pack density data from eleven nearby automated snow telemetry weather stations (Natural Resources Conservation Service SNOTEL data). Parameter values controlling response to spatial variables were either taken from the literature, or calibrated to spatial snow pack sampling data (Chapter 6 by Watson *et al.*, this volume). Final simulations were driven by temperature and precipitation data from spatially weighted combinations of eleven nearby SNOTEL stations as well as temperature data from the West Yellowstone NPS ranger station.

We used Langur output to characterize snowpack by mean SWE and snow heterogeneity (SNH), with each metric calculated at two spatial scales. Mean SWE was calculated as the average of all pixels within the scale of interest and represented the mean water content of the snow pack. SNH was calculated as the standard deviation of SWE for all pixels within the scale of interest and represented the spatial variability of the snow pack. All pixels within a 100 m radius of each elk and random location constituted the local scale (SWE_A, SNH_A) because we assumed elk responses to snow occurred at a scale larger than that of a single 28.5 m pixel. All pixels within the defined boundary of the study population constituted the landscape scale (SWE_L, SNH_L) because we assumed elk were capable of moving throughout the study area during the course of each winter. Landscape and local snow pack estimates were obtained daily, with local snow pack estimates assigned to each elk location and corresponding random locations on that day. In addition, we also calculated an annual index of snow pack severity (SWE_{acc}) by summing the mean daily SWE for the study area (SWE_L) from 1 October to 31 April (Garrott *et al.* 2003), which provided a single metric integrating snow pack depth, density, and duration.

B. Statistical Methods

We used log odds ratios to determine the likelihood of an elk occurring at a particular location depending upon local and landscape-scale snow pack conditions. The odds of an event is the ratio of the probability that the event occurs to the probability that it does not occur. An odds ratio is a ratio of two odds. Odds ratios are commonly used in epidemiology where, for example, the odds of the occurrence of a disease given the presence of a risk factor are compared to the odds of a disease when the risk factor is absent. Odds ratios have an asymmetrical distribution ranging from 0 to infinity with values >1 indicating increased odds of occurrence, values <1 indicating decreased odds and values of 1 indicating equal odds of occurrence. Log odds ratios, the natural log of odds ratios, are symmetrical about 0 and allow comparison of the strength of positive and negative relationships. The use and interpretation of odds ratios, including methods of statistical inference, can be found in many texts on categorical data analysis, *e.g.*, Agresti (2003).

All actual and random locations were first sorted into one of four categories depending upon the landscape SWE_L estimate on their date of collection (low 0.00–0.010 m, moderate >0.10–0.20 m, high >0.20–0.30 m, and very high >0.30). Locations were then sorted into local SWE_A levels, categorized into 0.05 m increments. This resulted in each location occurring in one local SWE_A level within one landscape SWE_L category. Odds ratios were then calculated for each local SWE_A level within each landscape SWE_L category. The odds of an elk location occurring in a particular local SWE_A level were calculated by dividing the probability of an elk location occurring in that level by the probability of an elk location not occurring in that level. After calculating the odds of a random location in the same manner, an odds ratio was obtained by dividing the odds of an elk location by the odds of a random location occurring in that level within each landscape category. Thus, we are comparing the odds of actual use to the odds of use expected under random selection. We obtained log odds ratios and calculated 95% confidence intervals when the proportion of locations occurring in a particular local SWE_A level exceeded 0.01. Using this approach, we also calculated log odds ratios of local SNH_A, categorized into 0.02 m increments, within the same four landscape SWE_L categories.

Predictions regarding the odds of elk occurrence were primarily based upon energetic costs of movement and foraging in snow. We predicted that the odds of elk occurrence would decrease with increasing local SWE_A because elk would be less likely to occupy energetically demanding areas of high snow mass compared to areas with low snow mass. We also predicted increasing odds of elk occurrence with increasing local SNH_A because high local SNH_A reflects a greater range of local snow conditions that provides elk more opportunities to locate areas of low SWE. Lastly, we anticipated that snow pack at the landscape level would influence elk responses to local conditions, with the odds of elk occurrence increasing across local SWE_A and local SNH_A levels as landscape SWE_L increased. Thus, the log odds curves would have shallower slopes with increasing landscape SWE_L because elk simply would be forced to occupy locations with higher local SWE_A at higher levels of landscape SWE_L.

To assess the contribution of SWE in explaining variation in winter elk distribution, we used matched case-control logistic regression in which each elk location (a case) was matched temporally with 20 random locations (controls). Matched case-control logistic regression is typically used to control for potential confounding variables, temporal variables in this analysis. Let $p_j(\mathbf{x}_{ij})$ be the probability that the ith location in the jth matched set of 21 locations is the used location. The vector $\mathbf{x}_{ij}$ contains the explanatory variables associated with the ith location, $i = 0, 1, \ldots, 20$ in the jth matched set $j = 1, \ldots, n$ ($\mathbf{x}_{0j}$ is the vector associated with the case), where j represents a unique date. Note that $p_j(\mathbf{x}_{ij})$ is the conditional probability that location i is the used location given (or conditional on) $\mathbf{x}_{0j}, \mathbf{x}_{1j}, \ldots, \mathbf{x}_{20j}$ being the explanatory vectors in the jth matched set. The probability is modeled as

$$\log\left(\frac{p_j(\mathbf{x}_{ij})}{1 - p_j(\mathbf{x}_{ij})}\right) = \alpha_j + \beta_1 x_{1ij} + \beta_2 x_{2ij} + \cdots + \beta_k x_{kij}$$

where x_{mij} denotes the mth explanatory variable ($m = 1, \ldots, k$) for the ith location in the jth matched set. The effect of the matching variables in the jth set is summarized by the constant or intercept term, α_j. Parameters are estimated by maximizing the conditional likelihood,

$$\prod_{j=1}^{n} \left[1 + \sum_{i=1}^{20} \exp\left\{\sum_{m=1}^{k} \beta_m(x_{mij} - x_{m0j})\right\}\right]^{-1}$$

Note that the constants (α_j, $j = 1, \ldots, n$, where n for our data set is 736) have canceled out of the likelihood implying that the effects of the matched variables are not estimable. Because our data were matched in time this means that we also matched on any variable that had the same study area wide value at a given time (*e.g.*, SWE_L, SNH_L). Thus, the impact of these variables on the response cannot be assessed. Including them as main effects in the model does not help because they simply cancel out of the likelihood. The effects of such variables can still be assessed by including interaction terms between matched variables and other explanatory variables. This is one exception to the general rule that one should not include interaction terms in a model without also including the main effect. Further details on matched case-control logistic regression can be found in Collett (2003).

We compared *a priori* hypotheses representing the potential effects of static landscape and dynamic snow pack covariates. Hypotheses were expressed as a set of candidate models with each model placed in one of three suites, depending upon which covariates it contained. The first suite contained seven models composed of static landscape covariates and represented the traditional approach to explaining winter distributions of large herbivores (Manly *et al.* 2002, Dussault *et al.* 2005, Sawyer *et al.* 2007). Within this suite, we hypothesized that lower elevations, higher SRI, and geothermal habitats would reflect lower SWE estimates and be selected for use due to lower energy requirements for foraging and movement. As large herbivores are known to respond to the quality and quantity of vegetation during winter, we hypothesized that geothermal and meadow habitats, with generally high vegetative productivity and potential forage biomass, would be selected over burned and unburned forests (Wallace *et al.* 1995). Though some geothermal habitats associated with active thermal features are devoid of vegetation, many provide higher quantities and/or quality of forage by supporting wet meadow areas with high vegetative productivity or plants that photosynthesize year round (Despain 1990).

We used a second suite of seven models composed of temporally dynamic snow pack covariates to directly examine the influence of snow pack on elk distribution. Due to the high correlation between SWE_L and SNH_L (Pearson correlation $= 0.92$), we chose to use only landscape SWE_L in model building. We predicted decreased selection with increasing local SWE_A, increased selection with increasing local SNH_A, and weaker responses to local SWE_A and SNH_A with increasing landscape SWE_L.

The third suite contained 13 models composed of both static landscape covariates and temporally dynamic snow pack covariates. Given the correlation between local SWE_A estimates and topographical covariates (Chapters 5 and 6 by Watson *et al.*, this volume), we assumed variation in elk distribution could be assessed with the most precision using local SWE_A estimates. Thus, elevation and SRI were only included in two simple additive models, while the remaining models examined variation in elk resource use using a combination of snow pack covariates and habitat. We fit models using PROC PHREG (SAS Institute 2000) in SAS. Models were ranked and the most parsimonious models supported by the data were selected using AIC (Burnham and Anderson 2002).

III. RESULTS

We collected 5929 locations from 45 instrumented elk over seven winter seasons. Group sizes ranged from 1 to 56 elk, with a mean of 6.2. Data from only one individual was used when multiple instrumented elk were located simultaneously in a single group. We retained 4869 locations (range 392–861 per season) after censoring dependent data. The area considered available to elk, using all observed elk location data, described a winter range of approximately 362 km^2 centered along the three main river valleys representing the Madison River headwaters and extending upslope to the edges of the high elevation plateaus.

The annual variation in snow pack experienced during the seven winters of this study was substantial and represented nearly the entire range of annual snow pack conditions evident in the historic data. Climatic data were available to parameterize the Langur snow pack model (Chapter 6 by Watson *et al.*, this volume) for the most recent 24 winters (1983–1984 through 2006–2007), allowing calculation of the SWE_{acc} index of annual snow pack severity. The median SWE_{acc} from this time series was 2093 with a range of 1042–5690 cm days. Winters during our study had SWE_{acc} measurements ranging from 1744 to 5690, with a median of 2643 cm days. The heaviest snow pack was experienced during the winter of 1996–1997 and represented the most severe snow pack conditions for the historic record. The lightest snow pack occurred during the winter of 1993–1994, with only five winters in the historic record experiencing lighter snow pack conditions. The seasonal and annual variation in snow pack characteristics experienced during this study is illustrated in Figure 8.2. While there was strong inter-annual variation in snow pack, there was also substantial variation in snow pack across the landscape, with an example of the variation in snow water equivalent estimates within and among drainages provided in Figure 8.3. Average and maximum landscape SWE_L estimates increase through winter until quickly decreasing at the beginning of spring thaw. A small portion of the study area always remained snow free due to geothermal activity (Chapter 4 by Watson *et al.*, this volume). Heterogeneity of the snow pack at the landscape scale (SNH_L) increased with increasing snow pack through most of the winter and dramatically decreased during spring thaw. These general seasonal trends were more pronounced during years of heavier snowfall, with higher maximum and average landscape SWE_L estimates and greater landscape SNH_L.

A comparison of the frequency distribution of all observed elk locations with the sample of random points drawn from the available winter range for all study years combined demonstrates that elk disproportionately selected areas of lower local snow pack and tended to avoid areas of higher local snow pack (Figure 8.4). Log odds ratios demonstrated that elk were less likely to occur in an area as local SWE_A increased with this trend consistent for all four categories of landscape SWE_L (Table 8.1, Figure 8.5A). The log odds ratios for any given local SWE_A, however, increased with increasing landscape SWE_L, indicating that as snow pack on the winter range became more severe elk were more likely to occupy areas of more severe local SWE_A than they were under less severe conditions. We did not, however, find a clear SWE_A threshold that precludes occupancy (Figures 8.4 and 8.5) with observed elk occupying areas of extremely severe snow pack (Figure 8.6). Local snow pack heterogeneity, as measured by SNH_A, also influenced elk occupancy. Across all four categories of landscape SWE_L, the likelihood of elk occurrence quickly increased and then stabilized as local SNH_A levels exceeded 0.06 (Table 8.1, Figure 8.5B). Comparing the log odds ratio plots among the four landscape SWE_L

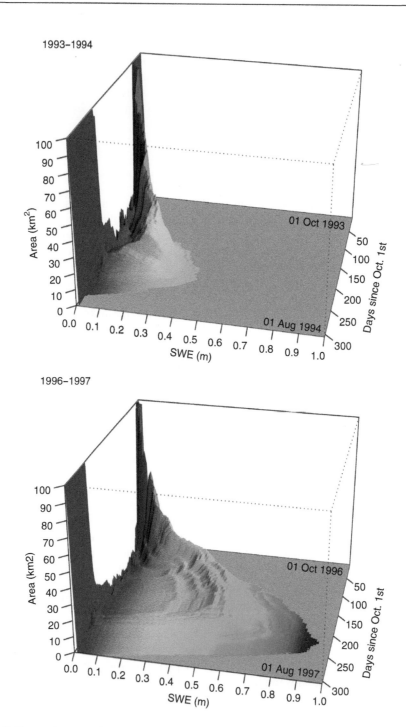

FIGURE 8.2 Temporal pattern of snow water equivalent estimates across the study area (362 km²) demonstrating strong variation in snow pack within and among winters. The heaviest and lightest snow pack accumulation occurred in winters 1993–1994 and 1996–1997, respectively. Snow typically began to accumulate in early October and quickly encompassed the entire study area, with the duration and accumulation of the snow pack varying among winters. The small portion of the study area that remained essentially snow-free throughout both winters (coded in red) represent the geothermal areas. A spatial depiction of this relationship is provided in (Figure 8.3).

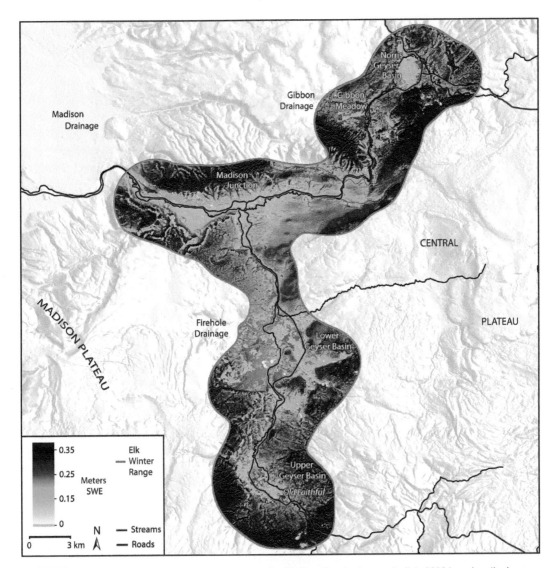

FIGURE 8.3 Spatial distribution of snowpack across the Madison headwaters on April 1, 2002 based on the Langur snow model (Chapter 6 by Watson *et al.*, this volume). Snow pack heterogeneity was highest in the Madison drainage with large continuous areas of lower snow pack, while higher and more uniform snow pack with distinct areas of lower snow were found in the Gibbon and Firehole.

categories indicated that elk increased their selection against occupying the most homogeneous snow pack areas as snow pack severity on the winter range increased. However, once local SNH_A increased modestly, elk were equally likely to occur regardless of increasing variation in local SNH_A or changing landscape SWE_L.

Results from model comparison techniques were consistent with results from interpreting log odds ratios and supported our predictions regarding the ability of covariates to explain variation in elk distributions (Table 8.2). The most-supported model in the landscape suite was an additive model containing all static landscape covariates: habitat, elevation, and SRI. The most-supported model from

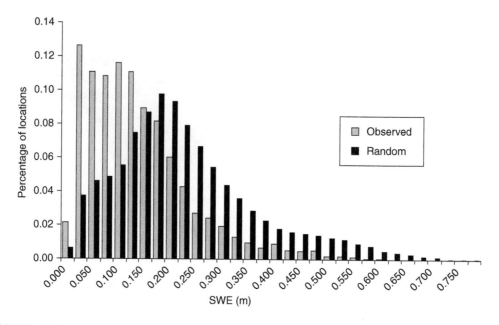

FIGURE 8.4 A comparison of the distribution of local-scale snow water equivalent, SWE$_A$, available within the study area (random) with the mean SWE$_A$ in the 100 m radius around observed elk locations (observed). Data are all independent elk groups for the seven pre-wolf years ($n = 4869$) and random points drawn for each elk observation matched to the date of the observation.

the snow suite was composed of the additive effects of local SWE$_A$, local SNH$_A$, and the interaction of each of the local-scale main effects and the landscape-scale mean SWE (SWE$_L$). Combined static landscape and dynamic snow covariates in the third suite resulted in the most-supported model across all suites, an additive model using all covariates: local SWE$_A$, local SNH$_A$, habitat, elevation, and SRI. The most parsimonious model from the landscape-snow suite also had an estimated AIC weight of 1.0 when compared across all competing models regardless of suite, suggesting no other competing models were well supported within these suites.

Coefficient estimates for the most-supported model in each of the three model suites supported predicted effects of covariates to explain variation in the distribution of elk (Table 8.3). All habitat coefficients in the most-supported model in the landscape suite were negative indicating that elk selection was strongest for geothermal environments, the reference habitat type. Coefficient confidence intervals overlapped for meadow and unburned forest suggesting similar selection, with the substantially larger burned forest coefficient indicating elk were least likely to use this habitat type. The elevation coefficient was negative with confidence limits that did not include zero, thus indicating elk selected for areas of lower elevation. The modest size of the SRI coefficient, combined with a confidence interval that nearly overlapped zero, provided only weak support for elk selection of areas with lower exposure to solar radiation. Coefficient estimates for the most-supported model in the snow suite indicated strong selection of elk for lower local-scale mean snow pack, SWE$_A$, and modest selection for increased local-scale snow heterogeneity, SNH$_A$, with selection weakening as mean landscape-scale snow pack, SWE$_L$, increased. The positive coefficient estimate for the SWE$_A \times$ SWE$_L$ interaction and the negative coefficient for the SNH$_A \times$ SWE$_L$ interaction terms indicate that elk selection for local SWE$_A$ and SNH$_A$ moderates as landscape-scale snow pack, SWE$_L$, increases.

To illustrate the influence of temporally dynamic snow pack on the seasonal distribution of elk we used our most-supported resource selection model to generate maps depicting the odds of elk

| TABLE 8.1 | Log odds ratios (LOR) and standard errors (SE) determining likelihood of elk occurrence under changing local- and landscape-scale snow conditions in the Madison headwaters area of Yellowstone National Park |

Landscape SWE categories

Local snow levels	Low (0.00–0.10 m)		Moderate (>0.10–0.20 m)		High (>0.20–0.30 m)		Very high (>0.30)	
	LOR	SE	LOR	SE	LOR	SE	LOR	SE
SWE								
0.00–0.05	0.62	0.09	1.63	0.05	1.85	0.08	1.64	0.14
0.05–0.10	0.02	0.09	1.21	0.05	1.46	0.08	1.81	0.13
0.10–0.15	−1.35	0.19	0.08	0.05	1.08	0.07	1.57	0.12
0.15–0.20	*		−1.20	0.07	0.46	0.07	1.42	0.11
0.20–0.25	*		−1.96	0.14	−0.60	0.08	0.51	0.12
0.25–0.30	*		−2.87	0.41	−1.25	0.10	0.39	0.12
0.30–0.35	*		−2.52	0.71	−1.83	0.18	−0.42	0.13
0.35–0.40	*		*		−2.84	0.45	−0.63	0.14
0.40–0.45	*		*		−3.53	1.00	−0.98	0.16
0.45–0.50	*		*		−2.63	1.00	−1.41	0.18
0.50–0.55	*		*		*		−1.86	0.25
0.55–0.60	*		*		*		−2.59	0.45
SNH								
0.00–0.02	−0.58	0.11	−1.08	0.05	−1.17	0.07	−1.71	0.13
0.02–0.04	0.60	0.12	0.26	0.05	−0.23	0.06	−0.69	0.09
0.04–0.06	0.36	0.30	0.94	0.05	0.67	0.07	0.49	0.09
0.06–0.08	*		1.01	0.08	0.96	0.08	0.77	0.10
0.08–0.10	1.92	1.16	0.70	0.16	1.04	0.10	1.12	0.11
0.10–0.12	*		0.31	0.46	0.82	0.15	1.15	0.13
0.12–0.14	*		*		0.13	0.33	1.17	0.15
0.14–0.16	*		*		0.23	0.73	1.09	0.18
0.16–0.18	*		*		*		0.99	0.24
0.18–0.20	*		*		3.01	1.00	0.97	0.32
0.20–0.22	*		*		*		1.09	0.44
0.22–0.24	*		*		*		0.61	0.74

* The proportion of elk locations in this level does not exceed 0.01.

occurrence in the Madison headwaters in early November 1996, when snow pack was light, and early April 1997, when snow pack was severe (Figure 8.7). In early winter the model predicts that elk would be widely distributed throughout all three drainages, as evidenced by relatively uniform odds of occurrence (Figure 8.7A). However, by late winter the model predicts that much of the study area would have very low odds of elk occurrence, with elk distributions restricted to differing extents in each of the three drainages of the study system (Figure 8.7B). The Madison drainage appeared to have higher, more contiguous odds of elk occurrence while areas of higher odds of occurrence in the Firehole were lower and more patchily distributed on the landscape. Odds of elk occurrence were even lower and substantially more localized in the Gibbon drainage. These differences were likely driven by differences in each drainage's respective snow pack accumulation, geothermal influence and topography (Figure 8.3). The Gibbon was characterized by heavy snow pack and relatively small and discrete areas of geothermal influence and accessible forage. While the Firehole was also characterized by heavy snow pack, it contains more abundant and contiguous geothermal areas. In contrast the only geothermal influence in the Madison drainage was riverbanks due to warming by the geothermally-influence

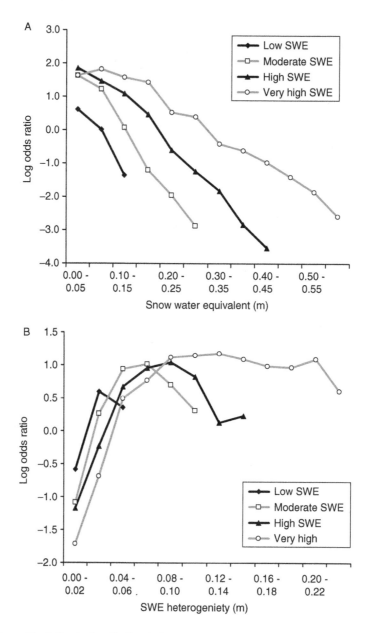

FIGURE 8.5 Log odds of elk selection (likelihood elk occurrence) across the range of snow water equivalent (A) and snow water equivalent heterogeneity (B) for four categories of average landscape snow water equivalent.

river waters, but this drainage had higher SNH due to the presence of steep south facing slopes running the length of the Madison canyon that provided a large contiguous area of lower snow and larger and more contiguous patches of accessible forage. Thus, it appeared that landscape characteristics resulted in differences in snow pack and that the distribution of animals was constrained by the interaction of static landscape attributes and temporally and spatially dynamic snow pack attributes.

FIGURE 8.6 While elk tended to occupy portions of the study area with reduced snow pack it was not uncommon to find elk occupying areas of extremely deep snow pack for extended periods during the winter as illustrated by this radio-collared cow and her calf (Photo by Robert Garrott).

IV. DISCUSSION

We used a validated snow model (Watson *et al.* 2006a,b, Chapter 6 by Watson *et al.*, this volume) to capture daily variations in snow pack at both the local and landscape spatial scales, and quantify elk responses to these changes. Our results demonstrated that elk exhibit decreased selection of areas with higher local SWE_A estimates, empirically supporting previous research that had observed decreasing utilization of deeper snow by large herbivores (Craighead *et al.* 1973, Houston 1982, Schaefer and Messier 1995, Ihl and Klein 2001, Boyce *et al.* 2003). Our results not only support this negative relationship, but also demonstrate that changing landscape-scale snow pack conditions strongly influenced the likelihood of elk occurrence within a particular local SWE_A level, a conclusion not documented in the existing literature. We documented an increased likelihood of elk locations occurring across the range of SWE levels present, as landscape SWE estimates increased. In addition, we noted that the range of local SWE_A estimates where elk had a greater likelihood of occurring increased as landscape SWE_L estimates increased. We interpreted this dynamic response to indicate that the study population had no choice but to occupy areas with higher local SWE_A as landscape SWE_L increased. In essence, elk were always most likely to select areas with the least amount of snow mass, but they began to occupy areas with higher snow mass as conditions on the winter range became more severe.

Numerous investigators of large herbivore behavior and spatial dynamics have suggested maximum foraging depths and upper occupancy thresholds (Dalke *et al.* 1965, Kelsall 1969, Rongstad and Tester 1969, Knight 1970, Sweeney and Sweeney 1984, Ball *et al.* 2001). Sweeney and Sweeney (1984) reported elk routinely occupying areas with <40 cm snow depth but >40 cm caused elk to move to areas with less snow. They reported a "critical depth" of 70 cm that physically impaired mobility and would preclude routine occupancy of an area. In this study we characterized snow pack using the snow water equivalent metric so our results are not directly comparable; however, qualitatively we found a similar

TABLE 8.2 Model selection results for *a priori* hypothesized models examining the effects of static landscape and dynamic snow covariates on elk resource selection

Model structure	k	AIC	Within suite ΔAIC	w_i	Among suite ΔAIC	w_i
Landscape models						
HBT	1	35368	3657	0.0	5100	0.0
SRI	1	36566	4855	0.0	6298	0.0
ELV	1	32663	952	0.0	2395	0.0
HBT + SRI	4	35363	3652	0.0	5095	0.0
HBT + ELV	4	31713	2	0.3	1445	0.0
SRI + ELV	2	32664	953	0.0	2396	0.0
HBT + SRI + ELV	5	31711	0	0.7	1443	0.0
Snow models						
SWE_A	1	32131	924	0.0	1863	0.0
SNH_A	1	35301	4094	0.0	5033	0.0
$SWE_A + SNH_A$	2	31756	549	0.0	1488	0.0
$SWE_A + SNH_A + (SWE_A*SNH_A)$	3	31656	449	0.0	1388	0.0
$SWE_A + (SWE_A*SWE_L)$	2	31658	451	0.0	1390	0.0
$SNH_A + (SNH_A*SWE_L)$	2	35192	3985	0.0	4924	0.0
$SWE_A + SNH_A + (SWE_A*SWE_L) + (SNH_A*SWE_L)$	4	31207	0	1.0	939	0.0
Landscape-snow models						
$SWE_A + HBT$	4	32081	1813	0.0	1813	0.0
$SWE_A + HBT + SRI + ELV$	6	30861	593	0.0	593	0.0
$SWE_A + SNH_A + HBT$	5	31736	1468	0.0	1468	0.0
$SWE_A + SNH_A + HBT + SRI + ELV$	7	30268	0	1.0	0	1.0
$SWE_A + HBT + (SWE_A*HBT)$	7	32020	1752	0.0	1752	0.0
$SWE_A + HBT + (SWE_L*HBT)$	7	32009	1741	0.0	1741	0.0
$SWE_A + SNH_A + HBT + (SWE_A*HBT)$	8	31728	1460	0.0	1460	0.0
$SWE_A + SNH_A + HBT + (SNH_A*HBT)$	8	31656	1388	0.0	1388	0.0
$SWE_A + SNH_A + HBT + (SNH_L*HBT)$	8	31656	1388	0.0	1388	0.0
$SWE_A + SNH_A + HBT + (SWE_A*SNH_A)$	6	31628	1360	0.0	1360	0.0
$SWE_A + SNH_A + HBT + (SWE_A*SNH_A) + (SWE_A*HBT)$	9	31628	1360	0.0	1360	0.0
$SWE_A + SNH_A + HBT + (SWE_A*SNH_A) + (SNH_A*HBT)$	9	31562	1294	0.0	1294	0.0
$SWE_A + SNH_A + HBT + (SWE_A*SNH_A) + (SWE_L*HBT)$	9	31552	1284	0.0	1284	0.0

HBT, habitat; SRI, solar radition index; ELV, elevation; SWE_A, local-scale snow water equivalent; SNH_A, local-scale snow heterogeneity; SWE_L, landscape-scale snow water equivalent.

pattern. Local snow pack of <0.15 m SWE was commonly occupied by elk with use of areas with >0.35 m substantially curtailed. However, we found that elk continued to use deeper snow locally as landscape snow pack deepened and were capable of occupying areas with local snow pack as severe as 0.50–0.60 m SWE, although few observations were recorded above 0.45 m SWE. Thus, we did not find a clear SWE threshold that precluded elk occupancy and suggest that the extreme heterogeneity of the snow pack due to the presence of geothermal features provided refuges from severe snow pack that allowed this population to continue occupying the high elevation upper Madison River drainages year-round when other elk populations seasonally migrated to lower elevation winter ranges (Figure 8.7).

Access to snow model data that provided metrics at a variety of scales also enabled us to demonstrate that elk were not likely to occur in areas where snow pack was relatively homogeneous. The importance of landscape heterogeneity on the distribution of organisms, and their utilization of resources, is well known (Weins 1989, Pearson *et al.* 1995, DuToit *et al.* 2003, Hobbs 2003) and this knowledge led to our prediction that elk occurrence was more likely in areas with greater heterogeneity. However, SNH has

| TABLE 8.3 | Coefficient values from the most-supported model for each of the three model suites identified through AIC model comparisons examining variation in winter elk distribution (Table 8.2) |

	Snow and landscape model		Snow model		Landscape model	
Covariate	β	SE	β	SE	β	SE
SWE_A	−6.507	0.271	−20.940	0.511		
SNH_A	14.066	0.571	21.300	1.462		
SWE_A*SNH_A						
SWE_A*SWE_L			34.534	1.596		
SNH_A*SWE_L			−36.375	4.512		
ELV	−0.00707	0.000194			−0.00917	0.000163
HBT						
Burned forest	−0.580	0.063			−1.704	0.053
Unburned forest	−0.383	0.054			−1.358	0.045
Meadow	−0.457	0.063			−1.266	0.056
SRI	−0.00043	0.000145			−0.00032	0.000155

HBT, habitat; SRI, solar radiation index; ELV, elevation; SWE_A, local-scale snow water equivalent; SNH_A, local-scale snow heterogeneity; SWE_L, landscape-scale snow water equivalent.

not been widely documented as a mechanism influencing the distribution of large herbivores (Schmidt 1993), likely because assessing spatial variation in snow pack is extremely difficult without detailed snow data. Thus, our finding of a strong positive relationship between elk occurrence and increasing local SNH, with a relatively equal likelihood of occurrence once snow mass variation increased slightly, represents new knowledge that should be confirmed by future investigations.

While our model selection results indicated that static landscape covariates also influenced the distribution of elk during winter, a comparison of the coefficient estimates from the landscape, snow, and combined landscape-snow models provided some interesting insights (Table 8.3). While the coefficient signs for SWE_A, SNH_A, ELV, and the HBT categories from the combined model remained the same as those estimated from the simpler landscape-only and snow-only model suites, the magnitude of the coefficients all decreased dramatically. Thus, the explanatory power of habitat type and elevation was reduced when snow covariates were included in the same model and, correspondingly, the explanatory power of snow covariates was reduced when habitat type and elevation were included in the same model. We interpret these changes as indicating that, in the absence of snow pack covariates, the habitat and elevation covariates captured some of the affect of snow pack on elk selection and, conversely, snow pack covariates considered in the absence of habitat and elevation covariates explained some of the affects of habitat and elevation on elk selection. While the magnitude of the habitat coefficients was reduced in the combined model, the relative ranking of elk selection among habitat types remained the same as estimated in the most-supported model from the landscape-only suite. This result was expected as we documented the strong influence of habitat and elevation on snow pack attributes and these static landscape effects were built into the Langur snow model used to develop the snow pack metrics considered in our elk resource selection models (Chapters 5 and 6 by Watson *et al.*, this volume).

In order to understand and predict a species' patterns of resource selection and hence their distributional dynamics, we need to be able to assess, as directly as possible, the resources required by that species and the constraints on acquiring those resources (Stephens and Krebs 1986, Morrison 2001, Austin 2002, Barry and Elith 2006). A demographic study of this elk herd, conducted concurrently with this resource selection study, clearly demonstrated that starvation was the dominant driver of variation in vital rates and that the magnitude of starvation mortality was strongly correlated with the severity of winter snow pack (Chapter 11 by Garrott *et al.*, this volume). Thus, the fundamental

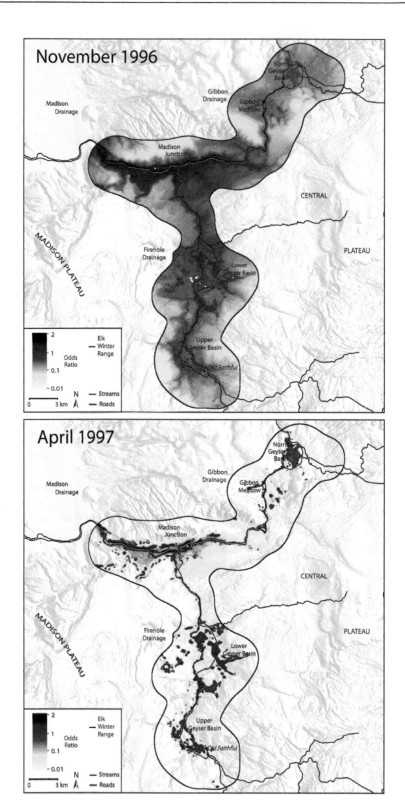

FIGURE 8.7 (Continued)

resource required was forage and the fundamental constraint was snow pack that reduced availability of forage and elevated the energetic cost of acquiring forage (Chapter 9 by White *et al.*, this volume). The development and validation of the Langur snow pack model for Yellowstone National Park allowed us to characterize the major constraint on this population of large herbivores at time and spatial scales relevant to the animals and our results clearly demonstrated that resource selection models that incorporated this constraint were far superior to models that did not include snow pack covariates.

We attempted to capture variation in the resource, forage, using surrogates such as habitat type, elevation, and SRI. These covariates certainly helped explain variation in elk resource selection so in that sense we can view them as successful, however, we suggest incorporating direct measures or estimates of the forage gradient into resource selection models at relevant spatial and temporal scales would substantially improve the predictive ability of such models. Thus we think it worthwhile to develop forage models similar to the Langur snow model that estimate dynamics of plant phenology (greenness) and hence quality, and plant biomass (quantity). Like the Langur model such forage models would incorporate data from the network of climate collection sites in the Yellowstone area, along with landscape attributes from current GIS data layers, remotely sensed data such as the normalized difference vegetation index, and ground measurements to calibrate and validate such a model. This is an active area of research and considerable progress has been made (Pettorelli *et al.* 2005a,b; Chapter 7 by Thein *et al.*, this volume). Successfully developing forage models could substantially simplify resource selection models and improve their predictive ability, thus extending our ecological understanding of large herbivore spatial and population dynamics in Yellowstone and contribute to sound policy and management decisions.

Most studies of habitat selection by ungulates reflect resource selection decisions by animals making tradeoffs between several factors limiting individual fitness, including predation, starvation, food availability, human activities, climate, disease, and parasites (Festa-Bianchet 1988, FitzGibbon and Lazarus 1995, Rowland *et al.* 2000, Dussault *et al.* 2005, Frair *et al.* 2005, Winnie *et al.* 2006). Disentangling the influence of these various factors is challenging because choices vary across scales often requiring considerable subjective assessments regarding causes and interactions. In contrast, this study provided a good understanding of how elk chose resources when the overwhelmingly dominant limiting factor was starvation due to chronic under-nutrition and energetic costs of winter snow pack, with other limiting factors such as predators or hunting were minimal or absent (Chapter 9 by White *et al.* and Chapter 11 by Garrott *et al.*, this volume). Thus, these findings provide an excellent opportunity to evaluate how elk resource selection changed after 1997–1998 when wolves recolonized the system and elk were confronted with tradeoffs between predation risk and starvation. We make such an assessment in Chapter 21 by White *et al.*, this volume.

V. SUMMARY

1. Previous studies investigating the distribution of ungulates in relation to snow pack have been limited by a lack of fine-scale metrics to characterize dynamic spatial and temporal changes in snow pack across the landscape.

FIGURE 8.7 Probability map developed from the best approximating *a priori* model examining the effects of temporally dynamic snow pack and static landscape attributes on elk resource selection within the Madison headwaters study area during early winter (low snow) (November) and late winter (peak snow) (April). Elk were more widely distributed in early winter prior to significant snowpack accumulation. With increasing snow pack, differences in landscape characteristics resulted in the odds of elk occurrence within and among the three drainages being considerably less uniform and elk became more concentrated in areas of high snow heterogeneity.

2. We collected 4869 locations from 45 radio-collared elk over seven winter seasons and used a validated snow model that captured daily spatial and temporal variations at both the local and landscape scales to quantify elk responses to dynamic changes in snow pack.

3. This novel approach enabled us to demonstrate that elk dynamically responded to local snow conditions as landscape-scale snow conditions changed. Elk generally occupied areas with low snow mass, but used areas with higher snow mass as conditions on the winter range became more severe.

4. Access to snow model estimates that provided metrics at a variety of scales also enabled us to detect a strong positive relationship between elk occurrence and increasing local SNH.

5. We did not detect a threshold of snow mass that precluded occupancy by elk and observed elk in areas with severe snow mass (0.55 m) that substantially exceeded the threshold previously estimated for Yellowstone elk.

6. This research highlights a significant effort to incorporate a mechanism of climate into our understanding of a species distribution, and as climatic modeling increases in prevalence and accuracy, researchers will have a greater ability to apply data, novel in both its' spatial and temporal detail, to a wide array of ecological questions.

7. This study provided a good understanding of how elk chose resources when the sole limiting factor was starvation. The findings provide an excellent opportunity to evaluate how elk resource selection changed after wolves recolonized the system and elk were confronted with tradeoffs between predation risk and starvation.

VI. REFERENCES

Ackerly, D. D. 2003. Community assembly, niche conservatism, and adaptive evolution in changing environments. *International Journal of Plant Sciences* **164**:S165–S184.

Agresti, A. 2003. *Categorical Data Analysis*. 2nd edn. Wiley, New York, NY.

Arthur, S. M., B. F. J. Manly, L. L. McDonald, and G. W. Garner. 1996. Assessing habitat selection when availability changes. *Ecology* **77**:215–227.

Austin, M. P. 2002. Spatial prediction of species distribution: An interface between ecological theory and statistical modeling. *Ecological Modelling* **157**:101–118.

Ball, J. P., C. Nordengren, and K. Wallin. 2001. Partial migration by large ungulates: Characteristics of seasonal moose *Alces alces* ranges in northern Sweden. *Wildlife Biology* **7**:39–47.

Barry, S., and J. Elith. 2006. Error and uncertainty in habitat models. *Journal of Applied Ecology* **43**:413–423.

Berthold, P. 2001. Bird migration: A novel theory for the evolution, the control and the adaptability of bird migration. *Journal Fur Ornithologie* **142**:148–159.

Boyce, M. S. J. S. Mao, E. H. Merrill, D. Fortin, M. G. Turner, J. Fryxell, and P. Turchin. 2003. Scale and heterogeneity in habitat selection by elk in Yellowstone National Park. *Ecoscience* **10**:421–431.

Burnham, K. P., and D. R. Anderson. 2002. *Model Selection and Inference*. Springer, New York, NY.

Collett, D. 2003. *Modelling Binary Data*. 2nd edn Chapman & Hall/CRC, Boca Raton, FL.

Cook, J. G., L. L. Irwin, D. Bryant, R. A. Riggs, and J. W. Thomas. 1998. Relations of forest cover and condition of elk: A test of the thermal cover hypothesis in summer and winter. *Wildlife Monographs* 141.

Cooper, A. B., and J. J. Millspaugh. 1999. The application of discrete choice models to wildlife resource selection studies. *Ecology* **80**:566–575.

Coughlan, J. C., and S. W. Running. 1997. Regional ecosystem simulation: A general model for simulating snow accumulation and melt in mountainous terrain. *Landscape Ecology* **12**:119–136.

Craighead, J. J., F. C. Craighead Jr, R. L. Ruff, and B. W. O'Gara. 1973. Home ranges and activity patterns of nonmigratory elk of the Madison Drainage herd as determined by biotelemetry. Wildlife Monographs 33.

Dalke, P. D., R. D. Beeman, F. J. Kindel, R. J. Robel, and T. R. Williams. 1965. Seasonal movements of elk in the Selway River drainage, Idaho. *Journal of Wildlife Management* **29**:333–338.

Despain, D. G. 1990. *Yellowstone Vegetation: Consequences of Environment and History in a Natural Setting*. Roberts Rinehart, Boulder, CO.

Doerr, J. G., E. J. Degayner, and G. Ith. 2005. Winter habitat selection by Sitka black-tailed deer. *Journal of Wildlife Management* **69**:322–331.

Dussault, C., J.-P. Ouellet, R. Courtois, J. Huot, L. Breton, and H. Jolicoeur. 2005. Linking moose habitat selection to limiting factors. *Ecography* **28**:619–628.

DuToit, J. T., K. H. Rogers, and H. C. Biggs. 2003. *The Kruger Experience: Ecology and Management of Savanna Heterogeneity.* Island Press, Washington, DC.

Festa-Bianchet, M. 1988. Seasonal range selection in bighorn sheep: Conflicts between forage quality, forage quantity, and predator avoidance. *Oecologia* **75**:580–586.

FitzGibbon, C. D., and J. Lazarus. 1995. Antipredator behavior of Serengeti ungulates: Individual differences and population consequences. Pages 274–296 in A. R. E. Sinclair and P. Arcese (Eds.) *Serengeti II: Dynamics, Management, and Conservation of an Ecosystem.* University of Chicago Press, Chicago, IL.

Fortin, D., H. L. Beyers, M. S. Boyce, D. W. Smith, T. Duchesne, and J. S. Mao. 2005. Wolves influence elk movements: Behavior shapes a trophic cascade in Yellowstone National Park. *Ecology* **86**:1320–1330.

Fretwell, S. D. 1972. *Populations in a Seasonal Environment.* Princeton University Press, Princeton, NJ.

Frair, J. L., E. H. Merrill, D. R. Visscher, D. Fortin, H. L. Beyer, and J. M. Morales. 2005. Scales of movement by elk (*Cervus elaphus*) in response to heterogeneity in forage resources and predation risk. *Landscape Ecology* **20**:273–287.

Gaillard, J.-M., M. Festa-Bianchet, N. G. Yoccoz, A. Loison, and C. Toïgo. 2000. Temporal variation in fitness components of population dynamics in large herbivores. *Annual Review of Ecology and Systematics* **31**:367–393.

Garrott, R. A., L. L. Eberhardt, P. J. White, and J. J. Rotella. 2003. Climate-induced variation in vital rates of an unharvested large-herbivore population. *Canadian Journal of Zoology* **81**:33–45.

Greene, E. M., G. E. Liston, and R. A. Pielke. 1999. Relationships between landscape, snowcover depletion, and regional weather and climate. *Hydrological Processes* **13**:2453–2466.

Grignolio, S., I. Rossi, B. Bassano, F. Parini, and M. Apollonio. 2004. Seasonal variations of spatial behaviour in female Alpine ibex (*Capra ibex ibex*) in relation to climatic conditions. *Ethology Ecology and Evolution* **16**:255–264.

Hebblewhite, M., E. H. Merrill, and T. L. McDonald. 2005. Spatial decomposition of predation risk using resource selection functions: An example in a wolf–elk predator–prey system. *Oikos* **111**:101–111.

Hobbs, N. T. 2003. Challenges and opportunities in integrating ecological knowledge across scales. *Forest Ecology and Management* **181**:223–238.

Houston, D. B. 1982. *The Northern Yellowstone Elk: Ecology and Management.* Macmillan, New York, NY.

Humphries, M. M., D. W. Thomas, and D. L. Kramer. 2003. The role of energy availability in mammalian hibernation: A cost–benefit approach. *Physiological and Biochemical Zoology* **76**:165–179.

Ihl, C., and D. R. Klein. 2001. Habitat and diet selection by muskoxen and reindeer in western Alaska. *Journal of Wildlife Management* **65**:964–972.

Iqbal, M. 1983. *An Introduction to Solar Radiation.* Academic Press, San Diego, CA.

Johnson, C. J., K. L. Parker, D. C. Heard, and M. P. Gillingham. 2002. A multiscale behavioral approach to understanding the movements of woodland caribou. *Ecological Applications* **12**:1840–1860.

Karl, J. W., P. J. Heglund, E. O. Garton, J. M. Scott, N. M. Wright, and R. L. Hutto. 2000. Sensitivity of species habitat-relationship model performance to factors of scale. *Ecological Applications* **10**:1690–1705.

Keating, K. A., P. J. P. Gogan, J. M. Vore, and I. R. Irby. 2007. A simple solar radiation index for wildlife habitat studies. *Journal of Wildlife Management* **71**:1344–1348.

Kelsall, J. P. 1969. Structural adaptations of moose and deer for snow. *Journal of Mammalogy* **50**:302–310.

Knight, R. R. 1970. *The Sun River elk herd.* Wildlife Mongraph 23.

Lele, S. R., and J. L. Keim. 2006. Weighted distributions and estimation of resource selection probabilities. *Ecology* **87**:3021–3028.

Levin, S. A. 1992. The problem of pattern and scale in ecology. *Ecology* **73**:1943–1967.

Loison, A., and R. Langvatn. 1998. Short- and long-term effects of winter and spring weather on growth and survival of red deer in Norway. *Oecologia* **116**:489–500.

Manly, B. F. J., L. L. McDonald, D. L. Thomas, T. L. McDonald, and W. P. Erickson. 2002. *Resource Selection by Animals: Statistical Design and Analysis for Field Studies.* Kluwer, Dordrecht, Netherlands.

Marsh, P. 1999. Snowcover formation and melt: Recent advances and future prospects. *Hydrological Processes* **13**:2117–2134.

Messer, M. A. 2003. Identifying large herbivore distribution mechanisms through application of fine-scale snow modeling. Montana State University, Bozeman, MT.

Morrison, M. L. 2001. A proposed research emphasis to overcome the limits of wildlife-habitat relationship studies. *Journal of Wildlife Management* **65**:613–623.

Nellemann, C. 1998. Habitat use by muskoxen (*Ovibos moschatus*) in winter in an alpine environment. *Canadian Journal of Zoology* **76**:110–116.

Parker, K. L., C. T. Robbins, and T. A. Hanley. 1984. Energy expenditures for locomotion by mule deer and elk. *Journal of Wildlife Management* **48**:474–488.

Pearson, S. M., M. G. Turner, L. L. Wallace, and W. H. Romme. 1995. Winter habitat use by large ungulates following fire in northern Yellowstone National Park. *Ecological Applications* **5**:744–755.

Pettorelli, N., A. Mysterud, N. G. Yoccoz, R. Langvatn, and N. C. Stenseth. 2005a. Importance of climatological downscaling and plant phenology for red deer in heterogeneous landscapes. *Proceedings Royal Society of London B* **272**:2357–2364.

Pettorelli, N., J. O. Vik, A. Mysterud, J. Gaillard, C. J. Tucker, and N. C. Stenseth. 2005b. Using the satellite-derived NDVI to assess ecological responses to environmental change. *Trends in Ecology and Evolution* **20**:503–510.

Post, E., and N. C. Stenseth. 1999. Climatic variability, plant phenology, and northern ungulates. *Ecology* **80**:1322–1339.

Robbins, C. T. 1993. *Wildlife Feeding and Nutrition.* Academic Press, New York, NY.

Rongstad, O. J., and J. R. Tester. 1969. Movements and habitat use of white-tailed deer in Minnesota. *Journal of Wildlife Management* **33**:366–379.

Root, T. 1988. Energy constraints on avian distributions and abundances. *Ecology* **69**:330–339.

Rowland, M. M., M. J. Wisdom, B. K. Johnson, and J. G. Kie. 2000. Elk distribution and modeling in relation to roads. *Journal of Wildlife Management* **64**:672–684.

Sæther, B. E. 1997. Environmental stochasticity and population dynamics of large herbivores: A search for mechanisms. *Trends in Ecology and Evolution* **12**:143–149.

SAS Institute Inc. 2000. *SAS/STAT User's Guide, Version 8.* SAS Institute, Cary, NC.

Sawyer, H., R. M. Nielson, F. G. Lindzey, L. Keith, J. H. Powell, and A. A. Abraham. 2007. Habitat selection of Rocky Mountain elk in a nonforested environment. *Journal of Wildlife Management* **71**:868–874.

Schaefer, J. A., and F. Messier. 1995. Habitat selection as a hierarchy: The spatial scales of winter foraging by muskoxen. *Ecography* **18**:333–344.

Schmidt, K. 1993. Winter ecology of nonmigratory alpine red deer. *Oecologia* **95**:226–233.

Stephens, D. W., and J. R. Krebs. 1986. *Foraging Theory.* Princeton University Press, Princeton, NJ.

Sweeney, J. M., and J. R. Sweeney. 1984. Snow depths influencing winter movements of elk. *Journal of Mammalogy* **65**:524–526.

Terry, E. L., B. N. Mclellan, and G. S. Watts. 2000. Winter habitat ecology of mountain caribou in relation to forest management. *Journal of Applied Ecology* **37**:589–602.

Turner, M. G., Y. Wu, L. L. Wallace, W. H. Romme, and A. Brenkert. 1994. Simulating winter interactions among ungulates, vegetation and fire in northern Yellowstone Park. *Ecological Applications* **4**:472–496.

Wallace, L. L., M. G. Turner, W. H. Romme, R. V. O'Neill, and Y. Wu. 1995. Scale of heterogeneity of forage production and winter foraging by elk and bison. *Landscape Ecology* **10**:75–83.

Watson, F. G. R., T. Anderson, W. Newman, S. Alexander, and R. A. Garrott. 2006a. Optimal sampling schemes for estimating mean snow water equivalents in stratified heterogeneous landscapes. *Journal of Hydrology* **328**:432–452.

Watson, F. G. R., W. Newman, J. C. Coughlan, and R. A. Garrott. 2006b. Testing a distributed simulation model against diverse observations. *Journal of Hydrology* **328**:453–466.

Watson, F. G. R., and J. M. Rahman. 2003. Tarsier: A practical software framework for model development, testing and deployment. *Environmental Modelling and Software* **19**:245–260.

Weins, J. A. 1989. Spatial scaling in ecology. *Functional Ecology* **3**:5–397.

White, G. C., and R. A. Garrott. 1990. *Analysis of Wildlife Radio-Tracking Data.* Academic Press, San Diego, CA.

Winnie, J.Jr, D. Christianson, S. Creel, and B. Maxwell. 2006. Elk decision-making rules are simplified in the presence of wolves. *Behavioral Ecology and Sociobiology* **61**:277–289.

CHAPTER 9

Diet and Nutrition of Central Yellowstone Elk During Winter

P. J. White,* Robert A. Garrott,[†] John J. Borkowski,[‡] James G. Berardinelli,[§] David R. Mertens,[¶] and Andrew C. Pils[†]

*National Park Service, Yellowstone National Park
[†]Fish and Wildlife Management Program, Department of Ecology, Montana State University
[‡]Department of Mathematical Sciences, Montana State University
[§]Department of Animal and Range Sciences, Montana State University
[¶]U.S. Department of Agriculture, Agricultural Research Service, U.S. Dairy Forage Research Center, University of Wisconsin

Contents

Theme

Winter for ungulates in the northern Rocky Mountains is a period of chronic under-nutrition and catabolism of lean muscle and body fat because forage availability and quality are low, while energetic demands are high (Moen 1976, Torbit et al. 1985, DelGuidice et al. 1990). The rate of ingestion of assimilable energy and nutrients, and the resulting state of body components, reflect the adequacy of forage quality and quantity to meet metabolic demands (Harder and Kirkpatrick 1994; Parker et al. 1999; Cook et al. 2004a,b). Furthermore, nutrition and nutritional condition strongly influence the probability of breeding, over-winter survival, lactation yields, and susceptibility to predation (Loudon et al. 1983, Hobbs 1989, Kohlmann 1999, Mech et al. 2001, Bender et al. 2002, Cook 2002). The heterogeneous environment inhabited by elk (Cervus elaphus) wintering in the Madison headwaters area of Yellowstone National Park provides a diversity of potential foods for elk with quite divergent energy, protein, secondary compound, and toxin characteristics (Craighead et al. 1973). We characterized the

The Ecology of Large Mammals in Central Yellowstone
R. Garrott, P. J. White and F. Watson
ISSN 1936-7961, DOI: 10.1016/S1936-7961(08)00209-1

Copyright © 2009, Elsevier Inc.
All rights reserved.

composition and quality of elk diets prior to wolf (*Canis lupus*) recolonization, evaluated the effects of diet selection and snow pack on nutrition using an index of metabolizable energy intake, and related annual variability in winter nutrition to demographic vital rates.

I. INTRODUCTION

Selection occurs when an animal uses resources disproportionately to their frequency of occurrence in the animal's environment (Johnson 1980). Resource selection by ungulates occurs in a hierarchical fashion. In northern temperate regions, animals choose a wintering range with a habitat mosaic that can satisfy their winter energetic requirements and then, prior to feeding, search for a habitat type within their range which offers a sufficient amount and quality of forage at an acceptable metabolic cost. When actually feeding, ungulates choose among plants of different species and adjacent plants of the same species. Also, they decide what size of bite to take from each plant or plant part, and how often to take bites. In making these choices, ungulates should attempt to maximize the intake and quality of forage while minimizing energetic costs (Stephens and Krebs 1986). However, the complexity of these choices often contributes to substantial variability among individuals (Pulliam 1989). Furthermore, ungulates may use different resource selection strategies yet achieve the same net nutritional gain (Hobbs *et al.* 1983). Some individuals may rely on relatively low quality forages that are abundant and require minimal energy expenditure to obtain, whereas others may choose higher quality forages that are less abundant and require higher energetic costs to obtain (Bunnell and Gillingham 1985; Owen-Smith 1993, 1994). In addition, different strategies may be favored under varying environmental conditions or by various sex or age groups (Maynard Smith 1982, Pulliam 1989).

Graminoids (grasses and grass-like plants) typically dominate elk diets and occur at higher proportions in the diet than in elk foraging habitats, indicating preference (Christianson and Creel 2007). However, there is likely strong selection for foraging strategies by elk during winter that increase energy and protein intake and reduce the catabolism of body mass (*i.e.*, weight loss; MacArthur and Pianka 1966, Schoener 1971, Charnov 1974, Stephens and Krebs 1986). Rocky Mountain elk choose diets from a wide variety of forage plants, the nutrient content of which varies among sites, species, parts, and phenological stage (Hobbs *et al.* 1979, Cook 2002). Elk can exploit this diversity of potential foods of varying quality by altering their foraging strategies to collect more energy or protein. Metabolizable energy is maximized by grazing (grasses, rushes, sedges), while dietary protein intake is generally maximized by browsing (conifers, deciduous trees, shrubs; Robbins 1993, Van Soest 1994). Also, intake of digestible dry matter will vary within and among winters as increased snow pack and decreased temperatures limit forage availability, reduce foraging efficiency, and increase energy expenditures (Gates and Hudson 1979, Skovlin 1982, Parker *et al.* 1984, Wickstrom *et al.* 1984).

The rugged mountainous landscape in the Madison headwaters area supports mature conifer forests, sedge-grass meadows, and aquatic communities that provide elk with a diversity of potential foods with quite divergent energy, protein, secondary compound, and toxin characteristics (Craighead *et al.* 1973). Numerous geyser basins provide warm, low-elevation meadows and rivers, thereby reducing snow cover and enabling some unique plant associations to continue photosynthesis through the winter (Despain 1990). Large-scale fires during 1988 burned 55% of this area, creating a complex mosaic of burned and unburned forests at different stages of succession (Turner *et al.* 1994). Climatic factors contribute to the variability of resources because snow depths frequently exceed one meter, thereby drastically reducing food availability and producing severe energetic problems for elk (Craighead *et al.* 1973).

We characterized the composition and quality of elk diets prior to wolf recolonization and used allantoin:creatinine (A:C) ratios from urine-saturated snow samples to index metabolizable energy intake and evaluate the effects of diet selection and snow pack on the nutrition of adult female elk wintering in the Madison headwaters area. Allantoin is a metabolite that reliably indexes rumen microbial flow in elk during a several-day period prior to the time urine was voided (Vagnoni *et al.* 1996, Garrott *et al.* 1997). Creatinine is a compound produced and excreted in relatively constant proportion to muscle mass, allowing comparisons to be made among animals of different body size and hydration and among samples of different snow dilution (Coles 1980). We predicted elk would selectively vary the forage class mix of their diet to maintain relatively stable dietary protein and energy levels through the winter (Hobbs *et al.* 1981, 1983). Also, we expected nutrition within and among winters would vary depending on snow accumulation patterns that affected the availability of forages (Garrott *et al.* 1996, 2003). In turn, we expected the extent nutrition was below maintenance levels each winter would be correlated with age and annual variation in over-winter survival (Cook *et al.* 2004a,b).

II. METHODS

A. Diet Composition

The elk population in the Madison headwaters area is nonmigratory and remains within the borders of the park through the year (Craighead *et al.* 1973). The protection afforded by the park and frequent contact with humans has resulted in these elk becoming relatively tolerant of humans. This tolerance greatly facilitates close observation and the collection of samples for indexing nutrition and demographic rates. We estimated the botanical composition of diets for radio-collared, adult, female elk during winters 1991–1992 through 1993–1994 using bite counts of forage intake by 15 elk (Spowart and Hobbs 1985) and microscopic examination of plant fragments in feces from five elk (Sparks and Malachek 1968), which reflected foraging decisions made over several hours and days, respectively. A single observer approached to within 5–30 m of feeding elk and recorded the number of bites and plant parts eaten by species or functional group (shrubs, graminoids, and forbs) during 5-min sampling intervals separated by five minutes (Figure 9.1). After observation sessions ($\leq$60 min), the observer collected 25 forage samples that mimicked the amounts and plant parts consumed by the elk. We determined bite weights by drying samples at 70 °C for 24 h and dividing total weight by 25 (Spowart and Hobbs 1985). We calculated the proportion by dry-weight of each forage class in the diet by multiplying bite frequency times mean bite weight and normalizing to sum to 100 (Hobbs *et al.* 1979). Because the number of bites observed was so large ($n = 125{,}636$), we calculated 90% confidence intervals on diet percentages as simple t intervals on the means of untransformed percentages across animals (Hobbs *et al.* 1979).

We also opportunistically collected fresh fecal pellets deposited by four radio-collared elk in the Old Faithful geyser basin and one elk in the Norris geyser basin. Samples were oven-dried at 70 °C and ground to 1-mm size in a Wiley Mill. We prepared two slides from each sample and examined 50 fields per slide to identify epidermal cell tissue fragments to at least the forage class level. We calculated percent relative density for each forage class.

B. Forage Quality

During January 1993, we collected 2–6 replicate samples of major forages eaten by elk from areas where they were frequently observed feeding in the Old Faithful geyser basin, Norris geyser basin, and upper Madison Valley. Samples included aquatic vegetation (*Myriophyllum, Potamogeton, Zannichellia*), bunchgrasses (*Agropyron, Bromus, Calamagrostis, Danthonia, Deschampsia, Elymus, Festuca, Phleum, Poa, Stipa*), burned bark, conifer tips and needles (*Pinus contorta*), elk sedge (*Carex geyeri*), leafy forbs (*Aster, Epilobium, Potentilla, Trifolium, Pteridium*), mixed grasses (bunchgrasses and other species; *e.g., Hordeum*), sedge shoots (*Carex sp.*), monocot mix (*Carex, Scirpus, Juncus*, grasses), spikerush

FIGURE 9.1 A biologist collecting bite count data from a radio-collared elk grazing in a geothermal basin in the Firehole River drainage of Yellowstone National Park, Wyoming, USA. Elk occupying the Madison headwaters area are non-migratory and remain within the park through the year. Thus, they are not subjected to human hunting and most animals are relatively tolerant of human presence (Photo by Robert Garrott).

(*Eleocharis rostellata*), and shrubs (*e.g.*, willow (*Salix* spp.)). Samples were lyophilized, ground to a 2-mm size in a Wiley mill, and stored in sealed vials at $-17\,^{\circ}C$. We did not wash plant samples prior to assay so that the values reflected dietary intake, including soil contamination, rather than simply plant constituents.

We analyzed samples for crude protein, acid detergent fiber (ADF), and mineral content following Schulte *et al.* (1987; Soil and Plant Analysis Laboratory, University of Wisconsin, Madison). Biogenic silica was measured by leaching ashed ADF residue with hydrobromic acid for two hours (Van Soest and Jones 1968, Wildlife Habitat Laboratory, Washington State University, Pullman). Fiber content was measured sequentially by neutral detergent fiber concentration, ADF, sulfuric acid lignin, cellulose, and insoluble ash (Goering and Van Soest 1970, Robbins *et al.* 1975). We determined 96-h *in vitro* dry-matter digestibility (IVDMD) following Goering and Van Soest (1970) and Tilley and Terry (1963), with modifications described in Jakubas *et al.* (1994; U.S. Dairy Forage Research Center, Madison, Wisconsin). The inoculum source was a nonlactating, fistulated, Holstein cow consuming an alfalfa hay diet. We calculated the crude protein content and diet *in vitro* digestibility as the sum of products of forage quality values of individual forage types multiplied by their dry-weight proportion (bite counts) or relative density (microhistology) in diets (Hobbs *et al.* 1979).

We modified procedures from Cork and Krockenberger (1991) to extract phenols by adding 3 ml of 70% aqueous acetone to darkened test tubes containing 0.1 g of ground, lyophilized plant material. The mixture was vortexed for 30 min and the supernatant was decanted. After repeating this process four times, we combined extracts and determined total phenols using the Prussian blue assay

(Price and Butler 1977). We tested for saponins (*i.e.*, triterpenoid glycosides) by adding 3 ml methanol (60 °C) to 0.1 g of plant material for one hour. After the mixture cooled, each sample was vigorously shaken to see if a persistent foam developed (Harborne 1984). We used a sequential extraction procedure (Lindroth *et al.* 1986) with 0.1 g of plant material and tested the final fraction for alkaloids by thin layer chromatography. Silica gel plates were developed in chloroform:ethanol:ethyl acetate: acetone (60:20:10:10) mobile phase and visualized with Dragendorff's reagent (Santavy 1969). We also tested for alkaloids by extracting plant material in 2 ml of chloroform for 24 h, after which the mixture was vortexed for 0.5 h. The supernatant was removed and analyzed by thin layer chromatography using silica gel plates developed in benzene:methanol (90:10) mobile phase.

C. Metabolizable Energy Intake

We estimated metabolizable energy intake (nutrition) of elk during 1991–1992 through 1997–1998 using molar ratios of allantoin and creatinine (A:C ratios) obtained from fresh urine deposited in snow. We attempted to collect samples from radio-collared, adult, female elk at ≤3-week intervals during December to April each winter using a restricted randomization design to determine the order and frequency of sampling (Garrott *et al.* 1996). Procedures for sample collection, handling, and storage before assays were described by DelGuidice *et al.* (1989). Frozen snow-urine samples were then thawed at room temperature and extraneous debris was eliminated by filtration from the snow-urine mixture. The filtrate was refrozen at −20 °C until assayed. Creatinine concentrations in filtered samples were assayed at the Department of Animal and Range Sciences, Montana State University, Bozeman, using a modification of the kinetic Jaffé reaction (Larsen 1972). Allantoin concentrations were determined using a microplate modification of the colorimetric procedures described by Young and Conway (1942). Allantoin and creatinine were assayed from the same sample at the same time to avoid additional freeze-thaw cycles that tend to degrade allantoin. Sensitivities of allantoin and creatinine assays were 0.027 and 0.079 μmol, respectively. Coefficients of variation for high and low elk snow-urine pools of allantoin and creatinine within each year were less than 15%.

We plotted A:C ratios for each elk by collection date and used the spatial power function of the SAS PROC MIXED procedure (SAS 2002) to generate models for predicting over-winter trends in A:C ratios (Pils *et al.* 1999). We then integrated, yielding the areas under the curves, to estimate the mean A:C ratio during December 4 through April 22 each winter. Garrott *et al.* (1997) estimated the winter maintenance requirements (550–725 kJ metabolizable energy/kg body mass$^{0.75}$) for free-ranging elk, and the A:C ratio (0.34–0.4) that represented this level of metabolizable energy intake, from experimental feeding trials with groups of captive elk. Thus, we integrated the area above the A:C curve for each year and below a threshold A:C level of 0.30 (*i.e.*, submaintenance index) to evaluate the relative extent elk nutrition was below maintenance each winter.

We compared monthly A:C ratios to indices of digestible energy and protein intake to provide field validation of previous captive feeding trials (Vagnoni *et al.* 1996) and evaluated the influence of diet selection on nutrition. We also compared trends in A:C ratios to cumulative snow pack estimates for the entire elk winter range during December through April. Mean daily snow pack estimates were derived from the dynamic Langur snow pack model developed and validated during five winters (Watson *et al.* 2006a,b; Chapters 5 and 6 by Watson *et al.*, this volume). The model estimated the number of centimeters of water in the snow pack (*i.e.*, snow water equivalent; SWE) at a 28.5×28.5-m pixel scale for the entire winter range and daily means were summed to obtain a cumulative estimate for the entire winter (Chapter 6 by Watson *et al.*, this volume). In addition, we regressed the submaintenance index each winter against adult and calf survival. Adult survival was estimated each year using the proportion of radio-collared females that survived (Evans *et al.* 2006). These estimates were likely confounded somewhat by the age distribution of radio-collared females each year because there was a strong effect of age on survival (*i.e.*, senescence; Chapter 10 by Garrott *et al.*, this volume).

Calf survival was estimated using the change-in-ratio method (Skalski *et al.* 2005) with autumn and spring calf:cow ratios calculated from groups associated with the first 100 and last 100 radio-collared animals observed during telemetry locations (Chapter 10 by Garrott *et al.*, this volume). These estimates were adjusted for adult female survival using a pooled estimate (0.94) for 1991–1992 through 1997–1998.

We used A:C ratios from snow-urine samples collected from radio-collared, adult, female elk during seven consecutive winter seasons (1991–1992 through 1997–1998) to fit a suite of 203 *a priori* mixed-models, where each model contained a subset of the following 12 model effects. The variable SWE was the daily snow water equivalent (*i.e.*, water content of a column of snow measured in cm) estimated as the average of all pixels in the study area by the Langur snow pack model (Chapter 6 by Watson *et al.*, this volume) with a lag of 2 days. We expected the coefficient of SWE would be negative, with increasing SWE associated with lower A:C values. The variable SWE^2 (*i.e.*, SWE*SWE) was also used to allow for non-linear (or curvature) effects with SWE. We expected the coefficient of SWE^2 would be positive, with the negative effect of SWE on A:C being dampened for larger values of SWE.

We defined a categorical variable DRAINAGE having three levels depending on where a given elk wintered in the Madison headwaters area: Firehole ($n = 745$), Gibbon ($n = 383$), or Madison ($n = 665$). We expected A:C would not be the same across drainages. Interactions of SWE*DRAINAGE and SWE^2*DRAINAGE were examined to determine if the effects of SWE or SWE^2 depended on the particular drainage. We also defined a categorical variable PERIOD indicating whether or not the date of a snow-urine sample collection was before (*i.e.*, pre) or after (*i.e.*, post) the date when SWE was at its maximum for that winter. We expected A:C values would gradually decrease as snow built-up and limited forage availability during winter, but increase more rapidly (*i.e.*, steeper slope) during spring melt-out as the snow pack became more heterogeneous and forage availability increased in snow-free patches; even though mean SWE may have been similar between periods. Interactions of SWE*PERIOD and SWE^2*PERIOD were examined to determine if the effects of SWE or SWE^2 depended on period, with a negative coefficient indicating a steeper slope for A:C values in the spring melt-out period.

We defined a variable AGE as the age category of the collared elk during the winter the A:C sample was collected, with two categories: prime (*i.e.*, ≤13 years old) and senescent (*i.e.*, >13 years old). We also used remotely sensed normalized differential vegetation index (NDVI) data as a measure of temporal and spatial variability in vegetation growth (Wittemyer *et al.* 2007). We selected the seven largest meadow complexes in the study area, ranging from 1 to 48 km² and encompassing a total of approximately 80 km² across the entire elevational gradient of the Firehole, Gibbon, and Madison River drainages. AVHRR satellite data at 1-km² resolution were processed by the U.S. Geological Survey Center for EROS (Eidenshink 1992, DeFelice *et al.* 2003). NDVI values were composited using the maximum value method over 14-day periods to eliminate spuriously low values caused by cloud contamination, haze, and other atmospheric effects. Values were then further smoothed temporally using the methods of Swets *et al.* (2000). Data included 26 composites per year and program TIMESAT with the Savitsky-Golay filter (Jönsson and Eklundh 2002, 2004) was used to extract NDVI metrics for each polygon and averaged across all polygons for each year. We used the L-integral, which is the integral of the seasonal NDVI curve over the growing season (Garel *et al.* 2006), to calculate a covariate, $NDVI_{int}$, to index summer range forage productivity. We also evaluated growing season length ($NDVI_{len}$), which indexes the time period when green, high-quality vegetation was available. We expected the coefficients of these NDVI metrics to be positive, with increasing NDVI values associated with higher A:C values during the following winter.

In addition, we used the Palmer Drought Severity Index (PDSI; Palmer 1965) as a single integrator of annual variability of regional warm season climate. The PDSI, a standard measure of drought severity in the United States, is based on a water balance model and uses a suite of inputs for its calculation, most significantly precipitation, evapotranspiration, soil moisture, and temperature. Negative values of the PDSI indicate drought, while positive values denote a wet period. We obtained monthly PDSI values for the Yellowstone region of Wyoming (region 1) from the National Climate Data Center (National Climate Data Center 2006). We defined a covariate, PDSI, as the average of monthly PDSI values during

May-July each year to index drought conditions during the growing season before the winter began. We expected the coefficient of PDSI would be positive, with increasing drought (*i.e.,* negative values of PDSI) associated with lower A:C values during the following winter.

Each *a priori* model had, at most, two of the $NDVI_{int}$, $NDVI_{len}$, and PDSI effects, and contained SWE^2 only if it also contained SWE. Also, each model contained an interaction only if both of the main effects associated with the interaction were in the model. The SAS PROC MIXED procedure was used to fit models. Because multiple A:C measurements were taken per radio-collared elk within each winter season and drainage, each model included random effects for individual animal effects to account for this lack of independence.

III. RESULTS

The overall proportions of forage classes in elk diets during winter based on 352 foraging bout observations were 59% $\pm$ 10% graminoids (grasses, rushes, sedges), 33% $\pm$ 7% browse (shrubs and trees), and 8% $\pm$ 3% forbs and aquatics. Similarly, the overall percent relative density for each forage class based on microscopic examination of 254 composite fecal samples were 50% $\pm$ 4% graminoids, 45% $\pm$ 5% browse, and 5% $\pm$ 1% forbs (Table 9.1). Estimates of browse consumption based on microscopic examination of feces were consistently higher than those derived from observations of forage intake, whereas the opposite was generally true for grasses.

The forage-class mix of elk diets changed through winter, with grasses initially dominating diets in October (68%), decreasing to 40% during December through January or February, and then increasing to 80% by April (Figure 9.2). The increased selection of grasses in the spring was primarily due to elk eating more sedges (Table 9.1). Conversely, the proportion of browse increased from 30% in October to 40–50% during December through January or February, and then decreased to 20% by April. The proportion of forbs in the diet increased from a small proportion (2%) of the diet in October to 16% during December and January before decreasing to <8% by March and April.

Portions of sedges, mixed grasses, and aquatic plants eaten by elk had relatively high dry-matter digestibility (39–51%), while shrubs, conifer tips, and burned bark were high in lignin (Table 9.2). Sedge shoots, bunch grasses, leafy forbs, shrubs, and aquatic plants were relatively high in protein (7–13%; Table 9.2). Shrubs and conifer tips contained relatively high levels of phenols (43–53 mg/g), while pine tips and burned bark contained saponins (Figure 9.3). No alkaloids were detected in any samples. Portions of spike-rush, bunch grasses, leafy forbs, and aquatic plants eaten by elk were high in silica (18–33%) and overall ash content (25–39%) which, in turn, reduced their digestibility (Tables 9.2 and 9.3). Spike-rush, sedge shoots, monocot mix (sedge, grass, rush), and aquatic plants were extremely high in arsenic (8–225 mg/kg) and/or fluoride (50–421 mg/kg; Table 9.3).

Winter diets contained mean proportions of 7% $\pm$ 0.4% crude protein and 34% $\pm$ 2% digestible dry matter (Figure 9.4). Estimates of diet digestibility based on microscopic examination of feces were consistently lower than those derived from observations of forage intake, whereas the opposite was generally true for crude protein (Figure 9.4). Over-winter changes in diet digestibility were similar to those of graminoid intake, with the digestibility of diets decreasing from 34% in October to 30% by January or February, and then increasing to 40% by March or April (Figure 9.4). In contrast, the protein content in elk diets remained relatively constant through winter (Figure 9.4).

Allantoin:creatinine ratios derived from 1793 urine-soaked snow samples collected from 45 radio-collared elk during 1991–92 through 1997–98 demonstrated strong temporal patterns within and among winters. The seasonal patterns in mean A:C ratios followed a quadratic, U-shaped pattern, with lowest values occurring in mid-winter and significantly higher values in December and April (Figure 9.5). The top 10 *a priori* models (Table 9.4) based on Akaike's Information Criterion corrected for small sample sizes (AIC_c) contained SWE, SWE^2, and PERIOD, and 8 of the 10 models

TABLE 9.1 Mean percent diet composition of radio-collared, adult, female elk in the Madison headwaters area of Yellowstone National Park during winters 1991–1992 through 1993–1994 based on bite counts of forage intake by 15 elk and microscopic examination of the relative density of epidermal cell tissue fragments in feces from five elk

	1992				1992–1993							1993–1994					
	Jan	Feb	Mar	Apr	Oct	Nov	Dec	Jan	Feb	Mar	Apr	Nov	Dec	Jan	Feb	Mar	Apr
Bite counts																	
Grasses	11	43	25	22	52	26	9	24	14	37	18						
Sedges	8	11	16	20	0	0	5	9	2	15	32						
Elk sedge	10	4	38	16	13	13	14	3	3	3	11						
Spike-rush	7	8	7	23	0	0	8	1	10	6	5						
Monocot mix	2	1	1	0	3	8	3	4	14	2	13						
Leafy forbs	11	1	0	0	2	10	13	10	1	3	3						
Shrubs	0	0	1	1	7	7	2	4	0	4	0						
Conifer tips	12	16	3	2	0	1	21	15	26	11	4						
Bark	32	16	9	16	22	34	21	29	21	15	8						
Aquatics	5	0	0	0	0	0	4	2	9	3	4						
Microhistology																	
Grasses	10	13	27	23		34	23	22	20	32	33	32	40	34	29	42	37
Sedges	7	12	21	20		21	10	4	2	5	15	28	13	9	2	5	9
Spike-rush	20	5	4	8		3	7	13	23	12	9	5	3	5	15	10	9
Monocot mix	4	1	2	2		2	3	0	1	0	1	0	1	0	0	0	1
Leafy forbs	2	1	3	3		1	2	3	6	10	3	7	2	4	6	9	10
Shrubs	2	3	5	1		1	2	2	1	2	0	11	1	2	2	1	1
Conifer tips	37	27	26	22		3	20	48	36	26	15	2	6	16	21	17	9
Bark	17	38	12	18		34	33	7	11	13	23	15	35	29	24	16	25
Aquatics	1	0	0	1		0	0	0	0	1	0	0	0	0	0	0	0

contained AGE. The main effect weights were 1.00 for SWE, SWE^2, and PERIOD, 0.99 for AGE, 0.20 for PDSI, 0.17 for NDVI$_{len}$, 0.01 for DRAINAGE, and <0.01 for NDVI$_{int}$. The interaction weights were 0.98 for SWE*PERIOD, 0.97 for SWE^2*PERIOD, and <0.01 for SWE*DRAINAGE and SWE^2*DRAINAGE. The coefficients for the top *a priori* model were 0.164 for the intercept (95% CI=0.129, 0.198), −0.265 for SWE (95% CI = −0.312, −0.217), 1.041 for SWE^2 (95% CI=0.739, 1.343), 0.022 for PERIOD = Post (95% CI = −0.079, 0.123; 0 otherwise), 0.029 for AGE=Prime (95% CI = −0.005, 0.064; 0 otherwise), −0.898 for SWE*PERIOD = Post (95% CI = −1.808, 0.012; 0 otherwise), and 6.533 for SWE^2*PERIOD=Post (95% CI=0.896, 12.169; 0 otherwise). SWE values were centered about their midpoint value of 0.237.

Allantoin:creatinine values decreased gradually during the snow build-up portion of winter, but then increased rapidly (*i.e.*, steeper slope) during the snow melt-out phase in spring (Figure 9.6). Using the centered values of SWE, the predictive equation for A:C values for prime-aged females at a given level of SWE during the snow build-up phase (*i.e.*, before maximum SWE was reached) was:

$$A : C = 0.19269 - 0.2646(SWE - 0.237) + 1.0410(SWE - 0.237)^2$$

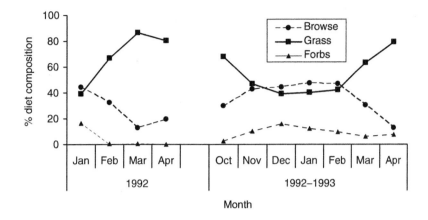

FIGURE 9.2 Mean proportion by dry-weight of forage classes in the winter diets of 15 radio-collared, adult, female elk during winters 1991–1992 and 1992–1993 in the Madison headwaters area of Yellowstone National Park, Wyoming, USA. Proportions were calculated as the product of bite frequency and mean bite weight, normalized to sum to 100 (Hobbs *et al.* 1979).

TABLE 9.2 Chemical composition of mid-winter forages used by elk during 1993 in the Madison headwaters area of Yellowstone National Park, Wyoming, USA

| Forage | Percent | | | | | IVDMD | Total phenols | Reaction to: | |
	Ash	Protein	ADF	NDF	Lignin			Saponins	Alkaloids
Spike-rush	24.5	5.4	33.5	60.2	4.2	26.6	2.0	No	No
Sedge shoots	14.4	7.0	30.2	60.1	2.8	51.4	4.8	No	No
Bunch grasses	38.9	10.8	18.7	38.5	3.1	34.0	2.3	No	No
Monocot mix	19.7	4.5	34.8	65.0	3.2	31.6	2.4	No	No
Elk sedge	9.7	5.2	33.4	63.6	2.6	47.1	4.5	No	No
Mixed grasses	8.6	3.3	39.1	72.6	2.6	38.6	3.1	No	No
Leafy forbs	39.2	11.8	13.6	25.2	3.5	34.4	3.5	No	No
Shrubs	3.3	9.0	33.0	47.9	14.1	34.9	43.4	Yes	No
Lodgepole pine	2.0	5.5	27.2	39.6	9.7	25.8	53.4	Yes	No
Burned bark	3.9	3.2	47.0	56.3	18.8	24.5	7.0	Yes	No
Aquatic plants	34.7	13.1	17.7	26.9	5.3	39.6	6.8	No	No

Acid detergent fiber (ADF), neutral detergent fiber (NDF), and lignin are reported as percent of organic matter (Goering and Van Soest 1970), while *in vitro* digestibility values (IVDMD) are reported in relation to total dry matter (Tilley and Terry 1963). Biogenic silica (%) assays were performed as described by Van Soest and Jones (1968). Total phenols (mg gallic acid/g) were determined using the Prussian blue assay (Price and Butler 1977). Saponins (*i.e.*, triterpenoid glycosides) were tested as described by Harborne (1984). Values are based on 2–6 aggregate samples collected from different sites within the study area and assayed in duplicate.

FIGURE 9.3 An instrumented cow elk browsing on the tips of lodgepole pine branches in the Madison headwaters area of Yellowstone National Park, Wyoming, USA, during mid-winter when deep snow pack covered grasses (Photo by Robert Garrott).

| TABLE 9.3 | Mineral concentration of mid-winter forages used by elk during 1993 in the Madison headwaters area of Yellowstone National Park, Wyoming, USA |

Forage	Parts per million												mg/kg dry		%
	Al	B	Ca	Cu	Fe	Mg	Mn	P	K	Na	S	Zn	As	Fl	Si
Spike-rush	2068	52	5057	4	631	439	806	505	1915	2589	1234	41	2	151	19
Sedge shoots	588	21	7256	3	714	703	481	1961	17696	3447	1971	30	8	50	9
Bunch grasses	2753	25	4909	7	2284	1116	651	1621	8400	625	1511	63	5	10	24
Monocot mix	801	18	2573	5	2173	541	340	469	1465	839	1619	45	15	78	10
Elk sedge	275	13	4088	5	304	927	511	787	4243	113	971	92	–	–	6
Mixed grasses	457	14	2938	5	460	704	214	645	2496	351	941	34	<1	27	6
Leafy forbs	2525	42	6724	10	4085	1353	252	1527	8559	2465	2176	93	–	–	33
Shrubs	303	17	2397	3	117	981	305	946	3449	<60	7867	51	–	–	–
Lodgepole	597	30	18840	<3	115	1698	662	691	1722	135	668	118	–	–	–
Burned bark	<40	21	9265	8	70	1282	110	1860	4988	<60	591	38	–	–	–
Aquatics	4733	71	14147	6	7840	1588	5950	2445	16405	13188	4295	125	225	421	18

Assays were performed as described by Schulte *et al.* (1987) and Van Soest and Jones (1968). All values are based on 2–18 aggregate samples collected from different sites within the study area. Al, aluminum; B, boron; Ca, calcium; Cu, copper; Fe, iron; Mg, magnesium; Mn, manganese; P, phosphorus; K, potassium; Na, sodium; S, sulfur; Zn, zinc; As, arsenic; Fl, fluoride and Si, silica.

The predictive equation for A:C values for prime-aged females at a given level of SWE during the snow melt-out phase (*i.e.*, after maximum SWE was reached) was:

$$A : C = 0.21446 - 0.2646(SWE - 0.237) + 1.0410(SWE - 0.237)^2 - 0.8982(SWE - 0.237)$$
$$+ 6.5325(SWE - 0.237)^2$$

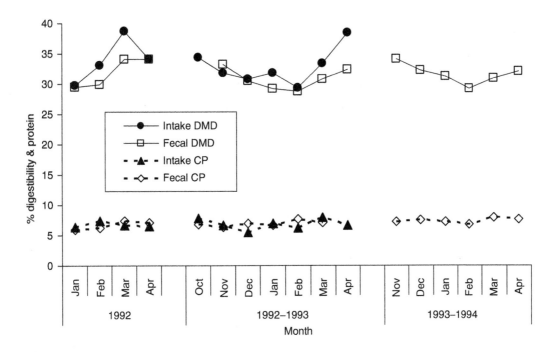

FIGURE 9.4 Mean percent diet digestibility (DMD) and crude protein (CP) content in the winter diets of radio-collared, adult, female elk during winters 1991–1992 through 1993–94 in the Madison headwaters area of Yellowstone National Park, Wyoming, USA, as determined from observations of forage intake (*n* = 15 elk) and microscopic examination of feces (*n* = 5 elk). Percentages were calculated as the sum of products of forage quality values of individual forage types multiplied by their dry-weight proportion (intake) or relative density (fecal) in diets (Hobbs *et al.* 1979).

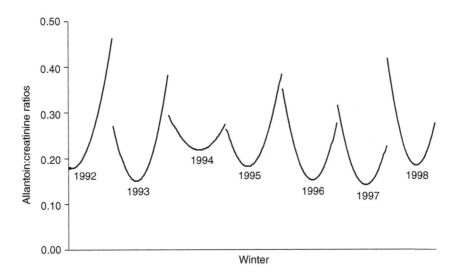

FIGURE 9.5 Allantoin:creatinine ratios of radio-collared, adult, female elk during 1991–1992 through 1997–1998 in the Madison headwaters area of Yellowstone National Park, Wyoming, USA.

TABLE 9.4	Top 10 *a priori* models for allantoin:creatinine (A:C) ratios for radio-collared, adult, female elk in Yellowstone National Park during seven consecutive winters from 1991–1992 through 1997–1998

Top 10 *a priori* models	K	ΔAIC_c	w_i
SWE + SWE^2 + PERIOD + AGE + SWE*PERIOD + SWE^2*PERIOD	7	0	0.610
SWE + SWE^2 + PERIOD + AGE + SWE* PERIOD + SWE^2* PERIOD + PDSI	8	2.4	0.184
SWE + SWE^2 + PERIOD + AGE + SWE* PERIOD + SWE^2* PERIOD + NDVI$_{len}$	8	2.7	0.158
SWE + SWE^2 + PERIOD + AGE	5	7.8	0.012
SWE + SWE^2 + PERIOD + AGE + SWE* PERIOD	6	9.1	0.006
SWE + SWE^2 + PERIOD + AGE + PDSI	6	9.7	0.005
SWE + SWE^2 + PERIOD + SWE* PERIOD + SWE^2* PERIOD + DRAINAGE	8	10.1	0.004
SWE + SWE^2 + PERIOD + SWE* PERIOD + SWE^2* PERIOD + DRAINAGE + SWE* DRAINAGE + SWE^2* DRAINAGE	12	10.3	0.004
SWE + SWE^2 + PERIOD + AGE + SWE* PERIOD + SWE^2* PERIOD + PDSI + NDVI$_{len}$	9	10.8	0.003
SWE + SWE^2 + PERIOD + AGE + NDVI$_{len}$	6	10.8	0.003

K (number of model parameters), ΔAIC_c (AIC corrected for small sample size), and w_k (AIC model weight). AIC_c was −3697.9 for the top *a priori* model.

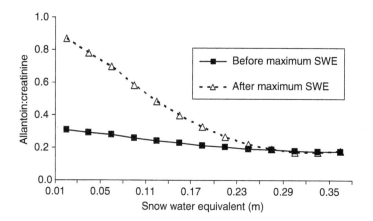

FIGURE 9.6 Predicted allantoin:creatinine values for a given snow water equivalent level before a maximum snow water equivalent of 0.35 m was reached during the snow build-up portion of winter and thereafter during the snow melt-out phase in spring.

Monthly mean A:C ratios were significantly correlated with the digestibility of diets estimated from microscopic examination of feces ($R^2 = 0.81$, $F_{1,12}=50.8$, $P = 0.0001$) and bite counts ($R^2 = 0.48$, $F_{1,7} = 6.5$, $P = 0.04$). Also, the relative extent elk nutrition was below maintenance each winter was significantly correlated with winter severity as indexed by measures of SWE$_{acc}$ ($R^2 = 0.59$, $F_{1,5} = 7.3$, $P = 0.04$). The primary limiting factor for elk in this population was over-winter mortality of both adults and calves due to starvation, with annual variation in mortality driven primarily by variation in snow pack severity (Chapter 10 by Garrott *et al.*, this volume). Estimates of adult survival were negatively correlated with the integrated area above the A:C curve for each winter and below a threshold A:C level of 0.30 (*i.e.*, extent nutrition was below maintenance), though the results were only marginally significant ($R^2 = 0.48$, $F_{1,5} = 4.6$, $P = 0.08$). Likewise, estimates of calf survival were negatively and significantly correlated with the extent nutrition was below maintenance levels each winter ($R^2 = 0.86$, $F_{1,5} = 24.9$, $P = 0.008$).

IV. DISCUSSION

Elk wintering in the Madison headwaters area selected diets dominated (>90%) by graminoids and browse. The consistently lower estimates of graminoid intake and diet digestibility based on microscopic examination of feces compared to bite counts of forage intake were likely due to the underestimation of highly digestible species by the microhistological method (Anthony and Smith 1974, Norbury 1988). The mean percentages of forage classes in elk diets during winter (*i.e.,* 59% grass, 33% browse, 8% forbs) were similar to those eaten by elk in Rocky Mountain National Park, Colorado, during diet trials (Hobbs *et al.* 1979). However, the proportions were quite different than the rumen contents of northern Yellowstone elk (82% grass, 11% browse, 7% forbs), except during a severe winter (1962) when browse comprised 28% of elk diets on the northern range (Greer *et al.* 1970). This result was not surprising given that the winter range for elk in northern Yellowstone receives far less snow and, as a result, typically has higher grass availability (Craighead *et al.* 1972, 1973). Also, the proportion of graminoids in the diet is significantly lower in elk populations experiencing severe winter conditions, with a concomitant increase in browse consumption (Christianson and Creel 2007).

Two atypical forages, aquatic plants (primarily *Myriophyllum*) and burned lodgepole pine bark, comprised significant portions of elk diets (Craighead *et al.* 1973, Jakubas *et al.* 1994). Elk were observed feeding extensively on these non-typical forages even when grasses and sedges were readily available. Burned bark had the lowest concentrations of inorganic substances and crude protein, the highest fiber values, and the lowest digestibility value of all forages assayed. In contrast, aquatic plants had nearly the highest concentrations of inorganic substances and crude protein, and very low ADF values; yet had digestibility values comparable to burned bark. Interestingly, bark from conifer trees was absent from the rumens of northern Yellowstone elk, except during severe winters (1962, 1965) when it was detected in up to 60% of rumens (Greer *et al.* 1970).

Sedges, mixed grasses, and aquatic plants contained more digestible dry matter than browses and were likely a better source of energy per unit of dry-matter intake because digestible dry matter and digestible energy are essentially equivalent (Moir 1961, Rittenhouse *et al.* 1971, Robbins *et al.* 1975, Milchunas *et al.* 1978, Mould and Robbins 1982). Conversely, shrubs were generally a better source of crude protein than grasses during winter, though bunchgrasses, leafy forbs, and sedges growing in thermal areas also had high protein content (Garrott *et al.* 2002). Graminoids contained low or insignificant levels of secondary compounds, which can deter feeding or reduce nutrition in ruminants by making certain plants unpalatable, negatively affecting rumen microflora, and causing adverse physiological effects if absorbed from the gut (Nagy *et al.* 1964, Oh *et al.* 1968, Mabry and Gill 1979, Hansson *et al.* 1986, Foley 1992). Conversely, woody plants had high levels of total phenols, similar to the findings of other studies (*e.g.,* McArthur *et al.* 1993). Elk consumption of woody plants with high phenols and saponins was limited to needles and twig tips and occurred primarily during periods of low protein levels in graminoids during early to mid-winter (Craighead *et al.* 1973, Hobbs *et al.* 1981). However, elk frequently consumed burned bark through the winter, likely because the 1988 fires substantially reduced the levels of secondary metabolites in lodgepole pine bark (Jakubas *et al.* 1994). Though bark was a low quality food compared to other winter forages (Table 9.2), it was readily available above the snow and consumption was likely related to harvesting efficiency rather than nutritional quality.

Geothermally-influenced plant communities were frequently used for foraging by elk in the Madison headwaters area during winter because the geothermal heat substantially reduced or eliminated snow pack in basins and along the banks of rivers draining these basins (Craighead *et al.* 1973). Also, plants such as sedges and spike-rush grew through the winter in geothermal run-off and along the banks of warm thermal streams, thereby providing relatively higher nutrition (Despain 1990). However, this foraging strategy also imposed a substantial cost because the chemical composition of plants generally reflects the chemical composition of the soil and water in which they grow (Kabata-Pendias 2001).

Soils, stream sediments, and surface waters in geothermally-influenced drainages of the Madison headwaters area contain anomalously high levels of arsenic, fluoride, and silica derived from magmatic sources (Thompson *et al.* 1975, Stauffer and Thompson 1984, Miller *et al.* 1997, Tuck *et al.* 1997). In turn, plants growing on soils derived from geothermally altered rocks or irrigated by geothermal waters absorb high levels of these elements or become contaminated on their surface by soil particles (Garrott *et al.* 2002). Silica reduces cell-wall digestibility, increases tooth wear, and promotes the formation of siliceous urinary calculi (Jones and Handreck 1967, Robbins 1993, Van Soest 1994). Also, excessive dietary ingestion of fluorine while permanent teeth are developing can result in compromised dentition due to fluoride toxicosis and, in turn, rapid tooth wear and reduced life span (Shupe *et al.* 1984, Garrott *et al.* 2002). Further, arsenic may be toxic at the concentrations present in many plants in the Madison headwaters area (Moxham and Coup 1968, Stauffer and Thompson 1984). Though Kocar *et al.* (2004) found some evidence that arsenic may have been detoxified by methylation reactions in the rumen, it is uncertain if this detoxification was adequate to ameliorate the potential for the high dietary arsenic concentrations to inhibit rumen microbial activity and result in lower digestibility. *In vitro* digestibility values suggest arsenic did reduce digestibility, but the inoculum for these assays was from a Holstein cow rather than an elk from our study area with rumen microflora that may have evolved the capability to detoxify arsenic. The high dietary intake of these elements by elk in geothermal areas was confirmed by bone samples from elk living in the Madison headwaters area containing 2-to 6-fold higher concentrations of arsenic and fluoride, respectively (Garrott *et al.* 2002, Kocar *et al.* 2004). Chronic exposure to these elements contributes to decreased life expectancy of elk in the Madison headwaters area relative to other populations in and near Yellowstone (Garrott *et al.* 2002, Chapter 10 by Garrott *et al.*, this volume).

The protein content of the winter diets of elk in the Madison headwaters area was high (7%) compared to other areas (4–6%; Craighead *et al.* 1973, Hobbs *et al.* 1981) and approximated the 6–7% crude protein needed by rumen microbes to sustain adequate fermentation rates with diets containing 40% digestible carbohydrate (McCullough 1969, Mertens 1973). Diet digestibility was lower (34%) than reports for elk wintering in other areas (35–50%) and substantially lower than the 50% digestible energy Ammann *et al.* (1973) suggested was necessary to maintain a ruminant (Hobbs *et al.* 1981). Thus, winter diets of elk in the Madison headwaters area contained levels of digestible dry matter that were below maintenance, but sufficient in protein. Christianson and Creel (2007) concluded that elk select graminoids almost exclusively in the absence of constraints, with the proportion of grass in the diet decreasing only as conditions restrict elk choice. Thus, elk nutrition and condition in winter would depend primarily on the nutritional profile of dormant grasses with their low protein content and digestibility. However, elk in the Madison headwaters area appeared to selectively vary their diet mix in response to seasonal changes in forage quality. Due to leaching of cell solubles from plant tissue, the crude protein content and digestibility of grasses and grass-like species decrease from autumn through winter, before increasing to a high-point in spring or early summer. In contrast, the protein content and digestibility of browse species does not change appreciably with advancing season (Hobbs *et al.* 1981). Thus, the relative value of grazing and browsing to meet energy and protein demands may change as winter progresses. As the difference in protein content of browse and grass increased during October through January, elk in the Madison headwaters area ate less grass and more browse with its higher protein content (Figure 9.2). Due to this change, the protein content in elk diets did not decrease through winter (Figure 9.4). This switch to browse intake was not without penalty, however, because diet digestibility decreased (Figure 9.4) owing to increased lignin and plant secondary compounds (phenols, saponins). As a result, elk switched their diet mix back to primarily grass and grass-like plants (*e.g.*, sedges) when the nutritional quality of these forages increased during March and April.

Hobbs *et al.* (1981) maintained similar seasonal changes in diet mix by tame elk foraging in Rocky Mountain National Park, Colorado, reflected a compromise, whereby relatively stable dietary protein was maintained through winter at the expense of decreased digestibility. Elk consumed primarily highly-digestible grasses when the protein content of grasses and browse were approximately equal, but switched to less digestible browse when protein in grasses decreased below critical levels. The observed selectivity for browse and other forages high in protein during mid-winter may have contributed to increased digestion of protein-deficient grasses and a more-favorable protein balance (Hobbs *et al.* 1981). Dietary nitrogen levels significantly affect rates of microbial protein synthesis and cellulose digestion in the rumen (Hume *et al.* 1970, Van Gylswyk 1970, Schwartz and Gilchrist 1975). Also, dietary protein deficiencies may be more costly to wintering elk than energy deficits because chronic negative protein balance may incur deleterious effects such as loss of muscle mass, reduced fetal viability, and susceptibility to disease (Harper *et al.* 1977, Robinson 1977, Swick and Benevenga 1977). Decreased protein intake results in diminished substrate available for protein synthesis, increased urea recycling, and decreased urinary urea in deer and elk when digestible energy intake is adequate for maintenance of positive nitrogen balance (Waterlow *et al.* 1977, Mould and Robbins, 1982, DelGuidice *et al.* 1991). However, as nutritional deprivation progresses energy becomes more limiting and animals continue to catabolize protein with increased renal filtration (Torbit *et al.* 1985, DelGuidice *et al.* 1991, Robbins 1993).

Vagnoni *et al.* (1996) demonstrated a positive, quadratic relation between digestible dry-matter intake and urinary A:C ratios in captive elk, similar to that observed in domestic ruminants (Chen *et al.* 1992*a,b*). Our data provided field validation of these captive feeding trials and indicated A:C ratios from urine deposited in snow were a useful non-invasive technique for evaluating digestible dry matter intake of individual animals and the population. Within-winter A:C ratios consistently decreased from the onset of sampling in December until middle to late winter, and then increased relatively rapidly in March or April. These changes were expected since diet digestibility decreased and then increased in a similar temporal pattern. Also, snow accumulation reduces forage availability and increases energetic costs during winter, but warm spring weather causes rapid melt-off and increases accessibility of forage in snow-free patches on the heterogeneous landscape. Thus, the level of elk nutrition for a given level of SWE was higher during the snow melt-out phase in spring than during the snow build-up portion of winter before maximum SWE was reached. Also, the rate of increase in nutrition was higher during spring compared to the rate of decrease during the snow build-up phase of winter.

Snow pack also had a significant influence on elk nutrition among winters and, in turn, nutrition was significantly related to adult female and calf survival. Though we did not directly assess nutrition in calves, we expected the annual variation in over-winter nutrition of adult female elk to be intensified in calves (Cook 2002). In addition, nutrition was higher in prime-aged animals, but lower in senescent animals with worn teeth. These findings support the results of demographic analyses (Chapter 10 by Garrott *et al.*, this volume) which indicated older animals had lower survival, especially during winters with increased snow pack. This decreased survival was due to starvation, which indicates poor nutrition (Figure 9.7). Our results demonstrate that elk in the Madison headwaters area have evolved fine-tuned strategies of diet selection influenced by snow pack, plant chemistry, and unique geothermal influences which, in turn, have allowed them to persist in one of the harshest winter environments in the continental United States. We suspect that elk would not winter in the Madison headwaters area without the spatial and temporal heterogeneity provided by the geothermal influence.

V. SUMMARY

1. We characterized the composition and quality of diets for adult, female elk in the Madison headwaters area during winter and prior to the establishment of resident wolf packs in the study area, evaluated the effects of diet selection and snow pack on nutrition using an index of metabolizable energy intake, and related nutritional costs to demography.

FIGURE 9.7 An instrumented cow elk photographed in early spring showing signs of severe nutritional deprivation. This animal wintered in one of the severest snow pack portions of the study area in the Gibbon Canyon. She survived this winter and eventually died of starvation at the age of 15, which is essentially the maximum longevity of elk in this population (Chapter 10 by Garrott *et al.*, this volume, Figure 10.2, Photo by Robert Garrott).

2. Wintering elk selected diets dominated by graminoids and browse (59% grass, 33% browse, 8% forbs). The protein content of winter diets was high (7%) compared to other areas (4–6%) and approximated the 6–7% crude protein needed by rumen microbes to sustain fermentation rates with diets containing 40% digestible carbohydrate.

3. Elk selectively varied their diet mix in response to seasonal changes in forage quality. As the difference in protein content of browse and grass increased during October-January, elk ate less grass and more browse with its higher protein content. This trend reversed in spring when the nutritional quality of grasses and grass-like plants increased.

4. Our data provided field validation of captive feeding trials in elk and indicated allantoin:creatinine ratios derived from urine deposited in snow were a useful non-invasive technique for evaluating digestible dry matter intake of individuals and populations.

5. The level of elk nutrition for a given level of snow water equivalent was higher during the snow melt-out phase in spring than during the snow build-up portion of winter before maximum snow water equivalent was reached.

6. Snow pack also had a significant influence on elk nutrition among winters and, in turn, nutrition was significantly related to survival. Nutrition was higher in prime-aged females, but lower in senescent animals with worn teeth.

VI. REFERENCES

Ammann, A. P., R. L. Cowan, C. L. Mothers-Head, and B. R. Baumgardt. 1973. Dry matter and energy intake in relation to digestibility in white-tailed deer. *Journal of Wildlife Management* **37**:195–201.

Anthony, R. G., and N. S. Smith. 1974. Comparison of rumen and fecal analysis to describe deer diets. *Journal of Wildlife Management* **38**:535–540.

Bender, L. C., P. E. Fowler, J. A. Bernatowicz, J. L. Musser, and L. E. Stream. 2002. Effects of open-entry spike bull, limited-entry branched bull harvesting on elk composition in Washington. *Wildlife Society Bulletin* **30**:1078–1084.

Bunnell, F. L., and M. P. Gillingham. 1985. Foraging behaviour: The dynamics of dining out. Pages 53–79 *in* R. J. Hudson and R. G. White (Eds.) *Bioenergetics of Wild Herbivores*. CRC Press, Boca Raton, FL.

Charnov, E. L. 1974. Optimal foraging, the marginal value theorem. *Theoretical Population Biology* **9**:129–136.

Chen, X. B., Y. K. Chen, M. F. Franklin, E. R. Orskov, and W. J. Shand. 1992a. The effect of feed intake and body weight on purine derivative excretion and microbial protein supply in sheep. *Journal of Animal Science* **70**:1534–1542.

Chen, X. B., G. Grubic, E. R. Ørskow, and P. Osuji. 1992b. Effect of feeding frequency on diurnal variation in plasma and urinary purine derivatives in steers. *Animal Production* **55**:185–191.

Christianson, D. A., and S. Creel. 2007. A review of environmental factors affecting elk winter diets. *Journal of Wildlife Management* **71**:164–176.

Coles, E. H. 1980. *Veterinary Clinical Pathology.* Saunders, Philadelphia, PA.

Cook, J. G. 2002. Nutrition and food. Pages 259–349 *in* D. E. Toweill and J. W. Thomas (Eds.) *North American Elk: Ecology and Management.* Smithsonian Institution Press, Washington, DC.

Cook, J. G., B. K. Johnson, R. C. Cook, R. A. Riggs, T. Delcurto, L. D. Bryant, and L. L. Irwin. 2004a. Effects of summer–autumn nutrition and parturition date on reproduction and survival of elk. *Wildlife Monographs* **155**:1–61.

Cook, R. C., J. G. Cook, and L. D. Mech. 2004b. Nutritional condition of northern Yellowstone elk. *Journal of Mammalogy* **85**:714–722.

Cork, S. J., and A. K. Krockenberger. 1991. Methods and pitfalls of extracting condensed tannins and other phenolics from plants: Insights from investigations on *Eucalyptus* leaves. *Journal of Chemical Ecology* **17**:123–134.

Craighead, J. J., G. Atwell, and B. W. O'Gara. 1972. Elk migrations in and near Yellowstone National Park. *Wildlife Monographs* **29**.

Craighead, J. J., F. C. Craighead Jr., R. L. Ruff, and B. W. O'Gara. 1973. Home ranges and activity patterns of nonmigratory elk of the Madison drainage herd as determined by biotelemetry. *Wildlife Monographs* **33**.

DeFelice, T. P., D. Lloyd, D. J. Meyer, T. T. Baltzer, and P. Piraino. 2003. Water vapour correction of the daily 1 km AVHRR global land dataset. I. Validation and use of the water vapour input field. *International Journal of Remote Sensing* **24**:2365–2375.

DelGuidice, G. D., L. D. Mech, and U. S. Seal. 1989. Physiological assessment of deer populations by analysis of urine in snow. *Journal of Wildlife Management* **53**:284–291.

DelGuidice, G. D., L. D. Mech, and U. S. Seal. 1990. Effects of winter under-nutrition on body composition and physiological profiles of white-tailed deer. *Journal of Wildlife Management* **54**:539–550.

DelGuidice, G. D., F. J. Singer, and U. S. Seal. 1991. Physiological assessment of winter nutritional deprivation in elk of Yellowstone National Park. *Journal of Wildlife Management* **55**:653–664.

Despain, D. 1990. *Yellowstone Vegetation: Consequences of Environment and History in a Natural Setting.* Roberts Rinehart, Boulder, CO.

Eidenshink, J. C. 1992. The 1990 conterminous U.S. AVHRR data set. *Photogrammetric Engineering and Remote Sensing* **58**:809–813.

Evans, S. B., L. D. Mech, P. J. White, and G. A. Sargeant. 2006. Survival of adult female elk in Yellowstone following wolf restoration. *Journal of Wildlife Management* **70**:1372–1378.

Foley, W. J. 1992. Nitrogen and energy retention and acid-base status in the common ringtail possum (*Pseudocheirus peregrinus*): Evidence of the effects of absorbed allelochemicals. *Physiological Zoology* **65**:403–421.

Garel, M., E. J. Solberg, B. Sæther, I. Herfindal, and K. Hogda. 2006. The length of growing season and adult sex ratio affect sexual size dimorphism in moose. *Ecology* **87**:745–758.

Garrott, R. A., J. G. Cook, J. G. Berardinelli, P. J. White, S. Cherry, and D. B. Vagnoni. 1997. Evaluation of the urinary allantoin: creatinine ratio as a nutritional index for elk. *Canadian Journal of Zoology* **75**:1519–1525.

Garrott, R. A., L. L. Eberhardt, J. K. Otton, P. J. White, and M. A. Chaffee. 2002. A geochemical trophic cascade in Yellowstone's geothermal environments. *Ecosystems* **5**:659–666.

Garrott, R. A., L. L. Eberhardt, P. J. White, and J. Rotella. 2003. Climate-induced variation in vital rates of an unharvested large-herbivore population. *Canadian Journal of Zoology* **81**:33–45.

Garrott, R. A., P. J. White, D. B. Vagnoni, and D. M. Heisey. 1996. Purine derivatives in snow-urine as a dietary index for free-ranging elk. *Journal of Wildlife Management* **60**:735–743.

Gates, C. C., and R. J. Hudson. 1979. Effects of posture and activity on metabolic responses of wapiti to cold. *Journal of Wildlife Management* **43**:564–567.

Goering, H. K., and P. J. Van Soest. 1970. Forage fiber analyses (apparatus, reagents, procedures, and some applications). Agricultural Handbook *379*. U.S. Department of Agriculture, Agricultural Research Service, Washington, DC.

Greer, K. R., J. B. Kirsch, and H. W. Yeager. 1970. Seasonal food habits of the northern Yellowstone elk (wapiti) herds during 1957 and 1962–1967 as determined from 793 rumen samples. Montana Fish, Wildlife & Parks, Kalispell, MT. Montana Department of Fish and Game Project W-83-R-12.

Hansson, L. H., R. Gref, L. Lundren, and O. Theander. 1986. Susceptibility to vole attacks due to bark phenols and terpenes in *Pinus contorta* provenances introduced into Sweden. *Journal of Chemical Ecology* **12**:1569–1578.

Harborne, J. B. 1984. *Phytochemical Methods: A Guide to Modern Techniques of Plant Analysis.* Chapman & Hall, London, United Kingdom.

Harder, J. D., and R. L. Kirkpatrick. 1994. Physiological methods in wildlife research. Pages 275–306 *in* T. A. Bookhout (Ed.) *Research and Management Techniques for Wildlife and Habitats.* The Wildlife Society, Bethesda, MD.

Harper, H. A., V. W. Rodwell, and P. A. Mayes. 1977. *Review of Physiological Chemistry.* Lange Medical Publications, Los Altos, CA.

Hobbs, N. T. 1989. Linking energy balance to survival in mule deer: Development and test of a simulation model. *Wildlife Monographs* **101:**1–31.

Hobbs, N. T., D. L. Baker, J. E. Ellis, and D. M. Swift. 1979. Composition and quality of elk diets during winter and summer: A preliminary analysis. Pages 47–53 *in* M. S. Boyce and L. D. Hayden-Wing (Eds.) *North American Elk: Ecology, Behavior, and Management.* University of Wyoming, Laramie, WY.

Hobbs, N. T., D. L. Baker, J. E. Ellis, and D. M. Swift. 1981. Composition and quality of elk winter diets in Colorado. *Journal of Wildlife Management* **45:**156–171.

Hobbs, N. T., D. L. Baker, and R. B. Gill. 1983. Comparative nutritional ecology of montane ungulates during winter. *Journal of Wildlife Management* **47:**1–16.

Hume, I. D., R. J. Moir, and M. Sommers. 1970. Synthesis of microbial protein in the rumen. I. Influence of the level of nitrogen intake. *Australian Journal of Agricultural Research* **21:**283–296.

Jakubas, W. J., R. A. Garrott, P. J. White, and D. R. Mertens. 1994. Fire-induced changes in the nutritional quality of lodgepole pine bark. *Journal of Wildlife Management* **58:**35–46.

Johnson, D. H. 1980. The comparison of usage and availability measurements for evaluating resource preference. *Ecology* **61:**65–71.

Jones, L. H. P., and K. A. Handreck. 1967. Silica in soils, plants and animals. *Advances in Agronomy* **19:**107–149.

Jönsson, P., and L. Eklundh. 2002. Seasonality extraction by function fitting to time-series of satellite sensor data. *IEEE Transactions on Geoscience and Remote Sensing* **40:**1824–1832.

Jönsson, P., and L. Eklundh. 2004. TIMESAT—A program for analyzing time-series of satellite sensor data. *Computers and Geosciences* **30:**833–845.

Kabata-Pendias, A. 2001. *Trace Elements in Soil and Plants.* CRC Press, Boca Raton, FL.

Kocar, B. D., R. A. Garrott, and W. P. Inskeep. 2004. Elk exposure to arsenic in geothermal watersheds of Yellowstone National Park, USA. *Environmental Toxicology and Chemistry* **23:**982–989.

Kohlmann, S. G. 1999. Adaptive fetal sex allocation in elk: Evidence and implications. *Journal of Wildlife Management* **63:**1109–1117.

Larsen, K. 1972. Creatinine assay by a reaction-kinetic approach. *Clinical Chemistry Acta* **41:**209–217.

Lindroth, R. L., J. M. Scriber, and M. T. S. Hsia. 1986. Differential responses of tiger swallowtail subspecies to secondary metabolites from tulip tree and quaking aspen. *Oecologia* **70:**13–19.

Loudon, A. S. I., A. S. McNeilly, and J. A. Milne. 1983. Nutrition and lactation control of fertility in red deer. *Nature* **302:**145–147.

Mabry, T. J., and J. E. Gill. 1979. Sesquiterpene lactones and other terpenoids. Pages 501–537 *in* G. A. Rosenthal and D. H. Janzen (Eds.) *Herbivores: Their Interaction with Secondary Plant Metabolites.* Academic Press, New York, NY.

MacArthur, R. H., and E. R. Pianka. 1966. On optimal use of a patchy environment. *American Naturalist* **100:**603–609.

Maynard Smith, J. 1982. *Evolution and the Theory of Games.* Cambridge University Press, New York, NY.

McArthur, C., C. T. Robbins, A. E. Hagerman, and T. A. Hanley. 1993. Diet selection by a ruminant generalist browser in relation to plant chemistry. *Canadian Journal of Zoology* **71:**2236–2243.

McCullough, D. R. 1969. *The Tule Elk: Its History, Behavior and Ecology.* University of California publication in Zoology, Vol. 88. University of California press Berkeley/Los Angeles. 209 pp.

Mech, L. D., D. W. Smith, K. M. Murphy, and D. R. MacNulty. 2001. Winter severity and wolf predation on a formerly wolf-free elk herd. *Journal of Wildlife Management* **65:**998–1003.

Mertens, D. R. 1973. Application of theoretical mathematical models to cell wall digestion and forage intake in ruminants. Dissertation. Cornell University, Ithaca, New York, NY.

Milchunas, D. G., M. I. Dyer, O. C. Wallmo, and D. E. Johnson. 1978. *In vivo/in vitro* relationships of Colorado mule deer forages (Colorado Division of Wildlife Special Report). Colorado Division of Wildlife, Fort Collins, CO.

Miller, W. R., A. L. Meier, and P. H. Briggs. 1997. *Geochemical Processes and Baselines for Stream Waters for Soda Butte-Lamar Basin and Firehole-Gibbon Basin, Yellowstone National Park.* Open-File Report No. 97–550. U.S. Geological Surrery. Reston, VA.

Moen, A. N. 1976. Energy conservation by white-tailed deer in the winter. *Ecology* **57:**192–198.

Moir, R. J. 1961. A note on the relationship between dry matter and digestible energy content of ruminant diets. *Australian Journal of Experimental Agriculture and Animal Husbandry* **1:**24–26.

Mould, E. D., and C. T. Robbins. 1982. Digestive capabilities in elk compared to white-tailed deer. *Journal of Wildlife Management* **46:**22–29.

Moxham, J. W., and M. R. Coup. 1968. Arsenic poisoning of cattle and other domestic animals. *New Zealand Veterinary Journal* **16:**161–165.

Nagy, J. G., H. W. Steinhoff, and G. M. Ward. 1964. Effects of essential oils of sagebrush on deer rumen microbial function. *Journal of Wildlife Management* **28:**785–790.

National Climate Data Center. 2006. Index of publications and data. http://www1.ndc.noaa.gov/pub/data/cirs/. Accessed November 2, 2006.

Norbury, G. L. 1988. A comparison of stomach and fecal samples for diet analysis of grey kangaroos. *Australian Wildlife Research* **15:**249–255.

Oh, H. K., M. B. Jones, and W. M. Longhurst. 1968. Comparison of rumen microbial inhibition resulting from various essential oils isolated from relatively unpalatable plant species. *Applied Microbiology* **16:**39–44.

Owen-Smith, N. 1993. Evaluating optimal diet models for an African browsing ruminant, the kudu: How constraining are the assumed constraints? *Evolutionary Ecology* **7:**499–524.

Owen-Smith, N. 1994. Foraging responses of kudus to seasonal changes in food resources: Elasticity in constraints. *Ecology* **75:**1050–1062.

Palmer, W. C. 1965. *Meteorological Drought*. Office of Climatology Research Paper 45, Weather Bureau, Washington, DC.

Parker, K. L., M. P. Gillingham, T. A. Hanley, and C. T. Robbins. 1999. Energy and protein balance of free-ranging black-tailed deer in a natural forest environment. *Wildlife Monographs* **143.**

Parker, K. L., C. T. Robbins, and T. A. Hanley. 1984. Energy expenditures for locomotion by mule deer and elk. *Journal of Wildlife Management* **48:**474–488.

Pils, A. C., R. A. Garrott, and J. J. Borkowski. 1999. Sampling and statistical analysis of snow-urine allantoin:creatinine ratios. *Journal of Wildlife Management* **63:**1118–1132.

Price, M. L., and L. G. Butler. 1977. Rapid visual estimation and spectrophotometric determination of tannin content of sorghum grain. *Journal of Agricultural and Food Chemistry* **25:**1268–1273.

Pulliam, H. R. 1989. Individual behavior and the procurement of essential resources. Pages 25–38 *in* J. Roughgarden, R. M. May, and S. A. Levin (Eds.) *Perspectives in Ecological Theory*. Princeton University Press, Princeton, NJ.

Rittenhouse, L. R., C. L. Streeter, and D. C. Clanton. 1971. Estimating digestible energy from digestible dry matter and organic matter in diets of grazing cattle. *Journal of Range Management* **24:**73–75.

Robbins, C. T. 1993. *Wildlife Feeding and Nutrition*. Academic Press, New York, NY.

Robbins, C. T., P. J. Van Soest, W. W. Mautz, and A. N. Moen. 1975. Feed analysis and digestion with reference to white-tailed deer. *Journal of Wildlife Management* **39:**67–79.

Robinson, J. J. 1977. The influence of maternal nutrition on ovine fetal growth. *Proceedings of the Nutrition Society* **36:**9–15.

Santavy, F. 1969. Alkaloids. Pages 421–470 *in* E. Stahl (Ed.) *Thin-Layer Chromatography*. Springer, Berlin, Germany.

SAS. 2002. *SAS Help and Documentation*. version 9.00. Cary, NC.

Schoener, T. W. 1971. Theory of feeding strategies. *Annual Review of Ecology and Systematics* **2:**369–404.

Schulte, E. E., J. B. Peters, and P. R. Hodgson. 1987. *Wisconsin Procedure for Soil Testing, Plant Analysis, and Feed and Forage Analysis*. Soil Fertility Serial 6, Department of Soil Science, University of Wisconsin, Madison, WI.

Schwartz, H. M., and F. M. C. Gilchrist. 1975. Microbial interactions with the diet and the host animal. Pages 165–179 *in* Digestion and Metabolism in the Ruminant. Proceedings of the IV International Symposium on Ruminant Physiology, Sydney, Australia.

Shupe, J. L., A. E. Olson, H. B. Peterson, and J. B. Low. 1984. Flouride toxicosis in wild ungulates. *Journal of the American Veterinary Medical Association* **185:**1295–1300.

Skalski, J. R., K. E. Ryding, and J. J. Millspaugh. 2005. *Wildlife Demography: Analysis of Sex, Age, and Count Data*. Elsevier, San Diego, CA.

Skovlin, J. M. 1982. Habitat requirements and evaluations. Pages 369–413 *in* J. W. Thomas and D. E. Toweill (Eds.) *Elk of North America: Ecology and Management*. Stackpole Books, Harrisburg, PA.

Sparks, D. R., and J. C. Malachek. 1968. Estimation percentage dry weight in diets using a microscope technique. *Journal of Range Management* **21:**264–265.

Spowart, R. A., and N. T. Hobbs. 1985. Effects of fire on diet overlap between mule deer and mountain sheep. *Journal of Wildlife Management* **49:**942–946.

Stauffer, R. E., and J. M. Thompson. 1984. Arsenic and antimony in geothermal waters of Yellowstone National Park, Wyoming, USA. *Geochimica et Cosmochimica Acta* **48:**2547–2561.

Stephens, D. W., and C. R. Krebs. 1986. *Foraging Theory*. Princeton University Press, Princeton, NJ.

Swets, D. L., B. C. Reed, J. D. Rowland, and S. E. Marko. 2000. *A Weighted Least-Squares Approach to Temporal NDVI Smoothing*. Proceedings of the 1999 ASPRS Annual Conference, From Image to Information, Portland, Oregon, May 17–21, 1999, American Society of Photogrammetry and Remote Sensing, Bethesda, MD.

Swick, R. W., and N. J. Benevenga. 1977. Labile protein reserves and protein turnover. *Journal of Dairy Science* **60:**505–515.

Thompson, J. M., T. S. Pressser, R. B. Barnes, and D. B. Bird. 1975. *Chemical Analysis of the Waters of Yellowstone National Park, Wyoming from 1965 to 1973*. Open-File Report No. 75–25. U.S. Geological Survey, Reston, VA.

Tilley, J. M. A., and R. A. Terry. 1963. A two stage technique for the *in vitro* digestion of forage crops. *Journal of the British Grassland Society* **18:**104–111.

Torbit, S. C., L. H. Carpenter, D. M. Swift, and A. W. Alldredge. 1985. Differential loss of fat and protein by mule deer during winter. *Journal of Wildlife Management* **49**:80–85.

Tuck, L. K., D. M. Dutton, and D. A. Nimick. 1997. *Hydrologic and Water-Quality Data Related to the Occurrence of Arsenic for Areas along the Madison and Upper Missouri Rivers, Southwestern and West-Central Montana.* Open-File Report No. 97–203. U.S. Geological Survey, Reston, VA.

Turner, M. G., W. H. Hargrove, R. H. Gardner, and W. H. Romme. 1994. Effects of fire on landscape heterogeneity in Yellowstone National Park, Wyoming. *Journal of Vegetation Science* **5**:731–742.

Vagnoni, D. B., R. A. Garrott, J. G. Cook, P. J. White, and M. K. Clayton. 1996. Urinary allantoin:creatinine ratios as a dietary index for elk. *Journal of Wildlife Management* **60**:728–734.

Van Gylswyk, N. O. 1970. The effect of supplementing a low-protein hay on the cellulolytic bacteria in the rumen of sheep and on the digestibility of cellulose and hemicellulose. *Journal of Agricultural Science* **74**:169–180.

Van Soest, P. J. 1994. *Nutritional Ecology of the Ruminant.* Cornell University Press. Ithaca, New York, NY.

Van Soest, P. J., and L. H. P. Jones. 1968. Effect of silica in forages upon digestibility. *Journal of Dairy Science* **51**:1644–1648.

Waterlow, J. C., M. Golden, and D. Picou. 1977. The measurement of rates of protein turnover, synthesis, and breakdown in man and the effects of the nutritional status and surgical injury. *American Journal of Clinical Nutrition* **30**:1333–1339.

Watson, F. G. R., T. N. Anderson, W. B. Newman, S. E. Alexander, and R. A. Garrott. 2006a. Optimal sampling schemes for estimating mean snow water equivalents in stratified heterogeneous landscapes. *Journal of Hydrology* **328**:432–452.

Watson, F. G. R., T. N. Anderson, W. B. Newman, S. E. Alexander, and R. A. Garrott. 2006b. Testing a distributed snowpack simulation model against spatial observations. *Journal of Hydrology* **328**:453–466.

Wickstrom, M. L., C. T. Robbins, T. A. Hanley, D. E. Spalinger, and S. M. Parish. 1984. Food intake and foraging energetics of elk and mule deer. *Journal of Wildlife Management* **48**:1285–1301.

Wittemyer, G., H. B. Rasmussen, and I. Douglas-Hamilton. 2007. Breeding phenology in relation to NDVI variability in free-ranging African elephant. *Ecography* **30**:42–50.

Young, E. G., and C. F. Conway. 1942. On the estimation of allantoin by the Rimini-Schryver reaction. *Journal of Biological Chemistry* **142**:839–853.

CHAPTER 10

Living in Yellowstone's Caldera: A Geochemical Trophic Cascade in Elk

Robert A. Garrott,* P. J. White,† James K. Otton,‡ and Maurice A. Chaffee‡

*Fish and Wildlife Management Program, Department of Ecology, Montana State University
†National Park Service, Yellowstone National Park
‡Geologic Division, U.S. Geological Survey, Denver Federal Center

Portions of this chapter were taken with kind permission from Springer Science + Business Media: Ecosystems, A geochemical trophic cascade in Yellowstone's geothermal environments, 5, 2002, 659–666, Garrott, R. A., L. L. Eberhardt, J. K. Otton, P. J. White, and M. A. Chaffee, ©2002, Springer-Verlag.

Contents

Theme

Though the geology of earth's rare geothermal environments and their associated microbial communities are intensely studied, less scientific attention has focused on their potential effects through the plant-herbivore-carnivore trophic chain. The west-central portion of Yellowstone National Park contains a 2000-km² volcanic caldera with 2- to 60-million-year-old, predominantly rhyolitic, rocks that produce relatively infertile soils. The caldera area contains one of the largest concentrations of active geothermal features in the world, including thousands of geysers, fumaroles, hot springs, and mud pots. In many other chapters of this book, we address the ecological impacts of this unique geology to the ecology of the large mammals that live within the caldera, primarily through the effects of geothermal heat on snow pack. In this chapter, we explore the unique geochemistry of the geothermal environments and the consequences of this geochemistry to elk and perhaps other herbivores that reside within the caldera. These studies were prompted by behavioral and demographic observations of elk during the initial years of our studies that we could not explain and, in turn, led us to formulate hypotheses about potential geochemical influences on elk ecology. Specifically, we contrast the concentrations of fluoride (F) and silica (SiO_2) found in the Madison headwaters area with areas on Yellowstone's

The Ecology of Large Mammals in Central Yellowstone
R. Garrott, P. J. White and F. Watson
ISSN 1936-7961, DOI: 10.1016/S1936-7961(08)00210-8

Copyright © 2009, Elsevier Inc.
All rights reserved.

northern range located approximately 50 km to the northwest, where geothermal features are rare or absent. We then trace the consequences of geochemical differences through abiotic and biotic linkages in the ecosystem.

I. INTRODUCTION

Though the high-elevation, Madison headwaters area of Yellowstone National Park routinely experiences deep snow pack during the winter months, geothermal heat dramatically reduces or eliminates snow pack in the geothermal basins and along the banks of the rivers draining these basins (Despain 1990, Chapter 4 by Watson *et al.*, this volume), thereby providing a refuge for elk and bison from deep snows (Figure 10.1). Thus, this geothermal influence essentially provides a winter range for bison and elk in west-central Yellowstone where one may otherwise not exist (Chapters 4 and 6 by Watson *et al.*, this volume). However, the geology of this area also results in unique geochemistry of the waters, soils, and sediments that provide the basis of the food web for these large herbivores. Despite this unique geochemistry, no work has explored the consequences of the elementally-enriched environments of Yellowstone and quantified the exposure of native ungulates using Yellowstone's geothermal areas.

FIGURE 10.1 A group of mature bull elk moving into the Firehole River in the vicinity of Biscuit Geyser Basin to forage on the bank vegetation immediately above the water line that was exposed due to the geothermal heat from the water. Geothermal areas provided a refuge for elk, bison, and other animals from the deep snow pack routinely experienced in the Madison headwaters area, but many radio collared elk predominantly occupied non-geothermal areas until snow pack became severe in late winter and early spring (Photo by Jeff Henry).

Given the presence of photosynthesizing plants and the minimal snow pack of geothermal areas it seemed obvious that these areas provided a substantial improvement in living conditions over the relatively deep snow pack and primarily senescent vegetation of the non-geothermal areas of the study system. By the third year of our telemetry studies, however, we were fundamentally puzzled by some of the behaviors we routinely observed. While the home range of nearly every radio collared elk in our sample of animals included geothermally-influenced areas, we found that many animals occupied and fed in these areas only rarely while some animals were never found far from geothermally-influenced areas. These observations led us to hypothesize that there were some sort of "costs" to animals that chose to occupy geothermal areas. We speculated that the costs were linked to the unique geochemistry of the geothermal areas (Thompson *et al.* 1975, Miller *et al.* 1997, Tuck *et al.* 1997). As we gradually instrumented additional animals for our studies, we also noticed that the age structure of the population seemed younger than expected based on the age structure of other elk populations reported in the literature. Initially, we made no link between this observation and the patterns of elk use of the geothermal areas. However, a major starvation mortality event caused by the severe snow pack experienced during the winter of 1997–1998 provided an opportunity to necropsy a sample of adult elk, which revealed a number of animals with very abnormal tooth wear patterns characteristic of clinical signs of fluoride toxicosis (Shupe *et al.* 1984). Combining these observations led us to hypothesize that elevated fluoride concentrations, perhaps combined with elevated silica, led to dental anomalies that, in turn, influenced the age structure of the Madison headwaters elk population. Thus, we embarked on a study to: (1) contrast the geochemistry of the Madison headwaters area, which has high concentrations of geothermal features, with nearby non-geothermal areas; (2) trace the consequences of geochemical differences through abiotic and biotic exposure pathways to higher trophic levels; and (3) evaluate the potential demographic consequences of geothermal chemistry on elk inhabiting the Madison headwaters.

II. METHODS

We collected control samples of stream sediments, plants, and elk mandibles from the northern winter range for elk in Yellowstone. The northern range is located along the northern boundary of Yellowstone and encompasses the watersheds of the Lamar and Yellowstone Rivers (Houston 1982, Chapter 1 by Garrott *et al.*, this volume). During our studies, approximately 200–500 bison and 12,000–20,000 elk wintered on the northern range (White and Garrott 2005). Northern range geology is composed mainly of andesitic rock with some metamorphic and sedimentary rocks, while the Madison headwaters area, which exists within the Yellowstone caldera, consists of Quaternary rhyolitic lavas and ash flow tuffs (Taylor *et al.* 1989, Chapter 2 by Newman *et al.*, this volume).

We collected samples of active stream sediment during 1996 and 1999 at 91 widely dispersed sites in the Lamar River watershed and at 46 similar sites in the Madison River watershed. Multiple subsamples were composited from several localities within a 30-m radius of each collection site to approximate the integrated chemistry of the material eroding from all rock exposures upstream of the sample sites and define the chemical environment of the area. A mean value calculated for all samples for each chemical variable in a given drainage area provides an estimate of the regional concentration level. Thus, the mean values can be used to identify broad chemical differences between areas.

We collected forage plant samples ($n = 83$) in late October 1997 from eight sites in the Lamar drainage and 10 sites in the Madison drainage. Collection sites were selected in non-forested valley bottoms where wintering elk were known to graze. At each site, we attempted to collect plant samples from four plant community types or microsites: (1) dry meadow sites dominated by *Poa* spp., *Festuca idahoensis*, *Stipa richardsonii*, and *J. balticus*; (2) transitional meadow sites dominated by *Deschampsia cespitosa*, *Poa pratensis*, *Phleum pratense*, and *J. balticus*; (3) wet sedge meadows dominated by *Carex rostrata*, *C. aquatilis*, and *C. nebrascensis*; and (4) riverbank sites within 1-m elevation of water surface dominated by *Carex* spp., *Poa palustris*, *D. cespitosa*, and *J. balticus*. We also sampled two additional unique plant community types associated with geothermal influences and heavily used by elk wintering in the Madison headwaters: (1) aquatic macrophytes, primarily *Myriophyllum* spp., *Ranunculus aquatilis*, and *Potamogeton* spp.; and (2) spike rush (*Eleocharis rostellata*) communities (Figure 10.2). Collections were made by clipping a 10–15-cm^2 area of plant material 1 cm above the ground to simulate elk cropping. Samples were not washed to determine total ingested fluoride and silica from both plants and soil contamination. We collected and composited a minimum of four subsamples within a 10-m radius at each site.

A. Elk Sampling

We collected location and demographic data from elk in the Madison headwaters during December 1991 through April 1998, which was prior to wolf (*Canis lupus*) colonization and, as a result, winter starvation was the primary form of mortality (Garrott *et al.* 2003, Chapter 11 by Garrott *et al.*, this volume). We maintained a sample of 25–32 female elk, 1–15 years old, with radio collars during this time period. The age of each elk was determined using cementum structures in the root of a canine tooth extracted at the time of capture (Hamlin *et al.* 2000). We used ground-based telemetry homing procedures to locate collared animals each winter between December and April. The order and frequency of animal locations were determined using a restricted randomized design and resulted in

FIGURE 10.2 Geothermal waters draining into the Gibbon and Firehole rivers warm water temperatures so that rivers in the Madison headwaters study area remain ice free though the winter and support robust growths of aquatic macrophytes that are high in protein and low in indigestible fiber. Elk forage on these plant communities during mid-winter and early spring when deep snow pack buries terrestrial forages but, in the process, are exposed to elevated levels of fluoride (Photo by Jeff Henry).

each animal being located 2–5 times per week (Chapter 8 by Messer *et al.*, this volume, $n = 7754$ locations). During winters 1991–1992 and 1992–1993, we recorded the activities of located animals for 30 min ($n = 1711$), including the amount of time animals were observed foraging on various types of plant communities during 830 foraging bouts (Chapter 9 by White *et al.*, this vloume). We used regression to explore snow pack-induced shifts in elk distribution by correlating the proportion of monthly animal locations recorded within geothermally-influenced areas with mean monthly snow pack, as indexed by snow water equivalent measurements recorded daily at an automated weather station. Mandibles were collected from all adult elk carcasses located during field activities in the Madison headwaters. Also, the National Park Service provided us with mandibles from adult elk that died on the northern range during winters 1994–1995 through 1996–1997. We evaluated these mandibles for clinical evidence of dental fluorosis (Shupe *et al.* 1984) and sectioned a sample of 12 mandibles from each herd across the diastema region to assay them for fluoride concentration.

B. Chemical Analyses

Stream sediment samples were air-dried, sieved through a 0.17-mm opening, and pulverized to less than 0.10-mm material prior to analysis. We analyzed samples for fluoride by a standard ion-selective electrode (Lachat Quik Chem Flow Injection Analyzer Lachat Instruments, Milwaukee, Wisconsin, USA) after fusing them with NaOH and dissolving the resulting material in dilute HNO_3. Dried plant and bone samples were crushed via mortar and pestle and ground in a Spex mill. We determined total fluoride concentration of plant samples by oxygen bomb combustion/ion-selective electrode method (American Society of Testing and Materials 1979). We determined total silica concentration by wet acid digestion and inductively coupled plasma-atomic emission spectroscopy (ICP-AES; Perkin-Elmer Optima, 3000 Perkin-Elmer Analytical Instruments, Shelton, Connecticut, USA). Bone samples were analyzed for total fluoride by the sodium hydroxide fusion/ion-selective electrode method (Orion Research Model 96-09 Thermo-Orion, Beverly, Massachusetts, USA).

C. Population Demography

We calculated an l_x survivorship schedule from the age-specific survival rate estimates derived from the logistic regression as described in Garrott *et al.* (2002) and fitted the resulting l_x estimates to the generalized survivorship model for large mammals (Eberhardt 1985) using nonlinear least squares (SAS/STAT version 6, SAS, Cary, North Carolina, USA). The l_x values from the fitted equation were converted to age-specific survival probabilities, $s_x = l_{x+1}/l_x$. The resulting s_x schedule for the Madison headwaters elk population was compared with a similar schedule for northern Yellowstone elk (Houston 1982). Average life expectancy for both elk populations were calculated directly from the l_x schedules (Pianka 1994).

III. RESULTS AND DISCUSSION

The chemistry of near-surface geothermal waters includes elements derived from magmatic sources at depth and elements leached from the rocks by circulating groundwater that mixes with the magmatic waters. The patterns of elemental enrichment in Yellowstone's geothermal waters vary from locale to locale; but fluoride and silica are enriched in virtually all of the geothermal waters of the Madison River drainage (Figure 10.3). These geothermal waters drain into the major streams of the area, including the Firehole, Gibbon, and Madison rivers and enrich these streams in many elements, including fluoride and silica, as compared to concentrations found in surface waters within the Lamar River drainage

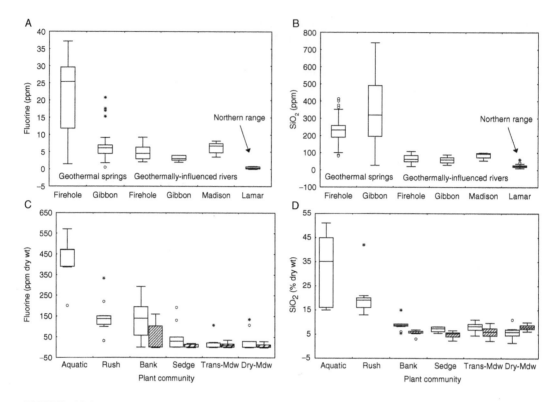

FIGURE 10.3 Contrast of water and plant chemistry between two major drainages of Yellowstone National Park, USA. The Madison headwaters area is within a volcanic caldera and contains thousands of geothermal features, while the Lamar drainage in the northern portion of Yellowstone is north of the caldera and has few geothermal features. (A, B), Fluorine and silica (SiO_2) concentrations in various sources of surface waters sampled in the two areas ($n = 200$). (C, D), Fluorine and silica concentrations in elk forage plants ($n = 83$) sampled along a gradient from plant communities growing in close association to surface waters to xeric communities more distant from surface waters. Open boxes represent samples collected from the upper Madison drainage and hatched boxes represent samples collected from the Lamar drainage.

(Thompson *et al.* 1975, Miller *et al.* 1997, Tuck *et al.* 1997, Figure 10.3). These differences in stream water chemistry are due partly to characteristics of the two dominant rock types in the park.

Rock and stream sediment samples also have contrasting chemistries that depend on the dominant rock type present in a given area. Representative mean values from the literature for rhyolite are 790 ppm fluoride and 75% silica, while values for andesite are 210 ppm fluoride and 59% silica (Fleischer and Robinson 1963, Taylor 1969). Similarly, the fluoride concentrations in active stream-sediment samples collected within these two areas reflect differences related to surface-rock chemistry (Madison River drainage: geometric mean = 792 ppm, $n = 46$; Lamar River drainage: geometric mean = 283 ppm, $n = 91$). In both areas, rocks altered by geothermal activity tend to have higher concentrations of chemical variables such as fluoride and silica than do their unaltered equivalents, regardless of rock type. Thus, the fluoride and silica chemistry of stream water and stream sediment is closely related to a combination of the chemistry of exposed rock material and chemical additions caused by geothermal activity.

The chemical composition of plants generally reflects the chemical composition of the soil and water in which they grow (Kabata-Pendias 2001). Plants growing in the Madison headwaters drainage on soils derived from geothermally altered rocks and/or irrigated by geothermal waters, as well as plants growing in the streams draining geothermal areas and along their banks, contained higher levels of fluoride and, in some environments, higher silica than plants collected at similar sites in the Lamar River drainage (Figure 10.3). The elevated concentrations of fluoride and silica are undoubtedly the result of a combination of absorption by the plant as well as surface contamination by soil particles (O'Reagain and Mentis 1989, Kabata-Pendias 2001). In particular, a pronounced biofilm was noted on both the aquatic macrophytes and spike rush specimens during collection. This film would have considerable potential to trap and accumulate suspended sediments in the wetland and river environments in which these plant communities are found. Thus, concentrations of fluoride and silica in the Madison headwaters were highest in aquatic plants and decreased as a function of distance from the primary surface waters.

All of the plant communities sampled were grazed by elk, but the degree to which the geothermally-influenced plant communities were used was dependent on the interaction of climate and geothermal heat as they affected snow pack. In high latitudes, the development of snow pack decreases the availability of forage plants and increases energetic expenditures by large herbivores such as elk for locomotion and displacement of snow to access buried forage (Parker *et al.* 1984). Winter diets are uniformly submaintenance due to low plant quality (Hobbs *et al.* 1981, Chapter 9 by White *et al.*, this volume). Thus, the additional energetic costs of snow pack increase the depletion of body reserves, which can lead to death (Garrott *et al.* 1997, Chapter 11 by Garrott *et al.*, this volume). Though the high-elevation, upper Madison River drainage has routinely deep snow pack, geothermal heat dramatically reduces or eliminates snow pack in the geothermal basins and along the banks of the rivers draining these basins (Despain 1990, Chapters 4 and 6 by Watson *et al.*, this volume), thereby providing a refuge for elk from deep snows. Data from instrumented animals indicates that elk tend to concentrate in these refugia as snow pack increases ($R^2 = 0.44$, $P < 0.001$, Figure 10.4, Chapter 8 by

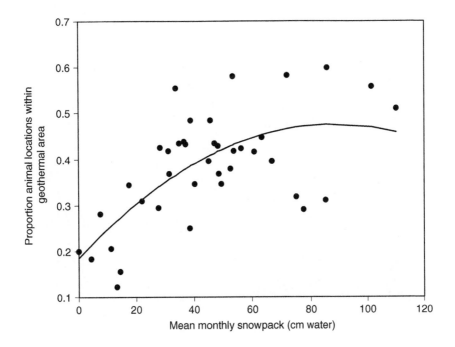

FIGURE 10.4 The interaction of winter climate and geothermal heat as it affects the probability of elk wintering in the Madison headwaters to congregate in the geothermally-influenced areas and feed on plants containing elevated levels of fluorine and silica. Increases in the proportion of monthly elk locations recorded within geothermally-influenced areas as snow pack increased through winter, 1991–1992 through 1997–1998.

Messer *et al.*, this volume) and feed extensively on plants containing high fluoride and silica concentrations (meadow grasses: 32% of total observed foraging minutes; sedges:14%; rushes: 6%; sedge-grass-rush mixture: 9%; aquatics: 6%).

Fluorine concentrates in mineralized tissues such as teeth and bones (Shupe *et al.* 1984). Thus, excessive dietary fluoride ingested while permanent teeth are developing in young animals interferes with matrix formation and mineralization of teeth, resulting in characteristic dental lesions and uneven and/or excessively rapid tooth wear (Shupe *et al.* 1984, Fejerskov *et al.* 1994, Kierdorf *et al.* 1996). This pathological consequence of fluoride toxicosis was evident in a comparison of mandibles collected from adult elk that died of natural causes on the Madison headwaters and northern Yellowstone winter ranges. Mandibles collected from elk in the Madison headwaters contained approximately six-fold higher levels of fluoride (1711 ± 64 ppm, $n = 12$) than mandibles collected from elk in northern Yellowstone (257 ± 30 ppm, $n = 12$), confirming high dietary exposure to fluoride due to ingestion of geothermal waters, soil, and plants. Though several mandibles collected from northern Yellowstone elk ($n = 34$) had minor dental anomalies, none showed signs of fluoride toxicosis. In contrast, 78% of the mandibles collected from carcasses in the Madison headwaters ($n = 74$) showed classic signs of fluoride toxicosis (Figure 10.5). Fifty-three percent of these affected mandibles were classified as severe, with extremely aberrant wear patterns, regression of alveolar processes, disfigurement, lost teeth, and localized bone apposition (Figure 10.5). Also, Becker et al. (Chapter 16 by Becker *et al.*, this volume) detected fluoride toxicosis and necrosis in approximately 50% of mandibles from adult bison and elk that were killed by wolves in the Madison headwaters. The compromised dentition of the fluorosed animals occupying the Madison headwaters was undoubtedly exacerbated by consumption of forages with high levels of abrasive silica (O'Reagain and Mentis 1989, Vicari and Basely 1993). Silica reduces cell-wall digestibility, increases tooth wear, and promotes the formation of siliceous urinary calculi (Jones and Handreck 1967, Robbins 1993, Van Soest 1994).

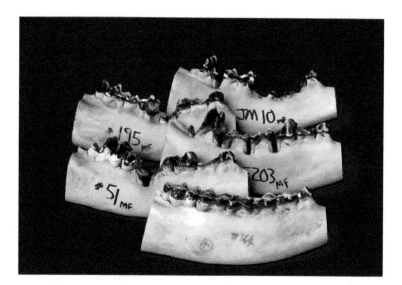

FIGURE 10.5 Elk mandibles contrasting the irregular and accelerated wear patterns and dental anomalies typical of fluoride toxicosis in four 12–15 year-old, adult, female elk from the geothermal Madison headwaters area that died of starvation (background) and a normal, relatively uniform wear pattern in the premolars and molars of a 15-year-old, adult, female elk killed by wolves in the non-geothermal Lamar River valley in northern Yellowstone (foreground). The substantial crown volume remaining in the Lamar sample suggests that, had this animal not been killed by wolves, she would have lived substantially longer before her teeth were sufficiently worn to cause terminal starvation.

Survival senescence in large herbivorous ungulates is dictated by tooth wear because as cheek teeth become severely abraded, proper mastication of forages is compromised (Laws 1981). The demographic consequences of fluoride toxicosis are readily apparent in a comparison of age-specific survival schedules for the northern Yellowstone (Houston 1982) and Madison headwaters elk populations (Figure 10.6). Onset of survival senescence in northern Yellowstone elk occurs at approximately 16 years of age, with the oldest animals surviving approximately 20–25 years (Houston 1982, White and Garrott 2005, Chapter 25 by Hamlin *et al.*, this volume). In contrast, the compromised dentition of Madison headwaters elk is responsible for a substantially abbreviated survival schedule, with onset of survival senescence occurring at 11 years of age and few animals surviving beyond 16 years (Chapter 11 by Garrott *et al.*, this volume). Interestingly, geothermal features represent the only major habitat difference between the northern Yellowstone and Madison headwaters areas, and age-specific fecundity schedules were essentially identical for the two populations through age 15 (Houston 1982, White and Garrott 2005, Chapter 11 by Garrott *et al.*, and Chapter 25 by Hamlin *et al.*, this volume). Consequently, the unique geochemistry of the Madison headwaters likely plays a pivotal role in decreased life expectancy.

The abbreviated survival schedule results in a compressed age structure, reduction of mean life span by approximately five years (northern Yellowstone: 17.9 years; Madison headwaters: 13.1 years; for animals surviving beyond first year), and an approximate doubling of the proportion of the adult population in the senescent age classes (northern Yellowstone: 5.8% Madison headwaters: 11.4%). Predators such as wolves preferentially prey on the young of the year and senescent animals in poorer nutritional condition (Peterson *et al.* 1984, Mech *et al.* 1998, Wright *et al.* 2006, Chapters 16 and 17 by Becker *et al.*, this volume). Thus, the geothermally-influenced age structure of the Madison headwaters population resulted in a higher proportion of the population being susceptible to predation by the colonizing wolf population, enhancing our ability to understand the interaction of wolf predation and age structure of a prey population (Chapter 23 by Garrott *et al.*, this volume). Thus, the "bottom-up" geochemical cascade we documented likely extends to predator-prey dynamics at the top of the trophic pyramid (Figure 10.7).

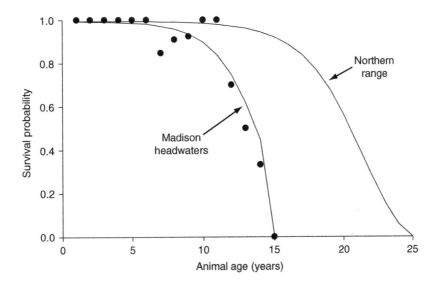

FIGURE 10.6 Differences in age-specific survival, s_x, between elk living in geothermal drainages of the Madison headwaters ($n = 185$ animal-years) and the northern portion of Yellowstone National Park where geothermal features are rare. Lines are the fitted generalized curves (Eberhardt, 1985) for each herd and points represent observed age-specific survival estimates for elk in the Madison headwaters.

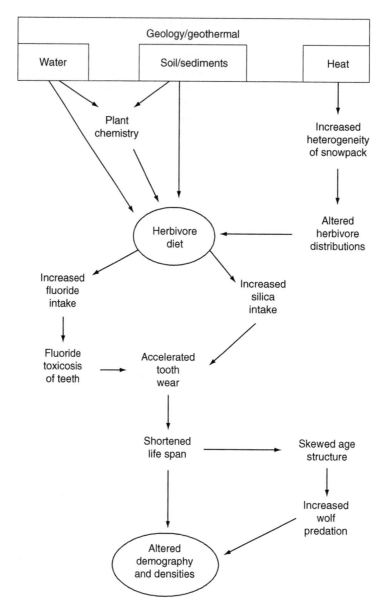

FIGURE 10.7 A conceptual model of the geochemical trophic cascade documented in the Madison headwaters area associated with the major geothermal landscapes of Yellowstone National Park.

A similar study of the Madison headwaters elk herd conducted by Kocar *et al.* (2004) focused on dietary exposure to arsenic, another element found in anomalously high concentrations in Yellowstone's geothermal waters. Unlike fluoride, which was found at high levels in a number of important elk forages, only aquatic plants contained high concentrations of arsenic. The ingested forms of arsenic were predominantly inorganic (Turpeinen *et al.* 1999, Koch *et al.* 2000, Kuehnelt *et al.* 2000), which can be toxic at relatively moderate concentrations. It is well documented that some microbial rumen communities are capable of detoxifying inorganic arsenic via methylation (Lasko and Peoples 1975,

Bentley and Chasteen 2002, Herbel *et al.* 2002). Kocar *et al.* (2004) found dimethyl arsonate in both rumen and fecal samples collected from elk in the Madison headwaters, suggesting the animals had at least some ability to detoxify inorganic arsenic. Such an adaptation would be advantageous as aquatic plants were high in protein and low in indigestible fiber fractions compared to more typical forages consumed by the study population (Chapter 9 by White *et al.*, this volume). Unfortunately, signs of chronic arsenic toxicosis, including depression, refusal of feed, and staggering during ambulation (Buck *et al.* 1973) are easily masked by the clinical signs of starvation, which often occurred during late winter for Madison headwaters elk in poor condition prior to wolf colonization (Chapter 11 by Garrott *et al.*, this volume). Thus, it is uncertain whether the unique geochemistry of the Yellowstone's geothermal areas affected the overall health of the Madison headwaters elk herd in any other way other than the early onset of survival senescence. Pregnancy rates, fall calf:cow ratios, and adult survival rates documented prior to wolf recolonization of our study system were similar to that observed in robust elk populations throughout western North America (Craighead *et al.* 1973, Aune 1981, Chapter 11 by Garrott *et al.*, this volume), suggesting effective adaptation of the elk to Yellowstone's geochemistry.

IV. SUMMARY

1. A combination of dominant rock type, geochemistry of geothermal waters, and geothermal heat in the Madison headwaters area of Yellowstone National Park affect plant chemistry and the winter distributions and diet of elk in the system.
2. Elk residing in the Madison headwaters are exposed to elevated levels of fluoride and silica through several major pathways, including aquatic and terrestrial plant material, soils, and stream sediments.
3. The high dietary intake of fluoride by elk in geothermal areas was confirmed by samples of bone from elk occupying the Madison headwaters area containing 6-fold higher concentrations than bone samples obtained from elk that occupied non-geothermal areas.
4. The pathological consequence of fluoride toxicosis was evident in a comparison of mandibles collected from adult elk that died of natural causes on the northern Yellowstone and Madison headwaters winter ranges. The compromised dentition of the fluorosed elk in the Madison headwaters was likely exacerbated by consumption of forages with high levels of abrasive silica.
5. The high dietary ingestion of fluoride and silica contributed to early onset of survival senescence and an abbreviated age structure in Madison headwaters elk compared to elk occupying the northern range of Yellowstone.
6. Wolves preferentially prey on old age class animals. Thus, the geochemical trophic cascade we documented likely influences predator-prey interactions at the top of the trophic chain.

V. REFERENCES

American Society of Testing and Materials. 1979. Standard test method for total fluorine in coal by the oxygen bomb combustion/ion selective electrode method. ASTM, Philadelphia, PA. ASTM standard D3761-D3779.

Aune, K. E. 1981. Impacts of winter recreationists on wildlife in a portion of Yellowstone National Park, Wyoming. M.S. Thesis. Montana State University, Bozeman, MT.

Bentley, R., and T. G. Chasteen. 2002. Microbial methylation of metalloids: Arsenic, antimony, and bismuth. *Microbiology and Molecular Biology Review* **66**:250–271.

Buck, W. B., G. D. Osweiler, and G. A. Van Gelder. 1973. *Clinical and Diagnostic Veterinary Toxicology.* Kendall/Hunt, Dubuque, IA.

Craighead, J. J., F. C. Craighead, Jr., R. L. Ruff, and B. W. O'Gara. 1973. Home ranges and activity patterns of nonmigratory elk of the Madison drainage herd as determined by biotelemetry. *Wildlife Monographs* **33**:1–50.

Despain, D. 1990. *Yellowstone Vegetation: Consequences of Environment and History in a Natural Setting.* Roberts Rinehart, Boulder, CO.

Eberhardt, L. L. 1985. Assessing the dynamics of wild populations. *Journal of Wildlife Management* **49**:997–1012.

Fejerskov, O., M. J. Larsen, A. Richards, and V. Baelum. 1994. Dental tissue effects of fluoride. *Advances in Dental Research* **8**:15–31.

Fleischer, M., and W. O. Robinson. 1963. Some problems of the geochemistry of fluorine. Pages 58–75 *in* D. M. Shaw (Ed.) *Studies in Analytical Geochemistry* (Royal Society Canada Special Publication 6). University of Toronto Press, Toronto, Ontario, Canada.

Garrott, R. A., J. G. Cook, J. G. Berardinelli, P. J. White, S. Cherry, and D. B. Vagnoni. 1997. Evaluation of urinary allantoin–creatinine ratio as a nutritional index for elk. *Canadian Journal of Zoology* **75**:1519–1525.

Garrott, R. A., L. L. Eberhardt, J. K. Otton, P. J. White, and M. A. Chaffee. 2002. A geochemical trophic cascade in Yellowstone's geothermal environments. *Ecosystems* **5**:659–666.

Garrott, R. A., L. L. Eberhardt, P. J. White, and J. Rotella. 2003. Climate-induced variation in vital rates of an unharvested large-herbivore population. *Canadian Journal of Zoology* **81**:33–45.

Hamlin, K. L., D. F. Pac, C. A. Sime, R. M. DeSimone, and G. L. Dusek. 2000. Evaluating the accuracy of ages obtained by two methods for Montana ungulates. *Journal of Wildlife Management* **64**:441–449.

Herbel, M. J., J.S Blum, S. E. Hoeft, S. M. Cohen, L. L. Arnold, J. Lisak, J. F. Stolz, and R. S. Oremland. 2002. Dissimilatory arsenate reductase activity and arsenate-respiring bacteria in bovine rumen fluid, hamster feces, and the termite hindgut. *FEMS Microbiology Ecology* **41**:59–67.

Hobbs, N. T., D. L. Baker, J. E. Ellis, and D. M. Swift. 1981. Composition and quality of elk winter diets in Colorado. *Journal of Wildlife Management* **45**:156–171.

Houston, D. B. 1982. *The Northern Yellowstone Elk.* MacMillan, New York, NY.

Jones, L. H. P., and K. A. Handreck. 1967. Silica in soils, plants and animals. *Advances in Agronomy* **19**:107–149.

Kabata-Pendias, A. 2001. *Trace Elements in Soil and Plants.* CRC, Boca Raton, FL.

Kierdorf, U., H. Kierdorf, F. Sedlacek, and O. Fejerskov. 1996. Structural changes in fluorosed dental enamel of red deer (*Cervus elaphus* L.) from a region with severe environmental pollution by fluorides. *Journal of Anatomy* **188**:183–195.

Kocar, B. D., R. A. Garrott, and W. P. Inskeep. 2004. Elk exposure to arsenic in geothermal watersheds of Yellowstone National Park, USA. *Environmental Toxicology and Chemistry* **23**:982–989.

Koch, I., L. Wang, C. A. Ollson, W. R. Cullen, and K. J. Reimer. 2000. The predominance of inorganic arsenic species in plants from Yellowknife, Northwest Territories, Canada. *Environmental Science and Technology* **34**:22–26.

Kuehnelt, D., J. Lintschinger, and W. Goessler. 2000. Arsenic compounds in terrrestrial organisms. IV. Green plants and lichens from an old arsenic smelter site in Austria. *Applied Organometallic Chemistry* **14**:411–420.

Lasko, J. U., and S. A. Peoples. 1975. Methylation of inorganic arsenic by mammals. *Journal of Agricultural and Food Chemistry* **23**:674–676.

Laws, R. M. 1981. Experiences in the study of large mammals. Pages 19–45 *in* C. W. Fowler and T. D. Smith (Eds.) *Dynamics of Large Mammal Populations.* Wiley, New York, NY.

Mech, L. D., L. G. Adams, T. J. Meier, J. W. Burch, and B. W. Dale. 1998. *The Wolves of Denali.* University of Minnesota Press, Minneapolis, MN.

Miller, W. R., A. L. Meier, and P. H. Briggs. 1997. *Geochemical Processes and Baselines for Stream Waters for Soda-Butte-Lamar Basin and Firehole-Gibbon Basin, Yellowstone National Park.* U.S. Geological Survey, Reston, VA. U.S. Geological Survey Open-File Report 97-550.

O'Reagain, P. J., and M. T. Mentis. 1989. Leaf silification in grasses—A review. *Journal of the Grassland Society of South Africa* **6**:37–43.

Parker, K. L., C. T. Robbins, and T. A. Hanley. 1984. Energy expenditures for locomotion by mule deer and elk. *Journal of Wildlife Management* **48**:474–488.

Peterson, R. O., J. D. Woolington, and T. N. Bailey. 1984. Wolves of the Kenai Peninsula, Alaska. *Wildlife Monographs* **88**.

Pianka, E. R. 1994. *Evolutionary Ecology.* HarperCollins, New York, NY.

Robbins, C. T. 1993. *Wildlife Feeding and Nutrition.* Academic Press, New York, NY.

Shupe, J. L., A. E. Olson, H. B. Peterson, and J. B. Low. 1984. Fluoride toxicosis in wild ungulates. *Journal of the American Veterinary Medical Association* **185**:1295–1300.

Taylor, R. L., J. M. Ashley, W. W. Locke, W. L. Hamilton, and J. B. Erickson. 1989. *Geological Map of Yellowstone National Park.* Department of Earth Sciences, Montana State University, Bozeman, MT.

Taylor, S. R. 1969. Trace element chemistry of andesites and associated calc-alkaline rocks. Pages 43–63 *in* A. R. McBirney (Ed.) *Proceedings of the Andesite Conference, Bulletin 65.* Oregon Department of Geology and Mining Industries, Portland, OR.

Thompson, J. M., T. S. Presser, R. B. Barnes, and D. B. Bird. 1975. *Chemical Analysis of the Waters of Yellowstone National Park, Wyoming from 1965 to 1973.* U.S. Geological Survey, Reston, VA. U.S. Geological Survey Open-File Report 75-25.

Tuck, L. K., D. M. Dutton, and D. A. Nimick. 1997. *Hydrologic and Water Quality Data Related to the Occurrence of Arsenic for Areas Along the Madison and Upper Missouri Rivers, Southwestern and West-Central Montana.* U.S. Geological Survey, Reston, VA. U.S. Geological Survey Open-File Report 97-203.

Turpeinen, R., M. Pantsar-Kallio, M. Haggblom, and T. Kairesalo. 1999. Influence of microbes on the mobilization, toxicity, and biomethylation of arsenic in soil. *Science of the Total Environment* **236:**173–180.

Van Soest, P. J. 1994. *Nutritional Ecology of the Ruminant*. Ithaca, New York, NY: Cornell University Press, Ithaca, New York, NY.

Vicari, M., and D. R. Basely. 1993. Do grasses fight back? The case for antiherbivore defenses. *Trends in Ecology and Evolution* **8:**137–141.

White, P. J., and R. A. Garrott. 2005. Northern Yellowstone elk after wolf restoration. *Wildlife Society Bulletin* **33:**942–955.

Wright, G. J., R. O. Peterson, D. W. Smith, and T. O. Lemke. 2006. Selection of northern Yellowstone elk by gray wolves and hunters. *Journal of Wildlife Management* **70:**1070–1078.

CHAPTER 11

The Madison Headwaters Elk Herd: Stability in an Inherently Variable Environment

Robert A. Garrott,* P. J. White,[†] and Jay J. Rotella*

*Fish and Wildlife Management Program, Department of Ecology, Montana State University
[†]National Park Service, Yellowstone National Park

Portions of this chapter were taken with kind permission from the Canadian National Research Council Press: Canadian Journal of Zoology, Climate-induced variation in vital rates of an unharvested large-herbivore population, 81, 2003, 33–45, Garrott, R. A., L. L. Eberhardt, P. J. White, and J. J. Rotella, ©2003, NRC Press.

Contents

Theme

While the elk herd that winters on the northern range of Yellowstone National Park has been the subject of almost continuous investigations since the inception of the park, the elk herd that occupies the Madison headwaters area in the central portion of Yellowstone has received much less scientific attention. This is a particularly interesting population for the study of regulatory processes because the elk remain within the confines of the park year-round and, thus, are not subjected to harvest by human hunters. Historic records also suggest that the herd was not targeted by market hunters during the 1800s nor managed by the National Park Service through intensive culling, as was common for ungulates occupying the northern range until 1968 when the Park Service adopted the natural regulation policy (Cole 1971, 1983). Hence, the dynamics of this herd have not been influenced to any appreciable extent by human manipulations other than the extirpation of wolves from the Park in the early 1900s. This portion of Yellowstone is also subjected to harsh winter conditions with periods of intense cold temperatures and deep snow pack in most years. These conditions provide an excellent opportunity to study the influence of density-dependent and density-independent factors and their interactions in regulating

The Ecology of Large Mammals in Central Yellowstone
R. Garrott, P. J. White and F. Watson
ISSN 1936-7961, DOI: 10.1016/S1936-7961(08)00211-X

Copyright © 2009, Elsevier Inc.
All rights reserved.

population processes in an ungulate herd. In this chapter, we present the results of an intensive demographic study of this population that was conducted just prior to the reestablishment of wolves.

I. INTRODUCTION

Describing the dynamics of populations and understanding the underlying processes that drive these dynamics is a prominent theme in population and community ecology (Cuppuccino 1995). Much of the classic work on population dynamics has focused on organisms that have relatively short generation times, are easily observed and enumerated, and effectively manipulated in experimental settings (Allee *et al.* 1949, Andrewartha and Birch 1954). Though large herbivores generally present considerable research challenges, a growing body of ecologists is conducting long-term demographic studies that are contributing significantly to our fundamental understanding of population regulation and limitation in complex age-structured populations (Gaillard *et al.* 2000). Large herbivores are prominent in many grassland, shrub-steppe, and tundra ecosystems and have important effects on nutrient cycling, energy transfer, fire regimes, soil erosion, hydrology, structuring of plant communities, and maintenance of species diversity (McNaughton 1985, Frank and McNaughton 1992, Hobbs 1996, Collins *et al.* 1998, Danell *et al.* 2006). In addition, landscape modifications and overexploitation have reduced the viability of many large herbivore populations to the point where sound conservation and management strategies are imperative. Knowledge of the mechanisms that limit or regulate large herbivore populations is, therefore, not only important to our theoretical understanding of population, community, and ecosystem patterns and processes, but also to effectively maintain assemblages of these organisms and the ecological processes they facilitate.

The central question in population dynamics has shifted from asking whether populations are regulated to finding the mechanisms that lead to regulation (Murdoch 1994). An initial step toward this goal is an expansion of research efforts from developing and analyzing time series of population estimates to estimating vital rates of populations with the aim of addressing three key topics: (1) the relative contribution of each vital rate to variability in population growth (λ), (2) the amount of temporal variability in each vital rate, and (3) the mechanisms responsible for variation in vital rates. Methodology for addressing the first two research topics, known as elasticity and sensitivity analyses and their variants, have become well developed and applied to a wide variety of species (Tuljapurkar and Caswell 1997, de Kroon *et al.* 2000). Analyses of the life history characteristics of long-lived large herbivores, which are very similar for many species, lead to the conclusion that λ is most sensitive to changes in adult female survival, followed by prime-age fecundity, fecundity of young-aged animals (age of first reproduction), and then juvenile survival. There is also a preponderance of data from many studies that demonstrate essentially the reverse pattern in temporal variation, with adult female survival remaining remarkably constant while juvenile survival varies substantially (Eberhardt 1977, 2002; Gaillard *et al.* 2000). Variation in large herbivore vital rates is undoubtedly influenced by both density-dependent and density-independent factors and their interactions. Though relatively rare, field studies that have developed time series of annual vital rates over a wide range of animal densities have supported Eberhardt's (1977) hypothesis of sequential changes in vital rates as animal density increases, with juvenile survival the most sensitive, followed by age of first reproduction, reproductive rate of prime-age animals, and adult survival (Gaillard *et al.* 1998). Such studies have also provided empirical evidence supporting nonlinear growth models (Ayala *et al.* 1973) with density-dependent changes in vital rates not expressed until the population is relatively close to K (Eberhardt 1977, Fowler 1981), a conclusion supported by theoretical work (Getz 1996).

The fundamental driver of density-independent changes in large herbivore vital rates is climate variation (Sæther 1997). Patterns or cycles of climatic variation can be studied at daily, seasonal, annual, decadal, multi-decadal, and longer temporal scales. However, studies of climatic influences on vital rates must focus on the annual scale because large herbivores have evolved life history patterns strongly aligned with the annual cycle of alternate periods of warm climatic conditions favorable for primary production, followed by a cold period of less favorable conditions when primary productivity is dramatically reduced (Stearns 1992). In contrast to this predictable seasonal climatic cycle, there is considerable stochastic climatic variation within the warm and cold seasons from year to year. Previous demographic research on large herbivores occupying mid to high latitudes provides a conceptual framework of the primary mechanisms by which warm and cold season climate variation may affect vital rates of species occupying these geographic regions (Figure 11.1). While there are strong mechanistic links between climatic variation and vital rates within the warm and cold season periods there are also cross-season links as well. Hence, the influence of unpredictable climatic variation and the relative strength of warm versus cold season variation on vital rates are key questions in the study of large herbivore population dynamics.

In a review of studies of large herbivore vital rates Gaillard *et al.* (2000) stated that "the greatest obstacle to better understanding of the population dynamics of large herbivores is the scarcity of data

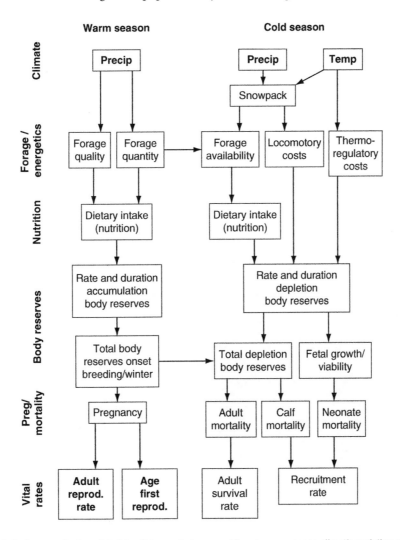

FIGURE 11.1 A conceptual model of the linkages between cold and warm season climatic variation and vital rates for ungulates occupying northern latitudes.

from long-term field studies of marked individuals." Here, we report on a study of variation in vital rates of an unharvested population of elk residing year-round within the protection of a national park. The study system experiences wide annual variation in both warm- and cold-season climate, with nearly the entire range of variation evident in historic climate records experienced during the 7 years that we collected data. We placed radio collars on known-aged animals (Figure 11.2) to enable the direct estimation of vital rates and eliminate the complication of estimating detection probability necessitated with capture–mark–recapture methodologies (Lebreton *et al.* 1992). Specific objectives were to (1) determine the magnitude of annual variation in adult survival, reproduction, and recruitment rates, (2) test *a priori* hypotheses regarding the influence of cold and warm season climatic variables on these vital rates, and (3) test potential interactions between age and climatic variables. We also integrated vital rates into a population projection model and compared model performance with the time series of population estimates available for the herd (Eberhardt 2002).

II. METHODS

A. Study Population

The approximately 600 elk residing in the Madison headwaters area of Yellowstone National Park (Eberhardt *et al.* 1998) are nonmigratory and remain within the borders of the park throughout the year, primarily in the Firehole, Gibbon, and Madison River drainages (Craighead *et al.* 1973; Chapter 18 by Gower *et al.*, this volume). A portion of the central Yellowstone bison (*Bison bison*) population also winters in the area with numbers fluctuating from approximately 300–1200 as animals migrate into and out of the area (Bjornlie and Garrott 2001). Neither the landscape nor the elk that inhabit the area have been manipulated or affected by humans to any significant extent since European settlement. This is

FIGURE 11.2 A sample of adult cow elk were captured via chemical immobilization from the ground and fitted with VHF radio collars in order to estimate vital rates and population demography, as well as resource selection (Chapter 8 by Messer *et al.*, this volume) and movements (Chapter 18 by Gower *et al.*, this volume). Here a collar with depleted batteries is being replaced in late spring in order for the animal to remain in the study until its death (Photo by Robert Garrott).

a particularly rare feature of the study system as nearly all other populations of large herbivores in North America are heavily harvested by humans. The only anthropogenic influence that may have appreciably affected the natural dynamics of the elk population was the extirpation of wolves (*Canis lupis*) from the park in the early 1900s (Cook 1993). Though wolves were reintroduced into the park in 1995 and 1996 (Fritts *et al.* 1997), only infrequent observations of 1–3 animals for a few days to several weeks were recorded during the last two years of this study. An unknown but relatively small number of grizzly bears (*Ursus arctos*), the only other predator capable of routinely killing adult elk, was present from early spring (late March–April) through late fall (October–November). Evidence of the presence of mountain lions (*Felix concolor*) on the study area was only detected during one winter when tracks were encountered in the Firehole drainage for a one-week period. Additional predators in the study area that were capable of killing calf elk up to several months old included black bears (*Ursus americanus*) and coyotes (*Canis latrans*). Black bears were rare, and coyotes were common.

B. Demographic Variables

Field studies were initiated in December 1991 and conducted for seven consecutive years through May 1998 with the objective of estimating annual adult female survival rates for summer and winter seasons, reproductive rates, recruitment of calves into the adult age class (≥ 1 year old), and overwinter calf survival rates. A total of 45 adult female elk were captured using ground-based delivery of immobilizing drugs via dart rifle and instrumented with collar-mounted VHF radio transmitters equipped with motion-sensitive "mortality" circuits. A vestigial canine tooth was extracted during handling to determine age via root cementum structures (Hamlin *et al.* 2000). A blood sample was also drawn for surveys of disease exposure and for developing and validating noninvasive pregnancy assays using fecal steroid concentrations. During each year of the study, 24–32 radio-marked females, ranging from 1 to 16 years of age, were monitored. Twenty-four animals were captured during the first year of the study, and 3–10 animals were added to the instrumented population in each of the subsequent years. We monitored 12 elk for 7 years, 6 elk for 6 years, 4 animals for both 5 and 4 years, and 18 animals for 1–3 years. Small numbers of calf elk (5–11) were also instrumented with mortality-sensing telemetry at the beginning of four of the seven winter seasons (1993–1994, 1994–1995, 1996–1997, and 1997–1998). Calves were immobilized when approximately six months of age and collared with units designed to drop off the animals during the following summer. Animals were arbitrarily selected for instrumentation in an attempt to maintain a relatively even distribution of collared individuals throughout the study area.

We used telemetry homing techniques (White and Garrott 1990) to locate and observe a sample of these elk during nearly every day throughout the winter field season (mid-November through early May). Instrumented animals were sampled via a restricted stratified sampling scheme, whereby we randomly selected one of three river drainages in the study area and then randomly determined the order elk residing in that drainage would be located. Located animals were excluded from the sampling pool in the days that followed until all instrumented animals were located, after which all animals were again placed in the sampling pool and the cycle repeated. This scheme resulted in each instrumented elk being visually located approximately 2–3 times per week during winter. Elk were not routinely relocated during the summer (mid-May through mid-November), though radio signals were monitored for mortality approximately monthly. Radio signals of instrumented calf elk were monitored a minimum of 3–4 times per week during winter to determine survival status. Deaths of instrumented animals were investigated immediately upon detection and the cause of death was determined by field necropsy and investigation of the immediate area for signs of predators, blood trails, and struggle.

The pregnancy status of each instrumented female was assessed each year by determining concentration of fecal progestagens (P_4). Located elk were observed for 5–30 min at a distance

(20–300 m) to avoid disturbing the animal's normal activities. If the animal defecated during the observation period, then the location of the excreta was noted and a sample was collected as soon as animals moved away from the area (White *et al.* 1995). All fecal P_4 concentrations were determined at the Conservation and Research Center (Smithsonian Institution, Front Royal, Virginia) using fecal hormone extraction and radioimmunoassay procedures described by Brown *et al.* (1994) and Wasser *et al.* (1994) and validated for elk (Garrott *et al.* 1998, Cook *et al.* 2002). If available, we used a fecal sample collected during March or April because White *et al.* (1995) reported that pregnancy determination using fecal P_4 was highest during late gestation. Animals with fecal P_4 concentrations $<1.00\ \mu g/g$ were considered nonpregnant, while animals with concentrations $>1.25\ \mu g/g$ were considered pregnant (Cook *et al.* 2002). Samples with concentrations between 1.0 and 1.25 $\mu g/g$ were considered inconclusive and an alternate sample was assayed, if available, to obtain a definitive fecal P_4 concentration.

To evaluate temporal trends in calf–cow ratios and provide an index of overwinter calf survival classifications of independent, randomly sampled elk groups were aggregated into five time periods (mid-November through December, January, February, March, April through early May) during each winter (Thompson 1992). The number of random groups used to calculate each ratio normally varied from 80 to 200, but occasionally varied considerably more during the first period each winter. Thus, we also calculated a "fall" and "spring" calf–cow ratio for each year using the first and last 100 random elk group records, respectively, to standardize the amount of data used to generate these ratios. The ratios were considered indices of calf "recruitment" to 6 months of age (fall ratio) and "recruitment" to yearling status (spring ratio). We used the two-sample change-in-ratio methods of Hanson (1963) and Paulik and Robson (1969) to estimate overwinter calf survival as described by Skalski *et al.* (2005, pp. 212–216), with the calf survival estimates corrected for adult cow survival using the winter survival estimates derived in this chapter.

Aerial surveys of the study area were used to develop Lincoln–Petersen mark–resight population estimates as described by Eberhardt *et al.* (1998). Complete surveys of the entire study area required 2–3 days of flying and were accomplished on four occasions during mid to late winter in 1993, 1994, 1996, and 1997.

C. Vital Rate Covariates

In order to evaluate hypotheses regarding variation in vital rates we developed a series of covariates representing animal attributes and climate variation. Adult survival and reproduction of long-lived large mammals may be strongly age-dependent. Hence, to properly evaluate the effect of density-independent covariates on these vital rates, we needed to first evaluate the effect of age. For adult females of long-lived large herbivores the general pattern of survival is consistently high annual survival rates through most of an animal's life, with a period of senescence where rates progressively decline for the oldest age classes (Siler 1979, Eberhardt 1985). A similar pattern has been documented for pregnancy; however, the extent of reproductive senescence appears to be more variable among species (Clutton-Brock *et al.* 1988). In addition, many large mammals demonstrate delayed sexual maturity such that the youngest age class adults have a lower probability of reproduction. Evidence from elk and red deer (*Cervus elaphus*) indicates lower pregnancy for the first several age classes (Clutton-Brock *et al.* 1982, Houston 1982). Based on these insights, age-related variation of adult female survival and reproduction was evaluated using a continuous age covariate (AGE) that allowed a progressive reduction in demographic performance indicative of senescence within the framework of logistic regression. Because we expected lower pregnancy in 1- and 2-year-old animals, and had very modest samples sizes in these young age classes; we evaluated a preliminary 2-category age covariate separating animals into 1–2-year-olds and animals >2 years old (AGE$_{Y/P}$). Inasmuch as the 2-category model was supported (see Section III), we excluded reproductive data from 1- to 2-year-olds when evaluating our reproductive models and considered a continuous AGE covariate when assessing reproductive senescence.

We chose to consider one warm season and one cold season covariate to capture climate-driven annual variation in density-independent factors that we suspected influenced elk vital rates. Temperature and precipitation regimes during the warm season are fundamental drivers of plant phenology and productivity which, in turn, affect the quality and quantity of forage available to large herbivores through the year (Figure 11.1). While it has been common to use temperature, precipitation, or some combination of these measures as covariates in ecological studies of large herbivores (Caughley and Gunn 1993, Loison and Langvatn 1998, Clutton-Brock and Coulson 2002, Vucetich *et al.* 2005, Wang *et al.* 2006), the recent availability of AVHRR satellite data (1989–2006) and algorithms for extracting seasonal normalized difference vegetation index (NDVI) parameters provide more direct measures of annual variation in vegetation canopy dynamics (Malingreau 1989, Reed *et al.* 1994; Chapter 7 by Thein *et al.*, this volume) that are biologically relevant to large herbivores. While forests dominate the study area (Chapter 2 by Newman and Watson, this volume), elk are primarily grazers (Christianson and Creel 2007). Thus, to best index annual variations in forage, we selected the seven largest meadow complexes in the study area, ranging from 1 to 48 km^2 and encompassing a total of approximately 80 km^2. The meadows were distributed across the entire elevational gradient and throughout the Firehole, Gibbon, and Madison River drainages. AVHRR satellite data at 1-km^2 resolution were processed by the USGS Center for EROS (Eidenshink 1992, DeFelice *et al.* 2003) with NDVI values composited using the maximum value method over 14-day periods to eliminate spuriously low values caused by cloud contamination, haze, and other atmospheric effects. These data were temporally smoothed using the methods of Swets *et al.* (2000). Data included 26 composites per year and program TIMESAT with the Savitsky–Golay filter (Jönsson and Eklundh 2002, 2004) was used to extract NDVI metrics for each polygon and averaged across all polygons for each year of the study. While 11 seasonal NDVI metrics were available, for *a priori* models we choose to evaluate only a single metric, growing season length (NDVI$_{len}$), which indexes the time period when green, high-quality vegetation was available. An alternate metric, NDVI$_{int}$ was also considered in exploratory analyses. NDVI$_{int}$ was the scaled integral of the seasonal NDVI curve and is strongly correlated with net primary productivity (Pettorelli *et al.* 2005); thus we considered this metric an index of annual variation in forage quantity.

Snow is a fundamental abiotic ecological force in temperate and high-latitude ecosystems, and we used the Langur snow pack model (Chapter 6 by Watson *et al.*, this volume) to generate a cold season covariate that indexed annual variations in snow pack. Langur generated daily estimates of snow water equivalent (SWE; the water content of a column of snow measured in cm) for the study area using temporally dynamic precipitation and temperature data, and spatially explicit slope, aspect, elevation, ground heat, vegetation cover type, and mean annual precipitation data. Mean daily SWE was calculated as the average of all pixels in the study area and represented the average water content of the snow pack for that day. An annual index of snow pack severity (SWE$_{acc}$) was calculated by summing the mean daily SWE values from 1 October through 31 April and thus provided a single metric integrating snow pack depth, density, and duration. Prior to analyses, SWE$_{acc}$ was converted from centimeter days by dividing by 1000.

D. Model Development and Evaluation

The five demographic response variables were adult female survival for the winter and summer periods, reproduction measured as late-term pregnancy, yearling recruitment indexed by spring calf–cow ratios, and winter calf survival. Our general analytical approach to assessing variation in vital rates was to form hypotheses that could be expressed as a suite of candidate models and to evaluate competing models using the data collected in this study and information-theoretic methods (Burnham and Anderson 2002).

We developed a restricted suite of candidate models that could be realistically evaluated given the quantity of data available from this study (Burnham and Anderson 2002). We hypothesized that the primary climatic drivers of annual variation in vital rates were snow pack during the cold season and $NDVI_{len}$ during the warm season. Thus, *a priori* models were limited to linear forms including only SWE_{acc} and $NDVI_{len}$ covariates. There was no indication of collinearity between SWE_{acc} and $NDVI_{len}$ during the historic record (1989–2007; $r^2 = 0.10$) or the period of this study (1992–1998; $r^2 = 0.04$). Thus, both covariates were considered in the same models. We expected SWE_{acc} to have a negative effect on vital rates because increasing snow pack reduces forage availability (Figure 11.1). Favorable temperature and precipitation regimes during the warm period extends the growing season, resulting in increased forage quality and quantity. Thus, we expected $NDVI_{len}$ to have a positive effect on vital rates (Figure 11.1). Most mortality of both adult and juvenile large herbivores occupying temperate and high latitudes occurs in the winter, especially in systems where effective predators are absent. Thus, we expected SWE_{acc} to have a stronger influence than $NDVI_{len}$ on adult female survival during the winter period, recruitment, and calf overwinter survival. Breeding occurs in the fall, and there is little direct evidence suggesting nutritionally related intra-uterine loss of fetuses during the winter gestation period in elk, red deer, and most other temperate cervids. Thus, we hypothesized that $NDVI_{len}$ had a stronger influence on reproductive rates than SWE_{acc}. Cross-seasonal effects are also possible (Figure 11.1). Thus, we evaluated models that included both SWE_{acc} and $NDVI_{len}$ covariates. Given that we expected strong age effects, we hypothesized that any effects of SWE_{acc} and $NDVI_{len}$ on adult female survival and reproduction would be more pronounced for senescent animals than for prime-aged animals. Thus we considered interactions in our candidate models. We limited *post priori* analyses to exploring the usefulness of the alternate NDVI metric. We used R version 2.5.0 (R Development Core Team 2006) to fit and evaluate regression models. Annual adult survival and reproduction were both dichotomous response variables as each instrumented animal was monitored each year to determine whether she lived or died and whether she was pregnant or not. Thus, we modeled data for these vital rates using generalized linear models with a logit link and binomial error structure (logistic regression). Standard regression was used to fit recruitment and winter calf survival models because these vital rates were derived from calf–cow ratios; thus, the response variables were continuous. Spring calf–cow ratios and calf survival estimates were logit transformed prior to analyses. We used Akaike's Information Criterion corrected for sample size, AIC_c, to compare the relative ability of each model to explain variation in the data and Akaike model weights (w_i) to address model-selection uncertainty (Burnham and Anderson 2002). Ninety-five percent confidence intervals were used to assess the degree to which signs and magnitudes of the estimated model parameters were reliably estimated. Violations of assumptions of independence and parameter homogeneity across individuals was a potential concern in our data because most animals were studied in more than one year. Thus, for logistic regression models, we (1) tested for evidence of overdispersion by dividing the residual deviance by the deviance degrees of freedom for the most general model (Burnham and Anderson 2002) and (2) formally evaluated goodness-of-fit using the le Cessie–van Houwelingen test (le Cessie and van Houwelingen 1991, Hosmer *et al.* 1997). This test is designed to assess goodness-of-fit for models with continuous covariates and binary responses based on nonparametric kernel methods. Goodness of fit of recruitment and calf survival regression models was evaluated using the adjusted coefficient of multiple determination (R^2).

E. Constructing Population Models

We used R software (R Development Core Team 2006) and code developed specifically for this study to execute stochastic modeling of a seasonal prebreeding age-structured population matrix model (Caswell 2001). Two matrices (**B1** and **B2**) per year were used to project the population through one annual cycle so that we could compare fall and spring calf–cow ratios produced by the matrix model with those observed in the field. **B1** was used to project the population from spring to fall and **B2** was used to project from fall to spring. Each matrix contained vital rate information for calves and females aged 1–17 years old. **B1** contained fecundity information for each age class and the probability that animals in each age class would survive from spring to fall. **B2** contained information on the probability that animals in each age class would survive from fall to spring and transition to the next age class. In the projection from spring to fall with **B1**, calves were created and some adult animals died, but no animals transitioned from one age to the next. In the projection from fall to spring with **B2**, no new calves were created and all animals either died or survived and transitioned to the next age with the constraint that all 17-year-old animals died during the winter. As a consequence, the population vector in spring contained only animals aged 1–17 years old, whereas the fall vector contained calves as well as 1–17-year-olds.

B1 was a deterministic matrix because cow elk survival was consistently high during summer and there was no indication that climate variation influenced pregnancy rates. We used the adult female summer survival estimate from data pooled across the six years of monitoring. Fecundity information was obtained based on our best estimates of pregnancy rates, litter size, and birth sex ratio. Birth rates (br) in **B1** were estimated using assessments of third trimester pregnancy rates during this study. Age-specific pregnancy rates were incremented one year to reflect the cow's age at time of parturition providing birth rate estimates of 0 for yearlings, 0.5 for 2-year-olds, 0.6 for 3-year-olds, and 0.9 for animals ≥4 years old. Litter size (ls) was assumed to be 1 and the sex ratio (sr) of offspring was assumed to be 0.5 (Raedeke *et al.* 2002). The probability that newborn female calves would survive from spring to fall (*Sjuv_sum*) was not estimated in this study. Therefore, we chose an estimate of summer calf survival (0.60) based on values reported in the literature (Myers *et al.* 1996, Singer *et al.* 1997, Smith and Anderson 1998, Zager *et al.* 2005, Hamlin 2006, Raithel *et al.* 2007). The age-specific fecundity values for **B1** were calculated as $br_{age} \times ls \times sr \times Sjuv\_sum$, where br_{age} was an age-specific birth rate.

Results of our logistic regression analyses indicated that survival was affected by both SWE and an animal's age during winter. Thus, **B2** was a stochastic matrix whose actual values depended on the value of SWE_{acc} in a given year. In each projection, a random value of SWE_{acc} was drawn with replacement from the values of SWE_{acc} recorded for the study area during winters from 1983–1984 to 2005–2006. We used the change in calf–cow ratios from fall to spring to estimate calf survival rates for winters 1992–1993 through 1997–1998. The log-odds of calf survival for each year were then modeled as a linear function of that year's observed SWE_{acc} value. The relationship between adult survival and SWE_{acc} was also negative, with estimated age-specific survival rates obtained from the logistic regression survival model that was most supported by the data (see Section III).

We conducted 10,000 simulations that projected a starting spring population of 375 females (N_0) forward for 10 years (N_{10}). A female population of 375 was chosen to approximate a fall elk population of 600 (adult sex ratio 0.18 M/F, 0.45 calves/adult female), the approximate size of the herd in the Madison headwaters area during 1965–1998. Rather than choosing a single age structure for the 375 animals in N_0, we obtained a set of 10,000 age structures by (1) estimating the stable age structure for the population that would result if the median observed value of SWE_{acc} recurred every year indefinitely, (2) applying that stable age structure to an N_0 of 375 animals and projecting the population forward for 10 years while using a random sample of SWE_{acc} (sampling with replacement from the observed values) in **B2** each year, (3) recording the age structure for N_{10}, and (4) repeating the procedure 10,000 times.

To conduct each of the 10,000 simulations, we first randomly chose a starting age structure for the N_0 of 375 animals from the set of 10,000 random age structures. We then used **B1** and **B2** to project that population forward for 10 years with random values of SWE_{acc} being used in each year. For each simulation, we recorded the annual values of SWE_{acc}, population growth rate (spring to spring), and fall and spring calf–cow ratios for each of the 10 projected years. We multiplied the matrix-derived calf–cow ratios by 1.67 to compare the field observations of fall and spring calf–cow ratios with matrix-derived ratios that reflected only female calves. Evidence from calf survival studies and sex ratios of hunter-killed calves in the region indicate females have higher survival and sex ratios are female biased (Smith and Anderson 1998, Hamlin and Ross 2002, Montana Fish Wildlife and Parks, unpublished reports). From the results of the 10,000 simulations, we estimated the arithmetic and geometric mean population growth rates (based on Tuljarpurkar's approximation using the observed arithmetic mean and variance for population growth rate), and the relationship between SWE_{acc} (winter of year t) and yearling–cow ratios (spring of year $t+1$). We calculated the asymptotic population growth rate of the deterministic versions of the prebreeding matrix product of **B1** and **B2** for comparison with stochastic results. We also constructed a post-breeding version of the deterministic matrix model in order to calculate the stable age distribution, reproductive values, elasticity, and sensitivity of vital rates for each age class from calves to the oldest age class adults, as well as net reproductive rate, and generation time. Because population growth rate is a nonlinear function of vital rates, we evaluated the robustness of the sensitivity values to changes in survival rates of various age classes of animals by examining the slope of the population growth rate across a range of absolute values for age-specific vital rates (de Kroon *et al.* 2000).

III. RESULTS

Variations in SWE_{acc} and $NDVI_{len}$ during our 7-year study captured nearly the entire range of annual variations in snow pack and growing season length described by available historic data. $NDVI_{len}$ during this study ranged from 13.88 (1996) to 17.91 (1997), with a mean of 15.51 (CV = 0.11), compared to a historic (1989–2006) range from 13.88 (1996) to 18.26 (2006) and a mean of 16.07 (CV = 0.08). Thus, the lowest $NDVI_{len}$ recorded occurred during this study, with only one year in the historic data having a substantially higher $NDVI_{len}$ than experience during this study. SWE_{acc} during this study ranged from 1744 to 5690 cm days, with a mean of 3008 (CV = 0.46). Langur-generated SWE_{acc} metrics were only available for the winters of 1989–1990 through 2006–2007. However, a SWE_{acc} metric derived from direct SWE measurements at the Madison Plateau automated SNOTEL site located west of the study area (Garrott *et al.* 2003) was strongly correlated with Langur-derived SWE_{acc} ($R^2 = 0.93$). These data provided a 40-year time series (winters 1967–1968 through 2006–2007) for comparison of historic variation in snow pack with that experienced during the seven winters of this study. SNOTEL SWE_{acc} recorded during the study period ranged from 4956 (1993–1994) to 12,404 cm days (1996–1997), with a mean of 7777, while the historic data ranged from 2975 (1976–1977) to 12,404 (1996–1997). Thus, the most severe snow pack year experienced in the historic record occurred during this study and lighter snow packs than the minimum recorded during our study period were recorded in only five winters.

We documented 12 deaths of radio-collared elk in 188 animal years of monitoring. Nine deaths were attributed to starvation (Figure 11.3), while one emaciated animal was killed in early spring by a grizzly bear, another animal was killed by a pair of transient wolves during the last year of the study, and the cause of one death was unknown. We accrued 151 animal years of survival monitoring during the summer period and documented only two mortalities. One animal was killed when hit by a recreational vehicle, while the cause of another death could not be reliably determined. We did not have adequate data to evaluate potential covariates during summer because only two instrumented cow elk died. Thus, we estimated a single summer survival probability of 0.987 (95% CI = 0.948 to 0.997) using the pooled data.

We evaluated 11 *a priori* models for adult female survival during winter and found strong support for both age and snow pack effects, with the additive model receiving 0.52 of the model weight and the interactive model receiving 0.26 of the model weight (Table 11.1). As predicted, both the AGE and SWE_{acc} coefficients were negative and confidence intervals did not encompass zero (additive model: $\hat{\beta}_{AGE} = -0.546$, 95% CI = -0.831 to -0.262; $\hat{\beta}_{SWE} = -0.845$, 95% CI = -1.376 to -0.314).

FIGURE 11.3 A radio-collared 15-year-old cow elk that died in early spring (April) of starvation after her body reserves were exhausted (Photo by Robert Garrott).

TABLE 11.1 Ranking of *a priori* hypothesized regression models concerning the effects of age (AGE), snowpack severity (SWE_{acc}), and growing season length ($NDVI_{len}$) on demographic vital rates of the Madison headwaters population

Model	K	AIC_c	ΔAIC_c	w_i
Adult female winter survival				
$AGE+SWE_{acc}$	3	60.78	0.00	0.52
$AGE+SWE_{acc}+(AGE \times SWE_{acc})$	4	62.18	1.40	0.26
$AGE+SWE_{acc}+NDVI_{len}$	4	62.74	1.96	0.20
$AGE+NDVI_{len}$	3	68.61	7.83	0.01
AGE	2	70.66	9.88	0.00
Constant	1	91.28	30.50	0.00
Pregnancy				
Constant	1	93.68	0.00	0.28
SWE_{acc}	2	94.67	0.99	0.17
$NDVI_{len}$	2	95.21	1.53	0.13
AGE	2	95.72	2.04	0.10
Calf recruitment (logit)				
SWE_{acc}	3	24.13	0.00	0.98
Constant	2	32.40	8.28	0.02
$NDVI_{len}$	3	38.13	14.01	0.00
Calf winter survival (logit)				
SWE_{acc}	3	27.48	0.00	0.94
Constant	2	32.95	5.46	0.06
$NDVI_{len}$	3	41.01	13.53	0.00

There was no evidence of overdispersion in the top-ranked model and the goodness-of-fit test supported the null hypothesis that the most general model fit the data ($P = 0.23$). A plot of the predicted age-specific winter survival probabilities indicated pronounced senescence with consistently high overwinter survival of young cow elk across the range of observed snow packs (Figure 11.4). In contrast, intermediate and older age elk were modestly and substantially affected by changes in snow

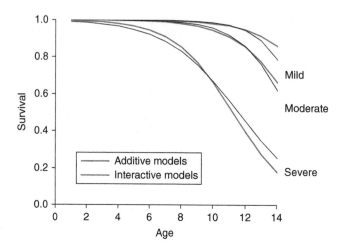

FIGURE 11.4 Predicted age-specific winter survival for cow elk based on the additive and interactive models most supported by the data collected during this study (Table 11.1). Survival curves are illustrated for the mildest ($SWE_{acc} = 1744$), moderate ($SWE_{acc} = 3091$), and severest ($SWE_{acc} = 5693$) snowpack winters experienced during the study. Confidence intervals for both SWE_{acc} and AGE covariates did not include zero; however, for clarity we did not include confidence bands about the plotted relationships.

pack severity, respectively. The third-ranked model which included $NDVI_{len}$ was nested within the top-ranked model and approximately two AIC_c units higher than the top model, indicating there was little useful information in this covariate. The alternate $NDVI_{int}$ metric we explored *post priori* also did not improve the top *a priori* models.

We assessed the pregnancy status of 21–28 instrumented elk per year based on fecal progestagen concentrations. Data from 1996–1997 were censored because field observations indicated substantial errors of commission. We suspect the extreme snow pack that winter resulted in a dietary shift to woody plants with high indigestible fiber fractions, effectively diluting steroid concentrations below normal levels and resulting in misclassification of pregnant animals as nonpregnant (Wasser *et al.* 1988). We used 157 animal years of pregnancy assessments to evaluate hypothesized effects of age, snow pack severity, and plant growing season length on reproduction (Table 11.2). As expected, the two age class model, $AGE_{Y/P}$, was most supported by the data ($AIC_c = 114.85$), being 6.22 AIC_c units lower than the AGE model and 6.91 AIC_c units lower than the null model. Thus, we excluded the pregnancy records of 1- and 2-year-old animals from the data set and evaluated the support for the *a priori* suite of models that included AGE, SWE_{acc}, and $NDVI_{len}$. No models were supported over the null (Table 11.1), with all estimated covariate coefficients of alternative models spanning zero. Therefore, we found no support for reproductive senescence or effects of snow pack severity or length of the plant growing season on the probability of pregnancy. The null model estimated a uniformly high pregnancy probability of 0.902 (95% CI = 0.841 to 0.941) for all females ≥ 3 years old. The probabilities of pregnancy for yearlings and 2-year-olds were estimated as 0.50 (2 of 4) and 0.60 (6 of 10), respectively.

We observed and classified 4965 groups of elk with radio-collared females during the seven winters of field work. The number of groups classified per time period ranged from 26 to 241 with mean group sizes for each period ranging from 3.7 to 10.0. We did not estimate calf–cow ratios during four periods (November–December 1991, 1993, 1994, and January 1992) when the minimum sample of >80 groups classified was not met. In general, calf–cow ratios progressively decreased throughout the winter (Figure 11.5). Because annual adult female survival rates were very high, we interpret the decline in calf–cow ratios as indicative of substantial overwinter mortality of calves. We collared 29 calves for mortality monitoring during four winter field seasons ($n = 5$, 1993–1994; $n = 6$, 1994–1995; $n = 11$,

TABLE 11.2 The age-specific and annual proportion of radio-collared cow elk assessed as pregnant using RIA techniques to measure fecal progestagens (P_4) from samples collected during the third trimester of gestation

Year	Number	1	2	3	4	5	6	7	8	9	10	11	12	13	14	15	16	Total	Proportion pregnant[a]
1992	Monitored	4	4	3	2	2	3	0	1	4	1	0	0	0	0	0	0	24	0.94
	Pregnant	2	3	2	2	2	3	0	1	4	1	0	0	0	0	0	0	20	
1993	Monitored	0	4	4	3	2	2	3	1	1	5	1	1	0	0	0	0	27	0.96
	Pregnant	0	3	4	3	2	1	3	1	1	5	1	1	0	0	0	0	25	
1994	Monitored	0	0	4	4	3	2	2	3	1	1	5	1	1	0	0	0	27	0.93
	Pregnant	0	0	4	3	3	2	2	3	1	1	4	1	1	0	0	0	25	
1995	Monitored	0	0	0	4	4	3	2	2	3	1	1	3	1	1	0	0	25	0.93
	Pregnant	0	0	0	4	4	3	2	2	3	1	1	3	1	0	0	0	21	
1996	Monitored	0	1	2	0	4	4	3	2	2	2	2	3	1	0	0	0	26	0.88
	Pregnant	0	0	2	0	4	4	2	0	2	2	2	3	1	0	0	0	22	
1997	Monitored	1	1	2	3	1	6	4	2	2	2	1	0	0	1	1	0	31	0.38
	Pregnant	0	0	2	3	0	1	3	2	1	2	1	0	0	1	0	0	11	
1998	Monitored	0	1	2	3	0	2	6	2	3	2	2	2	0	1	0	1	28	0.89
	Pregnant	0	0	2	3	0	1	5	2	3	2	2	2	0	1	0	1	24	
Total Monitored[b]		4	10	15	17	16	16	16	11	14	12	11	10	3	2	0	1	158	0.90
Total Pregnant[b]		2	6	14	15	15	14	14	8	12	12	10	10	3	1	0	1	137	
Proportion[b]		0.50	0.60	0.93	0.88	0.94	0.88	0.88	0.73	0.86	1.00	0.91	1.00	1.00	0.50	0.00	1.00	0.87	

[a] Proportions for cow ≥ 3 years old.

[b] Totals exclude 1997 data. Data from 1997 were considered unreliable due to known high levels of commission.

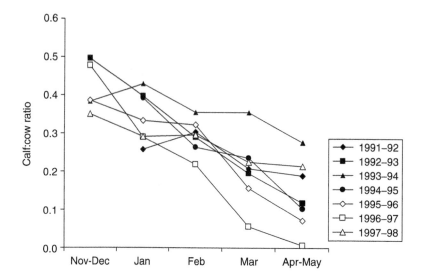

FIGURE 11.5 Overwinter changes in cow–calf ratios in the Madison headwaters elk population during seven consecutive years.

1996–1997; $n = 7$, 1997–1998), but three collars prematurely broke off before the end of the winter field season. Of the 26 remaining calves 13 calves died of starvation (0/5, 1993–1994; 2/6, 1994–1995; 9/9, 1996–1997; 1/6, 1997–1998) and the cause of death for one animal monitored the 1993–1994 season was undetermined. Starvation deaths occurred primarily in February ($n = 4$) and March ($n = 6$), with a single death in January and two deaths in April. Based on the limited sample of radio-collared calves monitored during this study, and data from carcasses discovered incidentally in the field, starvation was the dominant cause of calf mortality (Figure 11.6).

Annual recruitment, as indexed by spring calf–cow ratios, was highly variable among winters and ranged from 0.03 to 0.26 calves per cow (Table 11.3, Figure 11.7A). The data indicated a strong negative SWE_{acc} effect on annual recruitment ($\hat{\beta}_{SWE} = -1.419$, 95% CI=$-1.820$ to -1.018; $r^2 = 0.86$) with no evidence for an $NDVI_{len}$ effect (Table 11.1). Calf survival varied from 0.526 to 0.004 during the six winters (1992–1993 through 1997–1998) for which adequate fall and spring calf–cow ratios were available (Table 11.3). Similar to the results of the recruitment analysis, we found strong support for a snow pack effect ($\hat{\beta}_{SWE} = -1.339$; 95% CI $= -1.724$ to -0.953, $r^2 = 0.90$, Figure 11.7B) and no support for an $NDVI_{len}$ effect (Table 11.1).

Population estimates from four winter aerial surveys were 705 in 1993, 651 in 1994, 537 in 1996, and 821 in 1997. Because the number of marked elk groups was small, confidence limits on the annual point estimates were wide, ranging from ±150–220. Thus, we used the average of the four surveys to obtain our best estimate of recent population size (680) with bootstrapped 95% confidence limits on this mean estimate of 530–900 elk (Eberhardt *et al.* 1998).

The deterministic post-breeding matrix model yielded $\lambda = 1.053$, a net reproductive rate of 1.53, and a generation time of 8.25 years. The stochastic seasonal matrix model yielded an arithmetic mean value of $\lambda = 1.032$ (95% CI $= 0.869$ to 1.195) and a geometric mean value of $\lambda = 1.029$. Both annual population growth rate and the ratio between yearling and cow numbers in the spring were negatively related to winter snow pack (SWE_{acc}). Also, both fall and spring calf–cow ratios derived from the model mimicked the range of ratios observed in the field (Figure 11.8). The asymptotic properties of the deterministic post-breeding matrix model indicated that calves are the dominant age class, comprising approximately 27% of the population, while yearlings comprise approximately 7% of the population and each successive age containing a smaller proportion of the population. The highest reproductive

FIGURE 11.6 Calf elk that died on its bed during a mid-winter night when the air temperature had dropped to below −30 °C. The calf had depleted its body reserves and died of hypothermia (Photo by Robert Garrott).

TABLE 11.3 The annual changes in calf–cow ratios from fall to spring

	Fall		Spring		Uncorrected survival estimate[a]	Corrected survival estimate
Year	No. classified	Calf–cow ratio	No. classified	Calf–cow ratio		
1991–1992	–	–	566	0.177	–	–
1992–1993	577	0.476	591	0.113	0.24(0.17–0.30)	0.230
1993–1994	558	0.431	700	0.261	0.61(0.48–0.74)	0.595
1994–1995	386	0.485	405	0.122	0.25(0.17–0.33)	0.235
1995–1996	592	0.390	495	0.069	0.18(0.12–0.24)	0.169
1996–1997	682	0.526	688	0.003	0.01(0.00–0.01)	0.004
1997–1998	640	0.356	759	0.192	0.54(0.42–0.66)	0.526

[a] 90% confidence interval.
Ratios are calculated from the first and last 100 random elk groups classified at the beginning and end of each winter field season. Survival estimates are based on the two-sample change-in-ratio methods of Hanson (1963) and Paulik and Robson (1969) with the corrected estimate adjusted for annual variation in cow elk overwinter survival. Adequate sex-age composition data were not collected the fall of the first field season so a winter survival estimate could not be calculated.

values were for 1–8-year-old cows, with calves and cows ≥12 years old having the lowest reproductive values (Table 11.4)

Perturbing the age-specific mean survival rates revealed that the slope of the relationship between population growth and survival was relatively shallow and linear for old age animals; for younger adults slopes were steeper than those for older animals, especially when survival rates were set at lower values.

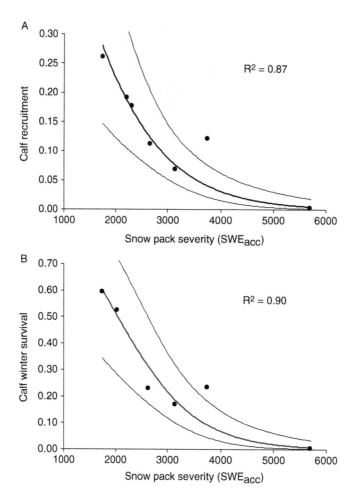

FIGURE 11.7 Panel A—The relationship between annual variation in an index of snowpack severity (SWE$_{acc}$) and spring calf–cow ratios, which can be considered an index of annual recruitment of calves to the yearling age class. Panel B—The relationship between annual variation in an index of snowpack severity (SWE$_{acc}$) for the Madison headwaters winter range and estimated overwinter calf survival based on changes in fall and spring calf–cow ratios (Table 11.3). The thick lines represent the regression line and the thin lines the 95% confidence intervals.

Overall, population growth was most sensitive to perturbations in calf survival with relatively modest departures from the mean survival estimated in this study resulting in major changes in λ (Figure 11.9).

IV. DISCUSSION

Snow pack was the primary driver of variation in vital rates, with the data supporting modest effects on adult female survival and strong effects on recruitment. We detected pronounced survival senescence in older animals, with snow pack effects differing in magnitude depending on age of the animal. Overwinter survival of older age class animals was highly variable and strongly correlated with annual variation in snow pack, while younger animals had relatively high survival typical of long-lived mammals that are not exploited by humans (Eberhardt 2002). While evidence of such climate-age

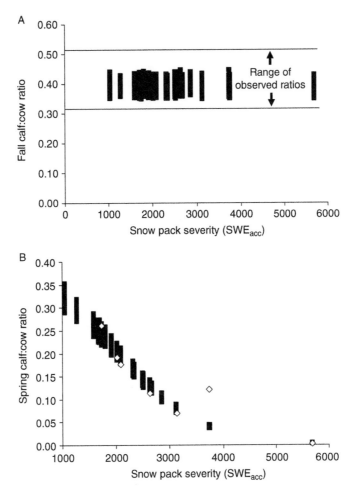

FIGURE 11.8 A comparison of the range of fall (panel A) and spring (panel B) calf–cow ratios observed in the field and those generated from a stochastic seasonal age-structured population matrix model that was parameterized with vital rates estimated from data collected during this study. The dark bars represent the range of calf–cow ratios derived from the matrix model and the open diamonds in panel B are the observed spring calf–cow ratios

interactions in large herbivores is rare (Gaillard *et al.* 2000), the mechanisms responsible for the differential affects of snow pack on different age classes of large herbivores are relatively intuitive. The proximate cause of senescence in large herbivores is believed to be due to tooth wear decreasing the ability of animals to efficiently crop forage and reduce plant material into small enough particles to facilitate efficient digestion by rumen microbes (Laws 1981, Van Soest 1994). Since nearly all plants used as forage by large herbivores in temperate and high latitudes are dormant during the cold season, the nutritional value of winter diets cannot meet maintenance requirements (Hobbs *et al.* 1981). Hence, sub-maintenance forage quality, combined with reduced forage availability and increased energetic costs due to snow pack (Parker and Robbins 1984, Wickstrom *et al.* 1984), resulted in the severe nutritional deprivation documented in this study (Garrott *et al.* 1997, Pils *et al.* 1999; Chapter 9 by White *et al.*, this volume). The compromised dentition of senescent animals exacerbates the level of nutritional deprivation and results in survival of these animals being much more sensitive to annual variation in snow pack than prime-aged animals possessing intact teeth.

TABLE 11.4	Asymptotic properties of the deterministic post-breeding matrix model parameterized using age-specific vital rates estimated from 7 years of field studies of the elk population occupying the Madison headwaters drainage of Yellowstone National Park		
Age	**Stable age distribution**	**Reproductive value**	**Sensitivity**
0	0.274	1.000	0.514
1	0.068	4.053	0.125
2	0.065	3.967	0.115
3	0.062	3.829	0.101
4	0.060	3.537	0.090
5	0.058	3.235	0.078
6	0.055	2.924	0.066
7	0.053	2.606	0.056
8	0.050	2.284	0.045
9	0.048	1.960	0.036
10	0.045	1.639	0.028
11	0.041	1.329	0.020
12	0.037	1.037	0.013
13	0.031	0.771	0.008
14	0.024	0.539	0.004
15	0.017	0.343	0.001
16	0.009	0.178	0.000
17	0.004	0.000	0.000

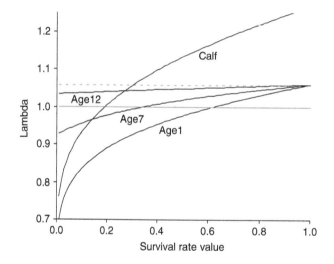

FIGURE 11.9 Changes in the population growth rate, λ, as a function of unit changes in survival rate for different age classes of female elk derived from a deterministic Leslie matrix model parameterized with vital rates estimated from a 7-year study of the Madison headwaters elk herd. The dashed horizontal line at $\lambda = 1.05$ represents the asymptotic population growth rate for the model using the mean values estimated from this study (calf–0.30, age 1–0.98, age 7–0.97, age 12–0.90).

Although survival senescence is a common trait of many long-lived large mammals, we found the onset of senescence at approximately 10–11 years of age (Figure 11.4) to be considerably earlier than for cow elk from the northern range of Yellowstone, where onset of senescence has been documented to occur at approximately 15–16 years of age (Houston 1982, White and Garrott 2005). An additional consequence of this early senescence was reduced longevity. We did not detect a cow older than 16 years of age, while 20–23-year-old animals are present in the northern range herd. These differences are apparently due to the unique regional geology of Yellowstone's volcanic caldera with its elementally-enriched geothermal waters and rhyolite soils. Elk in the Madison headwaters area ingest high levels of fluoride and highly-abrasive silica through consumption of water, plants, and soil particles (Garrott *et al.* 2002; Chapter 10 by Garrott *et al.*, this volume). Excessive dietary fluoride ingested while permanent teeth are developing in young animals interferes with matrix formation and mineralization of teeth, resulting in dental lesions and uneven and excessively rapid tooth wear (Shupe *et al.* 1984, Fejerskov *et al.* 1994, Kierdorf *et al.* 1996), which would undoubtedly be exacerbated by the abrasive action of silica. The pathological consequences of fluoride toxicosis were evident in many of the mandibles recovered from carcasses of adult elk in this area (Chapter 10 by Garrott *et al.*, this volume). These results provide a striking example of the influence of local geochemistry and the role of effectively functioning teeth in dictating survival schedules and longevity of large herbivores.

While annual variations in snow pack caused modest changes in adult female survival, the effects on calf recruitment were pronounced. Annual recruitment was highly variable, with strong evidence that stochastic fluctuations in snow pack directly influenced calf survival. All documented overwinter calf mortalities during this study were attributed to starvation, with an estimated 40% of the calves succumbing under even the mildest winter conditions. The most severe winter conditions (1996–1997) resulted in the virtual elimination of the juvenile cohort. These observations, combined with the strong correlation between SWE_{acc} and April calf–cow ratios and overwinter calf survival, indicate the predominant cause of the variable annual recruitment documented during this study was variability in snow pack. Though all sex and age classes of elk experience the same conditions each winter, the relatively large body size and long legs of adult females reduce energetic costs of thermoregulation and locomotion through snow (Gates and Hudson 1979, Robbins 1993). In addition, relatively large stores of body reserves in the form of fat and lean body mass buffer these animals from starvation mortality. In contrast, the small body size, short legs, and limited body reserves of calves make them more susceptible than adults to starvation mortality (Hudson and White 1985).

We expected climate variation during the warm season would have the most pronounced effects on reproduction, but found no support for this thesis. Pregnancy rates remained essentially constant and near their biological maxima (0.88–0.96) throughout the 7-year study. However, our assessment of variation in reproduction was limited to mature animals (≥ 3 years old). Variations in the fecundity of large herbivores are generally driven by annual changes in age at maturity, which have been linked to climatic variables and in some instances animal density (Sæther 1997). Lower body weight due to resource limitation during the growth and development period is believed to be the proximate mechanism for delayed age of maturity, as the onset of reproduction appears to be size-dependent in many ungulate species (Sinclair 1977, Gaillard *et al.* 1992, Sæther and Heim 1993, Jorgenson *et al.* 1993, Festa-Bianchet *et al.* 1994, Langvatn *et al.* 1996, Sand 1996, Sæther 1997, Adams and Dale 1998). Our limited data provide evidence of lower pregnancy rates in yearlings (0.50) and 2-year-olds (0.60), but we did not obtain adequate samples to document annual variation in these age classes.

Potential cross-seasonal effects of warm season climate variation on adult female survival and recruitment were also not supported by the data because we failed to find any reasonable evidence for NDVI effects. In recent years there have been numerous ecological studies that have explored correlations between annual variation in NDVI metrics and various aspects of large herbivore demographic performance, movements, and/or body mass (Pettorelli *et al.* 2005, 2007; Garel *et al.* 2006; Rasmussen *et al.* 2006; Tveraa *et al.* 2006; Wang *et al.* 2006; Wittemyer *et al.* 2007). Most of these studies detected what were interpreted as important correlations, so our failure to detect any warm season effects was unexpected. There are a number of possible reasons why we failed to find any correlation between NDVI covariates and annual estimates of vital rates in our study. There is no clear consensus in the literature on what metrics will be the most useful in ecological studies of large herbivores. NDVI metrics used have varied in both spatial and temporal scales, employing single NDVI measurements, averages over various time periods, peak values, slopes between consecutive measurements, the time period between baseline values, and integrated metrics. We only evaluated a few of the potential NDVI metrics that we felt were biologically the most reasonable and, as a result, may not have chosen the appropriate metric or temporal and spatial scale. More likely, however, we suspect that the annual variation in our cold-season snow pack metric (CV = 0.46) overwhelmed our ability to detect effects of the less variable warm season NDVI metrics (CV = 0.08).

Our field studies of temporal variations in vital rates corroborate the conclusions drawn from recent reviews of large-herbivore population dynamics (Sæther 1997; Gaillard *et al.* 1998, 2000). We found little annual variation in reproduction of animals >2 years old (0.88–0.96, CV = 0.06) and, though we found significant annual variation in adult female survival, the variation was limited primarily to older age class animals and small relative to variation in annual recruitment (<1–26 calves per 100 cows, CV = 0.64). Sæther (1997) concluded the population dynamics of ungulates are heavily influenced by annual recruitment, with mortality of juveniles outside the calving season affected by a combination of stochastic environmental variation and population density operating through a common effect on resource supply. Hence, when populations are near their stable attractor one would expect recruitment rates to be highly sensitive to climate variations that affect *per capita* resources. Though patterns of overwinter calf mortality due to resource limitation have been documented in populations of a wide variety of other ungulate species (Gaillard *et al.* 2000), the relationship between snow pack and recruitment documented in this study was unusually strong, suggesting the population was at or near its biological carrying capacity.

A comparison of population estimates and recruitment rates obtained during this investigation with historic data collected during monitoring and research initiatives conducted from 1965 to just prior to the initiation of this study (Craighead *et al.* 1973, Aune 1981, unpublished National Park Service files) support the premise that the population has been maintained in a dynamic equilibrium for at least three decades. Despite wide annual variability in recruitment (1.4–26 calves per 100 cows), the average population count from 1965 to 1988 was 520 (range 440–612, *n* = 8; Craighead *et al.* 1973, Aune 1981), which is only slightly lower than the average population estimate of 679 obtained during this study (range 537–821, *n* = 4, Eberhardt *et al.* 1998; Figure 11.10). The modest difference in the population estimates from the 1965–1988 and 1993–1997 periods may be attributed more to procedural differences in data collection than to actual differences in abundance. Earlier population data were simply

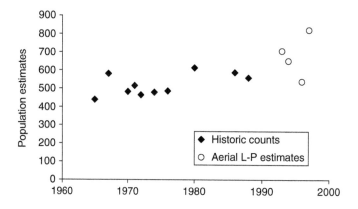

FIGURE 11.10 Population estimates for the elk herd occupying the Madison headwater drainages. Historic estimates are total counts based on aerial flights or a combination of aerial fights and ground surveys conducted during winter to early spring by NPS biologists, or university researchers (Craighead *et al.* 1973, Aune 1981, Cole 1983; NPS files). The most recent estimates are Lincoln–Petersen estimates based on aircraft mark–resight surveys of radio-collared elk groups in mid-winter (Eberhardt *et al.* 1998).

total counts of animals observed during ground or aerial surveys, while the recent population data are estimates based on mark–resight methodology. Using log-linear regression (Eberhardt 1987) provides an estimate of λ from these population estimates of 1.01 (95% CI = 1.00 to 1.02) which was quite similar to the mean estimate of 1.03 derived from the stochastic matrix model parameterized with the vital rate estimates obtained from this study. The close concordance of these two λ estimates obtained from independent data sources provides strong support for the validity of the estimates of demographic parameters for the population obtained during this study (Eberhardt 2002).

The relative constancy of population size, or fluctuations within relatively narrow limits, is notable because elk populations can increase at an annual rate of 20–28% (Eberhardt *et al.* 1996), indicating a biological potential for our study population to increase many fold over the 30-year period that data were available. During this same period, the migratory elk herd occupying the northern range of Yellowstone increased >440% (4305 to 19,045 elk) after intensive culling was ceased, demonstrating this potential for population growth (Coughenour and Singer 1996, Lemke *et al.* 1998, Taper and Gogan 2002). However, population increase was limited in the Madison headwaters elk population, as evidenced by the estimate of λ made directly from the population estimates and λ estimate obtained from the matrix model. We interpret the exceptionally strong correlation between snow pack and annual recruitment, and the evidence that the predominant source of mortality was starvation during the winter period, as strong evidence that the population was resource limited.

The matrix model results provided additional insights on growth and regulation in this population. Reproductive value is useful when considering contributions of individual animals to population growth. For example, the starvation loss of an animal with a low reproductive value such as a calf or senescent adult would have a lower impact on the population than removing a 6-year-old cow with a high reproductive value. However, when interpreting reproductive value in this manner there is an implicit assumption that removal of the animal does not affect the vital rates in the matrix. Thus, considering mortality agents such as starvation, predation, or harvest that kill many animals and, therefore, affect vital rates and their relative impact on the dynamics of a population, a different approach is needed. Here, it is useful to consider sensitivity metrics as these values provide an assessment of impacts of changes of age-specific vital rates on λ.

Sensitivity values are a function of the proportion of animals in an age class and the reproductive value of the following age class they would be recruited into if they were to survive. The general age-specific pattern illustrated in Table 11.4 is a gradual decline in sensitivity from the younger age classes to the oldest age classes. The notable exception to this pattern is calf sensitivity, which is approximately four times higher than the youngest age class of adults. The modest declines in sensitivity for adult age classes progressing from young to old is due to the fact that each successive adult age class is composed of only a slightly lower proportion of the population and reproductive values also decline quite gradually. In contrast, while calves have a low reproductive value, they make up a much larger proportion of the population, and recruit into the yearling age class that has the highest reproductive value. Thus, small changes in calf survival from year to year result in large changes in the number of animals moving into an age class that has very high reproductive value and hence, large changes in λ. This strong influence of changes in calf survival on population growth is most clearly illustrated with the results of the vital rate perturbation experiment (Figure 11.9) revealing the steeply curvilinear relationship that contrast markedly with sensitivities of older age class animals.

In this study, we documented very high annual variation in calf recruitment and demonstrated that this variation was strongly correlated with snow pack. We also demonstrated that while annual variation in adult survival was much more modest, it nevertheless covaried with calf survival. These results, combined with vital rate sensitivities derived from the matrix models, suggest that stochastic density-independent influences strongly affect the dynamics of this population. The time series of population estimates demonstrated substantial variability as well, which undoubtedly is at least partially a result of these density-independent influences in combination with sampling variability. However, the population has also shown a strong tendency over the last three decades to be bounded over a relatively narrow range, thus indicating density-dependent regulation. Direct assessments of density dependence are difficult, requiring long time series, reasonably precise demographic measurements (Shenk *et al.* 1998), and most importantly, measurements collected over a relatively wide range of densities. Thus, we could not directly investigate density dependence in this study; however, we propose that the regulatory mechanism involved an interaction of density dependence and density independence in that the influence of winter conditions on survival rates was weaker when the population was below the "stable attractor" and intensified when the population was above.

Sæther (1997) made a strong conclusion in his review paper that "In the absence of predation, a stable equilibrium is therefore unlikely to exist between an ungulate population and its food resources." We agree that stable equilibriums are unlikely because populations may be continually perturbed by temporary changes in limiting factors (e.g., severe winter or drought; mild winter or good rains) and rarely reach equilibrium (Caughley and Sinclair 1994). However, the population dynamics of elk in the Madison headwaters area are consistent with regulation through intraspecific competition for winter food resources since there were no significant predators of elk in the system (Caughley and Sinclair 1994). Similar results were observed for the northern Yellowstone elk after culling ceased and the population rebounded (Houston 1982). The relatively small range of population estimates over an approximately three-decade period (Figure 11.10) suggests density-dependent mortality in the Madison headwaters population was strong. Thus, we conclude that there was a mechanism in the population that slowed the rate of increase and regulated the population at a dynamic equilibrium through "bottom-up" processes.

Gaillard *et al.* (1998) noted that most studies of large herbivore population dynamics have been observational and suggested our understanding of ecological processes would be considerably enhanced through experimental manipulations. One line of investigation that is particularly contentious is the influence of large predators such as wolves on ungulate populations and the applicability of various theoretical predator–prey models that have lacked adequate evaluation due to a paucity of empirical data from field studies (Abrams and Ginzburg 2000). The fundamental questions are what proportion of wolf predation is additive and, if some proportion is additive, what will be the magnitude in change in the survival rates of the various age classes of animals. These are difficult questions to address with empirical data because most investigations of wolf–ungulate systems are confounded by human harvest of one or more components of the predator–prey system and, as a result, estimates of key vital rates of the prey population in the absence of predation are lacking. The establishment of resident wolf packs in the study system during the fall of 1998, combined with an understanding of the effects of climatic variation on vital rates and the sensitivity of population growth to changes in vital rates derived from this study, provided a rare opportunity for evaluating the strength, relative importance, and interactions of "bottom-up" and "top-down" forces (Matson and Hunter 1992) in a relatively intact and undisturbed system as this natural experiment unfolded (Chapter 23 by Garrott *et al.*, this volume).

V. SUMMARY

1. The dynamics of an elk (*C. elaphus*) population subjected to neither human hunting nor significant predation was studied using telemetry for seven consecutive years, 1991–1998.
2. We found pronounced senescence in survival rates, but no evidence for reproductive senescence. Prime-aged females experienced high annual survival with lower survival for senescent animals.
3. There was evidence that snowpack severity had little affect on survival of prime-age animals except during the most extreme winter while survival of senescent animals was progressively depressed as snowpack severity increased due primarily to overwinter starvation.

4. Reproductive rates remained essentially constant, near their biological maxima; however, annual recruitment was highly variable.

5. Snowpack had a pronounced effect on recruitment, with the most severe snowpack conditions resulting in the virtual elimination of a juvenile cohort due to overwinter starvation.

6. Population estimates and recruitment rates obtained during this investigation and historic data collected from 1965 to 1980 support the premise that the population was maintained in a dynamic equilibrium for at least three decades despite the stochastic affects of climate variation on vital rates.

7. We conclude that the population was "bottom-up" regulated due to resource limitation with variation about the equilibrium caused primarily by variable recruitment driven by stochastic annual snowpack.

VI. REFERENCES

Abrams, P. A., and L. R. Ginzburg. 2000. The nature of predation: Prey dependent, ratio dependent or neither? *Trends in Ecology and Evolution* **15**:237–241.

Adams, L. G., and B. W. Dale. 1998. Reproductive performance of female Alaskan caribou. *Journal of Wildlife Management* **62**:1184–1195.

Allee, W. C., A. E. Emerson, O. Park, T. Park, and K. P. Schmidt. 1949. *Principles of Animal Ecology*. W. B. Saunders, Philadelphia, PA.

Andrewartha, H. G., and L. C. Birch. 1954. *The Distribution and Abundance of Animals*. University of Chicago Press, Chicago, IL.

Aune, K. E. 1981. Impacts of winter recreationists on wildlife in a portion of Yellowstone National Park, Wyoming. M.S. Thesis, Montana State University, Bozeman, MT.

Ayala, F. J., M. E. Gilpin, and J. G. Ehrenfeld. 1973. Competition between species: Theoretical models and experimental tests. *Theoretical Population Biology* **4**:331–356.

Bjornlie, D. D., and R. A. Garrott. 2001. Effects of winter road grooming on bison in Yellowstone National Park. *Journal of Wildlife Management* **65**:423–435.

Brown, J. L., S. K. Wasser, D. E. Wildt, and L. H. Graham. 1994. Comparative aspects of steroid hormone metabolism and ovarian activity in felids, measured noninvasively in feces. *Biology of Reproduction* **51**:776–786.

Burnham, K. P., and D. R. Anderson. 2002. *Model Selection and Multimodel Inference: A Practical Information Theoretic Approach*. Springer, New York, NY.

Caswell, H. 2001. *Matrix Population Models: Construction, Analysis, and Interpretation*. 2nd edn. Sinauer Associates, Sunderland, MA.

Caughley, G., and A. Gunn. 1993. Dynamics of large herbivores in deserts: Kangaroos and caribou. *Oikos* **67**:47–55.

Caughley, G., and A. R. E. Sinclair. 1994. *Wildlife Ecology and Management*. Blackwell, Boston, MA.

le Cessie, S., and J. C. van Houwelingen. 1991. A goodness-of-fit test for binary regression models, based on smoothing methods. *Biometrics* **47**:1267–1282.

Christianson, D. A., and S. Creel. 2007. A review of environmental factors affecting elk winter diets. *Journal of Wildlife Management* **71**:164–176.

Clutton-Brock, T. H. 1988. *Reproductive Success: Studies of Individual Variation in Contrasting Breeding Systems*. University of Chicago Press, Chicago, IL.

Clutton-Brock, T. H., and T. N. Coulson. 2002. Comparative ungulate dynamics: The devil is in the detail. *Philosophical Transactions of the Royal Society B* **357**:1285–1298.

Clutton-Brock, T. H., F. E. Guinness, and S. D. Albon. 1982. *Red Deer: Behavior and Ecology of Two Sexes*. University of Chicago Press, Chicago, IL.

Clutton-Brock, T. H., M. Major, S. D. Albon, and F. E. Guinness. 1988. Reproductive success in male and female red deer. Pages 325–343 *in* T. H. Clutton-Brock (Ed.) *Reproductive Success: Studies of Individual Variation in Contrasting Breeding Systems*. University of Chicago Press, Chicago, IL.

Cole, G. F. 1971. An ecological rationale for the natural regulation or artificial regulation of native ungulates in national parks. *Transactions of the North American Wildlife Conference* **36**:417–425.

Cole, G. F. 1983. A naturally regulated elk population. Pages 62–81 *in* F. L. Bunnell, D. S. Eastman, and J.M Peek (Eds.) *Symposium on Natural Regulation of Wildlife Populations*. Proceedings no. 14, Forest Wildlife and Range Experiment Station, University of Idaho, Moscow, ID.

Collins, S. L., A. K. Knapp, J. M. Briggs, J. M. Blair, and E. M. Steinauer. 1998. Modulation of diversity by grazing and mowing in native tallgrass prairie. *Science* **280**:745–747.

Cook, R. C., J. G. Cook, R. A. Garrott, L. L. Irwin, and S. L. Monfort. 2002. Effects of diet and body condition on fecal progestagen excretion in elk. *Journal of Wildlife Diseases* **38**:558–565.

Cook, R. S. 1993. *Ecological Issues on Reintroducing Wolves into Yellowstone National Park.* Science Monograph 93/22, National Park Service, U.S. Government Printing Office, Washington, DC.

Coughenour, M. B., and F. J. Singer. 1996. Elk population processes in Yellowstone National Park under the policy of natural regulation. *Ecological Applications* **6**:573–583.

Craighead, J. J., F. C. Craighead Jr., R. L. Ruff, and B. W. O'Gara. 1973. Home ranges and activity patterns of nonmigratory elk of the Madison drainage herd as determined by biotelemetry. *Wildlife Monographs* **33**:1–50.

Cuppuccino, N. 1995. Novel approaches to the study of population dynamics. Pages 3–16 *in* N. Cupppuccino and P. W. Price (Eds.) *Population Dynamics: New Approaches and Synthesis.* Academic Press, San Diego, CA.

Danell, K., R. Bergström, P. Duncan, and J. Pastor. 2006. *Large Herbivore Ecology, Ecosystem Dynamics and Conservation.* Cambridge University Press, Cambridge, United Kingdom.

DeFelice, T. P., D. Lloyd, D. J. Meyer, T. T. Baltzer, and P. Piraino. 2003. Water vapour correction of the daily 1 km AVHRR global land dataset. I. Validation and use of the Water vapour input field. *International Journal of Remote Sensing* **24**:2365–2375.

Eberhardt, L. L. 1977. Optimal policies for conservation of large mammals with special reference to marine ecosystems. *Environmental Conservation* **4**:205–212.

Eberhardt, L. L. 1985. Assessing the dynamics of wild populations. *Journal of Wildlife Management* **49**:997–1012.

Eberhardt, L. L. 1987. Population projections from simple models. *Journal of Applied Ecology* **24**:103–118.

Eberhardt, L. L. 2002. A paradigm for population analysis of long-lived vertebrates. *Ecology* **83**:2841–2854.

Eberhardt, L. E., L. L. Eberhardt, B. L. Tiller, and L. L. Cadwell. 1996. Growth of an isolated elk population. *Journal of Wildlife Management* **60**:369–373.

Eberhardt, L. L., R. A. Garrott, P. J. White, and P. J. Gogan. 1998. Alternative approaches to aerial censusing of elk. *Journal of Wildlife Management* **62**:1046–1055.

Eidenshink, J. C. 1992. The 1990 conterminous U.S. AVHRR data set. *Photogrammetric Engineering and Remote Sensing* **58**:809–813.

Fejerskov, O., M. J. Larsen, A. Richards, and V. Baelum. 1994. Dental tissue effects of fluoride. *Advances in Dental Research* **8**:15–31.

Festa-Bianchet, M., M. Urquhart, and K. G. Smith. 1994. Mountain goat recruitment: Kid production and survival to breeding age. *Canadian Journal of Zoology* **72**:22–27.

Fowler, C. W. 1981. Density dependence as related to life history strategy. *Ecology* **62**:602–610.

Frank, D., and S. J. McNaughton. 1992. The ecology of plants, large mammalian herbivores, and drought in Yellowstone National Park. *Ecology* **73**:2043–2058.

Fritts, S. H., E. E. Bangs, J. A. Fontaine, M. R. Johnson, M. K. Phillips, E. D. Koch, and J. R. Gunson. 1997. Planning and implementing a reintroduction of wolves to Yellowstone National Park and central Idaho. *Restoration Ecology* **5**:7–27.

Gaillard, J.-M., M. Festa-Bianchet, and N. G. Yoccoz. 1998. Population dynamics of large herbivores: Variable recruitment with constant adult survival. *Trends in Ecology and Evolution* **13**:58–63.

Gaillard, J.-M., M. Festa-Bianchet, N. G. Yoccoz, A. Loison, and C. Toïgo. 2000. Temporal variation in fitness components of population dynamics in large herbivores. *Annual Review of Ecology and Systematics* **31**:367–393.

Gaillard, J.-M., A. J. Sempere, J. M. Boutin, C. VanLaere, and B. Boisaubert. 1992. Effects of age and body weight on the proportion of females breeding in a population of roe deer (*Capreolus capreolus*). *Canadian Journal of Zoology* **70**:1541–1545.

Garel, M., E. J. Solberg, B.-E. Sæther, I. Herdindal, and K.-A. Høgda. 2006. The length of growing season and adult sex ratio affect sexual size dimorphism in moose. *Ecology* **87**:745–758.

Garrott, R. A., J. G. Cook, J. G. Berardinelli, P. J. White, S. Cherry, and D. B. Vagnoni. 1997. Evaluation of urinary allantoin-creatinine ratio as a nutritional index for elk. *Canadian Journal of Zoology* **75**:1519–1525.

Garrott, R. A., L. L. Eberhardt, J. K. Otton, P. J. White, and M. A. Chaffee. 2002. A geochemical trophic cascade in Yellowstone's geothermal environments. *Ecosystems* **5**:659–666.

Garrott, R. A., L. L. Eberhardt, P. J. White, and J. Rotella. 2003. Climate-induced variation in vital rates of an unharvested large-herbivore population. *Canadian Journal of Zoology* **81**:33–45.

Garrott, R. A., S. L. Monfort, P. J. White, K. L. Mashburn, and J. G. Cook. 1998. One-sample pregnancy diagnosis in elk using fecal steroid metabolites. *Journal of Wildlife Diseases* **34**:126–131.

Gates, C. C., and R. J. Hudson. 1979. Effects of posture and activity on metabolic responses of wapiti to cold. *Journal of Wildlife Management* **43**:564–567.

Getz, W. M. 1996. A hypothesis regarding the abruptness of density dependence and the growth rate of populations. *Ecology* **77**:2014–2026.

Hamlin, K. L. 2006. *Monitoring and Assessment of Wolf-Ungulate Interactions and Population Trends Within the Greater Yellowstone Area, Southwestern Montana, and Montana Statewide.* Federal Aid Project W-120-R. Montana Fish, Wildlife and Parks, Wildlife Division, Helena, MT.

Hamlin, K. L., D. F. Pac, C. A. Sime, R. M. DeSimone, and G. L. Dusek. 2000. Evaluating the accuracy of ages obtained by two methods for Montana ungulates. *Journal of Wildlife Management* **64:**441–449.

Hamlin, K. L., and M. S. Ross. 2002. *Effects of Hunting Regulation Changes on Elk and Hunters in the Gravelly-Snowcrest Mountains, Montana.* Montana Fish Wildlife and Parks, Helena, MT.

Hanson, W. R. 1963. Calculation of productivity, survival, and abundance of selected vertebrates from sex and age ratios. *Wildlife Monographs* **9.**

Hobbs, N. T. 1996. Modification of ecosystems by ungulates. *Journal of Wildlife Management* **60:**695–713.

Hobbs, N. T., D. L. Baker, J. E. Ellis, and D. M. Swift. 1981. Composition and quality of elk winter diets in Colorado. *Journal of Wildlife Management* **45:**156–171.

Hosmer, D. W., T. Hosmer, S. le Cessie, and S. Lemeshow. 1997. A comparison of goodness-of-fit tests for the logistic regression model. *Statistics in Medicine* **16:**965–980.

Houston, D. B. 1982. *The Northern Yellowstone Elk.* MacMillan, New York, NY.

Hudson, R. J., and R. G. White. 1985. *Bioenergetics of Wild Herbivores.* CRC Press, Boca Raton, FL.

Jönsson, P., and L. Eklundh. 2002. Seasonality extraction by function fitting to time-series of satellite sensor data. *IEEE Transactions on Geoscience and Remote Sensing* **40:**1824–1832.

Jönsson, P., and L. Eklundh. 2004. TIMESAT—A program for analyzing time-series of satellite sensor data. *Computers and Geosciences* **30:**833–845.

Jorgenson, J. T., M. Festa-Bianchet, M. Lucherini, and W. D. Wishart. 1993. Effects of body size, population density, and maternal characteristics on age at first reproduction in bighorn ewes. *Canadian Journal of Zoology* **71:**2509–2517.

Kierdorf, U., H. Kierdorf, F. Sedlacek, and O. Fejerskov. 1996. Structural changes in fluorosed dental enamel of red deer (*Cervus elaphus*) from a region with severe environmental pollution by fluorides. *Journal of Anatomy* **188:**183–195.

de Kroon, H., J. van Groenendael, and J. Ehrlen. 2000. Elasticities: A review of methods and model limitations. *Ecology* **81:**607–618.

Langvatn, R., S. D. Albon, T. Burkey, and T. H. Clutton-Brock. 1996. Climate, plant phenology and variation in age of first reproduction in a temperate herbivore. *Journal of Animal Ecology* **65:**653–670.

Laws, R. M. 1981. Experiences in the study of large mammals. Pages 19–45 *in* C. W. Fowler and T. D. Smith (Eds.) *Dynamics of Large Mammal Populations.* Wiley, New York, NY.

Lebreton, J.-D., K. P. Burnham, J. Clobert, and D. R. Anderson. 1992. Modeling survival and testing biological hypotheses using marked animals: A unified approach with case studies. *Ecological Monographs* **62:**67–118.

Lemke, T. O., J. A. Mack, and D. B. Houston. 1998. Winter range expansion by the northern Yellowstone elk herd. *Intermountain Journal of Science* **4:**1–9.

Loison, A., and R. Langvatn. 1998. Short- and long-term effects of winter and spring weather on growth and survival of red deer in Norway. *Oecologia* **116:**489–500.

Malingreau, J. P. 1989. The vegetation index and the study of vegetation dynamics. Pages 285–303 *in* F. Toselli (Ed.) *Application of Remote Sensing to Agrometeorology.* ECSC, Brussels, Belgium and Luxembourg.

Matson, P. A., and M. D. Hunter. 1992. The relative contributions of top-down and bottom-up forces in population and community ecology. *Ecology* **73:**723.

McNaughton, S. J. 1985. Ecology of a grazing system: The Serengeti. *Ecological Monographs* **55:**259–294.

Murdoch, W. W. 1994. Population regulation in theory and practice. *Ecology* **75:**271–287.

Myers, W. L., B. Lyndaker, P. E. Fowler, and W. Moore. 1996. Investigations of calf elk mortalities in southeast Washington: A progress report 1992–1996. Washington Department of Wildlife, P-R Program Report, Olympia, WA.

Parker, K. L., and C. T. Robbins. 1984. Thermoregulation in mule deer and elk. *Canadian Journal of Zoology* **62:**1409–1422.

Paulik, G. J., and D. S. Robson. 1969. Statistical calculations for change-in-ratio estimators of population parameters. *Journal of Wildlife Management* **33:**1–27.

Pettorelli, N., A. Mysterud, N. G. Yoccoz, R. Langvatn, and N. C. Stenseth. 2005. Importance of climatological downscaling and plant phenology for red deer in heterogeneous landscapes. *Proceedings of the Royal Society B: Biological Sciences* **272:**2357–2364.

Pettorelli, N., F. Pelletier, A. von Hardenberg, M. Festa-Bianchet, and S. D. Côte. 2007. Early onset of vegetation growth vs. rapid green-up: Impact on juvenile mountain ungulates. *Ecology* **88:**381–390.

Pils, A. C., R. A. Garrott, and J. J. Borkowski. 1999. Sampling and statistical analysis of snow-urine allantoin:creatinine ratios. *Journal of Wildlife Management* **63:**1118–1132.

R Development Core Team. 2006. R: A language and environment for statistical computing. R Foundation for Statistical Computing, Vienna, Austria. http://www.R-project.org.

Raedeke, K. J., J. J. Millspaugh, and P. E. Clark. 2002. Population characteristics. Pages 449–491 *in* D. E. Toweill and J. W. Thomas (Eds.) *North American Elk: Ecology and Management.* Smithsonian Institution Press, Washington, DC.

Raithel, J. D., M. J. Kauffman, and D. H. Pletscher. 2007. Impact of spatial and temporal variation in calf survival on the growth of elk populations. *Journal of Wildlife Management* **71:**795-803.

Rasmussen, H. B., G. Wittemyer, and I. Douglas-Hamilton. 2006. Predicting time-specific changes in demographic processes using remote-sensing data. *Journal of Applied Ecology* **43:**366–376.

Reed, B. C., J. F. Brown, D. VanderZee, T. R. Loveland, J. W. Merchant, and D. O. Ohlen. 1994. Measuring phonological variability from satellite imagery. *Journal of Vegetation Science* **5**:703–714.

Robbins, C. T. 1993. *Wildlife Feeding and Nutrition*. Academic Press, San Diego, CA.

Sæther, B. E. 1997. Environmental stochasticity and population dynamics of large herbivores: A search for mechanisms. *Trends in Ecology and Evolution* **12**:143–149.

Sæther, B. E., and M. Heim. 1993. Ecological correlates of individual variation in age at maturity in female moose (*Alces alces*): The effects of environmental variability. *Journal of Animal Ecology* **62**:482–489.

Sand, H. 1996. Life history patterns in female moose (*Alces alces*): The relationship between age, body size, fecundity, and environmental conditions. *Oecologia* **106**:212–220.

Shenk, T. M., G. C. White, and K. P. Burnham. 1998. Sampling-variance effects on detecting density dependence from temporal trends in natural populations. *Ecological Monographs* **68**:445–463.

Shupe, J. L., A. E. Olson, H. B. Peterson, and J. B. Low. 1984. Fluoride toxicosis in wild ungulates. *Journal American Veterinary Medical Association* **185**:1295–1300.

Siler, W. 1979. A competing-risk model for animal mortality. *Ecology* **60**:750–757.

Sinclair, A. R. E. 1977. *The African Buffalo*. University of Chicago Press, Chicago, IL.

Singer, F. J., A. Harting, K. K. Symonds, and M. B. Coughenour. 1997. Density dependence, compensation, and environmental effects on elk calf mortality in Yellowstone National Park. *Journal of Wildlife Management* **61**:12–25.

Skalski, J. R., K. E. Ryding, and J. J. Millspaugh. 2005. *Wildlife Demography: Analysis of Sex, Age, and Count Data*. Elsevier, San Diego, CA.

Smith, B. L., and S. H. Anderson. 1998. Juvenile survival and population regulation of the Jackson Elk Herd. *Journal of Wildlife Management* **62**:1036–1045.

Stearns, S. C. 1992. *The Evolution of Life Histories*. Oxford University Press, Oxford, United Kingdom.

Swets, D. L., B. C. Reed, J. D. Rowland, and S. E. Marko. 2000. *A Weighted Least-Squares Approach to Temporal NDVI Smoothing* Proceedings of the 1999 ASPRS Annual Conference, From Image to Information. Portland, Oregon, May 17–21, 1999. American Society of Photogrammetry and Remote Sensing, Bethesda, MD.

Taper, M. L., and P. J. P. Gogan. 2002. The northern Yellowstone elk: Density dependence and climatic conditions. *Journal of Wildlife Management* **66**:106–122.

Thompson, S. K. 1992. *Sampling*. Wiley, New York, NY.

Tuljapurkar, S., and H. Caswell. 1997. *Structured-Population Models in Marine, Terrestrial, and Freshwater Systems*. Chapman and Hall, London, United Kingdom.

Tveraa, T., P. Fauchald, N. G. Yoccoz, R. A. Ims, R. Aanes, and K. A. Høgda. 2006. What regulate and limit reindeer populations in Norway? *Oikos* **116**:706–715.

Van Soest, P. J. 1994. *Nutritional Ecology of the Ruminant*. Cornell University Press, Ithaca, NY.

Vucetich, J. A., D. W. Smith, and D. R. Stahler. 2005. Influence of harvest, climate, and wolf predation on Yellowstone elk, 1961–2004. *Oikos* **111**:259–270.

Wang, G., N. T. Hobbs, R. B. Boone, A. W. Illius, I. J. Gordon, J. E. Gross, and K. L. Hamlin. 2006. Spatial and temporal variability modify density dependence in populations of large herbivores. *Ecology* **87**:95–102.

Wasser, S. K., S. L. Monfort, J. Souther, and D. E. Wildt. 1994. Excretion rates and metabolite of oestradiol and progesterone in baboons (*Papio cynocephalus cynocephalus*) faeces. *Journal of Reproduction and Fertility* **101**:213–220.

Wasser, S. K., L. Risler, and R. A. Steiner. 1988. Excreted steroids in primate feces over the menstrual cycle and pregnancy. *Biology of Reproduction* **39**:862–872.

White, G. C., and R. A. Garrott. 1990. *Analysis of Wildlife Radio-Tracking Data*. Academic Press, San Diego, CA.

White, P. J., and R. A. Garrott. 2005. Northern Yellowstone elk after wolf restoration. *Wildlife Society Bulletin* **33**:942–955.

White, P. J., R. A. Garrott, J. F. Kirkpatrick, and E. V. Berkeley. 1995. Diagnosing pregnancy in free-ranging elk using fecal steroid metabolites. *Journal of Wildlife Diseases* **31**:514–522.

Wickstrom, M. L., C. T. Robbins, T. A. Hanley, D. E. Spalinger, and S. M. Parish. 1984. Food intake and foraging energetics of elk and mule deer. *Journal of Wildlife Management* **48**:1285–1301.

Wittemyer, G., H. B. Rasmussen, and I. Douglas-Hamilton. 2007. Breeding phenology in relation to NDVI variability in free-ranging African elephant. *Ecography* **30**:42–50.

Zager, P., C. White, and G. Pauley. 2005. Study IV. Factors influencing elk calf recruitment. Job #s 1–3. Pregnancy rates and condition of cow elk. Calf mortality causes and rates. Predation effects on elk calf recruitment. Federal Aid in Wildlife Restoration, Job Progress Report, W-160-R-32, Subproject 31, Elk ecology, Idaho Department of Fish and Game, Boise, ID.

CHAPTER 12

Partial Migration in Central Yellowstone Bison

Jason E. Bruggeman,* P. J. White,[†] Robert A. Garrott,* and Fred G. R. Watson[‡]

*Fish and Wildlife Management Program, Department of Ecology, Montana State University
[†]National Park Service, Yellowstone National Park
[‡]Division of Science and Environmental Policy, California State University Monterey Bay

Contents

Theme

The conservation of bison (*Bison bison*) in Yellowstone National Park from near extinction to a high of 5000 animals has led to societal conflict regarding overabundance and potential transmission of brucellosis to cattle with widespread economic consequences. As abundance increased during 1971–1996, more bison migrated from the Hayden and Pelican valleys to the lower-elevation Madison headwaters area and, eventually, outside the park (Meagher 1998, Fuller *et al.* 2007). Meagher (1998) and others (Taper *et al.* 2000, Gates *et al.* 2005) concluded these migratory movements were stress-related responses to decreased food availability as bison fully occupied habitat in the Pelican and Hayden valleys and, subsequently, the Firehole and Madison river drainages. This hypothesis implies that these areas have a relatively fixed capacity for wintering bison and, as a

The Ecology of Large Mammals in Central Yellowstone
R. Garrott, P. J. White and F. Watson
ISSN 1936-7961, DOI: 10.1016/S1936-7961(08)00212-1

Copyright © 2009, Elsevier Inc.
All rights reserved.

result, a larger proportion of bison migrate to the Madison headwaters and elsewhere as bison numbers increase beyond this capacity. However, density-independent factors such as genetic predisposition, individual asymmetries (*e.g.*, age, sex), and stochastic variations in climate that influence food availability may also affect migration (Lundberg 1987). We used data collected during 1970–1971 through 2005–2006 to quantify annual variations in the magnitude and timing of migration by central herd bison, identify potential factors driving this variation, and evaluate if the proportion of migrants increased with abundance.

I. INTRODUCTION

On large spatial scales, migration serves to reduce the environmental heterogeneity experienced by an organism and place it under favorable conditions for survival (Dingle 1996). Migratory movements are often predicated on the need for resources, especially food, which may be affected by biotic and abiotic factors (*e.g.*, Whitehead 1996, Knight *et al.* 1999). A density-related reduction in per capita resources may lead to the establishment of migratory behavior for a population or affect the magnitude and timing of migration (Mahoney and Schaefer 2002, Whalen and Watts 2002, Marra *et al.* 2005). The effects of seasonality and climate on migratory patterns have also been documented in various studies (Kaňuščák *et al.* 2004), including effects of air temperature on birds (Gordo *et al.* 2005) and precipitation on insects (Dingle *et al.* 2000). These factors may modify the availability and quality of suitable habitat and forage. They can interact to influence an animal's choice of migratory decisions and destinations, as well as the rate of movement during migration (Ahola *et al.* 2004, Hulbert *et al.* 2005). Thus, quantifying the effects of density-dependent and independent mechanisms on migratory behavior is essential for comprehending population processes, evaluating habitat requirements, and designing management strategies for migratory species.

Variability in individual migratory behavior has been documented in various ungulate populations (Talbot and Talbot 1963, Morgantini and Hudson 1988, Bergerud *et al.* 1990, Nelson *et al.* 2004). White *et al.* (2007) suggested pronghorn (*Antilocapra americana*) assessed individual asymmetries, behaviors, and environmental conditions when deciding whether to migrate. Also, Ball *et al.* (2001) suggested differences in snow attributes between ranges of migrant and resident moose (*Alces alces*) may be a factor that influences behavior. Migrations in response to density and climate, which may ultimately affect population dynamics (Langvatn *et al.* 1996, Forchhammer *et al.* 1998, Post and Stenseth 1998, Jacobson *et al.* 2004), have been documented for assorted ungulates (Bergerud 1988, Fryxell and Sinclair 1988, Pettorelli *et al.* 2005). While wildebeest (*Connochaetes taurinus*) migration in response to rainfall (Maddock 1979) is a classic example, many other large herbivores follow forage productivity gradients and migrate in response to climate variation (Leimgruber *et al.* 2001, Mysterud *et al.* 2001). Mule deer (*Odocoileus hemionus*) migration between summer and winter ranges in response to snow has been detailed (Gilbert *et al.* 1970, D'Eon and Serrouya 2005). Further, sika deer (*Cervus nippon*) and elk (*Cervus elaphus*) have been found to migrate to areas of lesser snow pack in winter (Igota *et al.* 2004, White and Garrott 2005). Therefore, migratory movements may be influenced by weather conditions that affect vegetation quality and quantity, as well as forage availability.

After near extirpation in the early twentieth century, bison in the Lamar and Pelican valleys of Yellowstone National Park were subject to intense animal husbandry during 1902–1938 and reintroduced into the Hayden and Firehole Valleys of Yellowstone in 1936 (Meagher 1973). As numbers increased, movements between the Hayden and Firehole valleys were observed during all seasons and bison began using areas beyond the Firehole Valley and throughout the Madison headwaters area (Meagher 1973, Gates *et al.* 2005, Figure 12.1). Seasonal migrations are now the norm, with migration into the Madison headwaters area from the higher-elevation summer range in the Hayden and Pelican Valleys beginning in autumn and continuing through winter along the Mary Mountain trail until bison return to the summer range in June (Bjornlie and Garrott 2001, Chapter 27 by Bruggeman *et al.*, this volume). Increased use of areas beyond the Madison headwaters area has been occurring since the

FIGURE 12.1 After near extirpation a century ago Yellowstone's bison population has been fully restored and has reestablished migratory movements throughout the park and across its boundaries. These animals are crossing the Yellowstone River in the Hayden Valley where they summer, with many of the central herd animals migrating westward into the Madison headwater drainages to winter (Photo by Jeff Henry).

1990s. Meagher (1993, 1998) attributed these changes in distribution to stress-related dispersal in response to food limitations as bison fully occupied wintering areas and compensated for decreasing per capita resources by moving to additional areas. She concluded that movements by bison in the Pelican Valley westward had a domino effect on bison wintering in the Hayden Valley, resulting in larger movements of more bison earlier to the Madison headwaters area (Meagher 1998). This "domino effect" hypothesis, whereby increased abundance was coupled with cascading increases in distribution, was adopted by other scientists as a mechanism by which bison could maintain a relatively stable winter density. For example, Gates *et al.* (2005:115) referred to this hypothesized gradual and linear range expansion as "the density-equalization effect," whereby the area occupied by bison during winter expanded within available grassland and meadow habitats as the central and northern herds increased in abundance. Also, Taper *et al.* (2000) proposed range expansion as a mechanism explaining stable population growth in Yellowstone bison.

Our goals were to (1) test the predictions that significant migration by the central herd did not occur until bison had fully occupied the Hayden and Pelican valleys, and more animals migrated earlier as numbers increasingly exceeded this limit, and (2) assess the influence of biotic and abiotic mechanisms on the magnitude and timing of the annual migration to winter range in the Madison headwaters area. We analyzed one recent (1996–1997 through 2005–2006) data set from winter ground surveys and one historic (1970–1971 through 2005–2006) data set from winter aerial surveys using an information-theoretic approach to evaluate competing hypotheses regarding the relative influence of snow pack, forage productivity, and population size on the timing and magnitude of bison migration patterns.

II. METHODS

A. Development of Migration Response Variables

The number and distribution of bison wintering in the Madison headwaters area were determined by conducting comprehensive ground-based surveys every 10–14 days during November–May, 1996–1997 through 2005–2006 (Bjornlie and Garrott 2001, Bruggeman *et al.* 2006). Seventy-four sampling units were surveyed over two days using six distinct routes (Ferrari 1999, Bjornlie 2000) that afforded a nearly complete enumeration of bison in this area as determined from an aerial-ground double-sampling study (Hess 2002). A small portion of migratory bison remained uncounted because we did not survey winter range areas along the Mary Mountain trail and western park boundary. Observers using snowmobiles, trucks, or snowshoes started each route simultaneously to minimize missing or double counting bison (Bjornlie and Garrott 2001). We defined 2-week time intervals, i ($1 \leq i \leq 14$), from November–May for each winter, j ($1 \leq j \leq 10$), centered on the bimonthly ground surveys. To investigate the magnitude of migration, we defined a response variable, MAXIMUM, as the highest number of bison counted in the Madison headwaters area each winter during ground surveys. To investigate the timing of migration, we defined a response variable, TIMING, as the number of bison in the Madison headwaters area for the ijth period as determined from ground surveys.

We also obtained aerial estimates of abundance and distribution for central herd bison during winters 1970–1971 through 2005–2006, as well as removal data from the western boundary of the park, to investigate "historic" patterns in the magnitude of migration (Dobson and Meagher 1996, Hess 2002). We defined a response variable, HISTORIC$_{max}$, as the highest number of bison counted on the west-side winter range (Firehole, Gibbon, and Madison river drainages) during flights conducted between January and mid-April each year, plus removals that occurred before the flights. This definition of "west-side" winter range differs from Meagher (1993) who considered the west-side to be from Madison Junction along the Madison River to Hebgen Lake, excluding the Firehole drainage.

B. Development of Covariates

We anticipated snow would affect migration behavior by bison because it affects their foraging, movements, and distributions (Bruggeman 2006, Figure 12.2). We used a validated snow pack simulation model (Watson *et al.* 2006, Chapter 6 by Watson *et al.*, this volume) to compute daily estimates of average snow water equivalent (SWE; *i.e.*, amount of water in the snow) on the bison

FIGURE 12.2 Bison travelling and feeding at the mouth of Trout Creek in the Hayden Valley. As snow pack increases on the summer range of the central bison herd foraging and movement becomes difficult and provides an impetus for animals to migrate to the lower elevation winter range in the Madison headwater drainages (Photo by Jeff Henry).

summer range, encompassing all 28.5 × 28.5 m pixels within the Hayden and Pelican valleys. We defined a covariate, $SWES_{ij}$, as the average SWE on the bison summer range for each *ij*th period between November–May for each year to be used in the timing of migration analysis. We added daily SWE values from October 1–April 30 to calculate a covariate, SWE_{acc}, and obtain a measure of annual winter severity (Garrott *et al.* 2003) for use in both the recent and historic magnitude of migration analyses. We also calculated a covariate, $SWE_{acc,ij}$, as the sum of daily SWE estimates from October 1 through the *ij*th period for a temporal measure of winter severity for the timing of migration analysis.

Warm season climate variation, particularly with regard to precipitation, temperature, and the rate of evapotranspiration, influences plant growing conditions and, in turn, the quantity and quality of grasses available for bison (McNaughton 1985, Sala *et al.* 1988, Stephenson 1990). As a result, vegetation productivity may affect ungulate distribution, breeding phenology, and reproductive success (McNaughton 1985, Pettorelli *et al.* 2006, Wittemyer *et al.* 2007). Recent analyses of bison demographic rates and migration patterns in Yellowstone (*e.g.*, Bruggeman 2006, Fuller *et al.* 2007) have reported significant relationships with the Palmer Drought Severity Index (Palmer 1965), which was used as a single integrator of annual variability in regional warm season climate. Northwestern Wyoming experienced severe, sustained drought conditions during 1995–2006, with the mean Palmer Drought Severity Index during May–July decreasing from 0.9 to −9.0 compared to a mean index of −1.0 during 1969–1994 (range = −6.4 to 2.9). This index is cumulative, so the intensity of drought during a given month is derived from the current weather patterns plus the cumulative patterns of previous months. Certainly, food limitations during periods of drought are known to affect population dynamics (Sinclair *et al.* 1985, Mduma *et al.* 1999) and provide additional stimuli for animals to migrate (Polovina *et al.* 2001, Burtenshaw *et al.* 2004, Varpe *et al.* 2005). However, we did not use the Palmer Drought Severity Index in our current analyses because it was significantly and negatively correlated with summer counts of bison in the central herd during 1998–2006 ($R^2 = 0.80$, $F_{1,7} = 27.5$, $P = 0.001$). Also, we were unable to detect any correlation between this drought index and growing season precipitation in the Madison headwaters area during 1989–2005 ($R^2 = 0.03$, $F_{1,15} = 0.5$, $P = 0.51$).

Instead, we used remotely sensed, normalized differential vegetation index (NDVI) data from summers 1989–2005 for the Hayden and Pelican valleys as a measure of forage production. The NDVI is correlated with green biomass and serves as a measure of temporal and spatial variability in vegetation growth (Wittemyer *et al.* 2007). Specifically, we used the L-integral, which is the integral of the seasonal NDVI curve over the growing season (Garel *et al.* 2006), to calculate a covariate, $NDVI_{int}$, to index summer range forage productivity prior to winter. We also calculated a covariate, $NDVI_{len}$, as the length of the growing season as determined from NDVI data. Because NDVI data were available only from 1989–2005, we did not have complete data for the historic migration analysis. Therefore, we developed an index for the L-integral from 1971–1988 using L-integral data from 1989–2005 and assorted precipitation, temperature, and snow data obtained from the Canyon SNOTEL site (National Resource Conservation Service 2007). We conducted an all-subsets regression using PROC REG in SAS version 9.1.3 (SAS 2004) using the L-integral response variable and 78 climate covariates, and selected the model with the highest adjusted R^2 value (Neter *et al.* 1996). We then used data from the Yellowstone Lake CLIM site from 1970–1980 and Canyon SNOTEL site from 1981–1988 to estimate the $NDVI_{int}$ covariate for 1970–1988 for use in the historic migration analysis.

Aerial population estimates for the central herd were obtained during late July or August each year, prior to the migration (Dobson and Meagher 1996, Hess 2002). We used these annual estimates to define a covariate, BISON, which provided a measure of the effect of bison density on migration. We also defined a covariate, DATE, as the Julian date of the January–April winter survey used to calculate $HISTORIC_{max}$ because, in general, more migratory bison should be counted as winter progresses (*i.e.*, later survey dates; Bjornlie and Garrott 2001, Bruggeman *et al.* 2006).

C. 1996–1997 Through 2005–2006 Migration Patterns: Model Development and Statistical Analyses

We developed and compared *a priori* hypotheses, expressed as multiple regression models, in two modeling exercises to estimate the relative contributions of snow pack, forage productivity, and density on variations in the magnitude and timing of migration from 1996–1997 through 2005–2006 (Appendix 12A.1). We calculated variance inflation factors (VIFs; Neter *et al.* 1996) while forming our model list to quantify multicollinearity between model predictors, including interactions. Models containing predictors having a VIF > 5 were removed from our *a priori* list. Hypotheses for the timing of migration analysis were expressed as 11 regression equations consisting of additive main effects (SWES, BISON, $NDVI_{int}$) and interactions (SWES*BISON, BISON*$NDVI_{int}$) of covariates, while those for the magnitude of migration analysis were expressed as 11 equations of main effects (SWE_{acc}, BISON, $NDVI_{int}$) and interactions (SWE_{acc}*BISON, BISON*$NDVI_{int}$). In exploratory analyses, we systematically substituted covariates $SWE_{acc,ij}$ for SWES and $NDVI_{len}$ for $NDVI_{int}$.

For each covariate and interaction, we made a hypothesis about the sign of the effect on each response variable. We predicted the magnitude of migration (MAXIMUM) would be positively correlated with BISON because increasing population size would lead to a decrease in per capita resources on the summer range and result in more bison migrating to the winter range to find forage. Second, we hypothesized that MAXIMUM would be negatively correlated with $NDVI_{int}$ (or $NDVI_{len}$) since decreased primary productivity would decrease the quality and quantity of forage and bison would need to migrate to obtain adequate food resources. Third, we predicted increasing SWE_{acc} would result in increased bison migration (MAXIMUM) because greater snow pack on the summer range would provide more impetus for bison to migrate to lower elevations to find more accessible forage during winter. Fourth, we anticipated the influence of bison density would vary with snow pack in the form of a positive BISON*SWE_{acc} interaction effect because high population sizes at high SWE_{acc} would result in further increases in bison migration (MAXIMUM). Finally, we hypothesized that density would interact with primary productivity (BISON*$NDVI_{int}$) because the influence of population size would be accentuated during years of reduced productivity (negative sign), leading to increased bison migration (MAXIMUM).

We used similar rationale for the timing of migration analysis to hypothesize that TIMING would be positively correlated with BISON and negatively correlated with $NDVI_{int}$ (or $NDVI_{len}$). Second, we predicted that TIMING would be positively correlated with SWES (or $SWE_{acc,ij}$) because increasing snow pack on the summer range would provide an impetus for more bison to migrate to the lower-elevation, geothermally influenced winter range with easier access to forage. We anticipated the effect of population size would vary with climate and plant productivity in the form of BISON*SWES and BISON*$NDVI_{int}$ interactions because combinations of high population sizes and high snow levels, or high population sizes and reduced plant productivity, could further accelerate the timing of migration (TIMING).

We used regression techniques in R version 2.3.1 (R Development Core Team 2006) to fit models and estimate parameter coefficients. To facilitate comparisons of parameter coefficients, each continuous predictor was centered and scaled prior to analysis by subtracting the midpoint and dividing by half of the range, resulting in values between -1 and 1. We calculated a corrected Akaike's Information Criterion (AIC_c) value for each model, ranked and selected the best approximating models for the timing and magnitude analyses using ΔAIC_c values, and calculated Akaike weights (w_i) to obtain a measure of model selection uncertainty (Burnham and Anderson 2002).

D. 1970–1971 Through 2005–2006 Migration Patterns: Model Development and Statistical Analyses

Examination of central herd summer and winter count data from 1970–1971 through 2005–2006 revealed an unexpected relationship between the total number of bison enumerated from the July–August summer flight and the total count during the January to mid-April winter flight (Figure 12.3).

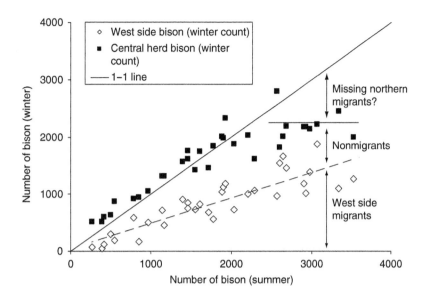

FIGURE 12.3 Relationships between the number of west-side migratory bison from the winter flight and central herd population size from the summer flight, and the central herd population size from the winter flight and central herd population size from the summer flight (1970–1971 through 2005–2006). The line depicting a 1–1 relationship is denoted for reference. Note the break point around 2350 bison for the central herd population size count comparisons.

We noted a change from an approximately 1–1 correlation, with fewer bison counted during winter compared to summer surveys as population size exceeded 2000 bison. We used piecewise regression methods (Neter *et al.* 1996) using PROC NLIN in SAS version 9.1.3 (SAS 2004) to determine this break point occurred at a count of approximately 2350 bison. Fuller *et al.* (2007) suggested a change in bison spatial dynamics, with increased emigration from the central herd and immigration into the northern herd, resulted in changes in herd dynamics beginning in 1981–1982. We suspected these movements onto the northern range caused this break point, but the data did not exist to further investigate the timing and magnitude of these movements. Thus, we censored summer counts when the central herd population was >2350 from the data set (1993–1997, 2002–2006) before beginning the analysis, and only examined migration to the west-side winter range.

We developed and compared *a priori* hypotheses to estimate the relative contributions of snow pack, forage productivity, density, and winter survey date on variations in the magnitude of migration for years when the central herd population size was <2350 bison during summer (*i.e.*, 1971–1992, 1998–2001; Appendix 12A.1). During 1982, 1983, and 1988, there was either no July–August flight or the count was deemed poor. Thus, no population size data were available to apply for the subsequent winters. We had 22 observations for the analysis after removing these years and a data point from 1987 owing to a high leverage $NDVI_{int}$ value. We calculated VIFs (Neter *et al.* 1996) while forming the model list and removed any predictors with VIF > 5. Hypotheses were expressed as 20 regression equations consisting of additive main effects (SWE_{acc}, BISON, $NDVI_{int}$, DATE) and interactions (SWE_{acc}*BISON, BISON*$NDVI_{int}$) of covariates. Our predictions were similar to the extent of migration analyses during 1996–1997 through 2005–2006. We used regression techniques in R version 2.3.1 (R Development Core Team 2006) to fit models and estimate parameter coefficients using centered and scaled continuous covariates as described above. We calculated AIC_c for each model, ranked and selected the best approximating models using AIC_c, and calculated w_i (Burnham and Anderson 2002).

E. Estimates of Nutritional Intake

To explore differences in nutrition of bison, which we hypothesize may be a primary driver of migration, we estimated metabolizable energy intake (nutrition) by bison in the Hayden Valley and Madison headwaters area during the winter of 2005–2006 using molar ratios of allantoin:creatinine (A:C) obtained from fresh urine deposited in snow. We opportunistically collected samples from mixed groups of adult and calf bison using the collection, storage, and assay procedures described by Pils *et al.* (1999) during eight 14-day intervals between December 18 and April 8. Frozen snow-urine samples were then thawed at room temperature and extraneous debris was eliminated by filtration from the snow-urine mixture. The filtrate was refrozen at 20°C until assayed. Creatinine concentrations in filtered samples were assayed at the Department of Animal and Range Sciences, Montana State University, Bozeman, using a modification of the kinetic Jaffé reaction (Larsen 1972). Creatinine is a compound produced and excreted in relatively constant proportion to muscle mass, allowing comparisons to be made among animals of different body size and hydration and among samples of different snow dilution (Coles 1980). Allantoin concentrations were determined using a microplate modification of the colorimetric procedures described by Young and Conway (1942). Allantoin is a metabolite that reliably indexes rumen microbial flow in domestic ruminants and free-ranging elk during a several-day period prior to the time urine was voided (Chen *et al.* 1992a,b; Vagnoni *et al.* 1996, Garrott *et al.* 1997). Allantoin and creatinine were assayed from the same sample at the same time to avoid additional freeze-thaw cycles that tend to degrade allantoin. Sensitivities of allantoin and creatinine assays were 0.027 and 0.079 μmol, respectively. Coefficients of variation for high and low snow-urine pools of allantoin and creatinine were less than 15%. The relationship between digestible dry matter intake and urinary A:C ratios has not been tested for bison during captive feeding trials.

Pils *et al.* (1999) reported that mean A:C ratios of samples collected from mixed groups of adult female and calf elk were highly variable and consistently overestimated the true mean A:C ratio of adult females. These problems were alleviated by trimming 15% off the right tail of the ordered sample distribution and 20% off the left tail and calculating the mean of the remaining samples. Trimmed estimates of mean A:C ratios were consistently less biased, with lower variability of sample means and smaller confidence intervals than untrimmed means (Pils *et al.* 1999). We calculated trimmed A:C ratios for bison in the Hayden Valley and Madison headwaters area during each collection period and used a one-tailed t-test to compare paired sample means. We predicted nutritional estimates for bison wintering in the Hayden Valley would be significantly lower than those of bison wintering in the lower-elevation and more heterogeneous Madison headwaters area.

III. RESULTS

A. Temporal Variation in Migratory Behavior (1996–1997 Through 2005–2006)

The number of bison observed in the Madison headwaters area (TIMING) increased with time each winter, usually peaking in late March or early April (see Figure 27.1a in Chapter 27 by Bruggeman *et al.*, this volume). During 109 ground distribution surveys, the number of bison counted in the Madison headwaters area ranged from 205–1538 bison (mean $\pm$ SE; 775 $\pm$ 30). The highest number of bison counted in the Madison headwaters area each year (MAXIMUM) varied between 888–1538 bison (1174 $\pm$ 64). The central herd population grew from 1473–3441 bison (2556 $\pm$ 250) during 1997–1998 through 2005–2006, after decreasing from 2928 bison in 1996–1997 owing to management-based removals at the park boundary and mortality during a severe winter. Snow accumulation in the Hayden and Pelican valleys began in October and built through the winter before generally peaking in early April, with annual peak SWES, SWE$_{\text{acc,ij}}$, and SWE$_{\text{acc}}$ ranging from 18.0–44.1 cm (26.1 $\pm$ 0.2), 66–5236 cm days (1367 ± 102), and 1750–5236 cm days (2843 ± 309), respectively. Covariates $NDVI_{int}$ and $NDVI_{len}$ ranged between 2071–2688 (2376 ± 56) and 12.7–17.1 weeks (14.6 ± 0.4), respectively.

One best approximating model structure was supported by the data for the magnitude of bison migration (Tables 12.1 and 12.2). This model ($AIC_c = 135.1$, $K = 2$, $w_i = 0.593$, adjusted $R^2 = 0.40$) contained a significant BISON effect (estimate: 190.5; 95% CI: 29.7, 351.3) and had a relative likelihood of 3.9 compared to the second best model ($\Delta AIC_c = 2.74$, $K = 3$, $w_i = 0.150$, adjusted $R^2 = 0.41$), which included BISON (estimate: 194.7; 95% CI: 35.3, 354.0) and SWE_{acc} (estimate: −100.6; 95% CI: −307.9, 106.8). There was no improvement in the top model when $NDVI_{len}$ was substituted for $NDVI_{int}$ in the exploratory analysis.

Three best approximating models were supported by the data for the timing of bison migration (Tables 12.1 and 12.3). The top model ($AIC_c = 1,518.2$, $K = 4$, $w_i = 0.426$, adjusted $R^2 = 0.35$) consisted of BISON (estimate: 208.0; 95% CI: 129.1, 286.9), SWES (estimate: 367.4; 95% CI: 243.3, 491.6; Figure 12.4), and $NDVI_{int}$ (estimate: 142.5; 95% CI: 27.6, 257.3) covariates. The second best model ($\Delta AIC_c = 0.94$, $K = 5$, $w_i = 0.266$, adjusted $R^2 = 0.35$) included a SWES*BISON interaction (estimate: −101.4; 95% CI: −284.3, 81.5) with the confidence interval overlapping zero, in addition to

TABLE 12.1 AIC model selection results for the best supported *a priori* models in the magnitude and timing of migration analyses

Model structure	k	ΔAIC_c	w_i
Magnitude of migration, 1996–1997 through 2005–2006			
$\beta_0 + \beta_1(BISON)$	2	0	0.59
$\beta_0 + \beta_1(SWE_{acc}) + \beta_2(BISON)$	3	2.74	0.15
$\beta_0 + \beta_1(BISON) + \beta_2(NDVI_{int})$	3	2.77	0.15
$\beta_0 + \beta_1(SWE_{acc})$	2	5.68	0.03
$\beta_0 + \beta_1(NDVI_{int})$	2	6.19	0.03
$\beta_0 + \beta_1(BISON) + \beta_2(NDVI_{int}) + \beta_3(BISON*NDVI_{int})$	4	6.79	0.02
$\beta_0 + \beta_1(SWE_{acc}) + \beta_2(BISON) + \beta_3(SWE_{acc}*BISON)$	4	7.87	0.01
$\beta_0 + \beta_1(SWE_{acc}) + \beta_2(BISON) + \beta_3(NDVI_{int})$	4	8.41	0.01
$\beta_0 + \beta_1(SWE_{acc}) + \beta_2(NDVI_{int})$	3	9.03	0.01
Timing of migration, 1996–1997 through 2005–2006			
$\beta_0 + \beta_1(SWES) + \beta_2(BISON) + \beta_3(NDVI_{int})$	4	0	0.43
$\beta_0 + \beta_1(SWES) + \beta_2(BISON) + \beta_3(NDVI_{int}) + \beta_4(SWES*BISON)$	5	0.94	0.27
$\beta_0 + \beta_1(SWES) + \beta_2(BISON) + \beta_3(NDVI_{int}) + \beta_4(BISON*NDVI_{int})$	5	1.61	0.19
$\beta_0 + \beta_1(SWES) + \beta_2(BISON)$	3	3.93	0.06
$\beta_0 + \beta_1(SWES) + \beta_2(BISON) + \beta_3(SWES*BISON)$	4	4.00	0.06
Historic magnitude of migration, 1970–1971 through 2005–2006			
$\beta_0 + \beta_1(BISON)$	2	0	0.33
$\beta_0 + \beta_1(BISON) + \beta_2(SWE_{acc})$	3	0.49	0.26
$\beta_0 + \beta_1(BISON) + \beta_2(NDVI_{int})$	3	2.69	0.08
$\beta_0 + \beta_1(SWE_{acc}) + \beta_2(BISON) + \beta_3(NDVI_{int})$	4	3.04	0.07
$\beta_0 + \beta_1(SWE_{acc}) + \beta_2(BISON) + \beta_3(SWE_{acc}*BISON)$	4	3.34	0.06
$\beta_0 + \beta_1(DATE) + \beta_2(SWE_{acc}) + \beta_3(BISON)$	4	3.46	0.06
$\beta_0 + \beta_1(BISON) + \beta_2(NDVI_{int}) + \beta_3(BISON*NDVI_{int})$	4	3.78	0.05
$\beta_0 + \beta_1(SWE_{acc}) + \beta_2(BISON) + \beta_3(NDVI_{int}) + \beta_4(BISON*NDVI_{int})$	5	5.28	0.02
$\beta_0 + \beta_1(DATE) + \beta_2(BISON) + \beta_3(NDVI_{int})$	4	5.71	0.02
$\beta_0 + \beta_1(SWE_{acc}) + \beta_2(BISON) + \beta_3(NDVI_{int}) + \beta_4(SWE_{acc}*BISON)$	5	6.07	0.02
$\beta_0 + \beta_1(DATE) + \beta_2(SWE_{acc}) + \beta_3(BISON) + \beta_4(NDVI_{int})$	5	6.43	0.01
$\beta_0 + \beta_1(DATE) + \beta_2(SWE_{acc}) + \beta_3(BISON) + \beta_4(SWE_{acc}*BISON)$	5	6.72	0.01
$\beta_0 + \beta_1(DATE) + \beta_2(BISON) + \beta_3(NDVI_{int}) + \beta_4(BISON*NDVI_{int})$	5	7.10	0.01

All models are ranked according to AIC_c and presented along with the number of parameters (k), the ΔAIC_c value (*i.e.*, change in AIC_c relative to the best model), and the Akaike weight (wi). The AIC_c values for the top models for the magnitude and timing of migration analyses during 1997–2006 were 135.1 and 1518.17, respectively. The AIC_c value for the top model for the historic magnitude of migration analyses during 1971–2006 was 289.18.

| TABLE 12.2 | Coefficient values (β_i) and 95% confidence limits for covariates from the best approximating models for the magnitude of migration analyses during 1996–1997 through 2005–2006 (recent) and 1970–1971 through 2005–2006 (historic) |

Covariate	$\beta_0 + \beta_1(\text{BISON})$	$\beta_0 + \beta_1(\text{SWE}_{acc}) + \beta_2(\text{BISON})$	$\beta_0 + \beta_1(\text{BISON}) + \beta_2(\text{NDVI}_{int})$
1996–1997 through 2005–2006 (Recent)			
BISON	**190.5 (29.7, 351.3)**	**194.7 (35.3, 354.0)**	**232.4 (50.8, 414.0)**
SWE$_{acc}$		−100.6 (−307.9, 106.8)	
NDVI$_{int}$			111.2 (−120.1, 342.5)
1970–1971 through 2005–2006 (Historic)			
BISON	**536.4 (417.1, 655.6)**	**554.3 (434.9, 673.8)**	**537.7 (410.5, 664.9)**
SWE$_{acc}$		95.0 (−48.0, 238.0)	
NDVI$_{int}$			5.07 (−130.3, 140.5)

Values in bold denotes significant coefficients at $\alpha = 0.05$.

| TABLE 12.3 | Coefficient values (β_i) and 95% confidence limits for covariates from the best approximating models for the timing of migration analyses during 1996–1997 through 2005–2006 |

Covariate	$\beta_0 + \beta_1(\text{SWES}) + \beta_2(\text{BISON}) + \beta_3(\text{NDVI}_{int})$	$\beta_0 + \beta_1(\text{SWES}) + \beta_2(\text{BISON}) + \beta_3(\text{NDVI}_{int}) + \beta_4(\text{SWES*BISON})$	$\beta_0 + \beta_1(\text{SWES}) + \beta_2(\text{BISON}) + \beta_3(\text{NDVI}_{int}) + \beta_4(\text{BISON* NDVI}_{int})$
SWES	**367.4 (243.3, 491.5)**	**371.2 (247.0, 495.3)**	**366.2 (241.7, 490.6)**
BISON	**208.0 (129.1, 286.9)**	**169.7 (64.9, 274.5)**	**208.8 (129.8, 287.9)**
NDVI$_{int}$	**142.5 (27.6, 257.3)**	**132.8 (16.8, 248.8)**	**125.8 (2.6, 249.0)**
SWES*BISON		−101.4 (−284.3, 81.5)	
BISON*NDVI$_{int}$			119.6 (−196.3, 435.5)

Values in bold denotes significant coefficients at $\alpha = 0.05$.

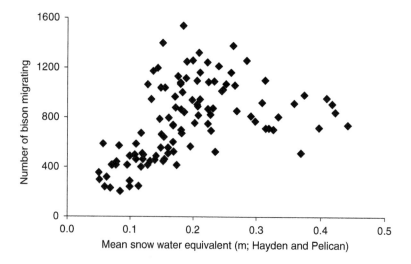

FIGURE 12.4 Data depicting the positive correlation between the number of bison in the Madison headwaters area for each two-week period (1996–1997 through 2005–2006) and the average summer range snow water equivalent covariate (SWES; in meters) for the timing of migration analysis in Yellowstone National Park, Wyoming, USA.

BISON (estimate: 169.7; 95% CI: 64.9, 274.5), SWES (estimate: 371.2; 95% CI: 247.0, 495.3), and NDVI$_{int}$ (estimate: 132.8; 95% CI: 16.8, 248.8). The third best model ($\Delta AIC_c = 1.61$, $K = 5$, $w_i = 0.190$, adjusted $R^2 = 0.35$) included a BISON* NDVI$_{int}$ interaction (estimate: 119.6; 95% CI: −196.3, 435.5), along with BISON (estimate: 208.8; 95% CI: 129.8, 287.9), SWES (estimate: 366.2; 95% CI: 241.8, 490.6), and NDVI$_{int}$ (estimate: 125.8; 95% CI: 2.6, 249.0). There was no improvement in the top models when substituting either SWE$_{acc,ij}$ for SWES or NDVI$_{len}$ for NDVI$_{int}$ in the exploratory analysis.

B. Historic Migration Patterns (1970–1971 Through 2005–2006)

Summer counts of the central herd increased from 261 to 3531 bison (1801 ± 161) during 1970–2005 with the maximum number of bison migrating westward to the Madison headwaters area (HISTORIC$_{max}$) ranging from 47 bison in winter 1971–1972 to 1877 in winter 1995–1996 (843 ± 75). There were no presurvey winter removals at the west boundary during 1970–1971 through 1984–1985 and removals thereafter were negligible (≤2% of migrating bison), except during winters 1992–1993 (78 bison removed), 1994–1995 (48 bison), and 1996–1997 (293 bison).

Summer and winter counts of the entire central herd were highly and positively correlated when counts were <2350 bison (slope of line = 0.81; 95% CI = 0.66, 0.95; $R^2 = 0.87$, df = 22). However, this relationship completely disappeared at higher counts (slope of line = −0.05; 95% CI = −0.72, 0.62; $R^2 = 0.003$, df = 9), with fewer bison counted during winter compared to summer surveys (Figure 12.3). Though the number of bison migrating west to the Madison headwaters area consistently increased with population size across the entire range of summer counts (slope of line = 0.44; 95% CI = 0.35, 0.53; $R^2 = 0.76$, df = 32; Figure 12.3), the proportion of migrants remained approximately constant though highly variable (slope of line = 0.002; 95% CI = −0.004, 0.008; $R^2 = 0.02$, df = 32). The maximum proportion of bison on the Madison headwaters range (*i.e.*, HISTORIC$_{max}$/BISON) varied between 0.12 in winter 1970–1971 and 0.75 in winter 1975–1976 (0.46 ± 0.03), but was relatively consistent after winter 1977–1978 with variation between 0.31–0.65 (0.49 ± 0.02; Figure 12.5). When summer population counts exceeded 2350, the total number of bison counted in the

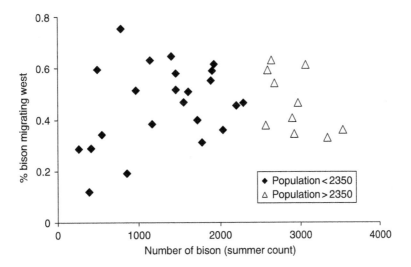

FIGURE 12.5 The proportion of bison migrating to the west-side winter range versus the total central herd population size as determined from the July–August summer count during 1970–1971 through 2005–2006 in Yellowstone National Park, Wyoming, USA.

west-central portion of the park (*i.e.*, Hayden and Pelican valleys, Madison headwaters area) during winter remained approximately constant (slope of line = −0.049; 95% CI = −0.721, 0.662; $R^2 = 0.004$, df = 9; Figure 12.3), even though summer counts of central bison continued to increase to >3500 bison by 2006.

During winters 1970–1971 through 2005–2006, SWE_{acc} in the Hayden and Pelican valleys ranged from 849–5236 cm days (2651 ± 155). The L-integral NDVI index developed for the $NDVI_{int}$ covariate had an $R^2 = 0.79$ and consisted of an intercept of 6116.6 and four covariates and their interactions: (1) average daily precipitation (inches) for April–August (coefficient = −59363.5), (2) average daily maximum temperature (°C) for April–May (coefficient = −4.53), (3) mean daily average temperature (°C) for May (coefficient = 113.64), (4) average daily maximum temperature (°C) for July (coefficient = −149.83), (5) interaction between average daily precipitation for April–August and the average daily maximum temperature for July (coefficient = 2265.97), (6) interaction between the average daily maximum temperature for April–May and the mean daily average temperature for May (coefficient = 24.45), and (7) interaction between the mean daily average temperature for May and the average daily maximum temperature for July (coefficient = 12.40). Predicted $NDVI_{int}$ for summers during 1970–1988 from this model varied between 2184.9 and 3610.2 (2486.0 ± 73.7), while $NDVI_{int}$ for 1989–2005 (obtained from actual NDVI data) varied between 2070.5 and 2687.5 (2341.6 ± 40.9).

Two top approximating models were supported by the data for the historic migration analysis when counts of the central herd were <2350 (Tables 12.1 and 12.2). The top model ($AIC_c = 289.2$, $K = 2$, $w_i = 0.325$, adjusted $R^2 = 0.81$) contained BISON (estimate: 536.4; 95% CI: 417.1, 655.6) and had a relative likelihood of 1.3 compared to the second best model ($\Delta AIC_c = 0.48$, $K = 3$, $w_i = 0.255$, adjusted $R^2 = 0.82$) that included BISON (estimate: 554.3; 95% CI: 434.9, 673.8) and SWE_{acc} (estimate: 95.0; 95% CI: −48.0, 238.0).

C. Estimates of Nutritional Intake

We collected 426 samples of urine-soaked snow from bison in the Hayden Valley ($n = 170$) and Madison headwaters area ($n = 256$; Table 12.4). Mean A:C ratios during five 14-day collection periods between January 1 and March 11 for which we had sufficient samples ($n > 20$) in each wintering area were consistently and significantly lower ($t = 5.12$, df = 4, $P = 0.003$) for bison in the Hayden Valley than bison in the Madison headwaters area (Figure 12.6).

TABLE 12.4 Allantoin:creatinine ratios from urine-soaked snow samples deposited by groups of adult and calf bison during 2005–2006 in the Madison headwaters area and the Hayden Valley of Yellowstone National Park, Wyoming, USA

			Madison headwaters			Hayden Valley		
Period	Start	End	n	Mean	Trimmed mean	n	Mean	Trimmed mean
1	18-Dec	31-Dec	17	0.623	0.641	0		
2	1-Jan	14-Jan	49	0.629	0.629	54	0.512	0.523
3	15-Jan	28-Jan	35	0.710	0.697	25	0.477	0.498
4	29-Jan	11-Feb	39	0.533	0.508	35	0.447	0.451
5	12-Feb	25-Feb	41	0.557	0.571	29	0.343	0.371
6	26-Feb	11-Mar	22	0.898	0.599	26	0.408	0.404
7	12-Mar	25-Mar	28	0.445	0.484	0		
8	26-Mar	8-Apr	25	0.530	0.524	1		

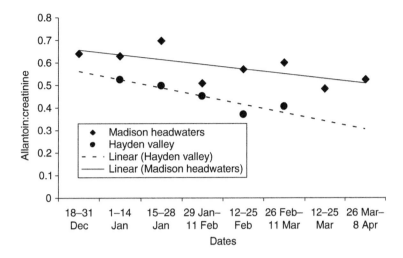

FIGURE 12.6 Allantoin:creatinine ratios from urine-soaked snow samples deposited by groups of adult and calf bison during 2005–2006 in the Madison headwaters area and Hayden Valley of Yellowstone National Park, Wyoming, USA.

IV. DISCUSSION

This study provided unequivocal evidence that bison from the central herd in Yellowstone National Park were partially migratory at least two decades before their abundance approached the estimated food-limiting carrying capacity of their range (Coughenour 2005). We detected substantial migration by central bison to the Madison headwaters area before the Hayden and Pelican valleys were fully occupied, as evidenced by 12–75% of the herd migrating each winter before counts reached 2000 bison. This partial migration was initially detected when bison abundance was low (<300 bison) and the Hayden and Pelican valleys should have provided ample resources for bison year-round. The number of bison migrating to, and remaining in, the Madison headwaters area increased consistently with abundance, but there was no evidence the proportion of migrants increased with population size—though there was substantial variability among years. As a result, the number of animals wintering in the Hayden and Pelican valleys (*i.e.*, nonmigrants) also increased with abundance. Thus, these findings do not support the "domino effect" hypothesis (Meagher 1993, 1998; Gates *et al.* 2005), whereby bison should have fully occupied the Pelican and Hayden valley wintering areas before compensating for decreasing per capita resources by moving westward over the Mary Mountain trail to the Madison headwaters area.

However, after the central herd exceeded 2350 animals the number of bison wintering in the Hayden and Pelican valleys appeared to stabilize, while bison continued to migrate to the Madison headwaters area along the west side of the park (Figure 12.3). These findings provide support for the "density-equalization effect," whereby the area occupied by bison during winter expanded as the central herd increased in abundance (Meagher 1998, Taper *et al.* 2000, Coughenour 2005, Gates *et al.* 2005). Also, the total number of wintering bison counted in the west-central portion of the park (*i.e.*, Hayden and Pelican valleys, Madison headwaters area) remained approximately constant, even though summer counts of central bison continued to increase. Given the high sightability of bison during aerial surveys in this system (Hess 2002), these results suggest some bison migrated outside the west-central portion of the park between the summer and winter counts, perhaps relocating to northern range as hypothesized by Meagher (1998) and Fuller *et al.* (2007). Thus, there may have been an increase in the proportion of migrants as habitat in the central and west-side ranges was occupied during winter and bison began migrating elsewhere. This speculation is supported by ground survey data dating back to winter 1996–1997 that indicate a relatively consistent maximum number of bison occupying the Madison headwaters area each winter, despite an increase in the central herd size by nearly 1000 bison during 1997–2006. Central herd bison have been detected making pre- or early-winter movements to the northern range through the Washburn Range, across the Solfatara Plateau, over the Mirror Plateau, and along the Norris to Mammoth road corridor before significant snow accumulates on the landscape (Meagher 1973, 1993; Bruggeman *et al.* 2007; Olexa and Gogan 2007; R. Wallen, National Park Service, unpublished data).

Population size had a significant, positive effect on the magnitude and timing of migration, with more bison migrating earlier to winter in the Madison headwaters area as density increased (as hypothesized by Meagher 1998). Decreased per capita resources due to increased abundance likely provided an impetus for bison to migrate. Coughenour (2005) reasoned that as the central bison herd grew following the cessation of management removals, they eventually reached a density where nutritional stress was high enough to elicit increased competition for key resources and subsequent behavioral responses to search for additional range. In turn, once new ranges were found, carrying capacity was increased in a positive feedback cycle. Estimates of metabolizable energy intake during the winter of 2005–2006 suggested bison in the Hayden Valley had consistently lower nutrition than bison in the Madison headwaters area, thereby providing an impetus for migration. Density-related movement responses, spurred by limitations in high quality resources, have been documented for bison in other regions (Gates and Larter 1990, Larter *et al.* 2000) and other ungulate populations (Heard and Calef 1986, Messier *et al.* 1988, Reynolds 1998, Ferguson and Messier 2000, Mahoney and Schaefer 2002).

Some of the annual variability in the proportion of bison migrating each winter was explained by density-independent climate covariates. Snow accumulation in the Hayden and Pelican valleys had a positive effect on the timing of migration with more bison moving to the lower-elevation Madison headwaters area as winter progressed and snow pack deepened. Also, the second best model in the historic magnitude of migration analysis included a positive annual winter severity (SWE_{acc}) effect, though its coefficient had confidence intervals slightly spanning zero. Ungulate migrations are affected by snow in many ecosystems owing to the need for easier access to forage (Sweeney and Sweeney 1984, Sabine *et al.* 2002, Doerr *et al.* 2005). Limitations on access to forage by deep or wet snow is the major factor influencing bison foraging behavior across the central winter range in Yellowstone (Bruggeman 2006), and migrations to lower elevations are likely a landscape-scale behavioral response to reduced forage availability caused by increasing snow pack. Also, during winter bison spend the majority of their time foraging, and displacing snow to access forage is a prominent energetic cost (Bjornlie and Garrott 2001, Bruggeman *et al.* 2006, Chapter 27 by Bruggeman *et al.*, this volume). The Madison headwaters area affords easier access to vegetation in winter because geothermally-warmed basins reduce snow accumulation, thereby providing a refuge for bison from nearby areas of deep snow (Figure 12.7). The influence of snow on the number of migrants suggests bison may employ a conditional migration strategy based on climate variability (Hazel *et al.* 2004).

Our findings are certainly not conclusive because analyses were performed on "snap-shots" of distributional patterns for the central bison herd obtained a few times per year. Only recently has intensive monitoring of movements by individual animals been initiated and, to date, sample sizes have been limited (Bruggeman *et al.* 2007, Olexa and Gogan 2007). There is much to be learned about the spatial dynamics of Yellowstone bison from continuing this work, including whether the central and northern herds are becoming more integrated, the differential effects of management removals on each herd given their respective movement dynamics, if individuals have strong fidelity to particular seasonal movement strategies and destinations, and factors influencing specific bison movement patterns and distributions. Also, limited assessments of the central herd from radio-collared animals suggest that the vast majority of bison have vacated the Hayden and Pelican valleys by the end of the winter in recent years—something not evident in the aerial survey data used in our analyses. In addition, we detected a significant influence of forage biomass (*i.e.*, $NDVI_{int}$) on the timing of migration, but the effect was the opposite (*i.e.*, earlier migration with higher forage) of what we predicted. Thus, monitoring and research of montane grasslands to link measurements of forage biomass and phenology with remote sensing indices is needed. The management of bison in and near

FIGURE 12.7 Adult bull bison foraging in reduced snow pack amidst steam in a geothermal area near Midway Geyser Basin in Yellowstone National Park, Wyoming, USA (Photo by Jason Bruggeman).

Yellowstone National Park is one of the most contentious and high-profile issues facing park managers. Thus, we encourage the continuation of research and monitoring efforts to gain insights into these key uncertainties and improve the knowledge base for making effective management decisions.

V. SUMMARY

1. We used aerial and ground data collected during 1970–1971 through 2005–2006 to quantify annual variations in the magnitude and timing of migration by central herd bison, identify potential factors driving this variation, and evaluate the "domino effect" hypothesis (Meagher 1998) that (a) significant migration to the Madison headwaters area did not occur until bison had fully occupied the Hayden and Pelican valleys, and (b) more animals migrated earlier as numbers increasingly exceeded this limit.

2. Bison from the central herd in Yellowstone National Park were partially migratory, with a portion of the animals migrating to the lower-elevation Madison headwaters area during winter while some remained year-round in or near the Hayden and Pelican valleys.

3. Contrary to the "domino effect" hypothesis, there was significant bison migration to the Madison headwaters area before the Hayden and Pelican valleys were fully occupied and abundance approached the food-limiting carrying capacity of these valleys.

4. However, after the central herd exceeded 2350 animals the number of bison wintering in the Hayden and Pelican valleys appeared to stabilize, while bison continued to migrate to the Madison headwaters area. Also, more bison migrated earlier as density increased (as hypothesized by Meagher 1998).

5. Our results suggest some bison migrated outside the west-central portion of the park between the summer and winter counts each year when the central herd exceeded 2350 bison, perhaps relocating to northern range as hypothesized by Meagher (1998) and Fuller *et al.* (2007).

6. Some of the annual variability in the proportion of bison migrating each winter was explained by density-independent climate covariates. The timing and magnitude of bison migration were accentuated during years of severe snow pack that limited access to food.

VI. REFERENCES

Ahola, M., T. Laaksonen, K. Sippola, T. Eeva, K. Rainio, and E. Lehikoinen. 2004. Variation in climate warming along the migration route uncouples arrival and breeding dates. *Global Change Biology* **10**:1610–1617.

Ball, J. P., C. Nordengren, and K. Wallin. 2001. Partial migration by large ungulates: Characteristics of seasonal moose *Alces alces* ranges in northern Sweden. *Wildlife Biology* **7**:39–47.

Bergerud, A. T. 1988. Caribou, wolves and man. *Trends in Ecology and Evolution* **3**:68–72.

Bergerud, A. T., R. Ferguson, and H. E. Butler. 1990. Spring migration and dispersion of woodland caribou at calving. *Animal Behavior* **39**:360–368.

Bjornlie, D. D. 2000. Ecological effects of winter road grooming on bison in Yellowstone National Park. Thesis. Montana State University, Bozeman, MT.

Bjornlie, D. D., and R. A. Garrott. 2001. Effects of winter road grooming on bison in Yellowstone National Park. *Journal of Wildlife Management* **65**:560–572.

Bruggeman, J. E. 2006. Spatio-temporal dynamics of the central bison herd in Yellowstone National Park. Dissertation. Montana State University, Bozeman, MT.

Bruggeman, J. E., R. A. Garrott, D. D. Bjornlie, P. J. White, F. G. R. Watson, and J. J. Borkowski. 2006. Temporal variability in winter travel patterns of Yellowstone bison: The effects of road grooming. *Ecological Applications* **16**:1539–1554.

Bruggeman, J. E., R. A. Garrott, P. J. White, F. G. R. Watson, and R. W. Wallen. 2007. Covariates affecting spatial variability in bison travel behavior in Yellowstone National Park. *Ecological Applications* **17**:1411–1423.

Burnham, K. P., and D. R. Anderson. 2002. *Model Selection and Multi-Model Inference.* Springer, New York, NY.

Burtenshaw, J. C., E. M. Oleson, J. A. Hildebrand, M. A. McDonald, R. K. Andrew, B. M. Howe, and J. A. Mercer. 2004. Acoustic and satellite remote sensing of blue whale seasonality and habitat in the Northeast Pacific. *Deep-Sea Research II* **51**:967–986.

Chen, X. B., Y. K. Chen, M. F. Franklin, E. R. Orskov, and W. J. Shand. 1992a. The effect of feed intake and body weight on purine derivative excretion and microbial protein supply in sheep. *Journal of Animal Science* **70**:1534–1542.

Chen, X. B., G. Grubic, E. R. Ørskow, and P. Osuji. 1992b. Effect of feeding frequency on diurnal variation in plasma and urinary purine derivatives in steers. *Animal Production* **55**:185–191.

Coles, E. H. 1980. *Veterinary Clinical Pathology.* W. B. Saunders, Philadelphia, PA.

Coughenour, M. B. 2005. *Spatial-Dynamic Modeling of Bison Carrying Capacity in the Greater Yellowstone Ecosystem: A Synthesis of Bison Movements, Population Dynamics, and Interactions with Vegetation.* Final report to U.S. Geological Survey Biological Resources Division, Bozeman, MT.

D'Eon, R. G., and R. Serrouya. 2005. Mule deer seasonal movements and multiscale resource selection using Global Positioning System radiotelemetry. *Journal of Mammalogy* **86**:736–744.

Dingle, H. 1996. *Migration: The Biology of Life on the Move.* Oxford University Press, New York, NY.

Dingle, H., W. A. Rochester, and M. P. Zalucki. 2000. Relationships among climate, latitude and migration: Australian butterflies are not temperate-zone birds. *Oecologia* **124**:196–207.

Dobson, A., and M. Meagher. 1996. The population dynamics of brucellosis in the Yellowstone National Park. *Ecology* **77**:1026–1036.

Doerr, J. G., E. J. DeGayner, and G. Ith. 2005. Winter habitat selection by Sitka black-tailed deer. *Journal of Wildlife Management* **69**:322–331.

Ferguson, M. A. D., and F. Messier. 2000. Mass emigration of arctic tundra caribou from a traditional winter range: Population dynamics and physical condition. *Journal of Wildlife Management* **64**:168–178.

Ferrari, M. J. 1999. An assessment of the risk of inter-specific transmission of *Brucella abortus* from bison to elk on the Madison-Firehole winter range. Thesis. Montana State University, Bozeman, MT.

Forchhammer, M. C., N. C. Stenseth, E. Post, and R. Langvatn. 1998. Population dynamics of Norwegian red deer: Density-dependence and climatic variation. *Proceedings of the Royal Society of London B* **265**:341–350.

Fryxell, J. M., and A. R. E. Sinclair. 1988. Seasonal migration by white-eared kob in relation to resources. *African Journal of Ecology* **26**:17–31.

Fuller, J. A., R. A. Garrott, and P. J. White. 2007. Emigration and density dependence in Yellowstone bison. *Journal of Wildlife Management* **71**:1924–1933.

Garel, M., E. J. Solberg, B. Saether, I. Herfindal, and K. Hogda. 2006. The length of growing season and adult sex ratio affect sexual size dimorphism in moose. *Ecology* **87**:745–758.

Garrott, R. A., J. G. Cook, J. G. Berardinelli, P. J. White, S. Cherry, and D. B. Vagnoni. 1997. Evaluation of the urinary allantoin: creatinine ratio as a nutritional index for elk. *Canadian Journal of Zoology* **75**:1519–1525.

Garrott, R. A., L. L. Eberhardt, P. J. White, and J. J. Rotella. 2003. Climate-induced variation in vital rates of an unharvested large-herbivore population. *Canadian Journal of Zoology* **81**:33–45.

Gates, C. C., and N. C. Larter. 1990. Growth and dispersal of an erupting large herbivore population in northern Canada: The Mackenzie wood bison (*Bison bison athabascae*). *Arctic* **43**:231–238.

Gates, C. C., B. Stelfox, T. Muhly, T. Chowns, and R. J. Hudson. 2005. *The Ecology of Bison Movements and Distribution in and Beyond Yellowstone National Park*. University of Calgary, Alberta, Canada.

Gilbert, P. F., O. C. Wallmo, and R. B. Gill. 1970. Effect of snow depth on mule deer in Middle Park, Colorado. *Journal of Wildlife Management* **34**:15–23.

Gordo, O., L. Brotons, X. Ferrer, and P. Comas. 2005. Do changes in climate patterns in wintering areas affect the timing of the spring arrival of trans-Saharan migrant birds? *Global Change Biology* **11**:12–21.

Hazel, W., R. Smock, and C. M. Lively. 2004. The ecological genetics of conditional strategies. *American Naturalist* **163**:888–900.

Heard, D. C., and G. W. Calef. 1986. Population dynamics of the Kaminuriak caribou herd, 1968–1985. *Rangifer Special Issue* **1**:159–166.

Hess, S. C. 2002. Aerial survey methodology for bison population estimation in Yellowstone National Park. Dissertation. Montana State University, Bozeman, MT.

Hulbert, L. B., A. M. Aires-da-Silva, V. F. Gallucci, and J. M. Rice. 2005. Seasonal foraging movements and migratory patterns of female *Lamna ditropis* tagged in Prince William Sound, Alaska. *Journal of Fish Biology* **67**:490–509.

Igota, H., M. Sakuragi, H. Uno, K. Kaji, M. Kaneko, R. Akamatsu, and K. Maekawa. 2004. Seasonal migration patterns of female sika deer in eastern Hokkaido, Japan. *Ecological Research* **19**:169–178.

Jacobson, A. R., A. Provenzale, A. von Hardenberg, B. Bassano, and M. Festa-Bianchet. 2004. Climate forcing and density dependence in a mountain ungulate population. *Ecology* **85**:1598–1610.

Kaňuščák, P., M. Hromada, P. Tryjanowski, and T. Sparks. 2004. Does climate at different scales influence the phenology and phenotype of the River Warbler *Locustella fluviatilis*? *Oecologia* **141**:158–163.

Knight, A., L. P. Brower, and E. H. Williams. 1999. Spring remigration of the monarch butterfly, *Danaus plexippus* (Lepidoptera: Nymphalidae) in north-central Florida: Estimating population parameters using mark-recapture. *Biological Journal of the Linnean Society* **68**:531–556.

Langvatn, R., S. D. Albon, T. Burkey, and T. H. Clutton-Brock. 1996. Climate, plant phenology and variation in age of first reproduction in a temperate herbivore. *Journal of Animal Ecology* **65**:653–670.

Larsen, K. 1972. Creatinine assay by a reaction-kinetic approach. *Clinical Chemistry Acta* **41**:209–217.

Larter, N. C., A. R. E. Sinclair, T. Ellsworth, J. Nishi, and C. C. Gates. 2000. Dynamics of reintroduction in an indigenous large ungulate: The wood bison of northern Canada. *Animal Conservation* **4**:299–309.

Leimgruber, P., W. J. McShea, C. J. Brookes, L. Bolor-Erdene, C. Wemmer, and C. Larson. 2001. Spatial patterns in relative primary productivity and gazelle migration in the eastern steppes of Mongolia. *Biological Conservation* **102**:205–212.

Lundberg, P. 1987. Partial bird migration and evolutionarily stable strategies. *Journal of Theoretical Biology* **125**:351–360.

Maddock, L. 1979. The "migration" and grazing succession. Pages 104–129 *in* A. R. E. Sinclair and M. Norton-Griffiths (Eds.) *Serengeti: Dynamics of an Ecosystem*. University of Chicago Press, Chicago, IL.

Mahoney, S. P., and J. A. Schaefer. 2002. Long-term changes in demography and migration of Newfoundland caribou. *Journal of Mammalogy* **83**:957–963.

Marra, P. P., C. M. Francis, R. S. Mulvihill, and F. R. Moore. 2005. The influence of climate on the timing and rate of spring bird migration. *Oecologia* **142**:307–315.

McNaughton, S. J. 1985. Ecology of a grazing ecosystem: The Serengeti. *Ecological Monographs* **55**:259–294.

Mduma, S. A. R., A. R. E. Sinclair, and R. Hilborn. 1999. Food regulates the Serengeti wildebeest: A 40-year record. *Journal of Animal Ecology* **68**:1101–1122.

Meagher, M. 1973. *The Bison of Yellowstone National Park*. National Park Service Scientific Monograph Series No. 1, National Park Service. US Government Printing, Office, Washington, DC.

Meagher, M. 1993. *Winter Recreation-Induced Changes in Bison Numbers and Distribution in Yellowstone National Park*. Yellowstone National Park, WY.

Meagher, M. 1998. Recent changes in Yellowstone bison numbers and distribution. Pages 107–112 *in* L. Irby and J. Knight (Eds.) *International Symposium on Bison Ecology and Management in North America*. Montana State University, Bozeman, MT.

Messier, F., J. Huot, D. LeHenaff, and S. Lettich. 1988. Demography of the George River caribou herd: Evidence of population regulation by forage exploitation and range expansion. *Arctic* **41**:279–287.

Morgantini, L. E., and R. J. Hudson. 1988. Migratory patterns of the wapiti, *Cervus elaphus*, in Banff National Park, Alberta. *Canadian Field-Naturalist* **102**:12–19.

Mysterud, A., R. Langvatn, N. G. Yoccoz, and N. C. Stenseth. 2001. Plant phenology, migration and geographic variation in body weight of a large herbivore: The effect of a variable topography. *Journal of Animal Ecology* **70**:915–923.

National Resource Conservation Service.2007. Canyon SNOTEL data. http://www.wcc.nrcs.usda.gov/snotel/snotel.pl?sitenum = 384&state = wy. Accessed April 1, 2007.

Nelson, M. E., L. D. Mech, and P. F. Frame. 2004. Tracking of white-tailed deer migration by Global Positioning System. *Journal of Mammalogy* **85**:505–510.

Neter, J., M. H. Kutner, C. J. Nachtsheim, and W. Wasserman. 1996. *Applied Linear Statistical Models*. McGraw-Hill, New York, NY.

Olexa, E. M., and P. J. P. Gogan. 2007. Spatial population structure of Yellowstone bison. *Journal of Wildlife Management* **71**:1531–1538.

Palmer, W. C. 1965. Meteorological Drought. Office of Climatology Research Paper 45, Weather Bureau, Washington, DC.

Pettorelli, N., J. Gaillard, A. Mysterud, P. Duncan, N. C. Stenseth, D. Delorme, G. Van Laere, C. Toigo, and F. Klein. 2006. Using a proxy of plant productivity (NDVI) to find key periods for animal performance: The case of roe deer. *Oikos* **112**:565–572.

Pettorelli, N., A. Mysterud, N. G. Yoccoz, R. Langvatn, and N. C. Stenseth. 2005. Importance of climatological downscaling and plant phenology for red deer in heterogeneous landscapes. *Proceedings of the Royal Society B* **272**:2357–2364.

Pils, A. C., R. A. Garrott, and J. J. Borkowski. 1999. Sampling and statistical analysis of snow-urine allantoin:creatinine ratios. *Journal of Wildlife Management* **63**:1118–1132.

Polovina, J. J., E. Howell, D. R. Kobayashi, and M. P. Seki. 2001. The transition zone chlorophyll front, a dynamic global feature defining migration and forage habitat for marine resources. *Progress in Oceanography* **49**:469–483.

Post, E., and N. C. Stenseth. 1998. Large-scale climatic fluctuation and population dynamics of moose and white-tailed deer. *Journal of Animal Ecology* **67**:537–543.

R Development Core Team. 2006. R: A language and environment for statistical computing. http://www.R-project.org. Accessed September 12, 2006.

Reynolds, P. E. 1998. Dynamics and range expansion of a reestablished muskox population. *Journal of Wildlife Management* **62**:734–744.

Sabine, D. L., S. F. Morrison, H. A. Whitlaw, W. B. Ballard, G. J. Forbes, and J. Bowman. 2002. Migration behavior of white-tailed deer under varying climate regimes in New Brunswick. *Journal of Wildlife Management* **66**:718–728.

Sala, O. E., W. J. Parton, L. A. Joyce, and W. K. Lauenroth. 1988. Primary production of the central grassland region of the Unites States. *Ecology* **69**:40–45.

SAS. 2004. SAS/STAT 9.1 User's Guide. Cary, NC.

Sinclair, A. R. E., H. Dublin, and M. Borner. 1985. Population regulation of Serengeti wildebeest: A test of the food hypothesis. *Oecologia* **65**:266–268.

Stephenson, N. L. 1990. Climatic control of vegetation distribution: The role of the water balance. *American Naturalist* **135**:649–670.

Sweeney, J. M., and J. R. Sweeney. 1984. Snow depths influencing winter movements of elk. *Journal of Mammalogy* **65**:524–526.

Talbot, L. M., and M. H. Talbot. 1963. The wildebeest in western Masailand, East Africa. *Wildlife Monographs* **12**:1–88.

Taper, M. L., M. Meagher, and C. L. Jerde. 2000. *The Phenology of Space: Spatial Aspects of Bison Density Dependence in Yellowstone National Park.* U.S. Geological Service, Biological Resources Division, Bozeman, MT.

Vagnoni, D. B., R. A. Garrott, J. G. Cook, P. J. White, and M. K. Clayton. 1996. Urinary allantoin:creatinine ratios as a dietary index for elk. *Journal of Wildlife Management* **60**:728–734.

Varpe, Ø., Ø. Fiksen, and A. Slotte. 2005. Meta-ecosystems and biological energy transport from ocean to coast: The ecological importance of herring migration. *Oecologia* **146**:443–451.

Watson, F. G. R., W. B. Newman, J. C. Coughlan, and R. A. Garrott. 2006. Testing a distributed snowpack simulation model against spatial observations. *Journal of Hydrology* **328**:453–466.

Whalen, D. M., and B. D. Watts. 2002. Annual migration density and stopover patterns of Northern Saw-whet Owls (*Aegolius acadicus*). *Auk* **119**:1154–1161.

White, P. J., T. L. Davis, K. K. Barnowe-Meyer, R. L. Crabtree, and R. A. Garrott. 2007. Partial migration and philopatry of Yellowstone pronghorn. *Biological Conservation* **135**:518–526.

White, P. J., and R. A. Garrott. 2005. Northern Yellowstone elk after wolf restoration. *Wildlife Society Bulletin* **33**:942–955.

Whitehead, H. 1996. Variation in the feeding success of sperm whales: Temporal scale, spatial scale and the relationship to migrations. *Journal of Animal Ecology* **65**:429–438.

Wittemyer, G., H. B. Rasmussen, and I. Douglas-Hamilton. 2007. Breeding phenology in relation to NDVI variability in free-ranging African elephant. *Ecography* **30**:42–50.

Young, E. G., and C. F. Conway. 1942. On the estimation of allantoin by the Rimini-Schryver reaction. *Journal of Biological Chemistry* **142**:839–853.

APPENDIX

APPENDIX 12A.1	AIC model selection results for the *a priori* models in the magnitudes and timing of migration analyses. All models are ranked according to AIC_c and presented along with the number of parameters (k), the $\triangle AIC_c$ value (*i.e.*, the change in $\triangle AIC_c$ value relative to the best model), and the Akaike weight (w_i)

Model structure	k	ΔAIC_c	w_i
Magnitude of migration, 1996–1997 through 2005–2006			
$\beta_0 + \beta_1(\text{BISON})$	2	0	0.59
$\beta_0 + \beta_1(\text{SWE}_{acc}) + \beta_2(\text{BISON})$	3	2.74	0.15
$\beta_0 + \beta_1(\text{BISON}) + \beta_2(\text{NDVI}_{int})$	3	2.77	0.15
$\beta_0 + \beta_1(\text{SWE}_{acc})$	2	5.68	0.03
$\beta_0 + \beta_1(\text{NDVI}_{int})$	2	6.19	0.03
$\beta_0 + \beta_1(\text{BISON}) + \beta_2(\text{NDVI}_{int}) + \beta_3(\text{BISON*NDVI}_{int})$	4	6.79	0.02
$\beta_0 + \beta_1(\text{SWE}_{acc}) + \beta_2(\text{BISON}) + \beta_3(\text{SWE}_{acc}\text{*BISON})$	4	7.87	0.01
$\beta_0 + \beta_1(\text{SWE}_{acc}) + \beta_2(\text{BISON}) + \beta_3(\text{NDVI}_{int})$	4	8.41	0.01
$\beta_0 + \beta_1(\text{SWE}_{acc}) + \beta_2(\text{NDVI}_{int})$	3	9.03	0.01
$\beta_0 + \beta_1(\text{SWE}_{acc}) + \beta_2(\text{BISON}) + \beta_3(\text{NDVI}_{int}) + \beta_4(\text{BISON*NDVI}_{int})$	5	15.30	0.00
$\beta_0 + \beta_1(\text{SWE}_{acc}) + \beta_2(\text{BISON}) + \beta_3(\text{NDVI}_{int}) + \beta_4(\text{SWE}_{acc}\text{*BISON})$	5	16.37	0.00
Timing of migration, 1996–1997 through 2005–2006			
$\beta_0 + \beta_1(\text{SWES}) + \beta_2(\text{BISON}) + \beta_3(\text{NDVI}_{int})$	4	0	0.43
$\beta_0 + \beta_1(\text{SWES}) + \beta_2(\text{BISON}) + \beta_3(\text{NDVI}_{int}) + \beta_4(\text{SWES*BISON})$	5	0.94	0.27
$\beta_0 + \beta_1(\text{SWES}) + \beta_2(\text{BISON}) + \beta_3(\text{NDVI}_{int}) + \beta_4(\text{BISON*NDVI}_{int})$	5	1.61	0.19
$\beta_0 + \beta_1(\text{SWES}) + \beta_2(\text{BISON})$	3	3.93	0.06
$\beta_0 + \beta_1(\text{SWES}) + \beta_2(\text{BISON}) + \beta_3(\text{SWES*BISON})$	4	4.00	0.06
$\beta_0 + \beta_1(\text{SWES})$	2	20.89	0.00
$\beta_0 + \beta_1(\text{SWES}) + \beta_2(\text{NDVI}_{int})$	3	22.99	0.00
$\beta_0 + \beta_1(\text{BISON})$	2	27.05	0.00
$\beta_0 + \beta_1(\text{BISON}) + \beta_2(\text{NDVI}_{int})$	3	28.69	0.00
$\beta_0 + \beta_1(\text{BISON}) + \beta_2(\text{NDVI}_{int}) + \beta_3(\text{BISON*NDVI}_{int})$	4	30.21	0.00
$\beta_0 + \beta_1(\text{NDVI}_{int})$	2	39.31	0.00
Historic magnitude of migration, 1970–1971 through 2005–2006			
$\beta_0 + \beta_1(\text{BISON})$	2	0	0.33
$\beta_0 + \beta_1(\text{SWE}_{acc}) + \beta_2(\text{BISON})$	3	0.49	0.26
$\beta_0 + \beta_1(\text{BISON}) + \beta_2(\text{NDVI}_{int})$	3	2.69	0.08
$\beta_0 + \beta_1(\text{SWE}_{acc}) + \beta_2(\text{BISON}) + \beta_3(\text{NDVI}_{int})$	4	3.04	0.07
$\beta_0 + \beta_1(\text{SWE}_{acc}) + \beta_2(\text{BISON}) + \beta_3(\text{SWE}_{acc}\text{*BISON})$	4	3.34	0.06
$\beta_0 + \beta_1(\text{DATE}) + \beta_2(\text{SWE}_{acc}) + \beta_3(\text{BISON})$	4	3.46	0.06
$\beta_0 + \beta_1(\text{BISON}) + \beta_2(\text{NDVI}_{int}) + \beta_3(\text{BISON*NDVI}_{int})$	4	3.78	0.05
$\beta_0 + \beta_1(\text{SWE}_{acc}) + \beta_2(\text{BISON}) + \beta_3(\text{NDVI}_{int}) + \beta_4(\text{BISON*NDVI}_{int})$	5	5.28	0.02
$\beta_0 + \beta_1(\text{DATE}) + \beta_2(\text{BISON}) + \beta_3(\text{NDVI}_{int})$	4	5.71	0.02
$\beta_0 + \beta_1(\text{SWE}_{acc}) + \beta_2(\text{BISON}) + \beta_3(\text{NDVI}_{int}) + \beta_4(\text{SWE}_{acc}\text{*BISON})$	5	6.07	0.02
$\beta_0 + \beta_1(\text{DATE}) + \beta_2(\text{SWE}_{acc}) + \beta_3(\text{BISON}) + \beta_4(\text{NDVI}_{int})$	5	6.43	0.01
$\beta_0 + \beta_1(\text{DATE}) + \beta_2(\text{SWE}_{acc}) + \beta_3(\text{BISON}) + \beta_4(\text{SWE}_{acc}\text{*BISON})$	5	6.72	0.01
$\beta_0 + \beta_1(\text{DATE}) + \beta_2(\text{BISON}) + \beta_3(\text{NDVI}_{int}) + \beta_4(\text{BISON*NDVI}_{int})$	5	7.10	0.01
$\beta_0 + \beta_1(\text{DATE}) + \beta_2(\text{SWE}_{acc}) + \beta_3(\text{BISON}) + \beta_4(\text{NDVI}_{int}) + \beta_5(\text{BISON*NDVI}_{int})$	6	9.05	0.00
$\beta_0 + \beta_1(\text{DATE}) + \beta_2(\text{SWE}_{acc}) + \beta_3(\text{BISON}) + \beta_4(\text{NDVI}_{int}) + \beta_5(\text{SWE}_{acc}\text{*BISON})$	6	9.90	0.00
$\beta_0 + \beta_1(\text{DATE})$	2	36.61	0.00
$\beta_0 + \beta_1(\text{NDVI}_{int})$	2	36.65	0.00
$\beta_0 + \beta_1(\text{SWE}_{acc})$	2	37.83	0.00
$\beta_0 + \beta_1(\text{SWE}_{acc}) + \beta_2(\text{NDVI}_{int})$	3	38.83	0.00
$\beta_0 + \beta_1(\text{DATE}) + \beta_2(\text{SWE}_{acc}) + \beta_3(\text{NDVI}_{int})$	4	41.14	0.00

The AIC_c values for the top models of the magnitude and timing of migration analyses during 1996–1997 through 2005–2006 were 135.1 and 1518.17, respectively. The AIC_c value for the top model of the historic magnitude of migration analyses during 1970–1971 through 2005–2006 was 289.18.

CHAPTER 13

Emigration and Density Dependence in Yellowstone Bison

Julie A. Fuller,* Robert A. Garrott,* and P. J. White[†]

*Fish and Wildlife Management Program, Department of Ecology, Montana State University
[†]National Park Service, Yellowstone National Park

Portions of this chapter were taken with kind permission from The Wildlife Society: Journal of Wildlife Management, Emigration and density dependence in Yellowstone bison, 71, 2007, 1924–1933, Fuller, J. A., R. A. Garrott, and P. J. White, ©2007, The Wildlife Society.

Contents

Theme

Density dependence is considered an overriding mechanism for limitation and regulation in large herbivores because increasing numbers of individuals reduce *per-capita* forage intake and affect the nutrition and physiological condition of animals which, in turn, affects survival, reproduction, and movement patterns. Density-independent mechanisms, primarily in the form of stochastic variations in climate, affect forage quantity, quality, and availability, which can also influence population processes. Understanding these processes is fundamental to crafting effective policies for the management of large herbivore communities in protected areas such as national parks. However, discerning the specific mechanisms and relative strengths of density-dependent and density-independent processes and their potential interactions is challenging due to the difficulty of obtaining adequate demographic data across a wide range of population densities. The successful restoration of bison in Yellowstone National Park from a few dozen animals in the remote interior of the park to the thousands of animals that now roam the entire park and beyond provides a unique time series of data to explore

The Ecology of Large Mammals in Central Yellowstone
R. Garrott, P. J. White and F. Watson
ISSN 1936-7961, DOI: 10.1016/S1936-7961(08)00213-3

Copyright © 2009, Elsevier Inc.
All rights reserved.

population processes in a large herbivore. We evaluated a 99-year record of changes in numbers of bison in the park's two herd units from 1902 through 2000, which includes periods of intensive husbandry, protection, and management culls, to gain insights into large herbivore population processes and management.

I. INTRODUCTION

A dominant paradigm in managing large herbivores is that populations increase to peak abundance following introduction to a new range, crash to a lower abundance, and then increase to a carrying capacity lower than peak abundance (Forsyth and Caley 2006). Increasing density regulates ungulate populations through declining per capita forage availability, negatively influencing nutrition and body condition, and decreasing survival and reproductive rates (Sinclair 1975, Caughley 1976, Eberhardt 2002). Stochastic effects of climate (*e.g.,* droughts, snows) can exacerbate these effects by further reducing the availability of forage and/or increasing energetic costs of foraging and locomotion (Clutton-Brock *et al.* 1985, Sæther 1997, Gaillard *et al.* 2000). Several recent reviews of large herbivore dynamics focused on density-related effects to survival and reproduction (Gaillard *et al.* 1998, 2000, Eberhardt 2002, Festa-Bianchet *et al.* 2003), but few studies considered the equally plausible possibility of spatial responses to increasing density (Sæther *et al.* 1999, Amarasekare 2004). Emigration and range expansion have been documented in several large ungulate populations when forage quantity or quality decreased due to density-dependent resource consumption (Lemke *et al.* 1998, Aanes *et al.* 2000, Larter *et al.* 2000, Ferguson *et al.* 2001).

The mechanisms underlying density-dependent feedbacks on population growth of bison (*Bison bison*) in Yellowstone National Park are of special interest to ecologists and park managers. As bison numbers increased from 46 animals in 1902 to nearly 5000 animals in 2005, bison expanded their range and began crossing the park boundary into adjacent areas of Montana (Gates *et al.* 2005). Range expansion was likely a natural response to increasing population density (Bjornlie and Garrott 2001, Gates *et al.* 2005, Chapters 27 and 28 by Bruggeman *et al.*, this volume), but may have been facilitated by the presence of mechanically snow-packed roads for snowmobiles in the central and western areas of Yellowstone National Park that provided energy-efficient travel routes to lower-elevation areas where forage was more readily available; thereby lessening winter mortality and resulting in increased population growth (Meagher 1993). Regardless, range expansion is of great interest because bison may be vectors of brucellosis (*Brucella abortus*) to cattle and a perceived threat to the brucellosis-free status of Montana (Cheville *et al.* 1998, National Park Service 2000).

Understanding the demography of Yellowstone National Park bison is essential for developing feasible conservation strategies and addressing controversies over how and why bison leave Yellowstone National Park. We analyzed multiple competing model formulations of population dynamics using a 99-year time series of bison count and removal data that spanned periods of intensive husbandry, protection, and management culls. Our objective was to evaluate the extent to which bison spatially and numerically respond to increasing density (Cole 1971, Meagher 1973, Dobson and Meagher 1996, Hess 2002). We expected population growth rates during intensive husbandry (1902–1954) would approximate the maximum because bison were not food-limited. We expected bison would respond spatially, rather than numerically, to increasing density during the period of protection within the park (1968–2000). As a result, we did not expect population growth rates to decrease with increasing population size.

II. YELLOWSTONE'S TWO BISON HERDS

Yellowstone's bison population existed almost entirely within the boundaries of the park and consisted of the central and northern herds. These herds were spatially distinct before the 1980s, but recent information suggests interchange may be occurring (Hess 2002, Gates *et al.* 2005). Present-day ranges of the central and northern herds were comparable in size (1200 km^2; Hess 2002); but the herds existed in areas with different plant communities, different precipitation patterns, and different numbers of wintering elk, potential competitors for forage. The range of the northern herd encompassed a decreasing elevation gradient extending approximately 90 km between Cooke City and Gardiner, Montana (Houston 1982, Barmore 2003). The northern range was drier and warmer than the rest of the park, with mean annual precipitation decreasing from 35 cm to 25 cm along the elevation gradient (Houston 1982, Farnes *et al.* 1999, Barmore 2003). Average snow water equivalents ranged from 29.5 cm to 2.0 cm in the higher- and lower-elevation portions of the range, respectively (Farnes *et al.* 1999). Upland grasses comprised the majority of forage in the northern range, followed by sedges (*Carex* spp.) and rushes (*Juncus* spp.; Barmore 2003). Bison shared this range with a large elk herd, which increased from 3200 to >19,000 individuals during 1968–1994, and then decreased to approximately 12,000 individuals by 2002 (White and Garrott 2005).

The range of the central herd extended from the Hayden and Pelican Valleys in the east to the lower-elevation Madison headwaters drainages in the west (Hess 2002). Winter conditions were severe, with snow water equivalents averaging 35.1 cm and temperatures reaching −42 °C (Meagher 1973, Farnes *et al.* 1999). Windswept areas in the upper portions of the Hayden Valley and snow-free geothermal areas throughout the range provided some relief from deep snows and facilitated access to forage (Kittams 1949, Craighead *et al.* 1973, Chapters 4, 5, and 6 by Watson *et al.*, this volume). The central range included a higher proportion of mesic meadows than the northern range, which contained grasses, sedges (*Carex* spp.), and willows (*Salix* spp.), with upland grasses in the drier areas (Craighead *et al.* 1973). The central herd coexisted with an average of 400–800 elk during winter (1965–1998; Craighead *et al.* 1973, Garrott *et al.* 2003, Chapter 11 by Garrott *et al.*, this volume).

Management actions to conserve Yellowstone National Park bison changed as their abundance increased. The northern herd was subject to intense animal husbandry during 1902–1938 to increase their remnant numbers. Park managers rounded up northern herd bison from their summer ranges, confined them, and fed them hay throughout winter in the Lamar Valley (Cahalane 1944, Figure 13.1). Roundups and confinement ceased in 1938, but bison were still baited into the northern range and fed through winter until 1952 (Meagher 1973). Periodic removals were implemented during 1925–1968 to

FIGURE 13.1 After near extirpation of bison in the late nineteenth century and remaining animals in Yellowstone National Park were captured and maintained under intensive husbandry at the "Buffalo Ranch" on the northern range in order to protect them from poaching and to increase population numbers to enhance restoration (Photo courtesy of Allison Backus, YNP Archives).

limit the growth of the bison population (Meagher 1973). The central herd was not subject to intense animal husbandry and remained <100 bison through the mid-1930s. To stimulate population growth, the central herd was augmented with 71 bison from the northern herd in 1936 (Cahalane 1944). Periodic culling was instituted to limit bison numbers in the central herd during 1954–1968 (Meagher 1973).

A new management policy was initiated by Yellowstone National Park in 1969 (Cole 1971). Bison herds were allowed to fluctuate without any direct manipulations (*e.g.*, culling) within the park; allowing a combination of weather, predators, and resource limitation to influence bison numbers. Between, 1984 and 2000, however, the State of Montana culled more than 3000 bison that left the park to prevent the possible transmission of brucellosis from bison to cattle (National Park Service 2000). A cooperative Bison Management Plan between the State of Montana and Yellowstone National Park (National Park Service 2000) allowed continued culling of bison leaving the park.

III. METHODS

A. Data Collection

The count and removal data for Yellowstone National Park bison consisted of two time series when bison were counted regularly: 1902–1954 and 1970–2000 (Figures 13.2 and 13.3). Counts during 1902–1954 were predominantly made from horseback, foot, or skis, although airplanes were also used after 1949 (Meagher 1973). The 1902–1954 counts on the northern herd were essentially population censuses, as bison were held captive in pens through winter until 1938, and supplemental feeding encouraged bison to stay centrally located during the winter until 1952. Little information regarding survey methods for the central herd over 1902–1969 existed, so the quality of these data cannot be determined. The 1954–1969 data were not considered because only nine counts occurred for each herd, counting methods were undocumented, and the 1965–1969 counts were more than 2.5 times lower than the 1970–1974 counts, suggesting that counting methods or areas surveyed were too different for sensible comparison. Counting methods were consistent throughout 1970–2000, when aerial counts of

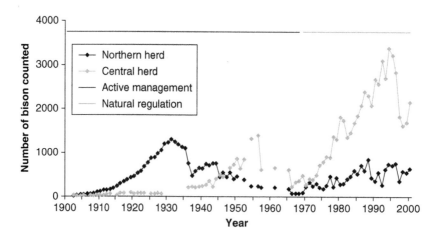

FIGURE 13.2 Population counts for the northern and central bison herds of Yellowstone National Park, Montana and Wyoming, during 1902–2000. Ground surveys were conducted during 1902–1949. Data were coalesced from Cahalane (1944), Kittams (1949), Barmore (1968), and Meagher (1973). Aerial surveys began in 1950 and multiple counts per year were conducted after 1970. We used the highest summertime (June–August) count each year from these aerial surveys (Dobson and Meagher 1996, Hess 2002).

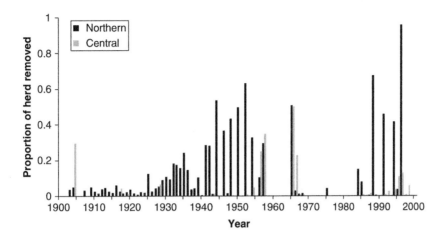

FIGURE 13.3 The proportion of bison removed each year from the central and northern herds of Yellowstone National Park, Montana and Wyoming, during 1902–2000. Data were coalesced from Cahalane (1944), Kittams (1949), Barmore (1968), Meagher (1973), and unpublished National Park Service reports.

FIGURE 13.4 In 1970, the National Park Service initiated aerial surveys of Yellowstone's bison populations using Piper Supercubs. The most accurate surveys were determined to be during the July–August period when bison aggregate for the rut (Hess 2002; Photo by Steven Hess).

all bison (calves and adults) occurred 2–18 times per year (Dobson and Meagher 1996, Hess 2002). However, survey effort was not recorded until 1997, precluding the use of some methods for population estimation.

For each year during 1970–2000, we used the bison count taken during summer months (June through August), a time when bison were highly detectable due to gregarious behavior during the rut (Hess 2002, Figure 13.4). If multiple counts occurred during a given summer, we used the single highest count during June through August. These counts occurred after the birth pulse in each year but before any management removals. Management removals of bison from the northern herd were sporadic and consisted of 1–6% of counted animals per year during 1902–1925. However, larger removals (up to 63% of counted bison) occurred every one or two years during 1926–1968 (Figure 13.3). There were no significant removals from the central herd until the mid-1950s, after which removals of 20–50% occurred sporadically. After 1968, the northern herd grew without removals until 1984, when the

State of Montana and the National Park Service began to cull bison calves and adults attempting to emigrate from the park along the northwestern boundary. The central herd began to emigrate out of the park at the west-central boundary after 1994, and several culling events were imposed on this herd as well. There was no bison hunting except during 1985–1989 when 668 bison were harvested outside the park by hunters and game wardens (State of Montana 1990).

We analyzed data from the central and northern herds separately because they were exposed to different habitat factors, environmental conditions, and management actions (Meagher 1973, Gates *et al.* 2005). We also analyzed the years 1902–1954 separately from 1970–2000 because of the different management paradigms and because herd sizes were large enough to expect to detect density dependence in 1970–2000.

B. Population Models—1902–1954

We did not evaluate density-dependent models for this period because density-related suppression of growth was highly unlikely for either herd. The northern herd received supplemental feeding throughout winter and was periodically culled to keep it at low abundance (Meagher 1973). The central herd began this period at 25 bison and only increased to 61 bison by 1928. There was an eight-year gap in the time series from 1928 to 1936, when the herd was augmented with 71 bison. After augmentation, the herd began to grow rapidly, but densities remained much lower than those eventually reached in the 1990s when the population exceeded 3000. Thus, we assumed bison were not resource limited during 1902–1954 and used exponential growth models to estimate the growth rates for each herd before and after significant management actions: before and after the augmentation of the central herd, and before and after the culling on the northern herd (Table 13.1a,b). We calculated the annual, relative change in the total size of each herd (r_t) as

$$r_t = \log_e(n_t) - \log_e(n_{t-1}) \tag{13.1}$$

TABLE 13.1A Time periods, management activity, best model used to estimate growth rates and associated 95% confidence intervals for central and northern bison herds in Yellowstone National Park, Montana and Wyoming, 1902–1954

Herd	Time	Management	Model	r(95% C.I.)
Northern	1902–1950	Feeding, removals 1926–1950	Exponential	0.16 (0.13, 0.20)
Central	1902–1928	No removals, pre-augment	$\log_e$-linear	0.06 (0.05, 0.07)
Central	1936–1954	No removals, post-augment	$\log_e$-linear	0.10 (0.08, 0.13)

TABLE 13.1B Time periods, management activity, best model used to estimate growth rates and associated equation to estimate growth rate given population size and period (P, an indicator variable), for central and northern bison herds in Yellowstone National Park, Montana and Wyoming, 1970–2000

Herd	Time	Management	Model	Equation
Northern	1970–1981; 1982–2000	Removals 1984 and after, period shift – winter 1982	2-period Ricker	$r_t = 1.157 - 0.004 (n_{t-1}) + 0.238(P) + 0.002(n_{t-1} \times P)$
Central	1970–2000	Removals 1985 and after	1-period Gompertz	$r_t = 1.04 - 0.13 (\log_e(n_{t-1}))$

where n refers to the number of counted individuals and the annual index $t = (1, 2, \ldots, N{-}1)$ (Eberhardt 1987). The 1926–1950 time series of northern herd data was complicated by frequent and substantial removals. We accounted for removals with the modification:

$$r_t = \log_e(n_t) - \log_e(n_{t-1} - RM_{t-1}) \qquad (13.2)$$

where RM_{t-1} represents removals taken after the count at n_{t-1} (Eberhardt 1987). We estimated growth rate and 95% confidence intervals using an equation describing perturbed exponential growth

$$r_t = a + \varepsilon \qquad (13.3)$$

where a represents the growth rate in the absence of density dependence and ε represents the stochastic contribution from noise and un-modeled processes (Zeng *et al.* 1998, Jacobson *et al.* 2004). For the northern herd, we compared the simple model estimating a single growth rate from the entire time series of r_t values (1902–1952) with a two-period model that estimated a separate growth rate for the pre-culling (1902–1925) and culling (1926–1952) periods by including an indicator variable in Eq. (13.3) to designate the two periods ($r_t = a_1 + a_2 P + \varepsilon$). We then compared one- and two-period models using corrected Akaike's Information Criterion (AIC_c) for model selection (Burnham and Anderson 2002).

We conducted the analysis for the central herd differently. Count data for the central herd were uncomplicated by removals over 1902–1954. Further, population sizes were different before (1902–1928) and after (1936–1954) augmentation, and population growth rates could have differed between the periods as well. We tested for differences between these periods using a piecewise $\log_e$-linear regression (Eberhardt 1987, Morris and Doak 2002) and AIC_c for model selection (Burnham and Anderson 2002). As with Eq. (13.2), the $\log_e$-linear model assumes exponential growth, but is advantageous because the linear regression allows residual analysis and uses the $\log_e$ of the population count as the response variable rather than a ratio of counts, which may be more susceptible to observer error. We evaluated three models: (1) a one-intercept, one-growth rate model (all 1902–1954); (2) a two-intercept, one-growth rate model (1902–1928 and 1936–1954); and (3) a two-intercept, two-growth rate model (1902–1928 and 1936–1954; Figure 13.5).

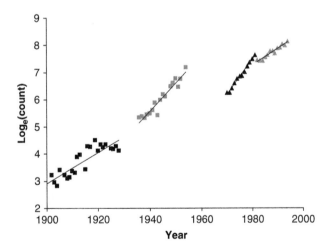

FIGURE 13.5 Piecewise $\log_e$-linear regressions of count data for the central bison herd in Yellowstone National Park, Montana and Wyoming, during 1902–1954 and 1970–1994 when there were few removals. Black squares indicate $\log_e$-counts during 1902–1928; gray squares indicate $\log_e$-counts during 1936–1954; black triangles indicate $\log_e$-counts during 1970–1981; and gray triangles indicate $\log_e$-counts during 1982–1994.

C. Population Models—1970–2000

We conducted preliminary analyses of all count data for the years before removals occurred (1970–1984 for the northern herd and 1970–1994 for the central herd) using $\log_e$-linear regression. This model regressed the $\log_e$ of count data against time, allowed inspection of residuals, and was used to detect abrupt changes in population growth rates (Piepho and Ogutu 2003). The time series of counts for the central herd during 1970–1994 revealed a significant breakpoint at 1982, with the two-intercept, two-slope piecewise regression model being more supported than a continuous model or a two-intercept model ($w_i = 1.0$; Figure 13.5). Based on this finding, we developed a suite of density-dependent and density-independent models for the entire time series (1970–2000) and for two-period models allowing different density-dependent or density-independent dynamics during 1970–1981 and 1982–2000.

We considered two density-dependent model formations to evaluate the relative annual change in total size for each herd over 1970–2000. We calculated r_t using Eq. (13.2), which accounted for time periods with removals and reduced to Eq. (13.1) for periods without removals. The Ricker model assumed linear density dependence,

$$r_t = a + bn_{t-1} + \varepsilon(\text{one period}) \tag{13.4}$$

$$r_t = a_1 + b_1 n_{t-1} + a_2 P + b_2 P n_{t-1} + \varepsilon(\text{two periods}) \tag{13.5}$$

whereas the Gompertz model assumed a decrease in growth rates with $\log_e$-counts.

$$r_t = a + b\big(\log_e(n_{t-1})\big) + \varepsilon(\text{one period}) \tag{13.6}$$

$$r_t = a_1 + b_1\big(\log_e(n_{t-1})\big) + a_2 P + b_2 P(\log_e(n_{t-1})) + \varepsilon(\text{two periods}) \tag{13.7}$$

In both these models, b represents the strength of density dependence and a population is said to exhibit a density-dependent response if b differs significantly from zero (Zeng *et al.* 1998, Jacobson *et al.* 2004). We also considered two density-independent models, including the stochastic growth equation describing perturbed exponential growth [Eq. (13.3)] and a random-walk model where population growth rate is uncorrelated with population size (Zeng *et al.* 1998, Jacobson *et al.* 2004):

$$r_t = \varepsilon \tag{13.8}$$

We explored the possibility that population changes depended on time-delayed dynamics using partial rate correlation functions (PRCF) for all periods without removals (Berryman and Turchin 2001). The results suggested we did not need to consider time lags (*i.e.*, delayed density dependence) >1 year in our analysis. Therefore, our final *a priori* model suite included Gompertz, Ricker, and exponential growth models calculated with and without the estimation of the first-order (AR1) autocorrelation parameter.

We used program R 2.0.0 (R Development Core Team 2004) to fit models and estimate parameter coefficients. We calculated AIC_c values for each model and then ranked and selected the best models using ΔAIC_c values (Burnham and Anderson 2002). Finally, we calculated Akaike weights (w_i) to obtain a measure of model selection uncertainty (Burnham and Anderson 2002). In an analysis such as this, measurement error inflates the variance around the estimated population growth parameter because counts are estimates and may not accurately reflect the true population size. In the case of density-dependent models, this type of variance may result in over-estimation of the strength of density dependence (Shenk *et al.* 1998, Viljugrein *et al.* 2005). We did not expect this would be problematic because bison in Yellowstone National Park are large, gregarious, and inhabit open landscapes, making count accuracy high relative to other herbivores (Hess 2002). To evaluate the level of sampling error, we used an approach recently developed by Staples *et al.* (2004), which uses a mixed-models approach to separating process and sampling error. This method was only available for the exponential model, and so was used for the time periods when exponential growth was thought to have occurred. We deemed this useful because we had no reason to expect sampling error to have changed. Results indicated that growth rate estimates and variances were only slightly inflated by sampling error, suggesting the influence of sampling variance on the interpretation of results was slight.

To explore the potential influence of annual climate variation on bison population dynamics, we evaluated one warm-season and one cold-season climate covariate. We used the Palmer Drought Severity Index (PDSI; Palmer 1968) from the National Climatic Data Center as our warm-season climate covariate because data were available for the length of our time series and this index incorporates multiple environmental factors used to gauge growing conditions across the USA (Alley 1985). We averaged PDSI over the growing season 1 May through 31 July across region one of Wyoming. We predicted a positive correlation between PDSI and relative population change because dry years (*i.e.*, low PDSI) would decrease plant production, thereby decreasing fat reserves for bison entering winter and resulting in lower calf survival. We lagged PDSI one year such that the drought index in $t-1$ affected the annual growth rate for year t. We used the accumulated daily value of snow water equivalent (SWE_{acc}) during 1 October to 30 April as our cold-season climate covariate because it integrates the depth, density, and duration of the snow pack (Garrott *et al.* 2003). SWE_{acc} estimates were available from 1949 to 2000 for the northern range from the Tower Falls CLIM site, and from 1981–2000 for the central range from the Canyon SNOTEL site (Farnes *et al.* 1999). We re-scaled the PDSI covariate by adding seven to each value to remove negative figures and allow a square-root transform, because we expected population growth rates to increase with increasing PDSI, but that growth rates could potentially plateau at higher values of PDSI. We re-scaled SWE_{acc} by dividing it by 1000 to enhance interpretability of coefficients and allow a quadratic transform, because we expected population growth rates to decrease with increasing SWE_{acc}, and that the higher values of SWE_{acc} could have stronger negative effects. We added combinations of our warm- and cold-season covariates to the top ranked models (*e.g.*, $r_t = a + \ldots + c(PDSI) + d(SWE_{acc}) + e(SWE_{acc} \times PDSI) + \varepsilon$) based on the AIC_c model selection results from the density-dependent and density-independent model suite. We used AIC_c to rank models and followed a stepwise model selection procedure to determine if the nonlinear forms of the covariates were more supported by the data than the linear forms (Borkowski *et al.* 2006).

IV. RESULTS

A. Population Models—1902–1954

The piecewise $\log_e$-linear regression model allowing two intercepts and the estimation of two growth rates (1902–1928 and 1936–1950) was the most supported model for the central herd during 1902–1954, receiving 98% of the Akaike model weight (Figure 13.5, Table 13.2). This model estimated the growth rate of the central herd at $\hat{r} = 0.06$ (95% C.I. = 0.05, 0.07; P < 0.01) during 1902–1928 and $\hat{r} = 0.10$ (95% C.I. = 0.08, 0.13; P < 0.01) after the herd was augmented with 71 bison in 1936 ($R^2 = 0.97$, $F_{1,36} = 339.2$; P < 0.01).

The northern herd one-period (1902–1950) and two-period (1902–1925, 1926–1950) models received similar support from the data (one-period: $\Delta AIC_c = 0.0$, $w_i = 0.55$; two-period: $\Delta AIC_c = 0.39$, $w_i = 0.45$). The population growth rate estimate for the one-period model (1902–1950) was $\hat{r} = 0.16$ (95% C.I. = 0.13, 0.20; P < 0.01). The population growth rate estimates for the two-period

TABLE 13.2 Model selection results for regression of factors influencing the $\log_e$ count ($n = 44$) of the central herd bison in Yellowstone National Park, Montana and Wyoming, during 1902–1954

Model	AIC$_c$	ΔAIC$_c$	w_i	R^2
$\log_e(n_t) = \beta_0 + \beta_1 Y + \beta_2 P + \beta_3 PY + \varepsilon$	7.44	0.00	0.98	0.97
$\log_e(n_t) = \beta_0 + \beta_1 Y + \varepsilon$	16.60	9.16	0.01	0.95
$\log_e(n_t) = \beta_0 + \beta_1 Y + \beta_2 P + \varepsilon$	17.78	10.34	0.01	0.95

The lowest AIC$_c$ is the most parsimonious model. Year is represented by Y, period is represented by indicator variable P ($0 = 1902$–1931, $1 = 1936$–1954). The β values represent coefficients estimated by least-squares regression and R^2 is the percent of variation in the data described by the regression model.

model were $\hat{r} = 0.19$ (95% C.I. $= 0.14, 0.23$; $P < 0.01$) for 1902–1925 and $\hat{r} = 0.14$ (95% C.I. $= 0.08, 0.21$; $P = 0.18$) for 1926–1954 ($R^2 = 0.04$, $F_{1,46} = 1.75$).

B. Population Models—1970–2000

The central herd showed evidence of a density-dependent response, with the one-period Gompertz and one-period Ricker models receiving high weights ($w_i = 0.40$ and 0.30, respectively). All other models had weights ≤ 0.08 (Table 13.3). The addition of autocorrelation parameters did not improve the fits of any of the top models. We also found evidence of density dependence in the northern herd during 1970–2000, with the two-period Ricker and two-period Gompertz models receiving nearly equal model weight (Table 13.4, $w_i = 0.45$ and 0.37). Residual analysis identified one influential point from the northern herd (r_{1997}) when growth rate was estimated at 2.44, substantially higher than biologically feasible given reproduction and survival alone (Chapter 14 by Geremia *et al.*, this volume). We censored this point because it was more than three standard deviations from the mean and was highly influential in the models.

For the northern herd, parameter estimates from the two-period Ricker equation were $\hat{a}_1 = 1.16$ (95% C.I. $= 0.63, 1.68$), $\hat{a}_2 = -0.48$ (95% C.I. $= -0.77, -0.19$), $\hat{b}_1 = -0.004$ (95% C.I. $= -0.006, -0.002$), and $\hat{b}_2 = 0.003$ (95% C.I. $= 0.001, 0.005$). The first period was associated with a rapid decrease in growth rates with increasing density, as indicated by the negative value of $\hat{b}_1$ and 95% C.I. that did not encompass zero. There was a lessening of density dependence in the second period, as indicated by the positive value of $\hat{b}_2$. The density dependence term for the second period ($\hat{b}_1 + \hat{b}_2$) was -0.001 (95% C.I. $= -0.002, -0.000$), indicating that density dependence had a stronger effect during 1970–1981 when population counts were lower (182–457), compared to 1982–2000 when population counts were higher (405–756; Figure 13.6).

The one-period Gompertz model was most supported for the central herd, whereas the two-period Ricker model was most supported for the northern herd. We added climate covariates to these models based on the availability of climate data. In both herds, transforming SWE$_{acc}$ to SWE$_{acc}^2$ resulted in a decrease in 1–2 AIC points per model. The square-root transformation was not supported for PDSI. For the central herd, the Gompertz one-period model received most of the model weight (0.79), but the SWE$_{acc}^2$ coefficient indicated a negative correlation with growth rate ($c = -0.007$; 95% C.I. $= -0.013, -0.002$; Table 13.5). For the northern herd, the two-period Ricker model without climate covariates was the top model, receiving 68% of the model weight. All four models containing climate covariates received 32% of total model weight (Table 13.5) and all climate covariates had coefficients overlapping zero, thus providing minimal support for effect of climate on population growth rates in the northern herd.

| TABLE 13.3 | Density-dependent and density-independent model selection results for the central bison herd in Yellowstone National Park, Montana and Wyoming, during 1970–2000 ($n = 30$) |

Model	AIC_c	ΔAIC_c	K	w_i
Gompertz (one-period)	−18.93	0.00	3	0.40
Ricker (one-period)	−18.32	0.61	3	0.30
Ricker (one-period AR1)	−15.65	3.29	4	0.08
Exponential (two-period)	−15.62	3.32	3	0.08
Exponential (two-period AR1)	−14.62	4.31	4	0.05
Exponential (one-period)	−13.59	5.34	2	0.03
Ricker (two-period)	−13.41	5.52	5	0.03
Gompertz (two-period)	−13.37	5.57	5	0.02
Exponential (one-period AR1)	−11.38	7.55	3	0.01
Ricker (two-period AR1)	−10.26	8.67	6	0.01
Random	−9.91	9.02	1	0.00
Gompertz (one-period AR1)	−5.24	13.69	4	0.00
Gompertz (two-period AR1)	−1.21	17.72	6	0.00

One-period models estimate a single growth rate for all data during 1970–2000, whereas two-period models estimate separate growth rates for the 1970–1981 and 1982–2000 periods. AR1 models include a parameter estimating the autocorrelation coefficient for a lag of one year.

| TABLE 13.4 | Density-dependent and density-independent model selection results for the northern bison herd in Yellowstone National Park, Montana and Wyoming, during 1970–2000, censoring 1997 ($n = 29$) |

Model	AIC_c	ΔAIC_c	K	w_i
Ricker (two-period)	13.25	0.00	5	0.45
Gompertz (two-period)	13.69	0.44	5	0.37
Ricker (two-period AR1)	16.33	3.07	6	0.10
Exponential (two-period AR1)	16.92	3.67	4	0.07
Exponential (one-period AR1)	21.41	8.15	3	0.01
Gompertz (one-period)	26.38	13.12	3	0.00
Exponential (one-period)	26.70	13.44	2	0.00
Ricker (one-period)	26.82	13.57	3	0.00
Exponential (two-period)	27.11	13.85	3	0.00
Random	29.33	16.08	1	0.00
Ricker (one-period AR1)	29.49	16.24	4	0.00
Gompertz (two-period AR1)	35.49	22.24	6	0.00
Gompertz (one-period AR1)	35.89	22.64	4	0.00

One-period models estimate a single growth rate for all data during 1970–2000, whereas two-period models estimate separate growth rates for the 1970–1981 and 1982–2000 periods. AR1 models include a parameter estimating the autocorrelation coefficient for a lag of one year.

V. DISCUSSION

Growth rates of the central and northern bison herds varied substantially over time in response to different management, climate, and competitive stressors. Intensive husbandry of the northern herd during 1902–1950 led to a growth rate ($r = 0.16$) similar to the eruptive phases of bison population growth ($r = 0.17$–0.22) in other areas of Montana and Canada (Roe 1951, Fredin 1984, Gates and Larter 1990). Likewise, the augmentation of the central herd with 71 bison in 1936 led to an increased

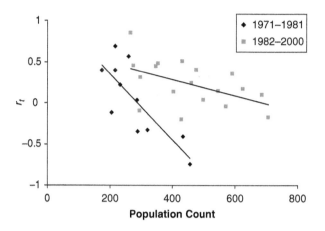

FIGURE 13.6 Graphical representation of the relative strengths of density dependence in the two-period Ricker model for the northern bison herd in Yellowstone National Park, Montana and Wyoming, during 1970–2000. The first period (1970–1981) is represented by black diamonds, while the second period (1982–2000) is represented by gray squares. The slopes of the lines indicate the strength of density dependence.

TABLE 13.5	Warm- and cold-season covariate models for the central bison herd (1981–2000) and the northern bison herd (1970–2000) in Yellowstone National Park, Montana and Wyoming. Weather covariates were added to the one-period Gompertz for the central herd and the two-period Ricker for the northern herd

	Model	AIC_c	ΔAIC_c	K	w_i	R^2
Central herd	One-period Gompertz; no covariates	−18.93	0.00	3	0.79	0.23
	SWE_{acc}^2	−15.89	3.04	4	0.17	0.34
	$SWE_{acc}^2 + PDSI$	−12.28	6.66	5	0.03	0.34
	PDSI	−8.88	10.05	4	0.01	0.07
	$SWE_{acc}^2 + PDSI + SWE_{acc}^2 \times PDSI$	−8.59	10.34	6	0.00	0.36
Northern herd	Two-period Ricker; no covariates	13.25	0.00	5	0.68	0.53
	SWE_{acc}^2	16.34	3.08	6	0.15	0.53
	PDSI	16.45	3.19	6	0.14	0.53
	$PDSI + SWE_{acc}^2$	19.85	6.60	7	0.03	0.53
	$PDSI + SWE_{acc}^2 + PDSI \times SWE_{acc}^2$	22.77	9.52	8	0.01	0.54

growth rate ($r = 0.10$) compared to 1902–1928 ($r = 0.06$). The depressed growth rate prior to augmentation may represent an Allee effect if intense poaching before the conservation period resulted in an unfavorable age or sex structure, or numbers too low for establishing social bonds necessary to facilitate population growth (Allee *et al.* 1949).

Counts for the northern herd during 1900–1930 and the central herd during 1969–1995 were indicative of an irruptive increase in population size following release to a new range or release from harvesting (Forsyth and Caley 2006). However, high-magnitude removals from both herds before they reached peak abundance curtailed natural density-dependent processes. Bison numbers increased rapidly after a moratorium on culling inside the park was instituted in 1969 and we detected evidence of density-dependent feedbacks on the dynamics of both herds. However, the structure of density dependence was different between the herds. Growth rates for the northern herd were strongly depressed by density-dependent processes during 1970–1981, but the influence of density dependence was substantially weaker during 1982–2000 when the number of animals in this herd exceeded that realized during the 1970–1981 period. Conversely, growth rates for the central herd were relatively high during 1970–1981, but decreased during 1982–2000.

These divergent dynamics in herds separated by only 30 km and at similar densities (0.2 bison/km^2) at the start of this period could be due to decreased survival and/or reproduction in the central herd, concurrent with increased survival and reproduction in the northern herd. This explanation seems unlikely, however, given that the only available herd-specific survival and reproductive rates for radio-marked cows during 1995–2001 did not differ significantly (Fuller 2006). Further, recruitment as indexed by calf–adult ratios during 1970–1997 did not significantly differ between herds or pre- and post-1981 periods (Fuller 2006). A more likely explanation for the divergent dynamics is that movement patterns changed, with bison from the central herd emigrating to the northern range. This emigration would have inflated population counts and growth rates of the northern herd, while resulting in opposite effects for the central herd. Winter conditions are known to cause large ungulates to disperse or migrate to find more accessible forage (Aanes *et al.* 2000). Winters are more severe in the central regions of Yellowstone National Park and the drier northern range would be a logical option for dispersing central herd bison. Range expansion in the central herd was documented in the 1980s as the central herd expanded westward into areas that were previously used rarely, if at all (Taper *et al.* 2000 in Gates *et al.* 2005). Central herd bison could also move to the northern range because no ecological barriers existed to the north, but high-elevation ridges and lack of foraging meadows likely blocked dispersal to the east and south. An influx of central herd bison onto the northern range would not have been easily detected because no individual bison in Yellowstone National Park were marked until 1995.

If our hypothesis of a spatial response to density dependence is correct, then growth rates of the central and northern herds should be negatively correlated during 1982–2000 because emigration would decrease central herd growth rate and increase northern herd growth rate. Conversely, we would only expect weak correlation in population growth rates of the two herds during 1970–1981 due to limited or no movements between the herds. These predictions were strongly supported when we examined the correlation using a linear regression model. There was no significant correlation between northern and central herd growth rates during 1970–1981 ($P = 0.17$, $F_{1,9} = 2.3$, $R^2 = 0.20$, slope = -1.09, 95% C.I. = -2.50, 0.32; Figure 13.7), but there was a strong negative correlation during 1982–2000 ($P < 0.01$, $F_{1,16} = 10.5$, $R^2 = 0.40$, slope = -1.13, 95% C.I. = -1.81, -0.45; Figure 13.8). The latter regression does not include the population growth rate estimate for 1997, which was 2.44 for the northern herd and -0.31 for the central herd. This influential point could be indicative of a large number of central herd bison moving to the northern herd in winter 1996–1997 and including this point in the regression increases the strength of the negative slope (-2.38, 95% C.I. = -3.55, -1.21) and the R^2 value (0.48). The hypothesized emigration of central herd animals to the northern herd after 1981 may have been triggered by the combination of relatively high bison densities during the winter of 1981–1982, combined with the most severe snow pack of the decade, resulting in strong resource limitation and starvation.

Other studies have documented strong interactive effects of environmental variation and density on dispersal, migration, vital rates, and population dynamics of ungulates (Sæther 1997, Gaillard *et al.* 1998, 2000, Aanes *et al.* 2000, Larter *et al.* 2000). The negative correlation of population growth rates and the divergent trends in density-dependent responses of the herds provide convincing indirect evidence of increasing emigration from the central herd to the northern range. This interpretation is also supported by count and removal data, which indicate the northern herd sustained the removal of >2000 bison during 1982–2000 even though counts never exceeded 900 bison. In contrast, the central herd sustained only half as many removals (1111 bison) even though it was three times larger (>3000 bison). We contend that the northern herd could not have sustained this high removal rate without immigrants from the central herd. For example, 877 bison were counted on the northern range during 1996 and 725 bison were subsequently removed that winter. The count of northern bison the following

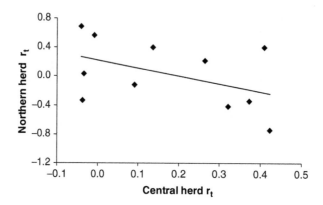

FIGURE 13.7 Correlations between estimated annual population growth rates of the northern and central bison herds in Yellowstone National Park, Montana and Wyoming, during 1970–1981.

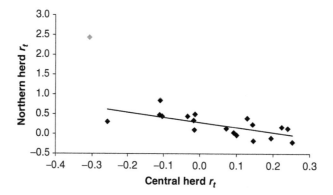

FIGURE 13.8 Correlations between estimated annual population growth rates of the northern and central bison herds in Yellowstone National Park, Montana and Wyoming, during 1982–2000. The 1997 value (gray) was excluded from the regression analysis.

year was 354, an increase of 230%. This increase could not have been realized solely from intrinsic productivity. Hence, substantial immigration must have occurred. Differential removals by sex could influence population growth rates, but the composition of approximately 1809 bison removed at the park boundaries during the winters of 1988–1989 and 1996–1997 indicated approximately equal proportions of males and females. If Yellowstone National Park bison herds have approximately equal sex ratios, as suggested in Shaw and Meagher (2000), then there was no overt bias in removals that would influence population growth rates.

Ungulate populations generally become more sensitive to density-independent factors that affect resource availability as they approach high densities (Sæther 1997, Gaillard *et al.* 1998, 2000). Therefore, we expected exogenous, density-independent processes such as drought and snow pack to have a major influence on the dynamics of both bison herds during 1970–2002. As predicted, the population growth rate of the central herd was negatively correlated with snow pack (SWE$_{acc}$), similar to the findings of numerous studies of large ungulates in relation to winter severity (Gaillard *et al.* 2000, Clutton-Brock and Coulson 2002, Jacobson *et al.* 2004, Wang *et al.* 2006, Chapter 11 by Garrott *et al.*, this volume). We did not observe a negative effect of snow pack on the northern herd, possibly due to influx from central herd bison during or immediately after severe winters. Spring precipitation has been shown to positively affect elk calf recruitment in Yellowstone National Park and surrounding areas (Merrill and Boyce 1991, Coughenour and Singer 1996, Taper and Gogan 2002, Lubow and Smith 2004), as well as ungulate population growth in other biomes (Sinclair 1975, Van Vuren and Bray 1986, Mduma *et al.* 1999, Gaillard *et al.* 2000). However, we found no strong evidence of warm-season drought effect on population growth rates of either the central or northern bison herds. It is possible that the effect of spring and summer precipitation on bison calf survival exists, but the overall population effect was too small to be detected. Further research into these effects would be useful.

Aggressive management intervention and protection during the twentieth century resulted in an approximately 100-fold increase in Yellowstone National Park's bison population from <50 individuals in 1902 to approximately 5000 animals in 2005. As numbers increased, an interaction between density and snow pack eventually led to resource limitation that fueled natural range expansion and the establishment of migratory patterns across much of the park. These same mechanisms also resulted in pulses of climate-induced dispersal from the central herd to the northern range, creating a source-sink dynamic that is contributing to the current controversy about management of bison when they leave the protection of the park and are culled to reduce the potential of brucellosis transmission to cattle (Cheville *et al.* 1998, National Park Service 2000).

VI. SUMMARY

1. Understanding the relative importance of density-dependent and density-independent feedback on population growth is essential for developing management strategies to conserve wildlife. We examined a 99-year time series of annual counts and removals for two bison (*B. bison*) herds occupying northern and central Yellowstone National Park.

2. Aggressive management intervention was effective at recovering bison from 46 animals in 1902 to >1500 animals in 1954. Supplemental feeding of the northern herd facilitated rapid growth ($r = 0.16$) during 1902–1952. Augmentation of the central herd with 71 animals also led to rapid growth over 1936–1954 ($r = 0.10$).

3. In 1969, manipulative management ceased in the park, and we detected evidence of density-dependent changes in population growth rates for both herds during 1970–2000 as numbers increased to >3000 animals. The central herd showed evidence of a constant density-dependent response over 1970–2000. In contrast, density dependence had a stronger effect on the northern herd's growth rate during 1970–1981 than during 1982–2000.

4. We found evidence to suggest that these trends resulted from pulses of emigration from the central herd to the northern range beginning in 1982 in response to resource limitation generated by an interaction between density and severe snow pack.

5. Density-related emigration from the central herd to the northern range has important implications for managers tasked with conserving bison, while reducing the potential risk of brucellosis transmission outside the park.

VII. REFERENCES

Aanes, R., B. E. Sæther, and N. A. Øritsland. 2000. Fluctuations of an introduced population of Svalbard reindeer: The effects of density dependence and climatic variation. *Ecography* **23**:437–443.

Allee, W. C., A. E. Emerson, O. Park, T. Park, and K. P. Schmidt. 1949. *Principles of Animal Ecology.* W.B. Saunders Company, Philadelphia, PA.

Alley, W. M. 1985. The Palmer Drought Severity Index as a measure of hydrologic drought. *Water Resources Bulletin* **21**:105–114.

Amarasekare, P. 2004. The role of density-dependent dispersal in source-sink dynamics. *Journal of Theoretical Biology* **226**:159–168.

Barmore, W. J. 1968. *Bison and Brucellosis in Yellowstone National Park.* July 1968, Yellowstone Archives, Yellowstone National Park, WY.

Barmore, W. J., Jr. 2003. *Ecology of Ungulates and their Winter Range in Northern Yellowstone National Park: Research and Synthesis, 1962–1970.* National Park Service, Mammoth Hot Springs, WY.

Berryman, A., and P. Turchin. 2001. Identifying the density-dependent structure underlying ecological time series. *Oikos* **92**:265–270.

Bjornlie, D. D., and R. A. Garrott. 2001. Effects of winter road grooming on bison in Yellowstone National Park. *Journal of Wildlife Management* **65**:560–572.

Borkowski, J. J., P. J. White, R. A. Garrott, T. Davis, A. R. Hardy, and D. J. Reinhart. 2006. Wildlife responses to motorized winter recreation in Yellowstone National Park. *Ecological Applications* **16**:1911–1925.

Burnham, K. P., and D. R. Anderson. 2002. *Model Selection and Multimodal Inference: A Practical Information-Theoretic Approach.* Springer, New York, NY.

Cahalane, V. H. 1944. Restoration of wild bison. *Transactions of the North American Wildlife Conference* **9**:135–143.

Caughley, G. 1976. Wildlife management and the dynamics of ungulate populations. Pages 183–246 *in* T. H. Coaker (Ed.) *Applied Biology 1.* Academic Press, London, United Kingdom.

Cheville, N. F., D. R. McCullough, and L. R. Paulson. 1998. *Brucellosis in the Greater Yellowstone Area.* National Research Council, Washington, DC.

Clutton-Brock, T. H., and T. Coulson. 2002. Comparative ungulate dynamics: The devil is in the detail. *Philosophical Transactions of the Royal Society of London Series B* **357**:1285–1298.

Clutton-Brock, T. H., M. Major, and F. E. Guinness. 1985. Population regulation in male and female red deer. *Journal of Animal Ecology* **54**:831–846.

Cole, G. F. 1971. An ecological rationale for the natural or artificial regulation of ungulates in parks. *Transactions of the North American Wildlife Conference* **36**:417–425.

Coughenour, M. B., and F. J. Singer. 1996. Elk population processes in Yellowstone National Park under the policy of natural regulation. *Ecological Applications* **6**:573–583.

Craighead, J. J., F. C. Craighead Jr., R. L. Ruff, and B. W. O'Gara. 1973. Home ranges and activity patterns of nonmigratory elk of the Madison drainage herd as determined by biotelemetry. *Wildlife Monographs* **33**.

Dobson, A., and M. Meagher. 1996. The population dynamics of brucellosis in the Yellowstone National Park. *Ecology* **77**:1026–1036.

Eberhardt, L. L. 1987. Population projections from simple models. *Journal of Applied Ecology* **24**:103–118.

Eberhardt, L. L. 2002. A paradigm for population analysis of long-lived vertebrates. *Ecology* **83**:2841–2854.

Farnes, P., C. Heydon, and K. Hansen. 1999. *Snowpack Distribution Across Yellowstone.* National Park. Department of Earth Sciences, Montana State University, Bozeman, MT.

Ferguson, M. A., D. L. Gauthier, and F. Messier. 2001. Range shift and winter foraging ecology of a population of Arctic tundra caribou. *Canadian Journal of Zoology* **79**:746–758.

Festa-Bianchet, M., J. M. Gaillard, and S. D. Côté. 2003. Variable age structure and apparent density dependence in survival of adult ungulates. *Journal of Animal Ecology* **72**:640–649.

Forsyth, D. M., and P. Caley. 2006. Testing the irruptive paradigm of large-herbivore dynamics. *Ecology* **87**:297–303.

Fredin, R. A. 1984. Levels of maximum net productivity in populations of large terrestrial animals. Pages 381–387 *in* W. F. Perrin, R. L. Brownell, Jr., and D. P. DeMaster (Eds.) *Reproduction in Whales, Dolphins, and Porpoises.* Reports of the International Whaling Commission, Special Issue 6, Cambridge, United Kingdom.

Fuller, J. A. 2006. Population demography of the Yellowstone National Park bison herds. Thesis. Montana State University, Bozeman, MT.

Gaillard, J. M., M. Festa-Bianchet, and N. G. Yoccoz. 1998. Population dynamics of large herbivores: Variable recruitment with constant adult survival. *Trends in Ecology and Evolution* **13**:58–63.

Gaillard, J. M., M. Festa-Bianchet, N. G. Yoccoz, A. Loison, and C. Toïgo. 2000. Temporal variation in fitness components and population dynamics of large herbivores. *Annual Review of Ecology and Systematics* **31**:367–393.

Garrott, R. A., L. L. Eberhardt, P. J. White, and J. Rotella. 2003. Climate-induced variation in vital rates of an unharvested large-herbivore population. *Canadian Journal of Zoology* **81**:33–45.

Gates, C. C., and N. C. Larter. 1990. Growth and dispersal of an erupting large herbivore population in northern Canada: The Mackenzie wood bison (*Bison bison athabascae*). *Arctic* **43**:231–238.

Gates, C. C., B. Stelfox, T. Muhley, T. Chowns, and R. J. Hudson. 2005. *The Ecology of Bison Movements and Distribution in and Beyond Yellowstone National Park.* University of Calgary, Alberta, Canada.

Hess, S. C. 2002. *Aerial Survey Methodology for Bison Population Estimation in Yellowstone National Park.* Dissertation. Montana State University, Bozeman, MT.

Houston, D. B. 1982. *The Northern Yellowstone Elk.* MacMillan, New York, NY.

Jacobson, A. R., A. P. Provenzale, A. von Hardenberg, B. Bassano, and M. Festa- Bianchet. 2004. Climate forcing and density dependence in a mountain ungulate population. *Ecology* **85**:1598–1610.

Kittams, W. H. 1949. *Preliminary Report on Hayden Valley Bison Range.* Document 720.04.1, July 1949, Yellowstone Archives, Yellowstone National Park, WY.

Larter, N. C., A. R. E. Sinclair, T. Ellsworth, J. Nishi, and C. C. Gates. 2000. Dynamics of reintroduction in an indigenous larger ungulate: The wood bison of northern Canada. *Animal Conservation* **4**:299–309.

Lemke, T. O., J. A. Mack, and D. B. Houston. 1998. Winter range expansion by the northern Yellowstone elk herd. *Intermountain Journal of Sciences* **4**:1–9.

Lubow, B. C., and B. L. Smith. 2004. Population dynamics of the Jackson elk herd. *Journal of Wildlife Management* **68**:810–829.

Mduma, S. A. R., A. R. E. Sinclair, and R. Hilborn. 1999. Food regulates the Serengeti wildebeest: A 40-year record. *Journal of Animal Ecology* **68**:1101–1122.

Meagher, M. M. 1973. *The Bison of Yellowstone National Park.* National Park Service Scientific Monograph Series Number 1, National Park Service, Washington, DC.

Meagher, M. M. 1993. *Winter Recreation-Induced Changes in Bison Numbers and Distribution in Yellowstone National Park.* National Park Service, Yellowstone National Park, WY.

Merrill, E. H., and M. S. Boyce. 1991. Summer range and elk population dynamics in Yellowstone National Park. Pages 263–273 *in* R. B. Keiter and M. S. Boyce (Eds.) *The Greater Yellowstone Ecosystem.* Yale University Press, New Haven, Connecticut.

Morris, W. F., and D. F. Doak. 2002. *Quantitative Conservation Biology.* Sinauer Press, Sunderland, MA.

National Park Service, U.S. Department of the Interior. 2000. *Winter Use Plans Final Environmental Impact Statement, Volume I, for the Yellowstone and Grand Teton National Parks and John D. Rockefeller, Jr., Memorial Parkway.* National Park Service Intermountain Regional Office, Lakewood, CO.

Palmer, W. C. 1968. Keeping track of crop moisture conditions, nationwide: The new crop moisture index. *Weatherwise* **21**:156–161.

Piepho, H. P., and J. O. Ogutu. 2003. Inference for the break point in segmented regression with application to longitudinal data. *Biometrical Journal* **45**:591–601.

R Development Core Team. 2004. *R: A language and environment for statistical computing.* R Foundation for Statistical Computing, Vienna, Austria. ISBN 3-900051-00-3, <http://www.R-project.org>.

Roe, F. G. 1951. *The North American Buffalo.* University of Toronto Press, Toronto, Ontario, Canada.

Sæther, B. E. 1997. Environmental stochasticity and population dynamics of large herbivores: A search for mechanisms. *Trends in Ecology and Evolution* **12**:143–149.

Sæther, B. E., S. Engen, and R. Lande. 1999. Finite metapopulation models with density-dependent migration and stochastic local dynamics. *Proceedings of the Royal Society of London Series B* **266**:113–118.

Shaw, J. H., and M. M. Meagher. 2000. Bison. Pages 447–464 *in* S. Demarais and P. R. Krausman (Eds.) *Ecology and Management of Large Mammals in North America.* Prentice Hall Press, NJ.

Shenk, T. M., G. C. White, and K. P. Burnham. 1998. Sampling-variance effects on detecting density dependence from temporal trends in natural populations. *Ecological Monographs* **68**:445–463.

Sinclair, A. R. E. 1975. The resource limitation of trophic levels in tropical grassland ecosystems. *Journal of Animal Ecology* **44**:497–520.

Staples, D. F., M. L. Taper, and B. Dennis. 2004. Estimating population trend and process variation for PVA in the presence of sampling error. *Ecology* **85**:923–929.

State of Montana, Montana Department of Fish, Wildlife and Parks, National Park Service and U.S. Forest Service. 1990. *Yellowstone Bison: Background and Issues.* Yellowstone National Park Library Archives, Yellowstone National Park, WY.

Taper, M. L., and P. J. Gogan. 2002. The northern Yellowstone elk: density dependence and climatic conditions. *Journal of Wildlife Management* **66**:106–122.

Van Vuren, D., and M. P. Bray. 1986. Population dynamics of bison in the Henry Mountains, Utah. *Journal of Mammalogy* **67**:503–511.

Viljugrein, H., N. C. Stenseth, G. W. Smith, and G. H. Steinbakk. 2005. Density dependence in North American ducks. *Ecology* **86**:245–254.

Wang, G., N. T. Hobbs, R. B. Boone, A. W. Illius, I. J. Gordon, J. E. Gross, and K. L. Hamlin. 2006. Spatial and temporal variability modify density dependence in populations of large herbivores. *Ecology* **87**:95–102.

White, P. J., and R. A. Garrott. 2005. Yellowstone's ungulates after wolves—Expectations, realizations and predictions. *Biological Conservation* **125**:141–152.

Zeng, Z., R. M. Nowierski, M. L. Taper, B. Dennis, and W. P. Kemp. 1998. Complex population dynamics in the real world: Modeling the influence of time-varying parameters and time lags. *Ecology* **79**:2193–2209.

CHAPTER 14

Demography of Central Yellowstone Bison: Effects of Climate, Density, and Disease

Chris Geremia,* P. J. White,* Robert A. Garrott,[†] Rick W. Wallen,* Keith E. Aune,[‡] John Treanor,* and Julie A. Fuller[†]

*National Park Service, Yellowstone National Park
[†]Fish and Wildlife Management Program, Department of Ecology, Montana State University
[‡]Montana Department of Fish, Wildlife and Parks

Contents

Theme

Over a century of concerted conservation recovered the bison population in Yellowstone National Park from 23 animals in 1901 to 5000 by 2005. This conservation success led to societal conflicts and disagreements among various management entities regarding classic issues of overabundance (Garrott *et al.* 1993), combined with concerns over the risk of brucellosis transmission to domestic livestock when bison migrate out of the park (Cheville *et al.* 1998). As a result, more than 6700 bison have been culled since 1983 as they attempted to leave the park (Gates *et al.* 2005). These large-scale removals are aimed at brucellosis risk management, but likely influence bison demographics and vital rates. The development of rigorously estimated vital rates that incorporate the effects of brucellosis and associated management actions is essential for formulating appropriate management strategies (*e.g.*, vaccination, culling) for long-term bison conservation. These estimates will also contribute to the growing scientific understanding of how climate, disease, and density affect managed ungulate populations. Fuller *et al.* (2007b) found high and consistent adult female survival and lower birth rates in

The Ecology of Large Mammals in Central Yellowstone
R. Garrott, P. J. White and F. Watson
ISSN 1936-7961, DOI: 10.1016/S1936-7961(08)00214-5

Copyright © 2009, Elsevier Inc.
All rights reserved.

brucellosis seropositive Yellowstone bison during 1995–2001. We focused on the central herd and incorporated additional information collected during 2002–2006 to extend those analyses by investigating density, climate, and brucellosis seroprevalence effects on age-specific survival and fecundity.

I. INTRODUCTION

It is widely accepted that increasing density regulates ungulate populations by decreasing per capita resources and, in turn, negatively influencing nutrition, condition, reproduction, and survival (Sinclair 1975; Caughley 1976; Eberhardt 1977, 2002). Stochastic climatic conditions such as drought or snow pack can exacerbate these effects by further limiting the availability of forage and increasing energetic costs (Clutton-Brock *et al.* 1985, Sæther 1997, Gaillard *et al.* 2000). An increase in energy demands during periods of forage limitation likely influences susceptibility to infectious disease. However, the role infectious diseases play in limiting ungulate populations is not as well developed as other density-dependent mechanisms. Disease effects are typically understood in terms of virulence or the degree of harm induced by a parasite, and focus on affecting survival (Ewald 1994). The result is an incomplete understanding of the impacts of diseases that minimally influence mortality rates, but largely affect reproduction (Joly and Messier 2005). Consequently, the limiting role of chronic diseases is often underestimated when evaluating the drivers of vital rates (Jolles *et al.* 2005).

Diseases such as brucellosis that cross the wildlife-livestock and wildlife-human interfaces are of particular interest due to their potential effects on public health and economic well-being (Cheville *et al.* 1998, Godfroid 2002). Brucellosis has largely been eradicated in cattle herds across the United States, but bison and elk in the Greater Yellowstone Ecosystem persist as one of the last reservoirs of infection (Gates *et al.* 2005). Brucellosis is a bacterial disease caused by *Brucella abortus* that may induce abortions or birth of nonviable calves in livestock and wildlife (Davis *et al.* 1990, 1991; Rhyan *et al.* 2001). When livestock are infected, it also results in economic loss from slaughtering infected cattle herds and imposed trade restrictions (Godfroid 2002). Because of the difficulty and cost of controlling disease in wildlife, managers often employ strategies such as aggressive culling that may unintentionally threaten the viability of otherwise healthy populations and are unpalatable to some constituencies (Plumb *et al.* 2007).

After intensively managing bison numbers for 60 years through husbandry and culling, Yellowstone National Park instituted a moratorium on culling ungulates inside the park in 1969 and allowed numbers to fluctuate in response to weather, predators, and resource limitations (Cole 1971). Bison numbers increased rapidly under this policy and since the mid-1980s increasing numbers have moved outside the park during winter where more than 3100 animals were culled by state and federal agencies during 1984–1997 (National Park Service 2000, Gates *et al.* 2005, Fuller *et al.* 2007b). These movements and removals led to claims that bison were overabundant and had degraded their range (Kay 1998, Wagner 2006). In turn, critics called for intensive management to limit the number and distribution of bison in the park, including fencing, fertility control, hunting, and brucellosis test-and-slaughter programs (Hagenbarth 2007, Kay 2007, Schweitzer 2007). In 2000, the federal government and state of Montana agreed to an Interagency Bison Management Plan that established guidelines for managing the risk of brucellosis transmission from bison to cattle by implementing hazing, test-and-slaughter, hunting, and other actions near the park boundary (National Park Service 2000).

Yellowstone bison provide an excellent opportunity to investigate how the demography of an ungulate population is affected by density, climate, and chronic disease. There is little information regarding bison demography due to the near eradication of free-ranging herds during the era of market hunting (Meagher 1973). Also, little is known about the effects of chronic infection with brucellosis on a bovid population since cattle herds in North America are generally destroyed after testing positive for the disease. Elucidating the processes that drive bison demography will help managers formulate appropriate management strategies for Yellowstone bison, including planning brucellosis vaccination strategies, estimating population abundance, and setting culling guidelines (National Park Service 2000).

Our objectives were to: (1) estimate adult female survival, pregnancy, and birthing rates; (2) evaluate the effects of climate, density, and brucellosis exposure on these rates; (3) explore the effects of management removals near the park boundary on adult female survival; and (4) estimate the population growth rate (λ) from these vital rates.

II. METHODS

A. Study Area and Population

After bison were nearly extirpated from the Greater Yellowstone Ecosystem in the early twentieth century, the population was restored through intensive husbandry, protection, and the reintroduction of bison into the Hayden and Firehole valleys (Meagher 1973). Today, Yellowstone bison function in two semi-distinct subpopulations that include the central and northern herds (Meagher 1993, Aune *et al.* 1998, Taper *et al.* 2000, Gates *et al.* 2005, Fuller *et al.* 2007a, Olexa and Gogan 2007). The central herd generally occupies the central plateau of Yellowstone, extending from the Pelican and Hayden valleys with a maximum elevation of 2400 m in the east to the lower-elevation and thermally-influenced Madison headwaters area in the west. Winters are often severe, with snow water equivalents (SWEs) (*i.e.*, mean water content of snow pack) averaging 35 cm and temperatures reaching −42 °C (Meagher 1973, Farnes *et al.* 1999). This area contains a high proportion of mesic meadows comprised of grasses, sedges, and willows, with upland grasses in drier areas (Craighead *et al.* 1973).

Central herd bison congregate in the Hayden valley for the breeding season (15 July–15 August). Most bison move between the Madison, Firehole, Hayden, and Pelican valleys during the remainder of the year. However, some animals travel to the northern portion of Yellowstone and commingle with the northern herd before returning to the Hayden Valley for the subsequent breeding season. Population counts of the central herd varied widely during 1995–2006 because bison that left the park in winter were subject to culling and up to 20% of the total population was removed annually. Counts decreased from 2593 to 1399 bison during 1996–1998 and varied between 2512 and 3531 animals during 2002–2006.

B. Demographic Variables

The Montana Department of Fish, Wildlife, and Parks and the U.S. Geological Survey conducted a study of adult female bison in Yellowstone National Park during 1995–2001 to evaluate survival, pregnancy, and birth rates. The National Park Service completed a similar study during 2002–2006. We coalesced data from these studies and differentiated central herd animals based on their breeding season distribution during mid-July through mid-August. We included bison proximal to the Hayden and Pelican valleys as central herd animals. Our designation of herd based on breeding distribution was different from other recent studies of Yellowstone bison demography that used autumn (Fuller *et al.* 2007b) and winter (Gates *et al.* 2005) distributions. Eighty bison were radio collared and monitored during 1995–2006, with animals entering the study throughout the duration of the project (Figure 14.1). One animal entered the study during the winter of October 1995 through April 1996 (*i.e.*, 1995–1996), three during 1996–1997, 21 during 1997–1998, and one during 1999–2000. The radio collars were removed from all surviving bison in 2001, after which another study began with 14 animals entering during 2001–2002, 17 during 2003–2004, 15 during 2004–2005, and eight during 2005–2006. Bison were captured by immobilization with carfentanil and xylazine (Aune *et al.* 1998) or at handling facilities near the boundary of the park (Figure 14.2; National Park Service 2000). Bison were fitted with mortality-sensing telemetry collars (Lotek, Newmarket, Ontario, Canada; Telonics, Mesa, Arizona, USA) and aged into three classes (3, 4–8, and ≥9-year-olds) by tooth eruption and wear patterns

FIGURE 14.1 Immobilization of an adult female bison in early winter near Slough Creek in Yellowstone National Park (National Park Service photo by Jenny Jones).

FIGURE 14.2 Bison held in a processing pen at the Stephen's Creek capture facility in Yellowstone National Park prior to brucellosis testing during February 2003. Additional bison are seen in the background moving towards the park boundary (National Park Service photo by Jim Peaco).

(Dimmick and Pelton 1996, Fuller 1959). The fourth incisor was collected from a sample of bison during capture to verify age through cementum annuli analysis (Moffitt 1998).

Bison were observed at least once per month to estimate survival, and mortalities were investigated to interpret the cause of death. We divided the year into summer (May–October) and winter (November–April) encounter periods. Bison <2 years old were excluded from analyses and animals captured during 1995–2001 did not enter the risk set until the start of the encounter period following initial collaring. Bison entered the risk set immediately after capture during 2002–2006 because collaring efforts overlapped the beginning of each encounter period. We censored (*i.e.,* excluded) animals that died within 30 days after handling from these analyses as potential capture-related deaths. We assumed the survival status of each bison was known at the beginning and end of every encounter period. Therefore, we right-censored the survival records for collar malfunctions during the encounter period in which they occurred.

Collared bison were captured each year to evaluate pregnancy rates. During 1995–2001, bison were handled during February and pregnancy status was determined through pregnancy-specific protein B serum assays (Haigh *et al.* 1991) and rectal palpation. Vaginal implant telemetry devices were inserted in animals palpated as pregnant. During 2002–2006, bison were captured during November and December and pregnancy status was determined using pregnancy-specific protein B serum assays.

Bison provide an excellent opportunity to evaluate birth rates because they predominantly calve near large groups and in open areas. Also, neonate calves are highly visible since they remain in close association with females (Green *et al.* 1989, Lott 1991). Brucellosis is believed to affect birth rates by inducing an abortion during the first and, possibly, second pregnancy following infection (Cheville *et al.* 1998, Rhyan *et al.* 2001). Consequently, bison fitted with vaginal implants were monitored through the calving period during 1995–2001 to evaluate birth rates. Biologists attempted to locate expelled transmitters and the affiliated female within 24 hours of release. Each transmitter expulsion site was examined for indications of an abortion event and cows were observed to determine if calves were nursing. Radio-collared bison determined to be pregnant during 2002–2006 were located 1–3 times per week beginning in April. Once females were observed with distended udders, they were observed daily to determine if a newborn calf was present. We considered birth successful if biologists observed a birthing event or a live calf in close association with the female. We considered birth unsuccessful if we observed an aborted fetus, stillborn calf, or repeatedly failed to detect a calf associated with the female. Field personnel generally confirmed births within 12–72 h.

C. Bison and Climate Covariates

We defined the annual covariate, BISON, to investigate the regulating effect of density as animals competed for diminishing per capita resources, such as forage. BISON was indexed as the maximum count of the central herd during aerial surveys of the Firehole, Madison, Hayden, and Pelican valleys between July and August (Hess 2002) each year.

We defined a continuous, individual covariate, AGE, using the incisor eruption patterns of bison <5 years old and cementum annuli analysis results for older animals. We also defined the categorical age class covariate, $AGE_{Y/P}$, whereby we differentiated 3-year-old bison from older animals. Bison ≥ 4 years old were coded with the reference value of 0.

It is generally accepted that brucellosis negatively affects reproduction in bison (Davis *et al.* 1990, 1991). However, detection is complicated since the bacteria can persist within lymphoid tissue until late pregnancy when it may replicate and infect the reproductive tract (Enright 1990). Serologic tests are the most cost-effective and reliable diagnostic tool available for assessing infection status in live animals (Sutherland and Searson 1990, Gall *et al.* 2000). However, these tests may be misleading since they provide indirect evidence of infection by detecting antibodies, which are a response to infection (Treanor *et al.* 2007). Roffe *et al.* (1999) were able to culture *Brucella* bacteria from 46% of seropositive Yellowstone bison removed during management culls, suggesting roughly one-half of seropositive bison are actively infected. We collected serum and tested for brucellosis exposure status when bison were captured during February (1995–2001) and November–December (2002–2006; Roffe *et al.* 1999, Rhyan *et al.* 2001). Animals were categorized as seropositive or seronegative based on the results of fluorescence polarization assay, card, buffered antigen plate agglutination, rivanol, complement fixation, standard plate, and standard tube tests performed by the Montana Department of Livestock Diagnostics Laboratory, Bozeman, Montana, USA (Gall *et al.* 2000). We created the categorical covariate, SERO, using these seroprevalence results. Bison testing seronegative were coded with the reference value of 0 and seropositive animals were coded as 1.

Snow is a fundamental limiting factor for ungulates occupying high-elevation, montane environments because it influences the energetic costs of foraging and locomotion. We used a validated snow pack simulation model (Watson *et al.* 2006; Chapter 6 by Watson *et al.*, this volume) to predict daily estimates of average SWE. We averaged SWE values across all 57×57 m pixels within a 99% kernel of bison use generated from year-round aerial surveys during 1997–2006 (Hess 2002). We generated a cold season covariate of accumulated SWE (SWE_{acc}) by adding daily SWE averages during 1 October through 31 April to index snow pack severity (Garrott *et al.* 2003).

We used the normalized differential vegetation index (NDVI) to generate a warm season covariate as a surrogate for primary productivity because it directly assesses the spatial and temporal variability in vegetation growth (Diallo *et al.* 1991, Rasmussen 1998; Chapter 7 by Thein *et al.*, this volume). This remote sensing value was derived from the ratio of red to near-infrared light reflected by the vegetation, and is highly correlated with green biomass (Goward and Prince 1995). Because bison are primarily grazers, we selected nine large meadows that were distributed across the elevation gradient of the central range. We integrated the 14-day composite NDVI values recorded by the AVHRR satellite at a spatial resolution of approximately 1 km^2. We chose to evaluate the absolute integral over each season ($NDVI_{L-int}$), averaged across all meadows.

D. Model Development and Evaluation

We evaluated the strength of evidence in the data for competing *a priori* models describing the variability of response variables for adult female survival, pregnancy, and birth rates. Our general analytical approach was to use logistic multiple regression to fit the *a priori* models to the data using the logit link and derive estimates of covariate coefficients. We scaled the year covariates BISON, SWE_{acc}, and $NDVI_{L-int}$ by dividing each by 1000. We calculated variance inflation factors (VIF), which measure multi-colinearity among variables, and retained those *a priori* models with annual covariate (*e.g.*, BISON, SWE_{acc}, and $NDVI_{L-int}$) combinations with VIF < 6 (Kutner *et al.* 2005). We also assessed the correlations between all pairs of annual covariates in the models that met the VIF criterion and removed those that had correlation coefficients of $R^2 > 0.50$. Akaike's Information Criteria corrected for small sample size (AIC_c) was used as our model selection criterion (Burnham and Anderson 2002). Since we repeatedly collected measurements from the same animals, we calculated the over-dispersion parameter, $\hat{c}$, as the model deviance divided by the degrees of freedom and adjusted AIC_c values to $QAIC_c$ if $\hat{c} >> 1.0$. Akaike model weights (w_i) were used to address model selection uncertainty. We calculated the predictor weight for each covariate as the sum of all model weights including that covariate to gain insight into the relative importance of each predictor. Model-averaged coefficient estimates for each covariate were calculated across all models receiving sufficient weight when there was no clear support for a single model (Burnham and Anderson 2002). We tested the goodness of fit for the most parameterized model from each *a priori* candidate suite using the le Cessie-van Houweilingen test (le Cessie and van Houwelingen 1991, Hosmer *et al.* 1997). All statistical analyses were performed using the R statistical package (R Core Development Team 2006).

While prior analyses of bison demography did not detect seasonal differences in adult female survival (Fuller *et al.* 2007b), the majority of deaths for many large ungulates living in high-latitude environments occur during winter (Sæther 1997, Gaillard *et al.* 2000; Chapter 11 by Garrott *et al.*, this volume). We analyzed winter and summer survival rates independently to evaluate seasonal differences given more years of data. We defined the response variable SSURV to evaluate summer survival during the 6-month period from April through September. We chose to evaluate two response variables for winter survival because analyses were complicated by brucellosis management actions whereby bison were frequently removed from the population after exiting the park. We defined WSURV to evaluate winter survival when records for culled bison were censored, and W_rSURV to evaluate winter survival when these records were treated as deaths.

We considered 22 *a priori* models for the WSURV and W_rSURV responses, including combinations of, and interactions between, the AGE, SWE_{acc}, BISON, and $NDVI_{L-int}$ covariates. When management culls were censored from analyses, we anticipated adult female winter survival (WSURV) would be dependent on age, with bison exhibiting consistently high survival through the majority of their life followed by a period of senescence. Thus, we expected a negative AGE coefficient. We predicted that forage availability would influence survival, with snow pack (SWE_{acc}) negatively affecting over-winter forage availability and forage productivity ($NDVI_{L-int}$) positively affecting forage availability during the following winter. We predicted that animal density (BISON) would negatively affect survival due to decreased per capita forage. In addition to considering all combinations of the main effects, we predicted the effects of snow pack, forage productivity, and density would be more pronounced for older animals. Thus, we considered single interaction models with either an AGE by snow pack (SWE_{acc}), density (BISON), or forage productivity ($NDVI_{L-int}$) term. When management culls were treated as deaths, we anticipated predictors of migration would best explain the variability in adult female survival (W_rSURV). The number of bison culled annually is a function of migration to lower-elevation areas during winter and spring because removals only occur when bison repeatedly attempt to leave the park (Chapter 12 by Bruggeman *et al.*, this volume). Hence, we expected snow pack (SWE_{acc}) and animal density (BISON) would negatively affect survival. We also considered a snow pack (SWE_{acc}) by density (BISON) interaction, predicting that the impetus to move would be exacerbated by increasing snow pack at higher densities.

We considered 11 *a priori* models to evaluate SSURV including combinations of, and interactions between, the AGE, BISON, and $NDVI_{L-int}$ covariates. These models were a subset of the models considered to evaluate WSURV and W_rSURV because all models with the SWE_{acc} covariate were not included. We predicted AGE, BISON, and $NDVI_{L-int}$ would have similar effects on summer survival as during winter.

We increased the age of each female at the time pregnancy status was determined to the next year to reflect her age at calving. We expected to detect delayed sexual maturity with 3-year-old bison having lower pregnancy rates than prime-aged adults. The extent of reproductive senescence in large mammals is variable among species (Eberhardt 2002) and has not been detected in Yellowstone bison (Fuller *et al.* 2007b). To consider both lower pregnancy rates in young and older bison and meet the assumptions of logistic regression, we used a two step approach to evaluate the response variable PREG. First, we used the 2-category $AGE_{Y/P}$ covariate to investigate pregnancy rates in 3-year-olds compared to adult (≥ 4 years old) animals. We excluded all annual covariates due to our limited number of 3-year-olds. We evaluated five *a priori* models that included all combinations of, and interactions between, $AGE_{Y/P}$ and SERO. Since the $AGE_{Y/P}$ covariate was supported (see Section III), we censored the records for 3-year-old animals and developed a second set of models to evaluate PREG for bison ≥ 4 years old. We considered a restricted suite of 17 *a priori* models given the limitations of our data, including all combinations of AGE, SERO, $NDVI_{L-int}$, and BISON. Fuller *et al.* (2007b) found weak evidence during their analysis of the 1995–2001 data that brucellosis exposure may reduce pregnancy rates. Thus, we predicted a negative SERO coefficient. We also considered an age by serology interaction because the preliminary analysis using the 2-category age covariate suggested the negative effects of serology may be exacerbated in bison ≥ 4 years old (see Section III). We assumed bison may experience reproductive senescence and predicted that the AGE coefficient would be negative. We expected that $NDVI_{L-int}$ and BISON would have similar effects on pregnancy as on survival.

We defined the response variable $BIRTH_P$ to evaluate the probability of known pregnant bison being observed with a live calf during the calving period. Three-year-old bison were included in the analysis set and we considered 32 *a priori* models, including all combinations of the AGE, SERO, BISON, SWE$_{acc}$, and NDVI$_{int}$ covariates. Brucellosis negatively influences birth rates because infection of the reproductive track may induce abortion or birth of a nonviable calf (Cheville *et al.* 1998, Rhyan *et al.* 2001). Thus, we anticipated a negative SERO coefficient. We expected AGE, BISON, SWE$_{acc}$, and NDVI$_{L-int}$ would have similar effects on birth as on pregnancy and survival. We considered an AGE by SERO interaction because most bison are exposed to brucellosis at a young age and the effects may wane with time (Cheville *et al.* 1998). Also, the effects of increasing snow pack and density may be more pronounced on older animals. Therefore, we considered AGE by SWE$_{acc}$ and AGE by BISON interactions. However, we decided not to include models incorporating an interaction term in our *a priori* suite because of the limited amount of available data. Instead, we considered these exploratory models *post priori*.

E. Matrix Model

We constructed a prebreeding, age-structured, deterministic Leslie matrix model for female bison using our vital rate estimates. We encountered 18- and 19-year-old animals during boundary management reductions occurring within the timeframe of this study (National Park Service, unpublished data). Thus, we constructed a 21×21 matrix model using age-specific survival estimates and a maximum age of 20 years. Survival estimates were generated using model-averaging techniques for the top models that evaluated adult female survival when management removals were censored. We used the predicted 2-year-old survival estimate for yearling survival. Calf survival was derived from ground-based composition surveys completed annually in July during 2003–2006 (National Park Service, unpublished data). We used the two-sample change-in-ratio methods of Hanson (1963) and Paulik and Robson (1969) to estimate survival as described by Skalski *et al.* (2005). However, the calf survival estimate (0.81) may be inflated because of some unknown mortality that occurs during the first few weeks of life.

Age-specific fecundity was derived as the product of pregnancy and birth estimates. These estimates were generated using model-averaging techniques for the top models evaluating pregnancy and birth for brucellosis seropositive and seronegative bison. Since age-specific pregnancy and birth was only estimated through 12 years, we assumed constant fecundity thereafter. We halved fecundity estimates because available evidence suggests equal sex ratios at birth (Pac and Frey 1991; National Park Service, unpublished data).

To explore the inherent population dynamics of the central herd and then consider the contribution of brucellosis we constructed two separate matrix models and estimates of the growth rate of the population (λ). First, we input fecundity parameters where model averaged estimates were weighted according to the proportion of seropositive (0.48) and seronegative (0.52) adult, female bison in the population as determined from seroprevalence information collected during winter reductions in 2003–2006 (National Park Service, unpublished data). Then, we only input fecundity parameters estimated from seronegative animals.

III. RESULTS

We collected 153 bison-years of age-specific survival observations during summer and 150 bison-years during winter. The average number of bison-years of data was 13.4 per age for bison ≤ 11 years old. For animals 12–15 years of age, we collected 19 bison-years of data during the summer and 16 during the winter. We documented 28 deaths of radio-collared bison during the study, including 19 (68%) during winter and 9 (32%) during summer. Management operations removed 3438 bison, including 370 during 1995–1996 when 26 were culled at the northern boundary and 344 were culled at the western boundary (*i.e.*, north = 26, west = 344), 1083 during 1996–1997 (north = 725, west = 358), 94 during 1998–1999 (west = 94), 6 during 2000–2001 (west = 6), 202 during 2001–2002 (west = 202), 244 during 2002–2003 (north = 231, west = 13), 280 during 2003–2004 (north = 266, west = 14), 115 during 2004–2005 (north = 1, west = 114), and 1044 during 2005–2006 (north = 979, west = 65). Thirteen study animals were removed as part of these management operations, with one (8%) occurring at the western boundary and 12 (92%) occurring at the northern boundary. Four additional bison were culled (north = 3, west = 1) but not included in these totals because the animals were initially handled just prior to removal and, as a result, did not enter the study. We documented six other winter mortalities, including two deaths due to motor vehicle accidents, two drownings, and two deaths due to unknown natural causes. The causes of death for the nine summer mortalities were three due to motor vehicle accidents, two due to late-winter malnutrition, one suspected bear predation, and four deaths due to unknown natural causes.

We collected 101 bison-years of age-specific pregnancy data, including 17 years from 3-year-olds and 79 years from 4- to 11-year-olds. There were 48 seropositive and 53 seronegative animals. We monitored 73 known pregnant animals through the parturition season, including 30 seropositive and 43 seronegative bison. There were five records for seropositive bison 12–15 years old collected from three individuals. Each of these animals was pregnant and later observed with a calf. We censored these records from both analyses due to the limited number of observations per age.

Variations in SWE_{acc} and $NDVI_{L-int}$ during the study period generally spanned the range of variability in the historic data. SWE_{acc} metrics generated using the snow pack simulation model (Chapter 6 by Watson *et al.*, this volume) ranged from 1181 (2000–2001) to 4774 (1996–1997), with a mean of 2448 cm (CV = 0.41) during the study period. The range of these metrics generated annually during 1970–2006 was from 643 (1976–1977) to 4774 (1996–1997), with a mean of 2461 cm (CV = 0.37). The minimum value observed during the study was the second smallest SWE_{acc} in the historic record. The maximum SWE_{acc} of 4774 was substantially greater than the second-highest value in the 37-year time series (*i.e.*, 3893 in 1976–1977). Therefore, we chose to use a SWE_{acc} value of 4000 cm to simulate severe snow pack severity in our model-averaged prediction exercises. $NDVI_{L-int}$ metrics were available since 1989 and the range for the historic data and study period was from 2178 to 2716. The mean values for the historic data and the study period were 2475 (CV = 0.07) and 2511 (CV = 0.07), respectively. The central herd increased in size after the conclusion of within-park reductions during 1969 through 1986 when sporadic management culls reduced the population by up to 20% annually. Population counts ranged from 261 (1970) to a maximum of 3531 (2005). The minimum count during the study period was 1399 bison in 1997.

The correlation coefficient of SWE_{acc} and $NDVI_{L-int}$ was $R^2 = -0.68$, even though each of the year covariates (BISON, SWE_{acc}, and $NDVI_{L-int}$) satisfied the criterion of VIF $\leq$ 6. This negative relationship was not entirely unexpected because higher $NDVI_{L-int}$ values indicate prolonged visible green vegetation, perhaps suggesting delayed establishment of the snow pack and lower SWE_{acc}. We removed models including SWE_{acc} and $NDVI_{L-int}$ as additive main effects from the *a priori* model set evaluating each response variable. Therefore, we considered 18 *a priori* models to evaluate winter survival and 24 models for birth rate. We used AIC_c to evaluate all model suites because we did not detect over-dispersion during any of the modeling exercises.

Evaluation of the *a priori* models for adult female survival when management removals were censored (WSURV) provided no single highly-supported model. There were four models within two AIC_c units and all models evaluated received some model weight (Table 14.1). The predictor weight of the AGE covariate was 0.88, suggesting it provided meaningful information in describing variability in the data. The AGE coefficient was negative, did not span zero ($\hat{\beta}_{AGE} = -0.28$, 95% CI = -0.55, -0.01) in the model only considering AGE, and was stable in other main effects models including the SWE_{acc}, $NDVI_{L-int}$, and BISON covariates. There was some evidence suggested by predictor weights that SWE_{acc} (0.52) and BISON (0.53) negatively affected survival, but coefficients estimates were unstable and confidence intervals spanned zero (Table 14.2). The most general model fit to the data based on the goodness of fit test ($P = 0.61$). To generate our best estimate of age-specific survival probabilities, we model averaged the eight models receiving ≥ 0.05 of w_i, fixing BISON, SWE$_{acc}$, and NDVI$_{L\text{-}int}$ covariates at the median values observed during the study. These results indicated senescence in survival, with consistently high survival for 2–11-year-old bison and progressively lower survival for older animals (Figure 14.3).

TABLE 14.1 Model selection results of *a priori* models for winter survival of adult female bison in the central herd when management culls were censored (WSURV)

Model	K	ΔAIC$_c$	w_i
AGE + SWE$_{acc}$ + BISON + (AGE × SWE$_{acc}$)	5	0.00	0.24
AGE	2	1.19	0.13
AGE + SWE$_{acc}$ + (AGE × SWE$_{acc}$)	4	1.68	0.10
AGE + BISON + (AGE × BISON)	4	1.96	0.09
AGE + SWE$_{acc}$	3	2.14	0.08
AGE + SWE$_{acc}$ + BISON	4	2.49	0.07
AGE + BISON	3	2.73	0.06
AGE + NDVI$_{L\text{-}int}$	3	3.19	0.05
Constant	1	4.07	0.03
AGE + NDVI$_{L\text{-}int}$ + (AGE × NDVI$_{L\text{-}int}$)	4	4.20	0.02
AGE + BISON + NDVI$_{L\text{-}int}$	4	4.60	0.02
BISON	2	5.43	0.02
SWE$_{acc}$	2	5.44	0.02
AGE + NDVI$_{L\text{-}int}$ + BISON + (AGE × NDVI$_{L\text{-}int}$)	5	5.63	0.01
BISON + SWE$_{acc}$	3	5.85	0.01
NDVI$_{L\text{-}int}$	2	5.87	0.01
NDVI$_{L\text{-}int}$ + BISON	3	7.05	0.01

The AIC$_c$ value for the top model was 47.23. K, the number of parameters; AIC$_c$, Akaike's Information Criteria corrected for small sample size; ΔAIC$_c$, the change in AIC$_c$ relative to the best model; and w_i, Akaike weight. Model covariates included the AGE, age of the animal in years; BISON, the maximum summer count of bison in the central herd; SWE$_{acc}$, an index of snow pack severity; and NDVI$_{L\text{-}int}$, an index of annual forage productivity.

TABLE 14.2 Coefficient estimates of all models within two AIC$_c$ units of the most-supported model evaluating the effects of age, snow pack, forage productivity, and bison density on adult female survival of central herd bison when management culls were censored (WSURV)

Model	AGE	BISON	SWE$_{acc}$	(AGE × SWE$_{acc}$)	(AGE × BISON)
AGE + SWE$_{acc}$ + BISON + (AGE × SWE$_{acc}$)	**−1.58 (−2.91, −0.24)**	−1.52 (−3.25, 0.20)	**−6.50 (−12.16, −0.85)**	**0.50 (0.00, 0.99)**	
AGE	**−0.28 (−0.55, −0.01)**				
AGE + SWE$_{acc}$ + (AGE × SWE$_{acc}$)	**−0.99 (−1.93, −0.06)**		−3.39 (−6.92, 0.15)	0.28 (−0.05, 0.61)	
AGE + BISON + (AGE × BISON)	0.49 (−0.42, 1.39)	2.45 (−1.25, 6.15)			−0.31 (−0.68, 0.06)

Values in bold indicate coefficients with 95% confidence intervals that do not span zero. Abbreviations are as in Table 14.1.

Evaluation of the *a priori* models for adult female survival when management removals were included as deaths (W_rSURV) indicated that the additive and interactive models including BISON and SWE_{acc} covariates were most supported and essentially equivalent with model weights of 0.31 and 0.32, respectively (Table 14.3). Predictor weights indicated that the BISON (0.97) and SWE_{acc} (0.79) covariates were the only predictors with reasonable support from the data, with ambiguous support for AGE (0.23) and $NDVI_{L\text{-}int}$ (0.12). The BISON and SWE_{acc} coefficients were negative and estimates and confidence intervals were stable among additive models (Table 14.4). The most general model provided the best fit to the data based on the goodness of fit test ($P = 0.50$). We model-averaged the eight models receiving >0.01 of w_i to illustrate the estimated affects of BISON and SWE_{acc}. We considered BISON values from 500 to 4000 and SWE_{acc} as the minimum (1182 cm), median (2145 cm), and near maximum (4000 cm) values observed during the study. AGE and $NDVI_{L\text{-}int}$ were held at the median of observed values during the study period. The resulting plots demonstrate the combined effects of

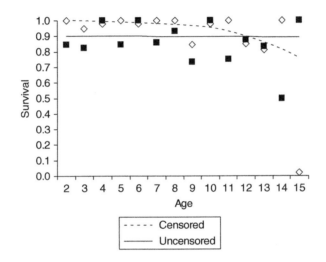

FIGURE 14.3 Age-specific survival estimates for adult female bison in the central herd when management culls were censored and treated as deaths (*i.e.*, uncensored) in Yellowstone National Park during 1995–2006. Estimates were generated using model-averaging techniques for the eight models receiving ≥0.05 and >0.01 of the Akaike model weight (*w_i*) for the censored and uncensored data sets, respectively. Diamonds and squares represent the observed proportion of animals surviving each age class when culls were censored and treated as deaths, respectively.

TABLE 14.3 Model selection results of *a priori* models for winter survival of adult female bison in the central herd when management culls were treated as deaths (W_rSURV)

Model	K	ΔAIC_c	w_i
$BISON + SWE_{acc} + (BISON \times SWE_{acc})$	4	0.00	0.32
$BISON + SWE_{acc}$	3	0.05	0.31
$AGE + SWE_{acc} + BISON$	4	1.94	0.12
$NDVI_{L\text{-}int} + BISON$	3	2.74	0.08
$AGE + SWE_{acc} + BISON + (AGE \times SWE_{acc})$	5	4.03	0.04
BISON	2	4.04	0.04
$AGE + BISON + NDVI_{L\text{-}int}$	4	5.81	0.03
$AGE + BISON$	3	5.95	0.02
$AGE + NDVI_{L\text{-}int} + BISON + (AGE \times NDVI_{L\text{-}int})$	5	6.83	0.01
$(AGE \times BISON)$	4	7.33	0.01

The AIC_c value for the top model was 107.21. Abbreviations are as in Table 14.1.

bison density and snow pack severity on survival when management culls were treated as deaths (Figure 14.4). Since management culls were the overwhelming source of mortality for this data set, these relationships can be interpreted as the influence of bison densities and snow pack severity on bison movements outside of the park boundaries.

We did not detect a significant main effect when evaluating summer survival rates. Thus, we used the null model to predict a single rate of 0.95 ± 0.02. The predicted survival rate from the null model for adult females during winter when culls were censored was 0.96 ± 0.03, compared to a rate of 0.87 ± 0.05 when culls were treated as deaths. We estimated the annual survival rate when removals were censored as 0.91 ± 0.05, compared to 0.83 ± 0.06 when removals were treated as deaths.

Evaluation of the *a priori* models for pregnancy using the 2-category age covariate indicated that model weight was distributed between the $AGE_{Y/P}$ (0.33; $\Delta AIC_c = 0.19$), additive (0.26; $\Delta AIC_c = 0.68$), and interactive (0.36; $AIC_c = 98.10$) models including the $AGE_{Y/P}$ and SERO covariates. There was considerable support that 3-year-old bison had lower pregnancy rates than bison ≥ 4 years old since the $AGE_{Y/P}$ covariate was negative with a confidence interval that did not span zero ($\beta_o + \beta_1(AGE1)$: $\hat{\beta}_{AGE_{Y/P}} = -1.51$, 95% CI $= -2.63, -0.39$). The top supported model included an age by serology interaction, suggesting that the negative effect of serology on pregnancy was most apparent in bison ≥ 4 years old ($\beta_o + \beta_1(AGE_{Y/P}) + \beta_2(SERO) + \beta_3(AGE_{Y/P} \times SERO)$: $\hat{\beta}_{AGE_{Y/P}} = -2.53$, 95% CI $= -4.20$, -0.86, $\hat{\beta}_{SEROLOGY} = -1.24$, 95% CI $= -2.51, 0.03$, $\hat{\beta}_{AGE_{Y/P} \times SEROLOGY} = -1.98$, 95% CI $= -0.36, 4.31$).

TABLE 14.4 Coefficient estimates of all models within two AIC_c units of the most-supported model evaluating the effects of age, snow pack, forage productivity, and bison density on adult female survival of central herd bison when management culls were treated as deaths (W_rSURV)

Model	AGE	BISON	SWE_{acc}	(BISON $\times$ SWE_{acc})
BISON + SWEacc + (BISON $\times$ SWE_{acc})		1.62 (-2.29, 5.53)	1.27 (-2.26, 4.80)	-1.14 (-2.74, 0.46)
BISON + SWE_{acc}		**-1.20** **(-2.00, -0.40)**	**-1.22** **(-2.24, -0.20)**	
AGE + SWE_{acc} + BISON	-0.04 (-0.19, 0.11)	**-1.20** **(-2.01, -0.39)**	**-1.23** **(-2.26, -0.20)**	

Values in bold indicate coefficients with 95% confidence intervals that do not span zero. Abbreviations are as in Table 14.1.

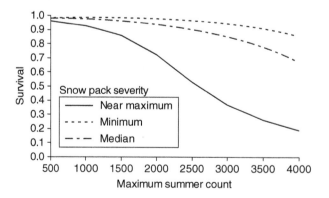

FIGURE 14.4 Survival estimates for adult female bison in the central herd when management culls were treated as deaths in Yellowstone National Park during 1995–2006. Predicted relationships were generated using model-averaging techniques for the eight models receiving >0.01 of the Akaike model weight (w_i). The three lines represent predictions assuming the minimum, median, and near maximum snow pack severities (SWE_{acc}) observed during the study.

For the data set that censored the 3-year-old animals, we found the additive and interactive models including the AGE and SERO covariates were most supported (Table 14.5). Predictor weights suggested that the SERO (0.81) and AGE (0.58) covariates provided useful information in describing the variability in the data, while the effects of BISON (0.35) and $NDVI_{L-int}$ (0.32) were ambiguous. The coefficient estimates and confidence intervals for both the SERO and AGE covariates were stable among additive models (Table 14.6). As predicted, the SERO coefficient was negative with a confidence interval that did not span zero, indicating that bison exposed to brucellosis had lower probability of pregnancy. The AGE covariate was positive with a confidence interval that marginally spanned zero, indicating that probability of pregnancy likely increased with age. Therefore, we found no evidence for reproductive senescence. The most general model that excluded the categorical seroprevalence covariate fit to the data ($P = 0.65$) based on the le Cessie-van Houweilingen test. We model-averaged the eight models within two AIC_c units of the most-supported model to illustrate the differences in age-specific pregnancy curves for seropositive and seronegative animals while holding $NDVI_{L-int}$ and BISON at the median observed values during the study (Figure 14.5).

Results of our evaluation of the probability of a successful birth given the female was known to be pregnant were ambiguous with the null model receiving the most model weight (0.13) and six models ranked within two AIC_c units of the top-supported model (Appendix 14A.1). All models received some model weight and the confidence intervals of all main effect regression coefficients spanned zero. However, there was weak support in the data for AGE and SERO effects. The predictor weight of the AGE covariate was 0.44 with a coefficient estimate that was positive, stable, and did not change sign across the main effects models. The SERO predictor weight was 0.42 with a negative coefficient estimate that also did not change sign across the models including main effect terms, despite a confidence interval that spanned zero. The most general model that excluded the categorical seroprevalence covariate fit to the data ($P = 0.92$). Our *post priori* analysis provided additional insight into the influence of AGE and SERO on birth rates because the interactive AGE × SERO model was 0.65 AIC_c units lower than the most-supported *a priori* model. The predictor weights of AGE and SERO

TABLE 14.5 Model selection results of *a priori* models for pregnancy rates of central herd bison ≥4 years old

Model	K	ΔAIC_c	w_i
AGE + SERO + (AGE × SERO)	4	0.00	0.17
AGE + SERO	3	0.61	0.13
SERO	2	0.87	0.11
SERO + BISON	3	1.70	0.07
AGE + SERO + BISON	4	1.83	0.07
SERO + BISON + $NDVI_{L-int}$	4	1.84	0.07
AGE + SERO + $NDVI_{L-int}$	4	1.84	0.07
SERO + $NDVI_{L-int}$	3	1.99	0.06
AGE + SERO + BISON + $NDVI_{L-int}$	5	2.21	0.06
Constant	1	2.70	0.05
BISON	2	3.36	0.03
AGE	2	3.55	0.03
$NDVI_{L-int}$	2	4.28	0.02
BISON + $NDVI_{L-int}$	3	4.28	0.02
AGE + BISON	3	4.48	0.02
AGE + $NDVI_{L-int}$	3	5.21	0.01
AGE + BISON + $NDVI_{L-int}$	4	5.57	0.01

The AIC_c value for the top model was 70.00. Model covariates included AGE, age of the animal in years; SERO, brucellosis serological status; BISON, the maximum summer count of bison in the central herd; SWE_{acc}, an index of snow pack severity; and $NDVI_{L-int}$, an index of annual forage productivity. Abbreviations are as in Table 14.1.

TABLE 14.6 Untransformed coefficient estimates of all models within two AIC$_c$ units of the top supported model evaluating the effects of age (AGE), seroprevalence (SERO), forage productivity (NDVI$_{L-int}$), and density (BISON) on pregnancy rates of central herd bison $\geq$4 years in Yellowstone National Park during 1995–2006

Model	AGE	SERO	BISON	NDVI$_{L-int}$	AGE × SEROLOGY
AGE + SERO + (AGE × SERO)	0.95 (−0.29, 2.20)	2.96 (−3.24, 9.16)			−0.85 (−2.13, 0.43)
AGE + SERO	0.21 (−0.07, 0.50)	**−1.46 (−2.78, −0.13)**			
SERO		−1.24 (−2.52, 0.04)			
SERO + BISON		−1.23 (−2.52, 0.05)	−0.45 (−1.21, 0.32)		
AGE + SERO + BISON	0.20 (−0.08, 0.48)	**−1.43 (−2.77, −0.10)**	−0.39 (−1.17, 0.38)		
SERO + BISON + NDVI$_{L-int}$		**−1.41 (−2.75, −0.06)**	−0.64 (−1.50, 0.21)	6.12 (−2.32, 14.55)	
AGE + SERO + NDVI$_{L-int}$	0.21 (−0.07, 0.49)	**−1.55 (−2.91, −0.20)**		3.95 (−3.80, 11.70)	
SERO + NDVI$_{L-int}$		**−1.35 (−2.66, −0.04)**		3.90 (−3.55, 11.36)	

Values in bold indicate coefficients with 95% confidence intervals that do not span zero. Abbreviations are as in Table 14.1.

increased to 0.52 and 0.50, respectively. We created model-averaged estimates of age-specific birth rate for seropositive and seronegative animals using the *post priori* AGE × SERO model and the top five *a priori* models receiving ≥0.05 of the model weight. SWE_{acc}, BISON, and $NDVI_{L-int}$ covariates were held at the median observed value during the study. These results indicated that birth rates were high and consistent for seronegative animals, but lower for 3–8-year-old seropositive bison (Figure 14.6). The combined effect of seroprevalence on pregnancy and birth rates resulted in lower calf production of seropositive bison across all ages (Figure 14.7).

The initial matrix model incorporating age-specific survival (0.34–0.99; Figure 14.3) and fecundity (0.18–0.33, Figure 14.7) weighted for seropositive (0.48) and seronegative (0.52) bison estimated $\lambda = 1.12$. The model using fecundity estimates (0.19–0.36) for just seronegative bison estimated

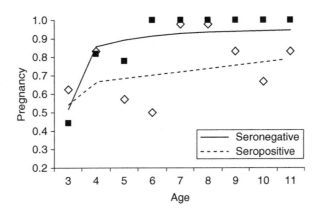

FIGURE 14.5 Age-specific pregnancy rate estimates for seronegative and seropositive bison in the central herd at Yellowstone National Park during 1995–2006. Estimates were generated using model-averaging techniques for three models receiving >0.03 and eight models receiving ≥0.06 of the Akaike model weight (w_i) for 3-year-old and ≥4-year-old animals. Diamonds and squares represent the observed proportion of seropositive and seronegative pregnant animals in each age class, respectively.

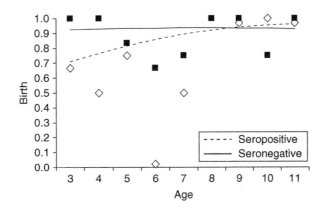

FIGURE 14.6 Age-specific birth rate estimates of known pregnant seronegative and seropositive bison in the central herd at Yellowstone National Park during 1995–2006. Estimates were generated using the AGE × SEROLOGY model and top five *a priori* models receiving ≥0.05 of the Akaike model weight (w_i). Diamonds and squares represent the observed proportion of seropositive and seronegative pregnant animals in each age class, respectively.

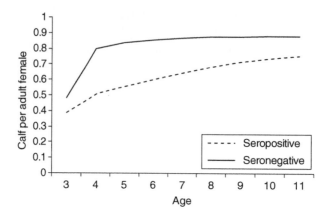

FIGURE 14.7 Age-specific calf production estimates of seronegative and seropositive bison ≥3 years of age generated from the product of model-averaged predictions of pregnancy and birth rates of central herd bison in Yellowstone National Park during 1995–2006.

$\lambda = 1.14$. Therefore, the growth rate of a brucellosis-free central herd could be >15% higher than the current population where approximately one-half of reproductively mature females exhibit exposure.

IV. DISCUSSION

Our analyses strongly support that large-scale removals aimed at managing brucellosis risk to cattle and chronic infection with the *Brucella* pathogen affect the demography of central herd bison. Adult female survival is highly age-dependent and management culls provide the overwhelming source of mortality, but were not age-specific. Likewise, our fecundity analyses provide intriguing and novel insights into the limiting effects of chronic brucellosis infection on the demography of Yellowstone bison. We detected what appear to be pronounced effects of chronic brucellosis infection on pregnancy rates. We found lesser effects on birth rates, and *Brucella*-induced abortions were uncommon.

We detected senescence in survival beginning after 12 years of age. In contrast, younger animals had relatively high survival typical of long-lived mammals that are not hunted (Eberhardt 2002). The proximate cause of senescence in large herbivores is believed to be tooth wear that decreases the ability of animals to efficiently crop forage and reduce plant material into small enough particles to facilitate efficient digestion by rumen microbes (Laws 1981, Van Soest 1994). The nutritional value of winter diets cannot meet maintenance requirements since nearly all plants used as forage by large herbivores in temperate and high latitudes are dormant during the cold season (Hobbs *et al.* 1981). Hence, sub-maintenance forage quality, combined with reduced forage availability and increased energetic costs due to snow pack (Parker and Robbins 1984, Wickstrom *et al.* 1984), contribute to prolonged nutritional deprivation (DelGuidice *et al.* 1994; Chapter 9 by White *et al.*, this volume). The compromised dentition of senescent animals exacerbates the level of nutritional deprivation and results in their survival being more sensitive to decreased food availability during the winter than prime-aged animals with intact teeth (Chapter 11 by Garrott *et al.*, this volume).

The onset of senescence in survival at approximately 12 years of age (Figure 14.3) was corroborated by age at death information collected during historic within-park culling operations (Meagher 1973) and recent boundary removals (Pac and Frey 1991). Senescence began at a surprisingly young age compared to the documented life span of >20 years for wild bison populations (McHugh 1958, Fuller 1966, Halloran 1968, Berger and Peacock 1988). The shortened longevity of central Yellowstone bison may result from the unique regional geology of Yellowstone's volcanic caldera, with its elementally-enriched geothermal waters and rhyolite soils. Similar to elk in this area, central Yellowstone bison likely ingest high levels of fluoride and highly-abrasive silica through consumption of water, plants, and soil particles (Garrott *et al.* 2002; Chapter 10 by Garrott *et al.*, this volume). Excessive dietary fluoride ingested while permanent teeth are developing in young animals interferes with matrix formation and mineralization of teeth, resulting in dental lesions and uneven and excessively rapid tooth wear (Shupe *et al.* 1984, Fejerskov *et al.* 1994, Kierdorf *et al.* 1996), which would undoubtedly be exacerbated by the abrasive action of silica. The pathological consequences of fluoride toxicosis were evident in many of the mandibles recovered from carcasses of adult bison in the Madison headwaters area (Chapter 16 by Becker *et al.*, this volume).

Brucellosis exposure indirectly lowered bison survival because more bison were culled due to concerns regarding transmission to cattle when bison attempted to migrate to lower-elevation areas outside the park. The magnitude and timing of migration by bison in the central herd from high-elevation summer range to lower-elevation wintering areas was driven by density and exacerbated by snow pack severity (Bruggeman 2006; Chapter 12 by Bruggeman *et al.*, this volume), similar to ungulates in many temperate areas (Maddock 1979, Bergerud 1988, Fryxell and Sinclair 1988). We detected a significant decrease in adult female survival when the number of bison in the central herd exceeded 2000–2500 animals and this decrease was exacerbated during winters with severe snow pack. Removals of Yellowstone bison were not age-specific because animals apparently migrated to the park boundary and were culled regardless of age. Thus, culls generally reflected the relative proportion of bison in each age class, similar to late-season elk hunts in the region (White and Garrott 2005).

Our finding that seropositive animals had significantly lower pregnancy rates across all age classes compared to seronegative bison was unexpected, but did not appear spurious because it was supported by two independent data sets (*i.e.*, animals radio-collared during 1995–2001 and 2002–2006; $n = 101$ bison-years), as well as animals captured in the Stephen's Creek pen near Gardiner, Montana ($n = 203$) during February–March 2004 (Appendix 14A.2). The majority of our understanding of brucellosis in bison is drawn from research on cattle with the assumption that the disease functions similarly in both species (Davis *et al.* 1990, Roffe *et al.* 1999, Rhyan *et al.* 2001). *B. abortus* is known to infect the reproductive tract, inducing abortion during the first and, sometimes, second pregnancies following infection (Enright 1990, Cheville *et al.* 1998). However, evidence supporting that *B. abortus* affects conception is lacking (Cheville *et al.* 1998). Thus, our findings do not directly implicate brucellosis infection as the causal mechanism for lower pregnancy rates in seropositive animals, but suggest this disease may play a role in reducing fecundity in chronically infected bison.

The potential mechanisms negatively affecting conception are difficult to ascertain since infectious birth tissues are believed to be the primary source of transmission. However, the effect of chronic diseases (Jolles *et al.* 2005, Joly and Messier 2005), parasite loading (Albon *et al.* 2002, Stien *et al.* 2002), and food availability (Caron *et al.* 2003) have been shown to decrease pregnancy rates in large ungulates. While *B. abortus* is less virulent than *Mycobacterium bovis* that causes bovine tuberculosis and decreases fecundity in wood buffalo (Joly and Messier 2005), morbidity from *Brucella* infection may be substantial enough to cause the diminished pregnancy rates we observed. Pregnancy rates for Yellowstone bison reaching reproductive maturity were similar for both seropositive and seronegative animals (Figure 14.5). The negative effect of seroprevalence was apparent starting at 4 years of age, suggesting that the effect of brucellosis infection became pronounced after the first and potentially failed pregnancy.

Uterine infections (*e.g.*, metritus—inflammation of the uterus) subsequent to a disruptive pregnancy or injury at the time of an abortion could reduce conception or induce early abortions (Neiland *et al.* 1968, Thorne *et al.* 1978). The postpartum reproductive tract is an ideal environment for bacterial growth and retained placental tissues may predispose the uterus to infection (Senger 1997).

Postpartum infection of the uterus by pathogenic bacteria can result in subfertility due to delayed ovulation and lower conception rates (Sheldon and Dobson 2004). Also, chronically infected individuals can develop carpal bursitis (*i.e.*, lameness) that may cause a substantial decrease in body condition (Neiland *et al.* 1968, Thorne *et al.* 1978, Cheville *et al.* 1998, Joly and Messier 2005), thereby further reducing fertility.

Our finding of lower birth rates in younger and seropositive bison was expected (Fuller *et al.* 2007b) and the difference in observed births between seropositive and seronegative bison provides insight into the effects of chronic brucellosis infection on this population (Figure 14.8). Abortions have rarely been observed in Yellowstone bison (Rhyan *et al.* 1994, Olsen *et al.* 1998) and, by definition, the term "chronic" implies a substantial degree of immunity for many individuals that retards the usual course of disease (Cheville *et al.* 1998). Age-specific seroprevalence rates of Yellowstone bison indicate that approximately 50% of bison are exposed prior to reproductive maturity (Treanor *et al.* 2007). This early exposure may allow immature bison to develop resistance to infection, thereby reducing the occurrence of *Brucella*-induced abortions (Meyer and Meagher 1997). Calves may acquire infections *in utero* or through ingesting contaminated milk (Nicoletti 1980). Vertical transmission of *B. abortus* from cow to calf through infected milk has been documented in cattle (Nicoletti and Gilsdorf 1997) and bison (Olsen *et al.* 2003). However, vertical transmission alone is unlikely to account for the differences in seroprevalence observed in adult male (75%) and female (49%) bison (National Park Service, unpublished data). Horizontal transmission through infected birth tissues likely plays an important role in transmission despite the lack of observed abortions. In fact, the degree of horizontal transmission occurring during live births may be substantial (National Park Service, unpublished data).

Exposure to *B. abortus* early in life may provide bison with some natural resistance to acute infection when reproductively mature. We observed 100% (2 of 2) of seropositive and pregnant 3-year-old bison producing viable calves and documented 80% (4 of 5) of similar births during a concurrent study of northern herd bison (Fuller *et al.* 2007b). However, Davis *et al.* (1990, 1991) reported that nearly all bison injected with a high dose of field strain *B. abortus* during mid-gestation aborted their first calf, suggesting that the timing and dose of exposure is important in determining the likelihood that an animal develops acute infection resulting in abortion. We found 63% (5 of 8) and 75% (3 of 4) of

FIGURE 14.8 Radio collared bison from the central herd of Yellowstone National Park licking her stillborn calf along the Madison River during May 2007 (National Park Service photo by Jenny Jones).

reproductively mature animals that were initially seronegative were not observed with calves during their first and second pregnancies, respectively, after converting to seropositive (Fuller *et al.* 2007b). These observations support that exposure during pregnancy, when internal conditions and hormonal signals are favorable for *B. abortus* infection, can result in fetal loss (Cheville *et al.* 1998). Seroconversion occurred in reproductively mature bison of ages 3, 4, 7, and 10, suggesting that all seronegative adult bison may be susceptible to a *Brucella*-induced abortion if exposed during gestation.

The Interagency Bison Management Plan that was established in 2000 assumed that bison culled at the northern boundary near Gardiner, Montana, would come from the northern herd, while bison culled at the western boundary near West Yellowstone, Montana, would come from the central herd (National Park Service 2000). Our findings based on radio-collared bison suggest that the vast majority of bison culled at both the northern and western boundary areas during 1995–2006 came from the central herd. The exception was during 1996–1997 when snow pack conditions on the northern range were more severe than during any other winter in the past 50 years and the entire northern range herd moved to the Gardiner basin (Taper *et al.* 2000). The finding that most bison culled at the northern boundary during our study came from the central herd is supported by time series and modeling analyses that suggest pulses of movement during winter from the central interior to the northern range in response to resource limitation began in approximately 1982 (Coughenour 2005, Fuller *et al.* 2007a). These movements and subsequent removals differentially affected the central herd during 1982–2000 by decreasing its growth rate while the growth rate of the northern herd increased (Fuller *et al.* 2007a).

The future management of Yellowstone bison is highly debated and contingent upon the management of brucellosis. Our findings suggest that the combined effect of brucellosis on pregnancy and birth rates resulted in lower fecundity across all ages. Consequently, chronic brucellosis infection may lower growth rate by more than 15%. Additionally, brucellosis exposure indirectly lowered bison survival because more bison were culled due to concerns regarding transmission to cattle when bison attempted to migrate to lower-elevation areas outside the park. These effects of brucellosis on the demography of Yellowstone bison are evident in time series data that indicate the population grew at a much slower rate than their biological potential during 1970–2000, even though densities were relatively low and per capita food resources should have been relatively high (Gates *et al.* 2005, Fuller *et al.* 2007a).

Bison have evolved to sustain robust population growth and are readily adapted to move when stressed by resource limitation (Gates *et al.* 2005; Chapter 12 by Bruggemaur *et al.*, this volume). Thus, the problems at the boundaries of Yellowstone are likely to continue, challenging society and policy makers with a bison population that exits the park in response to density and climate effects. If vaccination plans are implemented and successful at substantially reducing brucellosis, then population growth rates will likely increase and exacerbate bison dispersal attempts. Therefore, the future challenge of conserving Yellowstone bison involves developing a strategy that mitigates the societal conflict resulting from bison dispersing or migrating outside the park.

V. SUMMARY

1. We monitored 80 adult female bison from the central herd in Yellowstone National Park during 1995–2006 to estimate vital rates that incorporated the effects of brucellosis and could be used to formulate appropriate management strategies (*e.g.*, vaccination, culling).

2. Animals testing positive for exposure to brucellosis had significantly lower pregnancy rates across all age classes compared to seronegative bison. We do not understand the causal mechanism for this finding, which is difficult to ascertain since shedding through reproductive events is believed to be the primary route of brucellosis transmission.

3. Birth rates were high and consistent for seronegative animals, but lower for younger, seropositive bison. Seronegative bison that converted to seropositive while pregnant were likely to abort their first and second pregnancies. Thus, naïve seronegative adult bison may be highly susceptible compared to animals exposed before they were reproductively mature.

4. We detected pronounced senescence in survival for animals >12 years old. Also, brucellosis exposure indirectly lowered bison survival because more bison were culled over concerns about transmission to cattle when bison attempted to move to lower-elevation areas outside the park.

5. We detected a significant decrease in adult female survival when the number of bison in the central herd exceeded 2000–2500 animals, which was exacerbated during winters with severe snow pack because more bison moved outside the park. Except during 1996–1997, the vast majority of radio-collared bison culled at the northern and western boundaries during 1995–2006 came from the central herd.

6. Our findings suggest the combined effect of brucellosis on survival, pregnancy, and birth rates lowered the growth rate in the central herd. Thus, population growth rates will likely increase by more than 15% if vaccination plans are implemented and successful. Wildlife managers would then be challenged with greater numbers of disease-free bison dispersing or migrating outside of the park in response to density and climate effects.

VI. REFERENCES

Albon, S. D., A. Stien, R. J. Irvine, R. Langvatn, E. Ropstad, and O. Halvorsen. 2002. The role of parasites in the dynamics of a reindeer population. *Proceedings from the Royal Society of London B* **269**:1625–1632.

Aune, K. E., T. Roffe, J. Rhyan, J. Mack, and W. Clark. 1998. Preliminary results on home range movements, reproduction and behavior of female bison in northern Yellowstone National Park. Pages 61–70 *in* L. Irby and J. Knight (Eds.) *International Symposium on Bison Ecology and Management in North America*. Montana State University, Bozeman, MT.

Berger, J., and M. Peacock. 1988. Variability in size–weight relationships of *Bison bison*. *Journal of Mammalogy* **69**:618–624.

Bergerud, A. T. 1988. Caribou, wolves and man. *Trends in Ecology and Evolution* **3**:68–72.

Bruggeman, J. E. 2006. *Spatio-Temproal Dynamics of the Central Bison Herd of Yellowstone National Park*. Dissertation. Montana State University, Bozeman, MT.

Burnham, K. P., and D. R. Anderson. 2002. *Model Selection and Inference*. Springer, New York, NY.

Caron, A., P. C. Cross, and J. T. Du Toit. 2003. Ecological implications of bovine tuberculosis in African buffalo herds. *Ecological Applications* **13**:1338–1345.

Caughley, G. 1976. Wildlife management and the dynamics of ungulate populations. Pages 183–246 *in* T. H. Coaker (Ed.) *Applied Biology 1*. Academic Press, London, United Kingdom.

Cheville, N. F., D. R. McCullough, and L. R. Paulson. 1998. *Brucellosis in the Greater Yellowstone Area*. National Academy Press, Washington, DC.

Clutton-Brock, T. H., M. Major, S. D. Albon, and F. E. Guinness. 1985. Population regulation in male and female red deer. *Journal of Animal Ecology* **56**:53–67.

Cole, G. F. 1971. An ecological rationale for the natural regulation or artificial regulation of native ungulates in national parks. *Transactions of the North American Wildlife Conference* **36**:417–425.

Coughenour, M. B. 2005. *Spatial-Dynamic Modeling of Bison Carrying Capacity in the Greater Yellowstone Ecosystem: A Synthesis of Bison Movements, Population Dynamics, and Interactions with Vegetation*. Natural Resource Ecology Laboratory, Colorado State University, Fort Collins, CO.

Craighead, J. J., F. C. Craighead Jr., R. L. Ruff, and B. W. O'Gara. 1973. Home ranges and activity patterns of nonmigratory elk of the Madison drainage herd as determined by biotelemetry. *Ecological Applications* **6**:573–593.

Davis, D. S., J. W. Templeton, T. A. Ficht, J. D. Huber, R. D. Angus, and L. G. Adams. 1991. *Brucella abortus* in bison. II. Evaluation of strain 19 vaccination of pregnant cows. *Journal of Wildlife Diseases* **27**:258–264.

Davis, D. S., J. W. Templeton, T. A. Ficht, J. D. Williams, J. D. Kopec, and L. G. Adams. 1990. *Brucella abortus* in captive bison. I. Serology, bacteriology, pathogenesis, and transmission to cattle. *Journal of Wildlife Diseases* **26**:360–371.

DelGuidice, G. D., F. J. Singer, U. S. Seal, and G. Bowser. 1994. Physiological responses of Yellowstone bison to winter nutritional deprivation. *Journal of Wildlife Management* **58**:24–34.

Diallo, O., A. Diouf, N. P. Hanan, A. Ndiaye, and Y. Prevost. 1991. AVHRR monitoring of savanna primary production in Senegal, West Africa 1987–1988. *International Journal of Remote Sensing* **12**:1259–1279.

Dimmick, R. W., and M. R. Pelton. 1996. Criteria of sex and age. Pages 169–214 *in* T. A. Bookhout (Ed.) *Research and Management Techniques for Wildlife and Habitats.* The Wildlife Society, Bethesda, MD.

Eberhardt, L. L. 1977. Optimal policies for the conservation of large mammals, with special reference to marine ecosystems. *Environmental Conservation* 4:205–212.

Eberhardt, L. L. 2002. A paradigm for population analysis of long-lived vertebrates. *Ecology* 83:2841–2854.

Enright, F. M. 1990. *The Pathogenesis and Pathobiology of Brucella Infection in Domestic Animals.* CRC Press, Boca Raton, FL.

Ewald, P. W. 1994. *Evolution of Infectious Diseases.* Oxford University Press, Oxford, United Kingdom.

Farnes, P., C. Heydon, and K. Hansen. 1999. *Snowpack Distribution Across Yellowstone National Park.* Montana State University, Bozeman, MT.

Fejerskov, O., M. J. Larsen, A. Richards, and V. Baelum. 1994. Dental tissue effects of fluoride. *Advances in Dental Research* 8:15–31.

Fuller, J. A., R. A. Garrott, and P. J. White. 2007a. Emigration and density dependence in Yellowstone bison. *Journal of Wildlife Management* 71:1924–1933.

Fuller, J. A., R. A. Garrott, P. J. White, K. E. Aune, T. J. Roffe, and J. C. Rhyan. 2007b. Reproduction and survival of Yellowstone bison. *Journal of Wildlife Management* 71:2365–2372.

Fuller, W. A. 1959. The horns and teeth as indicators of age in bison. *Journal of Wildlife Management* 23:342–344.

Fuller, W. A. 1966. The biology and management of the bison of Wood Buffalo National Park. *Canadian Wildlife Service Wildlife Management Bulletin Series* 1:1–52.

Fryxell, J. M., and A. R. E. Sinclair. 1988. Seasonal migration by white-eared kob in relation to resources. *African Journal of Ecology* 26:17–31.

Gaillard, J. M., M. Festa-Bianchet, N. G. Yoccoz, A. Loison, and C. Toïgo. 2000. Temporal variation in fitness components and population dynamics of large herbivores. *Annual Review of Ecology and Systematics* 31:367–393.

Gall, D., K. Nielsen, L. Forbes, L. Davis, P. Elzer, S. Olsen, S. Balsevicius, L. Kelly, P. Smith, S. Tan, and D. Joly. 2000. Validation of the fluorescence polarization assay and comparison to other serological assays for the detection of serum antibodies to *Brucella abortus* in bison. *Journal of Wildlife Diseases* 36:469–476.

Garrott, R. A., L. L. Eberhardt, J. K. Otton, P. J. White, and M. A. Chaffee. 2002. A geochemical trophic cascade in Yellowstone's geothermal environments. *Ecosystems* 5:659–666.

Garrott, R. A., L. L. Eberhardt, P. J. White, and J. Rotella. 2003. Climate-induced variation in vital rates of an unharvested large-herbivore population. *Canadian Journal of Zoology* 81:33–45.

Garrott, R. A., P. J. White, and C. A. Vanderbilt White. 1993. Overabundance: An issue for conservation biologists? *Conservation Biology* 7:946–949.

Gates, C. C., B. Stelfox, T. Muhley, T. Chowns, and R. J. Hudson. 2005. *The Ecology of Bison Movements and Distribution in and Beyond Yellowstone National Park.* University of Calgary, Alberta, Canada.

Godfroid, J. 2002. Brucellosis in wildlife. *Revue Scientifique et Technique Office International des Epizooties* 21:277–286.

Goward, S. N., and S. D. Prince. 1995. Transient effects of climate on vegetation dynamics: Satellite observations. *Journal of Biogeography* 22:549–564.

Green, W., J. Griswold, and A. Rothstein. 1989. Post-weaning association among bison mothers and daughters. *Animal Behaviour* 38:847–858.

Hagenbarth, J. F. 2007. Testimony before the U.S. House of Representatives Subcommittee on National Parks, Forests, and Public Lands Oversight Hearing on Yellowstone National Park Bison. March 20, 2007, Washington, DC.

Haigh, J. C., C. Gates, A. Ruder, and R. Sasser. 1991. Diagnosis of pregnancy in wood bison using a bovine assay for pregnancy-specific protein B. *Theriogenology* 36:749–754.

Halloran, A. F. 1968. Bison (Bovidae) productivity on the Wichita Mountains Wildlife Refuge, Oklahoma. *Southwestern Naturalist* 13:23–26.

Hanson, W. R. 1963. Calculation of productivity, survival, and adundance of selected vertebrates from sex and age ratios. *Wildlife Monographs* 9.

Hess, S. C. 2002. *Aerial Survey Methodology for Bison Population Estimation in Yellowstone National Park.* Dissertation. Montana State University, Bozeman, MT.

Hobbs, N. T., D. L. Baker, J. E. Ellis, and D. M. Swift. 1981. Composition and quality of elk winter diets in Colorado. *Journal of Wildlife Management* 45:156–171.

Hosmer, D. W., T. Hosmer, S. Lemeshow, and S. le Cessie. 1997. A comparison of goodness of fit tests for the logistic regression model. *Statistics in Medicine* 16:965–980.

Jolles, A. E., D. V. Cooper, and S. A. Levin. 2005. Hidden effects of chronic tuberculosis in African buffalo. *Ecology* 86:2258–2264.

Joly, D. O., and F. Messier. 2005. The effect of bovine tuberculosis and brucellosis on reproduction and survival of wood bison in Wood Buffalo National Park. *Journal of Animal Ecology* 543–551.

Kay, C. E. 1998. Are ecosystems structured from the top–down or bottom–up: A new look at an old debate. *Wildlife Society Bulletin* 26:484–498.

Kay, C. E. 2007. Testimony before the U.S. House of Representatives Subcommittee on National Parks, Forests, and Public Lands Oversight Hearing on Yellowstone National Park Bison. March 20, 2007, Washington, DC.

Kierdorf, U., H. Kierdorf, F. Sedlacek, and O. Fejerskov. 1996. Structural changes in fluorosed dental enamel of red deer (*Cervus elaphus*) from a region with severe environmental pollution by fluorides. *Journal of Anatomy* **188**:183–195.

Kutner, M. H., C. J. Nachtsheiim, J. Neter, and W. Li. 2005. *Applied Linear Statistical Models*. McGraw-Hill, New York, NY.

Laws, R. M. 1981. Experiences in the study of large mammals. Pages 19–45 *in* C. W. Fowler and T. D. Smith (Eds.) *Dynamics of Large Mammal Populations*. Wiley, New York, NY.

le Cessie, S., and J. C. Houwelingen. 1991. A goodness-of-fit test for binary regression models, based on smoothing methods. *Biometrics* **47**:1267–1282.

Lott, D. 1991. American bison socioecology. *Applied Animal Behaviour Science* **29**:135–145.

Maddock, L. 1979. The migration and grazing succession. Pages 104–129 *in* A. R. E. Sinclair and M. Norton-Griffiths (Eds.) *Serengeti: Dynamics of an Ecosystem*. University of Chicago Press, Chicago, IL.

McHugh, T. 1958. Social behavior of the American buffalo (*Bison bison bison*). *Zoologica* **43**:1–40.

Meagher, M. 1973. *The Bison of Yellowstone National Park*. National Park Service/US Government Printing Office, Washington, DC. Scientific Monograph Series No. 1.

Meagher, M. 1993. *Winter Recreation-Induced Changes in Bison Numbers and Distribution in Yellowstone National Park*. Yellowstone National Park, WY.

Meyer, M. E., and M. Meagher. 1997. *Brucella abortus* infection in the free-ranging bison of Yellowstone National Park. Pages 20–32 *in* E. T. Thorne, M. S. Boyce, P. Nicoletti, and T. J. Kreeger (Eds.) *Brucellosis, Bison, Elk and Cattle in the Greater Yellowstone Area: Defining the Problem, Exploring Solutions*. Wyoming Game and Fish Department, Cheyenne, WY.

Moffitt, S. A. 1998. Aging bison by the incremental growth layers in teeth. *Journal of Wildlife Management* **62**:1276–1280.

National Park Service.2000. *Bison Management Plan for the State of Montana and Yellowstone National Park—Final Environmental Impact Statement*. U.S. Department of the Interior, Denver, CO.

Neiland, K. A., J. A. King, B. E. Huntley, and R. O. Skoog. 1968. The diseases and parasites of Alaskan wildlife populations, Part I. *Bulletin of Wildlife Disease Association* **4**:27–36.

Nicoletti, P. 1980. The epidemiology of bovine brucellosis. *Advances in Veterinary Science and Comparative Medicine* **24**:69–98.

Nicoletti, P., and M. J. Gilsdorf. 1997. Brucellosis—The disease in cattle. Pages 3–6 *in* E. T. Thorne, M. S. Boyce, P. Nicoletti, and T. J. Kreeger (Eds.) *Brucellosis, Bison, Elk, and Cattle in the Greater Yellowstone Area: Defining the Problem, Exploring Solutions*. Wyoming Game and Fish Department, Cheyenne, WY.

Olexa, E. M., and P. J. P. Gogan. 2007. Spatial population structure of Yellowstone bison. *Journal of Wildlife Management* **71**:1531–1538.

Olsen, S. C., A. E. Jensen, M. V. Palmer, and M. G. Stevens. 1998. Evaluation of serologic responses, lymphocyte proliferative responses and clearance from lymphatic organs after vaccination of bison with *Brucella abortus* strain RB51. *American Journal of Veterinary Research* **59**:410–415.

Olsen, S. C., A. E. Jensen, W. C. Stoffregen, and M. V. Palmer. 2003. Efficacy of calfhood vaccination with *Brucella abortus* strain RB51 in protecting bison against brucellois. *Research in Veterinary Science* **74**:17–22.

Pac, H. I., and K. Frey. 1991. *Some Population Characteristics of the Northern-Yellowstone Bison Herd During the Winter of 1988–89*. Montana Department of Fish, Wildlife and Parks, Bozeman, MT.

Parker, K. L., and C. T. Robbins. 1984. Thermoregulation in mule deer and elk. *Canadian Journal of Zoology* **62**:1409–1422.

Paulik, G. J., and D. S. Robson. 1969. Statistical calculations for change-in-ratio estimators of population parameters. *Journal of Wildlife Management* **33**:1–27.

Plumb, G., L. Babiuk, J. Mazet, S. Olsen, P.-P. Pastoret, C. Rupprecht, and D. Slate. 2007. Vaccination in conservation medicine. *Revue Scientifique et Technique Office International des Epizooties* **26**:229–241.

R Core Development Team. 2006. *R 2.0.5—A Language and Environment*. R Foundation for Statistical Computing, Vienna, Austria. < http://www.R-project.org>. Accessed June 1, 2007.

Rasmussen, M. S. 1998. Developing simple, operational, consistent NDVI-vegetation models by applying environmental and climatic information: part I. *Assessment of net primary production. International Journal of Remote Sensing* **19**:97–117.

Rhyan, J. C., T. Gidlewski, T. J. Roffe, K. Aune, L. M. Philo, and D. R. Ewalt. 2001. Pathology of brucellosis in bison from Yellowstone National Park. *Journal of Wildlife Diseases* **37**:101–109.

Rhyan, J. C., W. J. Quinn, L. S. Stackhouse, J. J. Henderson, D. R. Ewalt, J. B. Payeur, M. Johnson, and M. Meagher. 1994. Abortion caused by *Brucella abortus* biovar 1 in a free-ranging bison (*Bison bison*) from Yellowstone National Park. *Journal of Wildlife Disease* **30**:445–446.

Roffe, T. J., J. C. Rhyan, K. Aune, L. M. Philo, D. R. Ewalt, T. Gidlewski, and S. G. Hennager. 1999. Brucellosis in Yellowstone National Park bison: Quantitative serology and infection. *Journal of Wildlife Management* **63**:1132–1137.

Sæther, B.-E. 1997. Environmental stochasticity and population dynamics of large herbivores: A search for mechanisms. *Trends in Ecology and Evolution* **12**:143–149.

Schweitzer, B. 2007. Testimony Before the U.S. House of Representatives Subcommittee on National Parks, Forests, and Public Lands Oversight Hearing on Yellowstone National Park Bison. March 20, 2007, Washington, DC.

Senger, P. L. 1997. *Pathways to Pregnancy and Parturition*. Current Conceptions. Pullman, Washington, DC.

Sheldon, I., and H. Dobson. 2004. Postpartum uterine health in cattle. *Animal Reproduction Science* **82–83**:295–306.

Shupe, J. L., A. E. Olson, H. B. Peterson, and J. B. Low. 1984. Fluoride toxicosis in wild ungulates. *Journal American Veterinary Medical Association* **185**:1295–1300.

Sinclair, A. R. E. 1975. The resource limitation of trophic levels in tropical grassland ecosystems. *Journal of Animal Ecology* **44**:497–520.

Skalski, J. R., K. E. Ryding, and J. J. Millspaugh. 2005. *Wildlife Demography: Analysis of Sex, Age, and Count Data.* Elsevier, San Diego, CA.

Stien, A., R. J. Irvine, E. Ropstads, O. Halvorsen, R. Langvatn, and S. D. Albon. 2002. The impact of gastrointestinal nematodes on wild reindeer: Experimental and cross-sectional studies. *Ecology* **71**:937–945.

Sutherland, S. S., and J. Searson. 1990. The Immune response to *Brucella abortus*: The humoral response. *Animal Brucellosis.* CRC Press, Boca Raton, FL.

Taper, M. L., M. Meagher, and C. L. Jerde. 2000. *The Phenology of Space: Spatial Aspects of Bison Density Dependence in Yellowstone National Park.* Montana State University, Bozeman, MT.

Thorne, E. T., J. K. Morton, F. M. Blunt, and H. A. Dawson. 1978. Brucellosis in Elk. II. Clinical effects and means of transmission as determined through artificial infections. *Journal of Wildlife Diseases* **14**:280–291.

Treanor, J. J., R. L. Wallen, D. S. Maehr, and P. H. Crowley. 2007. Brucellosis in Yellowstone bison: Implications for conservation management. *Yellowstone Science* **15**:20–24.

Van Soest, P. J. 1994. *Nutritional Ecology of the Ruminant.* , New York, NY: Cornell University Press, Ithaca, New York, NY.

Wagner, F. H. 2006. *Yellowstone's Destabilized Ecosystem: Elk Effects, Science, and Policy Conflict.* Oxford University Press, New York, NY.

Watson, F. G. R., W. B. Newman, J. C. Coughlan, and R. A. Garrott. 2006. Testing a distributed snowpack simulation model against spatial observations. *Journal of Hydrology* **328**:728–734.

Wickstrom, M. L., C. T. Robbins, T. A. Hanley, D. E. Spalinger, and S. M. Parish. 1984. Food intake and foraging energetics of elk and mule deer. *Journal of Wildlife Management* **48**:1285–1301.

White, P. J., and R. A. Garrott. 2005. Northern Yellowstone elk after wolf restoration. *Wildlife Society Bulletin* **33**:942–955.

APPENDIX

Appendix 14A.1

Model selection results of *a priori* models for birth rates of known pregnant central herd bison ≥ 3 years old in Yellowstone National Park during 1995–2006 (Appendix 14A.1). The AIC_c value for the top model was 62.25. Model covariates included two individual covariates, age of the animal in years (AGE) and brucellosis serological status (SERO), and three annual covariates, the maximum summer count of bison in the central herd (BISON), an index of snow pack severity (SWE_{acc}), and an index of annual forage productivity ($NDVI_{L-int}$). Model abbreviations are the number of parameters (K), Akaike's Information Criteria corrected for small sample size (AIC_c), the change in AIC_c relative to the best model (ΔAIC_c), and Akaike model weight (w_i). *Post priori* we considered AGE by SERO, AGE by SWE_{acc}, and BISON by SERO interactions. The AGE $\times$ SERO model lowered AIC_c by 0.65 units below the top model.

APPENDIX 14A.1	Model selection results of *a priori* models for birth rates of known pregnant central herd bison ≥ 3 years old				

Model	K	AIC_c	ΔAIC_c	w_i
Constant	1	62.25	0.00	0.13
AGE	2	62.86	0.61	0.09
SERO	2	62.94	0.69	0.09
AGE + SERO	3	63.06	0.81	0.08
BISON	2	63.59	1.34	0.06
SERO + BISON	3	64.24	1.99	0.05
SWE_{acc}	2	64.26	2.00	0.05
AGE + BISON	3	64.26	2.01	0.05
$NDVI_{L\text{-}int}$	2	64.34	2.09	0.04
AGE + SERO + BISON	4	64.35	2.10	0.04
AGE + SWE_{acc}	3	64.92	2.67	0.03
AGE + $NDVI_{L\text{-}int}$	3	65.04	2.79	0.03
SERO + SWE_{acc}	3	65.11	2.85	0.03
SERO + $NDVI_{L\text{-}int}$	3	65.12	2.87	0.03
AGE + SERO + SWE_{acc}	4	65.30	3.05	0.03
AGE + SERO + $NDVI_{L\text{-}int}$	4	65.32	3.07	0.03
BISON + SWE_{acc}	3	65.40	3.15	0.03
BISON + $NDVI_{L\text{-}int}$	3	65.70	3.45	0.02
AGE + BISON + SWE_{acc}	4	66.14	3.89	0.02
SERO + BISON + SWE_{acc}	4	66.33	4.08	0.02
SERO + BISON + $NDVI_{L\text{-}int}$	4	66.48	4.23	0.02
AGE + BISON + $NDVI_{L\text{-}int}$	4	66.49	4.24	0.02
AGE + SERO + BISON + SWE_{acc}	5	66.52	4.27	0.01
AGE + SERO + BISON + $NDVI_{L\text{-}int}$	5	66.68	4.43	0.01
Post priori:				
AGE + SERO + (AGE × SERO)	4	61.60		
AGE + SWE_{acc} + (AGE × SWE_{acc})	4	67.04		
BISON + SERO + (BISON × SERO)	4	66.01		

Appendix 14A.2

Age class-specific pregnancy rate estimates of seronegative (dark) and seropositive (gray) central herd bison handled at the Stephens Creek capture pen in Yellowstone National Park during February and early March 2004.

Methods

We augmented the sample of radio-collared bison with pregnancy data collected from 176 adult female bison during a capture operation at the Stephens Creek pen near the northern boundary of Yellowstone National Park in February and early March 2004 (National Park Service 2000). Some of our study animals were in each of the captured groups, suggesting that many of the 176 bison were members of the central herd. Animals were aged into 3, 4, 5, 6, and older classes using incisor eruption patterns (Fuller 1959). We determined pregnancy status using pregnancy-specific protein B serum assays (Haigh *et al.* 1991). Bison were classified as seropositive or seronegative to brucellosis exposure status based on the results of the fluorescence polarization assay, card, buffered antigen plate agglutination, rivanol, complement fixation, standard plate, and standard tube tests performed by the Montana Department of Livestock Diagnostics Laboratory, Bozeman, Montana, USA (Roffe *et al.* 1999, Gall *et al.* 2000, Rhyan *et al.* 2001).

Results

We estimated the probability of pregnancy for seronegative and seropositive bison for each age class as the observed proportion pregnant for each age class. There were 21 3-year-old bison (seronegative = 5, seropositive = 16), 40 4-year-old bison (seronegative = 15, seropositive = 25), 16 5-year-old bison (seronegative = 5, seropositive = 11), and 99 ≥6-year-old bison (seronegative = 28, seropositive = 71). The proportion of pregnant seropositive bison ranged from 11% to 27% lower than seronegative animals (Figure 14A.1).

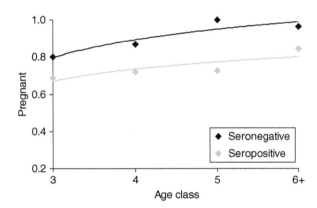

FIGURE 14A.1 Age class-specific pregnancy rate estimates of seronegative (dark) and seropositive (gray) central herd bison handled at the Stephens Creek capture pen during February and March 2004. Points represent the observed proportion of pregnant bison.

Section 4

Wolves: Re-Establishment and Predation

CHAPTER 15

Wolf Recolonization of the Madison Headwaters Area in Yellowstone

Douglas W. Smith,* Daniel R. Stahler,* and Matthew S. Becker[†]

*National Park Service, Yellowstone National Park
[†]Fish and Wildlife Management Program, Department of Ecology, Montana State University

Contents

Theme

After decades of absence, large carnivores are being restored or recolonizing substantial portions of their historical ranges worldwide. The reintroduction of wolves (*Canis lupus*) to Yellowstone National Park in 1995 is one such example of this larger trend. While the Greater Yellowstone Area was rich in prey abundance and diversity, how wolves would respond to the varying prey assemblages within the park was an open question. Wolves were expected to readily adapt to the elk-rich (*Cervus elaphus*) area of northern Yellowstone, but restoration success and subsequent wolf population ecology was less certain in the central portion of the park, where formidable bison outnumber elk and prey density was more patchy and seasonal due to severe winters. We describe wolf reintroduction efforts in the central Yellowstone area and the population ecology of wolf packs

The Ecology of Large Mammals in Central Yellowstone
R. Garrott, P. J. White and F. Watson
ISSN 1936-7961, DOI: 10.1016/S1936-7961(08)00215-7

Copyright © 2009, Elsevier Inc.
All rights reserved.

in this system as it transitioned from no wolves to an established population. We focused on pack dynamics in this relatively small area that supports dense ungulate prey in winter, but dispersed prey in summer.

I. INTRODUCTION

Following decades of active persecution and extermination by humans, the reintroduction of large predators is now a widely-employed conservation and management tool worldwide, with top predators recolonizing and affecting ecosystems from which they have been absent for generations (Maehr *et al.* 2001). Wolves were considered extirpated from Yellowstone National Park by 1930 and were reintroduced with considerable public support in 1995 (Bangs and Fritts 1996). While the success of the reintroduction quickly surpassed expectations for wolf recovery (Fritts *et al.* 2001), patterns of recovery and subsequent population ecology differed across systems within the Greater Yellowstone Area, likely due to differences in prey diversity, abundance and environmental and landscape variables. The central portion of Yellowstone is primarily a two-prey assemblage of elk and bison (*Bison bison*; Chapter 11 by Garrott *et al.*, Chapter 12 by Bruggeman *et al.*, and Chapter 13 by Fuller *et al.*, this volume). However, unlike most portions of the park, bison greatly outnumber elk. Also, prey abundance is highly patchy with elk primarily located in the Madison headwaters year-round, while bison migrate from the Hayden-Pelican valleys to the Madison headwaters area during winter (Chapter 12 by Bruggeman *et al.*, this volume).

Studies of predator population ecology are integral to understanding predator–prey dynamics because systems can behave in various manners, including stable equilibriums, oscillations, and even extirpation of prey or predator (Van Ballenberghe 1987, Boutin 1992). However, most studies of predator population ecology occur in established predator–prey systems. Thus, the dynamics that occur during the transition from a colonizing to an established predator population are rarely described. The reintroduction of wolves into central Yellowstone provided a rare opportunity to evaluate the dynamics of a large predator population from colonization to establishment. We describe efforts to reintroduce wolves during 1996–1998 and the subsequent population and spatial dynamics of the colonizing population.

A. Reintroduction History

The Madison headwaters area was the most difficult place in Yellowstone to establish wolves. An acclimation pen was constructed near Nez Perce Creek in September 1995 and used in January–March 1996 to acclimate a family group of wolves captured in British Columbia, Canada. Named after the nearby creek, the Nez Perce pack typified what we most feared—wide-ranging post-release movements. Of seven releases of wolves from pens in 1995 and 1996, only three were family groups comprised of a mated pair and offspring, and the Nez Perce pack was one of them. The Nez Perce pack's movements were ironic because we anticipated greater group cohesion and more site fidelity from family groups, yet this release was the most erratic.

The pack was released in March 1996, left their pen in two stages, and never reunited. The alpha female (#27) immediately left the pen with three female offspring (#26, #30, #37), while the alpha male (#28) and one male offspring (#29) left 24 h later. Female 27 traveled northward at a rapid pace with her female pups, averaging >30 km per day. Near Red Lodge, Montana, the pups began moving back towards Yellowstone, while female 27 continued north, stopping only at Interstate 90. She had a litter of pups in April along the ranching-wildland interface near Fishtail, Montana, which she raised alone. Despite summer trapping efforts and sporadic conflicts with domestic sheep, female 27 evaded capture until January 1997, when she and her only surviving pup (female 48) were captured via helicopter darting. After capture, both wolves were placed in the Nez Perce acclimation pen.

The alpha male (#28) never joined his mate and wandered widely in and out of Yellowstone. He was eventually discovered dead in the Madison River near Three Forks, Montana, in January 1997 from an apparent gunshot. Without the adults, all of the original offspring traveled widely following their release, sometimes together and other times as loners. Ultimately, these offspring formed pairs with other lone wolves and began packs (*e.g.*, Washakie, Thorofare, Teton packs) outside the Madison headwaters area.

Two other dispersing pups (male 29 and female 37) depredated livestock and were recaptured. They were initially placed in the Rose Creek pen on the northern range, but then were reunited with females 27 and 48 in the Nez Perce pen during February 1997. Around this same time, Yellowstone received 10 pups (#63–#72) born to the Sawtooth pack in Augusta, Montana. These pups were orphaned after their parents were killed due to livestock depredations. The orphaned pups were placed in the pen with the captured Nez Perce wolves in the hope that they would socialize and form a cohesive pack after release. This objective was not achieved because the Nez Perce and Sawtooth wolves split into three separate groups that did not unite. Only two of the Sawtooth pups were alive nine months after their release from the pen. The two surviving pups (males #70 and #72) both eventually became breeders and lived the majority of their lives in the Nez Perce pack.

Nez Perce siblings #29 and #37 were in the pen together during the breeding season and produced a litter of four pups in April 1997. This was the only documented case of close inbreeding in Yellowstone. Thereafter, these wolves were released with the remaining captive Nez Perce and Sawtooth wolves in two groups of eight during April and June. We hoped the first group would establish a territory near the pen and remain near the confined wolves, thereby inducing all the wolves to stay in the area after release. Instead, the wolves released in April traveled widely and were responsible for most of the livestock depredations by wolves that occurred in the Greater Yellowstone Area during 1997. In October 1997, these wolves left the park and depredated sheep near Dillon, Montana, prompting a capture effort and their return back to the Nez Perce pen. Within two weeks, they escaped the pen and depredated more sheep. Female 27 was culled due to her second depredation event, while the rest of the wolves were returned to the Nez Perce pen. It was not initially clear how the wolves escaped the pen because the inside had a chain-link ground apron that prevented digging where the fence met the ground and the pen was structurally intact. However, the overhang of the pen was only 10 ft above ground and we later learned that male #29 could scale the fence by running up the chain link, hanging by his teeth, and then maneuvering over the top. He then dug a hole under the fence and allowed all the wolves to escape.

Male 29 escaped the pen for a third time in February 1998 and became free-ranging, essentially ending the intensive management of this pack. The remaining wolves were released in June 1998 with a litter of pups, which caused them to localize around the pen before moving into the Hayden Valley. Eventually, these wolves united with male 29 and female 48 to form the Nez Perce pack. This pack became firmly established in the Madison headwaters during the winter of 1998–1999 and remained through 2005. They were the only pack in the area during 1996–2000, and the dominant pack until 2005. Their pack size ranged from 4–22 wolves (Table 15.1), and they had only one female breeder (#48; excluding #37 who had no surviving pups). Female 48 was killed at nine years of age in December 2005 by the Gibbon Meadows pack, which became the dominant pack in the area during 2005. Dispersing wolves from the Cougar

TABLE 15.1 Numbers of wolves and packs each year in the Madison headwaters area of Yellowstone National Park during 1995–2006

Pack	1995	1996	1997	1998	1999	2000	2001	2002	2003	2004	2005	2006
Biscuit basin										11		
Cougar creek							6	10	10	12	5	4
Gibbon meadows									5	8	9	12
Nez perce		6	5	7	13	22	18	20	15	14	4	
Totals		6	5	7	13	22	24	30	30	45	18	16
Mean		6.0	5.0	7.0	13.0	22.0	12.0	15.0	10.0	11.3	6	8
No. of packs	0	1	1	1	1	1	2	2	3	4	3	2

Creek pack formed the Gibbon Meadows pack in 2003 and their numbers increased to 12 wolves by 2006 (Table 15.1). The Gibbon Meadows pack essentially used the same territory as the Nez Perce wolves they displaced, with more-frequent use of the Hayden Valley (see below).

Two other packs used the Madison headwaters area. The Cougar Creek pack was formed in 2001 by a dispersing female (#151) from the Leopold pack in northern Yellowstone and dispersing wolves born in the Crystal Creek pack (Pelican Valley) or the Chief Joseph pack (Table 15.1). The Cougar Creek pack has persisted to present, with only one breeding female (#151). In April 2007, this female turned nine years of age and produced a litter of at least three pups. The Cougar Creek pack occupied a discrete territory away from conflicts between other Madison headwaters wolves and reached a peak of 12 wolves in 2004, after which the pack split. This second pack (Cougar Creek II) located northwest of Yellowstone, while the original Cougar Creek pack decreased to four wolves by 2007.

One other pack briefly occupied the Madison headwaters area. The Biscuit Basin pack formed in 2004 when dispersing Nez Perce wolves paired with at least one disperser of unknown origin. This pack never had an exclusive territory and spatially overlapped with the Nez Perce pack, to which they were clearly subordinate. Three Biscuit Basin wolves were captured and radio-collared in 2004. These wolves were in the worst condition of any wolves handled in Yellowstone and one had burns from a geothermal pool. These wolves emigrated from the system in early 2005, moving into southeastern Idaho where they still reside.

In addition to these known packs holding stable territories within the Madison headwaters area, both marked and unmarked wolves from outside this region of the park sporadically used the area. A pack of four wolves without radio-collars used the Gibbon and Madison drainages for much of winter 2002–2003, but several or all members were driven out with the formation of the Gibbon Meadows pack in winter 2003–2004. A distinctive male and female from that uncollared pack eventually formed the Hayden Valley pack and continued to use the Madison headwaters area infrequently from 2003–2004 to 2005–2006. Despite typically residing in areas well to the east of the Madison headwaters, they had a fairly frequent presence in the area during winter 2006–2007. Also, a pack of four wolves without radio-collars was detected in the Madison drainage during winter 2005–2006 and continued to be regularly detected in the Madison canyon area until the end of the study. Lastly, a pack of five wolves consisting of dispersers from the Gibbon Meadows and Nez Perce packs formed in late winter 2005–06 and occupied the Firehole drainage briefly before leaving the park.

II. METHODS

A. Wolf Reintroduction

The Nez Perce wolves were captured during early 1996 near Pink Mountain in British Columbia, Canada (U.S. Fish and Wildlife Service 1994). Their main prey was elk, deer (*Odocoileus hemionus*), and moose (*Alces alces*), though bison populations were within dispersal distances known for wolves and it was possible they fed on bison (Weaver and Haas 1998). The release strategy for these wolves differed from reintroductions in Central Idaho (Bangs and Fritts 1996, Fritts *et al.* 1997) because wolves were acclimated as family groups (*i.e.*, packs) in 1-acre pens and "soft" released 10 weeks later. As previously described, subsequent and longer acclimations in the pen were necessary for the Nez Perce wolves after their initial release. Yellowstone staff fed the penned wolves road-killed deer, elk, moose, and bison, twice a week.

B. Wolf Collaring and Handling

All wolves released into Yellowstone were initially radio-collared (Bangs and Fritts 1996). Post-reintroduction collaring efforts in the Madison headwaters area continued with the objective of maintaining contact with each pack, keeping the breeders in each pack marked, and marking up to

50% of the pups born each year. Wolves were captured via helicopter darting or net-gunning during November to March and fitted with a VHF radio collar with a built-in mortality sensor. Wolves were then measured and weighed, and approximately 10 ml of blood was drawn for genetic and disease studies. Captured individuals were classified as either pups (<12 months old from early April) or adults (>12 months old) based on tooth eruption, wear, and staining (Gipson *et al.* 2000), as well as body size, testicle size, and pelage markings compared against field observation of known individuals. Identification of pups was fairly certain, so collared pups were tracked as known-aged individuals through their life. Using these aging techniques, adult wolves were further categorized in age classes beyond yearling stage to better understand reproductive patterns, social status, and pack age structure.

C. Aerial Monitoring

All packs in the Madison headwaters area were tracked through the year using a fixed-wing aircraft (Supercub PA-18). Aerial locations of marked wolves averaged once per week for 10 months of the year, and almost daily during two 30-day periods in early (mid-November to mid-December) and late (March) winter (Smith *et al.* 2004). We recorded time of day and location coordinates for each radio-collared individual. When wolves were sighted (>90% of locations), we recorded group size, activity, pack age and sex composition (including unmarked individuals, if known), as well as an association with kills or den and rendezvous sites.

Changes in pack membership, new pack formation, and an individual's social status were determined via both aerial and ground monitoring of radio-collared and uncollared individuals. The intensity of observations and high proportion of marked animals in the population allowed us to reliably enumerate population size instead of relying on population estimators. We made total pack counts in early- and late-winter, when each pack was observed >10 times per month during our 30-day study periods and at least 5–10 times during the intervening months. The mode pack size was used to determine population size, plus the known loners. We did not estimate the number of loners that were not known, but we believe this to be <2% of the population (Smith *et al.* 2007). We report on early winter pack sizes (November–December), which were slightly larger than late winter (March) because early winter was the legally mandated reporting time by the U.S. Fish and Wildlife Service. In summer, packs were less cohesive relative to winter. Thus, aerial monitoring focused more on gathering reproductive data (*e.g.*, breeding female identification, den locations, and pup counts).

D. Ground Monitoring on the Ungulate Winter Range

The geothermally-influenced drainages of the Madison headwaters provided a more heterogeneous snow pack and easier access to forage for ungulates, resulting in very high densities of wintering elk and bison relative to the surrounding landscape (Chapter 4 by Watson *et al.*, Chapter 9 by White *et al.*, Chapter 11 by Garrott *et al.*, and Chapter 12 by Bruggeman *et al.*, this volume). Consequently, this relatively small winter range served as a highly valued hunting ground for most central Yellowstone wolf packs during the winter months. To evaluate wolf use of this shared winter range, ground-based winter studies were conducted on the approximately 31,000 ha Madison headwaters area during November 15–April 30 from 1996–2007, with intensive telemetry-based work beginning in 1998–1999. A linear drainage system with considerable road access (Chapter 2 by Newman *et al.*, this volume) allowed for rapid sampling of wolf presence in the study area each day. Sampling began at dawn with ground crews of 3–4 people covering all roads by snowmobile or vehicle and using strategic high points on the landscape to facilitate signal detection and observation. When collared packs were detected, locations were obtained by triangulations using a minimum of three bearings (White and Garrott 1990) or from observations with binoculars or spotting scopes. When possible, multiple locations were obtained in early morning and evening each day. Uncollared wolves opportunistically detected from tracks or observations were also recorded to better estimate the number of wolves using the study area. In addition, biologists studying elk and bison routinely covered backcountry areas to assist with wolf detection. We estimated wolf use of the Madison headwaters in wolf days and pack days, defined as one wolf or one pack in the study area for one day, respectively. The metric of multiple pack days was also defined to describe the number of days during the study period in which more than one pack was detected.

E. Population Demographic Parameters

Demographic parameters such as reproduction, mortality, survival, density, and population growth rate were determined using telemetry and observational data from both marked and unmarked individuals. Annual reproduction was assessed through the identification and monitoring of the breeding females in each pack. In the Madison headwaters area, the denning season typically began in mid-April when the breeding females would localize around a den. In cases where the breeding female was not marked during a denning season, monitoring of other marked wolves in the pack allowed for the location and monitoring of den sites. Assessment of annual pack reproduction was determined through observation of pups at dens and subsequent observations through the summer. Field-based parentage was based on observed dominant status of males and females, breeding behavior, morphological evidence of pregnancy prior to denning period, and individual denning behavior. Annual reproductive success and pup mortality were determined by comparing the difference between the high count of pups at the den in spring (May–June) to the number of pups in the pack in early winter (November–December).

Mortality and survivorship of adult wolves was determined via radio-tracking of collared individuals. Additional information on cause-specific mortality included data gathered opportunistically from uncollared individuals discovered by ground personnel. Upon discovery of a mortality beacon from a collared wolf, field investigation typically occurred within one week. Field necropsies were performed by investigating the body and surrounding area for evidence of trauma, injury, tracks, or site characteristics that would indicate cause of death. In many cases, the cause of death was obvious (*e.g.*, intra-specific strife, vehicle hit). However, some mortality was categorized as natural, unknown causes if we could rule out human influence but the period of time from death until carcass investigation precluded certainty due to decomposition or effects of scavengers. We estimated annual survival as the probability that a wolf alive at the beginning of a year survived all sources of mortality during that year, using the formula $(1-x/y)^{365}$, where x is the total number of deaths and y is the total number of transmitter-days (*i.e.*, number of days a radio-collared individual was alive) for all radio-collared individuals (Heisey and Fuller 1985).

Using annual census data, we determined the rate of population increase (λ) using the formula N_{t+1}/N_t, or as growth rate $r = \log_e \lambda$. To estimate the rate of increase when the wolf population was increasing in the Madison headwaters area, we took the natural logarithms of the population counts during 1997–2004 and fit a linear regression to the data points. We censored the count during 1996 from these analyses due to management actions (*i.e.*, restoration, depredation removals). We also censored the counts during 2005 and 2006 from these analyses because the number of wolves in the area decreased substantially, resulting in outlying residuals that strongly dampened the relationship (Figure 15.1). We estimated wolf density in the Madison headwaters area as the number of wolves per 1000 km², which is a common density measurement in other wolf systems (Fuller *et al.* 2003), and compared our results with similar measures from outside the study area.

Though a simple definition of dispersal can be difficult to assign to social species (Waser 1996), we defined dispersal as the departure of a wolf from its natal pack without return, regardless of whether or not it bred. Standard univariate statistics were used to compute differences between means and trends over time (Zar 1999). Two-tailed tests were used so assessments of differences would be conservative (Abelson 1995). Data were plotted and visually examined to determine appropriate groupings for analysis or biologically relevant groups.

F. Wolf Pack Territory Estimation

While there are numerous definitions of territory, we employed the home range definition of Kernohan *et al.* (2001:127) as "the extent of area with a defined probability of occurrence of an animal [in this case a wolf pack] during a specified time period." For each pack, we defined an annual territory describing a pack's movements through central Yellowstone and a winter territory describing its movements within the high-density ungulate winter ranges of the Madison headwaters area. Annual territories consisted of wolf pack locations collected aerially during April 1–March 31 from 1998–2007. We considered the dates defining a year to be biologically meaningful by encompassing the approximate time between the start of each year's denning period. Winter territories were comprised of ground locations collected from November 15–April 30 and were intended to evaluate pack spatial dynamics on a shared winter range that typically constituted only a portion of each wolf pack's territory, but was disproportionately used to varying degrees among packs.

To evaluate the best territory estimation technique, we performed goodness-of-fit simulations (Horne and Garton 2006a) on each pack's winter ground locations. Fixed and adaptive kernel techniques ranked as the top two competing estimators for each of the 14 wolf pack datasets, with fixed kernel chosen for the majority of datasets. We defined a territory as the area encompassing a 95% probability contour (White and Garrott 1990) and a core territory as the area encompassing a 50% contour.

To evaluate differences in winter territories between packs and among years, we determined the area-sample size asymptote by bootstrapping subsamples of locations from each dataset. Four-hundred bootstrap replications were performed for 21 different subsample sizes ranging from 10–217, with the optimized bandwidth for each determined from the bandwidth with the lowest average likelihood cross-validation score (Horne and Garton 2006b). Territory and core areas for each subsample size were then calculated at this bandwidth. Bandwidths calculated at 30 locations represented the sample size at which most resident packs were included and the effect of area and sample size was addressed. Winter territories could not be calculated for the Cougar Creek pack due to topography preventing adequate telemetry locations.

We were able to calculate winter territories for 14 pack years with a mean of 106 locations (range = 32–217). For each territory, we calculated the area of each contour, the proportion that was exclusive (*i.e.,* not overlapped by adjacent packs), and the proportion that was overlapped by another pack. Analyses were performed in Tarsier (Watson and Rahman 2003), ArcView 3.2, and ArcGIS 9. We evaluated wolf pack fidelity to winter territories among years by calculating the distance to consecutive centroids (Schoen and Kirchhoff 1985, Garrott *et al.* 1987, Brown 1992), estimated by the mean of all pack locations for a given winter. The Euclidean distance between consecutive centroids was then calculated for each pack present in the Madison headwaters for multiple winters.

While substantial spatial overlap can exist between wolf packs, temporal avoidance can decrease the probability of inter-pack strife, with adjacent and overlapping packs not present in the same area at the same time (Mech and Boitani 2003). To determine whether packs with overlapping territories exhibited temporal avoidance in their use of the Madison headwaters area, we first determined the number of days that adjacent packs were located in the study area on the same day to assess whether packs were avoiding the drainages altogether if other packs were present. When packs were located in the Madison headwaters on the same day and at approximately the same time, we calculated the Euclidean distance between locations. Our sampling methods rarely permitted simultaneous location times. Thus, we evaluated locations obtained during the same morning or evening sampling period (approximately three hours) to determine if there was a difference in separation distances as a function of time between locations. We did not detect a significant difference or pattern in separation distances and, as a result, used locations within the same early morning or evening sampling period in analyses. Only one paired location per day was used if multiple paired locations were available, with the minimum separation distance selected.

III. RESULTS

A. Population Status

A total of 37 wolves in four packs were captured and collared in the Madison headwaters area during 1996–2006. The number of wolves collared each year ranged from one (2002) to 11 (2005) and averaged five per year. Sixteen pups and 21 adults (0.8 pups per adult), comprised of 16 females and 21 males (0.8 females per male), were captured. The average weight of wolves captured in the Madison headwaters (41.8 kg $\pm$ 1.3) was slightly less than wolves captured in other areas of Yellowstone (45.1 kg $\pm$ 0.6), but age and sex structure of the captured sample was also different. The total number of wolves using the Madison headwaters area increased from five in 1997 (3 wolves per 1000 km^2) to 45 in 2004 (24 wolves per 1000 km^2), with an estimated mean rate of increase (r) of 0.30 (95% CI: 0.21, 0.39) during this period (Figure 15.1). The number of wolves decreased from 45 in 2004 to 18 in 2005 and 16 in 2006 ($r = -0.92$ and -0.12, respectively).

The number of packs ranged between one (1996–2000) and four (2004), with an early winter average pack size of 10.2 $\pm$ 5.3 wolves (range = 4–22). There was little overwinter loss of pack members. The Nez Perce pack was the only pack to occupy the area from 1996–2000 and ranged in size from 5–22 wolves (Figure 15.2, Table 15.1). Their numbers were stable until 2004, after which their numbers decreased until the pack dissolved after the death of the alpha female due to intraspecific killing by another pack. The Cougar Creek pack was formed in 2001 and included some immigrants from the Leopold pack (#151F). Pack sizes ranged from 4–12 wolves (Table 15.1). In 2005, the pack split permanently into two packs, with the Cougar Creek II pack emigrating north into Montana. Since the split, the Cougar Creek pack has aged, produced few pups, and remained a small pack. The Gibbon Meadows pack, which essentially usurped the territory of Nez Perce pack (see spatial analyses later in this chapter), formed in 2003 from dispersers that likely came from the Cougar Creek and Nez Perce packs (Figure 15.3). Pack sizes ranged from 5–12 wolves (Table 15.1), and the Gibbon Meadows pack is

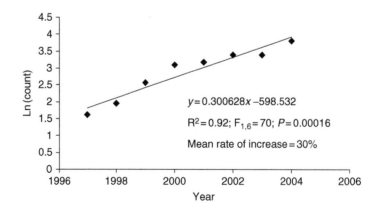

FIGURE 15.1 Rate of wolf population increase in the Madison headwaters area of Yellowstone National Park during 1997–2004.

FIGURE 15.2 Long-term alpha female (#48F) of the Nez Perce pack leads her mates across the lower geyser basin of the Madison headwaters area in Yellowstone National Park (National Park Service photo by Daniel Stahler).

FIGURE 15.3 The Gibbon Meadows pack howling in Yellowstone National Park. This pack eventually replaced the Nez Perce pack as the core pack in the Madison headwaters area. (National Park Service photo by Douglas Smith).

currently the dominant pack in the Madison headwaters area. The Biscuit Basin pack was present in 2004 and overlapped spatially with the Nez Perce pack before moving to Idaho in late winter, where the pack still resides.

B. Reproduction and Denning

A total of 67 pups were born to wolf packs in the Madison headwaters area, with 53 (79%; 4.8 surviving pups per year) surviving to early winter (Figure 15.4). This over-summer estimate of survival may be biased high because it was difficult to obtain early season observations of pups at dens to estimate litter size. However, pup mortality was high in 1996 and 1997, with only one pup surviving each year out of

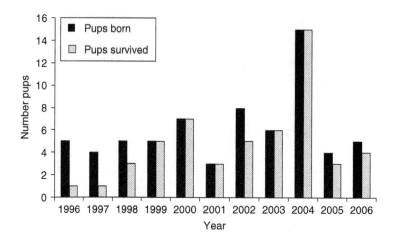

FIGURE 15.4 Numbers of wolf pups that were born and survived in the Madison headwaters area of Yellowstone National Park during 1996–2006.

five and four births, respectively. These mortalities were likely related to anthropogenic causes from management activities associated with reintroduction. We did not record more than one female breeding in any Madison headwaters pack, in contrast to other Yellowstone packs where two females may breed in certain packs when conditions are favorable (Smith *et al.* 2007).

Denning behavior of wolves in the Madison headwaters area was noticeably different compared to wolves occupying the northern portion of Yellowstone. During the denning season (April–August), breeders from the Gibbon Meadows and Nez Perce packs spent less time at the den than did northern range wolves from the Leopold and Druid Peak packs (females: $\chi^2 = 9.05$, $P = 0.003$, $n = 392$; males: $\chi^2 = 8.73$, $P = 0.003$, $n = 423$). The same was true for auxiliary females ($\chi^2 = 8.84$, $P = 0.003$, $n = 852$), but not males ($\chi^2 = 0.55$, $P = 0.46$, $n = 647$) where auxiliaries from both areas wandered equally from the den.

Other than the Biscuit Basin pack, which only denned in the area one year, few den sites were used. The Nez Perce pack used the same den site for five consecutive years, with only one additional site. The Gibbon Meadows pack usurped the den site of the Nez Perce pack one year, but also used two other sites. The Cougar Creek pack denned either in the same den or the same localized area each year.

C. Survival, Mortality and Dispersal

Nine of 29 radio-collared wolves died during 1996–2006 (20,826 radio days), for a pooled annual survival rate of 0.85. Overall, the annual mortality rate based only on radio-collared wolves averaged 15% and ranged from 0–57% (Figure 15.5). Mortality varied more year-to-year compared to annual mortality park-wide. In some years there were no mortalities (1999–2000, 2002–2003, 2006–2007), while eight of 14 radio-collared wolves died during 1997–1998. The leading cause of death was intraspecific strife (Figures 15.6 and 15.7).

A total of 24 wolves dispersed from the area, for an average annual dispersal rate of 2.2 wolves. The dispersal rate averaged 22% per year and ranged from 0–63% of the collared population (Figure 15.8). Combined with mortality, we calculated an average annual "disappearance rate" of 40% per year (mortality + dispersal = 18% + 22% = 40%). Nine females and 15 males dispersed, which did not significantly differ from parity ($\chi^2 = 0.17$, $P = 0.81$, $n = 95$).

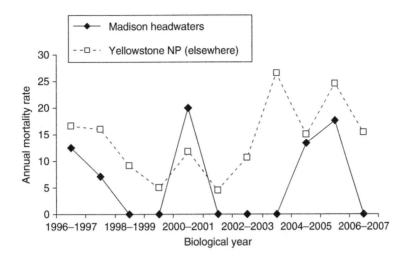

FIGURE 15.5 Annual mortality rates for wolves in the Madison Headwaters area during April 1 through March 31 each year during 1996–2007 compared to the mortality of wolves occupying other portions of Yellowstone National Park.

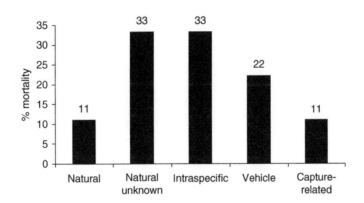

FIGURE 15.6 Causes of death for radio-collared, resident wolves in the Madison Headwaters area of Yellowstone National Park during 1996–2006.

D. Spatial and Temporal Territoriality

Wolf use of the Madison headwaters area during winter varied substantially during the study (Figure 15.9). The ungulate winter range received considerable wolf use, despite comprising a relatively small portion of each pack's annual territory (Figures 15.10 and 15.11). Wolves were detected in the study area on 1306 days of the 1837-day study period (71%), comprising a total of 16,801 wolf days, 1872 pack days, and 437 multiple pack days. Wolf presence during the 167-day winter field season ranged from 60–3964 wolf days, with pack days and multiple pack days ranging from 19–383 and 0–128, respectively. The percentage of days wolves were detected during the field season ranged from 19–96%. The Firehole drainage was used most extensively, with 78% of the total wolf days, followed by the Gibbon (11%) and Madison (11%) drainages. Wolf presence in the study area typically peaked in middle to late winter.

Similar to the overall wolf population trends in central Yellowstone, winter wolf use of the Madison headwaters steadily increased during 1996–2005, before decreasing substantially in subsequent years.

FIGURE 15.7 Scavengers feeding on the remains of a Biscuit Basin wolf that was killed by the Cougar Creek pack during a territorial intrusion. Intra-specific mortality is the leading cause of death for Yellowstone wolves (National Park Service photo by Daniel Stahler).

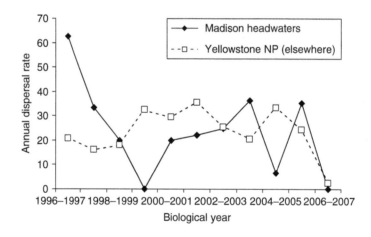

FIGURE 15.8 Annual dispersal rates for wolves in the Madison Headwaters area during April 1 through March 31 each year during 1996–2007 compared to dispersal rates of wolves from other portions of Yellowstone National Park.

Wolves were first detected in 1996–1997 when a few itinerant animals began using the study area and the Nez Perce pack was soft-released into the Firehole drainage. Once the Nez Perce pack became firmly established in winter 1998–1999, winter wolf use steadily increased as the pack increased in size and wolves were detected more frequently in the study area. The ungulate winter range began to transition from a single pack to a multiple pack system with the establishment of the Cougar Creek pack in winter 2001–2002, and an uncollared pack of four wolves in the Gibbon drainage during winter 2002–2003. However, the system was still primarily used by the Nez Perce pack. Multiple packs firmly established in

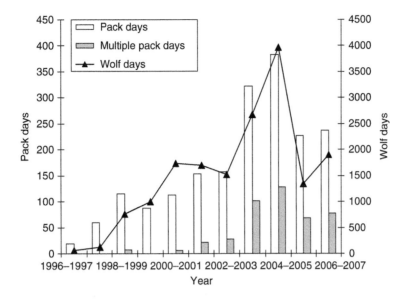

FIGURE 15.9 Variations in use of the Madison headwaters area of Yellowstone National Park by wolves during the winters of 1996–1997 through 2006–2007. Wolf use was estimated in wolf days and pack days, defined as one wolf or one pack in the study area for one day, respectively. The metric of multiple pack days described the number of days during the study period in which more than one pack was detected. The number of wolves, packs, and the frequency with which they used the Madison headwaters continued to increase until 2004–2005, after which it decreased substantially.

winter 2003–2004 with Nez Perce, Cougar Creek, Gibbon Meadows, and Biscuit Basin packs regularly using the Madison headwaters area. Wolf use peaked in winter 2004–2005, with an average of two packs detected in the study area each day, and as many as five packs routinely detected in middle to late winter. Wolf use dropped precipitously in winter 2005–2006, with the system being primarily used by the Gibbon Meadows pack and the Cougar Creek pack infrequently using the area. The Gibbon Meadows wolves continued as the primary pack in 2006–2007, but were joined by the Hayden Valley pack and an uncollared pack in the Madison drainage.

A total of 1369 telemetry locations and 534 visual locations were obtained from ground investigations of primarily five radio-collared packs (Nez Perce, Cougar Creek, Gibbon Meadows, Biscuit Basin, Hayden Valley). Mean winter territory size for wolf packs in the Madison headwaters area was 246 km^2 (95% CI = 196, 297) and ranged from 107 to 382 km^2. During winters 1998–1999 through 2002–2003, when the Nez Perce wolves were the primary pack using the Madison headwaters area, mean winter territory size for this pack was 266 km^2 (95% CI = 186, 346). However, the territory size for this pack decreased to a mean of 137 km^2 (95% CI = 111, 163) during the 2003–2004 through 2004–2005 winters, when peak wolf use occurred in the area. Mean winter territory size of all packs during this latter period was 236 km^2 (95% CI = 169, 302). The mean core area during the course of the study was 55 km^2 (95% CI = 44, 65) with a range of 26 to 81 km^2. Wolves spent the majority of their time near the center of their range, with the 50% contour covering an average territory proportion of 0.23 (95% CI = 0.22, 0.23). Mean pack size and the intensity of use by the wolf population (measured in wolf days) were poor predictors of winter territory size in the Madison headwaters ($r^2 = 0.04$ and 0.07, respectively).

Though territories changed in their annual configurations, and there was substantial variance in territory areas, winter territory fidelity was generally strong, with a mean distance between consecutive winter centroids of 3.4 km (95% CI = 0.9, 5.9; range = 0.8–14.3 km). The Nez Perce pack demonstrated remarkable territory fidelity over six winters, with a mean consecutive centroid distance of 1.4 km (95% CI = 0.8, 2.1) and a range of 0.8–3.1 km. The Gibbon Meadows pack exhibited the

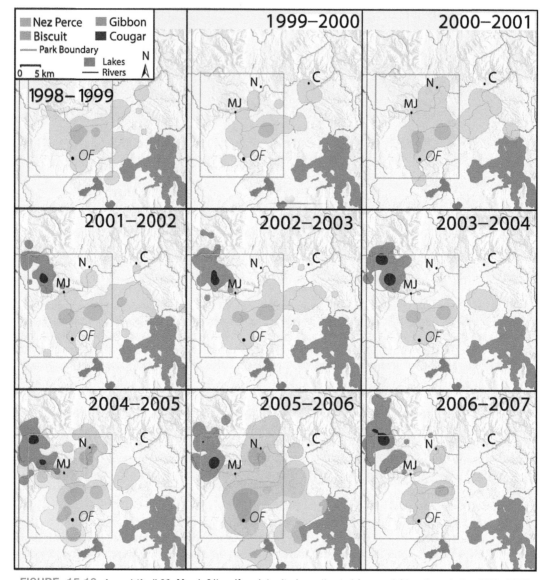

FIGURE 15.10 Annual (April 30–March 31) wolf pack territories estimated from aerial locations during 1998–2007. The territory of the Nez Perce pack during 2005–2006 was usurped by Gibbon Meadows pack in December 2005 and the Nez Perce pack disbanded shortly thereafter.

greatest territory change when they usurped the Nez Perce territory in winter 2005–2006 (distance = 14.3 km), but centroid distance changed little thereafter (distance = 2.9 km).

Wolf pack territories in the Madison headwaters area during winter exhibited substantial spatial overlap between packs. The mean proportion of a pack's winter territory that could be considered exclusive was 0.48 (95% CI = 0.32, 0.65) and ranged from 0.02 to 0.88. Several packs formed from dispersing members of resident packs that joined with unknown wolves and formed territories adjacent to the dispersing wolves' range, a process of pack formation known as budding (Fritts and Mech 1981). For example, the Biscuit Basin pack formed with dispersing Nez Perce wolves, while the Gibbon

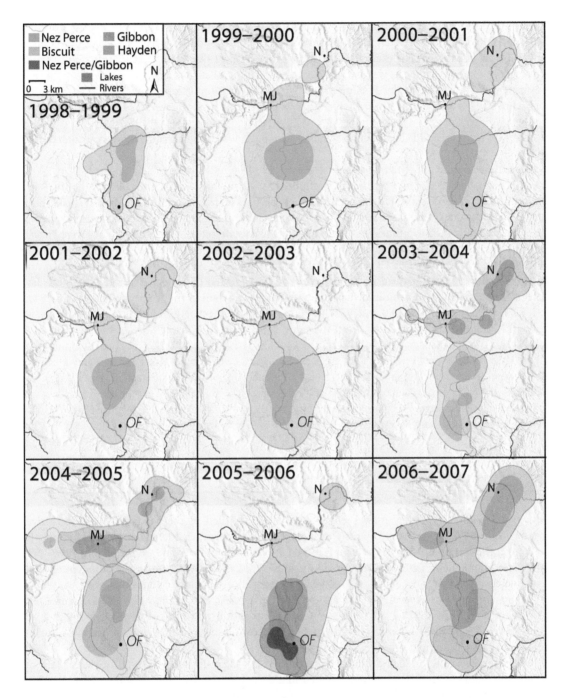

FIGURE 15.11 Winter (November 15–April 30) wolf pack territories in the Madison headwaters area estimated from ground locations during winters 1998–1999 to 2006–2007. The Nez Perce-Gibbon Meadows pack formed in winter 2005–2006 and was only present in late winter before leaving the area shortly thereafter. The Hayden Valley pack infrequently used the Madison headwaters area during the 2003–04 and 2005–06 winters, but substantially increased their use in 2006–2007.

Meadows pack formed with dispersing Cougar Creek wolves. A higher proportion of spatial overlap was apparently tolerated between the dispersers' pack and these budding packs for wolves overlapping in the Firehole drainage. For example, the mean proportion of overlap between territories of packs sharing the Firehole drainage was 0.64 (95% CI = 0.40, 0.88) and ranged from 0.23 to 1.00. For adjacent packs not sharing the Firehole drainage, the mean proportion of overlap was 0.23 (95% CI = 0.10, 0.35) with a range of 0.00–0.49. The mean proportion of core areas that were exclusive between adjacent packs was 0.86 (95% CI = 0.71, 1.00) and ranged from 0.43 to 1.00. All core area overlap was between budded packs in the Firehole drainage, with a mean proportion of overlap of 0.14 (95% CI = 0.00, 0.29).

Separation distances to evaluate temporal overlap between packs were calculated for winters 2003–2004 through 2006–2007 using eight pack combinations with adjacent territories. Despite the small size of the area and the large sizes of their territories, multiple packs were detected in the Madison headwaters on the same day on 374 days (56% of the time). The Biscuit Basin and Nez Perce packs were detected in the area on the same day 75 times (46% and 31% of the days detected, respectively). They were located during the same period on 58 days, with a mean Euclidean distance of 7.8 km (95% CI = 6.6, 9.0) in 2003–2004 and 12.1 km (95% CI = 10.1, 14.1) in 2004–2005. In contrast, the Gibbon Meadows and Biscuit Basin packs were detected in the study area on the same day 62 times (31% and 38% of the days detected respectively), with a mean distance between packs ($n = 46$) during the same period of 25.9 km (95% CI = 22.7, 29.2) and 19.0 km (95% CI = 16.5, 21.5) respectively. Similarly, the Nez Perce and Gibbon Meadows packs were detected on the same day 86 times (36% and 43% of the days detected, respectively) and distances ($n = 62$) averaged 16.0 km (95% CI = 12.9, 19.1) and 14.7 km (95% CI = 12.7, 16.7), respectively. The Gibbon Meadows and newly formed Nez Perce-Gibbon Meadows packs were detected on the same day 21 times (17% and 60% of the days detected, respectively) and distances ($n = 17$) in 2005–2006 averaged 8.1 km (95% CI = 5.8, 10.4). The Gibbon Meadows and Hayden Valley packs were detected in the study area on the same day 32 times (24% and 58% of the days detected, respectively) and distances ($n = 29$) in 2006–2007 averaged 15.6 km (95% CI = 13.0, 18.2).

IV. DISCUSSION

After some difficulty getting established, wolves became settled in the Madison headwaters river drainages in 1998 and have continuously occupied the area thereafter. Management interventions by staff from the National Park Service and Fish and Wildlife Service were necessary during 1996–1998 and curtailed population growth. The population grew rapidly after 1998, peaking at a moderately high wolf density of 24 wolves per 1000 km^2. In 2005, the population decreased precipitously to 8 wolves per 1000 km^2 or only 16 wolves in two packs. One of those packs comprised only three wolves. On a proximate level, this decrease was caused by poor pup production (three surviving pups in 2005 compared to 15 in 2004), the emigration of one pack (Biscuit Basin), and higher adult mortality (24%), some of which led to the demise of the Nez Perce pack.

The ultimate cause for such a dramatic decrease is less clear. Increased mortality, decreased pup production, and emigration could all be related to a decrease in available elk biomass (Chapter 23 by Garrott *et al.*, this volume). This decrease in elk availability presumably lowered the carrying capacity of the area for wolves. Also, wolves may have overshot the carrying capacity of the area in 2004 and, thereby, contributed to decreased elk numbers. We suggest that the availability of elk limited the number of wolves in this area because the wolf population decreased even though bison biomass was still abundant (Figure 15.12.; Chapter 12 by Bruggeman *et al.*, this volume). If the wolves in this area become more reliant on bison for prey, then the current low density and number of packs may increase again. However, we cannot rule out the effects of disease because wolves captured in the Madison headwaters area during 2005 tested positive for canine parvovirus, canine distemper, and canine

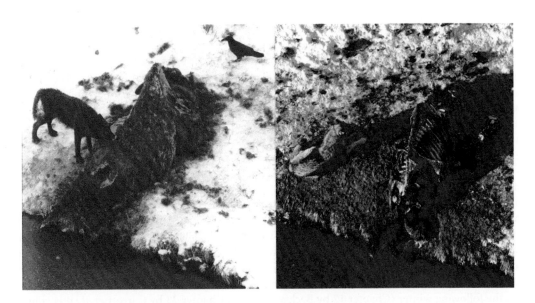

FIGURE 15.12 Members of the Gibbon Meadows pack feeding on a fresh bison carcass during consumption stages photographed two days apart. Bison represent a bounty of food when killed or scavenged by wolf packs (National Park Service photo by Douglas Smith).

adenovirus (Smith and Almberg 2007). High mortality in wolves occupying the northern portion of Yellowstone the same year was followed by rapid growth in the population thereafter, but this was not the case in wolves occupying the Madison headwaters area (Smith *et al.* 2007).

Other data also suggest that wolves in the Madison headwaters may need periodic augmentation by the immigration of wolves from other areas in Yellowstone. The average number of wolf pups surviving per year was low (4.8 pups per year) in the Madison headwaters area, and death and dispersal combined removed an average of 4.2 wolves per year. Thus, the loss of wolves from the system in most years was greater than 30%. Research from other North American wolf systems indicates that 30% or less is considered to be a sustainable mortality rate for many wolf populations (Keith 1983).

Data on pack formation supports the hypothesis that wolves in the Madison headwaters may need occasional immigrants to persist over time. The establishment of the two packs in the system (Cougar Creek and Gibbon Meadows packs) was dependent on immigrating dispersers from outside the study area. Likewise, the Biscuit Basin pack formed when dispersing Nez Perce wolves teamed with wolves from unknown origin. We also recorded wolves either passing through or temporarily inhabiting the area, without becoming permanently established.

The different ungulate densities and distributions in the Madison headwaters area compared to the rest of Yellowstone probably explain the notably different pack dynamics for this area. While wolf packs in central Yellowstone ranged over a considerable area, the dense winter concentrations of ungulates in the relatively small Madison headwaters area resulted in packs exhibiting a high degree of spatial and temporal overlap during winter as wolves used this area disproportionately. While one dominant pack (*i.e.,* Nez Perce or Gibbon Meadows) used the area, other smaller packs were also present in winter such that large proportions of each pack's winter territory were often used by at least one other pack. The presence of another pack may not have served as a deterrent from another pack using the drainage because multiple packs were frequently detected using areas at the same time. Packs tended to avoid direct encounters, but were often in close proximity and the mean separation distance between packs in the Firehole drainage was typically well within audible communication range (Harrington and Mech 1979).

The high degree of overlap between certain packs could be due to a variety of reasons. The highest amount of wolf use and territorial overlap occurred in the Firehole drainage, which was the most open drainage in terms of topography and vegetation and also typically had the highest overall ungulate density (Chapter 11 by Garrott *et al.*, this volume), though fewer elk than the Madison drainage (Chapter 21 by White *et al.*, this volume). The high concentration of prey in the Firehole drainage makes territory defense less feasible if a wolf spatial response to increased prey density occurred with bison migration. However, inter-pack strife resulting in wolf mortalities did occur between some packs, suggesting this was unlikely. Despite pack strife, elk redistribution and concentration at the borders of overlapping wolf pack territories did not occur in the Madison headwaters (Chapter 21 by White *et al.*, this volume) as has been observed in other studies (Mech 1977). This finding suggests wolves did not avoid areas of overlap due to risk of inter-pack strife.

It is plausible that tolerance of overlap between neighboring packs could also be due to a high degree of relatedness between packs, and genetic analyses to determine the degree of relatedness between packs could provide significant insights into observed patterns of territoriality. However, while relatedness may be associated with spatial and temporal overlap between packs, it is unlikely to supercede the importance of prey availability. For example, in winter 2003–2004 the Nez Perce and Biscuit Basin packs tolerated a strong degree of spatial and temporal overlap in the Firehole drainage, with both packs denning in the Firehole drainage and the Biscuit Basin pack exhibiting a relatively high kill rate (Chapter 17 by Becker *et al.*, this volume). However, coincident with a decrease in elk numbers and kill rates the following winter (Chapter 17 by Becker *et al.* and Chapter 23 by Garrott *et al.*, this volume), the Biscuit Basin pack was driven into the Madison drainage, encountered strife with the Cougar Creek pack that resulted in the death of a pack member, and eventually left the park entirely by the end of the winter. We documented extremely high carcass consumption in the winter of 2004–2005 (Chapter 17 by Becker *et al.*, this volume) and three wolves from the Biscuit Basin pack exhibited the worst body condition documented for Yellowstone wolves when they were darted mid-winter (Figure 15.13). Both of these findings suggest wolf packs in the Madison headwaters area were food stressed at high wolf densities following decreases in elk numbers.

Lastly, the Nez Perce pack demonstrated remarkable territory fidelity across the duration of seven winters. This pack primarily used the Firehole drainage despite the Madison and Gibbon drainages being equally available for the first five years, though their winter territory size decreased with increasing numbers of packs. Following the death of the alpha female and the subsequent dissolution of the Nez Perce pack in winter 2005–2006, the Gibbon Meadows pack quickly adopted the Nez Perce territory and demonstrated strikingly similar patterns of space use, possibly indicating that the Firehole was a preferred drainage. How patterns of wolf space use and territoriality will be affected with continued changes in prey abundance is an important question.

If current trends in prey biomass, distribution, and selection by wolves continue, then we predict the Madison headwaters area will only support one primary wolf pack. This prediction includes the area to the north of the Madison River, which is currently within the range of the Cougar Creek pack, and the use of Hayden Valley. The Nez Perce pack only occasionally used the Hayden Valley during the early years of our study, whereas in later years the Gibbon Meadows pack regularly used the Hayden Valley, likely because elk numbers in the Madison headwaters area became limiting. Therefore, future wolf occupation of the Madison headwaters area may depend on their contemporaneous use of other nearby areas (*e.g.*, Hayden Valley, Cougar Creek). The accessibility of these nearby areas is hard to predict because they are currently occupied by other territorial packs. The relatively small Hayden Valley pack (five wolves) has not deterred the Gibbon Meadows wolves from occasionally leaving the Madison headwaters area and using the Hayden Valley. However, use of Hayden Valley by Madison headwaters wolves may become limited if the Hayden Valley pack becomes larger, thereby indirectly affecting wolf occupation of the Madison headwaters area. Thus, managers should be conservative in their assessment of the future wolf population in central Yellowstone and expect only one pack to inhabit the Madison headwaters area, rather than the 2–4 packs observed during the past decade.

Another notable difference in wolves occupying the Madison headwaters area is the odd pattern of den attendance. Most other packs in Yellowstone showed localized behavior around den sites through

FIGURE 15.13 This 10-month old female from the Biscuit Basin pack, along with two other mates captured in January 2005, had the worst body condition of any wolves handled in Yellowstone, possibly indicating food stress in the Madison headwaters area that winter (National Park Service photo by Daniel Stahler).

the summer while they were tending pups. However, the dominant Nez Perce and then Gibbon Meadows packs in the Madison headwaters area spent less time tending their dens and were frequently widely scattered during summer. Interestingly, one den site was common to both of these packs. Perhaps Nez Perce-born individuals (*e.g.,* the breeding female) in the Gibbon Meadows pack returned to their natal area with prior knowledge of this den site that had suitable characteristics for raising offspring. Ungulate availability in the Madison headwaters area is more limited during summer than winter because the relatively small, nonmigratory, elk herd shifts to higher-elevation plateaus and most bison return to the Hayden Valley. Thus, wolves must range widely to find elk or other ungulate prey. The opposite appears to be true for wolf packs in other areas of Yellowstone because migratory elk and mule deer winter outside the park but return and increase local prey availability during summer; thereby creating a situation where wide-ranging movements by wolves are not as necessary during the main pup-rearing months (May through July).

Wolf population ecology and behavior in the Madison headwaters area is strikingly different than the rest of Yellowstone. Winter–summer prey densities and distributions led to patterns of landscape use and den attendance unique to wolves in Yellowstone. Decreasing elk numbers have strongly affected wolf numbers in the Madison headwaters area, but this could change if wolves increasingly prey on the numerous bison that migrate into the area during winter. How variations in winter weather and the associated effects of global warming will affect this system is an interesting subject for research in coming decades.

V. SUMMARY

1. We described wolf reintroduction efforts in the Madison headwaters area of Yellowstone National Park and changes in wolf population ecology as the system transitioned from no wolves to an established population.

2. Population growth of wolves was rapid from 1997 through 2004 and averaged 30% per year with a population beginning at five wolves in one pack and peaking at 45 wolves in four packs. Wolf numbers decreased during 2004 to 2006 to 16 wolves in two packs.

3. A combination of localized high density prey in winter and sparse, dispersed prey in summer likely contributes to the low number of wolves able to survive and reproduce in this system.

4. Intra-specific strife was the leading cause of death for wolves and caused the population to decrease in combination with emigration. A combination of mortality and emigration led to a "disappearance rate" of 4.2 wolves per year, while only 4.8 pups survived per year.

5. Wolf packs were territorial, but considerable overlap existed because packs used the relatively small, high-density winter range for ungulates disproportionate to the surrounding landscape. More spatial overlap was possibly tolerated between packs that were related.

6. No pack remained entirely in the Madison headwaters area through the winter, suggesting occupation of this area may depend on the contemporaneous use of other nearby areas (*e.g.*, Hayden Valley, Cougar Creek).

7. Compared to wolves occupying other areas of Yellowstone, den attendance was lower and summer movements were greater by wolves in the Madison headwaters area, likely due to local prey shortages around the den site.

VI. REFERENCES

Abelson, R. P. 1995. *Statistics as Principled Argument.* Erlbaum, Hillsdale, NJ.

Bangs, E. E., and S. H. Fritts. 1996. Reintroducing the gray wolf to central Idaho and Yellowstone National Park. *Wildlife Society Bulletin* **24:**402–413.

Boutin, S. 1992. Predation and moose population dynamics: A critique. *Journal of Wildlife Management* **56:**116–127.

Brown, C. G. 1992. Movement and migration patterns of mule deer in southeastern Idaho. *Journal of Wildlife Management* **56:**246–253.

Fritts, S. H., E. E. Bangs, J. A. Fontaine, M. R. Johnson, M. K. P, E. D. Koch, and J. R. G. 1997. Planning and implementing a reintroduction of wolves to Yellowstone National Park and central Idaho. *Restoration Ecology* **5:**7–27.

Fritts, S. H., C. M. Mack, D. W. Smith, K. M. Murphy, M. E. Phillips, M. D. Jimenez, E. E. Bangs, J. A. Fontaine, C. C. Niemeyer, W. G. Brewster, and T. J. Kaminski. 2001. Outcomes of hard and soft releases of reintroduced wolves in central Idaho and the greater Yellowstone area. Pages 125–147 *in* D. S. Maehr, R. F. Noss, and J. L. Larkin (Eds.) *Large Mammal Restoration: Ecological and Sociological Challenges in the Twenty First century.* Island Press, Washington, DC.

Fritts, S. H., and L. D. Mech. 1981. Dynamics, movements, and feeding ecology of a newly protected wolf population in northwestern Minnesota. *Wildlife Monographs* **80.**

Fuller, T. K., L. D. Mech, and J. F. Cochrane. 2003. Wolf population dynamics. Pages 161–191 *in* L. D. Mech and L. Boitani (Eds.) *Wolves: Behavior, Ecology, and Conservation.* University Chicago Press, Chicago, IL.

Garrott, R. A., G. C. White, R. M. Bartmann, L. H. Carpenter, and A. W. Alldredge. 1987. Movements of female mule deer in northwest Colorado. *Journal of Wildlife Management* **51:**634–643.

Gipson, P. S., W. B. Ballard, R. M. Nowak, and L. D. Mech. 2000. Accuracy and precision of estimating age of gray wolves by tooth wear. *Journal of Wildlife Management* **64:**752–758.

Harrington, F. H., and L. D. Mech. 1979. Wolf howling and its role in territory maintenance. *Behaviour* **68:**207–249.

Heisey, D. M., and T. K. Fuller. 1985. Evaluation of survival and cause-specific mortality rates using telemetry data. *Journal of Wildlife Management* **49:**668–674.

Horne, J. S., and E. O. Garton. 2006a. Selecting the best territory model: An information-theoretic approach. *Ecology* **87:**1146–1152.

Horne, J. S., and E. O. Garton. 2006b. Likelihood cross-validation versus least squares cross-validation for choosing the smoothing parameter in kernel home-range analysis. *Journal of Wildlife Management* **70:**641–648.

Keith, L. B. 1983. Population dynamics of wolves. Pages 66–77 *in* L. N. Carbyn (Ed.) *Wolves in Canada and Alaska: Their Status, Biology, and Management. Report Series, Number 45.* Canadian Wildlife Service, Edmonton, Alberta, Canada.

Kernohan, B. J., R. A. Gitzen, and J. J. Millspaugh. 2001. Analysis of animal space use and movements. Pages 125–166 *in* J. J. Millspaugh and J. M. Marzluff (Eds.) *Radio Tracking and Animal Populations.* Academic Press, San Diego, CA.

Maehr, D. S., R. F. Noss, and J. L. Larkin. 2001. *Large Mammal Restoration: Ecological and Sociological Challenges in the Twenty First Century.* Island Press, Washington, DC.

Mech, L. D. 1977. Wolf-pack buffer zones as prey reservoirs. *Science* **198**:320–321.

Mech, L. D., and L. Boitani. 2003. Wolf social ecology. Pages 1–34 *in* L. D. Mech and L. Boitani (Eds.) *Wolves: Behavior, Ecology, and Conservation.* University of Chicago Press, Chicago, IL.

Schoen, J. W., and M. D. Kirchoff. 1985. Seasonal distribution and home-range patterns of Sitka black-tailed deer on Admiralty Island, southeast Alaska. *Journal of Wildlife Management* **49**:96–103.

Smith, M. D., and E. Almberg. 2007. Wolf diseases in Yellowstone National Park. *Yellowstone Science* **15**:17–19.

Smith, E., T. D. Drummer, K. M. Murphy, D. S. Guernsey, and S. B. Evans. 2004. Winter prey selection and estimation of wolf kill rates in Yellowstone National Park, 1995–2000. *Journal of Wildlife Management* **68**:153–166.

Smith, S. B., D. R. Stahler, D. S. Guernsey, M. Metz, A. Nelson, E. Albers, and R. McIntyre. 2007. *Yellowstone Wolf Project Annual Report, 2006.* National Park Service, Yellowstone Center for Resources, Yellowstone National Park, WY.

U.S. Fish and Wildlife Service1994. *The Reintroduction of Gray Wolves to Yellowstone National Park and Central Idaho. Final Environmental Impact Statement.* U.S. Fish and Wildlife Service, Helena, MT.

Van Ballenberghe, V. 1987. Effects of predation on moose numbers: A review of recent North American studies. *Swedish Wildlife Research Supplement* **1**:431–460.

Waser, P. M. 1996. Patterns and consequences of dispersal in gregarious carnivores. Pages 267–295 *in* J. L. Gittleman (Ed.) *Carnivore Behaviour, Ecology, and Evolution.* Cornell University Press, Ithaca, NY.

Watson, F. G. R., and J. M. Rahman. 2003. Tarsier: A practical software framework for model development, testing and deployment. *Environmental Modeling and Software* **19**:245–260.

W, J. L., and G. T. H. 1998. Bison in the diet of wolves denning amidst high diversity of ungulates. Pages 141–144 *in* L. R. Irby and J. E. Knight (Eds.) *International Symposium on Bison Ecology and Management in North America.* Montana State University, Bozeman, MT.

White, G. C., and R. A. Garrott. 1990. *Analysis of Wildlife Radio-Tracking Data.* Academic Press, San Diego, CA.

Zar, J. H. 1999. *Biostatistical Analysis.* Prentice Hall, Upper Saddle River, NJ.

        CHAPTER 16

Wolf Prey Selection in an Elk-Bison System: Choice or Circumstance?

Matthew S. Becker,* Robert A. Garrott,* P. J. White,† Claire N. Gower,*
Eric J. Bergman,* and Rosemary Jaffe*

*Fish and Wildlife Management Program, Department of Ecology, Montana State University
†National Park Service, Yellowstone National Park

Contents

Theme

What a predator eats when given choices, and the subsequent effects of this behavior on ecosystem stability, has long been a topic of interest for ecologists. Prey selection is influenced by the absolute and relative abundances of prey types, the life history characteristics of predators and prey, and the attributes of the environment in which these interactions occur. Strong preference by a predator for a particular prey type can lead to ecosystem instability, while prey switching can lessen predation effects on the less abundant prey and enhance system stability. Evaluating prey selection in large mammal systems is difficult due to the broad spatial and temporal scales at which these predatory interactions occur, and investigations, particularly with wolf-ungulate systems, typically involve only the primary prey. Multiple prey species characterize most large mammal predator-prey systems, therefore research into predator-multiple prey dynamics has the potential to yield important ecological insights. We studied winter prey selection during 1996–1997 through 2006–2007 in a newly established wolf-elk-bison system where prey differed substantially in their vulnerability to wolf (*Canis lupus*) predation and wolves preyed primarily on elk (*Cervus elaphus*) but also used bison (*Bison bison*) to varying degrees within and

The Ecology of Large Mammals in Central Yellowstone
R. Garrott, P. J. White and F. Watson
ISSN 1936-7961, DOI: 10.1016/S1936-7961(08)00216-9

Copyright © 2009, Elsevier Inc.
All rights reserved.

among winters and packs. We analyzed the relative influences of prey abundance, predator abundance, and environmental variables on the selection of prey species and age classes and evaluated whether wolves exhibited prey switching from elk to bison.

I. INTRODUCTION

Predator-prey dynamics can be broadly classified by whether they are single prey or multiple prey systems. Predator diets can be relatively simple when only one prey type is present and few options exist. However, what a predator eats when given choices is a fundamental question germane to the multiple prey assemblages characteristic of most natural predator-prey systems. During the predatory sequence of encountering, attacking, and killing prey, most predators are assumed to select prey types based on abundance, thereby typically relying on encounter rates (Holling 1959). Once encountered, prey are frequently selected by sex, age, size, condition and behavior when individuals differ in their vulnerability to predation (Errington 1946, Morse 1980, Pastorok 1981, Greene 1986, Stephens and Krebs 1986, Quinn and Cresswell 2004). Selection across prey species can also differ based on their absolute or relative abundances and the life history characteristics of both predator and prey as manifested in morphology, defenses, and behavior. Thus, all of these variables have the potential to dramatically influence the dynamics of multiple-prey systems (Murdoch 1969, Fitzgibbon and Lazarus 1995, Moran *et al.* 1996, Denno and Peterson 2000, Denno *et al.* 2002, Rosenheim *et al.* 2004). While the physical vulnerability of a species or individual is of considerable importance in predatory interactions, environmental attributes can also influence vulnerability. Variables such as heterogeneity in climate, habitat structure, and landscape attributes can act alone or in concert with physical vulnerability to influence a predator's diet (Smuts 1978, Peckarsky and Penton 1989, Hunter and Price 1992, Langellotto and Denno 2004, Hopcraft *et al.* 2005, Chapter 24 by Garrott *et al.*, this volume).

Prey selection by predators in multiple-prey systems can have fundamental positive or negative effects on community stability and prey diversity (Oaten and Murdoch 1975, Murdoch and Bence 1987, Holt and Lawton 1994, Bonsall and Hassell 1997, Synder and Ives 2001). When a predator consumes a prey item disproportionately to its abundance it is said to exhibit a preference (Begon *et al.* 1996). Many predators have strong preferences for a certain prey type regardless of its abundance (*i.e.*, specialist), and this strong preference is typically viewed as destabilizing to a predator-prey system (Andersson and Erlinge 1977, Hanski *et al.* 1991, Turchin and Hanski 1997, Eubanks and Denno 2000). Conversely, other predators consume a wide variety of prey, with changes in prey availability strongly affecting their patterns of selection (*i.e.*, generalist). Prey switching behavior is typically associated with generalist predators and occurs when attacks are disproportionately frequent when a prey species is abundant and disproportionately infrequent when a prey species is rare (Murdoch 1969). Switching behavior by predators is generally viewed as stabilizing to a system because a predator can have a regulating influence through density-dependent predation on both prey species (Oaten and Murdoch 1975, Fryxell and Lundberg 1994). Thus, the patterns of prey selection a predator exhibits can have dramatically different ecological consequences (Paine 1966, Holt 1977, Caswell 1978, Hanski *et al.* 1991, Fryxell and Lundberg 1994, Krivan and Eisner 2003).

Studies of predation and its effects on ecosystem stability are difficult because, even in experimental settings, disentangling the myriad factors influencing prey selection is quite complicated. Nevertheless, decades of investigations have increased our understanding of these processes (Murdoch 1969, Oaten and Murdoch 1975, Post *et al.* 2000, Krebs *et al.* 2001, van Balaan *et al.* 2001, Prugh 2005). Relative to investigations of smaller taxa, intensive long-term studies of prey selection and stability in large mammal systems are hindered by the logistic and financial constraints imposed by the broad spatial and temporal scale of investigations. However, large mammal systems have the potential to yield significant insights because the life history characteristics of top predators and large herbivores with strong ecological influences differ substantially from those of smaller taxa typically used in prey selection studies (McNaughton 1985, Temple 1987, Frank and McNaughton 1992, Hobbs 1996, Terborgh *et al.* 2001, Garrott *et al.* 2007). Unlike systems of smaller taxa where prey typically rely on avoiding detection, large herbivore prey species are formidable and diverse in their array of defenses and behaviors they can employ once encountered and attacked. These defenses preclude large predators from killing all types of prey with equal effort and subject the predator to constant risk of severe injury and even death (Makacha and Schaller 1969, Mech 1970, Kruuk 1972, Schaller 1972, Carbyn and Trottier 1987, Creel and Creel 2002, Mech and Peterson 2003, Smith *et al.* 2003, MacNulty *et al.* 2007). As a result, substandard or vulnerable individuals are frequently selected. This vulnerability often depends on attributes of the prey (*e.g.*, age, size, physiological condition, behavior), environmental attributes, prey density, encounter rates, and the availability of alternative prey (see Mech and Peterson 2003 for review).

An impressive body of work has been compiled on wolf prey selection during the last several decades. Wolves are typically considered consummate generalists—opportunistic coursing predators taking advantage of whatever vulnerable prey are available within their territories (Mech 1970, Mech and Peterson 2003). However, virtually all wolf-ungulate investigations in multiple-prey systems have also demonstrated a clear selection for a particular prey species relative to other species in an assemblage (Carbyn 1983, Huggard 1993a, Dale *et al.* 1995, Jędrzejewski *et al.* 2000, Hebblewhite *et al.* 2003, Smith *et al.* 2004). Also, within a prey species wolves generally select certain age classes such as young-of-the-year (Mech 1970, Mech *et al.* 1995, Jaffe 2001, Smith *et al.* 2004) that could be considered different prey types due to differences in vulnerability. The dynamics of multiple-prey systems, and the mechanisms and conditions whereby wolf prey selection affects community stability and diversity have not been extensively investigated (Dale *et al.* 1995) and should yield insights into predation processes and the dynamics of large-mammal predator-prey systems.

Furthermore there is a need to formally evaluate prey switching for wolves (Dale *et al.* 1994), because switching has often been incorporated into models of wolf-ungulate dynamics and used to describe simple changes in predator diet composition rather than density-dependent predation (Garrott *et al.* 2007). Wolves exhibit many of the attributes common to predators that switch prey, including typically hunting by sight, cueing into the different areas where each prey species can be found (Bergman *et al.* 2006), and testing and evaluating individual prey (Murie 1944, Carbyn *et al.* 1993, MacNulty *et al.* 2007). Thus, it is feasible wolves could exhibit prey switching under some conditions. However, if prey preference for a particular species is strong, perhaps due to differences in vulnerability, then switching is unlikely to occur (Murdoch 1969, Murdoch and Marks 1973).

In addition to the ecological complexity inherent in studies of predation, comparative investigations are further complicated by the frequent use of terminology without consistent and explicit definitions and distinctions. Specifically, the concepts of prey selection, vulnerability, prey preference and prey switching are ubiquitously employed in predator-prey literature, but typically without concise definitions or differentiation. Prey selection is often considered to be what a predator eats when given choices, with no reference to the abundances of the various prey types available. Fundamental to this choice is the concept of differential vulnerability, or the factors that make an individual more susceptible to predation than other animals in a system. Vulnerability is considered to be of overriding importance in many predator-prey systems, especially those involving large mammals (Ivlev 1961, Mech 1970, Menge 1972, Power *et al.* 1992, Sinclair and Arcese 1995). However, precise definitions of vulnerability are infrequent and highly variable, ranging from the product of encounter rates and attack probabilities (Greene 1986, Pastorok 1981), "a combination of capture efficiency and profitability relative to risk" (Mech and Peterson 2003:140), to comprising "all of the behaviors that a prey can adopt to modify its risk of being targeted and caught when attacked" (Lind and Cresswell 2005:946). Disparities in definitions owe to the fact that quantifying vulnerability in most natural systems is extremely difficult because it is influenced by physical, behavioral, and environmental factors, and can vary among individuals, populations, species, and landscapes (Chapter 24 by Garrott *et al.*, this volume).

Consequently, defining and quantifying prey preference is equally fraught with difficulties and most ecologists employ what Taylor (1984) terms the "black box" definition of preference, defined as when a predator selects a prey type disproportionately to its occurrence in the environment. This interpretation is frequently employed in analyses because it serves as an umbrella for data that are rarely available in natural systems, encapsulating every decision a predator makes based on the myriad physical, behavioral, and environmental factors acting upon all stages of the predatory encounter, attack, and capture (Taylor 1984). Lastly, the definition of prey switching comes from Murdoch (1969) where "the number of attacks on a species is disproportionately large when the species is abundant relative to the other prey, and disproportionately small when the species is relatively rare." This indicates a preference that is not constant across all levels of abundances of the prey types but changes across a gradient of relative prey abundances, with a predator having a preference for the most abundant prey. Evaluations of dynamics in multiple-prey systems suffer from a lack of consistency in using the term "prey switching," with some investigators employing it to indicate a density-dependent change in predator preference (Murdoch 1969), while others use it to simply describe changes in predator diet composition.

We evaluated prey selection by wolves in a newly-colonized, bison-elk prey system in the Madison headwaters area of Yellowstone National Park during the winters of 1996–1997 through 2006–2007 (Figure 16.1). Wolf numbers varied between 2–50 wolves in 1–5 packs after they were reintroduced and colonized the area beginning in 1995–1996 (Chapter 15 by Smith *et al.*, this volume). Elk were resident throughout the year, but their numbers decreased from approximately 600 to 174 following wolf establishment (Chapters 11 and 23 by Garrott *et al.*, this volume). In contrast, bison were seasonally migratory with numbers increasing through each winter (200–1500) until they exceeded elk numbers by several orders of magnitude in late winter (Chapter 12 by Bruggeman *et al.*, this volume). The dramatic contrasts in life history characteristics, movements, and abundance between these two prey species, coupled with variations in snow pack and wolf abundance, offered a unique opportunity to evaluate prey selection. Elk are smaller in size and prone to flight as an anti-predator behavior, while bison are larger and tend to employ sophisticated group defenses (Carbyn 1974, Carbyn and Trottier 1987, Carbyn *et al.* 1993, MacNulty *et al.* 2007). Our objectives were to: (1) characterize

FIGURE 16.1 A member of the Hayden pack feeds on a freshly-killed bull elk in the Madison Canyon. Wolves exhibited a preference for elk but selected bison to varying degrees within and among winters (Photo by Shana Dunkley).

wolf prey selection over time and among packs, (2) evaluate the drivers of wolf prey selection at the species- and age-class levels, and (3) evaluate if wolves switched from elk to bison in this multiple-prey system.

II. METHODS

A. Detecting and Identifying Wolf Kills

We conducted intensive predation investigations in the primary winter ranges of bison and elk in the Madison headwaters area (31,000 ha), with concurrent investigations of these prey species allowing collection of wolf predation data in a tractable area with a well-described ungulate prey base. We documented prey selection by wolves during 15 November through 30 April each winter from 1996–1997 through 2006–2007. Our sampling unit was radio-collared wolf packs that used the study area as part of their territory. Wolves were aerially darted from helicopters by National Park Service biologists and equipped with VHF telemetry collars. A total of 37 wolves from four packs were collared during the course of the study (Chapter 15 by Smith *et al.*, this volume).

The number and sizes of wolf packs using the study area were dynamic within and among winters. Thus, we used ground observations, snow-tracking, and aerial counts during tracking flights by park biologists to estimate the wolf population. We defined two metrics, wolf days and pack days, as one wolf or one pack in the study area for one day, respectively (Chapter 15 by Smith *et al.*, this volume). We also defined multiple pack days as the number of days when more than one pack was present in the study area. We used roads traversing each river drainage in the study area (Chapter 2 by Newman *et al.*, this volume) to sample for wolf presence daily throughout the winter. Sampling began at dawn with ground crews of 3–4 people covering all roads by snowmobile or vehicle, and using strategic high points in the landscape to facilitate telemetry triangulations (White and Garrott 1990) and observations of wolves (Figure 16.2). When possible, multiple locations were obtained in early morning and evening each day. We also recorded any uncollared wolves detected opportunistically via tracks or

FIGURE 16.2 Using telemetry to detect, locate, and monitor radio-collared wolf packs. High points on the landscape were utilized for signal detection, scanning for avian scavengers, and observations (Photo by Shana Dunkley).

observations to aid in the estimation of the wolf population using the study area. In addition, biologists studying elk and bison routinely covered backcountry areas to assist with wolf detection.

When wolves were located, we used visual scans and monitoring of avian scavengers in the vicinity to detect kills. Ravens preferentially associate with wolves in winter, and an average of 28.6 ravens (*Corvus corax*) were present at fresh wolf kills in the northern range of Yellowstone (Stahler *et al.* 2002), with slightly lower averages in the Madison headwaters area (D. Stahler, National Park Service, personal communication). This association facilitated the detection of kills. We also conducted extensive snow-tracking after wolves departed the area to further facilitate kill detection (Huggard 1993a, Dale *et al.* 1995, Jędrzejewski *et al.* 2000, Jaffe 2001, Hebblewhite *et al.* 2003). We necropsied ungulate carcasses to determine cause of death, species, sex, age, and condition (Figure 16.3). Wolf kills were inferred from collective evidence of subcutaneous hemorrhaging indicative of injuries sustained before death, signs of struggle or chase at the kill site, blood trails, signs of predator presence, and our knowledge of wolf movements and activities. We documented frequent spring grizzly bear (*Ursus arctos*) predation on bison during the latter years of the study. Thus, when both bears and wolves were present on a kill, we classified it based on the patterns of injury and subcutaneous hemorrhaging. Bears typically attacked the head and spine, while wolves attacked the hindquarters and flanks. Similarly, mountain lion (*Puma concolor*) kills of elk were determined based on characteristics of the kill site and patterns of injury. Kills were sexed using the presence of genitalia, horns, antlers, or pedicels, and aged based on

FIGURE 16.3 Performing a necropsy on a wolf-killed calf elk in the Madison canyon. Extensive examinations were performed on every ungulate carcass to determine whether the animal was killed by predators or died of other causes, as well as to assess the condition of the animal and the attributes of the site in which it was killed. Wolves preferred elk calves over other prey types and the abundance of calves strongly influenced variation in wolf prey selection (Photo by Kevin Pietrzak).

size and patterns of tooth eruption and replacement (Fuller 1959, Hudson *et al.* 2002). When available, an incisor or canine was removed from adult ungulates and aged using cementum annuli (Moffitt 1998, Hamlin *et al.* 2000). Marrow fat from the femur or humerus was assessed visually based on color and consistency, and classified as: (1) solid and white, (2) 50–75% solid with red spots, (3) 25–50% solid, reddish, and (4) 0–25% solid, gelatinous and red (Cheatum 1949). We also classified the extent of fluoride toxicosis and necrosis (0 = none, 1 = mild, 2 = moderate, 3 = severe) in jaws from adult animals because these ailments were relatively common due to the strong geothermal influence (Shupe *et al.* 1984, Garrott *et al.* 2002, Chapter 10 by Garrott *et al.*, this volume).

Observed patterns of wolf predation can be biased by differing rates of detection for various prey types, with smaller prey such as calves consumed faster and, thus, potentially detected less frequently (Fuller and Keith 1980, Fuller 1989, Hebblewhite *et al.* 2003). While this bias is more likely in aircraft-based studies or studies that do not use snow-tracking (Fuller 1989, Dale *et al.* 1995), recent studies have considered kill detection efficiency in ground-tracking (Jędrzejewski *et al.* 2000, Jaffe 2001, Hebblewhite *et al.* 2003, Smith *et al.* 2004). We empirically evaluated our efficiency in detecting kills and concluded our methods provided accurate data on wolf prey selection patterns (Jaffe 2001).

B. Factors Influencing Wolf Prey Selection

The probability of a prey animal being consumed by a predator is the product of the probability of being encountered by the predator, the probability of the predator attacking the prey once encountered, and the probability that the attack is successful (Endler 1991). We did not assess encounter rates and attack rates, but recognized that bison and elk calves could be considered separate prey items given their dramatic differences in vulnerability compared to adults (Figure 16.4, Mech 1970, Carbyn and Trottier 1987, Mech *et al.* 1995). Thus, we identified four main prey types available to wolves (*i.e.*, elk calves, elk adults, bison calves, bison adults) in the Madison headwaters area.

FIGURE 16.4 An elk calf and bull bison feeding along the banks of the Firehole river. Elk calves and bull bison represent the extremes in prey sizes and defenses confronting wolves (Photo by Kevin Pietrzak).

Studies of prey selection in natural systems where encounter and attack rate data are unavailable typically employ selection indices (Lechowicz 1982), whereby the occurrence of a prey type in the predator's diet is compared to its abundance in the system. Thus, we calculated selection indices for the four prey types using Chesson's (1978) alpha method across early-, middle-, and late-winter periods from the establishment of resident packs in winter 1998–1999 through 2006–2007 to determine whether wolves selected any prey types disproportionately to their abundance.

However, selection indices are limited to prey abundance questions. Therefore, we employed a multinomial logit analysis (Menard 2002) to evaluate the relative importance of prey abundance, predator abundance, and environmental variables on selection across prey species and age classes. We modeled four response categories corresponding to the four main prey types available to resident wolf packs. Each kill comprised an observation and we used elk calves as the base model because wolves tended to select them when available (Jaffe 2001, Smith *et al.* 2004). Three logits were modeled as $L_a(x) = \log[\pi_a(x)/\pi_0(x)](a = 1, 2, 3)$, where $\pi_0(\mathbf{x})$, $\pi_1(\mathbf{x})$, $\pi_2(\mathbf{x})$, and $\pi_3(\mathbf{x})$ were the probabilities of a calf elk, adult elk, calf bison, and adult bison response, respectively. $\pi_0(\mathbf{x})$ was the denominator (*i.e.*, baseline response) of each odds and $\mathbf{x} = (x_1, x_2, \ldots, x_p)$ was a vector of model covariates. We developed a suite of covariates corresponding to prey, predator, and snow pack variables to evaluate factors influencing prey selection by wolves. These covariates were chosen based on our knowledge of the system and variables reported to significantly influence prey vulnerability and selection by wolves in other systems. We estimated covariates for the date that each wolf kill occurred.

Wolf population structure consisted of territorial packs, each occupying a particular territory (Chapter 15 by Smith *et al.*, this volume). Prey abundance also varied temporally and spatially, with bison prone to frequent large-scale movements between drainages (Chapter 12 by Bruggeman *et al.*, this volume). Thus, all prey in the study area were not equally available to all wolf packs, and such disproportionate abundance could affect wolf diets. Therefore, we estimated prey abundance at a scale contained within the river drainages (Madison, Gibbon, Firehole) of each pack's respective territory. We determined the drainages used by each pack each winter by constructing 95% fixed kernel territories from ground locations collected from 15 November through 30 April each winter (Chapter 15 by Smith *et al.*, this volume). We excluded temporary probes by packs into other drainages from these calculations.

We estimated abundance covariates (ELK_{calf}, ELK_{adt}, $BISON_{calf}$ and $BISON_{adt}$) for the four prey types within each drainage by decomposing the 167-day winter field season into three approximately 8-week periods corresponding to early, middle, and late winter. The non-migratory elk population (Craighead *et al.* 1973, Garrott *et al.* 2003) was not subject to the dramatic fluctuations in abundance characteristic of migratory populations, and typically only experienced decreases across winter due to starvation and predation, particularly of calves (Chapter 23 by Garrott *et al.*, this volume). Consequently, we estimated the abundance of adult elk and calves during early, middle, and late winter using mark-resight techniques and age composition data (Chapters 11 and 23 by Garrott *et al.*, this volume). We conducted multiple mark-resight surveys in late winter ($n = 10$–33) when elk were concentrated in meadow complexes. We also estimated calf:cow ratios during early and late winter using the respective first and last 100 random elk groups obtained from telemetry sampling (Chapters 11 and 23 by Garrott *et al.*, this volume). We then estimated adult elk abundance (ELK_{adt}) for the early-winter period by multiplying the previous spring's mark-resight population estimate by a pooled summer survival rate of 0.95 derived from telemetry data (Chapter 23 by Garrott *et al.*, this volume). We assumed 85% of the adult population was females (Chapter 11 and 23 by Garrott *et al.*, this volume) and multiplied the early-winter adult female estimate by the early-winter calf:cow ratio to obtain an early-winter calf (ELK_{calf}) estimate. Similarly, we multiplied the adult cow estimate by the late-winter calf:cow ratio to obtain the number of elk calves remaining in the late-winter period. A late-winter adult elk estimate was then calculated by subtracting the late-winter calf estimate from the total late-winter mark-resight population estimate. We averaged the respective means of the early- and late-winter estimates to approximate the abundance of both adult and calf elk for the mid-winter period. While elk distribution among the three drainages varied among winters (Chapter 21 by White *et al.*, this volume), there was little elk movement between drainages within winters (Chapter 18 by Gower *et al.*, this volume). Thus, we multiplied our estimates by the proportion of the elk population observed within each drainage during the spring mark-resight surveys to estimate the abundance of both prey types within each of the three drainages. Lastly, we estimated a covariate for the total abundance of elk (ELK) by summing the adult and calf estimates.

We estimated bison adult and calf abundance ($BISON_{adt}$ and $BISON_{calf}$, respectively) by conducting ground counts through the winter range every 10–16 days, with observers recording the number, location, sex, and age class of all observed bison. Each drainage was subdivided into discrete survey units and bison totals for each drainage were calculated by summing the respective unit totals. Because substantial changes in bison abundance could occur between surveys, we interpolated between estimates to derive the bison abundance estimates for the date of each wolf kill. We also estimated a total bison abundance covariate (BISON) by summing the adult and calf estimates. In addition, we calculated the ratio of bison to elk abundance (BISON.ELK) by dividing the bison estimate by the elk estimate.

The high density ungulate winter range of the Madison headwaters experienced considerable wolf use despite comprising a relatively small area. Following the establishment of multiple packs in the system there was a substantial increase in wolf abundance, spatial and temporal territory overlap, and inter-pack strife (Chapter 15 by Smith *et al.*, this volume). These dynamics, coupled with increases in elk anti-predator responses to increasing wolf numbers (Chapters 18, 19, and 20 by Gower *et al.*, Chapter 21 by White *et al.*, this volume) and decreasing elk numbers (Chapter 23 by Garrott *et al.*, this volume), were negatively related to kill rates for wolves using the system (Chapter 17 by Becker *et al.*, this volume) and likely affected prey selection. Thus we developed three covariates to index the strength of these competitive interactions: the wolf:ungulate ratio (WOLF:UNG); the wolf:elk ratio (WOLF:ELK); and multiple pack days ($PACK_{mult}$). We estimated each of these as population level indices for early-, middle-, and late-winter periods across the entire study area because wolf territories often overlapped extensively, pack territories often included more than one drainage, and adjacent packs likely influenced each others' movements and behaviors (Chapter 15 by Smith *et al.*, this volume). We calculated the wolf:ungulate and wolf:elk ratios by dividing the total wolf days estimated for the time period by the number of days in the period, and then dividing by the mean elk and bison estimates. We estimated multiple pack days by summing the total number of days in a period during which more than one pack was detected in the study area.

Snow pack substantially decreases ungulate mobility and increases their vulnerability to wolf predation (Peterson 1977, Parker *et al.* 1984, Nelson and Mech 1986, Huggard 1993b). Snow depth, density, and crusting can impede escape for ungulates that employ flight as an anti-predator tactic. Snow depth is not an accurate integrator of snow pack attributes due to differences in density, crust conditions, and layers. Thus, we described the temporal and spatial dynamics of snow pack using a validated snow model (Chapter 6 by Watson *et al.*, this volume) to estimate mean daily snow-water equivalents (*i.e.*, the amount of water in a column of snow; SWE_{mean}), for the Firehole, Gibbon, and Madison drainages during 1 October through 30 April. When a pack's territory encompassed more than one drainage, we calculated mean snow pack metrics across the two drainages. SWE_{mean} was estimated for the date of each wolf kill to provide an indirect measure of prey escape ability.

The nutrition and condition of ungulates in mid- to high-latitude systems decrease through the winter because most forage is senescent and animals must forage and travel through snow (Chapter 9 by White *et al.*, this volume). Consequently, the accumulation and duration of snow pack can have a long-term weakening influence on ungulate physiological condition that can ultimately be lethal in severe winters (Murie 1944, Severinghaus 1947, Jędrzejewski *et al.* 1992, Garrott *et al.* 2003). We estimated the sum of daily snow-water equivalent values beginning on 1 October each winter (SWE_{acc}; Garrott *et al.* 2003) for the date of each kill to provide an indirect measure of ungulate physiological condition.

We developed and evaluated *a priori* hypotheses to estimate the relative influence of prey abundance, snow pack, and wolf competition on wolf prey selection of ungulate species and age classes. Our *a priori* hypotheses were expressed in four main model structures incorporating prey abundance and other potential covariates of snow pack and wolf competition (Appendix). For each covariate, we then identified the metrics we believed were appropriate for estimation. We developed 84 candidate models in the form of multinomial logit equations to evaluate our hypotheses. Three logit equations were generated for each candidate model, describing: (1) the probability of an elk adult kill compared to an elk calf kill, (2) the probability of a bison calf kill compared to an elk calf kill, and (3) the probability of a bison adult kill compared to an elk calf kill.

For comparison of coefficient estimates, we scaled and centered each covariate prior to analysis by subtracting the dataset's midpoint from each covariate value and dividing them by the dataset's midrange. This restricts each covariate's values to fall within -1 and 1, inclusively. We assessed potential colinearity between covariates using variance inflation factors and did not use covariates with values >6 in the same model (Neter *et al.* 1996). Covariates that were not used in the same model due to strong colinearity were $BISON_{adt}$ and $BISON_{calf}$, SWE_{mean} and SWE_{acc}, and combinations of MULTPK, WOLF:UNG, or WOLF:ELK. We fitted all models in R version 2.4.1 using the function multinom in the nnet package (R Development Core Team 2006). Models were compared using Akaike's Information Criterion corrected for small samples (AIC_c; Burnham and Anderson 2002). We calculated Akaike weights and evaluated the importance of each covariate by its predictor weight (w_p), which we calculated by summing the Akaike weights for all models containing the covariate in the final model suite (Burnham and Anderson 2002). Our model selection followed a stepwise procedure within each suite, whereby we first fit all candidate models, calculated AIC_c and model weights and then determined for a given model structure which metric best estimated a given covariate. For example, if the three top models had identical structure and differed only by their inclusion of a different metric of wolf competition (WOLF:UNG, WOLF:ELK, MULTPK), then we determined which model was best-supported and removed the other two models from among suite comparisons. Among suites, we then recalculated model weights for the reduced set of models once we had determined the best metric for a given model structure.

Because elk are considerably more vulnerable to wolf predation than bison (MacNulty 2002), we predicted that covariates of bison abundance ($BISON_{adt}$, $BISON_{calf}$, BISON, BISON.ELK) would have no effect on the probability of wolves eating adult versus calf elk. We also predicted that elk abundance covariates (ELK_{adt}, ELK_{calf}, ELK) would be negatively related to the probability of wolves eating a bison adult or bison calf given that low elk abundance (absolute and relative to bison) would likely compel wolves to kill bison with increasing frequency. In addition, we predicted the probability that wolves would kill bison compared to elk calves would increase with the relative and absolute numbers of bison because the number of vulnerable individuals in the bison population would likely increase with the influx of migrating animals during winter.

We predicted that winters with more severe snow pack, as indexed by SWE_{acc}, would be positively correlated with wolves killing both bison age classes and adult elk because the larger and less vulnerable species and age classes would become weakened and relatively more vulnerable. Given that elk and bison differ in their responses to wolves, with elk typically employing flight and bison resorting to group defense (Carbyn 1974, Carbyn and Trottier 1987, MacNulty 2002), the two snow pack metrics could have different effects on vulnerability across species. Specifically SWE_{mean} could be more influential in predation of elk, as increasing values of SWE_{mean} equate to decreased mobility, while SWE_{acc} could be more influential in predation of bison by weakening their ability to defend themselves against attack. In addition, we predicted that competition imposed by multiple wolf packs (indexed by MULTPK, WOLF:UNG, and WOLF:ELK) would have a positive influence on the probability of wolves taking bison because packs competing for limited and decreasing elk resources would likely need to pursue other prey species to persist. Similarly, we predicted increasing wolf competition would result in increased selection for adult elk as wolves expanded through the study area.

Cooperative-hunting large carnivores often exhibit a positive relationship between group size and prey size (Rosenzweig 1966, Gittleman 1989, Creel and Creel 1995). However, the relationship between pack size and prey size for wolves is unclear (Mech and Boitani 2003). We did not expect pack size to be positively correlated with prey size (Chapter 17 by Becker *et al.*, this volume), but added a pack size covariate ($WOLF_{pk}$) to the best-supported *a priori* models to determine if the covariate improved model fit.

C. Prey Switching

Evaluating if wolves are capable of prey switching, or have a strong preference for elk regardless of bison abundance, cannot be determined by examining diet composition alone (Garrott *et al.* 2007). Thus, we evaluated wolf preference and potential prey-switching by relating the relative availability of bison and elk in the study system with the ratio of the two prey species in the wolves' diet. Murdoch (1969) provided the classic equation that relates the ratio of two prey types eaten by a predator (g_1/g_2) to the ratio of the prey types available to the predator (N_1/N_2). We evaluated the existence and extent of wolf prey switching by regressing the ratio of bison and elk in wolf diets to the ratio of bison and elk available in the population and evaluating the subsequent form (*i.e.*, linear or nonlinear) of the relationship (Murdoch 1969, Garrott *et al.* 2007). We used Murdoch's (1969) selection coefficient where the ratio of the two prey types eaten is denoted by:

$$\frac{g_{bison}}{g_{elk}} = c \frac{N_{bison}}{N_{elk}} \tag{16.1}$$

The left-hand side of Equation 1 is the ratio of bison to elk in wolf diets and N_{bison}/N_{elk} is the ratio of bison to elk in the population. The proportionality constant, c, measures "the bias in the predator's diet to one prey species" and relates the ratio of prey eaten to their relative abundance (Murdoch 1969:337). If wolves exhibit a high plasticity in their diet, then prey selection would likely change depending on the relative availability of the two prey types as determined by their abundance, vulnerability, and actual predator preference (Garrott *et al.* 2007). This dynamic nature of c can be incorporated by modifying the equation to allow changes in diet with changes in relative availability of elk and bison:

$$\frac{g_{bison}}{g_{elk}} = \left(c \frac{N_{bison}}{N_{elk}} \right)^b \tag{16.2}$$

The variable b is a measure of the extent of prey switching, with values greater than one denoting switching (Greenwood and Elton 1979, Elliot 2004). If wolves preferred elk proportional to relative abundance ratios of the two prey, then we would expect the relationship between diet and abundance ratios to be linear (Murdoch 1969, Garrott *et al.* 2007) because wolves would continue to prefer elk over all ranges of relative abundance ratios, even when elk were rare relative to bison (Figure 16.5). However, if wolves exhibit prey-switching then the relationship should be curvilinear and indicate a diet switch to the more abundant prey with increasing bison:elk ratios (Murdoch 1969, Garrott *et al.* 2007).

Obtaining sufficient data on the ratio of bison to elk in wolf diets required a time-scale of three winter periods of approximately eight weeks each, during which time bison abundance varied substantially among drainages (Chapter 12 by Bruggeman *et al.*, this volume). To account for this, we estimated prey abundance for the entire study system each winter period and calculated the ratio of the two prey species in wolf diets by pooling all wolf-killed elk and bison detected during the respective periods and deriving a ratio of bison to elk (g_{bison}/g_{elk}). Relative abundance ratios of prey (N_{bison}/N_{elk}) were estimated by calculating mean bison population estimates from surveys conducted during the early-, middle-, and late-winter periods and dividing by elk population estimates for these periods.

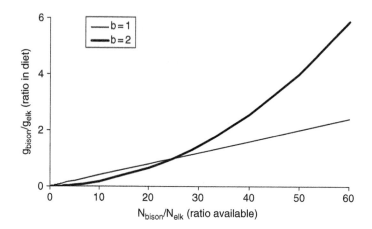

FIGURE 16.5 Theoretical relationship between the ratio of bison to elk in wolf diets versus the ratio of bison to elk available in the population. Curves for the scenarios for no prey switching ($b = 1$) and prey switching ($b = 2$) are depicted (Murdoch 1969, Garrott *et al.* 2007).

Elk estimates by winter period were calculated similarly to the multinomial model. Twenty-seven data points were generated corresponding to nine years of three winter periods each. To determine the form of the relationship between the wolf diet ratio and the ratio of prey abundance, we fit Equation 2 to these data and estimated parameter coefficients using the nls function from the nlme package in R version 2.4.1 (R Core Development Team 2006).

III. RESULTS

Wolves were detected in the study area on 1306 days of the 1837 day study period, comprising a total of 16,801 wolf days, 1872 pack days, and 437 multiple pack days. We obtained 1369 telemetry locations, 534 visual locations, and 4175 km of backtracking. Approximately 6600 person days were spent in the field, and an estimated 368,000 km were logged on snowmobiles and vehicles.

Wolf presence during the 167-day winter field season ranged from 60–3964 wolf days, with pack days and multiple pack days ranging from 19–383 and 0–128, respectively. Established packs ranged in size from 2 to 22 wolves (mean = 9.6; 95% CI = 9.4, 9.8), and the percentage of days wolves were detected during the field season ranged from 19% to 96%. Ten different wolf packs used the Madison headwaters area to varying degrees over the course of the study, with wolves first detected during the winter of 1996–1997 when several itinerant wolves used the area and the Nez Perce pack was soft released into the Firehole drainage (Chapter 15 by Smith *et al.*, this volume). The Nez Perce pack became established in the study area during 1998–1999, and was the only resident pack until the winter of 2002–2003 when the Cougar pack and another uncollared pack in the Gibbon drainage used portions of the study area. Two more packs became established in the study area during 2003–2004, and wolf presence peaked in 2004–2005 with up to five packs totaling approximately 45 wolves using the study area (Chapter 15 by Smith *et al.*, this volume). The wolf population decreased precipitously during winter 2005–2006 to primarily one pack, before increasingly to primarily three packs and an estimated 21 wolves the following winter (Chapter 15 by Smith *et al.*, this volume).

Elk population estimates for the study area ranged from 290–664 in autumn to 174–577 in spring, with the population decreasing 5–42% during winter. A progressive decrease in the elk population began in 2003–2004 and continued through spring 2007 when the population was estimated at 174 animals (Chapter 23 by Garrott *et al.*, this volume). Elk were equally distributed among the three drainages until the winter of 2000–2001 when the proportion of animals in the Madison drainage abruptly increased, accompanied by the virtual elimination of elk in the Gibbon drainage, and a gradual decrease in the Firehole drainage to a low of 28 animals (16% of the population) by the end of the study (Chapter 18 by Gower *et al.*, and Chapter 21 by White *et al.*, this volume). The proportion of elk in the Madison drainage gradually increased to 84% of the population in late-winter 2006–2007. Calf abundance also decreased 48–98% during each winter (Chapter 23 by Garrott *et al.*, this volume). Elk abundance estimates within pack territories ranged from 28–331 total animals, 0–84 calves, and 28–271 adults.

We conducted 114 ground distribution surveys of bison from 1997–1998 to 2006–2007, with abundance ranging from 205–1538 animals using the study area. Bison abundance generally increased as winter progressed and animals migrated into the study area from the Hayden and Pelican Valleys (Chapter 12 by Bruggeman *et al.*, this volume). Estimated bison abundance in pack territories ranged from 20–1108 animals, with 3–271 calves and 17–952 adults. With fluctuating populations of both prey and predator within and across seasons, we also documented considerable variation in the ratios of bison:elk, wolf:ungulate, and wolf:elk. Estimates ranged from 0.10–29.57 for bison:elk ratios, 0.003–0.038 for wolf:ungulate ratios, and 0.006–0.100 for wolf:elk ratios, respectively.

Snow pack accumulation in the study area typically began in late October (Chapter 6 by Watson *et al.*, this volume) and increased until late March when spring melt began, particularly in the lower-elevation meadows and drainages. Snow pack during the course of the study was below historical averages, with annual maximum SWE_{acc} values ranging from 1023–3612 cm days and averaging 2044 cm days (95% CI = 2038, 2050). Maximum SWE_{mean} values ranged from 9.8–31.9 cm and averaged 18.8 cm (95% CI = 14.0, 23.6).

A total of 759 wolf-killed ungulates and 21 canids were detected during the study period. Ungulate kills were comprised of elk (79.8%, $n = 606$), bison (19.9%, $n = 151$), moose (0.1%, $n = 1$), and mule deer (*Odocoileus hemionus*; 0.1%, $n = 1$), while wolf-killed canids consisted of coyotes (71.4%, $n = 15$), wolves (23.8%, $n = 5$), and red fox (4.8%, $n = 1$). Detected kills varied from 14–106 among winters, with a mean of 81.0 kills (sd = 16.8) following wolf establishment in 1998–1999. Elk were the primary prey species for wolves, with calves and adult females comprising 38% ($n = 292$) and 32% ($n = 241$) of total kills, respectively (Table 16.1). The percentage of bison in the pooled diets of resident wolf packs increased from zero soon after wolf recolonization to 53% ($n = 29$) in winter 2005–2006, when bison

TABLE 16.1 Numbers of bison and elk killed by wolves and detected in the Madison headwaters area of Yellowstone National Park during 1996–1997 through 2006–2007

Winter	Total elk	Total bison	Elk calves	Elk cows	Elk bulls	Bison calves	Bison cows	Bison bulls
1996–97	13	1	5	5	0	1	0	0
1997–98	15	0	11	4	0	0	0	0
1998–99	51	12	31	11	8	12	0	0
1999–00	49	3	28	19	2	1	1	0
2000–01	71	2	39	30	1	2	0	0
2001–02	75	16	32	33	10	9	5	2
2002–03	61	14	30	23	5	5	6	3
2003–04	82	24	30	41	11	12	9	3
2004–05	61	33	31	19	11	16	7	9
2005–06	47	37	22	21	4	26	11	0
2006–07	81	9	33	35	12	3	4	2
Totals	606	151	292	241	64	87	43	19

Age class and sex totals do not include kills that could not be categorized by age and sex.

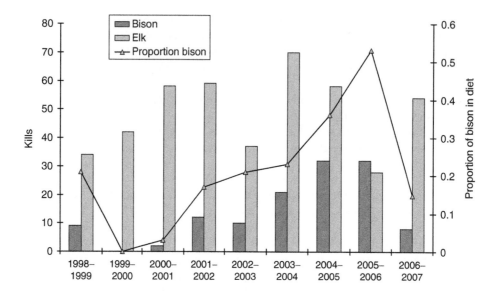

FIGURE 16.6 Wolf-killed elk and bison from resident packs in the Madison headwaters area of Yellowstone National Park during 1998–1999 through 2006–2007 ($n = 566$). Bison steadily increased in the diet until the mild winter of 2006–2007 when considerably more elk were killed despite a substantial decrease in elk abundance.

comprised the primary prey species (Figure 16.6). However, the proportion of bison in wolf diets decreased to 15% during winter 2006–2007 ($n = 9$). Prey selection by wolves was predictably variable within winters, with elk calves primarily killed in early-middle winter, bison killed during middle-late winter, and adult elk killed throughout winter (Figure 16.7). Selection indices calculated for early-, middle-, and late-winter periods from 1998–1999 through 2006–2007 demonstrated a strong preference only for elk calves (mean = 0.82, sd = 0.12) throughout every winter, with selection by winter period summarized in Table 16.2.

Mean ages of adult elk ($n = 280$) and bison ($n = 44$) killed by wolves were 8.3 years (95% CI = 7.8, 8.8) and 10.0 years (95% CI = 8.4, 11.6), respectively. Adult female elk killed by wolves were older than males, with a mean female age of 9.1 years ($n = 220$; 95% CI = 8.6, 9.7) and a mean male age of 5.6 years ($n = 58$; 95% CI = 4.6, 6.5). Ages of adult bison killed by wolves were older for females compared to males, with a mean age of 11.0 years ($n = 31$; 95% CI = 9.3, 12.6) and 7.7 years ($n = 13$; 95% CI = 4.3, 11.1) respectively, with considerably more variability for males. Wolves killed all age classes of adult bison and elk, but the highest proportion of kills was in the older age classes (Figures 16.8 and 16.9). The proportion of elk kills in the older age classes was higher during winters 2002–2003 through 2006–2007 when the number of resident packs and wolves increased compared to winters 1998–1999 through 2002–2003 when Nez Perce was the primary resident pack (Figure 16.9).

Marrow samples from 121 bison (52 adults, 69 calves) and 481 elk (275 adults, 206 calves) indicated condition decreased from early- to late-winter periods (Figure 16.10). Of the 120 elk jaws that were rated for necrosis, 50% ($n = 60$) showed no signs of necrosis, 18% ($n = 21$) were mild, 16% ($n = 19$) moderate, and 17% ($n = 20$) severe. Eleven bison jaws were rated, with 55% ($n = 6$) showing no signs of necrosis, 36% ($n = 4$) moderate, and 9% ($n = 1$) severe.

A. Factors Influencing Wolf Prey Selection

We fitted 84 models from three *a priori* suites to data from 564 kills (216 elk calves, 223 elk adults, 69 bison calves, 56 bison adults) by resident wolf packs during 1998–1999 through 2006–2007. Model selection results supported two top models with Akaike model weights (w_k) of 0.70 and 0.28,

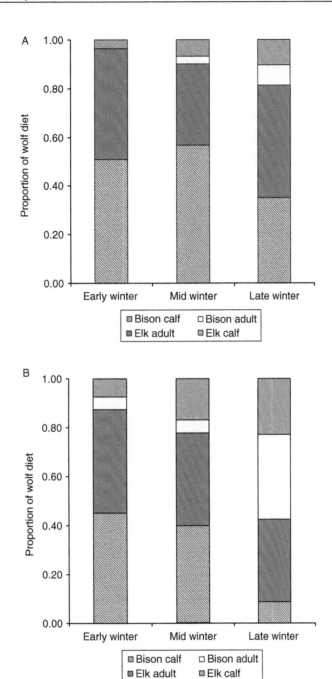

FIGURE 16.7 Winter trends in pooled diet composition from resident wolf packs in the Madison headwaters area of Yellowstone National Park during (A) 1998–1999 through 2002–2003 (*n* = 262) and (B) 2003–2004 through 2006–2007 (*n* = 302). Kills were classified by species, age class, and winter period (early = 15 November—10 January; middle = 11 January—6 March; late = 7 March—30 April).

TABLE 16.2	Selection indices (Chesson 1978) for four prey types from resident wolf pack kills during winters 1998–1999 through 2006–2007		
	Index of Selectivity		
Prey Type	**Early Winter**	**Middle Winter**	**Late Winter**
Elk Calf	0.752	0.827	0.816
Elk Adult	0.176	0.094	0.098
Bison Calf	0.059	0.073	0.070
Bison Adult	0.013	0.006	0.016

Kills were classified by species, age class, and winter period (early = November 15–January 10; middle = January 11–March 6; late = March 7–April 30). For m prey types a value greater than $1/m$ (i.e., 0.25) indicates a preference.

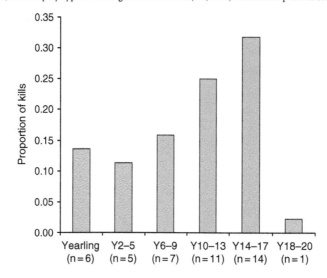

FIGURE 16.8 Age distribution of adult bison killed by wolves in the Madison headwaters area of Yellowstone National Park during 1996–1997 through 2006–2007 ($n = 41$).

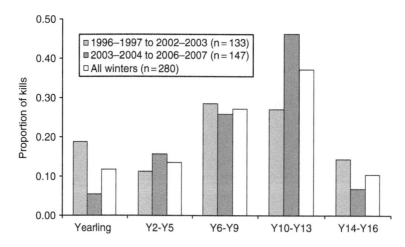

FIGURE 16.9 Age distribution of adult elk killed by wolves in the Madison headwaters area of Yellowstone National Park during 1996–1997 through 2002–2003 and 2003–2004 through 2006–2007. Reductions in younger age classes during the latter time period likely reflect lack of recruitment.

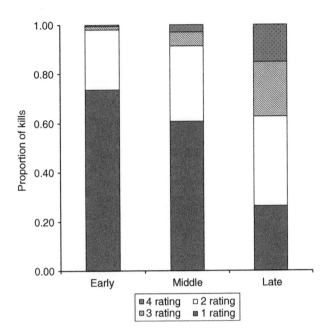

FIGURE 16.10 Marrow classifications for ungulates killed by wolves in the Madison headwaters area of Yellowstone National Park during early, middle, and late winter, 1996–1997 through 2006–2007 ($n = 602$).

respectively (Table 16.3). The covariates ELK_{calf}, $BISON_{calf}$, WOLF:UNG, and SWE_{acc} were included in each of these models, with predictor weights (w_p) of 0.99, 0.99, 0.99, and 1.00, respectively. The structure of the two models differed only in the inclusion of ELK_{adt} in the top model, with $w_p = 0.71$. All other models had $\Delta AIC_c > 8.5$ and differed in structure from the best-supported models by their lack of inclusion of the WOLF:UNG covariate. An exploratory analysis adding wolf pack size ($WOLF_{pk}$) to the two best-supported models did not improve the top model, but improved the second best model's AIC_c to 1259.56, with a resultant ΔAIC_c of 0.31.

Elk abundance was negatively related to the probability that wolves would kill bison of both age classes relative to elk calves and, in particular, the abundance of elk calves was strongly negatively correlated with predation of adult bison (Table 16.4). The abundance of bison, as estimated by the covariate $BISON_{calf}$, was positively related to the probability of wolves killing a bison calf relative to an elk calf. This relationship was similar for bison adults, though confidence intervals for the coefficient estimates spanned zero. In contrast to our hypotheses, bison abundance was negatively correlated with the probability that adult elk would be killed relative to calf elk. The wolf:ungulate ratio was positively correlated with the predation probability of all other prey types relative to elk calves, but was strongest for the bison adult logit (Table 16.4). As predicted, there was not a significant positive relationship between pack size and prey size, and there was a negative correlation with the probability of predation of bison calves relative to elk calves. The effect of SWE_{acc} was strongly positive for all logit equations, indicating that increasing snow pack resulted in increased probability of predation for all prey types relative to elk calves (Table 16.4).

There were significant increases in the odds of elk adult, bison calf, and bison adult kills with increases in bison calf abundance, wolf:ungulate ratios, and SWE_{acc} (Table 16.4). The odds of predation by wolves for adult elk were 0.57 lower for every 134 animal increase in bison calf abundance, 1.7 times greater for every 0.018 increase in wolf:ungulate ratios, and 6.5–7.6 times higher for each 1800 cm days

TABLE 16.3 *A priori* model structure and results from top models within and among suites for multinomial logit analyses of winter prey selection by resident wolf packs in a bison-elk system in the Madison headwaters area of Yellowstone National Park during 1998–99 through 2006–07. Covariate codes are the abundance of elk adults (ELK_{adt}, ELK_{calf}), bison calves ($BISON_{calf}$), accumulated snow pack (SWE_{acc}), wolf:elk ratio (WOLF:ELK), wolf:ungulate ratio (WOLF:UNG), and multiple pack days (MULTPK)

Model Structure	K	AIC$_c$	ΔAIC$_c$	w_k	ΔAIC$_c$	w_k
		Within Suite			**Among Suites**	
Prey Suite						
$ELK_{calf} + BISON_{calf}$	3	1284.6	0.00	0.74	25.35	0.00
$ELK_{calf} + ELK_{adt} + BISON_{calf}$	4	1286.8	2.14	0.26	27.49	0.00
ELK_{calf}	2	1298.9	14.29	0.00	39.65	0.00
Prey + Wolf Competition Suite						
$ELK_{calf} + BISON_{calf} + WOLF:ELK$	4	1283.4	0.00	0.43	24.15	0.00
$ELK_{calf} + BISON_{calf} + MULTPK$	4	1284.9	1.51	0.20	*a*	
$ELK_{calf} + ELK_{adt} + BISON_{calf} + WOLF:ELK$	5	1285.7	2.28	0.14	26.43	0.00
$ELK_{calf} + BISON_{calf} + WOLF:UNG$	4	1286.0	2.57	0.12	*a*	
$ELK_{calf} + ELK_{adt} + BISON_{calf} + WOLF:UNG$	5	1287.2	3.84	0.06	*a*	
$ELK_{calf} + ELK_{adt} + BISON_{calf} + MULTPK$	5	1287.6	4.16	0.05	*a*	
$ELK_{calf} + WOLF:UNG$	3	1300.2	16.77	0.00	40.91	0.00
Prey + Snow Pack Suite						
$ELK_{calf} + ELK_{adt} + BISON_{calf} + SWE_{acc}$	5	1267.8	0.00	0.65	8.54	0.01
$ELK_{calf} + BISON_{calf} + SWE_{acc}$	4	1269.1	1.33	0.34	9.87	0.01
$ELK_{calf} + ELK_{adt} + SWE_{acc}$	4	1277.5	9.67	0.01	18.21	0.00
$ELK_{calf} + SWE_{acc}$	3	1279.5	11.70	0.00	20.24	0.00
Prey + Wolf Competition + Snow Pack Suite						
$ELK_{calf} + ELK_{adt} + BISON_{calf} + WOLF:UNG + SWE_{acc}$	6	1259.3	0.00	0.56	0.00	0.70
$ELK_{calf} + BISON_{calf} + WOLF:UNG + SWE_{acc}$	5	1261.1	1.86	0.22	1.86	0.28
$ELK_{calf} + ELK_{adt} + BISON_{calf} + WOLF:ELK + SWE_{acc}$	6	1263.1	3.86	0.08	*a*	
$ELK_{calf} + BISON_{calf} + WOLF:ELK + SWE_{acc}$	5	1263.7	4.48	0.06	*a*	
$ELK_{calf} + ELK_{adt} + BISON_{calf} + MULTPK + SWE_{acc}$	6	1265.0	5.71	0.03	*a*	
$ELK_{calf} + BISON_{calf} + MULTPK + SWE_{acc}$	5	1265.0	5.76	0.03	*a*	
$ELK_{calf} + ELK_{adt} + WOLF:UNG + SWE_{acc}$	5	1269.5	10.20	0.00	10.20	0.00

a Values not included in among suite evaluations due to identical model structure differing only in wolf competition metric.

increase in SWE_{acc}. The odds of predation for bison calves decreased 0.32 and 0.38 times with every 42 animal and 122 animal increase in elk calf and elk adult abundance, respectively. The odds of predation for bison calves increased 2.21–2.67 times per 134 animal increase in bison calf abundance and 4.6–11.8 times per 1800 cm days increase in SWE_{acc}. The odds of predation for adult bison were 6.6–6.9 times greater for each 0.018 unit increase in wolf:ungulate ratio, 17.7–23.9 times greater for each 1800 cm days increase in SWE_{acc}, and 0.03–0.05 times greater for every 42 animal increase in elk calf abundance.

B. Wolf Prey Switching with Murdoch's Equation

There was a positive relationship between the ratios of bison to elk in wolf diets and the population (Figure 16.11). Fitting a non-linear model of Murdoch's equation to the data indicated a curvilinear relationship, with *c* and *b* values of 0.229 (95% CI = 0.203, 0.254) and 2.091 (95% CI = 1.175, 3.007),

TABLE 16.4 Coefficient values (B_j), lower and upper 95% confidence intervals (in parentheses), and odds ratios for the three best approximating models identified through AIC model comparison techniques for prey selection by resident wolf packs on bison and elk in the Madison headwaters area of Yellowstone National Park during 1998–1999 through 2006–2007

	Model	ELK$_{calf}$ w_p = 0.99	ELK$_{adt}$ w_p = 0.71	BISON$_{calf}$ w_p = 0.99	SWE$_{acc}$ w_p = 1.00	WOLF:UNG w_p = 0.99	WOLF$_{pack}$
Elk adult	PSW15	0.21 (−0.38, 0.80) 1.23	−0.21 (−0.68, 0.26) 0.81	**−0.56 (−1.10, −0.02)** **0.57**	**2.03 (1.13, 2.93)** **7.60**	**0.51 (0.01, 1.02)** **1.67**	
	PSW41	0.06 (−0.44, 0.56) 1.07		−0.51 (−1.04, 0.02) 0.60	**1.87 (1.03, 2.72)** 6.51	0.48 (−0.02, 0.98) 1.61	
	PSW41w	0.07 (−0.43, 0.57) 1.07		−0.52 (−1.08, 0.05) 0.60	**1.89 (1.01, 2.77)** 6.62	0.49 (−0.02, 0.99) 1.62	0.03 (−0.46, 0.51) 1.03
Bison calf	PSW15	−0.31 (−1.49, 0.87) 0.73		0.70 (−0.01, 1.42) 2.02	**2.47 (1.25, 3.68)** 11.76	0.55 (−0.41, 1.51) 1.73	
	PSW41	**−1.13 (−2.20, −0.06)** **0.32**	**−0.96 (−1.64, −0.27)** **0.38**	**0.79 (0.08, 1.50)** 2.21	**1.89 (0.75, 3.03)** 6.60	0.52 (−0.40, 1.44) 1.68	
	PSW41w	−1.06 (−2.13, 0.02) 0.35		**0.98 (0.24, 1.73)** 2.67	**1.52 (0.35, 2.70)** 4.59	0.55 (−0.36, 1.45) 1.73	**−0.85 (−1.57, −0.12)** **0.43**
Bison adult	PSW15	**−3.00 (−5.32, −0.69)** **0.05**	−0.41 (−1.14, 0.32) 0.67	0.37 (−0.43, 1.17) 1.45	**3.17 (1.83, 4.52)** 23.90	**1.93 (0.94, 2.93)** 6.91	
	PSW41	**−3.46 (−5.53, −1.39)** **0.03**		0.40 (−0.39, 1.19) 1.49	**2.87 (1.64, 4.11)** 17.70	**1.89 (.90, 2.89)** 6.65	
	PSW41w	**−3.34 (−5.38, −1.30)** **0.04**		0.36 (−0.47, 1.19) 1.43	**3.05 (1.71, 4.38)** 21.07	**1.89 (0.90, 2.88)** 6.61	0.29 (−0.62, 1.20) 1.34

Covariate codes are the abundance of elk calves and adults (ELK$_{calf}$, ELK$_{adt}$), bison calves (BISON$_{calf}$), accumulated snow pack (SWE$_{acc}$), wolf:ungulate ratio (WOLF:UNG), and wolf pack size (WOLF$_{pack}$). Values in bold indicates confidence intervals do not span zero.

respectively. The majority of this relationship was supported by data collected during the winters of 2001–2002 through 2006–2007, in particular two points corresponding to the late-winter periods of 2004–2005 and 2005–2006 appeared to have considerable influence on the shape of the relationship (Figure 16.11). To evaluate the effect of these points we removed them and refit the models. The estimated values of c and b decreased to 0.15 (95% CI = 0.08, 0.22) and 1.27 (95% CI = 0.60, 1.94) respectively, while the standard error of each estimate increased, and thus the confidence interval of b included values less than 1 (Table 16.5).

IV. DISCUSSION

We contributed new insights into wolf prey selection by investigating a temporally and spatially dynamic system that was unaffected by human harvests and contained two prey species, each with age classes differing in their size and defenses relative to adults. While the complexity of interacting

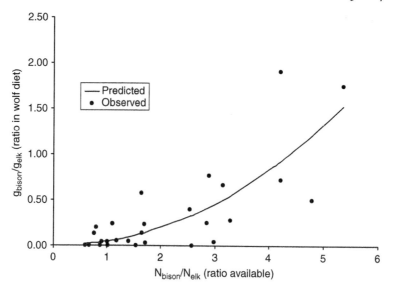

FIGURE 16.11 Observed versus predicted relationships between the ratio of bison:elk wintering in the Madison headwaters area of Yellowstone National Park and the ratio of bison:elk in wolf diets during 1998–1999 through 2006–2007. Predicted coefficients for fitted line are $c = 0.23$ and $b = 2.09$.

TABLE 16.5 Coefficient values and lower and upper 95% confidence intervals from analyses of prey switching by wolves in a bison-elk system in the Madison headwaters area of Yellowstone National Park during 1998–1999 through 2006–2007

	Parameter Estimates	
Model Structure	c	b
(c*BISON:ELK)[b]	0.23 (0.20, 0.25)	2.09 (1.18, 3.01)
Outliers Removed	0.15 (0.08, 0.22)	1.27 (0.60, 1.94)

Values indicate confidence intervals do not span zero. The constant c measures the bias in a predator's diet (Murdoch 1969), with values less than one indicating a preference for elk. The constant b measures the extent of prey-switching, with values greater than one indicating switching. The covariate code BISON:ELK measures the relative abundance of bison and elk in the system.

biotic and abiotic factors influencing prey selection in natural systems makes evaluations difficult, analyses across species and age classes demonstrated that selection was influenced by the absolute and relative abundance of prey types, the abundance of predators, and the duration of snow pack. Prey abundance, particularly elk calf and bison calf abundance, was important because wolves strongly preferred elk calves relative to all other prey types and elk calf abundance was inversely related to the occurrence of bison calves and adults in wolf diets. While wolves preferred elk to bison, their patterns of selection were also driven by the relative abundance of the two prey species with wolves killing disproportionately more bison at high bison:elk ratios, and the curvilinear form of Murdoch's equation indicating prey switching. In addition to prey abundance, the influence of predator numbers as reflected in the wolf:ungulate ratio resulted in a broadening of wolf prey selection from elk calves, with increasing probabilities of different prey types in the diet with increasing ratios. Lastly, we evaluated both the movement-inhibiting and weakening influence of snow pack on prey and demonstrated that the probability of predation on both bison age classes and adult elk increased dramatically with increasing snow pack duration and accumulation. The profound influence of snow pack illustrates the important role of environmental variables on prey selection in large mammal systems.

Preference for elk calves was strong within and among all years, presumably due to their small size and lack of defenses relative to other prey types. Unlike most small taxa, ungulates pose considerable injury risk to predators, and anti-predator defenses vary among species and age classes (Nelson and Mech 1981, Bergerud *et al.* 1984, Carbyn and Trottier 1987, Dale *et al.* 1995). Therefore, the ability of prey individuals to repel an attack can substantially influence large mammal predator-prey dynamics (Garrott *et al.* 2007). Elk typically employed flight as a primary anti-predator tactic and, as a result, elk calves did not benefit from group protection strategies such as those used by bison (Carbyn and Trottier 1987, Carbyn *et al.* 1993). Elk adults are more capable than calves of inflicting injury on wolves due to their larger size, strength, experience, and the presence of antlers on bulls. However, wolves also killed elk from all other age classes, with the proportion of older adult elk in wolf diets increasing in the latter years of the study due to the effects of consistently low recruitment on the age structure of the population (Chapter 23 by Garrott *et al.*, this volume).

While wolves preferred elk calves over adults they also preferred elk over bison due to differences in vulnerability, and selected both age classes of bison less than expected given their abundance. As in other studies (Carbyn *et al.* 1993, Smith *et al.* 2000, MacNulty *et al.* 2007), bison constituted an extremely formidable and dangerous prey to wolves due to their physical and behavioral defenses. We documented numerous instances of wolves seriously injuring bison and returning to kill and feed on them later (Carbyn *et al.* 1993). We also frequently witnessed bachelor and cow-calf herds continually defending injured or weak animals under attack by wolves (Carbyn and Trottier 1987, MacNulty *et al.* 2007). The majority of bison predation occurred in late winter when ungulates are likely in their most substandard physiological condition (Figure 16.12, Chapter 9 by White *et al.*, this volume) and in the latter years of the study when bison were most abundant relative to elk (Chapter 12 by Bruggeman *et al.*, this volume). The smallest bison prey with the fewest defenses, calves, comprised the majority of bison kills, followed by cows and bulls, respectively (Table 16.1). Bison adults of both sexes were likely weakened in late winter, but cows had increased energetic demands because they were at or nearing parturition (Chapter 14 by Geremia *et al.*, this volume). Bison were more abundant than elk during all years of our study, and were predictably found feeding in the open meadow complexes (Chapter 28 by Bruggeman *et al.*, this volume). Therefore encountering bison was unlikely to be a limiting factor influencing wolf predation on bison, even if a group was considered the unit of encounter (Huggard 1993a). Thus, while general models of predator behavior typically focus on encounter rates (Taylor 1984) we support the caution of Dale *et al.* (1995) against wolf predation models incorporating only encounter rates given the dramatic differences in vulnerability and risk of injury ungulates pose when attacked.

While absolute and relative prey abundance and life history characteristics were influential, the physiological effect of snow pack, as indexed by SWE_{acc}, had an overwhelming influence on predation

FIGURE 16.12 Winter-starved bull bison (A) and calf bison (B) in the Firehole river drainage. Late winter starvation was the primary source of mortality for both elk and bison prior to wolf recolonization and the weakening influence of snow pack made formidable prey such as bison considerably more vulnerable to wolf predation (Photos by Jeff Henry and Matt Becker).

by wolves among the four prey types. Though snow depth has been frequently attributed to increased vulnerability due to its inhibition of movement in a predation event and its longer-term weakening influence on ungulates (Nelson and Mech 1986, Huggard 1993b, Mech and Peterson 2003), the relative importance of each has not been analyzed for specific predator-prey systems and prey species. While both metrics have important and interacting influences, the mean snow pack on the ground at the time of a kill (SWE_{mean}) explained much less variation in prey selection than the duration and accumulation of snow up to that date (SWE_{acc}). Physical condition of wolf-killed ungulates decreased through each winter, consistent with nutritional profiles of the elk population (Chapter 9 by White *et al.*, this volume), and winter starvation mortalities were the predominant source of mortality for both elk and bison prior to wolf recolonization (Chapter 11 by Garrott *et al.*, this volume). The odds of wolf predation on bison increased many orders of magnitude with increasing accumulation and duration of snow pack, presumably weakening bison such that they were less able to defend themselves or their calves. While we did observe bison being killed in deep snow, observations of wolves attacking bison in late winter typically occurred in low snow meadow complexes and defense sometimes lasted several hours, as wolves continually attempted to isolate and injure vulnerable individuals. An animal in a weakened state is likely much less able to sustain such defense in the face of an attack. Snow pack is also highly influential in driving broad-scale movements of bison, such as their winter migrations into the Madison headwaters area and movements among drainages (Chapters 12 and 28 by Bruggeman *et al.*, this volume).

Snow pack also increased the vulnerability of adult elk to wolf predation both by weakening their condition and impeding their escape during flight. Because of their habitat selection and anti-predator defenses elk are likely more susceptible to environmental vulnerability in the form of hard habitat edges, structure, and changes in snow pack that can impede flight (Bergman *et al.* 2006). We frequently found wolf-killed elk that had been either encountered and killed in deep snow and complex forest structure or chased into it and killed. Thus, environmental vulnerability (Chapter 24 by Garrott *et al.*, this volume) can assume considerable importance in large mammal predator-prey interactions given the severe weakening influence of snow pack on prey animals, the potential for snow pack effects to differ among prey species depending on their life history characteristics, and the potentially negative effect of edge and the accompanying differences in habitat structure.

In addition to prey and snow pack variables, the ratio of wolves to ungulates was influential in wolf prey selection within and among species, likely due to a combination of wolf competition and elk anti-predator behavior. The transition from a one-wolf-pack system to a multiple-wolf-pack system resulted in wolves occupying the entire study area and overlapping extensively, with inter-pack strife the main cause of wolf mortality (Chapter 15 by Smith *et al.*, this volume) and wolf functional responses to elk best described as a Type II ratio-dependent response, indicating significant predator dependence (Chapter 17 by Becker *et al.*, this volume). Although the winter range comprised a relatively small area, packs did not exhibit temporal avoidance in their use of the system and were routinely detected in the same drainages, though they typically avoided direct encounters (Chapter 15 by Smith *et al.*, this volume). However, despite this intense use there was typically one dominant pack, and smaller packs were often displaced to more marginal areas of the study system. For example, the majority of the Biscuit Basin and Nez Perce pack territories overlapped with each other during winters 2003–2004 and 2004–2005; however when both packs were detected in the same drainage Nez Perce appeared to occupy the main hunting areas while Biscuit Basin was often displaced. Similar dynamics were apparent with the Hayden pack and dominant Gibbon pack in 2006–2007. Packs whose territories included large areas of marginal foraging based on the paucity of elk (*e.g.*, Gibbon drainage, Chapter 21 by White *et al.*, this volume) also preyed on adult elk and bison considerably more than packs occupying areas with relatively abundant elk, and the saturation of the system with wolves likely resulted in fewer places for adult elk to avoid encounters, such as bulls that often resided in minor drainages away from the main meadow complexes. This competition likely intensified as the elk population and the per-capita availability of calves decreased concurrent with a substantial distribution change within and among the three drainages (Chapter 18 by Gower *et al.*, and Chapter 21 by White *et al.*, this volume). While wolves killed elk calves throughout all winters, high wolf numbers in the system resulted in nearly all elk calves being killed by late winter in several of the latter years (Chapter 23 by Garrott *et al.*, this volume). While wolf pack size did not appear to affect selection, we are not aware of other studies that demonstrated the influence of wolf population numbers on prey selection. This may be because most of these analyses were confined to a single prey species or prey item, established wolf-ungulate systems do not typically undergo the dramatic changes in predator and prey populations that accompany newly-established systems, or the spatial and temporal dynamics of wolf territoriality in this system are unusual due to the high prey density of the Madison headwaters relative to the surrounding areas.

The significance of the wolf:ungulate ratio was also likely related to elk anti-predator responses. The elk population experienced a substantial decrease and re-distribution following wolf restoration, with 84% of the population residing in the Madison drainage by the end of the study period compared to approximately equal proportions distributed among the three major drainages before wolves (Chapter 21 by White *et al.*, this volume). In addition, elk increased the variability of their group sizes, changed their movements, and intensified their selection for habitats with high snow heterogeneity and possible escape terrain in response to increasing wolf numbers (Chapters 18, 19, and 20 by Gower *et al.*, Chapter 21 by White *et al.*, this volume). Under the behavioral resource depression hypothesis (Charnov *et al.* 1976), increasing predator presence should decrease the ability of individual predators, in this case packs, to capture prey due to increased wariness (and therefore decreased vulnerability) of prey (Chapter 17 by Becker *et al.*, this volume). Prey-switching is typically thought to result in prey persistence because the relative rarity of the primary prey results in lower encounter and kill rates and, as a result, the predator switches to the more abundant prey (Murdoch 1969). However, rarity alone is likely insufficient for prey persistence and has not been well-demonstrated (Matter and Mannan 2005). In systems where large predators are being restored, changes in prey selection and the potential for prey-switching may be driven in part by a shift in prey behaviors as species adopt more effective anti-predator strategies to reduce their vulnerability rather than changes in predator preference.

Virtually all studies of predation in large mammal multiple-prey systems report strong selection for certain prey species (Carbyn 1974, Potvin *et al.* 1988, Dale *et al.* 1995, Karanth and Sunquist 1995, Kunkel *et al.* 1999, Jędrzejewski *et al.* 2000, Creel and Creel 2002, Sinclair *et al.* 2003, Hayward *et al.* 2006). However, this selection is not consistent among studies or species assemblages. For example, wolves primarily select caribou in some systems and not in others (Dale *et al.* 1995, Wittmer *et al.* 2005). Likewise, our documentation of elk as the primary prey of wolves was consistent with some investigations in other multiple-prey systems containing elk (Carbyn 1974, 1983, Huggard 1993a,b, Weaver 1994, Hebblewhite *et al.* 2003, Husseman *et al.* 2003, Smith *et al.* 2004) and contrary to others (Kunkel *et al.* 1999). These apparent contrasts have led some investigators to conclude that use of the term "preference" to describe wolf prey selection is inappropriate because wolves select individuals of whatever species are most profitable with the least risk (*i.e.*, the most vulnerable; Mech and Peterson 2003).

Defining preference as what a predator eats when all prey types are equally abundant and available confines investigations to the sophisticated cafeteria feeding trial experiments used on smaller taxa (Rodgers 1990). Thus, it is admittedly infeasible to determine if wolves have an inherent preference for a particular prey species. However, investigations of prey selection patterns in the context of natural multiple-prey systems, where preference is defined relative to the prey species assemblage available and influenced by the backdrop of landscape and climate variables upon which these interactions occur, have the potential to significantly advance our understanding of wolf-prey dynamics and help explain the prey selection contrasts observed among different systems (Garrott *et al.* 2007). We used data on wolf diet composition and relative abundance ratios to evaluate whether wolves switched from preying primarily on elk to bison. Murdoch (1969) demonstrated that values of $c < 1$ are indicative of preference, while values of $b > 1$ indicate switching. Based on this equation our analyses indicate that wolves in the Madison headwaters area had a strong preference for elk relative to bison ($c = 0.229$), but switched to bison at high bison:elk ratios ($b = 2.091$). Garrott *et al.* (2007) estimated Murdoch's selection coefficient by decomposing c into vulnerability (v), preference (s), and biomass (m) to account for the profound differences in morphology and anti-predator defenses across ungulate prey species. Based on attack rate and attack success data on elk and bison in the northern portion of Yellowstone and the Pelican Valley (MacNulty 2002), Garrott *et al.* (2007) estimated the product of *svm* as 0.04, considerably lower than what we estimated by fitting Murdoch's equation to our data. This discrepancy may illustrate the difficulty of obtaining sufficient data on attack rates and success when decomposing c into svm, as well as the potential differences among wolf-elk-bison systems. Elk vastly outnumber bison on the northern range of Yellowstone, while the Pelican Valley prey base is primarily one bison herd of ≤ 150 animals (MacNulty 2002). Thus, it is reasonable to assume that differences in attack rates on the two prey species might differ as well among the three systems. While the utility of decomposing c into svm may primarily be confined to systems where such data are more readily collected (*e.g.*, Scheel 1993, Creel and Creel 2002, MacNulty *et al.* 2007), simply estimating c and b for a given system can be accomplished with data on wolf diet composition and relative prey abundance (Garrott *et al.* 2007).

The significance of distinguishing between Murdoch's (1969) definition of prey switching and using prey switching to simply describe changes in diet ultimately concerns the possible regulatory effects of a predator. By having a preference for the more abundant prey and thereby presumably lessoning predation on the less abundant prey at the same time, the predator can exert strong stabilizing density-dependent effects on the system (Murdoch and Oaten 1975, Oaten and Murdoch 1975). However, much of the experimental work on switching assumes constant predator and prey abundance (Murdoch 1969, Messier 1995). Thus theoretical and empirical treatments of switching that do not consider a numerical response (Messier 1995) or that assume a constant handling time do not address situations where predators respond numerically or can decrease their handling time under certain circumstances when prey become more available. When these equations are applied to natural systems with these characteristics a curvilinear relationship can be derived, yet a predator can take disproportionate amounts of the relatively more abundant prey without diminishing their take of the relatively less abundant prey. In this situation, predators will not exert a stabilizing influence on the less abundant and preferred prey species and consequently the switching evaluations recommended by Garrott *et al.* (2007) require further refinement to account for potentially common scenarios in natural settings.

Though Murdoch's (1969) equation suggested wolves in the Madison headwaters switched to bison at high bison:elk ratios, we did not detect a concurrent switch away from their preferred prey, elk, and we did not have constant abundances for either wolves or ungulates. Variations in wolf kill rates on elk were not negatively related to bison abundance and the effect of increasing bison abundance and increasing snow pack duration and accumulation was simply to increase the total wolf kill rate and the wolf kill rate on bison rather than reduce the kill rate on elk (Chapter 17 by Becker *et al.*, this volume). Furthermore, carcass consumption was negatively related to total kill rates and bison kill rate variation was best explained by snow pack or bison calf abundance (Chapter 17 by Becker *et al.*, this volume). The curvilinear relationship indicating switching was also heavily leveraged by two points comprising the late winter periods of 2004–2005 and 2005–2006 respectively. Disproportionate bison selection by wolves in late winter 2004–2005 occurred during the peak of wolf abundance in the study area, with calf elk abundance decreasing to an estimated six animals by winter's end. Peak snow pack accumulation was below average for the study period, but substantial numbers of bison relative to elk and high wolf abundance likely resulted in increased selection for bison. In contrast the winter of 2005–2006 followed a dramatic decrease in wolf abundance coupled with an above-average snow pack accumulation and high numbers of bison relative to elk, resulting in the highest proportion of bison killed during the study period. Nevertheless, elk continued to be preferred during the final winter of the study (2006–2007), when elk numbers were at their lowest and the bison.elk ratios were near their highest. This was further corroborated by selection indices indicating continued high preference for elk calves with declining elk availability both within and among winters. Thus, it appears that wolf prey selection in this system is driven by a strong preference for the most vulnerable prey items (*i.e.*, elk and elk calves in particular) and changes in prey selection are driven largely by circumstance (*i.e.*, high bison:elk ratios; high wolf abundances; severe winters) rather than by a density-dependent change in wolf preference. Consequently the ecological relevance of prey-switching, namely its density-dependent stabilizing effects, do not appear to be present in this system at this time, perhaps best evidenced by a continued decline of elk due to wolf predation (Chapters 23 and 24 by Garrott *et al.*, this volume). Most natural systems with wolves, whose abundance has a strong positive relationship to prey biomass (Fuller 1989), are unlikely to have constant predator or prey abundance and stochastic processes can contribute to variable handling time. Thus we suggest continued refinement of prey-switching evaluations to account for this variability, and that evaluations with the definition provided by Murdoch (1969) should also consider whether there is a concurrent decrease in predation on the formerly more abundant and preferred prey, as herein lies the density-dependent stabilizing effect that is of primary ecological interest.

Prior to reintroduction investigators predicted wolves would reduce the Yellowstone bison population by <15% (Boyce and Gaillard 1992, Boyce 1993). Bison predation park-wide was actually considerably less (<1%) during 1995–2000 (Smith *et al.* 2004). However Boyce (1995) did predict that prey switching from elk to bison could possibly occur in the Madison headwaters area in late winter. While we observed increased wolf predation on bison in late winter, it was unclear prior to our analyses whether this trend was driven by circumstance or by prey-switching to the increasing bison population and concurrently switching away from elk. Wolves are capable of subsisting almost exclusively on bison as evidenced in Wood Buffalo National Park (Carbyn *et al.* 1993). However, at this time wolves appear to primarily kill bison at high relative abundance ratios, particularly in severe winters and in times of high wolf abundances, with no indication that preference for elk has changed. Given this strong preference for elk it seems likely that elk numbers in the Madison headwaters area will continue to decrease to a low equilibrium, depending on their ability to escape predation via behavior or use of refuges (Creel *et al.* 2005, Hebblewhite *et al.* 2005, Chapters 18, 19, and 20 by Gower et al., Chapter 21 by White *et al.*, and Chapter 24 by Garrott *et al.*, this volume) that could produce pronounced switching away from elk and elk calves in particular. It is even possible that local extirpation of the Madison headwaters elk population could occur (Chapter 24 by Garrott *et al.*, this volume). However, the dynamics of wolves, elk, and bison in the Madison headwaters area are still those of a developing system, with wolves present for little over a decade. Understanding patterns of prey selection, preference, and the presence or absence of prey-switching and their effects on community stability and persistence will require subsequent years of study to distinguish between transitory phenomena and the myriad influences of predator, prey, and environment in a newly-established large mammal system.

V. SUMMARY

1. We contributed new insights into wolf prey selection by investigating a temporally and spatially dynamic system in the Madison headwaters area of Yellowstone National Park that was unaffected by human harvests and contained two prey species (bison, elk) with age classes differing in their sizes and defenses.
2. Prey selection by wolves was influenced by the absolute and relative abundance of prey types, the abundance of predators, and the duration of snow pack. Wolves strongly preferred elk calves relative to all other prey types, and elk calf abundance was inversely related to the occurrence of bison calves and adults in wolf diets.
3. An increase in predator numbers, reflected in the wolf:ungulate ratio, resulted in a broadening of wolf prey selection from elk calves, with increasing probabilities of different prey types in the diet with increasing ratios.
4. The probability of predation on both bison age classes and adult elk increased with increasing snow pack accumulation and duration, likely due to its long-term debilitating influence on ungulates that increased their vulnerability to wolves.

5. We evaluated whether wolves switched prey from elk to bison using Murdoch's (1969) equation and further evaluated potential changes in wolf preference using selection indices. While a curvilinear relationship existed between the ratio of bison to elk in wolf diets versus the ratio of bison to elk available in the population that suggested prey switching, confounding variability in wolf and prey numbers concurrent with no detected decrease in wolf preference away from elk did not support this stabilizing behavior.

6. Comparative investigations of prey selection by wolves are complicated by the frequent use of terminology without consistent and explicit definitions and distinctions. Further investigations into evaluating prey-switching and wolf preference are recommended and utilizing Murdoch's (1969) switching equation provides a rigorous evaluation of wolf preference and prey-switching, while eliminating inconsistencies in terminology. However, biologists should also consider whether there is a concurrent decrease in predation on the formerly more abundant and preferred prey to address confounding variability in predator and prey numbers common in natural systems.

VI. REFERENCES

Andersson, M., and S. Erlinge. 1977. Influence of predation on rodent populations. *Oikos* **29**:591–597.

Begon, M., J. L. Harper, and C. R. Townsend. 1996. *Ecology.* Blackwell Science, Malden, MA.

Bergerud, A. T., H. E. Butler, and D. R. Miller. 1984. Antipredator tactics of calving caribou: Dispersion in the mountains. *Canadian Journal of Zoology* **62**:1566–1575.

Bergman, E. J., R. A. Garrott, S. Creel, J. J. Borkowski, and R. M. Jaffe. 2006. Assessment of prey vulnerability through analysis of wolf movements and kill sites. *Ecological Applications* **16**:273–284.

Bonsall, M. B., and M. P. Hassell. 1997. Apparent competition structures ecological assemblages. *Nature* **338**:371–373.

Boyce, M. S. 1993. Predicting the consequences of wolf recovery to ungulates in Yellowstone National Park. Pages 234–269 in R. S. Cook (Ed.) *Ecological Issues on Reintroducing Wolves into Yellowstone National Park.* Scientific Monograph NPS/NRYELL/NRSM-93/22, USDI National Park Service, Denver, CO.

Boyce, M. S. 1995. Anticipating consequences of wolves in Yellowstone: Model validation. Pages 199–209 in L. N. Carbyn, S. H. Fritts, and D. R. Seip (Eds.) *Ecology and Conservation of Wolves in a Changing World.* Canadian Circumpolar Institute, Edmonton, Alberta, Canada.

Boyce, M. S., and J. -M. Gaillard. 1992. Wolves in Yellowstone, Jackson Hole, and the north fork of the Shoshone river: Simulating ungulate consequences of wolf recovery. Pages 4–71 to 4–115 in J. D. Varley and W. G. Brewster (Eds.) *Wolves for Yellowstone? A Report to the United States Congress, Research and Analysis,* Vol. 4. USDI National Park Service, Yellowstone National Park, WY.

Burnham, K. P., and D. R. Anderson. 2002. *Model Selection and Multimodel Inference: A Practical Information-Theoretic Approach.* Springer, New York, NY.

Carbyn, L. N. 1974. *Wolf Predation and Behavioral Interactions with Elk and other Ungulates in an Area of High Prey Diversity.* Canadian Wildlife Service, Edmonton, Alberta, Canada.

Carbyn, L. N. 1983. Wolf predation on elk in Riding Mountain National Park, Manitoba. *Journal of Wildlife Management* **47**:963–976.

Carbyn, L. N., and T. Trottier. 1987. Responses of bison on their calving grounds to predation by wolves in Wood Buffalo National Park. *Canadian Journal of Zoology* **65**:2072–2078.

Carbyn, L. N., S. M. Oosenbrug, and D. W. Anions. 1993. *Wolves, Bison and the Dynamics Related to the Peace-Athabasca Delta in Canada's Wood Buffalo National Park.* Canadian Circumpolar Institute, Edmonton, Alberta, Canada.

Caswell, H. 1978. Predator-mediated coexistence a nonequilibrium model. *American Naturalist* **110**:141–151.

Charnov, E. L., G. H. Orians, and K. Hyatt. 1976. Ecological implications of resource depression. *American Naturalist* **110**:247–259.

Cheatum, E. L. 1949. Bone marrow as an index of malnutrition in deer. *New York State Conservation* **3**:19–22.

Chesson, J. 1978. Measuring preference in selective predation. *Ecology* **59**:211–215.

Craighead, J. J., F. C. Craighead, Jr., R. L. Ruff, and B. W. O'Gara. 1973. Home ranges and activity patterns of nonmigratory elk of the Madison drainage herd as determined by biotelemetry. *Wildlife Monographs* **33**:1–50.

Creel, S., and N. M. Creel. 2002. *The African Wild Dog: Behavior, Ecology, and Conservation.* Princeton University Press, Princeton, NJ.

Creel, S., J. A. Winnie, Jr., B. Maxwell, K. Hamlin, and M. Creel. 2005. Elk alter habitat selection as an antipredator response to wolves. *Ecology* **86**:3387–3397.

Creel, S. R., and N. M. Creel. 1995. Communal hunting and pack size in African wild dogs, *Lycaon pictus*. *Animal Behaviour* **50**:1325–1339.

Dale, B. M., L. G. Adams, and R. T. Bowyer. 1994. Functional response of wolves preying on barren-ground caribou in a multiple-prey ecosystem. *Journal of Animal Ecology* **63**:644–652.

Dale, B. M., L. G. Adams, and R. T. Bowyer. 1995. Winter wolf predation in a multiple ungulate prey system: Gates of the Arctic National Park, Alaska. Pages 223–230 *in* L. N. Carbyn, S. H. Fritts, and D. R. Seip (Eds.) *Ecology and Conservation of Wolves in a Changing World.* Canadian Circumpolar Institute, Edmonton, Alberta, Canada.

Denno, R. F., C. Gratton, M. A. Peterson, G. A. Langellotto, D. L. Finke, and A. F. Huberty. 2002. Bottom-up forces mediate natural-enemy impact in a phytophagous insect community. *Ecology* **83**:1443–1458.

Denno, R. F., and M. A. Peterson. 2000. Caught between the devil and the deep blue sea: Mobile planthoppers elude natural enemies and deteriorating host plants. *American Entomology* **46**:95–109.

Elliott, J. M. 2004. Prey switching in four species of carnivorous stoneflies. *Freshwater Biology* **49**:709–720.

Endler, J. A. 1991. Interactions between predators and prey. Pages 169–196 *in* J. R. Krebs and N. B. Davies (Eds.) *Behavioural Ecology: An Evolutionary Approach.* Blackwell, Oxford, United Kingdom.

Errington, P. L. 1946. Predation and vertebrate populations. *Quarterly Review of Biology* **21**:145–177, 221–245.

Eubanks, M. D., and R. F. Denno. 2000. Health food versus fast food: The effects of prey quality and mobility on prey selection by a generalist predator and indirect interactions among prey species. *Ecological Entomology* **25**:140–146.

Fitzgibbon, C. D., and J. Lazarus. 1995. Antipredator behavior of Serengeti ungulates: Individual differences and population consequences. Pages 274–296 *in* A. R. E. Sinclair and P. Arcese (Eds.) *Serengeti II: Dynamics, Management, and Conservation of an Ecosystem.* University of Chicago Press, Chicago, IL.

Frank, D. A., and S. J. McNaughton. 1992. The ecology of plants, large mammalian herbivores, and drought in Yellowstone National Park. *Ecology* **73**:2043–2058.

Fryxell, J. M., and P. Lundberg. 1994. Diet choice and predator–prey dynamics. *Evolutionary Ecology* **7**:379–393.

Fuller, T. K. 1989. Population dynamics of wolves in north-central Minnesota. *Wildlife Monographs* **105**:1–41.

Fuller, T. K., and L. B. Keith. 1980. Wolf population dynamics and prey relationships in northeastern Alberta. *Journal of Wildlife Management* **44**:583–602.

Fuller, W. A. 1959. The horns and teeth as indicators of age in bison. *Journal of Wildlife Management* **23**:342–344.

Garrott, R. A., J. E. Bruggeman, M. S. Becker, S. T. Kalinowski, and P. J. White. 2007. Evaluating prey switching in wolf-ungulate systems. *Ecological Applications* **17**:1588–1597.

Garrott, R. A., L. L. Eberhardt, J. K. Otton, P. J. White, and M. A. Chaffee. 2002. A geochemical trophic cascade in Yellowstone's geothermal environments. *Ecosystems* **5**:659–666.

Garrott, R. A., L. L. Eberhardt, P. J. White, and J. J. Rotella. 2003. Climate-induced variation in vital rates of an unharvested large-herbivore population. *Canadian Journal of Zoology* **81**:33–45.

Gittleman, J. L. 1989. Carnivore group living: Comparative trends. Pages 183–207 *in* J. L. Gittleman (Ed.) *Carnivore Behavior, Ecology and Evolution.* Cornell University Press, Ithaca, NY.

Greene, C. H. 1986. Patterns of prey selection: Implications of predator foraging tactics. *American Naturalist* **128**:824–839.

Greenwood, J. J. D., and R. A. Elton. 1979. Analysing experiments on frequency-dependent selection in predators. *Journal of Animal Ecology* **48**:721–737.

Hamlin, K. L., D. F. Pac, C. A. Sime, R. M. DeSimone, and G. L. Dusek. 2000. Evaluating the accuracy of ages obtained by two methods for Montana ungulates. *Journal of Wildlife Management* **64**:441–449.

Hanski, I., L. Hansson, and H. Henttonen. 1991. Specialist predators, generalist predators, and the microtine rodent cycle. *Journal of Animal Ecology* **60**:353–367.

Hayward, M. W., P. Henschel, J. O'Brien, G. Hofmeyr, G. Balme, and G. I. H. Kerley. 2006. Prey preferences of the leopard (*Panthera pardus*). *Journal of Zoology* **270**:298–313.

Hebblewhite, M., P. C. Paquet, D. H. Pletscher, R. B. Lessard, and C. J. Callaghan. 2003. Development and application of a ratio estimator to estimate wolf kill rates and variance in a multiple-prey system. *Wildlife Society Bulletin* **31**:933–946.

Hebblewhite, M., C. A. White, C. G. Nietvelt, J. A. McKenzie, T. E. Hurd, J. M. Fryxell, S. E. Bayley, and P. C. Paquet. 2005. Human activity mediates a trophic cascade caused by wolves. *Ecology* **86**:2135–2144.

Hobbs, N. T. 1996. Modification of ecosystems by ungulates. *Journal of Wildlife Management* **60**:695–713.

Holling, C. S. 1959. The components of predation as revealed by a study of small mammal predation of European pine sawfly. *Canadian Entomology* **91**:293–320.

Holt, R. D. 1977. Predation, apparent competition, and the structure of prey communities. *Theoretical Population Biology* **12**:197–229.

Holt, R. D., and J. H. Lawton. 1994. The ecological consequences of shared natural enemies. *Annual Review of Ecology and Systematics* **25**:495–520.

Hopcraft, J. G. C., A. R. E. Sinclair, and C. Packer. 2005. Planning for success: Serengeti lions seek prey accessibility rather than abundance. *Journal of Animal Ecology* **74**:559–566.

Hudson, R. J., J. C. Haigh, and A. B. Bubenik. 2002. Physical and physiological adaptations. Pages 255–257 *in* D. E. Toweill and J. Ward Thomas (Eds.) *North American Elk: Ecology and Management.* Smithsonian Institution Press, Washington, DC.

Huggard, D. J. 1993a. Prey selectivity of wolves in Banff National Park I. *Prey species. Canadian Journal of Zoology* 71:130–139.

Huggard, D. J. 1993b. Effect of snow depth on predation and scavenging by gray wolves. *Journal of Wildlife Management* 57:382–388.

Hunter, M. D., and P. W. Price. 1992. Playing chutes and ladders: Heterogeneity and the relative roles of bottom-up and top-down forces in natural communities. *Ecology* 73:724–732.

Husseman, J. S., D. L. Murray, G. Power, C. Mack, C. R. Wenger, and H. Quigley. 2003. Assessing differential prey selection patterns between two sympatric large carnivores. *Oikos* 101:591–601.

Ivlev, V. S. 1961. Experimental ecology of the feeding of fishes. Yale University Press, New Haven, CT.

Jaffe, R. 2001. Winter wolf predation in an elk-bison system in Yellowstone National Park, Wyoming. Thesis. Montana State University, Bozeman, MT.

Jędrzejewski, W., B. Jędrzejewska, H. Okarma, and A. L. Ruprecht. 1992. Wolf predation and snow cover as mortality factors in the ungulate community of the Bialoweiza National Park, Poland. *Oecologia* 90:27–36.

Jędrzejewski, W., B. Jędrzejewska, H. Okarma, K. Schmidt, K. Zub, and M. Musiani. 2000. Prey selection and predation by wolves in Bialowieza Primeval Forest, Poland. *Journal of Mammalogy* 81:197–212.

Karanth, K. U., and M. E. Sunquist. 1995. Prey selection by tiger, leopard, and dhole in tropical forests. *Journal of Animal Ecology* 64:439–450.

Krivan, V., and J. Eisner. 2003. Optimal foraging and predator-prey dynamics III. *Theoretical Population Biology* 63:269–279.

Krebs, C. J., S. Boutin, and R. Boonstra (Eds.). 2001. *Ecosystem Dynamics of the Boreal Forest: The Kluane Project.* Oxford University Press, New York, NY.

Kruuk, H. 1972. *The Spotted Hyena.* University of Chicago Press, Chicago, IL.

Kunkel, K. E., T. K. Ruth, D. H. Pletscher, and M. G. Hornocker. 1999. Winter prey selection by wolves and cougars in and near Glacier National Park, Montana. *Journal of Wildlife Management* 63:901–910.

Langellotto, G. A., and R. F. Denno. 2004. Responses of invertebrate natural enemies to complex-structured habitats: A meta-analytical synthesis. *Oecologia* 139:1–10.

Lechowicz, M. J. 1982. The sampling characteristics of electivity indices. *Oecologia* 52:22–30.

Lind, J., and W. Cresswell. 2005. Determining the fitness consequences of anti-predation behavior. *Behavioral Ecology* 16:945–956.

MacNulty, D. R. 2002. The predatory sequence and the influence of injury risk on hunting behavior in the wolf. Thesis. University of Minnesota, Saint Paul, MN.

MacNulty, D. R., L. D. Mech, and D. W. Smith. 2007. A proposed ethogram of large-carnivore predatory behavior, exemplified by the wolf. *Journal of Mammalogy* 88:595–605.

Makacha, S., and G. B. Schaller. 1969. Observations on lions in Lake Manyara National Park, Tanzania. *East African Wildlife Journal* 7:99–103.

Matter, W. J., and R. W. Mannan. 2005. How do prey persist? *Journal of Wildlife Management* 69:1315–1320.

McNaughton, S. J. 1985. Ecology of a grazing system: The Serengeti. *Ecological Monographs* 55:259–295.

Mech, D. L. 1970. *The Wolf: The Ecology and Behavior of an Endangered Species.* University of Minnesota Press, Minneapolis, MN.

Mech, L. D., and L. Boitani. 2003. Wolf social ecology. Pages 1–34 *in* L. D. Mech and L. Boitani (Eds.) *Wolves: Behavior, Ecology, and Conservation.* University of Chicago Press, Chicago, IL.

Mech, L. D., T. J. Meir, J. W. Burch, and L. G. Adams. 1995. Patterns of prey selection by wolves in Denali National Park, Alaska. Pages 231–243 *in* L. N. Carbyn, S. H. Fritts, and D. R. Seip (Eds.) *Ecology and Conservation of Wolves in a Changing World.* Canadian Circumpolar Institute, Edmonton, Alberta, Canada.

Mech, L. D., and R. O. Peterson. 2003. Wolf-prey relations. Pages 131–160 *in* L. D. Mech and L. Boitani (Eds.) *Wolves: Behavior, Ecology, and Conservation.* University of Chicago Press, Chicago, IL.

Menard, S. 2002. *Applied Logistic Regression Analysis.* Sage Publications, Thousand Oaks, CA.

Menge, B. A. 1972. Foraging strategy of a starfish in relation to actual prey availability and environmental predictability. *Ecological Monographs* 42:25–50.

Messier, F. 1995. On the functional and numerical responses of wolves to changing prey density. Pages 187–197 *in* L. N. Carbyn, S. H. Fritts, and D. R. Seip (Eds.) *Ecology and Conservation of Wolves in a Changing World.* Canadian Circumpolar Institute, University of Alberta, Edmonton, Alberta, Canada.

Moffitt, S. A. 1998. Aging bison by the incremental cementum growth layers in teeth. *Journal of Wildlife Management* 62:1276–1280.

Moran, M. D., T. P. Rooney, and L. E. Hurd. 1996. Top-down cascade from a bitrophic predator in an old-field community. *Ecology* 77:2219–2227.

Morse, D. H. 1980. *Behavioral Mechanisms in Ecology.* Harvard University Press, Cambridge, MA.

Murdoch, W. W. 1969. Switching in general predators: Experiments on predator specificity and the stability of prey populations. *Ecological Monographs* 39:335–354.

Murdoch, W. W., and J. R. Bence. 1987. General predators and unstable prey populations. Pages 17–30 *in* W. C. Kerfoot and A. Sih (Eds.) *Predation: Direct and Indirect Impacts on Aquatic Communities.* University Press of England, Hanover, NH.

Murdoch, W. W., and J. R. Marks. 1973. Predation by coccinellid beetles: Experiments in switching. *Ecology* **54**:160–167.

Murdoch, W. W., and A. Oaten. 1975. Predation and population stability. *Advances in Ecological Research* **9**:1–131.

Murie, A. 1944. The wolves of Mount McKinley. Fauna of the national parks of the United States. *Fauna Series* **5**:1–238.

Nelson, M. E., and D. L. Mech. 1981. Deer social organization and wolf predation in northeastern Minnesota. *Wildlife Monographs Number 77.*

Nelson, M. E., and D. L. Mech. 1986. Relationship between snow depth and gray wolf predation on white-tailed deer. *Journal of Wildlife Management* **50**:471–474.

Neter, J., M. H. Kutner, C. J. Nachtsheim, and W. Wasserman. 1996. *Applied Linear Statistical Models.* McGraw-Hill, Boston, MA.

Oaten, A., and W. W. Murdoch. 1975. Switching, functional response, and stability in predator-prey systems. *American Naturalist* **109**:299–319.

Paine, R. T. 1966. Food web complexity and species diversity. *American Naturalist* **100**:65–75.

Parker, K. L., C. T. Robbins, and T.A Hanley. 1984. Energy expenditures for locomotion by mule deer and elk. *Journal of Wildlife Management* **48**:474–488.

Pastorok, R. A. 1981. Prey vulnerability and size selection by *Chaoborus* larvae. *Ecology* **62**:1311–1324.

Peckarsky, B., and M. A. Penton. 1989. Mechanisms of prey selection by stream-dwelling stoneflies. *Ecology* **70**:1203–1218.

Peterson, R. O. 1977. *Wolf Ecology and Prey Relationships on Isle Royale.* National Park Service Scientific Monograph Series 11, Washington, DC.

Post, D. M., M. E. Conners, and D. S. Goldberg. 2000. Prey preference by a top predator and the stability of linked food chains. *Ecology* **81**:8–14.

Potvin, F., H. Jolicoeur, and J. Huot. 1988. Wolf diet and prey selectivity during two periods for deer in Quebec: Decline versus expansion. *Canadian Journal of Zoology* **66**:1274–1279.

Power, M. E., J. C. Marks, and M. S. Parker. 1992. Variation in the vulnerability of prey to different predators: Community level consequences. *Ecology* **73**:2218–2223.

Prugh, L. 2005. Coyote prey selection and community stability during a decline in food supply. *Oikos* **110**:253–264.

Quinn, J. L., and W. Cresswell. 2004. Predator hunting behaviour and prey vulnerability. *Journal of Animal Ecology* **73**:143–154.

R Development Core Team. 2006. R: A language and environment for statistical computing. R Foundation for Statistical Computing, Vienna, Austria<http://www.R-project.org>.

Rodgers, A. R. 1990. Evaluating preference in laboratory studies of diet selection. *Canadian Journal of Zoology* **68**:188–190.

Rosenheim, J. A., T. E. Glik, R. E. Goeriz, and B. Rämert. 2004. Linking a predator's foraging behavior with its effects on herbivore population suppression. *Ecology* **85**:3362–3372.

Rosenzweig, M. L. 1966. Community structure in sympatric carnivores. *Journal of Mammalogy* **47**:602–612.

Schaller, G. B. 1972. *The Serengeti Lion: A Study of Predator–Prey Relations.* The University of Chicago Press, Chicago, IL.

Scheel, D. 1993. Profitability, encounter rates, and prey choice of African lions. *Behavioral Ecology* **4**:90–97.

Severinghaus, C. W. 1947. Relationship of weather to winter mortality and population levels among deer in the Adirondack region of New York. *Transactions North American Wildlife Conference* **12**:212–223.

Shupe, J. L., A. E. Olsen, H. B. Peterson, and J. B. Low. 1984. Flouride toxicosis in wild ungulates. *Journal of the American Veterinary Medical Association* **185**:1295–1300.

Sinclair, A. R. E., and P. Arcese. 1995. Population consequences of predation-sensitive foraging: The Serengeti wildebeest. *Ecology* **76**:882–891.

Sinclair, A. R. E., S. Mduma, and J. S. Brashares. 2003. Patterns of predation in a diverse predator-prey system. *Nature* **425**:288–290.

Smith, D. W., T. D. Drummer, K. M. Murphy, D. S. Guernsey, and S. B. Evans. 2004. Winter prey selection and estimation of wolf kill rates in Yellowstone National Park, 1995–2000. *Journal of Wildlife Management* **68**:153–166.

Smith, D. W., L. D. Mech, M. Meagher, W. E. Clark, R. Jaffe, M. K. Phillips, and J. A. Mack. 2000. Wolf-bison interactions in Yellowstone National Park. *Journal of Mammalogy* **81**:1128–1135.

Smith, D. W., R. O. Peterson, and D. B. Houston. 2003. Yellowstone after wolves. *BioScience* **53**:330–340.

Smuts, G. L. 1978. Interrelations between predators, prey, and their environment. *BioScience* **28**:316–320.

Stahler, D., B. Heinrich, and D. Smith. 2002. Common ravens, *Corvus corax*, preferentially associate with grey wolves, *Canis lupus*, as a foraging strategy in winter. *Animal Behaviour* **64**:283–290.

Stephens, D. W., and J. R. Krebs. 1986. *Foraging Theory.* Princeton University Press, Princeton, NJ.

Synder, W. E., and A. R. Ives. 2001. Generalist predators disrupt biological control by a specialist parasitoid. *Ecology* **82**:705–716.

Taylor, R. J. 1984. *Predation.* Chapman & Hall, New York, NY.

Temple, S. A. 1987. Do predators always capture substandard individuals disproportionately from prey populations? *Ecology* **68**:669–674.

Terborgh, J., L. Lopez, P. V. Nunez, M. Rao, G. Shahabuddin, G. Orihuela, M. Riveros, R. Ascanio, G. H. Adler, T. D. Lambert, and L. Balbas. 2001. Ecological meltdown in predator-free forest fragments. *Science* **294**:1923–1925.

Turchin, P., and I. Hanski. 1997. An empirically based model for latitudinal gradient in vole population dynamics. *American Naturalist* **149**:842–874.

van Baalen, M., V. Krivan, and P. C. J. van Rijn. 2001. Alternative food, switching predators, and the persistence of predator-prey systems. *American Naturalist* **157**:512–524.

Weaver, J. L. 1994. Ecology of wolf predation amidst high ungulate diversity in Jasper National Park, Alberta. Dissertation. University of Montana, Missoula, MT.

White, G. C., and R. A. Garrott. 1990. *Analysis of Wildlife Radio-Tracking Data.* Academic Press, San Diego, CA.

Wittmer, H. U., A. R. E. Sinclair, and B. N. McLellan. 2005. The role of predation in the decline and extirpation of woodland caribou. *Oecologia* **144**:257–267.

APPENDIX

Multinomial Model List

Model Suite 1: Prey Models

$P1 = PREY \sim ELK_{calf}$

$P2 = PREY \sim ELK_{calf} + ELK_{ad}$

$P3 = PREY \sim ELK_{calf} + ELK_{ad} + BISON_{calf}$

$P4 = PREY \sim ELK$

$P5 = PREY \sim ELK + BISON$

$P6 = PREY \sim BISON.ELK$

$P7 = PREY \sim ELK_{calf} + BISON_{calf}$

Model Suite 2: Prey + Snow Pack Models

$PS1 = PREY \sim ELK_{calf} + SWE_{acc}$

$PS2 = PREY \sim ELK_{calf} + ELK_{ad} + SWE_{acc}$

$PS3 = PREY \sim ELK_{calf} + ELK_{ad} + BISON_{calf} + SWE_{acc}$

$PS4 = PREY \sim ELK + SWE_{acc}$

$PS5 = PREY \sim ELK + BISON + SWE_{acc}$

$PS6 = PREY \sim BISON.ELK + SWE_{acc}$

$PS7 = PREY \sim ELK_{calf} + SWE_{mean}$

$PS8 = PREY \sim ELK_{calf} + ELK_{ad} + SWE_{mean}$

$PS9 = PREY \sim ELK_{calf} + ELK_{ad} + BISON_{calf} + SWE_{mean}$

$PS10 = PREY \sim ELK + SWE_{mean}$

$PS11 = PREY \sim ELK + BISON + SWE_{mean}$

$PS12 = PREY \sim BISON.ELK + SWE_{mean}$

$PS13 = PREY \sim ELK_{calf} + BISON_{calf} + SWE_{acc}$

$PS14 = PREY \sim ELK_{calf} + BISON_{calf} + SWE_{mean}$

Model Suite 3: Prey + Wolf Competition Models

$PW1 = PREY \sim ELK_{calf} + MULTPK$

$PW2 = PREY \sim ELK_{calf} + ELK_{ad} + MULTPK$

$PW3 = PREY \sim ELK_{calf} + ELK_{ad} + BISON_{calf} + MULTPK$

$PW4 = PREY \sim ELK + MULTPK$

$PW5 = PREY \sim ELK + BISON + MULTPK$

$PW6 = PREY \sim BISON.ELK + MULTPK$

$PW7 = PREY\sim ELK_{calf} + WOLF{:}ELK$

$PW8 = PREY\sim ELK_{calf} + ELK_{ad} + WOLF{:}ELK$

$PW9 = PREY\sim ELK_{calf} + ELK_{ad} + BISON_{calf} + WOLF{:}ELK$

$PW10 = PREY\sim ELK + WOLF{:}ELK$

$PW11 = PREY\sim ELK + BISON + WOLF{:}ELK$

$PW12 = PREY\sim BISON.ELK + WOLF{:}ELK$

$PW13 = PREY\sim ELK_{calf} + WOLF{:}UNG$

$PW14 = PREY\sim ELK_{calf} + ELK_{ad} + WOLF{:}UNG$

$PW15 = PREY\sim ELK_{calf} + ELK_{ad} + BISON_{calf} + WOLF{:}UNG$

$PW16 = PREY\sim ELK + WOLF{:}UNG$

$PW17 = PREY\sim ELK + BISON + WOLF{:}UNG$

$PW18 = PREY\sim BISON.ELK + WOLF{:}UNG$

$PW19 = PREY\sim ELK_{calf} + BISON_{calf} + MULTPK$

$PW20 = PREY\sim ELK_{calf} + BISON_{calf} + WOLF{:}ELK$

$PW21 = PREY\sim ELK_{calf} + BISON_{calf} + WOLF{:}UNG$

Model Suite 4: Prey + Snow Pack + Wolf Competition Models

$PSW1 = PREY\sim ELK_{calf} + MULTPK + SWE_{acc}$

$PSW2 = PREY\sim ELK_{calf} + ELK_{ad} + MULTPK + SWE_{acc}$

$PSW3 = PREY\sim ELK_{calf} + ELK_{ad} + BISON_{calf} + MULTPK + SWE_{acc}$

$PSW4 = PREY\sim ELK + MULTPK + SWE_{acc}$

$PSW5 = PREY\sim ELK + BISON + MULTPK + SWE_{acc}$

$PSW6 = PREY\sim BISON.ELK + MULTPK + SWE_{acc}$

$PSW7 = PREY\sim ELK_{calf} + WOLF{:}ELK + SWE_{acc}$

$PSW8 = PREY\sim ELK_{calf} + ELK_{ad} + WOLF{:}ELK + SWE_{acc}$

$PSW9 = PREY\sim ELK_{calf} + ELK_{ad} + BISON_{calf} + WOLF{:}ELK + SWE_{acc}$

$PSW10 = PREY\sim ELK + WOLF{:}ELK + SWE_{acc}$

$PSW11 = PREY\sim ELK + BISON + WOLF{:}ELK + SWE_{acc}$

$PSW12 = PREY\sim BISON.ELK + WOLF{:}ELK + SWE_{acc}$

$PSW13 = PREY\sim ELK_{calf} + WOLF{:}UNG + SWE_{acc}$

$PSW14 = PREY\sim ELK_{calf} + ELK_{ad} + WOLF{:}UNG + SWE_{acc}$

$PSW15 = PREY\sim ELK_{calf} + ELK_{ad} + BISON_{calf} + WOLF{:}UNG + SWE_{acc}$

$PSW16 = PREY\sim ELK + WOLF{:}UNG + SWE_{acc}$

$PSW17 = PREY\sim ELK + BISON + WOLF{:}UNG + SWE_{acc}$

$PSW18 = PREY\sim BISON.ELK + WOLF{:}UNG + SWE_{acc}$

$PSW19 = PREY\sim ELK_{calf} + MULTPK + SWE_{mean}$

$PSW20 = PREY\sim ELK_{calf} + ELK_{ad} + MULTPK + SWE_{mean}$

$PSW21 = PREY\sim ELK_{calf} + ELK_{ad} + BISON_{calf} + MULTPK + SWE_{mean}$

$PSW22 = PREY\sim ELK + MULTPK + SWE_{mean}$

$PSW23 = PREY\sim ELK + BISON + MULTPK + SWE_{mean}$

$PSW24 = PREY\sim BISON.ELK + MULTPK + SWE_{mean}$

$PSW25 = PREY\sim ELK_{calf} + WOLF{:}ELK + SWE_{mean}$

$PSW26 = PREY\sim ELK_{calf} + ELK_{ad} + WOLF{:}ELK + SWE_{mean}$

$PSW27 = PREY\sim ELK_{calf} + ELK_{ad} + BISON_{calf} + WOLF{:}ELK + SWE_{mean}$

$PSW28 = PREY\sim ELK + WOLF{:}ELK + SWE_{mean}$

$PSW29 = PREY\sim ELK + BISON + WOLF{:}ELK + SWE_{mean}$

$PSW30 = PREY\sim BISON.ELK + WOLF{:}ELK + SWE_{mean}$

$PSW31 = PREY\sim ELK_{calf} + WOLF{:}UNG + SWE_{mean}$

$PSW32 = PREY\sim ELK_{calf} + ELK_{ad} + WOLF{:}UNG + SWE_{mean}$

$\text{PSW33} = \text{PREY} \sim \text{ELK}_{\text{calf}} + \text{ELK}_{\text{ad}} + \text{BISON}_{\text{calf}} + \text{WOLF:UNG} + \text{SWE}_{\text{mean}}$

$\text{PSW34} = \text{PREY} \sim \text{ELK} + \text{WOLF:UNG} + \text{SWE}_{\text{mean}}$

$\text{PSW35} = \text{PREY} \sim \text{ELK} + \text{BISON} + \text{WOLF:UNG} + \text{SWE}_{\text{mean}}$

$\text{PSW36} = \text{PREY} \sim \text{BISON.ELK} + \text{WOLF:UNG} + \text{SWE}_{\text{mean}}$

$\text{PSW37} = \text{PREY} \sim \text{ELK}_{\text{calf}} + \text{BISON}_{\text{calf}} + \text{MULTPK} + \text{SWE}_{\text{acc}}$

$\text{PSW38} = \text{PREY} \sim \text{ELK}_{\text{calf}} + \text{BISON}_{\text{calf}} + \text{MULTPK} + \text{SWE}_{\text{mean}}$

$\text{PSW39} = \text{PREY} \sim \text{ELK}_{\text{calf}} + \text{BISON}_{\text{calf}} + \text{WOLF:ELK} + \text{SWE}_{\text{acc}}$

$\text{PSW40} = \text{PREY} \sim \text{ELK}_{\text{calf}} + \text{BISON}_{\text{calf}} + \text{WOLF:ELK} + \text{SWE}_{\text{mean}}$

$\text{PSW41} = \text{PREY} \sim \text{ELK}_{\text{calf}} + \text{BISON}_{\text{calf}} + \text{WOLF:UNG} + \text{SWE}_{\text{acc}}$

$\text{PSW42} = \text{PREY} \sim \text{ELK}_{\text{calf}} + \text{BISON}_{\text{calf}} + \text{WOLF:UNG} + \text{SWE}_{\text{mean}}$

CHAPTER 17

Wolf Kill Rates: Predictably Variable?

Matthew S. Becker,* Robert A. Garrott,* P. J. White,[†] Rosemary Jaffe,*
John J. Borkowski,[‡] Claire N. Gower,* and Eric J. Bergman*

*Fish and Wildlife Management Program, Department of Ecology, Montana State University
[†]National Park Service, Yellowstone National Park
[‡]Department of Mathematical Sciences, Montana State University

Contents

Theme

The ability of predators to successfully capture and kill prey is affected by the abundance and diversity of the prey assemblage, and such variation is a fundamental driver of ecosystem dynamics because *per capita* consumption rate strongly influences the stability and strength of community interactions. Descriptions of predatory behavior in this context typically include the functional response, specifically the kill rate of a predator as a function of prey density. Thus, a major objective in studying predator–prey interactions is to evaluate the strength of the numerous factors related to the kill rate of a predator, and to subsequently determine the forms of its functional response in natural systems because different forms have different consequences for ecosystem dynamics. Recent controversies over the nature of predation focus on the respective roles of prey and predator abundance in affecting the functional response. However, resolution requires more direct measures of kill rates in natural systems. We estimated wolf (*Canis lupus*) kill rates in a tractable and newly established wolf–elk (*Cervus elaphus*)–bison (*Bison bison*) system in the Madison headwaters area of Yellowstone National Park during winters 1998–1999 to 2006–2007 to document the transition from over seven decades without wolves to

The Ecology of Large Mammals in Central Yellowstone
R. Garrott, P. J. White and F. Watson
ISSN 1936-7961, DOI: 10.1016/S1936-7961(08)00217-0

Copyright © 2009, Elsevier Inc.
All rights reserved.

a well-established top predator population. Wolf abundance, distribution, and prey selection varied during the study, concurrent with variations in the demography, distribution, and behavior of elk and bison. These dynamics enabled us to evaluate factors influencing variations in wolf kill rates and the forms of their functional response.

I. INTRODUCTION

The role of a predator in regulating or destabilizing prey populations is widely believed to depend on the form of its functional response (Murdoch and Oaten 1975, Oaten and Murdoch 1975, Hassell 1978). There are three general forms of the functional response (Holling 1959). A predation rate increasing linearly with prey density is know as a Type I response, which is considered unrealistic for most predators due to lack of constraints in handling time or the time needed to capture and subdue a prey item. Consequently, Type I responses are most likely confined to predators with few handling constraints such as filter feeders (Jeschke *et al.* 2004). A more plausible description of predation behavior is the Type II response, which exhibits a decelerating predation rate with increasing prey density reflective of predator satiation (Holling 1959). This form of predatory behavior is thought to be common in natural systems consisting of a single prey species or a specialist predator (Peckarsky 1984) and is considered destabilizing because it is inversely density-dependent (Oaten and Murdoch 1975; Hassell 1978, 2000). However, Fryxell *et al.* (2007) argue that, while Type II responses can be destabilizing in the solitary prey and predator systems from which most predator–prey theory has been developed and refined, such a response can also be stabilizing if prey and predators aggregate in groups. The sigmoidal Type III functional response can be exhibited when more than one prey type exists and a predator is plastic in its foraging (Holling 1959). A Type III functional response can be generated by a variety of factors such as prey-switching (Murdoch 1969), selective foraging, and learning by the predator (Tinbergen 1960, Real 1979), or use of refuges by prey (Taylor 1984). Because a Type III response implies density-dependent predation, such behavior is considered to have a stabilizing influence on ecosystems (Oaten and Murdoch 1975).

Due largely to the dramatically different ecosystem trajectories that can ensue with different predator behaviors, an immense body of work has been performed on determining the drivers of kill rates. Kill rates are influenced by encounter rates and prey density (Holling 1959), the presence of alternative prey (Murdoch 1969), environmental factors (Thompson 1978, Anderson 2001), and prey distribution (Real 1979, Cosner *et al.* 1999, Pitt and Ritchie 2002). However, the causal factors driving the ultimate form of the functional response have been the subject of considerable debate. The long-standing belief that forms were driven by prey density alone ("prey-dependence") has been challenged by the idea that predator density can also appreciably influence *per capita* consumption ("predator dependence"). Strong predator dependence, typically denoted by the ratio of predators to prey ("ratio-dependence") has been vigorously debated as an alternative to the prey dependent functional response (Arditi and Ginzburg 1989, Berryman 1992, Abrams 1994, Akcakaya *et al.* 1995, Abrams and Ginzburg 2000). While this debate has not been resolved, it is likely that both prey and predator numbers influence predation and more empirical studies are needed (Abrams and Ginzburg 2000, Schenk *et al.* 2005).

Wolf–ungulate systems have received substantial attention in studies of kill rates, largely due to the strong scientific and societal interest in assessing the effects of wolves on prey populations. Kill rates of wolves are extremely variable and influenced by prey density, pack size, and snow pack (see reviews in Hebblewhite *et al.* 2003, Mech and Peterson 2003). There is considerable disagreement regarding the nature of wolf foraging behavior, but few studies of functional responses due to the difficulties inherent in data collection (Mech and Peterson 2003). Thus, various scientists have advocated that wolf kill rates were best described as constant (Eberhardt 1997), ratio-dependent (Vucetich *et al.* 2002, Jost *et al.* 2005), or prey-dependent (Messier 1994, 1995; Messier and Joly 2000; Varley and Boyce 2006). Recent analyses from the long-term wolf–moose dataset of Isle Royale indicated that ratio-dependence best described the nature of wolf predation in this system (Vucetich *et al.* 2002, Jost *et al.* 2005), and that the inclusion of a ratio-dependent functional response in kill rate analyses may provide significant insights into discrimination between types of predation.

Disagreement also stems, in part, from the inherent difficulty in accurately assessing kill rates in wolf–ungulate systems. These metrics are so difficult to measure that some scientists question the appropriateness and/or feasibility of estimating the functional response for describing wolf–ungulate systems (Eberhardt 1997, Marshal and Boutin 1999, Person *et al.* 2001, Mech and Peterson 2003) while others contend that distinguishing between functional response forms is not ecologically critical (Dale *et al.* 1994, Van Ballenberghe and Ballard 1994). Instead, some scientists suggest monitoring cause-specific mortality and recruitment rates of prey species (Kunkel *et al.* 2004) or changes in ungulate carrying capacity (Bowyer *et al.* 2005). In addition, the methods and metrics used to calculate kill rates vary widely. Thus, comparisons between systems and methods are often not possible (Hebblewhite *et al.* 2003). Lastly, the estimation of kill rates is subject to variable observer effort, weather conditions, and movement of wolves. Only recently have investigators attempted to account for these sources of variability (Jaffe 2001, Hebblewhite *et al.* 2003, Smith *et al.* 2004).

We evaluated drivers of kill rates and the forms of wolf functional response using long-term predation data collected during winters 1998–1999 through 2006–2007 from a tractable wolf–elk–bison system in the Madison headwaters area of Yellowstone National Park that experienced substantial seasonal and annual variation in prey abundance, predator abundance, and snow pack. Data on wolf numbers and kills were collected daily for each winter (15 November–21 April) of the study period (Figure 17.1). There are various ecological justifications for employing several different metrics to evaluate wolf kill rates. Metrics employing kills as the unit of measure are likely more appropriate than biomass for evaluating the effects of wolf predation on ungulate prey populations (Hayes and Harestad 2000, Hayes *et al.* 2000). Furthermore, it is useful to distinguish between kills per pack and kills per wolf because wolves are group hunting predators with a rigid social hierarchy and the pack is typically the hunting unit rather than individual wolves (Mech 1970). Questions concerning wolf population dynamics and their food acquisition are better addressed using a metric of biomass (Mech and Peterson 2003). Consequently, we calculated metrics of kills/pack/day, kills/wolf/day, and kg/wolf/day for each pack. Our objectives were to: (1) describe temporal trends in kill rates within and among winters and wolf packs; (2) determine the primary factors driving trends in total wolf kill rates, as well

FIGURE 17.1 Wolves from the Hayden pack feed on an elk kill in the Madison headwaters. Estimating the rate at which a predator kills prey is fundamental to understanding predator–prey dynamics (Photo by Shana Dunkley).

as kill rates for elk and bison for various metrics; and (3) assess the form of wolf functional response for elk. We predicted that kill rates would be positively influenced by the abundance of elk and bison and negatively influenced by wolf abundance. Also, we predicted that kill rates would be positively related to pack size when calculated per pack, and negatively related to pack size when calculated per wolf. Further, we predicted wolf functional responses would be best described by ratio dependence due to the strong potential for predator dependence.

II. METHODS

A. Wolf Tracking and Kill Detection

We conducted intensive kill rate investigations in the primary winter ranges of bison and elk in the Madison headwaters area (31,000 ha), with concurrent investigations of these prey species allowing collection of wolf predation data in a tractable area with a well-described ungulate prey base. We documented wolf kill rates during 15 November through 21 April each winter from the establishment of a resident pack in 1998–1999 (Chapter 15 by Smith *et al.*, this volume) through 2006–2007. Our sampling unit was radio-collared wolf packs that incorporated the study area as part of their territory. Wolves were aerially darted from helicopters by National Park Service biologists and fitted with VHF telemetry collars. A total of 37 wolves from four packs were collared during the course of the study (Chapter 15 by Smith *et al.*, this volume).

The number and sizes of wolf packs using the study area were dynamic within and among winters (Chapter 15 by Smith *et al.*, this volume). Thus, we used ground observations, snow-tracking, and counts during aerial tracking flights by park biologists to estimate the wolf population. We estimated the wolf population in wolf days, defined as one wolf in the study area for one day. We used roads traversing each river drainage in the study area (Chapter 2 by Newman and Watson, this volume) to rapidly sample for wolf presence daily through the winter. Sampling began at dawn with ground crews of three to four people covering all roads by snowmobile or vehicle and using strategic high points in the landscape to facilitate telemetry triangulations (White and Garrott 1990) and observations of wolves. When possible, multiple locations were obtained in early morning and evening each day. We also recorded any uncollared wolves detected opportunistically via tracks or observations to aid in the estimation of the wolf population using the study area. In addition, biologists studying elk and bison routinely covered backcountry areas to assist with wolf detection.

When wolves were located, we used visual scans and monitoring of avian scavengers in the vicinity to detect kills. Ravens preferentially associate with wolves in winter, and an average of 28 ravens (*Corvus corax*) were present at fresh wolf kills on the northern elk winter range in Yellowstone National Park (Stahler *et al.* 2002), with slightly lower averages in the Madison headwaters area (D. Stahler, National Park Service, personal communication). This association facilitated the detection of kills. We also conducted extensive snow-tracking after wolves departed the area to further facilitate kill detection (Huggard 1993, Dale *et al.* 1995, Jędrzejewski *et al.* 2000, Jaffe 2001, Hebblewhite *et al.* 2003, Bergman *et al.* 2006; Figure 17.2). We necropsied ungulate carcasses to determine cause of death, species, sex, age, condition, and percent consumed. Wolf kills were inferred from collective evidence of subcutaneous hemorrhaging indicative of injuries sustained before death, signs of struggle or chase at the kill site, blood trails, signs of predator presence, and our knowledge of wolf movements and activities. We documented frequent spring grizzly bear (*Ursus arctos*) predation on bison during the latter years of the study. Thus, when both bears and wolves were present on a kill, we classified it based on the patterns of injury and subcutaneous hemorrhaging. Bears typically attacked the head and spine, while wolves attacked the flanks, hindquarters, and underside of the neck. Similarly, mountain lion (*Puma concolor*) kills of elk were determined based on characteristics of the kill site and patterns of injury. Kills were

FIGURE 17.2 Snow tracking the Gibbon pack along Nez Perce Creek in the Firehole river drainage. A combination of telemetry, scanning, and tracking was employed on a daily basis to detect wolf kills in the study area (Photo by Shana Dunkley).

sexed using the presence of genitalia, horns, antlers, or pedicels, and aged based on size and patterns of tooth eruption and replacement (Fuller 1959, Hudson *et al.* 2002). When available, an incisor or canine was removed from adult ungulates and aged using cementum annuli (Moffitt 1998, Hamlin *et al.* 2000).

B. Kill Rate Estimation

Daily estimates of wolf numbers and kills detected for the wolf population at large and for each pack were used to estimate minimum kill rates each winter and for three winter periods of approximately eight weeks each that corresponded to early (15 November–6 January), middle (7 January–27 February), and late winter (28 February–21 April), ending near the mean pack denning date, after which packs were considerably less cohesive (Jaffe 2001; Chapter 15 by Smith *et al.*, this volume). The kills/pack/day metric was calculated by dividing the number of kills by the number of days in the sampling period in which the respective pack was detected, while the kills/wolf/day metric was calculated by dividing the number of kills by the estimated wolf days for a given pack for each period. Winter and winter period estimates of kills/wolf/day for the entire population were calculated from pooled estimates of wolf days and kills for a given period or winter. Estimates for kg/wolf/day were derived by summing the biomass of all kills for a given pack and dividing by the estimated wolf days for that period. We classified all kills into species, sex, and age, and used biomass estimates for elk and bison obtained from Murie (1951) and Meagher (1973), respectively. Elk bulls, cows, and calves were estimated at 287, 236, and 116 kg, respectively, while bison adults and calves were estimated at 500 and 136 kg, respectively. Bison adult age classes of both sexes can vary dramatically in weight depending on age (Berger and Peacock 1988). Thus, we did not use separate categories for males and females in biomass estimation. We assumed 75% edible biomass for each prey item (Peterson 1977), but did not account for scavenger loss or incomplete consumption of carcasses. Therefore, the kg/wolf/day metric was considered an index rather than an absolute measure of consumption per wolf. For each pack, we estimated total kill rates and separate kill rates for elk and bison using all three kill rate metrics for each winter period. We also estimated these kill rates using the same metrics for the entire wolf population during each winter and winter period.

C. Evaluating Kill Rate Variation

We used multiple linear regression techniques (Neter *et al.* 1996) to evaluate factors affecting variation in wolf kill rates within and among winters and packs. Because the Madison headwaters was a multiple-prey system with ungulate species differing substantially in abundance and defenses (Garrott *et al.* 2007), we evaluated variations in kill rates for each wolf pack using three separate response variables: total kill rates, kill rates for elk, and kill rates for bison. For each response variable, we calculated three kill rate metrics (kills/wolf/day, kills/pack/day, and kg/wolf/day) during a given winter period, comprising nine analyses in total. We developed eight covariates to evaluate the influences of prey abundance, wolf pack size, and snow pack. These covariates were judiciously selected from factors reported to be influential in the kill rate literature, as well as from our knowledge of the study system.

We used six covariates to describe wolf prey abundance, including elk abundance (ELK_{all}), bison abundance ($BISON_{all}$), and the respective abundances of elk adults and calves (ELK_{adt}, ELK_{calf}) and bison adults ($BISON_{adt}$) and calves ($BISON_{calf}$). We estimated prey abundance for the entire study area rather than just the drainages encompassed by a particular pack's territory because multiple packs overlapped spatially and temporally, larger packs were more dominant (Chapter 15 by Smith *et al.* and Chapter 16 by Becker *et al.*, this volume), and kill rates were estimated for each pack over nearly an 8-week period during which bison movement between wolf pack territories was considerable (Chapter 12 by Bruggeman *et al.*, this volume). We estimated the abundances of adult elk and calves during early, middle, and late winter using replicate mark–resight techniques and age composition data (Chapters 11 and 23 by Garrott *et al.* and Chapter 16 by Becker *et al.*, this volume). Estimates of bison abundance and age class were obtained via ground survey counts of the bison winter range in the study area. Surveys were conducted every 10–16 days during winter, with observers recording the number, location, sex, and age class of all bison sighted (Chapter 12 by Bruggeman *et al.* and Chapter 16 by Becker *et al.*, this volume). The effect of snow pack on prey vulnerability was estimated using a metric of accumulated snow-water equivalents (SWE_{acc}, Garrott *et al.* 2003). We used a validated model describing snow pack dynamics (Chapter 6 by Watson *et al.*, this volume) to estimate a mean daily SWE for the study area, and accumulated mean daily SWE values from the typical start of the first snowfall (1 October) until the end of a given winter period. By estimating the duration and severity of snow pack and its weakening effect on prey, we considered SWE_{acc} an indicator of prey physiological condition and SWE_{acc} explained substantially more variation in wolf prey selection in the Madison headwaters area than the mean SWE on the ground at the time of a kill (Chapter 16 by Becker *et al.*, this volume). Also, we calculated a mean wolf pack size ($WOLF_{pack}$) for each winter period from daily estimates of size for a given pack.

We developed and evaluated *a priori* hypotheses in the form of 12 candidate models fitted to each of the kill rate metrics (Appendix A). Thus, we performed three separate analyses for total kill rates and kill rates of elk and bison. To facilitate comparison of coefficient estimates, all covariates were centered and scaled prior to analysis by subtracting the midpoint and dividing by half of the range, resulting in values between −1 and 1. We assessed potential colinearity between covariates using variance inflation factors and did not use covariates with values >6 in the same model (Neter *et al.* 1996). Covariates that were not used in the same model due to strong colinearity were $BISON_{adt}$ and $BISON_{calf}$, and all bison covariates with SWE_{acc} because increasing snow pack resulted in increased bison migration into the study area (Chapter 12 by Bruggeman *et al.*, this volume). We fitted all models in R version 2.4.1 (R Development Core Team 2006). Models were compared using Akaike's Information Criterion corrected for small samples (AIC_c; Burnham and Anderson 2002). We calculated Akaike weights and evaluated the importance of each covariate by its predictor weight (w_p), which we calculated by summing the Akaike weights for all models containing the covariate (Burnham and Anderson 2002). Goodness of fit was evaluated using adjusted R-squared values for each model, and covariate coefficients were evaluated for direction (*i.e.*, positive or negative) and stability among different models. Next, we fit moderated and pseudo-threshold forms to all prey covariates to determine if the fit was improved. Lastly, we performed exploratory analyses fitting kill rates to a "per-pack" scale, wherein prey abundance was estimated for each pack's territory, to determine if covariate relationships and model selection results were affected.

Elk were considerably more vulnerable to wolf predation than bison in the Madison headwaters, though wolves increasingly selected bison in late winter at high bison:elk ratios (Chapter 16 by Becker *et al.*, this volume). Thus, we predicted elk abundance would be positively related to elk kill rates and total kill rates, but negatively related to bison kill rates, for all three metrics. Similarly, we predicted bison abundance covariates would be negatively correlated with elk kill rates of all three metrics because the migration of bison into the system would provide a large alternative food source that could decrease kill rates of elk. We predicted that increasing abundance of either or both prey species would be positively related to total kill rates. In social predators such as wolves, pack size has an important influence on kill rates because larger packs can make more kills, but acquire less food *per capita* (Thurber and Peterson 1993, Schmidt and Mech 1997). Thus, we predicted pack size would be positively related to kills/pack/day, but negatively related to kills/wolf/day and kg/wolf/day for total kill rates and kill rates of elk and bison. Lastly, snow pack has a considerable debilitating influence on prey, both in their ability to escape from predation and on their physiological condition (Chapter 16 by Becker *et al.*, this volume). Thus, we predicted that SWE_{acc} would be positively related to kill rates of all three metrics (Appendix A).

D. Evaluating the Functional Response for Elk

Multiple regression analyses evaluated the influence of various prey, predator, and environmental influences on the kill rates of wolf packs within the study area. To describe the average rate of elk consumption per wolf (*i.e.*, the functional response), however, we fit traditional functional response models (Holling 1959) to wolf kill rate data. Functional response curves were fit for the metric of elk kills/wolf/day estimated at a winter range scale, whereby a single elk kill rate for the study area during a given winter period was estimated by pooling all elk kills and dividing by the estimated wolf days in the period. Bison and elk abundance covariates were then estimated for the entire study area for each period as described for the kill rate variation analyses, and wolf abundance was estimated by dividing the total wolf days for a given winter period by the number of days in the period. While wolves killed bison to varying degrees within and among winters, this change in diet appeared to be heavily moderated by circumstances that increased bison vulnerability, such as severe winters and high wolf: ungulate ratios (Chapter 16 by Becker *et al.*, this volume) We also did not have sufficient kill rates on bison across a wide range of bison densities and therefore did not fit a wolf functional response for bison.

We fit elk kill rate data to seven *a priori* models to evaluate the form of wolf functional responses. The models were categorized into four groups, namely a null model of constant kill rate, prey-dependent Type II and Type III responses, ratio-dependent Type II and Type III responses, and prey- and ratio-dependent Type III responses with two prey. The generalized prey-dependent Type II and Type III equations (17.1a) and (17.1b), respectively from Holling's (1959) disk equation were:

$$\frac{\alpha N}{1 + \alpha h N} \tag{17.1a}$$

$$\frac{\alpha N^2}{1 + \alpha h N^2} \tag{17.1b}$$

where α is the elk attack rate, h is the handling time of a single prey item, and N is elk abundance. While prey abundance is certainly of essential importance in kill rates, there are a growing number of findings demonstrating the importance of predator dependence (Reeve 1997, Vucetich *et al.* 2002, Jost *et al.* 2005, Schenk *et al.* 2005, Tschanz *et al.* 2007). Thus, we used the Type II and Type III ratio-dependent models (17.2a) and (17.2b), respectively, from Arditi and Ginzburg (1989) denoted as:

$$\frac{\alpha N}{P + \alpha h N} \tag{17.2a}$$

$$\frac{\alpha N^2}{P + \alpha h N^2} \tag{17.2b}$$

where P is wolf abundance. Different functional responses can be exhibited for different species in multiple prey systems (Messier 1995), and indirect effects can exist between prey species sharing a predator (Holt 1977). Thus, we fit Type III prey-dependent and ratio-dependent functional responses that incorporated the abundances of both elk and bison. The structures of these models were adapted from Garrott *et al.* (2007), with Type III prey-dependent and ratio-dependent equations (17.3a) and (17.3b) as:

$$g_1 = \frac{\alpha N_1}{1 + \alpha\, N_2 h c^b m \left(N_2/N_1\right)^{b-1} + \alpha\, N_1 h} \tag{17.3a}$$

$$g_1 = \frac{\alpha N_1}{P + \alpha\, N_2 h c^b m \left(N_2/N_1\right)^{b-1} + \alpha\, N_1 h} \tag{17.3b}$$

Where subscripts 1 and 2 denote elk and bison respectively, the proportionality constant, c, measures "the bias in the predator's diet to one prey species" and relates the ratio of prey eaten to their relative abundance (Murdoch 1969, p. 337), and b is the extent of prey switching (Chapter 16 by Becker *et al.*, this volume). A value of c less than one indicates a preference for that prey, a value of b greater than one indicates prey switching (Elliott 2004, Greenwood and Elton 1979), and m is the biomass ratio between bison and elk. We estimated fixed quantities of c and b from prey selection data in the Madison headwaters to be 0.229 and 2.091, respectively (Chapter 16 by Becker *et al.*, this volume), with $m = 2$. A complete explanation of these equations is available in Appendix B.

Determining *a priori* the appropriate scale at which to evaluate functional responses can be difficult (Jost *et al.* 2005). However, we considered the effects of predator abundance and interference reflected in ratio dependence to be most pronounced on a population scale, where predator abundance was not pack size but rather wolf numbers. Thus, we fit models on a study area-wide scale. Elk populations typically declined modestly from fall to spring (Chapter 16 by Becker *et al.* and Chapter 23 by Garrott *et al.*, this volume) and therefore division into winter periods of approximately eight weeks each minimized the possibility of prey depletion bias (Jost *et al.* 2005). We fitted models and estimated parameter coefficients using the nls function from the nlme package in R version 2.4.1 (R Development Core Team 2006). We determined model and predictor weights and employed diagnostics similar to the multiple linear regression analysis. Based on previous empirical evaluations of wolf functional responses from Isle Royale (Vucetich *et al.* 2002, Jost *et al.* 2005), we predicted that ratio-dependent models would be more supported than prey-dependent models, and that the elk:wolf ratios would be positively related to wolf *per capita* kill rates on elk.

III. RESULTS

We detected and followed each radio-collared pack for an average of 35.1 days (95% CI = 31.6, 38.6) during each winter period, or ~66% of the time. Wolves were not in the Madison headwaters study area during the remainder of the time. We detected 688 ungulates (*i.e.*, 274 elk calves, 276 elk adults, 79 bison calves, 59 bison adults) killed by wolf packs during the winters of 1998–1999 through 2006–2007. The mean number of kills per period for each pack was 13.9 (range = 2–37, SD = 7.1) and mean pack size for which we estimated kill rates was 10.4 (range = 3.6–21.1, SD = 4.4). We censored kill rate estimates from four wolf packs for five predation periods that had inadequate tracking efficiencies due to wolves spending little time in the study area or poor tracking conditions.

Elk population estimates for the study area ranged from 290–664 in autumn to 174–577 in spring, with the population decreasing 5–42% during winter. Dramatic changes in elk distribution and abundance occurred over the course of the study (Chapter 21 by White *et al.* and Chapter 23 by Garrott *et al.*, this volume), with a progressive decrease beginning in 2003–2004 and continuing through 2006–2007. The distribution of elk also changed from approximately equal proportions in each drainage prior to 1997–1998 to 84% of the population residing in the Madison drainage by the end of 2006–2007 (Chapter 21 by White *et al.*, this volume). We detected no discernible trend in bison abundance over the 9-year study. However, pronounced seasonal trends were evident with bison numbers generally increasing as winter progressed as animals migrated into the study area from the Hayden and Pelican valleys (Chapter 12 by Bruggeman *et al.*, this volume). The mean numbers of bison recorded in the study area for the three periods used to estimate kill rates each year ranged from 234 to 1,356. The wolf population increased steadily from seven wolves during the first winter to a peak of ~45 animals in multiple packs during winter 2004–2005, followed by an abrupt decrease in wolf abundance in the ensuing winters (Chapter 15 by Smith *et al.*, this volume).

We collected kill rate data from four main radio-collared wolf packs during 26 winter predation periods, for a total of 36 measures of kill rate for each metric during 1998–1999 through 2006–2007. Total kills/wolf/day ranged from 0.020 to 0.083 (mean = 0.042; 95% CI = 0.037, 0.048), with elk kills/wolf/day ranging from 0.007 to 0.074 (mean = 0.033; 95% CI = 0.028, 0.039) and bison kills/wolf/day ranging from 0.000 to 0.045 (mean = 0.009; 95% CI = 0.013, 0.005). Total kills/pack/day ranged from 0.125 to 0.810 (mean = 0.400; 95% CI = 0.346, 0.450), while elk kills/pack/day ranged from 0.070 to 0.810 (mean = 0.320; 95% CI = 0.260, 0.381), and bison kills/pack/day ranged from 0.000 to 0.265 (mean = 0.076; 95% CI = 0.049, 0.103). Total kg/wolf/day ranged from 1.7 to 19.1 (mean = 6.6; 95% CI = 5.6, 7.7), while elk kg/wolf/day ranged from 1.1 to 9.4 (mean = 4.6; 95% CI = 3.9, 5.4), and bison kg/wolf/day ranged from 0.0 to 16.3 (mean = 2.0; 95% CI = 1.0, 3.0). Mean total kill rates did not significantly differ from mean elk kill rates across winter periods for all metrics except the late winter kg/wolf/day estimates, while mean bison kill rates increased from early to late winter (Figure 17.3). Mean elk kill rates were similar across all winter periods and were higher than mean bison kill rates for early and middle winter periods. However, the two kill rates had overlapping confidence intervals for the late winter period (Figure 17.3). We pooled kill rate data across all wolf packs to estimate total kills/wolf/day and elk kills/wolf/day for each winter period and winter, providing 26 and nine measures of kill rates, respectively. Winter range kill rate estimates for each winter exhibited steady decreases, with the lowest kill rate corresponding to peak wolf abundance in 2004–2005, before sharply increasing in 2005–2006 and decreasing again in 2006–2007 (Figure 17.4). There was also an inverse relationship between kill rates and consumption among winters, with the highest carcass consumption and lowest variance occurring in 2004–2005 when wolf numbers peaked and kill rates were at their lowest, and the lowest carcass consumption occurring when wolves first established in the system in 1998–1999 (Chapter 15 by Smith *et al.*, this volume; Figure 17.5). Aside from these two extremes, carcass consumption did not vary substantially among winters.

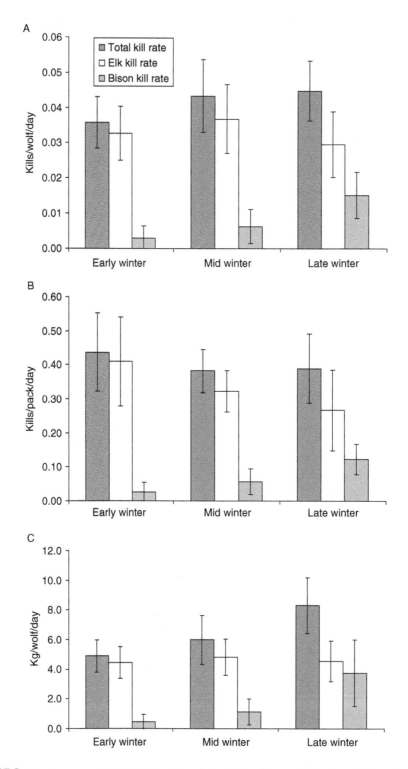

FIGURE 17.3 Kill rate summaries and 95% confidence intervals by winter period (early = 15 November–6 January; middle = 7 January–27 February; late = 28 February–21 April) for wolves in the Madison headwaters area of Yellowstone National Park during 1998–1999 through 2006–2007 using the following metrics: (A) kills/wolf/day, (B) kills/pack/day, and (C) kg/wolf/day.

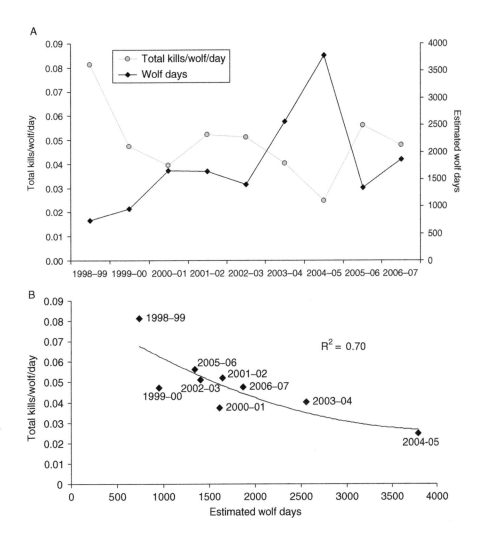

FIGURE 17.4 (A) Observed trends in estimated wolf days and winter kill rates (kills/wolf/day) in the Madison headwaters area of Yellowstone National Park during winters 1998–1999 through 2006–2007 and (B) the correlation between the two metrics.

A. Variations in Elk Kill Rate

We fitted 12 *a priori* models to 36 elk kill rate estimates of kills/wolf/day, kills/pack/day and kg/wolf/day, respectively, from resident wolf packs across nine winters. Model selection results supported one top model for both the kills/wolf/day and kills/pack/day metrics, with Akaike model weights (w_k) of 0.46 and 0.40, respectively (Table 17.1). The most-supported model structure was identical for both metrics, consisting of covariates for total elk abundance (ELK_{all}) and pack size ($WOLF_{pack}$). For both metrics, all other models had $\Delta AIC_c > 2$ and primarily differed in the substitution of elk age class covariates (ELK_{ad} and ELK_{calf}) for total elk abundance, though confidence intervals for elk calves overlapped zero (Table 17.3). Several models also included covariates for bison abundance, wolf population, and snow pack, but coefficient estimates overlapped zero. Predictor weight for total elk abundance (ELK_{all}) for kills/wolf/day and kills/pack/day was 0.70 and 0.68, respectively, while the predictor weight for wolf pack size ($WOLF_{pack}$) was 0.99 and 0.86, respectively. Elk abundance was

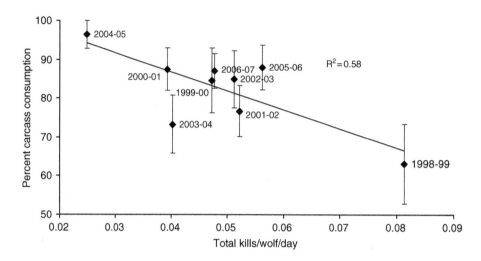

FIGURE 17.5 Relationship between total wolf kill rates and percent of carcass consumption with 95% confidence intervals for the Madison headwaters area of Yellowstone National Park during winters 1998–1999 through 2006–2007. Peak carcass consumption occurred in winter 2004–2005 and corresponded to peak wolf numbers, decreasing elk abundance, and low kill rates. The lowest carcass consumption occurred in winter 1998–1999 when wolves first became established in the Madison headwaters, elk were most abundant, and kill rates peaked.

significant in all models because models estimating elk abundance by age class accounted for the remaining predictor weights (0.29 and 0.30 for kills/wolf/day and kills/pack/day, respectively; Table 17.3).

Model results for kg/wolf/day were less clear, with three most-supported models differing primarily in whether total elk abundance or age class abundance was used (Table 17.1) and one model including snow pack (SWE_{acc}). Additional predictors aside from elk abundance and wolf pack size (predictor weights of 0.66 and 0.97, respectively) contributed little explanatory power (Tables17.1 and 17.3). Consistent with our predictions, the elk abundance covariates ELK_{all} and Elk_{ad} were positively related to kill rates of all three metrics. Wolf pack size was negatively related to kills/wolf/day and kg/wolf/day, but positively related to kills/pack/day (Table 17.3). Coefficients for significant top model covariates were stable and fitting modified and pseudo-threshold forms to prey abundance covariates did not improve model fits. Substantially more variation was explained by the most-supported model for elk kills/pack/day than the top models of kills/wolf/day and kg/wolf/day (r^2_{adj} of 0.51 *vs.* 0.37 and 0.36, respectively).

B. Variations in Bison and Total Kill Rates

We fitted 12 *a priori* models to 36 bison kill rate estimates of kills/wolf/day, kills/pack/day, and kg/wolf/day, respectively, from resident wolf packs across nine winters. Model selection results supported four top models for kills/wolf/day, one top model for kills/pack/day, and two top models for kg/wolf/day (Table 17.2). Akaike model weights (w_k) for the most-supported kills/wolf/day models were SWE_{acc} (0.29), $BISON_{calf}$ (0.20), $BISON_{calf}$ and $WOLF_{pack}$ (0.20), and Elk_{all} and SWE_{acc} (0.18; Table 17.2). Model weight for the most-supported kills/pack/day model with a single covariate structure, $BISON_{calf}$, was 0.58. Model weights for the most-supported models for kg/wolf/day were $BISON_{calf}$ (0.50) and $BISON_{calf}$ and $WOLF_{pack}$ (0.22; Table 17.2). For kills/wolf/day, confidence intervals for only the $WOLF_{pack}$ and $BISON_{calf}$ coefficients did not overlap zero. Similarly, coefficient estimates for kills/pack/day and kg/pack/day indicated that $BISON_{calf}$ was the only significant predictor (Table 17.4).

| **TABLE 17.1** | A priori model structure and results from top models for multiple linear regression analyses of *elk* kill rates by resident wolf packs in the Madison headwaters area of Yellowstone National Park during 1998–1999 through 2006–2007 |

Model structure and metric	ΔAIC_c	w_k	r^2_{adj}
Elk kills/wolf/day			
$ELK_{all} + WOLF_{pack}$	0.00	0.46	0.37
$ELK_{ad} + ELK_{calf} + WOLF_{pack}$	2.44	0.13	0.38
$ELK_{ad} + ELK_{calf} + BISON_{calf} + WOLF_{pack}$	2.53	0.13	0.35
$ELK_{all} + WOLF_{pack} + SWE_{acc}$	2.67	0.12	0.35
$ELK_{all} + BISON_{all} + WOLF_{pack}$	2.71	0.12	0.36
$ELK_{ad} + ELK_{calf} + WOLF_{pack} + SWE_{acc}$	5.31	0.03	0.33
ELK_{all}	10.22	0.00	0.12
Elk kills/pack/day			
$ELK_{all} + WOLF_{pack}$	0.00	0.40	0.51
$ELK_{ad} + ELK_{calf} + WOLF_{pack}$	2.33	0.12	0.50
$ELK_{all} + WOLF_{pack} + SWE_{acc}$	2.64	0.11	0.49
$ELK_{all} + BISON_{all} + WOLF_{pack}$	2.71	0.10	0.49
$ELK_{ad} + ELK_{calf} + BISON_{calf} + WOLF_{pack}$	2.87	0.09	0.51
$ELK_{ad} + ELK_{calf} + WOLF_{pack} + SWE_{acc}$	4.47	0.04	0.49
ELK_{all}	4.73	0.04	0.41
$ELK_{ad} + ELK_{calf} + SWE_{acc}$	5.58	0.02	0.45
$ELK_{ad} + ELK_{calf} + BISON_{calf}$	5.64	0.02	0.45
$ELK_{all} + SWE_{acc}$	5.87	0.02	0.42
$ELK_{ad} + ELK_{calf}$	6.81	0.01	0.40
$ELK_{all} + BISON_{all}$	7.20	0.01	0.40
Elk kg/wolf/day			
$ELK_{all} + WOLF_{pack}$	0.00	0.39	0.36
$ELK_{ad} + ELK_{calf} + WOLF_{pack}$	1.63	0.17	0.36
$ELK_{all} + WOLF_{pack} + SWE_{acc}$	1.74	0.16	0.36
$ELK_{all} + BISON_{all} + WOLF_{pack}$	2.53	0.11	0.34
$ELK_{ad} + ELK_{calf} + BISON_{calf} + WOLF_{pack}$	2.68	0.10	0.37
$ELK_{ad} + ELK_{calf} + WOLF_{pack} + SWE_{acc}$	4.36	0.04	0.34
$ELK_{all} + SWE_{acc}$	11.53	0.00	0.11

Covariate codes are total numbers of elk and bison (ELK_{all}, $BISON_{all}$), numbers of adult elk and bison (ELK_{adt}, $BISON_{adt}$), numbers of calf elk and bison (ELK_{calf}, $BISON_{calf}$), wolf pack size ($WOLF_{pack}$), and accumulated snow pack (SWE_{acc}).

Predictor weights for SWE_{acc} and $BISON_{calf}$ in kills/wolf/day analyses were 0.55 and 0.45, respectively, while the predictor weight for $BISON_{calf}$ was 0.88 and 0.92 for kills/pack/day and kg/wolf/day, respectively. Consistent with our predictions, bison calf abundance and increasing snow pack were positively related to kill rates of all three metrics (Table 17.4). Coefficients for significant top model covariates were stable and fitting modified and pseudo-threshold forms to prey abundance covariates did not improve model fits. The top models for kills/wolf/day and kills/pack/day explained similar amounts of variation relative to kg/wolf/day (r^2_{adj} of 0.40 and 0.39 *vs.* 0.34, respectively).

Top models for total kill rates primarily reflected the top elk kill rate models with the addition of SWE_{acc} (Appendix A). There was one most-supported model for kills/wolf/day with a weight of 0.73 and the structure of ELK_{all}, $WOLF_{pack}$, and SWE_{acc}. There were three top-ranking models for kills/pack/day that differed in their inclusion of SWE_{acc} and the covariates for elk abundance, with model weights of 0.32, 0.19, and 0.17, respectively. The kg/wolf/day analysis supported an identical top model to the analysis for kills/wolf/day, with a weight of 0.61. All predictors for these two analyses had coefficient estimates with confidence intervals that did not overlap zero, while confidence intervals for ELK_{calf} and SWE_{acc} coefficient estimates in the kills/pack/day analysis included zero (Appendix A). Elk abundance and SWE_{acc} were positively related to kill rates of all metrics. Pack size was positively related to kills/pack/day and negatively related to kills/wolf/day and kg/wolf/day (Appendix A).

TABLE 17.2 *A priori* model structure and results from top models for multiple linear regression analyses of *bison* kill rates by resident wolf packs in the Madison headwaters area of Yellowstone National Park during 1998–1999 through 2006–2007

Model structure and metric	ΔAIC_c	w_k	r^2_{adj}
Bison kills/wolf/day			
SWE_{acc}	0.00	0.29	0.39
$BISON_{calf}$	0.72	0.20	0.38
$BISON_{calf} + WOLF_{pack}$	0.73	0.20	0.40
$ELK_{all} + SWE_{acc}$	0.89	0.18	0.40
$ELK_{all} + SWE_{acc} + WOLF_{pack}$	3.52	0.05	0.38
$ELK_{ad} + ELK_{calf} + BISON_{calf}$	4.29	0.03	0.37
$ELK_{ad} + ELK_{calf} + SWE_{acc} + WOLF_{pack}$	4.87	0.03	0.39
$ELK_{ad} + ELK_{calf} + BISON_{calf} + WOLF_{pack}$	5.89	0.02	0.37
$BISON_{all} + WOLF_{pack}$	12.01	0.00	0.18
Bison kills/pack/day			
$BISON_{calf}$	0.00	0.58	0.39
$BISON_{calf} + WOLF_{pack}$	2.31	0.18	0.37
$ELK_{ad} + ELK_{calf} + BISON_{calf}$	3.88	0.08	0.37
SWE_{acc}	5.42	0.04	0.29
$ELK_{ad} + ELK_{calf} + BISON_{calf} + WOLF_{pack}$	5.53	0.04	0.37
$ELK_{all} + SWE_{acc} + WOLF_{pack}$	5.70	0.03	0.34
$ELK_{all} + SWE_{acc}$	6.26	0.03	0.30
$ELK_{ad} + ELK_{calf} + SWE_{acc} + WOLF_{pack}$	8.12	0.01	0.33
$BISON_{all}$	10.09	0.00	0.19
Bison kg/wolf/day			
$BISON_{calf}$	0.00	0.50	0.34
$BISON_{calf} + WOLF_{pack}$	1.59	0.22	0.34
$ELK_{ad} + ELK_{calf} + BISON_{calf}$	2.21	0.16	0.35
$ELK_{ad} + ELK_{calf} + BISON_{calf} + WOLF_{pack}$	5.06	0.04	0.33
$ELK_{all} + SWE_{acc}$	5.67	0.03	0.26
SWE_{acc}	5.92	0.03	0.22
$ELK_{all} + SWE_{acc} + WOLF_{pack}$	8.35	0.01	0.23
$BISON_{all}$	9.81	0.00	0.13

Covariate codes are defined in Table 17.1.

C. Functional Response

Fitting seven functional response models to 26 pooled wolf kill rate estimates for each winter period yielded overwhelming support for the Type II ratio-dependent model ($w_k = 0.70$), and ratio-dependent models comprised 0.92 of the model weights (Table 17.5). Coefficient values for estimated attack rate (α) and handling time (h) in the top model were 0.002 (95% CI = 0.001, 0.004) and 13.9 days (95% CI = 7.9, 19.9), respectively. The predicted functional response from the most-supported model increased rapidly at low elk:wolf ratios before gradually approaching an asymptote of approximately 0.058 kills/wolf/day (Figure 17.6A). One value in the data appeared to be an extreme outlier that could potentially influence the asymptotic value and model results (Figure 17.6A). Thus, we removed this data point and refit all models during an exploratory analysis. The asymptotic value decreased to 0.048 kills/wolf/day, coefficient values for attack rate and handling time changed to 0.004 (95% CI = 0.001, 0.007) and 18.79 (95% CI = 13.8, 23.8), respectively, and model selection results remained unchanged. The Type II functional response was also the most-supported prey-dependent model, but overall had little support and no clear asymptote as elk abundance increased (Figure 17.6B). The two-prey functional response models for both prey-dependent and ratio-dependent models were not supported by the data.

TABLE 17.3 Coefficient values (B_j), lower and upper 95% confidence intervals (in parentheses), and predictor weights (w_p) for the best approximating models for each kill rate metric identified through AIC model comparison techniques for *elk* kill rates by resident wolf packs in the Madison headwaters area of Yellowstone National Park during 1998–1999 through 2006–2007

Metric and model	ELK$_{all}$	ELK$_{adt}$	ELK$_{calf}$	BISON$_{all}$	BISON$_{calf}$	WOLF$_{pack}$	SWE$_{acc}$
Elk kills/wolf/day							
Predictor weight	0.70	0.29	0.29	0.12	0.13	0.99	0.15
ELK$_{all}$ + WOLF$_{pack}$	**0.020** **(0.011, 0.029)**					**−0.019** **(−0.028, −0.009)**	
Elk kills/pack/day							
Predictor weight	0.68	0.30	0.30	0.11	0.11	0.86	0.19
ELK$_{all}$ + WOLF$_{pack}$	**0.161** **(0.115, 0.207)**					**0.131 (0.037, 0.225)**	
Elk kg/wolf/day							
Predictor weight	0.66	0.31	0.31	0.11	0.10	0.97	0.20
ELK$_{all}$ + WOLF$_{pack}$	**2.19** **(0.97, 3.41)**					**−2.87** **(−4.18, −1.56)**	
ELK$_{adt}$ + ELK$_{calf}$ + WOLF$_{pack}$		**2.19** **(0.82, 3.56)**	0.06 (−1.57, 1.68)			**−2.91** **(−4.22, −1.60)**	
ELK$_{all}$ + WOLF$_{pack}$ + SWE$_{acc}$	**2.20** **(0.98, 3.40)**					**−2.62** **(−4.03, −1.21)**	0.67 (−0.73, 2.06)

Boldface type indicates confidence intervals do not span zero. Covariate codes are defined in Table 17.1.

TABLE 17.4 Coefficient values (B_i), lower and upper 95% confidence intervals (in parentheses), and predictor weights (w_p) for the best approximating models for each kill rate metric identified through AIC model comparison techniques for *bison* kill rates by resident wolf packs in the Madison headwaters area of Yellowstone National Park during 1998–1999 through 2006–2007

Metric and model	Covariate						
	ELK_{all}	ELK_{adt}	ELK_{calf}	$BISON_{all}$	$BISON_{calf}$	$WOLF_{pack}$	SWE_{acc}
Bison kills/wolf/day							
Predictor weight	0.23	0.08	0.08	0.00	0.45	0.30	0.55
SWE_{acc} + $BISON_{calf}$					**0.013** **(0.007, 0.019)**		**0.015 (0.009, 0.021)**
$BISON_{calf}$ + $WOLF_{pack}$					**0.011** **(0.005, 0.017)**	−0.005 (−0.011, 0.001)	
ELK_{all} + SWE_{acc}	−0.003 (−0.009, 0.003)						**0.015 (0.009, 0.021)**
Bison kills/pack/day							
Predictor weight	0.07	0.13	0.13	0.00	0.88	0.26	0.11
$BISON_{calf}$					**0.094** **(0.055, 0.133)**		
Bison kg/wolf/day							
Predictor weight	0.04	0.20	0.20	0.00	0.92	0.27	0.07
$BISON_{calf}$					**3.44** **(1.89, 4.99)**		
$BISON_{calf}$ + $WOLF_{pack}$					**3.23** **(1.62, 4.84)**	−0.85 (−2.61, 0.92)	

Boldface type indicates confidence intervals do not span zero. Covariate codes are defined in Appendix A, Table 1.

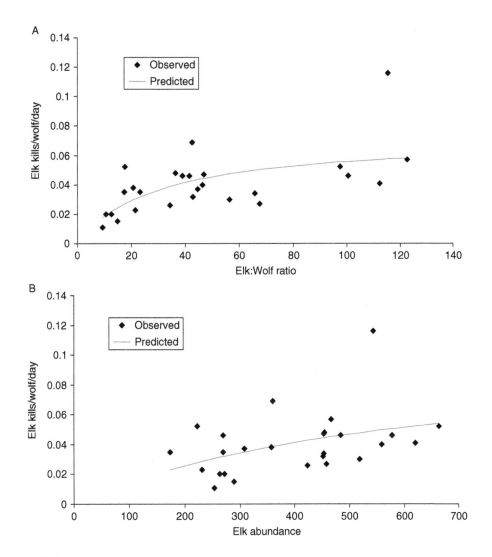

FIGURE 17.6 Predicted and observed functional response curves for elk from wolf predation data in the Madison headwaters area of Yellowstone National Park during winters 1998–1999 to 2006–2007, including (A) a Type II ratio-dependent curve and (B) a Type II prey-dependent curve.

TABLE 17.5	Results from functional response analyses of wolf kill rates on elk during 1998–1999 through 2006–2007 in the Madison headwaters area of Yellowstone National Park

Model	Structure	ΔAIC_c	w_k
Type II ratio-dependent	$\frac{\alpha N}{p + \alpha h N}$	0.00	0.70
Type III ratio-dependent	$\frac{\alpha N^2}{p + \alpha h N^2}$	2.49	0.20
Type II prey-dependent	$\frac{\alpha N}{1 + \alpha h N}$	6.47	0.03
Type III prey-dependent	$\frac{\alpha N^2}{1 + \alpha h N^2}$	6.94	0.02
Two-prey prey dependent Type III	$g_1 = \frac{\alpha N_1}{1 + \alpha N_2 h c^b m (N_2/N_1)^{b-1} + \alpha N_1 h}$	7.00	0.01
Two-prey ratio dependent Type III	$g_1 = \frac{\alpha N_1}{P + \alpha N_2 h c^b m (N_2/N_1)^{b-1} + \alpha N_1 h}$	7.88	0.01
Constant	α	8.27	0.01

IV. DISCUSSION

Adding a top predator to an ecosystem can result in profound demographic, spatial, and behavioral changes in prey and predator populations (Taylor 1984, Berger *et al.* 2001). Evaluating these dynamics requires descriptions of a predator's *per capita* consumption for a particular prey (Abrams and Ginzburg 2000). Thus, the functional response is a critical component embedded in virtually every predator–prey model. While prey abundance is essential for these descriptions, the influence of predator abundance on kill rates and functional responses is controversial because prey-dependent and predator-dependent models often make considerably different predictions about ecosystem dynamics (Abrams and Ginzburg 2000). We demonstrated that factors driving variation in wolf kill rates on elk and bison in the Madison headwaters area of Yellowstone differed between prey species. Kill rates on elk were primarily influenced by elk abundance and wolf pack size, while kill rates on bison were primarily influenced by the abundance of bison calves and snow pack severity. The form of the wolf functional response for elk was strongly Type II ratio-dependent, further supporting the importance of satiation and predator dependence in wolf–ungulate systems (Vucetich *et al.* 2002, Jost *et al.* 2005).

Elk were the preferred and primary prey for wolves in the Madison headwaters area, even though bison were more abundant during winter (Chapter 16 by Becker *et al.*, this volume). Thus, elk abundance significantly influenced variations in both total kill rates and kill rates on elk, similar to findings from other multiple-prey systems where elk were the primary prey (Hebblewhite *et al.* 2003). Discriminating between adult and calf elk abundance did not improve model fit relative to overall elk abundance, even though wolves preferred elk calves and calves were consumed faster due to their smaller size. The significance of total elk abundance and adult elk abundance in explaining kill rate variation on elk was likely due to an overall decrease in the elk numbers during the latter years of the study, which included a substantial increase in adult mortality (Chapter 23 by Garrott *et al.*, this volume). Calf survival was strongly affected by wolves (Chapter 23 by Garrott *et al.*, this volume) and typically decreased through winter due to starvation mortality prior to wolf recolonization (Chapter 11 by Garrott *et al.*, this volume), perhaps explaining why calf abundance was a poor predictor of kill rate variation on elk.

The Madison headwaters supported a two-prey system, making complex, indirect effects possible between prey that share a predator (Chapter 24 by Garrott *et al.*, this volume). However, the abundance of alternative bison prey explained little variation in wolf kill rates on elk. Rather than decreasing elk kill rates, increasing bison abundance, particularly calves, was correlated with increased bison kill rates and total kill rates (Table 17.1; Figure 17.1). Similarly, accumulated snow pack did not appear to explain variation in wolf kill rates on elk in our study system, contrary to findings from other studies (Mech *et al.* 2001, Hebblewhite *et al.* 2002).

We detected a significant relationship between pack size and kill rates across all metrics for elk kill rates, consistent with findings from other studies (Thurber and Peterson 1993, Schmidt and Mech 1997, Hayes *et al.* 2000). Larger packs killed more frequently than smaller packs and *per capita* kills and gross food availability decreased with increasing numbers of wolves in a pack. However, the importance of pack size on kill rates differs across systems. For example, in linear regression analyses, Hayes *et al.* (2000) found that variability in kill rates was best explained by pack size, while in multiple regression analyses Jędrzejewski *et al.* (2002) found it explained little variance relative to snow cover, though the size and range of pack sizes in their study was limited. In terms of total biomass acquired per wolf per day, our estimates were well above the minimum estimated intake of 1.6 kg/wolf/day (Mech 1970), particularly in late winter when kg/wolf/day was significantly higher due to the increased predation on bison. However, these estimates certainly reflect maximum intake because we did not account for scavenger loss or incomplete consumption (Figure 17.5). While smaller packs had a higher kg/wolf/day kill rate, the net disparity between large and small packs may not have been great given that larger packs are less prone to scavenger loss (Vucetich *et al.* 2004) and grizzly bears were common at wolf kills in spring and frequently usurped wolf kills (Ballard *et al.* 2003, Garrott unpublished data). Thus, while larger packs had lower gross *per capita* biomass at kills, they also were likely more effective at avoiding scavenger loss and increasing food acquisition (Vucetich *et al.* 2004) such that the difference in food intake was not as pronounced. Alternatively, smaller packs may have consumed less food at higher kill rates.

In contrast to kill rates on elk, wolf pack size poorly explained variation in kill rates on bison because larger packs did not kill bison more frequently. Kill rate variation on bison was best explained by the abundance of bison calves or by snow pack severity, but was not significantly affected by elk abundance. Due to the strong correlation between increasing snow pack and bison migration into the study system (Chapter 12 by Bruggeman *et al.*, this volume), we were unable to distinguish between the respective influences of snow pack and bison calf abundance. Wolves strongly selected for bison calves and predation coincided with increases in bison abundance and bison abundance relative to elk. Bison predation typically occurred in late winter when ungulates were likely in their worst physical condition due to prolonged nutritional deficits (Chapter 12 by Bruggeman *et al.* and Chapter 16 by Becker *et al.*, this volume). Bison are considerably more formidable prey than elk in both defenses and anti-predator behaviors (MacNulty *et al.* 2007; Chapter 16 by Becker *et al.*, this volume). Thus, wolf predation was largely opportunistic and primarily occurred when bison vulnerability increased late winter. Social carnivores typically take larger prey with larger foraging groups (Rosenzweig 1966, Gittleman 1989, Creel and Creel 1995). However, bison selection by wolves in the Madison headwaters was negatively related to pack size because the largest packs (22 wolves) selected primarily elk, while smaller packs (six wolves) killed the most bison (Chapter 16 by Becker *et al.*, this volume). This likely does not reflect an opposite trend from that observed in other systems so much as favorable conditions for killing bison occurred for these packs. In the more severe winters many bison were in very poor nutritional condition and therefore very vulnerable to wolf predation (Chapter 16 by Becker *et al.*, this volume; Figure 16.12), and this vulnerability was likely not substantially increased with more wolves in a pack. In addition several of the largest packs occurred in winters when elk were abundant and widely distributed throughout the system and wolf competition for them was low. An alternative explanation for this relationship is that particular packs learned how to kill bison more efficiently, which would also affect kill rates but would not be reflected in pack size. While we acknowledge that learning can assume an important role in predation, virtually all wolf packs had experience in killing bison and specific wolf packs did not exhibit a constant increase in bison kills as would be expected if they were simply improving their efficiency. For example, the Gibbon pack killed 63% bison in winter 2005–2006, but only 15% in 2006–2007 (Chapter 16 by Becker *et al.*, this volume).

Approximately 0.45–0.57 of the variation in total wolf kill rate was explained by the abundance of preferred prey, wolf pack size, and snow pack severity. While all of these factors have been identified as having important influences on wolf kill rates, we are not aware of other studies that conducted separate evaluations of factors affecting kill rates for multiple prey species. Nevertheless the most-supported models across all kill rate metrics for elk and bison explained 0.36–0.51 of the respective kill rate variation for each species (Tables 17.1 and 17.2), leaving a substantial amount of kill rate variation unexplained. This finding likely reflects the complexities of a multiple-prey system wherein species differed substantially in their relative vulnerability to wolves and vulnerability differed among age classes (Chapter 16 by Becker *et al.*, this volume). Also, there were complex interactions between heterogeneous landscapes, climate, and multiple, overlapping packs that undoubtedly contributed to variations in kill rate. In addition to these complexities, the appropriate scale at which to analyze kill rates and functional response is often unclear. Strong arguments can be made for using a per-pack scale, whereby kill rates and prey abundance are estimated for each pack's respective territory. However, kill rates could also be evaluated at a "mixed" scale, where kill rates are estimated for each pack but prey abundance is estimated for the entire system, or at a study area scale, where kill rates and prey abundance are estimated for the entire wolf and ungulate populations (Jost *et al.* 2005). In exploratory analyses, we evaluated kill rates on the per-pack scale and results were quite similar to the study-area scale analyses presented in this chapter. However, the per-pack scale models explained substantially less variation in kill rates with less distinction between models, possibly because much of the variation was related to decreases in elk abundance in the system during the latter years. Thus, estimating prey abundance at the study-area scale served as an umbrella for describing these influences. In addition, the study-area scale was appropriate for evaluating the shape of the functional response given that it is a description of the average *per capita* consumption rate of wolves. Nevertheless, the potential for differing results at differing scales should be heeded.

It is essential to avoid biases in the detection efficiency of kills when evaluating variation in kill rates and the subsequent shape of the functional response. For example, one potential reason for the effect of pack size on *per capita* kill rates is purely methodological in that larger packs can consume prey faster and potentially make kills more difficult to detect (Mech and Peterson 2003). Kill detection probability can also be subject to substantial variation in observer effort, weather conditions, and wolf movement, which can translate into substantial differences in kill detection both within and among studies (Jaffe 2001, Hebblewhite *et al.* 2003, Smith *et al.* 2004). We were able to obtain accurate estimates of wolf kill rates by evaluating wolf kill rates in a relatively small, tractable area defined by a high density ungulate winter range and employing intensive ground-based monitoring and tracking on a daily basis through each winter. Jaffe (2001) evaluated kill detection efficiency in the Madison headwaters and determined that the methods were effective in detecting at least 75% of the kills made by wolf packs in the study area, and subsequent estimates have improved efficiency to ~85% (Garrott unpublished data). No systematic biases were detected across prey types or pack sizes that would indicate inaccurate estimates.

Similar to other studies of functional responses, wolves exhibited a Type II curve for elk (Dale *et al.* 1994, Hayes and Harestad 2000, Vucetich *et al.* 2002). One of the primary ways an asymptotic, Type II functional response can arise is through predator satiation. However, this response may also be more likely to occur if a prey item is preferred (Holling 1959, Messier 1995). If a prey item is preferred relative to an alternative prey, then the functional response can be destabilizing (Eubanks and Denno 2000). Thus, the incorporation of a Type III functional response for elk in modeling wolf predation for Greater Yellowstone systems (Boyce 1993, 1995; Varley and Boyce 2006) may underestimate the effects of wolf predation on a preferred prey if alternative prey are considerably less vulnerable (Chapter 24 by Garrott *et al.*, this volume). Alternatively, recent investigations advocate that a Type II response can be stabilizing with social predators and prey (Fryxell *et al.* 2007). However, the dynamic nature of elk grouping strategies on fine and coarse temporal scales in response to variability in predation risk and habitat in this system (Chapter 19 by Gower *et al.*, this volume) make application of this idea difficult. Using data from Dale *et al.* (1994) and Messier (1991, 1994), Eberhardt (1997) demonstrated that wolf Type II functional response curves for moose and caribou increased rapidly before reaching asymptotic values at approximately 0.021 and 0.089 kills/wolf/day, respectively, across a wide range of prey densities.

The asymptotic value for elk functional response curves in the Madison headwaters was ~0.058 kills/wolf/day, similar to the mean total kill rates (90% of which were elk) reported by Smith *et al.* (2004) of 0.061 and 0.068 kills/wolf/day for wolf packs elsewhere in Yellowstone during winters 1995–1996 through 1999–2000. When the asymptotic value for wolf functional responses on elk in the Madison headwaters and caribou from Dale *et al.* (1994) are converted to moose equivalents (one moose equivalent to two elk or three caribou; Keith 1983), the resultant values are both 0.029 moose/wolf/day respectively, remarkably similar to the value calculated by Eberhardt (1997) for moose. Such consistency suggests a relatively uniform asymptotic wolf kill rate across a wide variety of wolf–ungulate systems and ungulate densities (Eberhardt 1997, Eberhardt *et al.* 2003). In the Madison headwaters, the functional response curve appears relatively constant across a wide range of values aside from an outlier (where wolves had a very high kill rate in the first winter of their establishment in the study area) and the three lowest values. High initial kill rates occurring on a naïve prey base and the extremely low elk:wolf ratios we recorded at peak wolf abundance (approximately one wolf per 10 elk) possibly represented transitory extremes in the system. Given the potential for a continuing decrease in elk abundance (Chapter 24 by Garrott *et al.*, this volume), additional estimates at very low elk:wolf ratios may be possible.

The multiple-species dependent models incorporating elk and bison abundance in describing elk functional response were not supported likely due to a variety of reasons. While it is possible that the model structure is inappropriate for these multiple-species interactions, bison did not appreciably influence variation in elk kill rates due to differences in vulnerability between the two prey species. Therefore, application of these models may be more appropriate in multiple prey systems where prey species do not differ so substantially (Chapter 24 by Garrott *et al.*, this volume). Alternatively, given that the Madison headwaters is still a developing two-prey system, the lack of fit may simply be due to a strong wolf preference for elk, with increased bison predation only under circumstances such as severe winters and high bison:elk ratios (Chapter 16 by Becker *et al.*, this volume); therefore additional estimates with continued decreases in elk abundance and increases in bison:elk ratios may provide better fits as wolves may increasingly kill bison (Chapter 24 by Garrott *et al.*, this volume). Regardless, we strongly advocate the continued development of multiple-species dependent functional response models to describe multiple-prey systems (Garrott *et al.* 2007, Tschanz *et al.* 2007).

Ratio-dependent models describe the functional response well for numerous predators and parasitoids (Arditi and Akcakaya 1990), and are supported by controlled experiments in natural settings (Reeve 1997) and field studies (Vucetich *et al.* 2002, Jost *et al.* 2005). The extensive long-term kill rate data from the Isle Royale wolf–moose system strongly indicated that the functional response for moose was best described by ratio dependence, and our investigations in a multiple prey system yielded similar results. Ratio dependence can arise through a variety of factors, including direct and hostile interference among predators, non-random foraging, the presence of prey refugia, changes in prey behavior resulting in less vulnerable prey with increasing predators, and differential vulnerability among the prey population (Charnov *et al.* 1976, Hassell 1978, Arditi and Ginzburg 1989, Abrams 1994). Elk were considerably more vulnerable than bison and preferred by wolves during our study. There was also differential vulnerability of prey across age classes (Chapter 16 by Becker *et al.*, this volume). In addition, wolves were territorial and inter-pack strife was the major cause of mortality (Chapter 15 by Smith *et al.*, this volume). Wolves had specific patterns of foraging and aggregated disproportionately in the Madison headwaters relative to the rest of their territories (Chapter 15 by Smith *et al.*, this volume) to target areas of high elk vulnerability (Bergman *et al.* 2006). Furthermore, landscape characteristics apparently created refugia in certain areas that contributed to a large-scale change in elk distribution within our study system (Chapter 21 by White *et al.*, this volume). Consequently, increases in wolf numbers and decreases in the elk:wolf ratios negatively affected elk kill rates by increasing competition and intra-specific strife between packs and increasing anti-predator behaviors by elk (Chapters 18–20 by Gower *et al.*, this volume).

While there is a vast literature on anti-predator behaviors in prey (Caro 2005), less is known about the effectiveness of this decision-making in actually reducing predation (Lima 2002), and elk demonstrate a variety of anti-predator responses to wolves (Hebblewhite and Pletscher 2002, Creel *et al.* 2005, Creel and Winnie 2005, Gude *et al.* 2006, Hebblewhite and Merrill 2007). We detected a strong correlation between mean winter group size for elk and wolf abundance, as well as an increase in elk group size variance with increasing wolf abundance, which we interpret as an elk behavioral response to increasing predation risk (Chapter 19 by Gower *et al.*, this volume). In addition, substantial changes in elk abundance, recruitment, and distribution occurred during the study period (Chapter 21 by White *et al.* and Chapter 23 by Garrott *et al.*, this volume), such that in the latter winters elk became more concentrated and predictable in areas that apparently provided refuge and escape habitat (Chapter 21 by White *et al.*, this volume). If an elk group is the unit of encounter rather than an individual (Huggard 1993), and herd sizes are variable on fine temporal scales as elk respond to immediate wolf threats by grouping and using escape terrain (Chapter 19 by Gower *et al.*, Chapter 21 by

White *et al.*, and Chapter 24 by Garrott *et al.*, this volume), then we would expect kill rates to decrease with increased wolf numbers due to prey depression, as fewer vulnerable individuals would be available despite being relatively predictable in their locations. There was a negative correlation between mean winter group size for elk and winter wolf kill rates on elk ($R^2 = 0.66$; Figure 17.7), indicating that these behavioral responses may have been effective at reducing predation risk. Whether these adjustments were part of the transitory dynamics of a newly-established system with prey adapting to the novel presence of a top predator or whether such plasticity in prey responses can be expected as the system continues to develop is unknown. Regardless, wolf kill rates on large herbivores in the Madison headwaters area were likely strongly dependent on the physical, behavioral, and environmental vulnerability of their prey (Chapter 24 by Garrott *et al.*, this volume), in addition to encounter rates.

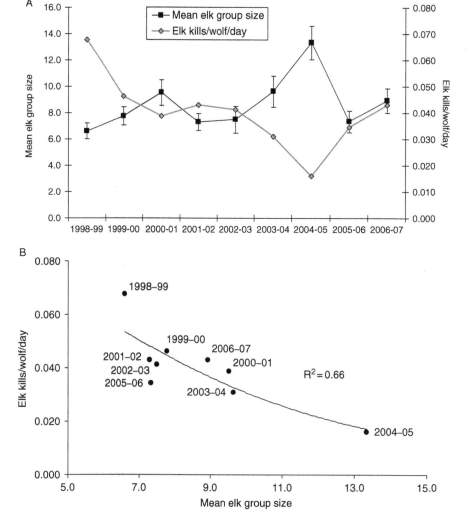

FIGURE 17.7 (A) Observed trends in mean winter elk group size (Chapter 19 by Gower *et al.*, this volume) and winter wolf kill rates (kills/wolf/day) on elk in the Madison headwaters area of Yellowstone National Park during 1998–1999 through 2006–2007 and (B) the correlation between the two metrics.

FIGURE 17.8 Members of the Hayden pack scavenging on an adult, female bison carcass that was repeatedly revisited by wolves. Increased wolf use of the Madison headwaters area was strongly correlated with decreased kill rates and increased carcass consumption. Scavenging and revisitation of old kills also appeared to increase with decreased elk abundance and increased wolf use of the system (Photo by Shana Dunkley).

Though it is extremely difficult, if not impossible, to disentangle the respective influences of all these different factors into models of predator–prey interactions, we concur with other investigators that ratio-dependence is a parsimonious means of describing the effects of predator density on *per capita* consumption rates (Jost *et al.* 2005; Figure 17.8). Precise ratio-dependence or prey dependence is likely rare in nature, and can change within systems (Abrams and Ginzburg 2000, Schenk *et al.* 2005, Tschanz *et al.* 2007). Thus, further studies are necessary before generalizations can be made. However, our evaluation of a newly-established large mammal predator–prey system further corroborates the importance of considering predator population density in understanding the nature of predation.

V. SUMMARY

1. We estimated wolf kill rates in a tractable and newly established wolf–elk–bison system in the Madison headwaters area of Yellowstone National Park during winters 1998–1999 to 2006–2007 to document the transition from more than seven decades without wolves to a well-established, top predator population.
2. Multiple regression analyses of 501 elk and bison kills made by the four primary packs in the study system indicated that variations in kill rates (kills/wolf/day, kills/pack/day, and kg/wolf/day) of elk, the preferred prey, were positively related to elk abundance, negatively related to pack size for *per capita* kill rates, and positively related to pack size for kills/pack/day.
3. Variations in kill rates of bison were positively related to snow pack and bison calf abundance. Increases in bison kill rates were not related to reductions in elk kill rates, but simply served to increase total kill rates.

4. Elk abundance was a poor single predictor of variations in wolf kill rates and describing the shape of the functional response. Also, the duration and accumulation of snow pack did not significantly influence wolf kill rates on elk due to a concurrent, progressive increase in the numbers of nutritionally-deprived bison migrating into the area from higher elevations.

5. The functional response of wolves for elk was best described by a Type II ratio-dependent model that rapidly increased at low elk:wolf ratios before approaching an asymptote at approximately 0.058 kills/wolf/day across a wide range of elk:wolf ratios.

6. Potential mechanisms generating ratio-dependence in the Madison headwaters were likely substantial increases in wolf abundance concurrent with elk responses resulting in distribution, abundance, and behavioral changes.

7. Our findings further suggest the influence of predator abundance as described by ratio-dependence is important for understanding the nature of predation.

VI. REFERENCES

Abrams, P. A. 1994. The fallacies of ratio-dependent predation. *Ecology* **75**:1842–1850.

Abrams, P. A., and L. R. Ginzburg. 2000. The nature of predation: Prey dependent, ratio dependent or neither? *Trends in Ecology and Evolution* **15**:337–341.

Akcakaya, H. R., R. Arditi, and L. R. Ginzburg. 1995. Ratio-dependent predation: An abstraction that works. *Ecology* **76**:995–1004.

Anderson, T. W. 2001. Predator responses, prey refuges, and density-dependent mortality of a marine fish. *Ecology* **82**:245–257.

Arditi, R., and H. R. Akcakaya. 1990. Underestimation of mutual interference of predators. *Oecologia* **83**:358–361.

Arditi, R., and L. R. Ginzburg. 1989. Coupling in predator–prey dynamics: Ratio dependence. *Journal of Theoretical Biology* **139**:311–326.

Ballard, W. B., L. N. Carbyn, and D. W. Smith. 2003. Wolf interactions with non-prey. Pages 259–271 *in* L. D. Mech and L. Boitani (Eds.) *Wolves: Behavior, Ecology, and Conservation.* University of Chicago Press, Chicago, IL.

Berger, J., and M. Peacock. 1988. Variability in size–weight relationships of *Bison bison*. *Journal of Mammalogy* **69**:618–624.

Berger, J., J. E. Swenson, and I.-L. Persson. 2001. Recolonizing carnivores and naïve prey: Conservation lessons from Pleistocene extinctions. *Science* **291**:1036–1039.

Bergman, E. J., R. A. Garrott, S. Creel, J. J. Borkowski, and R. M. Jaffe. 2006. Assessment of prey vulnerability through analysis of wolf movements and kill sites. *Ecological Applications* **16**:273–284.

Berryman, A. A. 1992. The origins and evolution of predator–prey theory. *Ecology* **73**:1530–1535.

Bowyer, R. T., D. K. Person, and B. M. Pierce. 2005. Detecting top–down versus bottom–up regulation of ungulates by large carnivores: Implications for conservation of biodiversity. Pages 342–361 *in* J. C. Ray, K. H. Redford, R. S. Steneck, and J. Berger (Eds.) *Large Carnivores and the Conservation of Biodiversity.* Island Press, Washington, DC.

Boyce, M. S. 1993. Predicting the consequences of wolf recovery to ungulates in Yellowstone National Park. Pages 234–269 *in* R. S. Cook (Ed.) *Ecological Issues on Reintroducing Wolves into Yellowstone National Park.* USDI National Park Service, Denver, CO. Scientific Monograph NPS/NRYELL/NRSM-93/22.

Boyce, M. S. 1995. Anticipating consequences of wolves in Yellowstone: Model validation. Pages 199–209 *in* L. N. Carbyn, S. H. Fritts, and D. R. Seip (Eds.) *Ecology and Conservation of Wolves in a Changing World.* Canadian Circumpolar Institute, Edmonton, Alberta, Canada.

Burnham, K. P., and D. R. Anderson. 2002. *Model Selection and Multimodel Inference: A Practical Information-Theoretic Approach.* Springer, New York, NY.

Caro, T. M. 2005. *Antipredator Defenses in Birds and Mammals.* The University of Chicago Press, Chicago, IL.

Charnov, E. L., G. H. Orians, and K. Hyatt. 1976. Ecological implications of resource depression. *American Naturalist* **110**:247–259.

Cosner, C., D. L. DeAngelis, J. S. Ault, and D. B. Olson. 1999. Effects of spatial grouping on the functional response of predators. *Theoretical Population Biology* **56**:65–75.

Creel, S., and J. A. Winnie Jr. 2005. Response of elk herd size to fine-scale and temporal variations in the risk of predation by wolves. *Animal Behavior* **69**:1181–1189.

Creel, S., J. A. Winnie Jr., B. Maxwell, K. Hamlin, and M. Creel. 2005. Elk alter habitat selection as an antipredator response to wolves. *Ecology* **86**:3387–3397.

Creel, S. R., and N. M. Creel. 1995. Communal hunting and pack size in African wild dogs, *Lycaon pictus. Animal Behaviour* **50:**1325–1339.

Dale, B. M., L. G. Adams, and R. T. Bowyer. 1994. Functional response of wolves preying on barren-ground caribou in a multiple-prey ecosystem. *Journal of Animal Ecology* **63:**644–652.

Dale, B. M., L. G. Adams, and R. T. Bowyer. 1995. Winter wolf predation in a multiple ungulate prey system: Gates of the Arctic National Park, Alaska. Pages 223–230 *in* L. N. Carbyn, S. H. Fritts, and D. R. Seip (Eds.) *Ecology and Conservation of Wolves in a Changing World.* Canadian Circumpolar Institute, Edmonton, Alberta, Canada.

Eberhardt, L. L. 1997. Is wolf predation ration-dependent? *Canadian Journal of Zoology* **75:**1940–1944.

Eberhardt, L. L., R. A. Garrott, D. W. Smith, P. J. White, and R. O. Peterson. 2003. Assessing the impact of wolves on ungulate prey. *Ecological Applications* **13:**776–783.

Elliott, J. M. 2004. Prey switching in four species of carnivorous stoneflies. *Freshwater Biology* **49:**709–720.

Elton, C. S. 1927. *Animal ecology.* Sidgwick and Jackson, London, United Kingdom.

Eubanks, M. D., and R. F. Denno. 2000. Health food versus fast food: The effects of prey quality and mobility on prey selection by a generalist predator and indirect interactions among prey species. *Ecological Entomology* **25:**140–146.

Fryxell, J. M., A. Mosser, A. R. E. Sinclair, and C. Packer. 2007. Group formation stabilizes predator–prey dynamics. *Nature* **449:**1041–1044.

Fuller, W. A. 1959. The horns and teeth as indicators of age in bison. *Journal of Wildlife Management.* **23:**342–344.

Garrott, R. A., J. E. Bruggeman, M. S. Becker, S. T. Kalinowski, and P. J. White. 2007. Evaluating prey switching in wolf–ungulate systems. *Ecological Applications* **17:**1588–1597.

Garrott, R. A., L. L. Eberhardt, P. J. White, and J. J. Rotella. 2003. Climate-induced variation in vital rates of an unharvested large-herbivore population. *Canadian Journal of Zoology* **81:**33–45.

Gittleman, J. L. 1989. Carnivore group living: Comparative trends. Pages 183–207 *in* J. L. Gittleman (Ed.) *Carnivore Behavior, Ecology and Evolution.* Cornell University Press, Ithaca, New York, NY.

Greenwood, J. J. D., and R. A. Elton. 1979. Analysing experiments on frequency-dependent selection in predators. *Journal of Animal Ecology* **48:**721–737.

Gude, J. A., R. A. Garrott, J. J. Borkowski, and F. King. 2006. Prey risk allocation in a grazing environment. *Ecological Applications* **16:**285–298.

Hamlin, K. L., D. F. Pac, C. A. Sime, R. M. DeSimone, and G. L. Dusek. 2000. Evaluating the accuracy of ages obtained by two methods for Montana ungulates. *Journal of Wildlife Management* **64:**441–449.

Hassell, M. P. 1978. *The Dynamics of Arthropod Predator–Prey Systems.* Princeton University Press, Princeton, NJ.

Hassell, M. P. 2000. *The Spatial and Temporal Dynamics of Host Parasitoid Interactions.* Oxford University Press, Oxford, United Kingdom.

Hayes, R. D., A. M. Baer, U. Wotschikowsky, and A. S. Harestad. 2000. Kill rate by wolves on moose in the Yukon. *Canadian Journal of Zoology* **78:**49–59.

Hayes, R. D., and A. S. Harestad. 2000. Wolf functional response and regulation of moose in the Yukon. *Canadian Journal of Zoology* **78:**60–66.

Hebblewhite, M., and E. H. Merrill. 2007. Multiscale wolf predation risk for elk: Does migration reduce risk? *Oecologia* **152:**377–387.

Hebblewhite, M., P. C. Paquet, D. H. Pletscher, R. B. Lessard, and C. J. Callaghan. 2003. Development and application of a ratio estimator to estimate wolf kill rates and variance in a multiple-prey system. *Wildlife Society Bulletin* **31:**933–946.

Hebblewhite, M., and D. Pletscher. 2002. Effects of elk group size on predation by wolves. *Canadian Journal of Zoology* **80:**800–809.

Hebblewhite, M., D. H. Pletscher, and P. C. Paquet. 2002. Elk population dynamics in areas with and without predation by recolonizing wolves in Banff National Park, Alberta. *Canadian Journal of Zoology* **80:**789–799.

Holling, C. S. 1959. The components of predation as revealed by a study of small mammal predation of European pine sawfly. *Canadian Entomology* **91:**293–320.

Holt, R. D. 1977. Predation, apparent competition, and the structure of prey communities. *Theoretical Population Biology* **12:**197–229.

Hudson, R. J., J. C. Haigh, and A. B. Bubenik. 2002. Physical and physiological adaptations. Pages 255–257 *in* D. E. Toweill and J. Ward Thomas (Eds.) *North American Elk: Ecology and Management.* Smithsonian Institution Press, Washington, DC.

Huggard, D. J. 1993. Prey selectivity of wolves in Banff National Park. Prey species. *Canadian Journal of Zoology* **71:**130–139.

Jaffe, R. 2001. Winter wolf predation in an elk–bison system in Yellowstone National Park, Wyoming. Thesis. Montana State University, Bozeman, MT.

Jędrzejewski, W., B. Jędrzejewska, H. Okarma, K. Schmidt, K. Zub, and M. Musiani. 2000. Prey selection and predation by wolves in Bialowieza Primeval Forest, Poland. *Journal of Mammalogy* **81:**197–212.

Jędrzejewski, W., K. Schmidt, J. Theuerkauf, B. Jędrzejewska, N. Selva, K. Zub, and L. Szymura. 2002. Kill rates and predation by wolves on ungulate populations in Bialowieza Primeval Forest (Poland). *Ecology* **83:**1341–1356.

Jeschke, J. M., M. Kopp, and R. Tollrian. 2004. Consumer-food systems: Why type I functional responses are exclusive to filter feeders. *Biological Review* 79:337–349.

Jost, C., G. DeVulder, J. A. Vucetich, R. O. Peterson, and R. Arditi. 2005. The wolves of Isle Royale display scale-invariant satiation and ratio-dependent predation on moose. *Journal of Animal Ecology* 74:809–816.

Keith, L. B. 1983. Population dynamics of wolves. *Canadian Wildlife Service Report Series* 45:66–77.

Kunkel, K. E., D. H. Pletscher, D. K. Boyd, R. R. Ream, and M. W. Fairchild. 2004. Factors correlated with foraging behavior of wolves in and near Glacier National Park, Montana. *Journal of Wildlife Management* 68:167–178.

Lima, S. L. 2002. Putting predators back into behavioral predator–prey interactions. *Trends in Ecology and Evolution* 17:70–75.

MacNulty, D. R., L. D. Mech, and D. W. Smith. 2007. A proposed ethogram of large-carnivore predatory behavior, exemplified by the wolf. *Journal of Mammalogy* 88:595–605.

Marshal, J. P., and S. Boutin. 1999. Power analysis of wolf–moose functional response. *Journal of Wildlife Management* 63:396–402.

Meagher, M. M. 1973. *The Bison of Yellowstone National Park.* National Park Service, Washington, DC. Scientific Monograph Series Number 1.

Mech, D. L. 1970. *The Wolf: The Ecology and Behavior of an Endangered Species.* University of Minnesota Press, Minneapolis, MN.

Mech, L. D., and R. O. Peterson. 2003. Wolf–prey relations. Pages 131–160 *in* L. D. Mech and L. Boitani (Eds.) *Wolves: Behavior, Ecology, and Conservation.* University of Chicago Press, Chicago, IL.

Mech, L. D., D. W. Smith, K. M. Murphy, and D. R. MacNulty. 2001. Winter severity and wolf predation on a formerly wolf-free elk herd. *Journal of Wildlife Management* 65:998–1003.

Messier, F. 1991. The significance of limiting and regulating factors on the demography of moose and white-tailed deer. *Journal of Animal Ecology* 60:377–393.

Messier, F. 1994. Ungulate population models with predation: A case study with the North American moose. *Ecology* 75:478–488.

Messier, F. 1995. On the functional and numerical responses of wolves to changing prey density. Pages 187–197 *in* L. N. Carbyn, S. H. Fritts, and D. R. Seip, editors. Ecology and conservation of wolves in a changing world. Canadian Circumpolar Institute, University of Alberta, Edmonton, Alberta, Canada.

Messier, F., and D. O. Joly. 2000. Comment: Regulation of moose populations by wolf predation. *Canadian Journal of Zoology* 78:506–510.

Moffitt, S. A. 1998. Aging bison by the incremental cementum growth layers in teeth. *Journal of Wildlife Management* 62:1276–1280.

Murdoch, W. W. 1969. Switching in general predators: Experiments on predator specificity and stability of prey populations. *Ecological Monographs* 39:335–354.

Murdoch, W. W., and J. R. Marks. 1973. Predation by coccinellid beetles: experiments in switching. Ecology 54:160–167.

Murdoch, W. W., and A. Oaten. 1975. Predation and population stability. *Advances in Ecological Research* 9:1–131.

Murie, O. J. 1951. *The Elk of North America.* Stackpole, Harrisburg, PA.

Neter, J., M. H. Kutner, C. J. Nachtsheim, and W. Wasserman. 1996. *Applied Linear Statistical Models.* McGraw-Hill, Boston, MA.

Oaten, A., and W. W. Murdoch. 1975. Switching, functional response, and stability in predator–prey systems. *American Naturalist* 109:299–319.

Peckarsky, B. L. 1984. Predator–prey interactions among aquatic insects. Pages 196–254 *in* V. H. Resh and D. M. Rosenberg (Eds.) *The Ecology of Aquatic Insects.* Praeger Publishers, New York, NY.

Person, D. K., R. T. Bowyer, and V. Ballenberghe. 2001. Density dependence of ungulates and functional responses of wolves: Effects of predator–prey ratios. *Alces* 37:253–273.

Peterson, R. O. 1977. *Wolf Ecology and Prey Relationships on Isle Royale.* National Park Service, Washington, DC. Scientific Monograph Series 11.

Pitt, W. C., and M. E. Ritchie. 2002. Influence of prey distribution on the functional response of lizards. *Oikos* 96:157–163.

R Development Core Team. 2006. R: A language and environment for statistical computing. R Foundation for Statistical Computing, Vienna, Austria. <http://www.R-project.org>.

Real, L. A. 1979. Ecological determinants of functional response. *Ecology* 60:481–485.

Reeve, J. D. 1997. Predation and bark beetle dynamics. *Oecologia* 112:48–54.

Rozenzweig, M. L. 1966. Community structure in sympatric carnivores. *Journal of Mammalogy* 47:602–612.

Schenk, D., L. F. Bersier, and S. Bacher. 2005. An experimental test of the nature of predation: Neither prey-nor ratio-dependent. *Journal of Animal Ecology* 74:86–91.

Schmidt, P. A., and L. D. Mech. 1997. Wolf pack size and food acquisition. *American Naturalist* 150:513–517.

Smith, D. W., T. D. Drummer, K. M. Murphy, D. S. Guernsey, and S. B. Evans. 2004. Winter prey selection and estimation of wolf kill rates in Yellowstone National Park, 1995–2000. *Journal of Wildlife Management* 68:153–166.

Stahler, D., B. Heinrich, and D. Smith. 2002. Common ravens, *Corvus corax*, preferentially associate with grey wolves, *Canis lupus*, as a foraging strategy in winter. *Animal Behaviour* 64:283–290.

Taylor, R. J. 1984. *Predation.* Chapman and Hall, New york, NY.

Thompson, D. J. 1978. Towards a realistic predator–prey model-the effect of temperature on functional response and life history of larvae of damselfly, *Ischnura elegans. Journal of Animal Ecology* 47:757–767.

Thurber, J. M., and R. O. Peterson. 1993. Effects of population density and pack size on the foraging ecology of gray wolves. *Journal of Mammalogy* **74**:879–889.

Tinbergen, L. 1960. The natural control of insects in pine woods. Factors influencing intensity of predation by song birds. *Archives Neerlandaises de Zoologie* **13**:265–343.

Tschanz, B., L. F. Bersier, and S. Bacher. 2007. Functional responses: A question of alternative prey and predator density. *Ecology* **88**:1300–1308.

Van Ballenberghe, V., and W. B. Ballard. 1994. Limitation and regulation of moose populations: The role of predation. *Canadian Journal of Zoology* **72**:2071–2077.

Varley, N., and M. S. Boyce. 2006. Adaptive management for reintroductions: Updating a wolf recovery model for Yellowstone National Park. *Ecological Modelling* **193**:315–339.

Vucetich, J. A., R. O. Peterson, and C. L. Schaefer. 2002. The effect of prey and predator density on wolf predation. *Ecology* **83**:3003–3013.

Vucetich, J. A., R. O. Peterson, and T. A. Waite. 2004. Raven scavenging favours group foraging in wolves. *Animal Behaviour* **67**:1117–1126.

White, G. C., and R. A. Garrott. 1990. *Analysis of Wildlife Rradio-Tracking Data.* Academic Press, San Diego, CA.

APPENDIX

Appendix A. *A priori* Model Lists for Evaluating Kill Rate Variation

Total and Elk Kill Rates (Kill Rate is Kills/Wolf/Day, Kills/Pack/Day, or kg/Wolf/Day)

S1. Kill Rates $\sim$ ELK_{all}
S2. Kill Rate $\sim$ ELK_{all} + SWE_{acc}
S3. Kill Rate $\sim$ ELK_{all} + $BISON_{all}$
S4. Kill Rate $\sim$ ELK_{all} + $WOLF_{pack}$
S5. Kill Rate $\sim$ ELK_{all} + $WOLF_{pack}$ + SWE_{acc}
S6. Kill Rate $\sim$ ELK_{all} + $BISON_{all}$ + $WOLF_{pack}$
S7. Kill Rate $\sim$ ELK_{adt} + ELK_{calf}
S8. Kill Rate $\sim$ ELK_{adt} + ELK_{calf} + SWE_{acc}
S9. Kill Rate $\sim$ ELK_{adt} + ELK_{calf} + $BISON_{calf}$
S10. Kill Rate $\sim$ ELK_{adt} + ELK_{calf} + $WOLF_{pack}$
S11. Kill Rate $\sim$ ELK_{adt} + ELK_{calf} + $WOLF_{pack}$ + SWE_{acc}
S12. Kill Rate $\sim$ ELK_{adt} + ELK_{calf} + $BISON_{calf}$ + $WOLF_{pack}$

Bison Kill Rates (Kill Rate is Kills/Wolf/Day, Kills/Pack/Day, or kg/Wolf/Day)

S1. Kill Rate $\sim$ $BISON_{all}$
S2. Kill Rate $\sim$ SWE_{acc}
S3. Kill Rate $\sim$ ELK_{all} + $BISON_{all}$
S4. Kill Rate $\sim$ $BISON_{all}$ + $WOLF_{pack}$
S5. Kill Rate $\sim$ ELK_{all} + $BISON_{all}$ + $WOLF_{pack}$
S6. Kill Rate $\sim$ ELK_{all} + SWE_{acc} + $WOLF_{pack}$
S7. Kill Rate $\sim$ ELK_{all} + SWE_{acc}
S8. Kill Rate $\sim$ $BISON_{calf}$
S9. Kill Rate $\sim$ ELK_{ad} + ELK_{calf} + $BISON_{calf}$
S10. Kill Rate $\sim$ $BISON_{calf}$ + $WOLF_{pack}$
S11. Kill Rate $\sim$ ELK_{adt} + ELK_{calf} + $BISON_{calf}$ + $WOLF_{pack}$
S12. Kill Rate $\sim$ ELK_{adt} + ELK_{calf} + SWE_{acc} + $WOLF_{pack}$

APPENDIX 17A.1 Predictive table for multiple regression analyses of wolf kill rate variation

Kill rate and metric	Covariate							
	ELK_{all}	ELK_{adt}	ELK_{calf}	$BISON_{all}$	$BISON_{adt}$	$BISON_{calf}$	$WOLF_{pack}$	SWE_{acc}
Elk								
Kills/wolf/day	+	+	+	−	−	−	−	+
Kills/pack/day	+	+	+	−	−	−	+	+
kg/wolf/day	+	+	+	−	−	−	−	+
Bison								
Kills/wolf/day	−	−	−	+	+	+	−	+
Kills/pack/day	−	−	−	+	+	+	+	+
kg/wolf/day	−	−	−	+	+	+	−	+
Total								
Kills/wolf/day	+	+	+	+	+	+	−	+
Kills/pack/day	+	+	+	+	+	+	+	+
kg/wolf/day	+	+	+	+	+	+	−	+

Covariate codes are total elk and bison abundance (ELK_{all}, $BISON_{all}$), abundance of elk and bison adults (ELK_{adt}, $BISON_{adt}$) and calves (ELK_{calf}, $BISON_{calf}$), wolf pack size ($WOLF_{pack}$), and accumulated snowpack (SWE_{acc}).

APPENDIX 17A.2 *A priori* model structure and results from top models for multiple linear regression analyses of *total* wolf kill rates by resident wolf packs in the Madison headwaters area of Yellowstone National Park during 1998–1999 through 2006–2007

Model Structure and Metric	ΔAIC_c	w_k	r^2_{adj}
Total kills/wolf/day			
$ELK_{all} + WOLF_{pack} + SWE_{acc}$	0.00	0.73	0.57
$ELK_{ad} + ELK_{calf} + WOLF_{pack} + SWE_{acc}$	2.95	0.17	0.55
$ELK_{ad} + ELK_{calf} + WOLF_{pack}$	5.88	0.04	0.49
$ELK_{all} + WOLF_{pack}$	6.08	0.03	0.47
$ELK_{all} + BISON_{all} + WOLF_{pack}$	7.18	0.02	0.47
$ELK_{ad} + ELK_{calf} + BISON_{calf} + WOLF_{pack}$	7.93	0.01	0.49
$ELK_{all} + SWE_{acc}$	15.14	0.00	0.31
Total kills/pack/day			
$ELK_{ad} + ELK_{calf} + WOLF_{pack}$	0.00	0.32	0.53
$ELK_{all} + WOLF_{pack}$	1.05	0.19	0.50
$ELK_{all} + WOLF_{pack} + SWE_{acc}$	1.22	0.17	0.52
$ELK_{all} + BISON_{all} + WOLF_{pack}$	2.09	0.11	0.50
$ELK_{ad} + ELK_{calf} + WOLF_{pack} + SWE_{acc}$	2.64	0.08	0.52
$ELK_{ad} + ELK_{calf} + BISON_{calf} + WOLF_{pack}$	2.71	0.08	0.52
$ELK_{ad} + ELK_{calf}$	6.01	0.02	0.42
ELK_{all}	6.95	0.01	0.38
$ELK_{ad} + ELK_{calf} + SWE_{acc}$	7.76	0.01	0.42
$ELK_{ad} + ELK_{calf} + BISON_{calf}$	8.64	0.00	0.40
Total kg/wolf/day			
$ELK_{all} + WOLF_{pack} + SWE_{acc}$	0.00	0.61	0.45
$ELK_{ad} + ELK_{calf} + WOLF_{pack} + SWE_{acc}$	2.66	0.16	0.44
$ELK_{ad} + ELK_{calf} + BISON_{calf} + WOLF_{pack}$	3.99	0.08	0.42
$ELK_{all} + SWE_{acc}$	4.35	0.07	0.35
$ELK_{ad} + ELK_{calf} + SWE_{acc}$	5.34	0.04	0.37
$ELK_{all} + BISON_{all} + WOLF_{pack}$	7.52	0.01	0.33
$ELK_{ad} + ELK_{calf} + WOLF_{pack}$	8.91	0.01	0.30
$ELK_{all} + WOLF_{pack}$	9.08	0.01	0.26
$ELK_{ad} + ELK_{calf} + BISON_{calf}$	12.89	0.00	0.22

Covariate codes are defined in Table 17.1.

APPENDIX 17A.3 Coefficient values (B_j), lower and upper 95% confidence intervals (in parentheses), and predictor weights (w_p) for the best approximating models for each kill rate metric identified through AIC model comparison techniques for *total* kill rates by resident wolf packs in the Madison headwaters area of Yellowstone National Park during 1998–1999 through 2006–2007

Metric and model	ELK_{all}	ELK_{adt}	ELK_{calf}	$BISON_{all}$	$BISON_{calf}$	$WOLF_{pack}$	SWE_{acc}
Total kills/wolf/day							
Predictor weight	0.78	0.22	0.22	0.02	0.01	1.00	0.90
$ELK_{all} + WOLF_{pack} + SWE_{acc}$	**0.018** (0.012, 0.023)					**-0.020** (-0.029, -0.012)	**0.013** (0.005, 0.022)
Total kills/pack/day							
Predictor weight	0.48	0.51	0.51	0.11	0.08	0.95	0.26
$ELK_{adt} + ELK_{calf} + WOLF_{pack}$		**0.160** (0.074, 0.246)	-0.043 (-0.145, 0.059)			**0.124** (0.042, 0.205)	
$ELK_{all} + WOLF_{pack}$	**0.130** (0.051, 0.209)					**0.128** (0.043, 0.213)	
$ELK_{all} + WOLF_{pack} + SWE_{acc}$	**0.131** (0.054, 0.209)					**0.154** (0.064, 0.243)	0.069 (-0.019, 0.158)
Total kg/wolf/day							
Predictor weight	0.70	0.29	0.29	0.01	0.08	0.88	0.88
$ELK_{all} + WOLF_{pack} + SWE_{acc}$	0.85 (-0.86, 2.56)					**-2.65** (-4.62, -0.68)	**3.51 (1.56, 5.47)**

Boldface type indicates confidence intervals do not span zero. Covariate codes are defined in Table 17.1.

Appendix B. Description of Multiple Species Functional Response Models

We evaluate the relative availability of the bison and elk in the study system with the ratio of the two prey species in the wolves' diet. Murdoch (1969) provides the classic equation that relates the ratio of two prey types eaten by a predator (g_1/g_2) to the ratio of the prey types available to the predator (N_1/N_2):

$$\frac{g_1}{g_2} = \left(c\frac{N_1}{N_2} \right)^b \tag{A.1}$$

where subscripts 1 and 2 correspond to prey types 1 and 2, respectively; g is the functional response (prey killed per predator per day); N is the number of prey available, and c is a selection coefficient (Murdoch's "proportionality constant") that measures the "bias in the predator's diet to one prey species." If $c = 1$, then there is no bias and the predator kills the two prey types in proportion to their availability. If $c > 1$, the predator kills prey type 1 disproportionately, and if $c < 1$ prey type 2 is killed disproportionately. The bias of a predator's diet could be quite malleable and, thus, c may not remain constant but change depending on the relatively availability of the two prey types and perhaps other factors (Elton 1927). The coefficient b is a measure of the extent of prey switching with values of $b > 1$ denoting prey switching (Greenwood and Elton 1979). Various relationships between the ratio of available prey and the ratio of prey types in the predator's diet can occur using this equation. If both c and b equal 1 (*i.e.*, no bias in the wolves' diet and no switching), then wolves simply consume prey in proportion to their availability and the relationship is linear with a slope of 1. If there is a bias in the wolves' diet ($c \neq$ and there is no prey switching ($b = 1$), the relationship will still be linear but the slope of the line will be <1. If switching occurs ($b > 1$), the relationship will be curvilinear.

The selection coefficient for two-prey systems can be influenced by a variety of factors, including differences in ungulate abundance, body size, anti-predator behaviors and defenses, and vulnerability, as well as variability in wolf preference for the two prey types. Thus, to capture inherent differences between bison and elk, we decompose c into three 3 components such that $c = svm$, where s is the differential preference for a predator to attack prey type 1 compared to type 2, v is the differential vulnerability of prey type 1 compared to type 2, and m is the relative nourishment of prey type 1 to type 2. Estimates of c and b were derived from switching analyses in Chapter 16 by Becker *et al.*, this volume, while m estimates were based upon body mass, with bison being much larger than elk and providing approximately twice as much nourishment than elk to wolves when killed ($m = 2$; Murie 1951, Meagher 1973).

The first step toward this goal is development of functional response equations that incorporate two prey types and the potential for switching. For prey-dependence, following the structure proposed by Murdoch (1973), the functional response model for two prey types is:

$$g_1 = \frac{\alpha_1 N_1}{1 + \alpha_1 N_1 h_1 + \alpha_2 N_2 h_2}; \; g_2 = \frac{\alpha_2 N_2}{1 + \alpha_1 N_1 h_1 + \alpha_2 N_2 h_2} \tag{A.2}$$

For ratio-dependence, the functional response model for two prey types is:

$$g_1 = \frac{\alpha_1 N_1}{P + \alpha_1 N_1 h_1 + \alpha_2 N_2 h_2}; \; g_2 = \frac{\alpha_2 N_2}{P + \alpha_1 N_1 h_1 + \alpha_2 N_2 h_2} \tag{A.3}$$

where subscripts 1 and 2 correspond to prey types 1 and 2, respectively; g and N are defined as above; P is the number of predators; α is the "attack rate" (*i.e.*, instantaneous rate of discovering prey by one predator) in days^{-1}, and h is the "handling time" (days per predator per prey killed) taken by one predator for each prey killed. Switching can be incorporated into Eqs. (A.3) and (A.4) by defining $m = h_1/h_2$ and using Eqs. (A.1)–(A.3) to derive an expression for α_1. It can be shown that Eq. (A.2) becomes:

$$g_1 = \frac{\alpha_2 N_1 (svm)^b (N_1/N_2)^{b-1}}{1 + \alpha_2 N_1 h_2 (sv)^b m^{b+1} (N_1/N_2)^{b-1} + \alpha_2 N_2 h_2}; \; g_2 = \frac{\alpha_2 N_2}{1 + \alpha_2 N_1 h_2 (sv)^b m^{b+1} (N_1/N_2)^{b-1} + \alpha_2 N_2 h_2}$$

$$(A.4)$$

and Eq. (A.3), for a ratio-dependent functional response, becomes:

$$g_1 = \frac{\alpha_2 N_1 (svm)^b (N_1/N_2)^{b-1}}{P + \alpha_2 N_1 h_2 (sv)^b m^{b+1} (N_1/N_2)^{b-1} + \alpha_2 N_2 h_2}; \; g_2 = \frac{\alpha_2 N_2}{P + \alpha_2 N_1 h_2 (sv)^b m^{b+1} (N_1/N_2)^{b-1} + \alpha_2 N_2 h_2}$$

$$(A.5)$$

From Chapter 16 we estimated variables b and c, and estimated m above. Thus rearranging the equations and substituting $c = svm$ as well as values of b and m the new equations for two-prey functional for prey-dependent and ratio-dependent functional response models respectively are:

$$g_1 = \frac{\alpha N_1}{1 + \alpha N_2 h c^b m (N_2/N_1)^{b-1} + \alpha N_1 h}$$

$$g_1 = \frac{\alpha N_1}{P + \alpha N_2 h c^b m (N_2/N_1)^{b-1} + \alpha N_1 h}$$

Section 5

Wolf-Ungulate Dynamics

Spatial Responses of Elk to Wolf Predation Risk: Using the Landscape to Balance Multiple Demands

Claire N. Gower,* Robert A. Garrott,* P. J. White,[†] Fred G. R. Watson,[‡] Simon S. Cornish,[‡] and Matthew S. Becker*

*Fish and Wildlife Management Program, Department of Ecology, Montana State University
[†]National Park Service, Yellowstone National Park
[‡]Division of Science and Environmental Policy, California State University Monterey Bay

Contents

Theme

In the absence of an effective predator, spatial patterns of large herbivores in northern temperate regions are largely influenced by food acquisition and energy conservation during winter, when resources are limited and the energetic cost of movement is high. In these circumstances animals would be expected to minimize movement to avoid unnecessary energy expenditures. With the addition of a top predator such a strategy may not be compatible with avoiding predation risk, therefore animals may increase their movement to avoid detection or escape capture. Such increased movement may occur on relatively fine scales within an animal's home range or

The Ecology of Large Mammals in Central Yellowstone
R. Garrott, P. J. White and F. Watson
ISSN 1936-7961, DOI: 10.1016/S1936-7961(08)00218-2

Copyright © 2009, Elsevier Inc.
All rights reserved.

may be manifested in broader scale responses such as dispersal or migration. With the reintroduction of wolves (*Canis lupus*) to Yellowstone National Park in 1995 and 1996, much attention has focused on the behavioral responses of large herbivores to wolf predation risk and the implications of these responses on ecosystem structure and function. Investigations are numerous but evaluations are complicated due partly to a paucity of data on elk (*Cervus elaphus*) spatial patterns prior to wolf reintroduction. We quantified winter movement patterns of a non-migratory elk herd in the Madison headwaters prior to the reintroduction of wolves, when animals were constrained only by nutritional restrictions, and following wolf colonization and establishment, when elk experienced significant wolf predation. We evaluated changes in home range size and fidelity and described broader scale elk movement patterns such as dispersal and migration. We predicted elk would shift from a winter strategy of being relatively sedentary and occupying areas based on energetic considerations in the absence of wolves, to a more spatially dynamic strategy as elk responded to increased predation risk from wolves. Furthermore, we incorporated our knowledge of landscape characteristics of the system to provide possible explanations for any changes in elk spatial responses.

I. INTRODUCTION

Animal movement is a consequence of evolutionary pressures which contribute significantly to the long-term survival and reproductive success of the individual (Gadgil 1971, MacArthur 1972). Ecological requirements dictate spatial and temporal changes in location for most taxonomic groups, with movements often driven by seasonal and changing environmental conditions (Sinclair 1983), resource distribution and habitat quality, forage and prey availability (Pyke 1983), distribution of mates (Greenwood 1983, Ostfeld 1986, Ims 1988), territory maintenance (Ostfeld 1985, 1986), and thermal regulation. Interactions within communities such as intra- and inter-specific competition (Denno and Roderick 1992), and predation (Fryxell *et al.* 1988, Lima and Dill 1990, Mitchell and Lima 2002) also influence how and why animals move.

Hence animal movement is influenced by a combination of top-down and bottom-up forces, with the spatial dynamics of large herbivores often driven by quantity and quality of food in the absence

FIGURE 18.1 A small group of elk traveling past Old Faithful geyser in the upper geyser basin of Yellowstone National Park. Prior to wolf reintroduction movements were made over relatively short distances and elk predictably traveled between small pockets of available forage (Photo by Jeff Henry).

of any other constraints (McNaughton 1985, Senft *et al.* 1987, Fryxell *et al.* 2004). This is especially true when forage is patchily distributed across the landscape and predictably governed by fixed features such as soil, slope, aspect, and elevation (Bailey *et al.* 1996, Frair *et al.* 2005, Figure 18.1). While landscape attributes affect the occurrence and distribution of resources, seasonal interactions with temperature and precipitation, and thus changes in the phenology of plant communities, can also produce substantial spatial and temporal variation in the availability and quality of forage (Chapter 7 by Thein *et al.*, this volume). This heterogeneity strongly influences how large herbivores move (McNaughton and Banyikwa 1995, Pastor *et al.* 1997, Anderson *et al.* 2005, Fortin *et al.* 2005, Frair *et al.* 2005, Saïd and Servanty 2005), and the spatial distribution of limiting resources can define the location and size of an animal's home range (Ford 1983, Mitchell and Powell 2004).

Deep snow considerably reduces the availability of food resources for large herbivores in northern temperate regions during many months of the year (Jenkins and Wright 1987), making forage less accessible, and at a lower quality and higher energetic cost. Thus, large herbivores rapidly lose fat reserves and become chronically undernourished for a large portion of the year (Hobbs *et al.* 1981, Cook 2002, Parker *et al.* 2005). Also, energy required for thermoregulation during periods of intense cold contributes to their negative energetic state (Gates and Hudson 1979, Cook 2002, Parker *et al.* 2005). Variation in reproduction has been correlated with poor body condition and nutrition of large herbivores (Clutton-Brock *et al.* 1983, Testa and Adams 1998, Cook *et al.* 2004), and snow exacerbates these conditions due to the high energetic costs of locomotion and cratering for unexposed vegetation (Parker *et al.* 1984, Sweeney and Sweeney 1984, Hudson and White 1985, Fancy and White 1987). Thus, chronic nutritional deprivation and high energetic costs during winter strongly influence the demography of large herbivore populations in mid- to high-latitude environments due to starvation mortality (Clutton-Brock *et al.* 1985, Gaillard *et al.* 1998, Chapter 11 by Garrott *et al.*, this volume). Whilst migration is considered a strategy to lesson or avoid these costs (Garrott *et al.* 1987, Nelson 1995), many large herbivores do not migrate entirely out of their winter range but attempt to limit the loss of body reserves and potential for starvation during winter by adopting an energy conservation strategy of minimizing movement, including white-tailed (*Odocoileus virginianus*; Ozoga and Gysel 1972, Verme 1973, Moen 1976), moose (*Alces alces*; Miguelle *et al.* 1992, Renecker and Schwartz 1998, Dussault *et al.* 2005), woodland caribou (*Rangifer tarandus caribou*; Johnson *et al.* 2001), and red deer and elk (*Cervus elaphus*; Craighead *et al.* 1973, Georgii and Schröder 1983, McCorquodale 1993).

While energy conservation strategies are effective at reducing starvation risk, minimal movement can also make animals predictable in their locations, and thus easily encountered and potentially attacked by predators (Mitchell and Lima 2002). Relying on minimal movement to avoid detection in the presence of predators would appear to be an effective strategy in an environment where predator and prey are in close proximity to one another, or where predators require visual cues to detect their prey (Sih 1982, 1992, 1994; Lima and Dill 1990, Azevedo-Ramos *et al.* 1992, Lima 1998, Sih and McCarthy 2002). Under these circumstances movement by prey would greatly enhance the probability of being detected by predators, and thus a dangerous response (Lima 1998). But, in large scale predator-prey systems where predator and prey are both mobile and predators use multiple senses to detect prey and have a well-developed spatial memory, it may be advantageous for prey to increase movement in an attempt to remain unpredictable, avoid encounters, and/or move away from areas of high predation risk (Sih 1984, Sonerud 1985, Formanowicz and Bobka 1989, Mitchell and Lima 2002, Hammond *et al.* 2007). Additionally, prey could predictably move across the landscape by utilizing refuges or escape terrain (Brown 1988), or spatially segregate from predators by dispersing or migrating (Fryxell *et al.* 1988).

Large herbivores often adopt strategies of moving away from predators at a coarse spatial scale via migration (Bergerud *et al.* 1984, 2008; Fryxell *et al.* 1988; Seip 1992), but non-migratory animals that do not spatially separate themselves from predation must employ an effective movement strategy that addresses predation risk (Rettie and Messier 2001). Elk in particular have reacted to wolves by exhibiting dynamic local movements within their range (Creel *et al.* 2005, Fortin *et al.* 2005, Frair *et al.* 2005, Gude *et al.* 2006, Winnie and Creel 2007). In addition elk utilize refuges, or adopt long-distance movements away from high-density wolf areas (Hebblewhite and Merrill 2007). Only recently has the effectiveness of these different movement strategies been evaluated (Hebblewhite and Merrill 2007), but how large herbivores modify their movements to address predation risk while still meeting nutritional needs remains unclear. To evaluate the spatial responses of non-migratory elk it is necessary to know how much area is used to meet biological needs; estimation of this requirement is typically employed using the concept of home range, while the stability of this range can be addressed using the metric of site fidelity.

We evaluated the drivers of elk spatial dynamics on multiple spatial and temporal scales to determine whether spatial changes occurred as elk balanced resource acquisition and energy conservation with avoiding wolf predation. We conducted intensive long-term investigations on a non-migratory elk herd in the Madison headwaters of Yellowstone National Park, documenting elk spatial patterns prior to the recolonization of wolves in the system, during the recolonizing period, and following the establishment of an abundant top predator population. We estimated home range size and fidelity for individual radio-collared elk and described broader scale movements such as dispersal and migration for elk that did not remain in the system following wolf recolonization. Prior to wolf colonization, we predicted that elk constrained their winter movements to minimize energy costs and reduce the rate of depletion of body reserves that was their primary cause of mortality (Chapter 11 by Garrott *et al.*, this volume). Thus, we expected home range sizes would be small and site fidelity would be high as elk predictably used the same foraging sites within and among years. After wolf colonization, we predicted elk movements would be more dynamic and fluid as they moved to areas of lower immediate risk due to continual encounters with wolves. Thus, we expected home range sizes would be larger and site fidelity would be lower as elk used more spatially diverse landscape attributes to reduce predation risk and cope with nutritional constraints (Kie 1999).

II. METHODS

We examined the spatial use of elk in the Madison headwaters area during November 15 through April 30 of 16 consecutive winters (1991–1992 through 2006–2007) but also included long distance movements detected during the 17th year of study while this manuscript was in preparation. Each winter, we repeatedly sampled 20–35 adult, female elk fitted with VHF radio-collars (*i.e.*, approximately 4–10% of the population). Collared elk were typically monitored for multiple winters (median = 4, range = 1–16 years) until death, collar failure or detachment, or dispersal from the study area (Fig. 18.2). Each animal was located approximately two to three times per week following a stratified random sampling regime that ensured sampling times were distributed throughout daylight hours to capture variation in space use. Hand-held telemetry equipment and homing procedures were employed to visually locate radio-collared animals. We calculated home range area and home range (site) fidelity for use as response variables in our analyses and to describe the dispersal behavior of elk in the Madison headwaters.

A. Elk Winter Home Ranges

Home range has been defined as the extent of the area an animal occupies during a specified time frame and is an important indicator of the area used by an animal to meet its biological needs (Burt 1943, White and Garrott 1990). We used kernel density estimation to estimate the utilization distribution (UD) (Silverman 1986, Worton 1989), and defined the home range as the area encompassing 95% of the cumulative UD. We selected fixed rather than adaptive kernel based on goodness-of-fit simulations

FIGURE 18.2 Adult female elk remained instrumented with collars until they died, permanently left the study area, or the collar mechanism dropped-off. Therefore, individuals were part of the sampling regime for multiple years. One elk that was collared as a yearling in 1991 is still being tracked within the study area (Photo by Claire Gower).

(Horne and Garton 2006a) on a sample of our elk location data. Fixed and adaptive kernel techniques were ranked as the two top competing estimators, with fixed kernel chosen for the majority of the sampled datasets. We used the Likelihood Cross Validation method (Silverman 1986:53, Horne and Garton 2006b) to objectively estimate the kernel bandwidth smoothing parameter. Likelihood Cross Validation method was selected due to its ability to perform well with small sample sizes (Horne and Garton 2006b) compared to the commonly used Least Squares Cross Validation and reference bandwidth (Sain *et al.* 1994, Seaman *et al.* 1999, Blundell *et al.* 2001, Hemson *et al.* 2005). We estimated home ranges for each radio-collared elk that we had obtained ≥ 20 locations over the period November 15 through April 30, but restricted the inclusion of animal years to those spanning a minimum time interval of January 1–March 25 each year. Home range estimators are generally sensitive to the number of locations used in the calculation (Swihart and Slade 1985, Seaman and Powell 1996, Seaman *et al.* 1999) and because there was a disparity in the number of locations obtained for each animal during each winter of monitoring (N varied from 20 to 59, mean = 32.2, sd = 8.37) we predicted that biological comparisons between animal years would clearly be confounded given this sample size effect. We confirmed experimentally that the resulting kernel home range area was biased by sample size, and similarly confirmed that the Likelihood Cross Validation optimal bandwidth would also be influenced by samples size, particularly when sample size is small (F. G. Watson and S. Cornish unpublished data). Thus we developed a sample size correction procedure to remove these confounding effects. Specifically, from a sample of N locations from a given animal year (when N was ≥ 20) we selected a standard sub-sample size of n = 20. We then randomly selected (without replacement) 200 replicate sub-samples of size n. We used Likelihood Cross Validation to estimate the maximum likelihood UD and corresponding home range area for each replicate as recommended by Kernohan *et al.* (2001:147). We used the mean of the replicates as our estimate of home range area for that particular animal year; this estimate likely incorporated all locations for a given animal year. We then made our comparisons between animal years in terms of the standardized areas.

Home range areas can be inflated by one or a few outlying points that are a result of occasional sallies (Burt 1943). In the latter half of the study, we observed elk making such movements for one to three days in a winter, representing one to two locations. These locations disproportionately inflated the home range and therefore, we censored a total of ten locations from six independent animal years prior to generating the home range replicates to obtain estimates that more accurately represented the typical area used. Censoring these locations was a conservative approach so that increased movement did not just reflect these outlying points. In addition, we censored three complete sampling years when an insufficient amount of data were collected. This included the first winter (1991–1992) when the study was being established and many animals were not collared until after January 1. Also, we did not collect sufficient locations during winters 1998–1999 and 1999–2000 to generate home range estimates for a large enough sample of animals.

B. Site Fidelity

Site fidelity is the "tendency of an animal to either return to an area previously occupied, or to remain within the same area for an extended period of time" (White and Garrott 1990:133). Thus, we evaluated if radio-collared elk remained in the same approximate area each winter by calculating the centroid of each elk's winter distribution as the arithmetic mean of all the X coordinates and the arithmetic mean of all the Y coordinates. We used the Euclidean distance (km) between centroids from consecutive winter seasons as a metric of fidelity (White and Garrott 1990:134). A large Euclidean distance represented animals that shifted the area of use between years, whereas a small Euclidean distance reflected high fidelity and little shift in the center of the distribution of locations from year to year. We only included animals in the analysis if we had ≥ 20 locations spanning a minimum time interval of January 1–March 25 each year. Estimates of centroids did not appear to vary with sample size, so we included all the locations for each animal year when calculating the centroid, provided an animal was located at least 20 times during the specified period. However, we did censor single locations that were >10 km from the next nearest point ($n = 9$) to avoid distortion of the centroid due to a single non-typical movement. Also, the insufficient number of locations collected in the middle of the study precluded us from comparing estimates of site fidelity between 1998 and 1999.

C. Long-Distance Movements

Long-distance movements by an animal outside of its normal area were considered by Burt (1943:351) as occasional sallies that were "perhaps exploratory in nature and should not be considered as part of the home range." Dispersal has been defined as the "one way movement of individuals from their natal site, or an area that has been occupied for a period of time" (White and Garrott 1990:121), while migration is "a round trip movement of individuals between two or more areas or seasonal ranges" (White and Garrott 1990:121). We collectively termed sallies, dispersal, and migration as long-distance movements by elk away from their original winter ranges. We defined occasional sallies as movements lasting one to four days and outside of the individuals' normal range to a location where the home range had not previously encompassed. Dispersal was defined as a long-distance movement (>15 km) away from the area where the animal had remained sedentary for one or more prior winters, or if the radio signal was not located anywhere within the boundaries of the study area. Monitoring radio-collared animals over consecutive winter seasons allowed us to categorize dispersal behavior into one of two states: (1) permanently leaving the drainage where the elk had remained sedentary for one or more prior winters, and (2) temporarily leaving the drainage where the elk had remained sedentary for one or more prior winters, but returning after >1 year and residing in that drainage during subsequent winters. Lastly, we defined a migratory animal as an individual that vacated the drainage during winter where it had remained sedentary for one or more prior years, moving >15 km away from its former range, or if the radio signal was not located anywhere within the boundaries of the study area. These individuals returned and were relocated within the study area during the snow-free period. Although we were unsure of the location where a few of the temporary dispersers moved to, they all returned to the study area with effectively working collars. Thus, we were confident that animals which we could not detect within the boundaries of the study area had left, and we were not mistaking dispersal events with premature collar malfunctions. We aerially detected the locations of all permanent dispersers, and visually observed the individuals from the ground, and retrieved the collar from dead animals. We defined an annual metric of elk dispersal (*i.e.*, dispersal plus migration) as the percentage of the animals that chose to reside in a new location for at least one winter. This metric did not include animals that conducted exploratory fine temporal movements (sallies) within a winter. We determined that animals had dispersed after failing to detect them at the start of each winter field season. The percentage "dispersed" was based on the total number of collared animals that we were monitoring at the start of the winter season. We monitored the telemetry signals of dispersing and migratory animals during subsequent summers and winters, using ground-based telemetry and occasional aerial flights to determine locations and fates.

D. Non-Wolf Predictors of Elk Movement

We used a validated snow pack simulation model for the central Yellowstone region (Chapter 6 by Watson *et al.*, this volume) to estimate the mean daily snow water equivalent (SWE, water content of snow in meters) on the elk winter range (landscape SWE). Daily values were averaged over the entire winter season to provide the annual snow pack covariate of mean daily SWE (SWE_{mean}). We felt this metric would most appropriately capture the physical constraints that impede movements and foraging by large herbivores during winter because the sinking depth of an animal is a function of snow depth, snow density, and hardness (Parker *et al.* 1984) which, in turn, influences energetic costs (Parker *et al.* 1984, Sweeney and Sweeney 1984).

The spatial arrangement of habitat and terrain features are important for determining the distribution of resources, patterns of ungulate movement (Anderson *et al.* 2005, Frair *et al.* 2005), and home range sizes and shapes (Kie *et al.* 2002, 2005; Mitchell and Powell 2004; Saïd and Servanty 2005; Forester *et al.* 2007). We suspected landscape differences between the three drainages in the Madison headwaters area would influence spatial responses. Thus, we defined a categorical covariate DRAINAGE which classified elk into the Madison, Gibbon, or Firehole drainages. Drainages were assigned based on the telemetry locations for a given animal year and, on the rare occasion where animals used multiple drainages, assignment was based on where the majority of the locations occurred. For the site fidelity analysis, we assigned DRAINAGE for the drainage used during the first of the pair of consecutive years.

We suspected that elk density may influence home range size (Tufto *et al.* 1996, Kilpatrick *et al.* 2001, Kjellander *et al.* 2004), but excluded this potential covariate from analyses because the annual estimate of population size by drainage was highly correlated with the wolf covariates described in subsequent paragraphs (Pearson's Correlation Coefficient = -0.53 and -0.56 with $WOLF_{days}$ and KILLS, respectively). We recognized that age of an animal could affect movements, particularly the tendency to remain faithful to a site. Exploratory movements of large herbivores tend to be made by males (Cederlund and Sands 1994), which had no specific relevance on our analyses, or by first time female breeders (*i.e.*, 2-year-olds) (Garrott *et al.* 1987). Treating age as a categorical covariate (<2 and ≥ 2) would have allowed us to incorporate the age effect into our models, but our data had a limited number of the young age class animals in which to effectively incorporate this covariate.

E. Covariates of Predation Risk

We detected and quantified the presence of wolves in the Madison headwaters from November 15 to April 30 during the winters of 1996–1997 through 2006–2007. National Park Service biologists captured wolves and fitted radio-collars on animals in each pack (Smith 2005, Chapter 15 by Smith *et al.*, this volume). We monitored wolf presence and their locations using ground-based telemetry, snow tracking, and visual observations of collared and un-collared individuals. We intensively monitored each drainage for wolf presence daily with crews of three to four people using snowmobiles, vehicles, snowshoes, and high points in the landscape to facilitate telemetry and observations. When packs containing radio-collared wolves were detected, we estimated multiple locations of all detected animals through the day using triangulation (White and Garrott 1990). We used snow tracking, visual observations, and counts during aerial monitoring by National Park Service biologists to estimate the number of animals per pack and aid our daily assessments of wolf presence or absence (Chapter 15 by Smith *et al.*, and Chapter 16 by Becker *et al.*, this volume). These methods also allowed us to obtain a variety of data on wolf landscape use patterns, and the frequency and distribution of kills (Bergman *et al.* 2006, Chapter 16 by Becker *et al.*, this volume). Detection of un-collared wolves was facilitated by opportunistic observations of tracks and wolves by field personnel working on concurrent elk and bison investigations (Chapter 11 by Garrott *et al.*, Chapter 12 by Bruggeman *et al.*, and Chapter 21 by White *et al.*, this volume). The total numbers of individual wolves known to be present in each drainage each day were estimated based on the information obtained from these various wolf monitoring techniques and quantified as wolf days.

We developed three wolf covariates to assess our hypotheses about the influence of predation risk on space use of elk in the Madison headwaters. The covariate $WOLF_{period}$ categorized the 16-year data set into three periods: before, during, and after wolf colonization. This covariate was drainage-specific because wolves became established in different drainages during different winters (Firehole: 1997–1998; Gibbon: 2000–2001; Madison: 2001–2002; Chapter 15 by Smith *et al.*, this volume). No wolves were present in the study area during 1991–1992 through 1995–1996. To account for the potential transitory behavioral dynamics due to the initial naïveté of the prey (Berger *et al.* 2001), we defined a colonizing period immediately following wolf reintroduction, when elk were initially exposed to wolf predation risk, but no wolf pack (≥ 2 animals) was routinely detected in the drainage. We considered wolves to be established in a drainage during the first winter that a pack was consistently detected in that drainage. The level of wolf predation risk for elk occupying each drainage was highly variable among winters, and we suspected the magnitude or frequency of any behavioral responses of elk to predation risk may also have been scaled to the number of wolves in the immediate area. Thus, we constructed a continuous covariate, $WOLF_{days}$, which described our estimate of the number of wolves in each drainage each day. For the home range analysis, this covariate was the estimated number of wolves in a drainage for a given year. We also developed a third wolf metric, KILLS, that represented the number of wolf-killed ungulates discovered in each drainage in a given year (Chapter 15 by Smith *et al.* and Chapter 16 by Becker *et al.*, this volume). For the site fidelity analysis, we used $WOLF_{period}$, $WOLF_{days}$, and KILLS made in that drainage during the first of the two consecutive years because the Euclidean distance was a metric based on two consecutive years.

F. Statistical Analyses

We used Analysis of Variance (ANOVA) and Tukey multiple comparisons with unequal sample size (Kutner *et al.* 2005:750) to evaluate if elk winter home range size and fidelity changed with varying intensities of wolf predation risk. Therefore, we compared differences in the mean home range sizes and mean Euclidean distances between consecutive annual home range centroids among the pre-wolf, colonizing, and established wolf periods ($WOLF_{period}$). We *ln*-transformed both home range size and Euclidean distance and used diagnostic residual plots to evaluate the assumption of constant variance and normality of residuals. We used 95% confidence intervals to quantify uncertainty in parameter estimates.

We also used mixed-effects linear models (Pinheiro and Bates 2000) to evaluate competing *a priori* models for assessing change in home range size and the Euclidean distance between years. These multiple regression models treated SWE_{mean}, DRAINAGE, and WOLF covariates as fixed effects in the home range and fidelity analyses. Because these datasets included observations of the same individuals through time we treated individual animal identity as a random effect (*i.e.,* intercept-only) to account for individual variability in space use among animals within and between years. For both response variables, this covariate allowed us to partition the variation between individuals (intercept) and within an individual (residual) and determine how much of the variation in home range size and site fidelity was accounted for by the fixed and random effects. Combinations of the covariates were included in the additive form for all analyses. We predicted a fixed rate of change in the response variable per unit change in the predictor variable (*i.e.,* linear form). We evaluated diagnostic plots to assess the assumptions of normality and constant variance of residuals, and transformed both response variables using the natural log to conform to linear models assumptions. We centered and scaled all continuous covariates to facilitate comparisons and interpretations of covariate coefficients. Variance inflation factors (VIF), which measure multi-colinearity among variables, were calculated for all combinations of predictors. Those models that included predictor combinations with VIF <6 were retained in the model list. This was a conservative approach because VIF in excess of 10 implies multi-colinearity (Kutner *et al.* 2005: 409). Correlation coefficients were also calculated to further check for multi-colinearity between the predictor variables. We used Akaike's Information Criteria corrected for small sample size (AIC$_c$) to rank models given the data and compare the relative ability of each model to explain variation in the data (Burnham and Anderson 2002, 2004). Akaike model weights (wi_c) were used to address model selection uncertainty (Burnham and Anderson 2002) and evidence ratios (ratio of wi_c/wj_c) were used to measure the relative likelihood of model pairs (Burnham and Anderson 2002). Covariate coefficients and variance of the random effects were estimated using restricted maximum likelihood. Comparable AIC$_c$ values were calculated using maximum likelihood estimation (Pinheiro and Bates 2000). All statistical analyses were performed using the R statistical package (R Development Core Team 2006).

G. *A Priori* Model Suites and Predictions

We expressed our hypotheses on elk spatial behavior as two suites of candidate *a priori* models constructed for home range size (15 models) and site fidelity (7 models; Appendices 18.A1, 18.A2). These models were constructed based on literature and our field knowledge of the study system and included one or more of the climate or landscape covariates independently or with the inclusion of a single wolf covariate. Every model in both suites also included the individual random effect. We also constructed a null model for each suite which hypothesized constant home range area and site fidelity.

Snow reduces forage availability and creates a wider dispersion of forage over the landscape. Thus, large herbivores may range over greater areas when critical resources are scarce (Geist 2002), resulting in a positive correlation between home range size and snow (Kjellander *et al.* 2004, Anderson *et al.* 2005). However, snow conditions in the Madison headwaters routinely reach a depth and density that would impose major constraints to the locomotion of most large herbivores (Craighead *et al.* 1973, Parker *et al.* 1984, Fancy and White 1985, Chapter 8 by Messer *et al.,* this volume). Under these conditions, we predicted that elk would constrain their movements and home range size would be inversely related to SWE$_{mean}$ (Georgii and Schröder 1983, Schmidt 1993, Krasińska *et al.* 2000, Grignolio *et al.* 2004, Dussault *et al.* 2005).

We also hypothesized that the distribution of habitat types, patch size, and topography would influence the way large herbivores moved. Specifically, we predicted that elk residing in drainages that contained a high density of geothermally-heated foraging sites (*i.e.,* the Firehole, and to a lesser degree the Gibbon) would have smaller home ranges and higher site fidelity as they made localized movements within and between patches (Craighead *et al.* 1973, Kie *et al.* 2002, Chapter 8 by Messer *et al.,* this volume). Conversely, mountainous terrain could facilitate movements around landscape features (Kie *et al.* 2005). Therefore, we predicted elk residing in drainages that were more constrained by local topography (*i.e.*, the Madison and to a lesser degree the Gibbon) would move linearly around topographic features and, as a result, have a larger area of use.

In addition, we predicted that elk movement patterns within and between winters would change as elk responded to variations in predation risk. There is strong evidence among wolf-elk systems that elk move between habitats in response to predation threat or vacate areas of high wolf use (Creel *et al.* 2005, Fortin *et al.* 2005, Frair *et al.* 2005, Gude *et al.* 2006, Hebblewhite and Merrill 2007, Winnie and Creel 2007). Consequently, we predicted home range sizes of elk would become larger with less site fidelity within and among years as elk were increasingly encountered, attacked, and displaced by wolves, resulting in elk utilizing multiple habitat components as they attempted to balance predation risk and nutritional demands. We were uncertain which wolf covariate would be the best predictor of these responses. Therefore, we considered *a priori* models with identical non-wolf covariate structures and each of the three wolf covariates (WOLF$_{period}$, WOLF$_{days}$, KILLS).

Based on the outcome of the *a priori* model-selection results, we conducted *post hoc* exploratory analyses to generate hypotheses for future work. We included a WOLF*DRAINGE and WOLF*SWE$_{mean}$ interaction in the top model from the home range model suite because we predicted the strength of the wolf effect would differ by drainage and/or with increasing snow pack. We also evaluated different non-linear forms of the continuous covariates (β*(quadratic $(x+x^2)$)), exponential (β*(exp(x))), and negative exponential (β*(exp($-$x)))). For the site fidelity analysis, we included a WOLF*DRAINGE interaction and evaluated different non-linear forms of the continuous covariates. We used changes in AIC$_c$ scores to evaluate if models were improved with the covariate additions and substitutions.

III. RESULTS

We obtained 11,908 randomly collected elk locations from repeated sampling of 115 radio-collared, adult, female elk during the winters 1991–1992 through 2006–2007. We collected 3927 locations before wolf colonization, 2501 locations during wolf colonization, and 5480 locations after wolf packs became established in the study area. We used these data to generate 277 home range estimates for 91 radio-collared elk (95 Pre, 50 Col., 132 Est.), and 221 Euclidean distances between consecutive winter centroids for 66 radio collared elk (100 Pre, 39 Col., 82 Est.). Wolves were absent from the Madison headwaters area prior to reintroduction in 1995–1996. Thereafter, the total number of wolf days increased from 55 in 1996–1997 to a peak of 3657 in 2004–2005. The Firehole drainage experienced high wolf use immediately following reintroduction, while the Madison and Gibbon drainages did not experience wolf use until several years later and never reached the magnitude of use observed in the Firehole. Pack size and number of packs varied both spatially and temporally (Chapter 15 by Smith *et al.*, this volume).

Home range sizes of radio-collared elk ranged from 2.8 km^2 to 86.0 km^2 ($n = 277$, mean = 21.2 km^2, sd = 14.3), with an increase of 53% ($P = 0.001$) in mean home range size as the system transitioned from no wolves ($n = 95$, mean = 16.8 km^2, sd = 10.8), through wolf colonization ($n = 50$, mean = 17.4 km^2, sd = 9.5), to established wolf packs ($n = 132$, mean = 25.7 km^2, sd = 16.8; Figure 18.3A). *Ln*-transformed home ranges were significantly larger after wolf packs became established in the system compared to the pre-wolf ($\hat{D}_{\ln(established)-\ln(pre-wolf)} = 0.372$; 95% CI = 0.162, 0.583; $P < 0.001$) and colonizing periods ($\hat{D}_{\ln(established)-\ln(colonizing)} = 0.279$; 95% CI = 0.019, 0.538; $P = 0.032$). No significant differences in home range size were detected between the pre-wolf and colonization periods ($\hat{D}_{\ln(colonizing)-\ln(pre-wolf)} = 0.094$; 95% CI = -0.180, 0.367; $P = 0.699$). $\hat{D}$ defines the difference in transformed mean home range size between two wolf periods. An advantage of log transformations is that the results can be interpreted on the original scale of the variable (Ramsey and Schafer 2002). So by taking the exponent of the difference in the mean log (Y) between two time periods we can obtain a multiplicative value for the change in the median (Y) between the time periods. Consequently the median home range size for the established period was exp (0.372) = 1.451 times bigger than the median

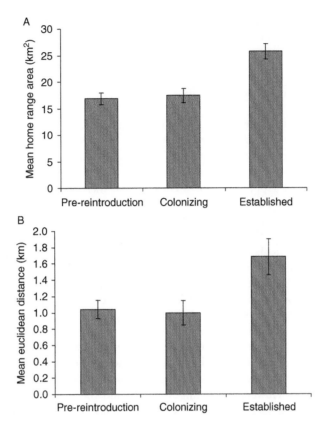

FIGURE 18.3 Mean 95% fixed kernel home range areas (A) and Euclidean distances (B) by wolf period. Plotted values are based on the untransformed data, with the error bars representing standard errors.

home range size for the pre-wolf period (95% CI = 1.176, 1.791), and 1.322 times bigger than the median home range size for the colonizing period (95% CI = 1.019, 1.712). Median home range size for the colonizing period was only 1.099 times bigger (95% CI = 0.835, 1.443) than that of the pre-wolf period.

Elk were generally faithful to their winter range within a specific drainage from year to year, with the Euclidean distance between consecutive centroids ranging from 0.02 to 10.15 km ($n = 221$, mean = 1.3 km, sd = 1.5). However, we did observe a 69% increase ($P = 0.001$) in mean Euclidean distance as the system transitioned from no wolves ($n = 100$, mean = 1.0 km, sd = 1.1), through wolf colonization ($n = 39$, mean = 1.0 km, sd = 0.9), to established wolf packs ($n = 82$, mean = 1.7 km, sd = 2.0; Figure 18.3B). $\hat{D}$ defines the difference in transformed mean Euclidean distance between two wolf periods and *in*-transformed Euclidean distances were significantly larger after wolf packs became established in the system compared to the pre-wolf period ($\hat{D}_{\ln(\text{established})-\ln(\text{pre-wolf})} = 0.438$; 95% CI = 0.081, 0.795; $P = 0.012$). No significant differences in Euclidean distances were detected on the log scale between pre-reintroduction and colonization ($\hat{D}_{\ln(\text{colonizing})-\ln(\text{pre-wolf})} = 0.180$; 95% CI = −0.273, 0.632; $P = 0.617$) or between colonization and establishment ($\hat{D}_{\ln(\text{established})-\ln(\text{colonizing})} = 0.258$; 95% CI = −0.208, 0.724; $P = 0.394$). The lack of a significant difference in Euclidean distance on a natural log scale between the colonizing and established periods likely reflects higher variance in Euclidean distance during the established wolf period compared to the pre-wolf and colonization periods. Median Euclidean distance for the established period was 1.550 times bigger than the median Euclidean distance for the pre-wolf period (95% CI = 1.084, 2.214), and 1.294 times bigger than the median Euclidean distance for the colonizing period (95% CI = 0.812, 2.063). Median Euclidean distance for the colonizing period was 1.197 times bigger (95% CI = 0.761, 1.881) than that of the pre-wolf period.

A. Mixed Effects Linear Regression Results

Model selection results for the home range size analysis supported two top models that received all of the model weight and contained the covariates SWE_{mean}, DRAINAGE, and $WOLF_{days}$ (Table 18.1). Snow had a negative effect on *ln* home range size ($\hat{\beta}_{SWEmean} = -0.065$; 95% CI = -0.138, 0.008), but SWE_{mean} only appeared in one of the top ranked models and confidence limits spanned zero. Thus, we did not find substantial support for our prediction that there was a linear relationship between SWE_{mean} and elk home range size. DRAINAGE and $WOLF_{days}$ received all of the predictor weight and appeared in both of the top ranked models. As predicted, the drainage where the animals resided played an influential role in the size of the area used by elk during winter. In the absence of wolves, elk in the Madison ($\hat{\beta}_{DRAINAGE(Madison)} = 3.092$; 95% CI = 2.936, 3.248) and Gibbon ($\hat{\beta}_{DRAINAGE(Gibbon)} = 3.017$; 95% CI = 2.802, 3.231) drainages used larger areas than elk in the Firehole drainage ($\hat{\beta}_{DRAINAGE(Firehole)} =$ 2.581; 95% CI = 2.429, 2.732; coefficient estimates and 95% CI are provided in the natural-log scale). As predicted, there was a positive correlation between *ln* home range size and the presence of wolves in all three drainages ($\hat{\beta}_{WOLFdays} = 0.265$; 95% CI = 0.168, 0.362; Figures 18.4 A and B). Compared to the other wolf covariates $WOLF_{days}$, received all the predictor weight, while $WOLF_{period}$ and KILLS received no predictor weight. Results of the random effect from the mixed modeling indicated that within-animal variability ($\hat{\sigma}_W^2 = 0.29$) was substantially higher than between-animal variability ($\hat{\sigma}_B^2 = 0.09$). The interactions included in *post hoc* exploratory analyses did not provide any improvement over the most supported *a priori* models. However, including SWE_{mean} into the model as a quadratic or negative exponential improved the model (6.22 and 4.45 ΔAICc units, respectively) over the linear form. The estimated coefficient for $SWE_{mean} + SWE_{mean}^2$ was -0.189 (95% CI: = -0.299, -0.078) + 0.082

TABLE 18.1 Model selection results for the most supported models examining the natural logarithms of home range size and site fidelity (*i.e.*, Euclidean distance) of elk in the Madison headwaters area of Yellowstone National Park during 1991–1992 through 2006–2007

Model structure	k^*	ΔAIC_c	w_{ic}
***ln* (home range area)**			
SWE_{mean} + DRAINAGE + $WOLF_{days}$	7	0.00	0.62
DRAINAGE + $WOLF_{days}$	6	1.03	0.38
SWE_{mean} + DRAINAGE + KILLS	7	12.25	0.00
DRAINAGE + KILLS	6	14.76	0.00
SWE_{mean} + $WOLF_{days}$	5	15.15	0.00
***ln* (Euclidean distance)**			
DRAINAGE + KILLS	6	0.00	0.59
DRAINAGE + $WOLF_{days}$	6	2.62	0.16
DRAINAGE + $WOLF_{period}$	7	2.79	0.15
DRAINAGE	5	4.62	0.06
$WOLF_{period}$	5	7.08	0.02

* Residual error from the mixed modeling accounts for one parameter value.
All models are ranked according to AIC$_c$ values, and presented along with the number of parameters (k), ΔAIC$_c$ value (*i.e.*, change in AIC$_c$ value relative to the best model), and the Akaike weight (w_i). AIC$_c$ values for the top home range and site fidelity models were 514.4 and 594.6, respectively. *Abbreviations:* SWE_{mean} (mean daily snow water equivalent), DRAINAGE (Madison, Firehole, Gibbon), $WOLF_{period}$ (pre-reintroduction, colonizing, established), $WOLF_{days}$ (total number of wolf days per drainage per year), and KILLS (total number of wolf-killed ungulates per drainage per year). For the site fidelity analysis the wolf covariates describe wolf use in the particular drainage during the first of the two consecutive years.

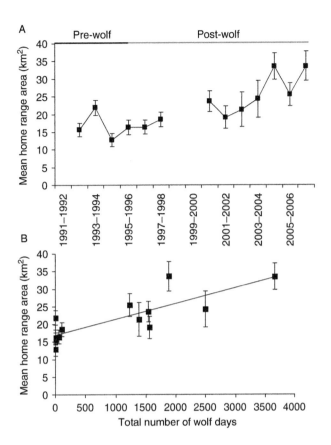

FIGURE 18.4 Mean annual 95% kernel home range areas for the study area with error bars representing standard error (gaps represent insufficient data collected for 1991, 1998, and 1999) (A), and the relationship ($R^2_{adj} = 0.67$, $F_{1,11} = 22.3$, $P < 0.001$) between mean annual 95% fixed kernel home range areas and the total number of wolf days for the study area, with error bars representing standard errors (B).

(95% CI: $= 0.027, 0.139$), while exp ($-SWE_{mean}$) was 0.112 (95% CI $= 0.032, 0.194$). Because the confidence intervals did not span zero, SWE_{mean} appeared to be a strong predictor of elk home range size when included in the model in a non-linear form. Changing the functional form changed the coefficients for the best *a priori* model only slightly. Re-fitting the model with the wolf covariates as different functional forms did not make any additional improvements.

Model selection results for the site fidelity analysis supported one top model that received a model weight of 0.59 and contained the covariates DRAINAGE and KILLS (Table 18.1). DRAINAGE was a highly influential predictor of Euclidean distance, receiving a predictor weight of 0.96. In the absence of wolves, the distance between the centers of consecutive winter distributions was greater in the Madison ($\hat{\beta}_{DRAINAGE(Madison)} = 0.016$; 95% CI $= -0.259, 0.292$) and the Gibbon ($\hat{\beta}_{DRAINAGE(Gibbon)} = 0.132$; 95% CI $= -0.266, 0.530$) compared to the Firehole drainage ($\hat{\beta}_{DRAINAGE(Firehole)} = -0.692$; 95% CI $= -0.988, -0.395$; coefficient estimates and 95% CI are shown in the natural-log scale). As predicted, the distance between consecutive winter centroids was positively correlated with the wolf covariate (Figure 18.5A and B). The most influential of the wolf covariates on the *ln* Euclidean distance was the number of kills in a given drainage the prior year ($\hat{\beta}_{KILLS} = 0.229$; 95% CI $= 0.064, 0.394$). KILLS received a predictor weight of 0.60 compared to predictor weights of 0.17 for both the WOLF$_{period}$ and WOLF$_{days}$

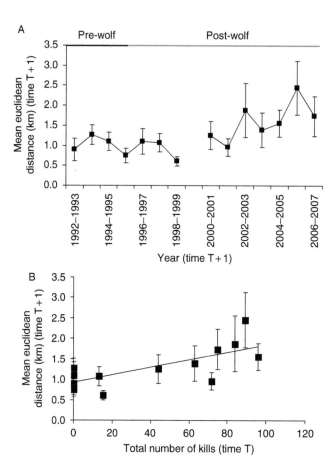

FIGURE 18.5 Mean annual Euclidean distance for the study area at time T + 1 with error bars representing standard error (gaps represent insufficient data collected to compare estimates of site fidelity between 1998 and 1999) (A), and the relationship ($R^2_{adj} = 0.54$, $F_{1,12} = 13.87$, $P = 0.002$) between mean annual Euclidean distance at time T + 1 and the total number of wolf-killed ungulates for the study area at time T, with error bars representing standard errors (B).

covariates. On the log scale, variance of Euclidean distance between years was relatively high within ($\hat{\sigma}^2_W = 0.62$) and among ($\hat{\sigma}^2_B = 0.31$) elk. This finding suggested some animals remained very faithful to a site from year to year, while others demonstrated less philopatric behavior. *Post hoc* exploratory analyses provided no improvement over the *a priori* models.

B. Dispersal and Migratory Movements

A total of 115 adult, female elk were radio-collared and monitored during the 17-year study period (mean elk per year = 27, range = 22–35), constituting 426 animal years of movement monitoring from 1991–1992 through 2007–2008. During the first half of the study (1991–1992 through 1998–1999), when wolves were absent or initially colonizing the study area, we accrued 215 animal years of monitoring and all radio-collared elk remained within the Madison headwaters area during each winter. After wolf packs became established in the system (1999–2000 through 2007–2008), we accrued 211 animal years of monitoring during which 19 individuals displayed some form of long-distance movement away from the study area and were absent from the Madison headwaters during one or more winters (Figure 18.6). All of these individuals were monitored for 1–6 years prior to their long-distance movement, with data indicating they had been sedentary and part of the non-migratory Madison headwaters population. Of the 19 long-distance movements we recorded, four collared females that had a history of association left the Gibbon drainage and relocated approximately 21 km away. Three permanently left the drainage, while one became migratory, residing at this new location (Canyon) during the winters and migrating back to the Gibbon drainage during the summer. Nine other collared females (five independently and four jointly) dispersed from the Firehole drainage and were discovered between 18 and 63 km (mean = 43 km, sd = 16.7) from their traditional ranges within the Madison headwaters. Four other instrumented females dispersed from the study area to unknown locations for one winter and returned to their former home ranges in the Madison headwaters the following year. These animals subsequently remained in the study area for 2–4 years until death, collar detachment, or the end of this study. Two additional individuals became migratory, leaving the study area in autumn and returning to the study area for the summer. These individuals were relocated aerially approximately 55 km to the north in the Gardiner basin area of the northern range. The three migratory animals were sedentary for two, four, and five years prior to changing their behavior, and remained migratory for six, six, and two subsequent winters, respectively. All animals that conducted long-distance movements resided in the Gibbon ($n = 5$) and Firehole ($n = 14$) drainages, where their home ranges the previous year had overlapped wolf use areas (Figure 18.7). No collared animals in the Madison drainage performed any long-distance movements during the study. Of these 19 animals that moved long distances from the study area six were killed by wolves in their new locations or upon returning the Madison headwaters. In total 15 of the 79 collared elk which experienced at least one winter with wolf predation risk permanently dispersed or became migratory and wintered outside of the Madison headwaters. Thus approximately 19% of the collared population altered their traditional patterns of range occupancy in such a way that they were no longer classified as part of the Madison headwaters sedentary herd following wolf reintroduction.

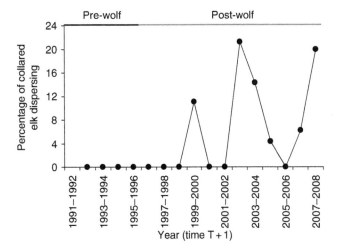

FIGURE 18.6 The percentages of radio-collared elk dispersing from the Madison headwaters area of Yellowstone National Park during 1992–1993 through 2007–2008. The percentage "dispersed" each year was estimated as the number of collared animals missing from the Madison headwaters in November divided by the number of collared animals that we were monitoring at the start of the winter season. All movements were conducted away from home range areas that overlapped wolf use areas the prior year (Figure 18.7).

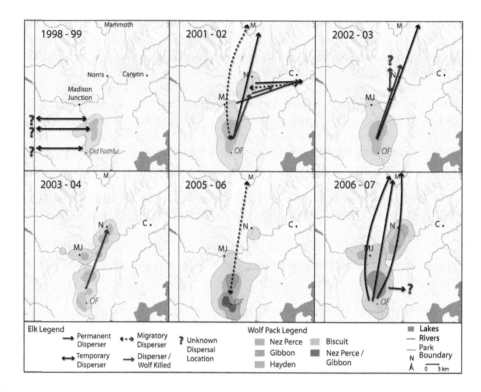

FIGURE 18.7 Radio-collared elk that dispersed from the Madison headwaters of Yellowstone National Park after wolf colonization appeared to be vacating core wolf use areas (50% and 95% kernel isopleths—darker shading represents 50% isopleth) during the previous winter. One-way arrows depict permanent dispersers, two-way arrows depict animals that returned to the study area after a temporary dispersal event (solid line), or became seasonally migratory (broken line). Red arrows represent animals that dispersed and were subsequently killed by wolves at their new location. Question marks indicate that the dispersal destination was unknown. Arrows start at the centroid of the distribution of locations for a given year.

IV. DISCUSSION

How predators influence space use by large herbivores has become of increasing interest to ecologists over the past decade due to the implications these spatial changes may have on ecosystem structure and functions (Fortin *et al.* 2005). For example, wolves in Yellowstone have been credited with inducing a trophic cascade of alterations in vegetation structure and composition by a combination of changing elk space use and decreasing elk numbers (Ripple *et al.* 2001, Beschta 2003, Ripple and Beschta 2003, 2004, Smith *et al.* 2003, Creel *et al.* 2005, Fortin *et al.* 2005). We detected changes in elk movement patterns that supported findings from other wolf-elk systems that showed elk spatially responded to wolves at a localized scale (Creel *et al.* 2005, Fortin *et al.* 2005, Frair *et al.* 2005, Gude *et al.* 2006, Winnie and Creel 2007) and over larger spatial scales as well (Hebblewhite and Merrill 2007). We demonstrated modest increases in home range and reduced fidelity, and we documented animals dispersing from the study area or becoming migratory following wolf reintroduction. Unlike other studies, however, we also obtained direct information on elk spatial behavior in the absence of wolves because data collection spanned a period prior to and following wolf reintroduction. Thus, our analyses provided a direct empirical evaluation of the mechanisms driving ungulate movement behavior in the presence and absence of wolves when elk where exposed to various ecological constraints.

Elk occupied smaller ranges before wolf reintroduction or when winter wolf abundance in the system was low. Smaller home ranges in the absence of wolves appear typical of large herbivores that were minimizing movement and using small areas for food acquisition through winter. White-tailed deer congregate for thermal shelter and relief from deep snow (Ozoga and Gysel 1972, Verme 1973, Moen 1976), and likewise moose (Miquelle *et al.* 1992, Renecker and Schwarz 1998, Dussault *et al.* 2005), woodland caribou (Johnson *et al.* 2001), red deer (Georgii and Schröder 1983), and elk (McCorquodale 1993) restricted foraging activities to small isolated patches, or where snow conditions are locally reduced during winter. Our findings also corroborate the work of Craighead *et al.* (1973) who observed similar-sized home ranges of elk in the Madison headwaters more than two decades before wolves were introduced into the system. Movements in the absence of predators should simply reflect nutritional constraints, and home range sizes should strongly reflect the spatial distribution of forage resources (Ford 1983, Geist 2002, Mitchell and Powell 2004, Anderson *et al.* 2005).

Our results suggest that elk in the Madison headwaters ranged over a wider area, thus, occupying larger home ranges as wolf presence and activity within a given drainage increased. Mitchell and Lima (2002) suggested a predator-prey "shell game" would be established in the presence of a highly mobile predator that has a good spatial memory and the ability to seek out prey locations. Consequently, prey movement would be favored in an attempt to remain unpredictable and elusive in space. However, elk in the Madison headwaters appeared to become more concentrated in their use of preferred habitats following reintroduction (Chapter 21 by White *et al.*, this volume) so it is unlikely that movement to elude detection was being adopted within this system. More likely, elk were moving within their range following a predation event to avoid any further imminent threat. We frequently observed groups of elk with a collared animal using foraging areas when wolves were absent, but shifting from one end of the collared animal's home range to the other immediately after wolves were active in the vicinity. These observations were similar to those of Gude *et al.* (2006) who observed elk leaving wolf encounter sites before wolves returned to hunt again, and would corroborate the suggestions of Bergman *et al.* (2006) that elk would dynamically move among patches over short periods and at a fine spatial scale. Therefore, these ideas support the game-theoretical notion proposed by Mitchell and Lima (2002) that increased prey movement and larger home ranges in the presence of wolves are consistent with patterns of fine-scale predator avoidance within an animal's range. Sih (1986) distinguished between avoidance and escape, defining avoidance as behavior occurring prior to an attack, and the escape as behavior occurring following an attack. Behavioral responses of elk in the Madison headwaters that resulted in increased movement with wolf presence reflected more of an escape strategy than a pre-meditated attempt to avoid wolf encounters. It is doubtful that elk had the capacity to remain elusive in a landscape with a predator that has such well-developed spatial memory as wolves (Mech 1970). However, we adopted the terminology "avoidance" because we suspect elk moved to areas of lower immediate predation risk to avoid further attack and/or relocated to less vulnerable locations before further encounters could occur. Because elk are working within the confines of a winter environment, and predation risk is spatially and temporally dynamic (Creel and Winnie 2005, Chapter 15 by Smith *et al.* and Chapter 19 by Gower *et al.*, this volume), a pre-emptive increase in movement to avoid being detected by wolves would not be an efficient or advantageous approach.

Changes in the fidelity of animals to a winter home range were also observed with the addition of wolves to the system. Non-migratory elk in the Madison headwaters were extremely philopatric to their range year after year, particularly prior to the establishment of wolves. This is similar to other investigations of spatial fidelity on sedentary large herbivores (Franklin and Lieb 1979, Irwin and Peek 1983, Edge *et al.* 1986, Cederlund and Sands 1992, Krasińska *et al.* 2000, Millspaugh *et al.* 2004) with a high degree of philopatry associated with a strong social structure with conspecifics (Edge *et al.* 1986), and familiarity with resource availability and distribution (Franklin and Lieb 1979, Irwin and Peek 1983, Linnell and Anderson 1995). Geothermal areas and meadow complexes in the Madison headwaters provide access to reduced or snow-free foraging year-round (Craighead *et al.* 1972, 1973; Chapters 4 and 6 by Watson *et al.*, this volume). Therefore, forage availability remains relatively constant within these habitats from year to year (Chapter 8 by Messer *et al.*, this volume). Location centroids for elk before wolf reintroduction appear to reflect the spatial arrangement of these foraging sites and the use of the same areas over consecutive years, suggesting high site fidelity during this period, reflects an energetically efficient foraging strategy.

After wolf packs were established in the Madison headwaters, we observed an increasing trend in the Euclidean distance between centroids of consecutive winter ranges, implying that elk became less faithful to a site between successive years. Elk shifted their location of activity when a greater number of wolf kills were detected in the drainage the previous winter. Bergman *et al.* (2006) reported that elk in the Madison headwaters were more vulnerable in geothermal sites and wolves selected these areas non-randomly as they coursed the landscape. Because wolves have well-developed learning abilities (Mech 1970) and periodically return to the same area within their range (Jędrzejewski *et al.* 2001; R. Garrott, unpublished data) to presumably re-encounter unsuspecting prey (Lima and Steury 2005, Roth and Lima 2007), we expected elk to modify their space use from the pre-wolf strategy of predictably foraging in geothermal areas where the majority of kills were made (Bergman *et al.* 2006). Surprisingly elk did not re-locate permanently from these areas, as evidenced by the continued use of vulnerable geothermal sites (Chapter 21 by White *et al.*, this volume). Landscape-level variability in risk has been discussed in ungulate-wolf systems (Kunkel and Pletscher 2000, Hebblewhite *et al.* 2005, Kauffman *et al.* 2007), with different habitat and physical attributes of the landscape found to offer different levels of security or vulnerability (Kauffman *et al.* 2007). Habitats that facilitate hiding or escape may not necessarily provide the best foraging opportunities. Conversely, areas that are attractive for foraging may be accompanied with increased levels of threat (Fortin *et al.* 2005, Bergman *et al.* 2006). The continued selection of geothermal sites and other habitats with high snow heterogeneity by elk even after wolf reintroduction (Chapter 21 by White *et al.*, this volume) was probably because these sites offered the highest availability of forage at the lowest energetic costs during nutritionally critical months. It is likely that this continued use of preferred habitats was possible because many of these sites were adjacent to other habitat components such as meadows (Hebblewhite *et al.* 2005, Chapter 21 by White *et al.*, this volume), forests (Wolff and Van Horn 2003, Creel and Winnie 2005, Creel *et al.* 2005, Fortin *et al.* 2005), deep rivers (Chapter 21 by White *et al.*, this volume), and areas with high human activity (Hebblewhite *et al.* 2005, Hebblewhite and Merrill 2007, Chapter 21 by White *et al.*, this volume) which collectively provided an area that could facilitate escape or provide protection when predation threat was high. The increase in mean Euclidean distance does not imply that elk made long-distance shifts in their use areas or switched drainages between years. Rather, we interpret this decrease in site fidelity as a more variable distribution of locations within the elk's winter range, and selecting a home range that encompassed a wider diversity of landscape components would allow elk to mediate the risk and starvation by moving between areas of relative risk and safety as the level of perceived risk changed (Kie 1999). This suggests that predictable behavior may not be detrimental if landscape attributes serve as refuges, but would appear hazardous if landscape features contribute to the vulnerability of prey, as observed in our system (Chapter 21 by White *et al.*, this volume). This broader use of landscape characteristics explains the reduced site fidelity and increased home range sizes for elk following wolf colonization.

Large-scale movements such as dispersal and migration have been widely discussed as mechanisms to avoid predation risk by spatially segregating from predators at a much coarser scale (Bergerud *et al.* 1984, 2008; Fryxell *et al.* 1988; Seip 1992; Fryxell 1995; Hebblewhite and Merrill 2007). We observed long-distance movements by elk after wolf packs became established in the system. These movements were never observed when wolves were absent from the system. At the finest temporal scale, we observed long-distance sallies by elk from their traditional home ranges following sequences of successful predation attacks by wolves. During the established wolf phase, it was not uncommon to observe elk temporarily leaving their range and moving an average of 10–15 km, but returning several days later. On one occasion, most of the elk in the Old Faithful sub-herd were observed moving north from their typical range after multiple kills were made by wolves. These elk were subsequently located approximately 20 km away (Figure 18.8). This movement event only lasted for three to four days, after which the elk returned to their traditional range and were again attacked by wolves.

We also observed elk temporarily (>1 year) or permanently dispersing from the study area the year following intensive wolf use of their traditional range. These movements ranged from between 18–63 km away from their former range in the Madison headwaters. Dispersal would only appear to be a profitable strategy if environmental conditions varied spatially and temporally (Stenseth 1983) and the probability of survival had the potential to be higher in the new location. In our system, this assumes predation risk is not evenly distributed across the landscape and there are relatively safe places for elk to occupy. This assumption was likely correct immediately following wolf reintroduction when packs were unevenly distributed through the park. However, wolf packs continued to expand their range and few areas of the park were devoid of wolves during the latter years of our studies (Chapter 15 by Smith *et al.*, this volume). Five of the nineteen collared elk that we defined as dispersers were subsequently killed by wolves away from the Madison headwaters, suggesting limited effectiveness of long-distance movements for increasing survival. Dispersing could be considered more perilous than staying within the former home range if unfavorable conditions occur from place to place (Gadgil 1971). Conversely, a 74% chance of survival for the animals that vacated the study area may have been a better outcome than staying in their former ranges where high predation rates by wolves have essentially depleted elk from the Gibbon and Firehole drainages (Chapter 21 by White *et al.* and Chapter 23 by Garrott *et al.*, this volume).

We also detected a switch from sedentary to migratory behavior in three of the radio-collared, female elk. These animals were monitored as part of the non-migratory herd during the early colonization period, but changed their behavior after wolf packs became established in the system, suggesting that wolf activity was the probable cause for the abandonment of their tradition winter ranges. While these observations involved only a small number of animals, they allowed us to speculate that while an environment without predation may favor year-round sedentary behavior, migratory movements may be evolving as the environment changes with the addition of wolves. In African

FIGURE 18.8 Most of the Old Faithful sub-herd fled north from their typical range after multiple, successful attacks by wolves during January 2005. This sub-herd included several radio-collared animals and traveled the entire day until they were approximately 20 km north-west of Old Faithful along the Mary Mountain trail between the Hayden Valley and Firehole River drainage. Madison headwaters elk were not documented in this area previously. These elk returned to Old Faithful 3–4 days later and several more herd members were attacked and killed by wolves (Photo by Claire Gower).

systems migration has been suggested as a way to enhance survivorship (Fryxell 1995). Theoretical modeling of migration in the Serengeti ecosystem suggests that population regulation by predators may affect non-migratory animals, while migratory species are more commonly regulated by food (Fryxell *et al.* 1988). This implies that the top-down effect of predation would dominate in a non-migratory herd such as the Madison headwaters. Thus, it is not surprising that high wolf numbers have contributed to low rates of over-winter adult survival, low calf recruitment, and a significant population decrease (Chapter 21 by White *et al.* and Chapter 23 by Garrott *et al.*, this volume). In the Madison headwaters, winter is a time when deep snow exacerbates the vulnerability of large herbivores to wolves due to reduced mobility and potential for escape (Mech and Peterson 2003; Chapter 16 by Becker *et al.*, this volume). It is also the time when wolves have an almost continual presence within the Madison headwaters (Chapter 15 by Smith *et al.*, this volume). Under these conditions, seasonally escaping predators during winter when vulnerability reaches a peak, and returning in summer when vulnerability is reduced may be more profitable. Interestingly, all long-distance movements that we documented following reintroduction occurred from areas of intensive wolf activity (Firehole, Gibbon). No collared animals vacated the Madison drainage, which is the area wolves frequented the least (Chapter 15 by Smith *et al.*, this volume). These data thus allow us to speculate that animals that have displayed strong fidelity to a range can actually "make a decision" that their traditional range has changed in such a fundamental way that it is no longer conducive to remain in this area. Thus, while it has been documented that density dependent factors such as crowding and resource limitation would promote animals to relocate in search of more profitable surroundings (Lambin *et al.* 2001, Clobert *et al.* 2004, Ims and Andreassen 2005), our data suggest that the risk of predation can promote a similar response. These results also indicate that while we attributed the majority of the decline of the Madison headwaters elk population to direct predator mortality (Chapter 23 by Garrott *et al.*, this volume), permanent dispersal and animals switching from non-migratory to migratory seasonal movement strategies also contributed to the population decline.

Our results show that elk in the Madison headwaters made modest adjustments in their use of space during winter, presumably to reduce their vulnerability from predators at a fine-scale within their range (Fortin *et al.* 2005). We suspect that these changes were modest because elk were constrained by other attributes of the system and the fundamental energetic limitations of surviving in a severe snow-pack environment. This limited flexibility could explain why elk have been reduced to low densities in most areas of the Madison headwaters (Chapter 21 by White *et al.* and Chapter 23 by Garrott *et al.*, this volume) and, also, why a relatively large proportion of elk left two of the drainages where wolf predation pressure was exceptionally high. We expected one optimal movement strategy to be manifested throughout the Madison headwaters study area, but our results illustrate the contrary. Different animals appeared to adopt different strategies within a given system, which we propose is characteristic of a heterogeneous landscape with considerable variation in predation risk.

In the absence of predation the spatial patterns of elk in the Madison headwaters were heavily influenced by the characteristics of the landscape, as evidenced by the importance of the DRAINAGE covariate in both the home range and site fidelity analyses. Independent of wolves, smaller home ranges and higher site fidelity were observed in the Firehole drainage compared to the other two drainages. This finding is likely a consequence of the evenly distributed thermal areas through much of the Firehole drainage, providing an essentially contiguous snow-free foraging area where elk did not need to travel far between forage sites (Craighead *et al.* 1973). The Gibbon drainage also contains a high proportion of geothermal areas, but their configuration is quite different (Chapter 4 by Watson *et al.*, this volume). The Gibbon drainage has a few large geothermal complexes (*e.g.,* Norris, Primrose) and numerous smaller geothermal sites arranged linearly along river corridors. Therefore, the larger home ranges for elk in the Gibbon likely reflected animals feeding within and between these geothermal areas, with more movements due to the spatial arrangement of these snow-free sites. These finding support the work of Kie *et al.* (2002) who suggested that when patches were clumped rather than evenly distributed, home range size of mule deer (*Odocoileus hemionus*) were larger because animals had to travel longer distances from one feeding patch to another. The Madison drainage differs from the Gibbon and the Firehole drainages in that it does not contain any geothermal areas. However, movement patterns within this drainage could still be explained by the spatial arrangement of snow-free sites because the nearly continuous, steep, south-facing slope along this drainage provides another topographic feature with reduced snow pack. This slope provided a contiguous area elk used extensively for locomotion and foraging. Movements in the Madison drainage also reflected the topographic relief (Georgii and Schröder 1983, Kie *et al.* 2005) and linear arrangement of large meadow complexes. The spatial arrangement of snow-free geothermal areas had a pronounced effect on the way elk moved within the system and we strongly suspect that the assemblage and distribution of these geothermal sites also contributed to the vulnerability of elk in the Madison headwaters (Bergman *et al.* 2006, Chapter 21 by White *et al.*, this volume).

Elk in the Madison headwaters appeared to adopt several risk avoidance behaviors over multiple temporal and spatial scales, including: (1) increased use of the landscape at the local scale, (2) long-distance sallies that lasted several days, (3) temporary (>1 year) and permanent long-distance movements away from a former home range, and (4) transitions from sedentary to migratory behavior. While these findings suggest elk were using the landscape in an attempt to mitigate predation risk, we were limited in our ability to discern if and how effective these spatial responses were at reducing risk (Hebblewhite and Merrill 2007). However, the spatial responses observed in the Madison headwaters elk population suggest animals were making movement decisions under the competing environmental constraints of managing predation risk in addition to starvation risk. Familiarity with foraging resources and knowledge of habitat attributes that help facilitate avoidance and escape will favor spatial responses that occur within the home range (Linnell and Anderson 1995). Spatially avoiding predators is costly, so nutritionally vulnerable prey should have a high tolerance of disturbance by wolves before they completely abandon their range. Also, adopting an integrative approach that couples spatial responses within the range with other anti-predator behaviors such as grouping (Chapter 19 by Gower *et al.*, this volume), alterations in foraging behavior (Chapter 20 by Gower *et al.*, this volume), and effective use of landscape features that facilitate escape (Chapter 21 by White *et al.*, this volume) may be more effective at mitigating predation risk than trying to completely remain elusive in a high predation risk environment.

Our results suggest that elk will manifest a more dynamic movement behavior as wolves course the landscape looking for prey. These results are consistent with the predictions of Fortin *et al.* (2005) and Gude *et al.* (2006) that predation may facilitate the dilution of foraging pressure on plant communities if elk are constantly moving to reduce further encounters with wolves. This response could combine with predation-induced changes in the abundance and distribution of large herbivores to cause changes in ecosystem dynamics (Ripple *et al.* 2001, Beschta 2003, Smith *et al.* 2003, Fortin *et al.* 2005). However, our results also suggest that landscape characteristics substantially moderate the magnitude of the spatial response by elk to wolves, which could potentially affect the intensity of the trophic cascade. The spatial distribution of foraging areas in relation to the spatial distribution of safe or vulnerable attributes of the landscape may also determine the degree to which predators inflict direct and indirect costs on their prey.

V. SUMMARY

1. We evaluated how elk changed their spatial use of the landscape in the presence of wolves by collecting 11,908 locations from 115 radio-collared, adult, female elk in the Madison headwaters area of Yellowstone National Park during winters 1991–1992 through 2006–2007.
2. Prior to wolf colonization, winter movements of elk were constrained and predictable as elk attempted to conserve energy and decrease starvation risk. Home ranges were small and elk

displayed strong spatial fidelity. After wolf colonization, elk movements were more dynamic as elk moved more over the landscape as they were increasingly encountered, attacked, and displaced by wolves. Home range sizes were larger, with modest decreases in philopatry.

3. We documented 19 long-distance (>15 km, range = 18–63 km) dispersal movements by radio-collared elk away from high-density wolf areas in the Gibbon and Firehole drainages. These apparent predator-avoidance movements were not observed prior to wolf colonization or from areas where the risk of predation was lower.

4. Elk in the Madison headwaters are constrained by deep snow and vulnerable environmental conditions. This provides limited flexibility to mitigate predation risk by modifications in movement within the traditional range. This limited flexibility could explain why elk have been reduced to low densities in most areas of the Madison headwaters and, also, why a relatively large proportion of elk left two of the drainages where wolf predation pressure was exceptionally high.

5. Dynamic movements by prey to mitigate predation risk have important implications for encounter rates, search time, and the functional responses of predators. Likewise, predation-induced changes in the abundance and distribution of large herbivores may have substantial effects through the plant-herbivore-carnivore trophic chain, as well as important implications for managing elk at the landscape level.

VI. REFERENCES

Anderson, D. P., J. D. Forester, M. G. Turner, J. L. Frair, E. H. Merrill, D. Fortin, J. S. Mao, and M. S. Boyce. 2005. Factors influencing female home range size in elk (*Cervus elaphus*) in North America. *Landscape Ecology* **20**:257–271.

Azevedo-Ramos, C., M. V. Sluys, J. M. Hero, and W. E. Magnusson. 1992. Influence of tadpole movement on predation by Odonate naiads. *Journal of Herpetology* **26**:335–338.

Bailey, D. W., J. E. Gross, E. A. Laca, L. R. Rittenhouse, M. B. Coughenour, D. M. Swift, and P. L. Sims. 1996. Mechanisms that result in large herbivore grazing distribution patterns. *Journal of Range Management* **49**:386–400.

Berger, J., J. E. Swenson, and I.-L. Persson. 2001. Recolonizing carnivores and naive prey: Conservation lessons from Pleistocene extinctions. *Science* **291**:1036–1039.

Bergerud, A. T., H. E. Butler, and D. R. Miller. 1984. Anti-predator tactics of calving caribou—Dispersion in mountains. *Canadian Journal of Zoology* **62**:1566–1575.

Bergerud, A. T., S. N. Luttich, and L. Camps. 2008. The return of the caribou to Ungava. McGill-Queen's University Press, Montreal, Canada.

Bergman, E. J., R. A. Garrott, S. Creel, J. J. Borkowski, R. Jaffe, and F. G. R. Watson. 2006. Assessment of prey vulnerability through analysis of wolf movements and kill sites. *Ecological Applications* **16**:273–284.

Beschta, R. L. 2003. Cottonwoods, elk, and wolves in the Lamar valley of Yellowstone National Park. *Ecological Applications* **13**:1295–1309.

Blundell, G. M., J. A. K. Maier, and E. M. Debevec. 2001. Linear home ranges: Effects of smoothing, sample size, and autocorrelation on kernel estimates. *Ecological Monographs* **71**.

Brown, J. S. 1988. Patch use as an indicator of habitat preferences, predation risk, and competition. *Behavioral Ecology and Sociobiology* **22**:37–47.

Burnham, K. P., and D. R. Anderson. 2002. *Model Selection and Multimodel Inference: A Practical Information-Theoretic Approach.* Springer, New York, NY.

Burnham, K. P., and D. R. Anderson. 2004. Multimodel inference—Understanding AIC and BIC in model selection. *Sociological Methods and Research* **33**:261–304.

Burt, W. H. 1943. Territoriality and home range concepts as applied to mammals. *Journal of Mammalogy* **24**:346–352.

Cederlund, G. N., and H. K. G. Sands. 1992. Dispersal of subadult moose (*Alces alces*) in a nonmigratory population. *Canadian Journal of Zoology* **70**:1309–1314.

Cederlund, G. N., and H. K. G. Sands. 1994. Home-range size in relation to age and sex in moose. *Journal of Mammalogy* **75**:1005–1012.

Clobert, J., R. A. Ims, and F. Rousset. 2004. Causes, mechanisms, and consequences of dispersal. Pages 307–336 *in* I. Hanski and O. Gaggiotti (Eds.) *Ecology, Genetics, and Evolution of Metapopulations.* Academic Press, London, United Kingdom.

Clutton-Brock, T. H., F. E. Guiness, and S. D. Albon. 1983. The cost of reproduction to red deer hinds. *Journal of Animal Ecology* **52**:367–383.

Clutton-Brock, T. H., M. Major, and F. E. Guiness. 1985. Population regulation in male and female red deer. *Journal of Animal Ecology* **54**:831–846.

Cook, J. G. 2002. Nutrition and food. Pages 259–349 *in* D. E. Toweill and J. W. Thomas (Eds.) *North American Elk: Ecology and Management.* Smithsonian Institution Press, Washington, DC.

Cook, J. G., B. K. Johnson, R. C. Cook, R. A. Riggs, T. Delcurto, L. D. Bryant, and L. L. Irwin. 2004. Effects of summer–autumn nutrition and parturition date on reproduction and survival of elk. *Wildlife Monographs* **155**.

Craighead, J. J., G. Atwell, and B. W. O'Gara. 1972. Elk migrations in and near Yellowstone National Park. *Wildlife Monographs* **29**.

Craighead, J. J., F. C. J. Craighead, R. L. Ruff, and B. W. O'Gara. 1973. Home ranges and activity patterns of non-migratory elk of the Madison drainage herd as determined by biotelemetry. *Wildlife Monographs* **33**.

Creel, S., and J. A. WinnieJr. 2005. Response of elk herd size to fine-scale and temporal variations in the risk of predation by wolves. *Animal Behavior* **69**:1181–1189.

Creel, S., J. A. WinnieJr, B. Maxwell, K. Hamlin, and M. Creel. 2005. Elk alter habitat selection as an antipredator response to wolves. *Ecology* **86**:3387–3397.

Denno, R. F., and G. K. Roderick. 1992. Density-related dispersal in planthoppers: Effects of interspecific crowding. *Ecology* **73**:1323–1334.

Dussault, C., R. Courtois, J.-P. Ouellet, and I. Girard. 2005. Space use of moose in relation to food availability. *Canadian Journal of Zoology* **83**:1431–1437.

Edge, D. W., C. L. Marcum, S. L. Olson, and J. F. Lehmkuhl. 1986. Nonmigratory cow elk herd ranges as management units. *Journal of Wildlife Management* **50**:660–663.

Fancy, S. G., and R. G. White. 1985. Incremental costs of activity. Pages 143–160 *in* R. J. Hudson and R. G. White (Eds.) *Bioenergetics of Wild Herbivores.* CRC Press, Boca Raton, FL.

Fancy, S. G., and S. G. White. 1987. Energy expenditure for locomotion by barren-ground caribou. *Canadian Journal of Zoology* **65**:122–128.

Ford, G. L. 1983. Home range in a patchy environment: Optimal foraging predictions. *American Zoologist* **23**:315–326.

Forester, J. D., A. I. Ives, M. G. Turner, D. R. Anderson, D. Fortin, H. L. Beyers, D. W. Smith, and M. S. Boyce. 2007. State-spaced models link elk movement patterns to landscape characteristics in Yellowstone National Park. *Ecological Monographs* **77**:285–299.

Formanowicz, D. R., and M. S. Bobka. 1989. Predation risk and microhabitat preference: An experimental study of the behavioral response of prey and predator. *American Midland Naturalist* **121**:379–386.

Fortin, D., H. L. Beyers, M. S. Boyce, D. W. Smith, T. Duchesne, and J. S. Mao. 2005. Wolves influence elk movements: Behavior shapes a trophic cascade in Yellowstone National Park. *Ecology* **86**:1320–1330.

Frair, J. L., E. H. Merrill, D. R. Visscher, D. Fortin, H. L. Beyers, and J. M. Morales. 2005. Scales of movement by elk (*Cervus elaphus*) in response to heterogeneity in forage resource and predation risk. *Landscape Ecology* **20**:273–287.

Franklin, W. L., and J. W. Lieb. 1979. The social organization of a sedentary population of North American elk: A model for understanding other populations. Pages 185–195 *in* M. S. Boyce and L. D. Hayden-Wing (Eds.) *North American Elk: Ecology, Behavior, and Management.* University of Wyoming, Laramie, WY.

Fryxell, J. M. 1995. Aggregation and migration by grazing ungulates in relation to resources and predation. Pages 257–273 *in* A. R. E. Sinclair and P. Arcese (Eds.) *Serengeti II: Dynamics Management, and Conservation of an Ecosystem.* University of Chicago Press, Chicago, IL.

Fryxell, J. M., J. Greever, and A. R. E. Sinclair. 1988. Why are migratory ungulates so abundant? *American Naturalist* **131**:781–798.

Fryxell, J. M., J. F. Wilmshurst, and A. R. E. Sinclair. 2004. Predictive models of movement by Serengeti grazers. *Ecology* **85**:2429–2435.

Gadgil, M. 1971. Dispersal: Population consequences and evolution. *Ecology* **52**:253–261.

Gaillard, J.-M., M. Festa-Bianchet, and N. G. Yoccoz. 1998. Population dynamics of large herbivores: Variable recruitment with constant adult survival. *Trends in Ecology and Evolution* **13**:58–63.

Garrott, R. A., G. C. White, R. M. Bartmann, L. H. Carpenter, and W. A. Alldredge. 1987. Movements of female mule deer in northwest Colorado. *Journal of Wildlife Management* **51**:634–643.

Gates, C. C., and R. J. Hudson. 1979. Effects of posture and activity on metabolic responses of wapiti to cold. *Journal of Wildlife Management* **43**:564–567.

Geist, V. 2002. Adaptive behavioral strategies. Pages 389–433 *in* D. E. Toweill and J. W. Thomas (Eds.) *North American Elk: Ecology and Management.* Smithsonian Institute Press, Washington, DC.

Georgii, B., and W. Schröder. 1983. Home range and activity patterns of male red deer (*Cervus elaphus L.*) in the Alps. *Oecologia* **58**:238–248.

Greenwood, P. J. 1983. Mating systems and the evolutionary consequences of dispersal. Pages 116–131 *in* I. R. Swingland and P. J. Greenwood (Eds.) *The Ecology of Animal Movement.* Clarendon Press, Oxford, United Kingdom.

Grignolio, S., I. Rossi, B. Bassano, F. Parrini, and M. Appolinio. 2004. Seasonal variation of spatial behavior in female alpine ibex (*Capra ibex ibex*) in relation to climate conditions and age. *Ethology Ecology and Evolution* **16**:255–264.

Gude, J. A., R. A. Garrott, J. J. Borkowski, and F. King. 2006. Prey risk allocation in a grazing environment. *Ecological Applications* **16**:285–298.

Hammond, J. I., B. Luttberg, and A. Sih. 2007. Predator and prey space use: Dragonflies and tadpole in an interactive game. *Ecology* **88**:1525–1535.

Hebblewhite, M., and E. H. Merrill. 2007. Multiscale wolf predation risk for elk: Does migration reduce risk? *Oecologia* **152**:377–387.

Hebblewhite, M., E. H. Merrill, and T. L. McDonald. 2005. Spatial decomposition of predation risk using resource selection functions: An example in a wolf-elk predator-prey system. *Oikos* **111**:101–111.

Hemson, G., P. Johnson, A. South, R. E. Kenward, R. Ripley, and D. MacDonald. 2005. Are kernels the mustard? Data from global positioning system (GPS) collars suggest problems for kernel home range analyses with least squares cross-validation. *Journal of Animal Ecology* **74**:455–463.

Hobbs, N. T., D. L. Baker, J. E. Ellis, and D. M. Swift. 1981. Composition and quality of elk winter diets on Colorado. *Journal of Wildlife Management* **45**:156–171.

Horne, J. S., and E. O. Garton. 2006a. Selecting the best home range model: An information-theoretic approach. *Ecology* **87**:1146–1152.

Horne, J. S., and E. O. Garton. 2006b. Likelihood cross-validation versus least squares cross validation for choosing the smoothing parameter in kernel home-range analyses. *Journal of Wildlife Management* **70**:641–648.

Hudson, R. J., and R. G. White. 1985. *Bioenergetics of Wild Herbivores*. CRC Press, Boca Raton, FL.

Ims, R. A. 1988. Spatial clumping of sexually receptive females induces space sharing among male voles. *Nature* **335**:541–543.

Ims, R. A., and H. P. Andreassen. 2005. Density-dependent dispersal and spatial population dynamics. *Proceedings of the Royal Society B—Biological Science* **272**:913–918.

Irwin, L. P., and J. M. Peek. 1983. Elk habitat use relative to forest succession in Idaho. *Journal of Wildlife Management* **47**:664–672.

Jędrzejewski, W., K. Schmidt, J. Theuerkauf, B. Jędrzejewska, and H. Okarma. 2001. Daily movements and territory use by radio-collared wolves (*Canis lupus*) in Białowieża Primeaval Forest in Poland. *Canadian Journal of Zoology* **79**:1993–2004.

Jenkins, K. J., and R. G. Wright. 1987. Dietary niche relationships among cervids relative to winter snowpack in northwestern Montana. *Canadian Journal of Zoology* **65**:1397–1401.

Johnson, C. J., K. L. Parker, and D. C. Heard. 2001. Foraging across a variable landscape: Behavioural decisions made by woodland caribou. *Oecologia* **127**:590–602.

Kauffman, M. J., N. Varley, D. W. Smith, D. R. Stahler, D. R. MacNulty, and M. S. Boyce. 2007. Landscape heterogeneity shapes predation in a newly restored predator–prey system. *Ecology Letters* **10**:690–700.

Kernohan, B. J., R. A. Gitzen, and J. J. Millspaugh. 2001. Analysis of animal space use and movement. Pages 125–166 *in* J. Millspaugh and J. Marzluff (Eds.) *Radio Tracking and Animal Populations*. Academic Press, San Diego, CA.

Kie, J. G. 1999. Optimal foraging and risk of predation: Effects on behavior and social structure in ungulates. *Journal of Mammalogy* **80**:1114–1129.

Kie, J. G., A. A. Ager, and T. R. Bowyer. 2005. Landscape-level movements of North American elk (*Cervus elaphus*): Effects of habitat patch structure and topography. *Landscape Ecology* **20**:289–300.

Kie, J. G., T. R. Bowyer, M. C. Nicholson, B. B. Boroski, and E. R. Loft. 2002. Landscape heterogeneity at differing scales: Effects on spatial distribution of mule deer. *Ecology* **83**:530–544.

Kilpatrick, H. J., S. M. Spohr, and K. K. Lima. 2001. Effects of population reduction on home ranges of female white-tailed deer at high densities. *Canadian Journal of Zoology* **79**:949–954.

Kjellander, P., A. J. M. Hewison, O. Liberg, J.-M. Angibault, E. Bideau, and B. Cargnelutti. 2004. Experimental evidence for density dependence of home range size in roe deer (*Capreolus capreolus*, L.): A comparison of two long term studies. *Oecologia* **139**:478–485.

Krasińska, M., Z. Krasiński, and A. N. Bunevich. 2000. Factors affecting the variability in home range and distribution in European bison in the Polish and Belarussian parts of the Białowieża Forest. *Acta Theriologica* **45**:321–334.

Kunkel, K. E., and D. H. Pletscher. 2000. Habitat factors affecting vulnerability of moose to predation by wolves in southeastern British Columbia. *Canadian Journal of Zoology* **78**:150–157.

Kutner, M. H., C. J. Nachtsheim, J. Neter, and W. Li. 2005. *Applied Linear Statistical Models* 5th edn, McGraw-Hill, New York, NY.

Lambin, X., J. Aars, and S. B. Piertnet. 2001. Dispersal, intraspecific competition, kin competition, and kin facilitation: A review of the empirical evidence. Pages 261–272 *in* J. Clobert, E. Danchin, A. A. Dhondt, and J. D. Nichols (Eds.) *Dispersal*. Oxford University Press, New York, NY.

Lima, S. L. 1998. Stress and decision making under the risk of predation: Recent developments from behavioral, reproductive and ecological perspectives. *Advances in the Study of Behavior* **27**:215–290.

Lima, S. L., and L. M. Dill. 1990. Behavioral decisions made under the risk of predation: A review and prospectus. *Canadian Journal of Zoology* **68**:619–640.

Lima, S. L., and T. D. Steury. 2005. Perception of predation risk. Pages 166–188 *in* P. Barbosa and I. Castellanos (Eds.) *Ecology of Predator–Prey Interactions*. Oxford University Press, Oxford, United Kingdom.

Linnell, J. D., and R. Anderson. 1995. Site tenacity in roe deer: Short-term effects of logging. *Wildlife Society Bulletin* **12**:31–35.

MacArthur, R. H. 1972. *Geographical Ecology*. Harper and Row, New York, NY.

McCorquodale, S. M. 1993. Winter foraging behavior of elk in the shrub-steppe of Washington. *Journal of Wildlife Management* **57**:881–890.

McNaughton, S. J. 1985. Ecology of a grazing ecosystem: The Serengeti. *Ecological Monographs* **55**:259–294.

McNaughton, S. J., and F. F. Banyikwa. 1995. Plant communities and herbivory. Pages 49–70 *in* A. R. E. Sinclair and P. Arcese (Eds.) *Serengeti II: Dynamics, Management, and Conservation of an Ecosystem*. University of Chicago Press, Chicago, IL.

Mech, L. D. 1970. *The Wolf: The Ecology of an Endangered Species*. Natural History Press, New York, NY.

Mech, L. D., and R. O. Peterson. 2003. Wolf-prey relations. Pages 131–160 *in* L. D. Mech and L. Boitani (Eds.) *Wolves: Behavior, Ecology, and Conservation*. University of Chicago Press, Chicago, IL.

Millspaugh, J. J., G. C. Brundige, R. A. Gitzen, and K. J. Raedeke. 2004. Herd organization of cow elk in Custer State Park, South Dakota. *Wildlife Society Bulletin* **32**:506–514.

Miquelle, D. G., J. M. Peek, and V. Van Ballenberghe. 1992. Sexual segregation in Alaskan moose. *Wildlife Monographs* **122**.

Mitchell, M. S., and R. A. Powell. 2004. A mechanistic home range model for optimal use of spatially distributed resources. *Ecological Modelling* **177**:209–232.

Mitchell, W. A., and S. L. Lima. 2002. Predator–prey shell games: Large-scale movement and its implications for decision making prey. *Oikos* **99**:249–259.

Moen, A. N. 1976. Energy conservation by white-tailed deer in the winter. *Ecology* **57**:192–198.

Nelson, E. H. 1995. Winter range arrival and departure of White-tailed deer in Northern Minnesota. *Canadian Journal of Zoology* **73**:1069–1076.

Ostfeld, R. S. 1985. Limiting resources and territoriality in microtine rodents. *American Naturalist* **126**:1–15.

Ostfeld, R. S. 1986. Territoriality and mating systems of California voles. *Journal of Animal Ecology* **55**:691–706.

Ozoga, J. J., and L. W. Gysel. 1972. Responses of white-tailed deer to winter weather. *Journal of Wildlife Management* **36**:892–896.

Parker, K. L., P. S. Barboza, and T. R. Stephenson. 2005. Protein conservation in female caribou (*Rangifer tarandus*): Effects of decreasing diet quality during winter. *Journal of Mammalogy* **86**:610–622.

Parker, K. L., C. T. Robbins, and T. A. Hanley. 1984. Energy expenditure for locomotion by mule deer and elk. *Journal of Wildlife Management* **48**:474–488.

Pastor, J., R. Moen, and Y. Cohen. 1997. Spatial heterogeneities, carrying capacity, and feedbacks in animal-landscape interactions. *Journal of Mammalogy* **78**:1040–1052.

Pinheiro, J. C., and D. M. Bates. 2000. *Mixed-Effects Models in S and S-Plus*. Springer, New York, NY.

Pyke, G. H. 1983. Animal movements: An optimal foraging approach. Pages 7–31 *in* I. R. Swingland and P. J. Greenwood (Eds.) *The Ecology of Animal Behavior*. Clarendon Press, Oxford, United Kingdom.

Ramsey, F., and D. Schafer. 2002. *The Statistical Sleuth: A Course in Methods of Data Analysis*. 2nd edn. Duxbury Press, Pacific Grove, CA.

R Development Core Team 2006. *R: A language and environment for statistical computing*. R Foundation for Statistical Computing, Vienna, Austria. ISBN 3-900051-07-0, URLhttp://www.R-project.org.

Renecker, L. A., and C. C. Schwarz. 1998. Food habits and feeding behavior. Pages 403–439 *in* A. W. Franzmann and C. C. Schwarz (Eds.) *Ecology and Management of the North American Moose*. Smithsonian Institution Press, Washington DC.

Rettie, W. J., and F. Messier. 2001. Range use and movement rates of woodland caribou in Saskatchewan. *Canadian Journal of Zoology* **79**:1933–1941.

Ripple, W. J., and R. L. Beschta. 2003. Wolf reintroduction, predation risk, and cottonwood recovery in Yellowstone National Park. *Forest Ecology and Management* **184**:299–313.

Ripple, W. J., and R. L. Beschta. 2004. Wolves and the ecology of fear: Can predation risk structure ecosystems? *BioScience* **54**:755–766.

Ripple, W. J., E. J. Larsen, R. A. Renkin, and D. W. Smith. 2001. Trophic cascades among wolves, elk and aspen on Yellowstone National Park's northern range. *Biological Conservation* **102**:227–234.

Roth, T. C., and S. L. Lima. 2007. Use of prey hotspots by an avian predator: Purposeful unpredictability. *American Naturalist* **169**:264–273.

Saïd, S., and S. Servanty. 2005. The influence of landscape structure on female roe deer home range size. *Landscape Ecology* **20**:1003–1012.

Sain, S. R., K. A. Baggerly, and D. W. Scott. 1994. Cross-validation of multivariate densities. *Journal of the American Statistical Association* **89**:807–817.

Schmidt, K. 1993. Winter ecology of non-migratory alpine red deer. *Oecologia* **95**:226–233.

Seaman, D. E., J. J. Millspaugh, B. J. Kernohan, G. C. Brundige, K. J. Raedeke, and R. A. Gitzen. 1999. Effects of sample size on kernel home range estimates. *Journal of Wildlife Management* **63**:739–747.

Seaman, D. E., and R. A. Powell. 1996. An evaluation of the accuracy of kernel density estimates for home range analysis. *Ecology* **77**:2075–2085.

Seip, D. R. 1992. Factors limiting woodland caribou populations and their interrelationship with wolves and moose in southeastern British Columbia. *Canadian Journal of Zoology* **70**:1494–1503.

Senft, R. L., M. B. Coughenour, D. W. Bailey, L. R. Rittenhouse, E. O. Sala, and D. M. Swift. 1987. Large herbivore foraging and ecological hierarchies. *BioScience* **37**:789–795798–799.

Sih, A. 1982. Foraging strategies and the avoidance of predation by an aquatic insect *Notonecta hoffmanni*. *Ecology* **63**:786–796.

Sih, A. 1984. The behavioral response race between predator and prey. *American Naturalist* **123**:143–150.

Sih, A. 1986. Antipredator responses and the perception of danger by mosquito larvae. *Ecology* **67**:434–441.

Sih, A. 1992. Prey uncertainty and the balancing of antipredator and feeding needs. *American Naturalist* **139**:1052–1069.

Sih, A. 1994. Predation risk and the evolutionary ecology of reproductive behavior. *Journal of Fish Biology* **45A**:111–130.

Sih, A., and T. M. McCarthy. 2002. Prey responses to pulses of risk and safety: Testing the risk allocation hypothesis. *Animal Behaviour* **63**:437–443.

Silverman, B. W. 1986. *Density Estimation for Statistics and Data Analysis.* Chapman and Hall, London, United Kingdom.

Sinclair, A. R. E. 1983. The influence of distance movements in vertebrates. Pages 240–258 *in* I. R. Swingland and P. J. Greenwood (Eds.) *The Ecology of Animal Movement.* Clarendon Press, Oxford, United Kingdom.

Smith, D. W. 2005. Ten years of Yellowstone wolves. *Yellowstone Science* **13**:7–33.

Smith, D. W., R. O. Peterson, and D. B. Houston. 2003. Yellowstone after wolves. *BioScience* **53**:330–340.

Sonerud, G. A. 1985. Brood movements in grouse and waders as defense against win-stay search in their predators. *Oikos* **44**:287–300.

Stenseth, N. C. 1983. Causes and consequences of dispersal in small mammals. Pages 63–101 *in* I. R. Swingland and P. J. Greenwood (Eds.) *The Ecology of Animal Movement.* Clarendon Press, Oxford, United Kingdom.

Sweeney, J. M., and J. R. Sweeney. 1984. Snow depths influencing winter movements of elk. *Journal of Mammalogy* **65**:524–526.

Swihart, R. K., and N. A. Slade. 1985. Influence of sampling interval on estimates of home range size. *Journal of Wildlife Management* **49**:1019–1025.

Testa, J. W., and G. P. Adams. 1998. Body condition and adjustments to reproductive effort in female moose (*Alces alces*). *Journal of Mammalogy* **79**:1345–1354.

Tufto, J., R. Andersen, and J. Linnell. 1996. Habitat use and ecological correlates of home range size in a small cervid: The roe deer. *Journal of Animal Ecology* **65**:715–724.

Verme, L. J. 1973. Movements of white-tailed deer in upper Michigan. *Journal of Wildlife Management* **37**:545–552.

White, G. C., and R. A. Garrott. 1990. *Analysis of Wildlife Radio-Tracking Data.* Academic Press, San Diego, CA.

Winnie, J. A.Jr, and S. Creel. 2007. Sex-specific behavioural responses of elk to spatial and temporal variation in the threat of wolf predation. *Animal Behaviour* **73**:215–225.

Wolff, J. O., and T. Van Horn. 2003. Vigilance and foraging patterns of American elk during the rut in habitats with and without predators. *Canadian Journal of Zoology* **81**:266–271.

Worton, B. J. 1989. Kernel methods for estimating the utilization distribution in home range studies. *Ecology* **70**:164–168.

APPENDIX

APPENDIX 18A. 1	Complete model list used in the mixed effects linear regression analysis to evaluate factors influencing home range size of elk in the Madison headwaters area of Yellowstone National Park during 1991–1992 through 2006–2007

Model number	Model structure
0	NULL
1	SWE_{mean}
2	DRAINAGE
3	$WOLF_{period}$
4	SWE_{mean} + DRAINAGE
5	SWE_{mean} + $WOLF_{period}$
6	DRAINAGE + $WOLF_{period}$
7	SWE_{mean} + DRAINAGE + $WOLF_{period}$
8	$WOLF_{days}$
9	SWE_{mean} + $WOLF_{days}$
10	DRAINAGE + $WOLF_{days}$
11	SWE_{mean} + DRAINAGE + $WOLF_{days}$
12	KILLS
13	SWE_{mean} + KILLS
14	DRAINAGE + KILLS
15	SWE_{mean} + DRAINAGE + KILLS

Abbreviations: SWE_{mean} (mean daily snow water equivalent), DRAINAGE (Madison, Firehole, Gibbon), $WOLF_{period}$ (pre-reintroduction, colonizing, established), $WOLF_{days}$ (total number of wolf days per drainage per year), and KILLS (total number of wolf-killed ungulates per drainage per year).

APPENDIX 18A. 2	Complete model list used in the mixed effects linear regression analysis to evaluate factors influencing site fidelity of elk in the Madison headwaters area of Yellowstone National Park during 1991–1992 through 2006–2007

Model number	Model structure
0	NULL
1	$WOLF_{period}$
2	DRAINAGE
3	$DRAINAGE + WOLF_{period}$
4	$WOLF_{days}$
5	$DRAINAGE + WOLF_{days}$
6	KILLS
7	DRAINAGE + KILLS

Abbreviations are as in Appendix 18A.1. Wolf covariates describe wolf use in the particular drainage during the first of the two consecutive years.

CHAPTER 19

Elk Group Size and Wolf Predation: A Flexible Strategy when Faced with Variable Risk

Claire N. Gower,* Robert A. Garrott,* P. J. White,[†] Steve Cherry,[‡] and Nigel G. Yoccoz[§]

*Fish and Wildlife Management Program, Ecology Department, Montana State University
[†]National Park Service, Yellowstone National Park
[‡]Department of Mathematical Sciences, Montana State University
[§]Department of Biology, University of Tromsø, Norway

Contents

Theme

When predators and prey are in close proximity, prey may change how they aggregate to reduce their probability of being attacked and killed. However, the best strategy to reduce the risk of predation may not be the best strategy to acquire the resources needed to meet physiological demands for body maintenance and reproduction. How animals balance the competing risks of starvation and predation mortality is an active area of research in a wide variety of predator–prey systems. The non-migratory elk (*Cervus elaphus*) population in the Madison headwaters area of Yellowstone National Park provided an excellent opportunity to evaluate prey grouping tendencies in the absence of predators and, following wolf (*Canis lupus*) reestablishment, when faced with trade-offs between avoiding predation and acquiring adequate resources. We collected long-term data on elk grouping behavior prior to wolf reintroduction, during the colonization stage, and after wolves became fully established in the system. Wolf presence varied spatially and temporally within and among years, and there was a wide range of snow conditions during the 16-year study. Our objectives were to (1) evaluate competing

The Ecology of Large Mammals in Central Yellowstone
R. Garrott, P. J. White and F. Watson
ISSN 1936-7961, DOI: 10.1016/S1936-7961(08)00219-4

Copyright © 2009, Elsevier Inc.
All rights reserved.

hypotheses on whether elk formed larger or smaller aggregations when wolves were present, (2) determine if grouping strategies changed under different abiotic conditions, and (3) determine if variability in group size increased following wolf reintroduction as elk responded to fine-scale temporal variation in predation risk.

I. INTRODUCTION

It is widespread in the animal kingdom that animals aggregate (*i.e.*, flocks, schools, groups, and herds) (Figure 19.1) and has long been considered an evolutionary adaptation whereby members derive benefits of communal foraging, mating, thermal regulation, reduced costs of movement, and protection against predators (Allee 1927). Animal grouping is a complex behavioral process driven by a combination of environmental, population, and community processes that change with ecological conditions over relatively short temporal and spatial scales (Chapman and Chapman 2000, Krause and Ruxton 2002). Thus, decisions by individuals to remain solitary or aggregate, and how large an aggregation to tolerate, are clearly influenced by abiotic and biotic factors that individuals confront.

Many drivers of aggregation patterns have been suggested, including that predation risk is reduced when individuals join with conspecifics. Aggregating into larger groups has the benefit of enhancing predator detection and reducing individual vigilance (Pulliam 1973, Elgar 1989, Roberts 1996), individual protection and numerical dilution (Hamilton 1971, Bertram 1978, Dehn 1990), confusion effects, and group cooperative defense (Caro 2005). However, predation may be facilitated by larger aggregations of animals because they are more conspicuous to predators and easier to detect at a distance (Treisman 1975). Other potential countervailing consequences of aggregating into larger groups include increased intra-specific competition for resources (Chapman and Chapman 2000), increased foraging costs due to local food depletion (Buckel and Strona 2004), higher levels of aggression (Clutton-Brock *et al.* 1982, Molvar and Bowyer 1994), and higher exposures to diseases and parasites. Eventually, the costs of grouping may exceed the benefits, and often these physiological and competitive constraints limit group size (Pulliam and Caraco 1984, Giraldeau and Caraco 2000, Krause and Ruxton 2002).

FIGURE 19.1 A small group of cow and calf elk in the Upper Geyser Basin of Yellowstone National Park. Bison and elk are commonly observed foraging in geothermally heated areas, or benefiting from the warmth on cold frosty winter days (Photo by Jeff Henry).

Forming smaller groups or remaining solitary may lessen detection by predators and is an advantageous strategy if predators preferentially target larger groups of prey (Hebblewhite and Pletscher 2002, Krause and Ruxton 2002). Under these circumstances, individuals may form small groups, or adopt a solitary existence as an alternative response to predation. If animals succeed in remaining elusive, then they may escape the attention of a predator and, thus, the probability of encounter with a predator may be lower (Jarman 1974, Semeniuk and Dill 2004, Creel and Winnie 2005). Solitary prey can exploit resources in the absence of competition, but may also incur foraging costs by not aggregating in higher-quality foraging patches. Group sizes in many taxa appear to be determined by trade-offs or interactions between food availability, intra-specific competition for food, and predation risk (Heard 1992, Molvar and Bowyer 1994, Grand and Dill 1999, Parrish and Edelstein-Keshet 1999, Buckel and Strona 2004, Semeniuk and Dill 2004), with the most advantageous group size reflecting the balance between the costs and benefits of the competing demands for each group member. Theoretically, this optimal size would offer maximum fitness to each group member and may vary for any given individual, environment, or situation (Sibly 1983, Pulliam and Caraco 1984, Heard 1992, Higashi and Yamamura 1993, Krause and Ruxton 2002).

Determining if an optimal group size exists for large herbivores is difficult because available forage and predation risk are often highly variable in both space and time and difficult to quantify. Variability in forage biomass due to heterogeneous landscapes, changes in plant phenology (and thus plant quality), and changes in availability of forage due to herbivory and snow pack conditions all result in group size varying predictably over seasons and among different types of plant communities (Heard 1992, Jędrzejewski *et al.* 1992, Borkowski and Furubayashi 1998). In a seasonal environment where the main predator is absent and snow limits access and availability to forage for a large proportion of the year, individuals should adopt an efficient and predictable grouping strategy that minimizes physiological costs (Chapter 8 by Messer *et al.*, this volume). If a system includes highly mobile, coursing predators such as wolves, then predation risk could be highly variable at a fine temporal scale because predators could frequently enter and leave areas occupied by prey. Considering the constraints imposed during winter, staying in a large group for a sustained period is costly to the individual. However, remaining solitary or in a small aggregation provides limited protection in the event the group is detected by predators. Thus, it is highly unlikely that one single group size is optimal to balance the demands of energy conservation while minimizing predation risk. Instead, individuals may aggregate into groups to reap the anti-predator behavior benefits during times of high predation risk, but disperse when there is not an imminent threat or as the level of individual risk diminishes (Lima and Bednekoff 1999, Sih *et al.* 2000, Sih and McCarthy 2002). If risk changes over the course of hours or even minutes, then a sustained response may not occur. Rather, a more ephemeral response may be evident, with frequent fragmenting and combining of groups as the level of risk changes.

It is generally accepted that animals reduce their risk of predation by associating with other herd members and, as a result, larger aggregations of large herbivores should be observed as predator abundance increases (Fryxell 1991, 1995, Jędrzejewski *et al.* 1992). However, there is only limited support for this prediction from recent studies of elk–wolf systems in the northern Rocky Mountains (Hebblewhite and Pletscher 2002, Creel and Winnie 2005, Gude *et al.* 2006). Varied grouping responses by elk in these different systems suggest predation alone is not wholly responsible for changes in group sizes and precludes generalizing one optimal grouping strategy by large herbivores in the presence of predators. These findings also indicate additional research is needed to evaluate competing hypotheses concerning the mechanisms driving grouping behavior in large herbivores.

Our goal was to improve understanding of the mechanisms driving grouping behavior in large herbivores by initially evaluating how elk in the Madison headwaters area of Yellowstone National Park responded to environmental conditions in the absence of predators and, subsequently, evaluating how their behavior changed when wolves were added to the system in 1995 and 1996. Data were collected for 5 years prior to 1995 when wolves were entirely absent from the system and for 11 years after wolf reintroduction. We used information-theoretic model selection techniques to evaluate the strength of evidence in the data for the competing predictions that elk would either aggregate in response to increasing wolf presence or form smaller groups or become more solitary once wolves re-colonized the area. We also evaluated the prediction that elk would exhibit greater variation in group size as the level of predation risk varied.

II. METHODS

We recorded the grouping behavior of elk in the Madison headwaters area during 15 November through 30 April for 16 consecutive winters (1991–1992 through 2006–2007) by repeatedly sampling 20–35 adult female elk per year fitted with radio collars (*i.e.*, approximately 4–10% of the population). Sampling of collared individuals followed a stratified random sampling regime, which ensured that sampling times were randomly distributed during daylight hours to capture daily variation in group size and composition. Hand-held telemetry equipment and homing procedures were employed to obtain direct observations of collared individuals (Chapter 11 by Garrott *et al.*, this volume) and obtain group locations, total group counts and compositions (*i.e.*, calf, cow, yearling bull, adult bull), and identify the primary habitat occupied (meadow, thermal, unburned forest, burned forest, riparian).

These data were used to develop three response variables related to different elements of elk grouping behavior: group size, group size variation, and typical group size. Group size was defined as a single animal or individuals "that remain together for a period of time while interacting with one another to a distinctly greater degree than with other conspecifics" (Wilson 1975:585). Group size variation was derived by calculating the absolute difference between a given group size and the mean group size for that particular year, and was used to assess the frequency of aggregation and dispersion of groups. Typical group size was defined as the size of the group in which the average animal finds itself and calculated as $\sum G_i^2 / \sum G_i$, where G_i is the size of the *i*th group (Jarman 1974). This metric of individual behavior is a descriptive statistic which, in addition to mean group size, could provide more information for discerning the effects of predation on prey behavior and the consequences of prey aggregation on the predator. Mean group size is often sensitive to the number of records of solitary animals and consequently can underestimate the size of the group that is experienced by the individual animal. In contrast, typical group size is less sensitive to the number of records of solitary animals and less sensitive to the commonly observed mean group size frequency distribution (a lot of small groups at the far left of the distribution, with few larger groups representing the right tail) (Heard 1992, Lingle 2003). Mean group size is therefore an "observer-centered" measure that provides useful information about how an outsider (*i.e.*, an observer or a predator) views the group (Lingle 2003), whereas the typical group size is an "animal-centered" measure that provides insight into the group size than an individual chooses to occupy (Jarman 1974). Typical group size has been used in other studies that have evaluated large herbivore aggregation patterns (Heard 1992, Heard and Ouellet 1994, Lingle 2003, Mao 2003, Festa-Bianchet and Côte 2008) and would appear particularly useful when assessing grouping behavior in a predator–prey framework. Because selection acts upon the individual, not the group (Heard 1992), this metric was used to describe the evolving behavior between predator and prey. Because typical group size is derived from the sum of an assemblage of groups over a determined time period, we calculated this metric using 10-day sampling intervals through the winter season at the study area level (16 periods per winter).

A. Non-Wolf Predictors of Grouping Behavior

We explored the influence of landscape attributes on elk grouping tendencies using estimates of snow pack and habitat type. Snow pack and habitat type influence resource availability which, in turn, could play a vital role governing how large herbivores aggregate (Heard 1992, Jędrzejewski *et al.* 1992,

Borkowski and Furubayashi 1998). We used a validated snow pack simulation model for the central Yellowstone region (Chapter 6 by Watson *et al.*, this volume) to construct a covariate that represented the average snow water equivalent value (SWE_L) on the elk winter range (landscape SWE) for each day that radio-collared elk were observed. This metric described the mean water content of the snow (Farnes *et al.* 1999) and provided an index of resource availability because higher values reflected decreased forage accessibility and higher energetic costs (Parker *et al.* 1984). We also developed a snow heterogeneity metric (SNH_L) to index the continuity of snow cover across the landscape. This metric was used as a surrogate for snow melt, which creates patches of available forage during spring. In addition, we defined a categorical habitat covariate (HBT) and classified each elk observation into meadow, burned forest, unburned forest, thermal, or riparian categories based on field observations of the plant communities the animals occupied at the time of the observation.

We considered SWE_L and time period (SEASON) as our non-wolf covariates in the typical group size analyses. SWE_L represented the average daily snow water equivalent value for the 10-day period. SEASON was a continuous variable from 1 to 16 representing the consecutive 10-day intervals within each winter season. Foraging dynamics likely reflect the seasonal variation in the quality and quantity of available forage (Craighead *et al.* 1973, Green and Bear 1990, Ager *et al.* 2003), so we used this covariate to evaluate seasonal trends in typical group size because larger groups of elk occurred in openings during spring green up.

B. Covariates of Predation Risk

As described in Chapter 15 by Smith *et al.*, and Chapter 16 by Becker *et al.*, this volume, we detected and quantified the presence of wolves from 15 November to 30 April during 1996–1997 to 2006–2007. National Park Service biologists captured wolves to maintain radio-collared animals in each pack (Smith 2005; Chapter 15 by Smith *et al.*, this volume), and we monitored wolf presence and their location using ground-based telemetry, snow tracking, and visual observations of collared and uncollared individuals. We intensively monitored each drainage for wolf presence daily with crews of 3–4 people using snowmobiles, vehicles, snowshoes, and high points in the landscape to facilitate telemetry and observations. When packs with radio-collared wolves were detected, we estimated telemetry locations using triangulation (White and Garrott 1990) and obtained multiple locations throughout the day. We used snow tracking, visual observations, and counts during aerial monitoring by National Park Service biologists to provide estimates of the number of animals per pack and aid our daily assessments of wolf presence or absence (Chapter 15 by Smith *et al.* and Chapter 16 by Becker *et al.*, this volume). Detection of uncollared wolves was facilitated by opportunistic observations of tracks and wolves by field personnel that were working throughout the study area on elk and bison investigations (Chapter 11 by Garrott *et al.*, Chapter 12 by Bruggeman *et al.*, and Chapter 21 by White *et al.*, this volume). The total number of wolves present in each drainage each day was estimated based on the information obtained from the various wolf monitoring techniques and was quantified in the form of wolf days (*i.e.*, the number of individual wolves known to be present in each of the drainages each day).

Using these data on wolf presence, we developed three wolf covariates to assess our hypotheses about the influence of predation risk on aggregation behavior by elk. The covariate $WOLF_{period}$ categorized wolf presence within each drainage into three periods: before wolf reintroduction, during wolf colonization, and after wolf establishment. Pre-reintroduction occurred during the initial years of the study (1991–1992 through 1995–1996), prior to wolf reintroduction when no wolves were present in the study area. To account for the potential transitory behavioral dynamics due to the initial naïveté of the prey (Berger *et al.* 2001), we defined a colonizing period immediately following wolf reintroduction, when elk were initially exposed to wolf predation risk, but no wolf pack (≥ 2 animals) was routinely detected in the drainage. The established wolf presence period began the first winter that a pack was consistently detected in a drainage. Wolves became established in different drainages during different winters (Firehole: 1997–1998, Gibbon: 2000–2001, Madison: 2001–2002; Chapter 15 by Smith *et al.*, this volume); therefore, the WOLF$_{period}$ covariate was drainage-specific. The level of wolf predation risk for elk occupying each drainage was highly variable among winters, and we suspected the magnitude or frequency of any behavioral responses would be scaled to predation risk. Thus, we used wolf presence data to construct a continuous wolf covariate, WOLF$_{days}$, that indexed the prevalence of wolf activity in each drainage each day. For the group size and group size variation analyses, WOLF$_{days}$ was calculated as the estimated number of wolves within that drainage on any particular day. Wolf days were assigned to multiple drainages if wolves traveled between drainages during any one day. For the typical group size analysis, WOLF$_{days}$ was calculated as the total number of wolf days in the study area during the 10-day period. We could not use the covariate WOLF$_{period}$ for this analysis because the study area encompassed drainages at different stages of wolf colonization or establishment. For all three analyses, we developed a third wolf metric, KILLS, that represented the number of wolf-killed ungulates discovered in each drainage on a given day (for methodology see Chapter 15 by Smith *et al.* and Chapter 16 by Becker *et al.*, this volume), or in the case of the typical group size analyses, the total number of wolf-killed ungulates detected in the study area over the 10-day period.

C. Statistical Analyses

We used Analysis of Variance (ANOVA) and Tukey multiple comparisons with unequal sample size (Kutner *et al.* 2005:750) to evaluate the change in mean elk group size and identify differences in the means between the three time periods defined by the predictor variable wolf presence (WOLF$_{period}$). We *ln*-transformed the data and used diagnostic residual plots to evaluate the assumption of constant variance and normality of residuals. We used 95% confidence intervals to quantify uncertainty in parameter estimates.

We used mixed-effects linear models (Pinheiro and Bates 2000) to evaluate competing *a priori* models for assessing group size and group size variation, which treated the covariates SWE$_L$, HBT, and WOLF$_{period}$ as fixed effects. We treated individual animal identity (ID) as a random effect (*i.e.*, intercept-only) because these datasets included observations of the same individual through time, and we suspected individuals tended to repeatedly choose similar size groups. This covariate allowed us to partition the variation in group size preference between individuals (intercept) and within an individual (residual), and determine how much of the variation in group size was accounted for by the fixed and random effects. We used multiple regression analyses with the covariates SWE$_L$, SEASON and WOLF$_{days}$ to evaluate competing *a priori* models for assessing typical group size because this dataset did not include repeated measures sampling. Combinations of the covariates were included in the additive form or as an interaction for all analyses. We predicted a fixed rate of change in the response variable per unit change in the predictor variable (*i.e.*, linear form). We evaluated diagnostic plots to assess the assumptions of normality and constant variance of residuals for all three response variables, and transformed the group size and typical group size response variables using the natural log to conform to linear models assumptions. We centered and scaled all continuous covariates to facilitate comparisons and interpretations of covariate coefficients. Variance inflation factors (VIF), which measures multi-colinearity among variables, were calculated for all additive and interactive combinations of predictors. Those models that included predictor combinations with VIF <6 were retained in the model list. This was a conservative approach because VIF in excess of 10 implies multi-colinearity (Kutner *et al.* 2005:409). Correlation coefficients were also calculated to further check for multi-colinearity between the predictor variables. We used Akaike's Information Criteria (AIC) to rank models given the data and compare the relative ability of each model to explain variation in the data (Burnham and Anderson 2002, 2004). Akaike model weights (w_i) were used to address model selection uncertainty, and model-averaged coefficient estimates for each covariate were computed across all models (Burnham and Anderson 2002) when no clear support for a single model was found. Covariate coefficients and variance of the random effects were estimated using restricted maximum likelihood. Comparable AIC values were calculated using maximum likelihood estimation (Pinheiro and Bates 2000). All statistical analyses were performed using the R statistical package (R Development Core Team 2006).

D. *A Priori* Model Suites and Predictions

Hypotheses representing the relationships between elk grouping behavior and covariates were expressed as suites of candidate *a priori* models for group size (6 models), group size variation (6 models), and typical group size (6 models). Every model in the group size and group size variation suites contained the covariates snow and habitat (Hirth 1977, Heard 1992, Bender and Haufler 1995), as well as the individual random effect. We also included a null model into each suite which hypothesized constant grouping behavior for the respective response variable.

We predicted the covariate estimates for the response variables group size and group size variation would be negatively correlated with the landscape snow covariate and positively correlated with open habitat types. While there is empirical evidence that group size in large herbivores is positively correlated with snow depth (Heard 1992, Jędrzejewski *et al.* 1992), we predicted the opposite would be observed for elk in the heterogeneous landscape of the Madison headwaters area. Group size is positively influenced by the size of the food patch (Berger *et al.* 1983) and if snow limits forage availability (Telfer and Kelsall 1984, Jenkins and Wright 1987) by reducing the number and size of available food patches, then the resulting small foraging areas would be unlikely to support large aggregations of elk. In addition, we predicted open habitat types (*i.e.*, meadow and thermal) would promote a foraging response and have a positive effect on group size. Social ungulates regularly form larger groups in more open areas that contain more abundant, high-quality forage than closed habitat types (Jarman 1974, Hirth 1977, Clutton-Brock *et al.* 1982, Creel and Winnie 2005, Gude *et al.* 2006). Thus, habitat type often serves as a surrogate for relative forage biomass that is available in different plant communities.

There is strong evidence that group sizes of large herbivores increase with predator density or predation pressure (Fryxell 1991, Heard 1992, Jędrzejewski *et al.* 1992, Molvar and Bowyer 1994, Mao *et al.* 2005). Thus, we predicted that elk group size would increase from pre-wolf to colonizing to established wolf periods if elk adopted the "many eyes" strategy (Pulliam 1973) in an attempt to benefit from risk dilution and better detection of predators (Bertram 1978, Dehn 1990, Lima 1995). However, groups of large herbivores have also been empirically shown to disperse in response to predation presence (Creel and Winnie 2005). Thus, a competing hypothesis predicted that a higher frequency of smaller aggregations and solitary elk would be observed as the system transitioned from pre-wolf through to the established period if elk tried to remain elusive during times of high predation risk (Creel and Winnie 2005).

We predicted the influence and strength of the response of elk group size to snow would vary by habitat type (*e.g.*, a $SWE_L \times HBT$ interaction), with more elk congregating in geothermally warmed areas as snow pack increased in other areas (Chapter 8 by Messer *et al.*, this volume). We also included models with a $WOLF_{period} \times HBT$ interaction in our *a priori* suite because anti-predator strategies may be determined by habitat characteristics (Molvar and Bowyer 1994, White and Berger 2001). If elk aggregate in open areas as an anti-predator strategy, then groups should be even larger in the presence of predators. However, if elk aggregate in open areas as a foraging response (rather than a direct response to predation; Clutton-Brock *et al.* 1982, Creel and Winnie 2005), then aggregations should become smaller in open areas and larger in closed habitat types as predation risk increases. If elk are reacting to wolf presence differently in different habitat types, then the effect of wolf predation pressure on elk group size should be intensified in particular habitat types. We predicted a larger response (either positive or negative) in thermal areas and meadows because wolf kills often occur in these habitat types (Bergman *et al.* 2006).

We used the same suite of *a priori* candidate models to evaluate the importance of specific ecological factors causing elk groups to vary in size. We predicted increasing SWE_L would result in more stable groups because snow reduces variation in food resources, while making movement between patches more difficult (Sweeney and Sweeney 1984). We also predicted there would be more frequent aggregation and dispersion of groups as elk adjusted to spatial and temporal variations in predation risk (Lima and Bednekoff 1999, Sih *et al.* 2000). For the effects of wolves on typical group size, we adopted the competing hypotheses described previously for the group size response variable. We also predicted that typical group size would be negatively correlated with snow pack, but positively correlated with SEASON because elk aggregate in larger groups during spring green up (Craighead *et al.* 1973, Ager *et al.* 2003).

Based on the outcome of the *a priori* model selection results, we conducted *post hoc* exploratory analyses to generate hypotheses for future work. We substituted $WOLF_{period}$ with $WOLF_{days}$ in the group size and group size variation analyses to evaluate if a finer-temporal scale metric indexing the frequency of encounter with wolves described grouping behavior better than the defined multi-year wolf presence covariate. We also substituted SWE_L with snow heterogeneity (SNH_L), to evaluate if this covariate made an improvement over the *a priori* model, and we hypothesized that elk would aggregate into larger groups as the snow pack became more heterogeneous. For all three of the analyses, we added the covariate KILLS to the top-ranked model to evaluate if elk responded to successful predation events by wolves. We used changes in AIC scores to evaluate if models were improved with the covariate additions and substitutions.

III. RESULTS

A. Temporal Trends in Elk Group Size

We obtained 8373 random elk group sizes from repeated sampling of a total of 115 radio-collared, adult, female elk during the winters 1991–1992 through 2006–2007. Group size ranged from 1 to 128 (mean = 7.2, se = 0.09). There was a 56% increase ($P = 0.001$) in mean group size as the system transitioned from no wolves ($n = 3103$, mean = 5.58, se = 0.17) to wolf colonization ($n = 1810$, mean = 7.55, se = 0.09) and, eventually, to established wolf packs ($n = 3460$, mean = 8.70, se = 0.17). Transformed group sizes differed between the pre-wolf and colonization ($\hat{D}_{ln(colonizing)-ln(pre-wolf)} = 0.297$; 95% CI = 0.230, 0.364; $P < 0.001$) and established wolf periods ($\hat{D}_{ln(established)-ln(pre-wolf)} = 0.299$; 95% CI = 0.243, 0.355; $P < 0.001$), but were similar during the colonization and established wolf periods ($\hat{D}_{ln(established)-ln(colonizing)} = 0.002$; 95% CI = -0.064, 0.068; $P = 0.998$). However, untransformed mean group sizes were significantly different ($\hat{D}_{established-colonizing} = 1.152$; 95% CI = 0.602, 1.701; $P < 0.001$) between the colonization and established wolf periods (Figure 19.2) (Note: $\hat{D}$ defines the difference in transformed or untransformed mean group size between two wolf periods). There was also a shift in group size distribution between the colonization and established wolf periods (Figure 19.3). Thus, the lack of a significant difference in group size on a natural log scale between the colonizing and established periods likely reflects higher variance in group sizes during the established wolf period compared to the pre-wolf and colonization periods. An advantage of log transformations is that the results can be interpreted on the original scale of the variable (Ramsey and Schafer 2002). So by taking the exponent of the difference in the mean log (Y) between two time periods we can obtain a multiplicative value for the change in the median (Y) between the time periods. Consequently the median group size for the established period was exp (exp (0.299) = 1.349 times bigger than the median group size for the pre-wolf period (95% CI = 1.275, 1.426). Similarly, the median group size during the colonizing period was

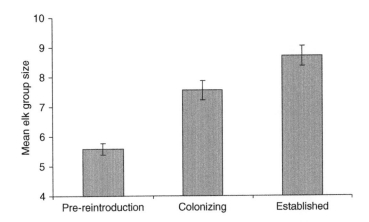

FIGURE 19.2 Changes in mean elk group size before wolf reintroduction, during wolf colonization, and after wolf establishment in the Madison headwaters area of Yellowstone National Park during the winters of 1991–1992 through 2006–2007. The plotted values are based on the untransformed data, with error bars representing standard errors.

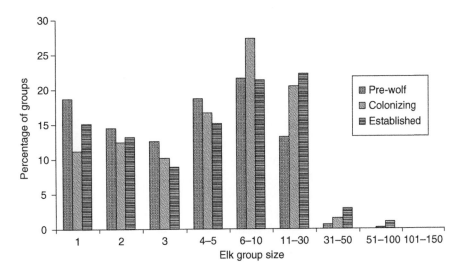

FIGURE 19.3 Changes in the distribution of elk group size (percentage of groups) before wolf reintroduction, during wolf colonization, and after wolf establishment in the Madison headwaters area of Yellowstone National Park during the winters of 1991–1992 through 2006–2007.

1.346 times bigger than the median group size for the pre-wolf period (95% CI = 1.259, 1.439), while the median group size for the established period and colonizing period were quite similar (1.000, 95% CI = 0.938, 1.073).

B. Model Selection Results

Wolves were absent from the Madison headwaters area prior to reintroduction in 1995–1996. Thereafter, the total number of wolf days increased from 55 in 1996–1997 to a peak of 3657 in 2004–2005. The Firehole drainage experienced high wolf use immediately following reintroduction, while the Madison and Gibbon drainages did not experience wolf use until several years later and never reached the magnitude of use observed in the Firehole. Pack size and number of packs also varied both spatially and temporally (Chapter 15 by Smith *et al.*, this volume).

Model selection results for the group size analysis supported one top model that received all of the model weight and contained the covariates SWE_L, HBT, $WOLF_{period}$, and a HBT $\times$ $WOLF_{period}$ interaction (Table 19.1). As predicted, snow had a negative effect on elk group size and, in the absence of wolves, larger groups were evident in meadow and thermal areas compared to other habitat types (Figure 19.4). Elk group size was positively correlated with wolf presence, but its influence was dependent on habitat type. Elk in meadows and riparian habitats were aggregated into larger groups during the colonization and established wolf periods compared to the pre-wolf period. Elk in forested habitats aggregated into larger groups during the colonization period than the pre-wolf period, but did not appear to differ in the established period. Elk groups in the thermal habitat remained similar or became slightly smaller as wolf presence increased (Table 19.2).

On the log scale, heterogeneity of group size preference between ($\hat{\sigma}_B^2$) and within ($\hat{\sigma}_W^2$) individual females was relatively high ($\hat{\sigma}_B^2 = 0.18$ and $\hat{\sigma}_W^2 = 0.70$, respectively). Individual elk did not routinely seek larger or smaller groups, and their choice of group sizes was quite variable (*i.e.*, there was considerable variation between different elk, and each elk displayed considerable variation in their

TABLE 19.1 Model selection results for elk group size, group size variation, and typical group size multiple regression analyses

Model structure	k^a	ΔAIC	w_i
ln (Group size)			
SWE_L + HBT + $WOLF_{period}$ + (HBT $\times$ $WOLF_{period}$)	18	0.00	1.00
SWE_L + HBT + $WOLF_{period}$ + ($SWE_L \times$ HBT)	14	20.16	0.00
SWE_L + HBT + $WOLF_{period}$	10	35.26	0.00
SWE_L + HBT + ($SWE_L \times$ HBT)	12	35.80	0.00
SWE_L + HBT + $WOLF_{period}$ + ($SWE_L \times WOLF_{period}$)	12	38.18	0.00
SWE_L + HBT	8	50.16	0.00
NULL	3	564.17	0.00
Group size variation			
SWE_L + HBT + $WOLF_{period}$ + ($SWE_L \times$ HBT)	14	0.00	0.98
SWE_L + HBT + $WOLF_{period}$ + ($SWE_L \times WOLF_{period}$)	12	7.71	0.02
SWE_L + HBT + $WOLF_{period}$	10	13.30	0.00
SWE_L + HBT + $WOLF_{period}$ + (HBT $\times$ $WOLF_{period}$)	18	16.50	0.00
SWE_L + HBT + ($SWE_L \times$ HBT)	12	130.00	0.00
SWE_L + HBT	8	141.00	0.00
NULL	3	353.63	0.00
ln (Typical group size)			
SEASON + $WOLF_{days}$	4	0.00	0.60
$Wolf_{days}$	3	1.59	0.27
SWE_L + $WOLF_{days}$	4	3.04	0.13
SWE_L + SEASON	4	52.16	0.00
SWE_L	3	71.61	0.00
SEASON	3	73.40	0.00
NULL	2	77.27	0.00

[a] Residual error from the mixed modeling accounts for 1 parameter value for the *ln* (group size) and group size variation analyses.

All models are ranked according to AIC and presented along with the number of parameters (*k*), the ΔAIC value (the change in AIC value relative to the best model), and the Akaike weight (w_i). The AIC value for the top model for the *ln* (group size), group size variation, and *ln* (typical group size) were 21120, 52631, and 213, respectively. Abbreviations are (group size and group size variation): SWE_L (landscape-scale snow water equivalent), HBT (habitat), $WOLF_{period}$ (Pre-reintroduction, Colonizing, and Established); (typical group size): SEASON (time period), SWE_L (average landscape-scale snow water equivalent for the 10-day period), $WOLF_{days}$ (total number of wolf days within the study area for the 10-day period).

FIGURE 19.4 Elk frequently aggregate in meadows and geothermal areas due to reduced snow accumulation and exposed vegetation that is often present within these habitat types (Photo by Jeff Henry).

TABLE 19.2 Coefficient estimates and 95% confidence limits for the factors influencing *In* (group size) and group size variation

	In (Group size)			Group size variation		
Covariate	Estimate	Lower CI	Upper CI	Estimate	Lower CI	Upper CI
Intercept	**1.631**	**1.510**	**1.752**	**4.684**	**4.215**	**5.152**
SWE_L	**−0.261**	**−0.312**	**−0.210**	**−2.388**	**−2.963**	**−1.813**
HBT-Riparian	**−0.499**	**−0.657**	**−0.341**	**−1.837**	**−2.426**	**−1.249**
HBT—BF	**−0.292**	**−0.381**	**−0.203**	**−1.440**	**−1.805**	**−1.075**
HBT—UF	**−0.326**	**−0.422**	**−0.230**	**−1.584**	**−1.982**	**−1.186**
HBT—TH	−0.040	−0.150	0.070	**−1.194**	**−1.712**	**−0.676**
$WOLF_{period}$—Col	**0.213**	**0.100**	**0.326**	**1.717**	**1.304**	**2.130**
$WOLF_{period}$—Est	**0.245**	**0.130**	**0.360**	**2.488**	**2.061**	**2.914**
Riparian × Col	0.077	−0.158	0.313			
BF × Col	−0.062	−0.192	0.069			
UF × Col	−0.132	−0.283	0.019			
TH × Col	**−0.254**	**−0.447**	**−0.060**			
Riparian × Est	−0.199	−0.400	0.001			
BF × Est	**−0.302**	**−0.418**	**−0.187**			
UF × Est	**−0.358**	**−0.489**	**−0.227**			
TH × Est	−0.125	−0.279	0.029			
SWE_L × Riparian				0.818	−0.428	2.063
SWE_L × BF				**1.486**	**0.778**	**2.194**
SWE_L × UF				**1.483**	**0.658**	**2.307**
SWE_L × TH				0.355	−0.774	1.485

All covariate levels are compared to the intercept reference level which is the meadow habitat type in the pre-reintroduction $WOLF_{period}$. Bold font denotes coefficient estimates with 95% confidence limits that do not span zero.

group size choice). *Post hoc* exploratory analyses revealed that substituting $WOLF_{days}$ for $WOLF_{period}$, and SNH_L (snow heterogeneity) for SWE_L, did not improve models. However, the addition of KILLS to the original top model improved the model by 17.96 AIC units. The estimated coefficient for KILLS was

0.339 (95% CI = 0.191, 0.489) and was a strong predictor of elk group size. Adding KILLS changed the coefficients for the best *a priori* model only slightly.

Model selection results for the group size variation analysis supported one model that included the covariates SWE_L, HBT, and $WOLF_{period}$, and a $SWE_L \times$ HBT interaction which received an Akaike weight of 0.98 (Table 19.1). Overall, models that contained the $WOLF_{period}$ covariate accounted for all of the model weights. Similar to the group size analysis, snow had a negative affect on group size variation. Variation in group size also appeared to be habitat specific and, for an average snow pack, groups were more stable in riparian, thermal, and forested habitats compared to meadows. After wolf reintroduction, grouping behavior by elk became more dynamic and as predicted, more group size variation was observed during the colonization and established wolf periods (Table 19.2). The interaction term $SWE_L \times$ HBT indicated that more variation in grouping behavior was observed under deep snow conditions in the forested habitat types. No females had a specific tendency to aggregate and disperse more frequently than others ($\hat{\sigma}_B^2 = 1.25$), and variation in grouping behavior was very inconsistent within individual elk ($\hat{\sigma}_W^2 = 30.84$). *Post hoc* exploratory analyses revealed that adding KILLS to the original top model improved the model by 19.6 AIC units. The estimated coefficient for KILLS was 2.33 (95% CI = 1.35, 3.31) and was a strong predictor of elk group size variation. Adding KILLS changed the coefficients for the best *a priori* model only slightly. Substituting $WOLF_{days}$ and SNH_L for $WOLF_{period}$ and SWE_L, respectively, did not improve models. The sizes of elk aggregations and associated variation matched the trend in wolf days among years (Figure 19.5), suggesting a response to predation risk at an intermediate temporal scale.

Typical group size was substantially larger than mean group size (Figure 19.6), which is always the case when the variance of mean group size is larger than zero (Heard 1992). Model selection results for the typical group size analysis supported two top models (Table 19.1). The most supported model included the covariates SEASON and $WOLF_{days}$ and received an Akaike weight of 0.60 ($R_{adj}^2 = 0.30$, $F_{2,217} = 48.48$, $P < 0.001$). The second-ranked model included only $WOLF_{days}$ and received an Akaike weight of 0.27 ($R_{adj}^2 = 0.29$, $F_{1,218} = 92.31$, $P < 0.001$). Models with the $WOLF_{days}$ covariate accounted for all the model weight, indicating that models incorporating the effect of wolf activity received more support from the data than the non-wolf models when explaining differences in typical group size. We found support for our predictions of larger typical group sizes as wolf activity increased and, also, that a slight increase in typical group size would occur throughout winter (Table 19.3). Adding KILLS to the top model during *post hoc* exploratory analyses provided no improvement to *a priori* models.

IV. DISCUSSION

Several investigations of the relationship between large herbivore behavior and predation risk have been conducted in the Greater Yellowstone Ecosystem, focusing on the behavior of naïve elk after wolves became established on the landscape (Laundré *et al.* 2001, Childress and Lung 2003, Wolff and Van Horn 2003). These studies documented elk behavior in accordance with general theories of prey responses to predation risk, forming larger group sizes and increasing vigilance as predation risk increased. However, empirical data documenting pre-wolf behavior was lacking and prevented comparisons between behaviors with and without a top predator, and in at least one of the previously published studies the investigators assumed low wolf presence in a drainage within our study area where we documented substantial wolf activity. We gained additional insights by incorporating empirical data on elk grouping behavior before wolf reintroduction, during the colonizing stage, and when wolf packs were resident in the study area. Elk in the Madison headwaters area demonstrated changes in aggregation patterns that coincided with the reintroduction of wolves and provided strong support for the effects of wolves on elk grouping behavior. Larger aggregations were positively correlated with wolf presence, activity, and the number of kills in a drainage each day. Groups were

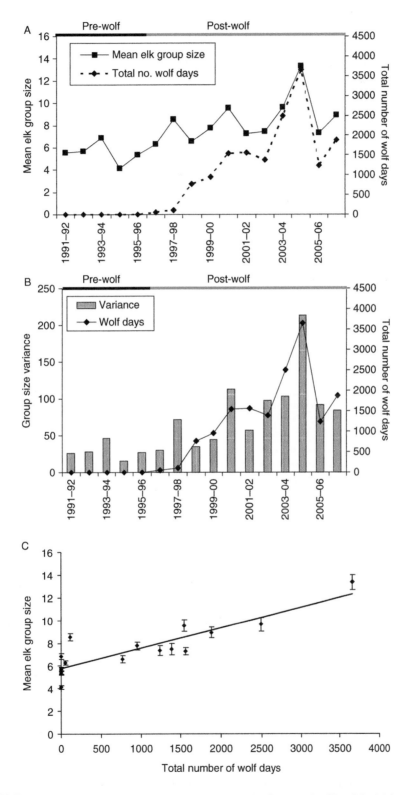

FIGURE 19.5 Changes in mean elk group size (A) or the variance in elk group size (B) and the total number of wolf days in the Madison headwaters area of Yellowstone National Park during the winters of 1991–1992 through 2006–2007. Mean annual group size showed a strong correlation with the total number of wolf days for the study area ($R_{adj}^2 = 0.75$, $F_{1,14} = 47.86$, $P < 0.001$). Mean annual group size is shown with error bars representing standard errors (C).

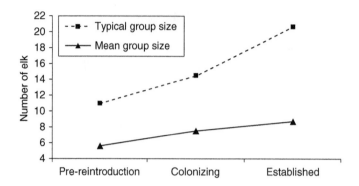

FIGURE 19.6 Temporal trends in mean elk group size and typical group size before wolf reintroduction, during wolf colonization, and after wolf establishment in the Madison headwaters area of Yellowstone National Park during the winters of 1991–1992 through 2006–2007.

TABLE 19.3	Coefficient estimates and 95% confidence limits for covariates affecting *ln* (typical group size)

	ln (Typical group size)		
Covariate	**Estimate**	**Lower CI**	**Upper CI**
SWE$_L$	−0.049	−0.181	0.082
WOLF$_{days}$	**0.517**	**0.409**	**0.625**
SEASON	0.090	−0.003	0.183

Bold font denotes coefficient estimates with 95% confidence limits that do not span zero. Coefficients for the typical group size are based on model-averaged results.

also more variable in the presence of wolves relative to the pre-wolf period, and on days that wolves had been actively hunting the drainage and successfully killed prey. Models that contained only non-wolf covariates received little support from the data when compared to the models that incorporated a wolf covariate.

These results are similar to those of other studies reporting larger aggregations of large herbivores during times of high predation risk (Heard 1992, Jędrzejewski *et al.* 1992, Molvar and Bowyer 1994, Lingle 2001, Mao *et al.* 2005). However, our findings are in sharp contrast to two other recent studies of elk behavioral responses to wolf predation risk conducted in the Greater Yellowstone Ecosystem. Creel and Winnie (2005) studied elk–wolf interactions along the western border of Yellowstone National Park (approximately 75 km north of our study area) and concluded elk formed smaller groups when wolves were in the same drainage compared to when wolves were absent. Also, Gude *et al.* (2006) reported that wolves had no detectable effect on the size of elk groups in the lower Madison Valley (40 km west of the Madison headwaters area). These equivocal results from studies in close geographic proximity suggest wolf predation may not be the only factor influencing elk group size and variation. The top model from our group size analyses contained a habitat × wolf presence interaction, by which grouping behavior in the presence of wolves was amplified in certain habitat types. However, the response was not uniform across all habitat types in the entire study area. Likewise, Kunkel and Pletscher (2000) reported that certain habitat and landscape features affected vulnerability to predation, and Bergman *et al.* (2006) found that predators are capable of selecting such vulnerable conditions, which may explain this habitat specific response. Thus, differences in habitat, snow pack, and landscape features among sites in close proximity may influence elk grouping behavior in the presence of wolves. The Madison headwaters area is a highly heterogeneous landscape with burned and unburned forest patches, snow free thermal areas, steep rocky canyons, and small open meadow complexes along major river corridors. In contrast, the Gallatin Valley is comprised of large expanses of sage scrub and grassland, with relatively continuous unburned stands of coniferous forest and some small riparian zones (Creel and Winnie 2005). The lower Madison Valley consists of expansive flat grassland and sage brush on low-elevation benches along the Madison River, with small riparian communities and west-facing slopes with coniferous forest and intermixed stands of aspen (Gude *et al.* 2006). Snow conditions also differ among the sites. Snow depths commonly exceeding 90 cm are typical for the Madison headwaters (Eberhardt *et al.* 1998), whereas in the lower Madison Valley snow may reach depths of 40 cm in wooded areas but rarely exceeds 10 cm in the grassland. Typically these large grassland/shrub areas are windblown and many areas remain snow free (Gude *et al.* 2006). The Gallatin Valley experiences intermediate snow conditions (Figure 19.7).

FIGURE 19.7 (Continued)

FIGURE 19.7 Characteristic landscape and topography of the Madison headwaters (A) Gallatin Valley (B) and lower Madison Valley (C) The grouping behavior of elk in response to wolf predation risk was studied at each of these sites which are in close geographic proximity to one another, but quite divergent results were found. The large group of elk in the lower Madison Valley photo is typical of this windswept grassland/shrub winter range that contrasts with the much smaller groups that are typical of the Madison headwaters study site (Photos by Matthew Becker—A, David Christianson—B, Justin Gude—C).

The heterogeneous nature of the landscape in the Madison headwaters area imposes numerous hard edges that can impede animal movement away from predators (Bergman *et al.* 2006). If animals occupy habitats where escape is not an effective anti-predator tactic or conditions such as deep snow make movement difficult, then forming groups rather than fleeing may be an effective alternative (Lima 1992). Elk in the Gallatin and lower Madison valleys were not as severely constrained by such landscape heterogeneity or deep snow conditions, which may explain why these elk adopted a strategy of moving instead of aggregating (Gude *et al.* 2006). The smaller aggregations of elk observed in forested areas of the Gallatin Valley when wolf predation risk was high may reduce the probability of being detected by predators (Creel and Winnie 2005) or, alternatively, may be a result of a bias due to the difficulty of accurately determining group size in dense forested environments. Regardless, landscape disparities among areas may strongly influence the behavioral responses of elk to wolf presence, and the degree to which these behaviors are manifested. As Hebblewhite and Pletscher (2002) point out, prey grouping behavior could be a fundamental component affecting the detection and hunting success by wolves. As a result, grouping behavior could potentially influence the functional response of wolves and, consequently, elk population dynamics. We have evidence from our studies that this may have occurred as indicated by the negative correlation ($R^2 = 0.66$) between mean annual elk group size and mean annual wolf kill rates on elk (Chapter 17 by Becker *et al.*, this volume). Thus, if landscape differences dictate behavioral responses, then they could also alter the strength of top-down effects by predators on their prey and therefore sociality among prey may have important implications for stability in predator–prey dynamics (Fryxell *et al.* 2007).

Elk in the Madison headwaters area also responded to the presence of wolves at short time scales, when wolves were actively hunting and killing in a given drainage. These findings suggest elk may adopt risk allocation in their grouping behavior to some degree, raising the baseline level of anti-predator

FIGURE 19.8 Elk aggregating in response to wolves in the Madison headwaters. Elk remained clumped and vigilant as a single wolf approached near Biscuit basin in the Firehole drainage (A) and were often observed fleeing into deep spots in the river when wolves were close (B). Elevated behavioral alertness and tight aggregation were maintained when a predation threat was imminent, but elk often resumed feeding just a short period of time after wolves left the immediate vicinity (Photos by Claire Gower).

behavior in the presence of predation risk and demonstrating the greatest level of anti-predator behavior during brief, but infrequent, high risk situations (Lima and Dill 1990, Lima and Bednekoff 1999, Sih *et al.* 2000, Sih and McCarthy 2002, Gude *et al.* 2006; Figure 19.8). After wolves became established in the Madison headwaters area, the moderately stable aggregations of elk observed during the pre-wolf period changed to more dynamic aggregations with frequent variations in the aggregation and dispersion of groups. Such variations in group size may be an effective anti-predator strategy because unpredictable behavior can reduce predator efficiency (Bowyer *et al.* 1999). Also, because predator–prey feedback mechanisms are governed by the way an individual (rather than a group) responds, these results may better explain evolutionary forces that shape prey and predator behavior over time.

Another explanation for dynamic grouping behavior in the presence of predators could be that animals are trying to balance competing demands. Costly anti-predator behavior is adopted during bouts of elevated predation threat and relaxed when the immediate threat of predation subsides (White and Berger 2001, Wolff and Van Horn 2003). Our data support this interpretation, particularly considering the extreme temporal and spatial variation in wolf presence exhibited within our study tract. The relatively modest variation in elk group size before wolf reintroduction may reflect elk adopting a single feeding strategy solely based on energy conservation and food acquisition. During the colonizing and established wolf periods, however, elk aggregated ephemerally when wolf predation risk was high, and then reverted to smaller aggregations similar to the pre-wolf period as the immediate risk subsided. An alternative interpretation for the higher variation observed during the colonization and established wolf periods is that the threat of predation makes the baseline group size larger, which then fragments and disperses when the group is attacked. This would also explain the increased variation observed on days when wolves had been successful in making a kill. Clearly, the response is temporally dependent and the temporal scale to which the response is observed may yield different insights.

In the absence of wolves, elk were less aggregated with increasing snow conditions. The Madison headwaters area is a mosaic of small foraging patches during winter that likely cannot support large aggregations of elk for long periods of time. Thus, the small group sizes we observed when snow pack was high likely reflect elk using smaller foraging patches with reduced snow. Deep snow also restricts elk movements and increases the effective distance between suitable feeding areas (Heard 1992), thereby inducing elk to conserve energy and remain in stable groups. Because snow pack severity has been related to the risk of predation for large herbivores in numerous studies (Mech 1970, Jędrzejewski *et al.* 1992, Huggard 1993) including our own research (Chapters 16 and 17 by Becker *et al.*, this volume), we were surprised to find no support from the data for an interaction between snow and wolf presence in our analyses.

A positive correlation between group size and population density has been reported for several large ungulate species (Clutton-Brock *et al.* 1982, Hebblewhite and Pletscher 2002, Krause and Ruxton 2002). Population estimates for elk in the Madison headwaters area indicate a decreasing trend in abundance since 1998 (Chapter 23 by Garrott *et al.*, this volume). Though we did not include population size as a covariate in any of the models due to correlations with the wolf covariates, an obvious expectation would be that decreasing population size should produce smaller aggregations of elk. This was not the case because group size increased as population size decreased. Similar findings were reported for red deer in response to human hunting (Jędrzejewski *et al.* 2006) and Gude *et al.* (2006) observed the large elk herds in the lower Madison valley fragmented as a response to human hunters, suggesting that predation in any form can have an effect on elk grouping behavior regardless of density.

Garrott *et al.* (2005) stressed that the effects of wolves on the demographics of ungulate prey populations should not be generalized between areas and, based on our findings, we extend this caution to the behavior of prey at different sites. Variations in the behavior of subpopulations in close proximity have been previously documented (Brashares and Arcese 2002), and geographic variation in behavior has been widely discussed in relation to evolutionary theory (Foster 1999, Foster and Endler 1999). Behavior is an important component of the way animals adapt to local conditions (Foster and Endler 1999), so it is not unexpected that differentiation in anti-predator behavior between locations occurs. As a result, ecologists are becoming increasingly aware that behavioral and physiological changes resulting from anti-predator constraints should be considered in predator–prey models (Beckerman *et al.* 1997, Schmitz 1997, 1998, Brown *et al.* 1999, Hebblewhite and Pletscher 2002, Denno *et al.* 2003, Nelson *et al.* 2004, Preisser *et al.* 2005, Eshel *et al.* 2006, Mchich *et al.* 2006).

While it could be argued that animals aggregate to form stable optimal groups (Giraldeau and Gillis 1985), results from this study suggest the contrary. Optimal group size is what would be expected under stable conditions, whereby every individual member's fitness is maximized. However, our study highlights that the theoretically optimal group size is unlikely to be met under most natural conditions because of temporal and spatial variation, uncertainty in prey knowledge of predation risk, and the ephemeral nature of animal groups (Sibly 1983, Pulliam and Caraco 1984). Individuals must aggregate in a manner that maximizes their individual fitness at a specific location or circumstance. They do this by individually selecting to be part of the typical group size (Jarman 1974) which could essentially be defined as a composite of individual decisions made to maximize individual fitness. Therefore, typical group size consists of an aggregation of animals where every member in the group strives for, but potentially falls short of maximizing their fitness because of variation in the environment. As such "typical group size" would appear to be a more realistic definition to describe the most advantageous group to be a part of rather than the "optimal group size." Typical group size seems to be an underused metric in most behavioral studies; however, it would appear to be a valuable and more realistic metric to use when variable environmental conditions prevail.

We documented changes in elk behavior that we interpret as responses to the reestablishment of predation risk due to the reintroduction of wolves into the Yellowstone Ecosystem. Specifically, elk have adapted to coping with predation risk from wolves by individually increasing their typical group size. With these prey aggregation responses, how wolves perceive the vulnerability of their prey must also differ, and predators must now face the challenge of adjusting their behavior accordingly. In summary, we have observed how elk respond to the newly established wolf population and the consequent risk of predation, what we now expect to see is an adjustment in wolf behavior to compensate for this prey behavioral response, and we expect that optimality and stability in prey behaviors will not occur as this continuous process of predator and prey responding to one another continues to evolve in our study system.

V. SUMMARY

1. Since wolves were reintroduced to Yellowstone in 1995–1996 a considerable amount of research has contributed to our understanding of prey behavior in response to this newly established top carnivore. We extended this work by adding a pre-wolf component which allowed us to evaluate how these behaviors have changed following reintroduction. Collectively these studies provide a comprehensive synthesis of elk grouping behavior over a wide ecological range. This allowed for possible disparities to be identified, and the mechanisms responsible could be explored.

2. We sampled 8373 randomly selected elk groups during 16 consecutive winters between 1991–1992 and 2006–2007 to compare group sizes before wolf reintroduction, during colonization, and after wolf establishment. With these data we evaluated non-wolf and wolf covariates influencing elk group size, group size variation, and typical group size.

3. We determined that group size and group size variation were negatively correlated with snow pack severity and positively correlated to the presence of wolves under certain habitat conditions. Elk altered their grouping behavior at multiple temporal scales, including among years, within winters, and daily depending on wolf presence and if a kill had occurred in a given drainage.

4. We interpreted the relatively modest variation in group size before wolf restoration as a consequence of elk adopting a single feeding strategy based solely on energy conservation and food acquisition. During the wolf colonization and establishment periods, we documented a substantial increase in group size variation that we suggest was due to animals trying to balance the conflicting demands of minimizing predation risk and maximizing food acquisition.

5. While we have documented elk aggregating into larger groups in response to wolf predation risk, studies similar to our own, and in close geographic proximity to our study site, have documented both no detectable change in elk group size and elk aggregating into smaller groups in the presence of wolf predation risk. We attribute these differences in prey behavioral responses to substantial differences in landscapes attributes such as snow pack severity and habitat types, complexity, and patch size that influence predation risk and dictate different prey behavioral responses.

6. Prey individuals can adjust their behavior rapidly to the presence or absence of predators and site-specific situations. Ultimately, how individual animals aggregate and the influence of these aggregation responses on the risk of predation, could influence predator–prey dynamics at the community level.

VI. REFERENCES

Ager, A. A., B. K. Johnson, J. W. Kern, and J. G. Kie. 2003. Daily and seasonal movements and habitat use by female Rocky Mountain elk and mule deer. *Journal of Mammalogy* **84**:1076–1088.

Allee, W. C. 1927. Animal aggregations. *The Quarterly Review of Biology* **2**:367–398.

Beckerman, A. P., M. Uriarte, and O. J. Schmitz. 1997. Experimental evidence for a behavior-mediated trophic cascade in terrestrial food chain. *Proceedings Natural Academy of Science* **94**:10735–10738.

Bender, L. C., and J. B. Haufler. 1995. Relationships between social group size of elk (*Cervus elaphus*) and habitat cover in Michigan. *American Midland Naturalist* **135**:261–265.

Berger, J., D. Daneke, J. Johnson, and S. H. Berwick. 1983. Pronghorn foraging economy and predator avoidance in a desert ecosystem: Implications for the conservation of large mammalian herbivores. *Biological Conservation* **25**:193–208.

Berger, J., J. E. Swenson, and I. -L. Persson. 2001. Recolonizing carnivores and naive prey: Conservation lessons from Pleistocene extinctions. *Science* **291**:1036–1039.

Bergman, E. J., R. A. Garrott, S. Creel, J. J. Borkowski, R. Jaffe, and F. G. R. Watson. 2006. Assessment of prey vulnerability through analysis of wolf movements and kill sites. *Ecological Applications* **16**:273–284.

Bertram, B. C. R. 1978. Living in groups: Predator and prey. Pages 64–96 *in* J. R. Krebs and N. B. Davies (Eds.) *Behavioural Ecology: An Evolutionary Approach*. Blackwell Scientific, Oxford, United Kingdom.

Borkowski, J. J., and K. Furubayashi. 1998. Seasonal and diel variation in group size among Japanese sika deer in different habitats. *Journal of Zoology* **245**:29–34.

Bowyer, T. R., V. Van Ballenberghe, J. G. Kie, and J. A. K. Maier. 1999. Birth-site selection by Alaskan moose: Maternal strategies for coping with a risky environment. *Journal of Mammalogy* **80**:1070–1083.

Brashares, J. S., and P. Arcese. 2002. Role of forage, habitat and predation in the behavioural plasticity of a small African antelope. *Journal of Animal Ecology* **71**:626–638.

Brown, J. S., J. W. Laundre, and M. Gurung. 1999. The ecology of fear: Optimal foraging, game theory, and trophic interactions. *Journal of Mammalogy* **80**:385–399.

Buckel, J. A., and A. W. Strona. 2004. Negative effects of increasing group size on foraging in two estuarine Piscivores. *Journal of Experimental Marine Biology and Ecology* **307**:183–196.

Burnham, K. P., and D. R. Anderson. 2002. *Model Selection and Multimodel Inference: A Practical Information-Theoretic Approach.* 2nd edn Springer, New York, NY.

Burnham, K. P., and D. R. Anderson. 2004. Multimodel inference—Understanding AIC and BIC in model selection. *Sociological Methods and Research* **33**:261–304.

Caro, T. M. 2005. *Antipredator Defenses in Birds and Mammals.* The University of Chicago Press, Chicago, IL.

Chapman, C. A., and L. J. Chapman. 2000. Constraints on group size in red Colobus and red-tailed Guenons: Examining the generality of the ecological constraints model. *International Journal of Primatology* **21**:565–585.

Childress, M. J., and M. A. Lung. 2003. Predation risk, gender and the group size effect: Does elk vigilance depend upon the behavior of conspecifics? *Animal Behaviour* **66**:389–398.

Clutton-Brock, T. H., F. E. Guinness, and S. D. Albon. 1982. *Red Deer: Behaviour and Ecology of Two Sexes.* University of Chicago Press, Chicago, IL.

Craighead, J. J., F. C. J. Craighead, R. L. Ruff, and B. W. O'Gara. 1973. Home ranges and activity patterns of non-migratory elk of the Madison drainage herd as determined by biotelemetry. *Wildlife Monographs* **33**:6–50.

Creel, S., and J. A. Winnie, Jr. 2005. Response of elk herd size to fine-scale and temporal variations in the risk of predation by wolves. *Animal Behaviour* **69**:1181–1189.

Dehn, M. M. 1990. Vigilance for predators: Detection and dilution effects. *Behavioral Ecology and Sociobiology* **26**:337–342.

Denno, R. F., C. Gratton, D. Hartmut, and D. L. Finke. 2003. Predation risk affects relative strength of top-down and bottom-up impacts on insect herbivory. *Ecology* **84**:1032–1044.

Eberhardt, L. L., R. A. Garrott, P. J. White, and P. J. Gogan. 1998. Alternative approaches to aerial censusing of elk. *Journal of Wildlife Management* **62**:1046–1055.

Elgar, M. A. 1989. Predator vigilance and group size in mammals and birds: A critical review of the empirical evidence. *Biological Review* **64**:13–33.

Eshel, I., E. Sansone, and A. Shaked. 2006. Gregarious behaviour of evasive prey. *Journal of Mathematical Biology* **52**:595–612.

Farnes, P., C. Heydon, and K. Hansen. 1999. *Snow Pack Distribution Across Yellowstone National Park, Wyoming.* Final report cooperative agreement number CA 1268-1-9014. Montana State University, Bozeman, MT.

Festa-Bianchet, M., and S. D. Côte. 2008. *Mountain Goats: Ecology, Behavior, and Conservation of an Alpine Ungulate.* Island Press, Washington, DC.

Foster, S. A. 1999. The geography of behavior: An evolutionary perspective. *Trends in Ecology and Evolution* **14**:190–195.

Foster, S. A., and J. A. Endler. 1999. *Geographic Variation in Behavior: Perspectives on Evolutionary Mechanisms.* Oxford University Press, New York, NY.

Fryxell, J. M. 1991. Forage quality and aggregation by large herbivores. *American Naturalist* **138**:478–498.

Fryxell, J. M. 1995. Aggregation and migration by grazing ungulates in relation to resources and predators. Pages 257–273 *in* A. R. E. Sinclair and P. Arcese (Eds.) *Serengeti II: Dynamics, Management, and Conservation of an Ecosystem.* University of Chicago Press, Chicago, IL.

Fryxell, J. M., A. Mosser, A. R. E. Sinclair, and C. Packer. 2007. Group formation stabilizes predator-prey dynamics. *Nature* **449**:1041–1044.

Garrott, R. A., J. A. Gude, E. J. Bergman, C. N. Gower, P. J. White, and K. L. Hamlin. 2005. Generalizing wolf effects across the Greater Yellowstone Area: A cautionary note. *Wildlife Society Bulletin* **33**:1245–1255.

Giraldeau, L.-A., and T. Caraco. 2000. *Social Foraging Theory.* Princeton University Press, Princeton, NJ.

Giraldeau, L.-A., and D. Gillis. 1985. Optimal group sizes can be stable: A reply to Sibly. *Animal Behaviour* **33**:666–667.

Grand, T. C., and L. M. Dill. 1999. The effects of group size on the foraging behavior of juvenile coho salmon: Reduction of predation risk or increased competition. *Animal Behaviour* **58**:443–451.

Green, R. A., and G. D. Bear. 1990. Seasonal cycles and daily activity patterns of Rocky Mountain elk. *Journal of Wildlife Management* **54**:272–279.

Gude, J. A., R. A. Garrott, J. J. Borkowski, and F. King. 2006. Prey risk allocation in a grazing environment. *Ecological Applications* **16**:285–298.

Hamilton, W. D. 1971. Geometry for the selfish herd. *Journal of Theoretical Biology* **31**:295–311.

Heard, D. C. 1992. The effect of wolf predation and snow cover on musk-ox group size. *American Naturalist* **139**:190–204.

Heard, D. C., and J. -P. Ouellet. 1994. Dynamics of an introduced caribou population. *Arctic* **47**:88–95.

Hebblewhite, M., and D. Pletscher. 2002. Effects of elk group size on predation by wolves. *Canadian Journal of Zoology* **80**:800–809.

Higashi, M., and N. Yamamura. 1993. What determines animal group size? Insider–outsider conflict and its resolution. *American Naturalist* **142**:553–563.

Hirth, D. H. 1977. Social behavior in white-tailed deer in relation to habitat. *Wildlife Monographs* **53**:1–55.

Huggard, D. J. 1993. Effects of snow depth on predation and scavenging by gray wolves. *Journal of Wildlife Management* **57**:382–388.

Jarman, P. J. 1974. Social organization of antelope in relation to their ecology. *Behaviour* **48**:215–267.

Jędrzejewski, W., B. Jędrzejewska, H. Okarma, and A. L. Ruprecht. 1992. Wolf predation and snow cover as mortality factors in the ungulate community of the Białowieza National Park, Poland. *Oecologia* **90**:27–36.

Jędrzejewski, W., H. Spaedtke, J. F. Kamler, B. Jędrzejewska, and U. Stenkewitz. 2006. Group size dynamics of red deer in Białowieza primeval forest, Poland. *Journal of Wildlife Management* **70**:1054–1059.

Jenkins, K. J., and R. G. Wright. 1987. Dietary niche relationships among cervids relative to winter snowpack in northwestern Montana. *Canadian Journal of Zoology* **65**:1397–1401.

Krause, J., and G. D. Ruxton. 2002. *Living in Groups.* Oxford University Press, Oxford, United Kingdom.

Kunkel, K. E., and D. H. Pletscher. 2000. Habitat factors affecting vulnerability of moose to predation by wolves in southeastern British Columbia. *Canadian Journal of Zoology* **78**:150–157.

Kutner, M. H., C. J. Nachtsheim, J. Neter, and W. Li. 2005. *Applied Linear Statistical Models.* 5th edn. McGraw-Hill, New York, NY.

Laundré, J. W., L. Hernández, and K. B. Altendorf. 2001. Wolves, elk, and bison: Reestablishing the "landscape of fear" in Yellowstone National Park, USA. *Canadian Journal of Zoology* **79**:1401–1409.

Lima, S. L. 1992. Strong preference for apparently dangerous habitats? A consequence of differential escape from predators. *Oikos* **64**:597–599.

Lima, S. L. 1995. Back to the basics of anti-predatory vigilance: The group size effect. *Animal Behaviour* **48**:734–736.

Lima, S. L., and P. A. Bednekoff. 1999. Temporal variations in danger drives anti predator behavior: The predation risk allocation hypothesis. *American Naturalist* **153**:650–659.

Lima, S. L., and L. M. Dill. 1990. Behavioral decisions made under the risk of predation: A review and prospectus. *Canadian Journal of Zoology* **68:**619–640.

Lingle, S. 2001. Anti-predator strategies and grouping patterns in white-tailed deer and mule deer. *Ethology* **107:**295–314.

Lingle, S. 2003. Group composition and cohesion in sympatric white-tailed deer and mule deer. *Canadian Journal of Zoology* **81:**1119–1130.

Mao, J. S. 2003. *Habitat Selection by Elk Before and After Wolf Reintroduction in Yellowstone National Park, Wyoming. Master of Science.* University of Alberta, Edmonton, Canada.

Mao, J. S., M. S. Boyce, D. W. Smith, F. J. Singer, D. J. Vales, J. M. Vore, and E. H. Merrill. 2005. Habitat selection by elk before and after wolf reintroduction in Yellowstone National Park. *Journal of Wildlife Management* **69:**1691–1707.

Mchich, R., P. Auger, and C. Lett. 2006. Effects of aggregative and solitary individual behaviors on the dynamics of predator–prey game models. *Ecological Modelling* **197:**281–289.

Mech, L. D. 1970. *The Wolf: The Ecology of an Endangered Species.* Natural History Press, Garden City, NY.

Molvar, E. M., and T. R. Bowyer. 1994. Costs and benefits of group living in a recently social ungulate: The Alaskan moose. *Journal of Mammalogy* **75:**621–630.

Nelson, E. H., C. E. Matthews, and J. A. Rosenheim. 2004. Predator reduced population growth by inducing changes in prey behavior. *Ecology* **85:**1853–1858.

Parker, K. L., C. T. Robbins, and T. A. Hanley. 1984. Energy expenditure for locomotion by mule deer and elk. *Journal of Wildlife Management* **48:**474–488.

Parrish, J. K., and L. Edelstein-Keshet. 1999. Complexity, pattern, and evolutionary trade-off's in animal aggregation. *Science* **284:**99–101.

Pinheiro, J. C., and D. M. Bates. 2000. *Mixed-Effects Models in S and S-Plus.* Springer, New York, NY.

Preisser, E. L., D. I. Bolnick, and M. F. Benard. 2005. Scared to death? The effects of intimidation and consumption in predator–prey interactions. *Ecology* **86:**501–509.

Pulliam, H. R. 1973. On the advantage of flocking. *Journal of Theoretical Biology* **38:**419–422.

Pulliam, H. R., and T. Caraco. 1984. Living in groups: Is there an optimal group size? Pages 122–147 *in* J. R. Krebs and N. B. Davies (Eds.) *Behavioural Ecology: An Evolutionary Approach.* Blackwell Scientific, Oxford, United Kingdom.

R Development Core Team 2006. R: A language and environment for statistical computing. R Foundation for Statistical Computing, Vienna, Austria. ISBN 3-900051-07-0, URL http://www.R-project.org.

Ramsey, F., and D. Schafer. 2002. *The Statistical Sleuth: A Course in Methods of Data Analysis.* 2nd edn. Duxbury Press, Pacific Grove, CA.

Roberts, G. 1996. Why individual vigilance declines as group size increases. *Animal Behavior* **51:**1077–1086.

Schmitz, O. J. 1997. Behaviorally mediated trophic cascades: Effects of predation risk on food web interactions. *Ecology* **78:**1388–1399.

Schmitz, O. J. 1998. Direct and indirect effects of predation and predation risk in old-field interaction webs. *American Naturalist* **151:**327–342.

Semeniuk, C. A. D., and L. M. Dill. 2004. Cost/benefit analysis of group and solitary resting in the cowtail stingray, *Pasinachus sephen. Behavioral Ecology* **16:**417–426.

Sibly, R. M. 1983. Optimal group size is unstable. *Animal Behaviour* **31:**947–948.

Sih, A., and T. M. McCarthy. 2002. Prey responses to pulses of risk and safety: Testing the risk allocation hypothesis. *Animal Behaviour* **63:**437–443.

Sih, A., R. Ziemba, and K. Harding. 2000. New insights on how temporal variation in predation risk shapes prey behavior. *Trends in Ecology and Evolution* **15:**3–4.

Smith, D. W. 2005. Ten Years of Yellowstone Wolves. *Yellowstone Science* **13:**7–33.

Sweeney, J. M., and J. R. Sweeney. 1984. Snow depths influencing winter movements of elk. *Journal of Mammalogy* **65:**524–526.

Telfer, E. S., and J. P. Kelsall. 1984. Adaptions of some large North American mammals for survival in snow. *Ecology* **65:**1828–1834.

Treisman, M. 1975. Predation and the evolution of gregariousness. I. Models for concealment and evasion. *Animal Behaviour* **23:**799–800.

White, G. C., and R. A. Garrott. 1990. *Analysis of Wildlife Radio-Tracking Data.* Academic Press Inc., San Diego, CA.

White, K. S., and J. Berger. 2001. Antipredator strategies of Alaskan moose: Are maternal trade-offs influenced by offspring activity? *Canadian Journal of Zoology* **79:**2055–2062.

Wilson, O. E. 1975. *Sociobiology: The New Synthesis.* Harvard University Press, Cambridge, MA.

Wolff, J. O., and T. Van Horn. 2003. Vigilance and foraging patterns of American elk during the rut in habitats with and without predators. *Canadian Journal of Zoology* **81:**266–271.

CHAPTER 20

Elk Foraging Behavior: Does Predation Risk Reduce Time for Food Acquisition?

Claire N. Gower,* Robert A. Garrott,* and P. J. White[†]

*Fish and Wildlife Management Program, Department of Ecology, Montana State University
[†]National Park Service, Yellowstone National Park

Contents

Theme

Large herbivores that inhabit seasonal environments face considerable challenges when trying to balance the conflicting demands of satisfying physiological requirements and avoiding predation. The harsh environment imposes strong nutritional constraints which influence activity patterns and foraging strategies. In addition, the need to evade predators strongly determines the way in which prey behave (Lima and Dill 1990). The traditional view is that prey manage food acquisition and predation risk by trading one behavior at the expense of the other. However, if a range of anti-predator behaviors can collectively be adopted at appropriate temporal scales, then this trade-off may be somewhat reduced (Lind and Cresswell 2005). Plasticity in prey behaviors may balance the need to forage and minimize predation risk, thereby allowing prey to remain in an environment that both provides adequate forage resources but also poses a high threat of predation. We observed the behavior of elk (*Cervus elaphus*) in the Madison headwaters area of Yellowstone National Park during 15 consecutive winter field seasons, accruing approximately 3100 h of behavioral data before and after wolf (*Canis lupus*) reintroduction to Yellowstone, when elk were exposed to varying degrees of wolf predation risk. We hypothesized that environmental and temporal factors would play a fundamental role governing the likelihood of winter foraging by elk.

The Ecology of Large Mammals in Central Yellowstone
R. Garrott, P. J. White and F. Watson
ISSN 1936-7961, DOI: 10.1016/S1936-7961(08)00220-0

Copyright © 2009, Elsevier Inc.
All rights reserved.

We proposed competing hypotheses regarding the effect of wolves. Based on the traditional view of the foraging-vigilance trade-off, we suspected that the addition of wolves to the system would manifest a higher degree of behavioral alertness in elk and reduce the likelihood of foraging. Alternatively, increasing behavioral alertness when wolves were actively hunting at night would manifest increased diurnal foraging. Furthermore, we incorporated our knowledge of the range of other behavioral responses exhibited by elk to determine if they behaviorally compensated to mediate any fitness costs associated with behaviors that reduced predation risk.

I. INTRODUCTION

Animals should make decisions regarding their foraging behavior by selecting strategies that ultimately maximize long-term fitness (Pyke *et al.* 1977, Pyke 1984, Kie 1999). For animals that are not constrained by predation this strategy would theoretically appear quite simple and decisions would be based predominately on energetic considerations (Parker *et al.* 2005, Loe *et al.* 2007; Figure 20.1). When predation risk becomes an additional selective force, an individual must attempt to minimize their vulnerability to predators while still obtaining the necessary food resources for survival. Optimization theory predicts that animals will maximize their fitness by modulating their behavior to balance energy gain and predator avoidance (Abrams 1984). The response often observed reflects an individual's nutritional state (Stephens and Krebs 1986, Lima 1998a, Winnie and Creel 2007), the relative frequency of predation risk (Lima and Bednekoff 1999a, Sih *et al.* 2000, Sih and McCarthy 2002, Laurila *et al.* 2004), and the range of possible compensatory behaviors (Lind and Cresswell 2005, Ajie *et al.* 2007, Watson *et al.* 2007).

Increased visual awareness by prey, commonly referred to as vigilance, is a universal behavioral response among many different taxa (Elgar 1989, Treves 2000). In the presence of predators, prey will increase their level of behavioral alertness to facilitate early detection of predators which may reduce

FIGURE 20.1 A calf elk forages in the harsh winter conditions of the Madison headwaters area of Yellowstone National Park. Foraging strategies before wolf reintroduction were constrained only by energetic considerations. Elk in the presence of wolves must negotiate the additional threat of predation risk (Photo by Claire Gower).

their probability of being attacked and ultimately killed (Elgar 1989, Quenette 1990, Hunter and Skinner 1998, Lima and Bednekoff 1999b, Treves 2000, Caro 2005). Probability of survival may be enhanced if early predator detection allows prey to find an effective escape route, seek protection from landscape features in close proximity that serve as refuges (Lima 1992, Berryman and Hawkins 2006), and/or congregate with conspecifics for defense (Caro 2005) and dilution of risk (Hamilton 1971, Bertram 1978, Dehn 1990). However, anti-predator responses aimed at reducing predation risk may have measurable fitness costs, particularly if the responses interrupt other fitness-enhancing activities such as feeding (Fitzgibbon and Lazarus 1995). Scanning the environment for predators may reduce nutrition which, in turn, could translate into nonlethal costs, or even indirect lethal effects if predator presence is severe enough to cause starvation in prey (Lima and Dill 1990, Lima 1998b).

The threat of predation affects a greater number of individuals than those that are killed. Therefore, there is increasing interest to more fully understand predator intimidation and quantify the nonconsumptive effects predators inflict on their prey (Nelson *et al.* 2004, Bolnick and Preisser 2005, Preisser *et al.* 2005, Creel *et al.* 2007). Experimental manipulations in a laboratory or field setting have revealed behavioral shifts in foraging as a result of predation threat (Beckerman *et al.* 1997, Schmitz 1997, 1998), but trying to empirically quantify individual-level foraging responses in large mammal systems presents a far more challenging task. In most large herbivore systems, it is notoriously hard to study the effects of predation threat on prey behavior, particularly if there is a heavy human influence that causes confounding effects. Protected areas in Africa provide conducive systems to study behavioral alertness under varying levels of risk due to the lack of any significant human disturbance and tolerance of human observers by both predators and prey (Underwood 1982, Hunter and Skinner 1998). However, few systems exist in North America where prey are tractable and/or the system has an intact predator guild and human influence is negligible. Recently, vigilance levels of elk have been compared in areas with and without the presence of wolves, or following reintroduction of wolves into Yellowstone National Park in 1995 and 1996 (Laundré *et al.* 2001, Childress and Lung 2003, Wolff and Van Horn 2003, Lung and Childress 2006, Winnie and Creel 2007, Liley and Creel 2008). These studies empirically documented noticeable changes in vigilance levels as the threat of predation risk by wolves increased. While there were differences among the studies, they demonstrated strong relationships between variation in vigilance with group size, sex and age class, and maternal state. In a few cases, a corresponding decrease in the proportion of time foraging with increasing vigilance was also observed (Laundré *et al.* 2001, Childress and Lung 2003, Wolff and Van Horn 2003, Winnie and Creel 2007). However, whether increased vigilance actually results in any detrimental consequence for the prey remains uncertain.

Vigilance is normally addressed by measuring the proportion of time prey spend scanning the environment for predators. This is often considered time that would otherwise be invested in searching for and consuming food (Quenette 1990). While this is a reasonable assumption when foraging and scanning are mutually exclusive behaviors, it is not always the case for prey that can simultaneously conduct both activities. The extent to which foraging opportunities are compromised and, thus, the cost of vigilance, consequently depends on how much visual attention is required when searching for and handling food (Studd *et al.* 1983; Lima 1988, Cowlishaw *et al.* 2003, Cresswell *et al.* 2003, Fortin *et al.* 2004a,b, Caro 2005, Figure 20.2). Therefore, it is more difficult to directly assign a cost of vigilance for prey animals that can process their food and survey their surroundings at the same time because vigilance does not necessarily imply that feeding activities are interrupted. Large herbivores often perform additional activities such as scanning their immediate surroundings, while simultaneously searching for, chewing, and physiologically processing food (Owen-Smith and Novellie 1982). Illius and Fitzgibbon (1994) theoretically proposed that this efficient use of time can enable large herbivores to essentially forage cost free, and Fortin *et al.* (2004b) empirically showed that if the two behaviors somewhat overlap, the foraging costs can be reduced.

While there have been many investigations addressing the anti-predator behavior of prey, few can draw unequivocal conclusions about the fitness consequences associated with the behavioral response

FIGURE 20.2 Scanning for predators is precluded while searching for food in deep snow, but can be conducted while chewing and digesting food (Photo by Claire Gower).

(Lind and Cresswell 2005). This is particularly true for studies of vigilance in large mammal predator–prey systems that assumed vigilance was conducted at the exclusion of foraging. While many of these studies have been able to correlate changes in vigilance behaviors with perceived variation in predation risk, assessments of potential reductions in foraging as a consequence of increased vigilance could only be suggested or inferred.

To avoid the inherent and subjective difficulty of discerning if foraging is actually interrupted while being vigilant, we focused on quantifying foraging behavior directly and correlating variation in the time devoted to foraging with varying levels of predation risk. We evaluated the likelihood of changes in foraging activity by elk associated with varying levels of perceived predation risk before, during, and after wolf colonization of the Madison headwaters area in Yellowstone National Park. Due to the tractable nature of the elk occupying this area, we were able to accurately quantify elk behavior and wolf predation risk simultaneously at relatively fine temporal and spatial scales. Working with a nonhunted, nonmigratory herd provided a rare opportunity to isolate the effects of wolf predation on elk foraging behavior where the confounding effects of human hunting were absent. We predicted that elk would respond to wolves by altering their level of behavioral alertness and modifying the time devoted to foraging activities. We also predicted that foraging behavior would be strongly influenced by environmental conditions that also vary substantially at relatively fine spatial and temporal scales.

II. METHODS

We recorded winter foraging behavior of nonmigratory elk in the Madison headwaters area over 15 consecutive winters from 1991–1992 through 2005–2006, during which time we maintained a VHF radio-collared population of 20–35 adult cow elk per year (*i.e.*, ~4–10% of the herd). We repeatedly sampled collared individuals annually from 15 November to 30 April using a stratified random sampling regime to select the focal collared cow to be monitored for behavioral data (Chapter 11 by Garrott *et al.*, this volume). This sampling design also ensured that daily variation in elk behavior was captured because sampling times were randomly distributed through the daylight hours.

Hand-held telemetry equipment and homing procedures were employed to locate the selected animal. When the focal animal was observed, we recorded a Universal Transverse Mercator (UTM) location, total group count and composition of the herd (calf, cow, yearling bull, adult bull), the predominant habitat type used through the majority of the observation (meadow, thermal, unburned forest, and burned forest), and if the focal animal was positively identified to have a calf. We then conducted a 30-min continuous behavioral observation (Altmann 1974; Figure 20.3). Focal animal watches were typically carried out from a distance of 20–300 m to avoid disturbing the animal's normal activities, with the observation bout terminated if the animal walked out of view or if its behavior was altered by an anthropogenic disturbance. Observation bouts with <15 min of data were censored from all analyses. Often observations were conducted on more than one collared elk in the same group, but to retain independence of the data we only used one observation per group in all analyses.

For each observation period, we quantified the time the focal animal spent in each of five discrete behavioral states: foraging; bedded; standing; traveling; and grooming/socializing. We included a number of specific behaviors into our classification of a foraging bout including cropping plant material, displacing snow to access forage, traveling short distances at a slow pace with lowered head, and brief periods when the animal would raise its head apparently surveying its surroundings. These ephemeral look-resume behaviors were often performed while the animal masticated plant material prior to swallowing and, thus, we considered this behavior part of a foraging bout. However, we also recorded

FIGURE 20.3 Behavioral observations on radio-collared, adult, female elk were conducted during 15 November though 30 April for 15 consecutive winters (1991–1992 through 2005–2006) to evaluate the foraging behavior of elk in the Madison headwaters of Yellowstone National Park (Photo by Thor Anderson).

how many times this behavior occurred during the observation. If the head-up posture was maintained more than momentarily, then we recorded the behavior as standing. These data were used to develop two response variables, including the proportion of time foraging and the total number of look-resume behavioral scans that an animal made during a foraging bout. The proportion of time spent foraging was the product of time spent searching for and consuming food divided by the total length of the observation. This metric provided an estimate of the proportion of the daytime activity allocated to maintaining physiological condition during winter. The total number of look-resume behavioral scans that the animal made during a foraging bout was not mutually exclusive from processing food (masticating or swallowing plant material), so it merely provided an index of the amount of time extracted from searching for and cropping plant material. The number of look-resume behavioral scans was only evaluated for 30-min behavioral observations that were exclusively composed of foraging behaviors. We intentionally avoided using the term "vigilance" because it is often associated with the assumption that other ongoing activities are interrupted (Quenette 1990) or that surveillance is directed specifically towards predator detection. Due to the uncertainty surrounding these two assumptions, we refrained from using this terminology and contend the term "behavioral scan" was more appropriate. There is the potential for an observer effect and confounding variables of other human presence during any behavioral study. However, elk in our study area exhibited habituation to humans (Chapter 26 by White *et al.*, this volume), so we had no reason to suspect this would be confounded with any wolf influence.

A. Landscape, Temporal, and Social Predictors of Elk Foraging Behavior

We considered three landscape, three temporal, and one social covariate that we suspected influenced large herbivore foraging behavior during winter. Snow is a fundamental factor limiting the availability of forage in temperate and high latitude environments. Thus, we used a validated snow pack simulation model for the central Yellowstone region (Chapter 6 by Watson *et al.*, this volume) to construct two covariates that captured important attributes of snow pack. We calculated the mean snow water equivalent (SWE, water content of snow) of all 28.5×28.5 m pixels within a 100-m radius of the elk location (SWE_A (m)) specific to the day the radio-collared elk was observed. The standard deviation of all pixels within a 100-m radius of each elk location was also calculated as a metric of snow heterogeneity across the local landscape (SNH_A). Different habitat types offer different feeding opportunities for large herbivores (Craighead *et al.* 1973, Hobbs *et al.* 1981). Thus, we defined a categorical habitat covariate (HBT) and classified each elk observation into meadow, burned forest, unburned forest, or thermal based on field observations of the plant communities the animals were predominately using at the time of the observation. Meadow habitat type was used as the reference category with which to compare foraging behavior in other habitat types.

We generated covariates at three temporal scales that literature suggested may be biologically meaningful with respect to foraging behavior. We constructed the covariate YEAR, which was a continuous variable from 1 to 15 representing the successive winters from 1991–1992 through 2005–2006. Extensive wildfires burned approximately 48% of the study area in 1988 (Chapter 2 by Newman and Watson, this volume). Thus, this metric was used to index the temporal trend in forest succession and potential changes in forage availability for large herbivores post-fire (Houston 1973, Knight and Wallace 1989, Pearson *et al.* 1995). We hypothesized that the effect of fires would only influence foraging behavior in the burned forest habitat type. Thus, we only included YEAR as a YEAR × BF interaction (BFYR). Foraging behavior is also likely influenced by seasonal variation in the quality and quantity of available plant material (Craighead *et al.* 1973, Green and Bear 1990, Ager *et al.* 2003) and changes in body condition and physiological requirements of elk. To capture this within season variability, we generated the covariate SEASON, which was a continuous variable from 1 to 167 (168 in leap years) and represented the day within the season starting from 15 November. Because large herbivores have daily behavioral cycles or rhythms (Green and Bear 1990, Ager *et al.* 2003), we developed the covariate $TIME_{day}$. This metric was a continuous variable, calculated as the number of hours and proportion of the hour between 6:00 am and the time the behavioral observation was initiated. Finally, we included the covariate GROUP to define the size of the group the focal animal occupied during the time of the observation. Foraging behavior may be affected by intra-specific competition for resources (Ranta *et al.* 1993, Fritze and De Garne-Wichatitsky 1996, Rita *et al.* 1996), attraction towards conspecifics in foraging patches (Clark and Mangel 1984, Valone 1989, Ruxton *et al.* 1995), and anti-predator behaviors such as vigilance (Elgar 1989, Kie 1999). While others have reported adult, female elk showing elevated anti-predator responses if they are accompanied by young (Hunter and Skinner 1998, Laundré *et al.* 2001, Childress and Lung 2003, Wolff and Van Horn 2003), we did not include maternal status in our analyses because it was often difficult to discern how long a female retained a calf. Also, once calves become part of the herd they must learn to avoid danger independently and there is often little assistance provided by the mother (Geist 2002). Adult, female elk in the Madison headwaters were often observed fleeing when approached by wolves, without waiting for their calves to join them. Thus, we had no reason to suspect maternal status would be an important covariate in our analyses.

B. Covariates of Predation Risk

We detected and quantified the presence of wolves from 15 November to 30 April during 1996–1997 through 2005–2006. We intensively monitored each drainage daily for wolf presence using ground-based telemetry, snow tracking, and visual observations of collared and un-collared individuals. When packs containing radio-collared wolves were detected, we estimated telemetry locations using triangulation (White and Garrott 1990) and obtained multiple locations through the day. We used snow tracking, visual observations, and counts during aerial monitoring by National Park Service biologists to provide estimates of the number of animals per pack and aid our daily assessments of wolf presence or absence (Chapter 15 by Smith *et al.* and Chapter 16 by Becker *et al.*, this volume). Detection of un-collared wolves was facilitated by opportunistic observations of tracks and wolves by field personnel that were working throughout the study area on elk and bison investigations (Chapter 11 by Garrott *et al.*, Chapter 12 by Bruggeman *et al.*, and Chapter 21 by White *et al.*, this volume). The total number of wolves present in each drainage each day was estimated based on the information obtained from these various wolf monitoring techniques.

We used these data on wolf presence to develop three wolf covariates, reflecting different temporal scales, and assess our hypotheses about the influence of predation risk on elk foraging behavior. The covariate $WOLF_{period}$ categorized the 15-year data set into three periods: before wolf reintroduction, during wolf colonization, and after wolf establishment. The $WOLF_{period}$ covariate was drainage-specific because wolves established in different drainages during different winters (Firehole: 1997–1998, Gibbon: 2000–2001, Madison: 2001–2002; Chapter 15 by Smith *et al.*, this volume). No wolves were present in the study area during the initial years of the study (1991–1992 through 1995–1996). To account for the potential transitory behavioral dynamics due to the initial naïveté of the prey (Berger *et al.* 2001), we defined a colonizing period immediately following wolf reintroduction when elk were initially exposed to wolf predation risk, but no wolf pack (≥ 2 animals) was routinely detected in the drainage. The established wolf presence period began during the first winter that a pack was consistently detected in a drainage. Wolves were wide-ranging, routinely moving among drainages and in and out of the study area (Chapter 15 by Smith *et al.* and Chapter 16 by Becker *et al.*, this volume). Thus, we also developed a dichotomous covariate, $WOLF_{presence}$, to indicate whether or not wolves were present in a drainage on a given day (0 = absent, 1 = present). To evaluate if surviving elk responded to successful hunting and killing by wolves, we developed a third wolf metric, KILLS, that represented the number of wolf-killed ungulates discovered in each drainage on a given day (Chapter 15 by Smith *et al.* and Chapter 16 by Becker *et al.*, this volume).

C. Statistical Analyses

We used Analysis of Variance (ANOVA) to evaluate the change in mean proportion of time foraging by elk during winter and Tukey multiple comparisons with unequal sample size (Kutner *et al.* 2005:750) to identify differences in the means between the pre-wolf, colonizing, and established wolf periods (WOLF$_{period}$). We transformed the response variable, the proportion of an observation bout that an elk was engaged in foraging behaviors, using the logit transformation: $\text{logit}(P) = \log_e \frac{P}{1-P}$. Because logit transformations cannot be applied to proportions of exactly zero or one, we adjusted the proportions using the following equation; $P' = \frac{F + 0.5}{N + 1}$ (Fox 1997:80), where F is the frequency of the focal category (*e.g.*, the number of minutes foraging) and N is the total number of minutes included in the observation bout. We evaluated the adequacy of the logit transformation using diagnostic residual plots to assess the assumption of constant variance. The diagnostic plots showed that assumption of normality of the residuals was not met. However, ANOVA is robust to the assumption of normality, particularly when evaluating average values and the sample size is large (Gelman and Hill 2007:46). We used 95% confidence intervals to quantify uncertainty in parameter estimates.

We complimented the ANOVA analysis that assessed differences in the mean proportion of time elk spent engaged in foraging behaviors among the three time periods with a regression analysis to gain further insight into the potential influences of wolf predation risk on foraging behavior. A large number of observations were comprised of animals either feeding through the entire observation period or not feeding at all. These long sequences of a single behavior prevented us from using a generalized linear model with the logit link and binomial error structure on the proportions directly because they violated the assumptions of independence. This lack of independence was consistent with the extremely large over-dispersion parameter ($\hat{c} = 29$). Therefore, we created a dichotomous response variable from these data by coding observation bouts where the elk spent 0–25% of the observation time in foraging behaviors as zero, and bouts where the elk spent 75–100% of the observation time engaged in foraging behaviors as one. We censored 15% of the observation bouts with intermediate foraging proportions (0.26–0.74) (Figure 20.4). Thus, each observation bout included in the analysis was classified as either a foraging or non-foraging bout which allowed us to use logistic regression (logit link, assuming binomial error structure) to estimate the log odds. These odds were then used to generate the odds ratios (exp(log odds)) which were used to define the ratio of probability of the event occurring to the probability that it does not occur (Agresti 1996), thereby providing a ratio of the likelihood of foraging under certain conditions. Odds ratios lie between zero and ∞, with values larger than one indicating higher probability of an event occurring, values less than one indicating lower probability of an event occurring, and values of one indicating equal probability of an event occurring. Log odds lie between $-\infty$ and $+\infty$, so the coefficient values from the logistic regression provided an interpretation of the strength and direction of the relationship (Agresti 1996).

We conducted a subsidiary analysis on a subset of the data (1998–1999 through 2005–2006) to evaluate if the frequency of look-resume behavioral scans was influenced by the daily presence or absence of wolf predation threat. This data set included only 30-min behavioral observations that were entirely composed of foraging and occurred during the colonizing and established wolf periods. We did not feel comfortable using the complete 15-year data set due to potential inconsistencies with how we recorded look-resume behavioral scans in the field prior to 1998–1999. The data set was based on counts (*i.e.*, total number of scans in 30 min), and made up of a large proportion of zero's (53%). While the distribution showed a large proportion of the observations at the left tail and few high counts representing the right tail, it was not Poisson distributed. The dispersion of the data was larger than expected for a Poisson distribution, but histograms of the expected versus observed showed a good fit

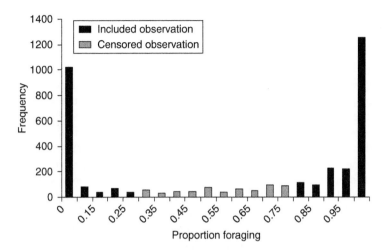

FIGURE 20.4 Frequency distribution of the proportion of an observation bout that radio-collared, adult, female elk in the Madison headwaters of Yellowstone National Park engaged in foraging behaviors. Note the bimodal distribution of these data.

to the negative binomial distribution. Therefore, we used a model assuming a negative binomial distribution and a log link (Venables and Ripley 2000) to evaluate the scanning behavior of elk during and after wolf colonization. Such a model assumes an over-dispersion compared to a Poisson distribution described by the relationship variance = mean + mean$^2/\theta$, rather than variance = mean, as would be expected with the Poisson distribution.

We developed competing hypotheses that were expressed as suites of *a priori* candidate models for both the logistic and negative binomial regression and used Akaike Information Criteria (AIC) to rank models given the data and compare the relative ability of each model to explain variation in the data (Burnham and Anderson 2002, 2004). Akaike model weights (w_i) were used to address model selection uncertainty, and evidence ratios (ratio of w_i/w_j) were used to measured the relative likelihood of model pairs (Burnham and Anderson 2002).

We repeatedly sampled the same collared elk through a winter and frequently over numerous consecutive winters. Thus, for the logistic regression analysis we calculated the over-dispersion parameter, *c*, to assess potential violations of the assumption of independence in the data (Burnham and Anderson 2002). We calculated *c* by dividing the residual deviance of the most general model by the deviance degrees of freedom (McCullagh and Nelder 1989). Because this is a biased-high estimate of the true over-dispersion (White 2002), we also estimated *c* using the Pearson's residuals (Faraway 2006). We evaluated goodness of fit for the most general model using the le Cessie's test for binary response variables (le Cessie and van Houwelingen 1991, Hosmer *et al.* 1997). This was conducted using the lrm function in the design library in R (R Development Core Team 2006). The negative binomial takes into account over-dispersion through the extra parameter θ. Therefore, we did not need to evaluate over-dispersion for this analysis. All continuous variables were centered and scaled prior to analyses to facilitate interpretation of the coefficient estimates and to alleviate problems with correlation among covariates. Variance inflation factors (VIF), which measure multi-colinearity among variables, were calculated for all combinations of predictors. Those models that included predictor combinations with VIF < 6 were retained in the model list. This was a conservative approach because VIF in excess of 10 implies multi-colinearity (Kutner *et al.* 2005:409). Correlation coefficients were also calculated to further check for multi-colinearity between the predictor variables. All statistical analyses were performed using the R statistical package (R Development Core Team 2006).

D. *A Priori* Model Suites and Predictions

To evaluate the likelihood of a foraging bout, we formulated two restricted *a priori* model suites of additive models which were based on literature and our field knowledge of the study system. One suite of 14 models represented our non-wolf hypotheses and, also, included our null model (constant likelihood of a foraging bout), while the second suite of 13 models had the same structure, but with the addition of the $WOLF_{period}$ covariate (Appendix 20A.1). Every model contained the covariate indexing mean snow pack in the vicinity of the elk, SWE_A, and the habitat covariate, HBT due to their importance in the foraging ecology of large herbivores during winter. We proposed competing hypotheses regarding the effects of snow on the probability of foraging by elk. We hypothesized that the probability of foraging would be negatively correlated with SWE_A because increasing snow pack diminishes forage availability (Jenkins and Wright 1987) and increases energetic costs associated with locomotion and displacing snow to expose forage (Fancy and White 1985). Alternatively, we predicted that elk would need to spend more time searching and displacing snow for forage when snow conditions are high. Because we are defining foraging as the product of time invested into searching and actually consuming forage, the absolute time required to obtain the baseline level of food would be amplified when elk foraged under high snow pack conditions. Feeding and other activities are often related to specific habitat types (Craighead *et al.* 1973, Collins and Urness 1983, Green and Bear 1990), with grasses and shrubs typically constituting the winter diet of elk (Kufeld 1973, Hobbs *et al.* 1981, 1983, Christianson and Creel 2007; Chapter 9 by White *et al.*, this volume). Because elk intensively use areas of high herbaceous biomass while foraging, we predicted that the probability of observing a foraging bout would not be uniform across all habitat types, but observations of elk in open habitats (meadow and thermal) would reflect a higher probability of foraging than in the less productive forested habitats. We suspected that in the initial years after the extensive 1988 wildfires high quantity and quality of forage may have been available in the burned forests (Hobbs and Spowart 1984, Pearson *et al.* 1995), but that the reestablishment of lodgepole pine saplings gradually reduced forage quantity and quality over the duration of our study. Since this potential effect would be limited to the burned forest habitat type, we considered a YEAR × BF interaction and predicted the probability of a foraging bout in burned forest habitat would gradually decrease over the period of our study.

Foraging dynamics reflect seasonal fluctuations in the quantity and quality of available forage (Hobbs *et al.* 1981, Green and Bear 1990), and changes in physiological demand (Clutton-Brock *et al.* 1989) that occur over winter. Therefore, we expressed SEASON as a quadratic function predicting that the odds of foraging would be highest during the late autumn and early spring periods of reduced snow pack and higher forage availability (Craighead *et al.* 1973, Georgii and Schröder 1983, Green and Bear 1990, Ager *et al.* 2003). We hypothesized that the odds of foraging would not be uniform across the daytime hours, but feeding bouts would coincide with sunrise and sunset and extended periods of rest would occur during the day (Georgii and Schröder 1983, Green and Bear 1990, Ager *et al.* 2003). Therefore, $TIME_{day}$ was also expressed as a quadratic function to capture crepuscular foraging activity. We also hypothesized that if larger groups are formed as a foraging response (Creel and Winnie 2005), then the odds of foraging would be positively correlated with GROUP.

Finally, we proposed competing hypotheses regarding the effects of wolf period on the likelihood of a foraging bout. If elk in the Madison headwaters cannot simultaneously forage and scan for predators (McNamara and Houston 1987), then anti-predator vigilance should carry a cost of reduced foraging time. Thus, the odds of a foraging bout would be lower during wolf colonization and establishment than the pre-wolf period. Alternatively, if a large proportion of foraging time is at night or during the crepuscular hours (Green and Bear 1990, Ager *et al.* 2003), coinciding with the most active hunting time for wolves (Mech 1970, Peterson and Ciucci 2003; Chapter 16 by Becker *et al.*, this volume), then we predicted that elk would trade-off foraging when wolves are most active and adopt a strategy of increased foraging during the daytime hours when predation risk was lower. Therefore, the odds of a daytime foraging bout would increase from the pre-reintroduction to the established wolf period. Lastly, we proposed there would be no effect of the covariate $WOLF_{period}$ on the probability of a foraging bout by elk due to digestive system constraints and the ability of large herbivores to remain vigilant while still foraging (Illius and Fitzgibbon 1994, Fortin *et al.* 2004a,b).

Based on the outcome of the *a priori* model-selection results, we conducted *post hoc* exploratory analyses to generate hypotheses for future work. In the top model for both the non-wolf and wolf suite, we replaced the mean snow water equivalent covariate, SWE_A, with the snow heterogeneity covariate, SNH_A, hypothesizing that elk would spend a greater proportion of their time foraging with increased heterogeneity of snow pack. We replaced the $WOLF_{period}$ with the $WOLF_{presence}$ covariate to evaluate if a metric indexing daily presence or absence described variation in foraging behavior better than the coarse temporal scale of $WOLF_{period}$. We also explored the possibility that elk behavioral responses to daily wolf presence differed when predation risk was a relatively novel phenomenon during wolf colonizing period compared to latter years after wolves had been established in the system. To accomplish this we included both $WOLF_{presence}$ and $WOLF_{period}$ main effects and their interaction. We added the covariate KILLS to the top wolf model to evaluate if the likelihood of foraging by surviving elk was influenced by successful predation events by wolves. Though we suspected SWE_A and HBT would contribute heavily to the best supporting model because of the importance these covariates play in the winter ecology of large herbivores, we removed them and their associated interactions from the top models to verify their importance. Similarly, we removed the other covariates from the top model to verify their significance.

To evaluate scan behavior, we formulated an *a priori* model suite with a null model (hypothesizing constant number of scans) and 23 models that included the covariates GROUP, HBT, TIME, $WOLF_{presence}$, and KILLS independently or in the additive form (Appendix 20A.2). We proposed competing hypotheses for the effect of group size on the scanning behavior of elk. We first predicted that we would observe a decrease in the frequency of scans within a 30-min foraging bout with larger groups. This would support the assumption of reduced individual vigilance (Pulliam 1973, Elgar 1989), which is suspected to be one of the principal benefits of group living. Alternatively, larger groups would facilitate more scans which were directed towards conspecifics within the group (Quenette 1990, Beauchamp 2001). We suspected that the number of scans would increase in closed habitats where visibility is impaired (Underwood 1982). We also predicted an increase in the number of scans in the morning and evening to coincide with the main hunting times for wolves. Thus, we included $TIME_{day}$ in the model in the quadratic form. Finally, we predicted that elk would increase their level of behavioral alertness on a given day that wolves were present within the same drainage as the elk (Liley and Creel 2008), or if a kill had been made within the drainage. Thus, the number of scans would be positively related to the $WOLF_{presence}$ and KILLS covariates.

III. RESULTS

We accrued 186,823 min (3114 h) of behavioral data (6428 observation bouts) through repeated sampling of a total of 108 radio-collared, adult, female elk during the winters 1991–1992 through 2005–2006. To retain independence of the data for analyses only one radio-collared animal was included per observation bout, thus reducing data to 113,806 min (1897 h) of observations accrued during 3749 independent behavioral observation bouts (mean = 249 observations annually; range = 115–553, se = 26.6; range 15–180 min per observation, mean = 30.4 min, se = 0.11). We conducted 1700 independent observation bouts prior to wolf reintroduction, 657 bouts during wolf colonization, and 1392 bouts after wolf packs were established in the area. Feeding comprised a mean proportion of 0.58 (se = 0.007) of elk activity budgets, compared to bedding (0.33; se = 0.007), standing (0.05; se = 0.002), traveling (0.03; se = 0.001), and grooming/socializing (0.01; se = 0.001). We recorded only modest variation in the proportions of each of these behaviors among the pre-reintroduction, colonizing, and established time periods defined by WOLF$_{period}$ (Figure 20.5). The proportion of time spent foraging was the only behavioral category that demonstrated a consistent trend across the three periods, with indications of a slight increase in time devoted to foraging from the pre-reintroduction to the established periods (Figure 20.6). Results of the ANOVA analysis of the logit transformed adjusted proportion of time foraging (logit (P′)), indicated no significant difference in the proportion of time spent foraging between the pre-reintroduction and wolf colonization period ($\hat{D}_{colonizing-prewolf} = 0.292$; 95% CI $= -0.056, 0641$; $P = 0.12$) or between the colonizing and established wolf periods ($\hat{D}_{established-colonizing} = 0.115$; 95% CI $= -0.244, 0.474$; $P = 0.73$). However,

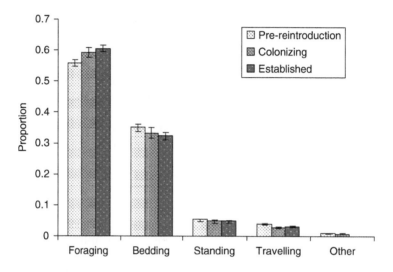

FIGURE 20.5 Variation in daytime activity budgets for elk in the Madison headwaters of Yellowstone National Park during periods before, during, and after wolf colonization. Data were collected over 15 consecutive winters (1991–1992 through 2005–2006) from 15 November to 30 April annually. Proportion represents the mean proportion of all observation time, with error bars representing standard errors.

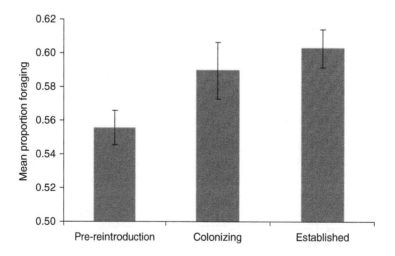

FIGURE 20.6 Changes in the mean proportion of time radio-collared, adult, female elk in the Madison headwaters of Yellowstone National Park engaged in foraging behaviors before, during, and after wolf colonization. The plotted values are based on the untransformed data with error bars representing standard errors.

a significant difference was detected between the pre-reintroduction and established wolf periods ($\hat{D}_{\text{established}-\text{prewolf}} = 0.408$; 95% CI $= 0.133, 0.682$; $P < 0.001$). The lack of a significant difference in the proportion of time foraging on a logit scale between the colonizing and the established periods, or between the colonizing and pre-wolf period likely reflects higher variance during the colonizing wolf period compared to the other two periods. $\hat{D}$ defines the difference between the logit transformed mean proportion of time foraging between two wolf periods.

A. Model Selection Results

Wolves were absent from the Madison headwaters area prior to reintroduction in 1995–1996. Thereafter, the total number of wolf days increased from 55 in 1996–1997 to a peak of 3657 in 2004–2005. The Firehole drainage experienced high wolf use immediately following reintroduction, while the Madison and Gibbon drainages did not experience wolf use until several years later and never reached the magnitude of use observed in the Firehole. Pack size and number of packs also varied both spatially and temporally (Chapter 15 by Smith *et al.*, this volume).

Conversion of the observation data to a dichotomous response variable resulted in classifying 1925 foraging observation bouts (foraging proportion 0.75–1.0), 1248 non-foraging bouts (foraging proportion 0–0.25), and 576 observation bouts with intermediate foraging proportions (0.26–0.74) that were censored. Thus, we used a total of 3173 behavioral observations of ≥ 15 min in length to evaluate our *a priori* logistic regression model suites. The residual deviance/degrees of freedom for the most general model suggested slight over-dispersion ($\hat{c} = 1.32$), but the Pearson's test indicated no evidence of over-dispersion ($\hat{c} = 1.01$). Because the deviance/degrees of freedom method yields estimates of c which are usually biased high, and because this method is as well approximated as the Pearson's method for large sample sizes (White 2002), we had no reason to suspect over-dispersion was a concern. The goodness of fit tests revealed the most general model from the wolf suite fit the data only reasonably well ($P = 0.042$), but because of the large sample size, we were comfortable that this was a satisfactory fit.

Model selection results for the non-wolf suite supported two top models that received approximately 0.85 of the model weight (Table 20.1). As predicted, the SWE_A coefficient was negative, with the odds of a foraging bout being less likely with increasing SWE_A (Table 20.2A, Figure 20.7). Also as predicted, the odds of a foraging bout in burned forest and unburned forest were 0.79 and 0.61 times, respectively, lower than the odds of a foraging bout in meadow habitats, while the odds of foraging in geothermal was not significantly different from meadow. These results suggest that open habitats rather than closed forested habitats types offered better foraging opportunities and hence, higher probability of elk foraging bouts. We found strong support for our hypothesis that elk in the Madison headwaters exhibited crepuscular foraging behavior as illustrated by the predicted temporal curves from the top-ranked model when other coefficients were fixed at their means (Figure 20.8). Also, the likelihood of a foraging bout decreased with increasing elk group size. While the seasonal and year by burned forest interaction covariates were included in top-ranked models, confidence intervals of coefficient estimates spanned zero. Thus, there was no strong evidence supporting an effect of season or forest succession in predicting the probability of a foraging bout.

Within and between model suite comparisons allowed us to evaluate if models with the wolf covariate were more supported by the data. The inclusion of the wolf period covariate improved the model performance for almost all model pairs, with evidence ratios suggesting that the top model from the wolf suite was approximately seven times more supported than the top non-wolf model. Model selection results for the wolf suite supported two top models that received approximately 0.86 of the model weight (Table 20.1). The most parsimonious models for the wolf suite consisted of SWE_A, HBT, $TIME_{day}$ and $TIME_{day}^2$, SEASON and $SEASON^2$, BFYR, GROUP, and $WOLF_{period}$, with coefficient values and confident intervals changing little for the landscape and temporal covariates between the

TABLE 20.1	Model selection results for the most-supported logistic regression models (non-wolf and wolf suites) examining the likelihood of a foraging bout by elk in the Madison headwaters area of Yellowstone National Park during 1991–1992 through 2005–2006. All models are ranked according to AIC values, and presented along with the number of parameters (k), ΔAIC value (change in AIC value relative to the best model), and the Akaike weight (w_i). AIC values for the top non-wolf model and top wolf model were 4195.03 and 4191.22, respectively. Among-suite ΔAIC values were calculated based on the difference in AIC value from the top wolf model (AIC value of 4191.22)

		Within suite		Among suite	
Model structure	k	ΔAIC	w_i	ΔAIC	w_i
Non-wolf models					
SWE_A + HBT + $TIME_{day}$ + $TIME_{day}^2$ + SEASON + $SEASON^2$ + GROUP	10	0.00	0.55	3.81	0.08
SWE_A + HBT + BFYR + $TIME_{day}$ + $TIME_{day}^2$ + SEASON + $SEASON^2$ + GROUP	11	1.23	0.30	5.05	0.04
SWE_A + HBT + BFYR + $TIME_{day}$ + $TIME_{day}^2$ + GROUP	9	3.56	0.09	7.38	0.01
SWE_A + HBT + BFYR + $TIME_{day}$ + $TIME_{day}^2$ + SEASON + $SEASON^2$	10	5.60	0.03	9.41	0.00
SWE_A + HBT + $TIME_{day}$ + $TIME_{day}^2$	7	7.03	0.02	10.84	0.00
Wolf-models					
SWE_A + HBT + $TIME_{day}$ + $TIME_{day}^2$ + SEASON + $SEASON^2$ + GROUP + $WOLF_{period}$	12	0.00	0.60	0.00	0.52
SWE_A + HBT + BFYR + $TIME_{day}$ + $TIME_{day}^2$ + SEASON + $SEASON^2$ + GROUP + $WOLF_{period}$	13	1.72	0.26	1.72	0.22
SWE_A + HBT + BFYR + $TIME_{day}$ + $TIME_{day}^2$ + GROUP + $WOLF_{period}$	11	3.24	0.12	3.24	0.10
SWE_A + HBT + BFYR + $TIME_{day}$ + $TIME_{day}^2$ + SEASON + $SEASON^2$ + $WOLF_{period}$	12	8.52	0.01	8.52	0.01
SWE_A + HBT + $TIME_{day}$ + $TIME_{day}^2$ + $WOLF_{period}$	9	8.68	0.01	8.69	0.01

SWE_A, local-scale snow water equivalent; HBT, habitat; BFYR, burned forest × year interaction; SEASON, day within the season; $TIME_{day}$, time of day; GROUP, elk group size; and $WOLF_{period}$, pre-reintroduction, colonizing, and established.

TABLE 20.2A	Coefficient estimates (log odds) and 95% confidence limits for the best supported logistic regression model from the *a priori* non-wolf model suite examining the factors affecting the probability of a foraging bout by elk in the Madison headwaters area of Yellowstone National Park during 1991–1992 through 2005–2006

Covariate	Estimate	L.CI	U.CI	Odds ratio
SWE_A	−0.099	−0.175	−0.024	0.905
HBT—BF	−0.239	−0.431	−0.048	0.787
HBT—UF	−0.499	−0.719	−0.279	0.607
HBT—TH	0.190	−0.050	0.430	1.209
GROUP	−0.095	−0.171	−0.020	0.909
$TIME_{day}$	**0.119**	**0.045**	**0.193**	**1.126**
$TIME_{day}^2$	**0.117**	**0.037**	**0.197**	**1.124**
SEASON	−0.001	−0.003	0.001	0.999
$SEASON^2$	**0.075**	**0.002**	**0.148**	**1.078**

All covariate levels are compared to the Meadow habitat type. Bold font denotes coefficient estimates with 95% confidence limits that do not include zero. Abbreviations are explained in Table 20.1.

FIGURE 20.7 Elk bedded and ruminating along the bank of the Madison River. Deep snow impedes access to forage and induces high energetic costs of searching for and moving between foraging patches, thus elk were often observed bedding down for long periods during or shortly after severe weather events, and this reduced activity is typical of animals attempting to minimize energy expenditure. Also, during times of high predation risk, such activities were often conducted in close proximity to escape terrain such as rivers which further suggest the ability of elk to meet maintenance requirements even in the presence of predation (Photo by Claire Gower).

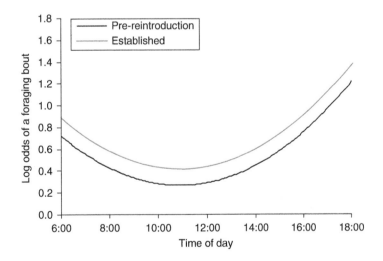

FIGURE 20.8 The estimated log odds of a foraging bout during the day for radio-collared, adult, female elk in the Madison headwaters of Yellowstone National Park before and after wolf colonization. Predicted values are based on the top wolf model (Tables 20.1 and 20.2B) for elk foraging in meadow habitat type.

non-wolf and wolf suite (Table 20.2B). The coefficients for the wolf covariate indicated that the likelihood of a foraging bout was slightly higher in the colonizing and established period compared to the pre-wolf period, thus the baseline level of the likelihood of a foraging bout was raised following reintroduction (Figure 20.8). None of the *post hoc* exploratory analyses provided improvements over

| TABLE 20.2B | Coefficient estimates (log odds) and 95% confidence limits for the best supported logistic regression model from the *a priori* wolf model suite (and best among-suite model) examining the factors affecting the probability of a foraging bout by elk in the Madison headwaters area of Yellowstone National Park during 1991–1992 through 2005–2006 |

Covariate	Estimate	L.CI	U.CI	Odds ratio
SWE_A	**−0.108**	**−0.188**	**−0.029**	**0.897**
HBT—BF	**−0.223**	**−0.415**	**−0.031**	**0.800**
HBT—UF	**−0.468**	**−0.689**	**−0.247**	**0.627**
HBT—TH	0.229	−0.012	0.471	1.258
GROUP	**−0.116**	**−0.193**	**−0.039**	**0.891**
$TIME_{day}$	**0.103**	**0.027**	**0.178**	**1.108**
$TIME_{day}^2$	**0.136**	**0.054**	**0.218**	**1.146**
SEASON	−0.040	−0.116	0.035	0.960
$SEASON^2$	0.070	−0.003	0.143	1.073
$WOLF_{PERIOD}$—COL	**0.256**	**0.044**	**0.469**	**1.292**
$WOLF_{PERIOD}$—EST	**0.196**	**0.024**	**0.369**	**1.217**

All covariate levels are compared to the Meadow habitat type, pre-reintroduction. Bold font denotes coefficient estimates with 95% confidence limits that do not include zero. Abbreviations are explained in Table 20.1.

the top *a priori* models and our decision to include both SWE_A and the HBT covariates in all *a priori* models was supported because dropping either of these covariates singly or in combination resulted in increases of 5.1–34.6 AIC units. Dropping TIME from the top model resulted in an increase of 15.3 AIC units, which further highlighted the importance of time of day in the foraging behavior of elk.

From 1998–1999 through 2005–2006 (*i.e.*, the post-wolf reintroduction period), we classified 435 independent observations of 30 min in length which were made up entirely of foraging (217.5 h of observation time). Overall, the number of behavioral scans performed by an individual while feeding was extremely low, and in the majority of the observations (53%), the individuals did not look up from foraging once ($n = 435$, mean = 1.6, se = 0.15, range = 0–23). Though our data set did not incorporate observations from the entire colonization period, we still had a sufficient sample of observations that could be compared between wolf colonization ($n = 75$) and establishment ($n = 360$). Using the negative binomial distribution, we observed a significant difference between the two time periods ($P < 0.001$) with a mean number of scans increasing from 0.47 (se = 0.15, range = 0–10) to 1.92 (se = 0.17, range = 0–23) as the system transitioned from colonizing to establishment. There was also a significant difference in the number of scans made when wolves were detected in the same drainage as the elk compared to days where there was a high probability that no wolves were using the drainage ($P = 0.03$). When there was no detection of wolves in the drainage, elk scanned the environment on average 1.4 times during a 30-min foraging bout ($n = 255$, se = 0.18, range = 0–21), with 58% of the observations not including a single scan response. Even when wolves were present in the same drainage as the collared elk, 46% of the observations showed that an animal did not actively scan the surroundings once, and the mean number of look-resume scans was only 2.05 which is still low for a 30-min foraging bout ($n = 180$, se = 0.25, range = 0–23).

Model selection results from the negative binomial regression analysis supported one top model that contained the covariates GROUP and $WOLF_{presence}$, receiving approximately 0.57 of the model weight (Table 20.3). As predicted, GROUP was negatively correlated with the behavioral scan response ($\hat{\beta}_{GROUP} = -0.234$, 95% CI = −0.416, −0.053) and there was an increase in the number of scans when wolves were present in a drainage ($\hat{\beta}_{WOLFpresence} = 0.397$; 95% CI = 0.054, 0.741). Predictor weights showed that the most important predictors were GROUP and $WOLF_{presence}$, receiving a weight of 0.95 and 0.75, respectively. The covariates HBT, $TIME_{day}$, and KILLS were not influential predictors and received predictor weights of 0.09, 0.13, and 0.07, respectively.

TABLE 20.3 Model selection results for the best-supported negative binomial linear regression models to evaluate factors affecting the scanning behavior of elk in the Madison headwaters area of Yellowstone National Park during 1998–1999 through 2005–2006

Model structure	k^*	ΔAIC	w_i
GROUP + WOLF$_{presence}$	4	0.00	0.57
GROUP	3	3.15	0.12
GROUP + TIME$_{day}$ + TIME$_{day}^2$ + WOLF$_{presence}$	6	3.87	0.08
GROUP + HBT + WOLF$_{presence}$	7	4.60	0.06
GROUP + KILL	4	5.13	0.04

* Extra parameter to account for the θ distribution.
All models are ranked according to AIC values and presented along with the number of parameters (k), ΔAIC value (change in AIC value relative to the best model), and the Akaike weight (w_i). AIC value for the top model was 1447.8. GROUP, elk group size; HBT, habitat; TIME$_{day}$, time of day; WOLF$_{presence}$, presence or absence of wolves within a drainage on a given day; and KILL, number of wolf-killed ungulates within a drainage on a given day.

IV. DISCUSSION

A prediction of foraging theory is that animals will alter their feeding behavior under the risk of predation (Lima and Dill 1990), which has been demonstrated empirically in a wide variety of predator–prey systems (Sih 1980, Edwards 1983, Brown *et al.* 1988, Kohler and McPeek 1989, Schmitz 1997, Abramsky *et al.* 2002). One of the predominant behavioral responses that is enhanced during times of high predation threat is vigilance (Lima 1987). This response has lead to the assumption that increased vigilance will decrease foraging consumption (Lima and Dill 1990, Brown 1999) which, in turn, could lead to reduced growth and decreased survival and reproduction (Lima 1998a). In recent years, there has been a substantial amount of research investigating vigilance levels of elk following the reintroduction of wolves to Yellowstone National Park (Laundré *et al.* 2001, Childress and Lung 2003, Wolff and Van Horn 2003, Lung and Childress 2006, Winnie and Creel 2007, Liley and Creel 2008). As expected, these studies documented that elk expressed a high level of awareness when wolves were present in the system. Several of these studies also demonstrated less time foraging in areas that contained wolves compared to wolf-free sites. While these studies support the majority of theoretical predator–prey models of vigilance, which assume that foraging by prey will be sacrificed at the expense of remaining behaviorally aware for predators, we present results indicating the risk of predation may not reduce time for food acquisition to any detrimental degree. Because recent literature questions the traditional foraging-vigilance trade-off, particularly in large herbivores where these two behaviors can be conducted simultaneously (Caro 2005:117), we proposed a different method that used a direct measure of foraging (*i.e.*, likelihood of a foraging bout) to provide additional insights regarding the effects predators have on large herbivore feeding strategies.

Logistic regression analysis provided strong evidence that the likelihood of a foraging bout was influenced by the presence of wolves. The direction of this response supported the *a priori* prediction that elk would manifest a strategy of increasing the frequency and length of foraging bouts in the presence of wolves. These analyses also corroborated the results of the ANOVA analysis in which the proportion of time foraging marginally increased as wolves colonized and eventually became established in the Madison headwaters. In addition, our results suggested that elk did not excessively allocate time to scanning the environment during a foraging bout when wolves were colonizing the Madison headwaters area or after they were well established in the system. In fact, there was a high prevalence of complete observation periods where elk never looked up from feeding. In accordance with our predictions, the level of scanning subtly increased when wolves were present within the drainage, but we suspect this modest increase did not inflict any foraging costs.

We observed that wolves in the Madison headwaters were most often found on a kill at dawn and spent a large proportion of the diurnal period resting, a pattern documented by others (Mech 1970, Peterson and Ciucci 2003; Chapter 16 by Becker *et al.*, this volume). This implied that the majority of hunting occurred at night or during early morning. Thus, one explanation for the increase in the likelihood of a foraging bout following wolf reintroduction could be that elk were forced to compromise feeding during the night when wolves were actively hunting, but increased forage during the diurnal period to compensate for this loss. Elk likely remained acutely aware during the main hunting period and increased the intensity or number of foraging bouts during the diurnal period when wolves were typically not hunting and, as a result, relative risk of predation was lower.

The modest increase in the likelihood of a foraging bout that was associated with increased wolf predation risk would initially imply that our results were in sharp contrast to other studies that observed increased vigilance when wolves were associated with the observed elk (Laundré *et al.* 2001, Childress and Lung 2003, Wolff and Van Horn 2003, Lung and Childress 2006, Winnie and Creel 2007, Liley and Creel 2008). However, we evaluated whether or not a foraging bout occurred during each observation. Thus, although it would appear that elk in the Madison headwaters were not trading-off foraging time for anti-predator awareness, the way we recorded "foraging" meant that short duration vigilance was incorporated into our "foraging" category. Even if vigilance levels were elevated in the presence of wolves, there would not necessarily have been any observable change in the likelihood of a foraging bout in our study. This limitation in our methodology precludes making a strong conclusion regarding whether changes in the likelihood of a foraging bout came with any nutritional costs. Similarly the inherent ability of ruminants to routinely overlap behaviors (Illius and Fitzgibbon 1994, Fortin *et al.* 2004a,b), which would allow a large herbivore to essentially continue foraging without interruption while other activities are simultaneously being carried out (Figure 20.9), highlights that using vigilance as a metric comes with its own set of limitations. Thus, it is very challenging for observational studies of this kind to provide a quantitative measure that associates predation and foraging, and this becomes more of a task when behavioral observations are being conducted at a time when the threat of predation on prey is relatively reduced (*i.e.*, daylight hours).

Our results suggest that there is no obvious reduction in diurnal foraging by elk, but we can speculate that wolves may still have induced a cost which we were unable to detect. For example, costs could have been derived if foraging and scanning did not completely overlap and consequently animals would be less effective at gaining adequate nutrition (Fortin *et al.* 2004a). Similarly, elk may have been scanning for predators through the observation while at the same time masticating forage. While they may have partially compensated by increasing the length or number of foraging bouts in a day, when wolves were in the system, if interruptions were abundant, bouts may have been less efficient and elk could have sustained a foraging cost. These costs would have most likely occurred due to a reduction of the harvest rate of forage, which was too fine a behavioral response to capture with our methodology. However, we suspect these costs were inconsequential because the minimal number of look-resume scans we recorded during foraging bouts would not have equated to any significant cost.

If elk were trading vigilance behavior at the expense of reduced foraging at night when relative risk was high, then they may have incurred costs which were only partially offset by increased foraging during the day. Heightened behavioral alertness is an obvious behavioral response when the threat of attack by a predator is imminent (Chapter 19 by Gower *et al.*, this volume; Figures19.8 and 20.9A). Loss of foraging would be expected during or immediately following a direct encounter between a predator and its prey. However, elk are in poor nutritional condition over winter (Cook 2002) and we suspect this interruption of feeding only was sustained for a short duration following a predation event. The main hunting period for wolves occurred outside the sampling period of this study, which could explain why we were unable to empirically or statistically detect any apparent change in the likelihood of a foraging bout by elk.

We also speculate that foraging costs could have been incurred by elk if they were forced into safer but poorer quality habitats in the presence of wolves. This is a commonly adopted strategy by prey in many aquatic and terrestrial systems (Sih 1980, 1982, Brown *et al.* 1988, Werner and Hall 1988,

Beckerman *et al.* 1997, Schmitz 1997, 1998, Lima 1998b). In some wolf–ungulate systems, prey have shifted habitats to reduce their level of predation risk (Creel *et al.* 2005) and sustained a decrease in diet quality as a consequence (Edwards 1983, Hernández and Laundré 2005). However, Mao *et al.* (2005) and Kauffman *et al.* (2007) did not show any significant changes in resource selection by northern Yellowstone elk due to wolves. Similarly, we have no evidence to indicate that elk were moving to nutritionally less profitable habitats or that costs were being accrued via this mechanism. On the contrary, resource selection was relatively stable before and after wolf colonization (Chapter 8 by Messer *et al.* and Chapter 21 by White *et al.*, this volume), with no definitive changes in the nutritional status of elk during this time (Chapters 9 and 22 by White *et al.*, this volume).

FIGURE 20.9 (Continued)

FIGURE 20.9 The typical definition of vigilance assumes mutually exclusive behaviors. Thus, all other behaviors are interrupted while animals respond to an external stimulus (A). However, large herbivores have the capacity to conduct multiple behaviors simultaneously. This elk calf in the Madison headwaters of Yellowstone National Park continues to process forage while remaining aware of its surroundings (B). Also, a bedded bull elk surveys the area while masticating a bolus of regurgitated forage (C). Further, an adult, female elk masticates a bolus of regurgitated forage while walking and scanning in the direction of her travel (D) (Photos A, B, and D by Kevin Pietrzak; photo C by Claire Gower).

Our results suggest that elk maintain the same level of foraging time, and retain a relatively constant level of nutrition, by adopting other behaviors to mitigate indirect predation costs. Results from this system, and other wolf–elk systems, demonstrated many responses to predation risk such as changes in aggregation patterns (Hebblewhite and Pletcher 2002, Creel and Winnie 2005; Chapter 19 by Gower *et al.*, this volume), habitat shifts (Creel *et al.* 2005), and distribution and movement (Fortin *et al.* 2005, Gude *et al.* 2006, Hebblewhite and Merrill 2007; Chapter 18 by Gower *et al.* and Chapter 21 by White *et al.*, this volume). Increased group size by elk after wolf colonization of the Madison headwaters was accompanied by a decrease in the number of behavioral scans. Thus, these additional behavioral modifications may be balancing trade-offs between resource acquisition and vigilance. If plasticity in behavioral responses allowed elk to forage efficiently in the presence of wolves, then our findings support the work of Lind and Cresswell (2005), Ajie *et al.* (2007), and Watson *et al.* (2007) who suggest that predation costs can be mitigated if prey integrate multiple behavioral responses collectively. Most studies correlate a single behavior with the apparent fitness cost of predation, but addressing collective responses simultaneously may provide a more realistic understanding of the fitness costs that predators inflict on their prey (Lind and Cresswell 2005, Ajie *et al.* 2007).

Integrating anti-predator behaviors in a multiplicative fashion, rather than considering these behaviors as additive could provide another explanation for the ability of prey to maintain adequate foraging activities in the presence of predation risk. Frid (1997) explored the possibility that vigilance will increase as group sizes decrease, but this relationship will not be as strong if prey are close to a refuge. Vigilance in Dall's sheep (*Ovis dalli dalli*) was negatively correlated with group size, but the magnitude of the response decreased as they foraged closer to steep cliffs. This implied that animals did not redundantly invest in anti-predator behavior when risk was perceived as low (*i.e.*, close to a refuge), even if they were in a small group. Alternatively, animals did not need to employ strong vigilance with increasing distance to a refuge when they were in larger groups. Following wolf reintroduction, we observed a redistribution of elk in the Madison headwaters, which we attribute to differences in vulnerability within the heterogeneous landscape of our study system (Chapter 21 by White *et al.*, this volume). If elk select safe areas on the landscape to enhance their survival (*i.e.*, refuge habitat such as rivers; Chapter 21 by White *et al.* and Chapter 24 by Garrott *et al.*, this volume), then the integration of anti-predator behaviors could effectively reduce predation risk while maintaining foraging activities in the presence of wolves. Thus, remaining close to refugia may be more profitable and less costly than sustaining vigilance for long periods of time. In the Madison headwaters, this would appear even more attractive because the majority of the meadows are adjacent to rivers and increased vigilance is likely an unnecessary adjustment if elk can continue to feed, or conduct other daily activities, and simply flee into the river upon detection of wolves (Figure 20.7 and Chapter 21 by White *et al.* this volume; Figure 21.10).

In addition to adopting behavioral compensation to mediate the effects of wolves, it appears elk adaptively manage their foraging strategy to cope with environmental and temporal conditions—both in the presence and absence of wolves. Our results supported predictions that the decision to forage was heavily influenced by depth and density of the local snow. On days when levels of local SWE_A at a specific location were high, the likelihood of foraging by elk was strongly reduced. Deep snow impedes access to forage and induces high energetic costs of searching for and moving between foraging patches (Sweeney and Sweeney 1984, Fancy and White 1985, Jenkins and Wright 1987). Thus, periods of reduced activity would be typical of animals attempting to minimize energy expenditure (Craighead *et al.* 1973; Figure 20.7). Due to the duration and the severity of the winter, however, high levels of local SWE_A persist for long periods of time and elk cannot give up feeding for extended periods. Thus, it seems elk attempted to minimize the detrimental effects of snow by selecting sites with reduced local SWE_A (Chapter 8 by Messer *et al.*, this volume). In the absence of predation, elk generally selected low elevation meadows and geothermal areas (Chapter 8 by Messer *et al.*, this volume), which coincided with the areas that a foraging bout was most likely to occur. This supported our prediction that these open habitat types were selected because they permit easier locomotion and access to relatively high quality forage than the closed habitat types (Craighead *et al.* 1973). The strength of selection for these sites increased after wolf reintroduction (Chapter 21 by White *et al.*, this volume), implying elk did not adopt the strategy of permanently moving to different habitats in the presence of predators (Mao *et al.* 2005, Kauffman *et al.* 2007). Rather, elk continued to forage in these areas at similar rates to that documented during the pre-wolf period. Even though these areas were heavily selected for by wolves (Bergman *et al.* 2006), they clearly possessed foraging attributes which would appear essential to elk during this time of nutritional hardship.

Surprisingly, we found no evidence to support the prediction that the effect of forest succession would alter foraging by elk in the burned forest. Rather the consistently low likelihood of foraging among years in this habitat type may have been because sapling regeneration curbed the benefits of foraging several years after the initiation of this study. Also, a large proportion of the burned forests were on slopes or higher elevation plateaus, offering very little opportunity to forage without substantial energetic costs. Contrary to our predictions, the probability of foraging did not appear to coincide with periods of higher quantity and quality of forage, low snow pack during autumn and spring, or times of high nutritional demand (Craighead *et al.* 1973, Georgii and Schröder 1983, Green and Bear 1990). Our results suggested that the likelihood of foraging remained constant through winter. This is not surprising in an environment like the Madison headwater where both the quantity and quality of forage biomass were lowest during winter and elk need to forage consistently through winter to avoid starvation. Time of day also played a significant role determining the likelihood of foraging by elk, with the likelihood of foraging coincided with dawn and dusk (Craighead *et al.* 1973, Green and Bear 1990, Ager *et al.* 2003). It has been suggested that these feeding patterns may be linked to maximizing energy intake rates, but could also be in response to predators (Leuthold 1977, Loe *et al.* 2007). If crepuscular foraging was linked solely to predation it would not explain why we observed crepuscular foraging pre-wolf reintroduction, though this inherent behavior evolved in the system with an intact predator guild and likely may not be lost over the short evolutionary time span since anthropogenic activities have altered large predator abundance.

Our studies provided insights into the adaptive behavioral strategies of elk to accommodate harsh environmental constraints during winter. Complementary components of this study (Chapters 18 and 19 by Gower *et al.*, this volume and Chapters 21 and 22 by White *et al.*, this volume) provide additional evidence that wolf colonization of the system may not necessarily be contributing detrimentally to these constraints. We did not detect evidence that foraging bouts were decreasing with wolves, and during the colonizing and the established periods elk did not scan the environment at a level that would contribute to any significant decrease of forage intake. The level of scanning did not substantially increase when wolves were present, and did not increase on days when a kill had been made within the drainage. Also, we had no evidence that elk selected poorer quality food in the presence of wolves (Chapter 21 by White *et al.*, this volume) or that nutrition was substantially lower when wolves occupied the system (Chapter 22 by White *et al.*, this volume). We interpret these results as evidence large herbivores have evolved to live and forage efficiently in the presence of predators. We suspect this ability to apparently minimize nutritional costs of predator detection may at least partially be due to numerous senses elk may employ for predator detection (Hudson and Haigh 2002, Mech and Peterson 2003). Sight is just one of several senses that large herbivores employ to detect predators, as their acute auditory and olfactory senses would also allow them to forage and remain aware at the same time. In addition, the ability of large herbivores to simultaneously process food and be visually aware (Illius and Fitzgibbon 1994, Fortin *et al.* 2004a,b), and the development of a complex and sophisticated range of possible compensatory behaviors exhibited by large herbivores (Chapters 18 and 19 by Gower *et al.*, this volume and Chapter 21 by White *et al.*, this volume), would provide elk with the capacity to cope with environmental constraints and lessen the effects of predation risk simultaneously.

While we did not directly observe any indication that nutritional costs are being derived via predation in the Madison headwaters, we do not doubt that the addition of wolves to central Yellowstone is an added complication to an already strained foraging strategy of elk. Contrary to other studies that have inferred quite significant reductions in foraging time due to wolves (Laundré *et al.* 2001, Childress and Lung 2003, Wolff and Van Horn 2003, Winnie and Creel 2007), we concluded that potential detriments of predation pressure on foraging and nutrition were not substantially realized in Madison headwaters elk. These unequivocal results between studies likely reflect emphasis on differing behavioral activities, the subjective nature in classifying behaviors such as vigilance, and monitoring of elk that are in different physiological states. Also, the potential for several behaviors to occur simultaneously, the range of possible compensatory behaviors, and the types of sampling designs and statistical analyses employed by various studies will contribute to these disparities. These complicating factors demonstrate that employing observational studies to assess the indirect effects of predation in wolf–ungulate systems and obtain insights about this potentially important component of predator–prey interactions is fundamentally challenging.

V. SUMMARY

1. We collected 113,806 min of independent behavioral observations from 108 individual radio-collared, adult, female elk during daytime in the Madison headwaters of Yellowstone National Park, to evaluate changes in the proportion of time elk devoted to foraging before, during, and after wolf colonization.
2. We found the likelihood of a foraging bout by elk was marginally higher in the presence of wolves. This finding may reflect that wolves actively hunted primarily during the crepuscular and nighttime periods, but were relatively inactive during the day. Thus, elk could sacrifice foraging for predator vigilance or avoidance during the high-risk nighttime period, but compensate by increasing foraging during the relatively low-risk daytime hours.
3. Elk never looked up during a high proportion of observation periods, even after wolves were established in the system. Behavioral scans by elk were short in duration and often occurred while the animals were chewing food prior to swallowing. Thus, there was little apparent reduction in foraging efficiency.
4. Elk can likely mitigate predation risk with only minimal effects on food acquisition due to their highly acute senses, ability to simultaneously scan the environment for predators and process food, and high plasticity in their behaviors.
5. In contrast to other studies, we found little evidence that vigilance during foraging significantly reduced foraging efficiency. However, the literature is conflicting because different studies emphasize different behavioral activities, classifying vigilance behavior is subjective, foraging and vigilance behavior can occur simultaneously, and prey could compensate by employing other behaviors. Thus, studies of anti-predator behaviors come with limitations when trying to assess the foraging costs that wolves inflict on their ungulate prey.

VI. REFERENCES

Abrams, P. A. 1984. Foraging time optimization and interactions in food webs. *American Naturalist* **124**:80–96.
Abramsky, Z., M. L. Rosenzweig, and A. Subach. 2002. The cost of apprehensive foraging. *Ecology* **83**:1330–1340.
Ager, A. A., B. K. Johnson, J. W. Kern, and J. G. Kie. 2003. Daily and seasonal movements and habitat use by female Rocky Mountain elk and mule deer. *Journal of Mammalogy* **84**:1076–1088.
Agresti, A. 1996. *An Introduction to Categorical Data Analysis*. Wiley, New York, NY.
Ajie, B. C., L. M. Pintor, J. Watters, J. L. Kerby, J. I. Hammond, and A. Sih. 2007. A framework for determining the fitness consequences of anti-predator behavior. *Behavioral Ecology* **18**:267–270.
Altmann, J. 1974. Observational study of behaviour: Sampling methods. *Behaviour* **49**:227–267.
Beauchamp, G. 2001. Should vigilance always decrease with group size? *Behavioral Ecology and Sociobiology* **51**:47–52.
Beckerman, A. P., M. Uriarte, and O. J. Schmitz. 1997. Experimental evidence for a behavior-mediated trophic cascade in terrestrial food chain. *Proceedings Natural Academy of Science* **94**:10735–10738.
Berger, J., J. E. Swenson, and I.-L. Persson. 2001. Recolonizing carnivores and naive prey: Conservation lessons from Pleistocene extinctions. *Science* **291**:1036–1039.
Bergman, E. J., R. A. Garrott, S. Creel, J. J. Borkowski, R. Jaffe, and F. G. R. Watson. 2006. Assessment of prey vulnerability through analysis of wolf movements and kill sites. *Ecological Applications* **16**:273–284.
Berryman, A. A., and B. A. Hawkins. 2006. The refuge as an integrating concept in ecology and evolution. *Oikos* **115**:192–196.

Bertram, B. C. R. 1978. Living in groups: Predator and prey. Pages 64–96 *in* J. R. Krebs and N. B. Davies (Eds.) *Behavioural Ecology: An Evolutionary Approach.* Blackwell, Oxford, United Kingdom.

Bolnick, D. I., and E. L. Preisser. 2005. Resource competition modifies the strength of the trait-mediated predator–prey interaction: A meta-analysis. *Ecology* **86**:2771–2779.

Brown, J. S. 1999. Vigilance, patch use and habitat selection: Foraging under predation risk. *Evolutionary Ecology Research* **1**:49–71.

Brown, J. S., B. P. Kotler, R. J. Smith, and W. O. Wirtz. 1988. The effects of owl predation on the foraging behavior of heteromyid rodents. *Oecologia* **76**:408–415.

Burnham, K. P., and D. R. Anderson. 2002. *Model Selection and Multimodel Inference: A Practical Information-Theoretic Approach.* Springer, New York, NY.

Burnham, K. P., and D. R. Anderson. 2004. Multimodel inference—Understanding AIC and BIC in model selection. *Sociological Methods and Research* **33**:261–304.

Caro, T. M. 2005. *Antipredator Defenses in Birds and Mammals.* University of Chicago Press, Chicago, IL.

le Cessie, S., and J. C. van Houwelingen. 1991. A goodness-of-fit test for binary regression models based on smoothing methods. *Biometrics* **47**:1267–1282.

Childress, M. J., and M. A. Lung. 2003. Predation risk, gender and the group size effect: Does elk vigilance depend upon the behavior of conspecifics? *Animal Behavior* **66**:389–398.

Christianson, D. A., and S. Creel. 2007. A review of environmental factors affecting elk winter diets. *Journal of Wildlife Management* **71**:164–176.

Clark, C. W., and M. Mangel. 1984. Foraging and flocking strategies: Information in an uncertain environment. *American Naturalist* **123**:626–641.

Clutton-Brock, T. H., S. D. Albon, and F. E. Guinness. 1989. Fitness costs of gestation and lactation in wild mammals. *Nature* **337**:260–262.

Collins, W. B., and P. J. Urness. 1983. Feeding behavior and habitat selection of mule deer and elk on northern Utah summer range. *Journal of Wildlife Management* **47**:646–663.

Cook, J. G. 2002. Nutrition and food. Pages 259–349 *in* D. Toweill and J. Thomas (Eds.) *North American Elk: Ecology and Management.* Stackpole Books, Harrisburg, PA.

Cowlishaw, G., M. J. Lawes, M. Lightbody, A. Martin, R. Pettifor, and J. M. Rowcliffe. 2003. A simple rule for the cost of vigilance: Empirical evidence from a social forager. *Proceedings Royal Society of London* **271**:27–33.

Craighead, J. J., F. C. J. Craighead, R. L. Ruff, and B. W. O'Gara. 1973. Home ranges and activity patterns of non-migratory elk of the Madison drainage herd as determined by biotelemetry. *Wildlife Monographs* **33**.

Creel, S., D. Christianson, S. G. Liley, and J. A. Winnie, Jr. 2007. Predation risk affects reproductive physiology and demography of elk. *Science* **315**:960.

Creel, S., and J. A. Winnie, Jr. 2005. Response of elk herd size to fine-scale and temporal variations in the risk of predation by wolves. *Animal Behavior* **69**:1181–1189.

Creel, S., J. A. Winnie, Jr, B. Maxwell, K. Hamlin, and M. Creel. 2005. Elk alter habitat selection as an antipredator response to wolves. *Ecology* **86**:3387–3397.

Cresswell, W., J. L. Quinn, M. J. Whittingham, and S. Butler. 2003. Good foragers can also be good at detecting predators. *Proceedings Royal Society of London* **270**:1069–1076.

Dehn, M. M. 1990. Vigilance for predators: Detection and dilution effects. *Behavioral Ecology and Sociobiology* **26**:337–342.

Edwards, J. 1983. Diet shifts in moose due to predator avoidance. *Oecologia* **60**:185–189.

Elgar, M. A. 1989. Predator vigilance and group size in mammals and birds: A critical review of the empirical evidence. *Biological Review* **64**:13–33.

Fancy, S. G., and R. G. White. 1985. Incremental costs of activity. Pages 143–160 *in* R. J. Hudson and R. G. White (Eds.) *Bioenergetics of Wild Herbivores.* CRC Press, Boca Raton, FL.

Faraway, J. L. 2006. *Generalized Linear, Mixed Effects and Nonparametric Regression Models.* Chapman and Hall/CRC, Boca Raton, FL.

Fitzgibbon, C. D., and J. Lazarus. 1995. Antipredator behavior of Serengeti ungulates: Individual differences and population consequences. Pages 274–296 *in* A. R. E. Sinclair and P. Arcese (Eds.) *Serengeti II: Dynamics, Management, and Conservation of an Ecosystem.* University of Chicago Press, Chicago, IL.

Fortin, D., H. L. Beyers, M. S. Boyce, D. W. Smith, T. Duchesne, and J. S. Mao. 2005. Wolves influence elk movements: Behavior shapes a trophic cascade in Yellowstone National Park. *Ecology* **86**:1320–1330.

Fortin, D., M. S. Boyce, and E. H. Merrill. 2004a. Multi-tasking by mammalian herbivores: Overlapping processes during foraging. *Ecology* **85**:2312–2322.

Fortin, D., M. S. Boyce, E. H. Merrill, and J. M. Fryxell. 2004b. Foraging costs of vigilance in large mammalian herbivores. *Oikos* **107**:172–180.

Fox, J. 1997. *Applied Regression Analysis, Linear Models, and Related Methods.* Sage Publications, Thousand Oaks, CA.

Frid, A. 1997. Vigilance by female Dall's sheep: Interactions between predation risk factors. *Animal Behavior* **53**:799–808.

Fritze, H., and M. De Garne-Wichatitsky. 1996. Foraging in social antelope: Effects of group size on foraging choices and resource perception in impala. *Journal of Animal Ecology* **65**:736–742.

Geist, V. 2002. Adaptive behavioral strategies. Pages 389–433 *in* D. E. Toweill and J. W. Thomas (Eds.) *North American Elk: Ecology and Management*. Smithsonian Institute Press, Washington, DC.

Gelman, A., and J. Hill. 2007. *Data Analysis Using Regression and Multilevel/Hierarchical Models*. Cambridge University Press, Cambridge, United Kingdom.

Georgii, B., and W. Schröder. 1983. Home range and activity patterns of male red deer (*Cervus elaphus L.*) in the Alps. *Oecologia* **58**:238–248.

Green, R. A., and G. D. Bear. 1990. Seasonal cycles and daily activity patterns of Rocky Mountain elk. *Journal of Wildlife Management* **54**:272–279.

Gude, J. A., R. A. Garrott, J. J. Borkowski, and F. King. 2006. Prey risk allocation in a grazing environment. *Ecological Applications* **16**:285–298.

Hamilton, W. D. 1971. Geometry for the selfish herd. *Journal of Theoretical Biology* **31**:295–311.

Hebblewhite, M., and E. H. Merrill. 2007. Multiscale wolf predation risk for elk: Does migration reduce risk? *Oecologia* **152**:377–387.

Hebblewhite, M., and D. Pletscher. 2002. Effects of elk group size on predation by wolves. *Canadian Journal of Zoology* **80**:800–809.

Hernández, L., and J. W. Laundré. 2005. Foraging in the "landscape of fear" and its implications for habitat use and diet quality of elk *Cervus elaphus* and bison Bison bison. *Wildlife Biology* **11**:215–220.

Hobbs, N. T., D. L. Baker, J. E. Ellis, and D. M. Swift. 1981. Composition and quality of elk winter diets on Colorado. *Journal of Wildlife Management* **45**:156–171.

Hobbs, N. T., D. L. Baker, and R. B. Gill. 1983. Comparative nutritional ecology of montane ungulates during winter. *Journal of Wildlife Management* **47**:1–16.

Hobbs, N. T., and R. A. Spowart. 1984. Effects of prescribed fire on nutrition of mountain sheep and mule deer during winter and spring. *Journal of Wildlife Management* **48**:551–560.

Hosmer, D. W., T. Hosmer, S. le Cessie, and S. Lemeshow. 1997. A comparison of goodness-of-fit tests for the logistic regression model. *Statistics in Medicine* **16**:965–980.

Houston, D. B. 1973. Wildfires in northern Yellowstone National Park. *Ecology* **54**:1111–1117.

Hudson, R. J., and J. C. Haigh. 2002. Physical and physiological adaptations. Pages 199–257 *in* D. E. Toweill and W. J. Thomas (Eds.) *North American Elk: Ecology and Management*. Smithsonian Institution Press, Washington, DC.

Hunter, L. T. B., and J. D. Skinner. 1998. Vigilance behavior in African ungulates: The role of predation pressure. *Behaviour* **135**:195–211.

Illius, A. W., and C. Fitzgibbon. 1994. Cost of vigilance in foraging ungulates. *Animal Behaviour* **47**:481–484.

Jenkins, K. J., and R. G. Wright. 1987. Dietary niche relationships among cervids relative to winter snowpack in northwestern Montana. *Canadian Journal of Zoology* **65**:1397–1401.

Kauffman, M. J., N. Varley, D. W. Smith, D. R. Stahler, D. R. MacNulty, and M. S. Boyce. 2007. Landscape heterogeneity shapes predation in a newly restored predator–prey system. *Ecology Letters* **10**:690–700.

Kie, J. G. 1999. Optimal foraging and risk of predation: Effects on behavior and social structure in ungulates. *Journal of Mammalogy* **80**:1114–1129.

Knight, D. H., and L. L. Wallace. 1989. The Yellowstone fires: Issues in landscape ecology. *BioScience* **39**:700–706.

Kohler, S. L., and M. A. McPeek. 1989. Predation risk and the foraging behavior of competing stream insects. *Ecology* **70**:1811–1825.

Kufeld, R. C. 1973. Foods eaten by the Rocky Mountain elk. *Journal of Range Management* **26**:106–112.

Kutner, M. H., C. J. Nachtsheim, J. Neter, and W. Li. 2005. *Applied Linear Statistical Models*. McGraw-Hill, New York, NY.

Laundré, J. W., L. Hernández, and K. B. Altendorf. 2001. Wolves, elk, and bison: Reestablishing the "landscape of fear" in Yellowstone National Park, USA. *Canadian Journal of Zoology* **79**:1401–1409.

Laurila, A., M. Jarvi-Laturi, S. Pakkasmaa, and J. Merila. 2004. Temporal variation in predation risk: Stage-dependency, graded responses and fitness costs in tadpole antipredator defenses. *Oikos* **107**:90–99.

Leuthold, W. 1977. African ungulates. *Zoophysiology and Ecology* **8**:1–307.

Liley, S., and S. Creel. 2008. What best explains vigilance in elk: Characteristics of prey, predator, or the environment. *Behavioral Ecology* **19**:245–254.

Lima, S. L. 1987. Vigilance while feeding and its relation to the risk of predation. *Journal of Theoretical Biology* **124**:303–316.

Lima, S. L. 1988. Vigilance and diet selection: The classical diet model revisited. *Journal of Theoretical Biology* **132**:127–143.

Lima, S. L. 1992. Strong preferences for apparently dangerous habitats? A consequence of differential escape from predators. *Oikos* **64**:597–600.

Lima, S. L. 1998a. Stress and decision making under the risk of predation: Recent developments from behavioral, reproductive and ecological perspectives. *Advances in the Study of Behavior* **27**:215–290.

Lima, S. L. 1998b. Non-lethal effects in the ecology of predator–prey interactions: What are the ecological effects of anti-predator decision making? *BioScience* **48**:25.

Lima, S. L., and P. A. Bednekoff. 1999a. Temporal variations in danger drives anti predator behavior: The predation risk allocation hypothesis. *American Naturalist* **153**:650–659.

Lima, S. L., and P. A. Bednekoff. 1999b. Back to the basics of anti-predatory vigilance: Can non-vigilant animals detect attack? *Animal Behaviour* **58**:537–543.

Lima, S. L., and L. M. Dill. 1990. Behavioral decisions made under the risk of predation: A review and prospectus. *Canadian Journal of Zoology* **68**:619–640.

Lind, J., and W. Cresswell. 2005. Determining the fitness consequences of antipredation behavior. *Behavioral Ecology* **16**:945–955.

Loe, L. E., C. Bonenfant, A. Mysterud, T. Severinsen, N. A. Øritsland, R. Langvatn, A. Stien, R. J. Irvine, and N. C. Stenseth. 2007. Activity patterns of arctic reindeer in a predator free environment: No need to keep a daily rhythm. *Oecologia* **152**:617–624.

Lung, M. A., and M. J. Childress. 2006. The influence of conspecifics and predation risk on the vigilance of elk (*Cervus elaphus*) in Yellowstone National Park. *Behavioral Ecology* **18**:12–20.

Mao, J. S., M. S. Boyce, D. W. Smith, F. J. Singer, D. J. Vales, J. M. Vore, and E. H. Merrill. 2005. Habitat selection by elk before and after wolf reintroduction in Yellowstone National Park. *Journal of Wildlife Management* **69**:1691–1707.

McCullagh, P., and J. A. Nelder. 1989. *Generalized Linear Models.* Chapman and Hall, New York, NY.

McNamara, J. M., and A. I. Houston. 1987. Starvation and predation as factors limiting population sizes. *Ecology* **68**:1515–1519.

Mech, L. D. 1970. *The Wolf: The Ecology of an Endangered Species.* Natural History Press, Garden City, New York, NY.

Mech, L. D., and R. O. Peterson. 2003. Wolf–prey relations. Pages 131–160 *in* L. D. Mech and L. Boitani (Eds.) *Wolves: Behavior, Ecology, and Conservation.* University of Chicago Press, Chicago, IL.

Nelson, E. H., C. E. Matthews, and J. A. Rosenheim. 2004. Predator reduced population growth by inducing changes in prey behavior. *Ecology* **85**:1853–1858.

Owen-Smith, N., and P. Novellie. 1982. What should a clever ungulate eat? *American Naturalist* **119**:151–178.

Parker, K. L., P. S. Barboza, and T. R. Stephenson. 2005. Protein conservation in female caribou (*Rangifer tarandus*): Effects of decreasing diet quality during winter. *Journal of Mammalogy* **86**:610–622.

Pearson, S. M., M. G. Turner, L. L. Wallace, and W. H. Romme. 1995. Winter habitat use by large ungulates following fire in northern Yellowstone National Park. *Ecological Applications* **5**:744–755.

Peterson, R. O., and P. Ciucci. 2003. The wolf as a carnivore. Pages 104–130 *in* L. D. Mech and L. Boitani (Eds.) *Wolves: Behavior, Ecology, and Conservation.* University of Chicago Press, Chicago, IL.

Preisser, E. L., D. I. Bolnick, and M. F. Benard. 2005. Scared to death? The effects of intimidation and consumption in predator–prey interactions. *Ecology* **86**:501–509.

Pulliam, H. R. 1973. On the advantage of flocking. *Journal of Theoretical Biology* **38**:419–422.

Pyke, G. H. 1984. Optimal foraging theory: A critical review. *Annual Review of Ecology and Systematics* **15**:523–575.

Pyke, G. H., H. R. Pulliam, and E. L. Charnov. 1977. Optimal foraging: A selective review of theory and tests. *The Quarterly Review of Biology* **52**:137–154.

Quenette, P.-Y. 1990. Functions of vigilance behaviour in mammals: A review. *Acta Oecologica* **11**:801–818.

Ranta, E., R. Hannu, and K. Lindstrom. 1993. Competition versus cooperation: Success of individuals foraging alone or in groups. *American Naturalist* **142**:42–58.

R Development Core Team. 2006. R: A language and environment for statistical computing. R Foundation for Statistical Computing, Vienna, Austria, ISBN 3-900051-07-0. URLhttp://www.R-project.org.

Rita, H., E. Ranta, and N. Peuhkuri. 1996. Competition in foraging groups. *Oikos* **76**:583–586.

Ruxton, G. D., S. J. Hall, and S. C. Gurney. 1995. Attraction towards feeding conspecifics when food patches are exhaustible. *American Naturalist* **145**:653–660.

Schmitz, O. J. 1997. Behaviorally mediated trophic cascades: Effects of predation risk on food web interactions. *Ecology* **78**:1388–1399.

Schmitz, O. J. 1998. Direct and indirect effects of predation and predation risk in old-field interaction webs. *American Naturalist* **151**:327–342.

Sih, A. 1980. Optimal behavior: Can foragers balance two conflicting demands? *Science* **210**:1041–1043.

Sih, A. 1982. Foraging strategies and the avoidance of predation by an aquatic insect *Notonecta hoffmanni*. *Ecology* **63**:786–796.

Sih, A., and T. M. McCarthy. 2002. Prey responses to pulses of risk and safety: Testing the risk allocation hypothesis. *Animal Behaviour* **63**:437–443.

Sih, A., R. Ziemba, and K. Harding. 2000. New insights on how temporal variation in predation risk shapes prey behavior. *Trends in Ecology and Evolution* **15**:3–4.

Stephens, D. W., and J. R. Krebs. 1986. *Foraging Theory.* Princeton University Press, Princeton, NJ.

Studd, M., R. D. Montgomerie, and R. J. Robertson. 1983. Group size and predator surveillance in foraging house sparrows (*Passer domesticus*). *Canadian Journal of Zoology* **61**:226–231.

Sweeney, J. M., and J. R. Sweeney. 1984. Snow depths influencing winter movements of elk. *Journal of Mammalogy* **65**:524–526.

Treves, A. 2000. Theory and method in studies of vigilance and aggregation. *Animal Behaviour* **60**:711–722.

Underwood, R. 1982. Vigilance behaviour in grazing African antelopes. *Behaviour* **79**:81–107.

Valone, T. J. 1989. Group foraging, public information, and patch estimation. *Oikos* **56**:357–363.

Venables, W. N., and B. D. Ripley. 2000. *Modern Applied Statistics with S.* Springer, New York, NY.

Watson, M., N. J. Aebischer, and W. Cresswell. 2007. Vigilance and fitness in grey partridges *Perdix perdix*: The effects of group size and foraging-vigilance trade-offs on predation mortality. *Journal of Animal Ecology* **76**:211–221.

Werner, E. E., and D. J. Hall. 1988. Ontogenetic habitat shifts in bluegill: The foraging rate-predation risk trade off. *Ecology* **69**:1352–1366.

White, G. C. 2002. Discussion comments on: The auxiliary variables in capture–recapture modeling. An overview. *Journal of Applied Statistics* **29**:103–106.

White, G. C., and R. A. Garrott. 1990. *Analysis of Wildlife Radio-Tracking Data.* Academic Press, San Diego, CA.

Winnie, J. A. Jr., and S. Creel. 2007. Sex-specific behavioural responses of elk to spatial and temporal variation in the threat of wolf predation. *Animal Behaviour* **73**:215–225.

Wolff, J. O., and T. Van Horn. 2003. Vigilance and foraging patterns of American elk during the rut in habitats with and without predators. *Canadian Journal of Zoology* **81**:266–271.

APPENDIX

APPENDIX 20A.1	Complete Model Suite for the Non-Wolf and Wolf Models Used in the Logistic Regression Analysis to Evaluate the Factors Influencing Elk Foraging Behavior in the Madison Headwaters Area of Yellowstone National Park During 1991–1992 Through 2005–2006

Model number[*]	Model structure—non-wolf models
0	NULL
1	$SWE_A + HBT$
2	$SWE_A + HBT + BFYR$
3	$SWE_A + HBT + SEASON + SEASON^2$
4	$SWE_A + HBT + TIME_{day} + TIME_{day}^2$
5	$SWE_A + HBT + GROUP$
6	$SWE_A + HBT + BFYR + SEASON + SEASON^2$
7	$SWE_A + HBT + BFYR + TIME_{day} + TIME_{day}^2$
8	$SWE_A + HBT + BFYR + GROUP$
9	$SWE_A + HBT + BFYR + TIME_{day} + TIME_{day}^2 + SEASON + SEASON^2$
10	$SWE_A + HBT + BFYR + TIME_{day} + TIME_{day}^2 + GROUP$
11	$SWE_A + HBT + BFYR + SEASON + SEASON^2 + GROUP$
12	$SWE_A + HBT + BFYR + TIME_{day} + TIME_{day}^2 + SEASON + SEASON^2 + GROUP$
13	$SWE_A + HBT + TIME_{day} + TIME_{day}^2 + SEASON + SEASON^2 + GROUP$

Model structure—wolf models

14 (1)	$SWE_A + HBT + WOLF_{period}$
15 (2)	$SWE_A + HBT + BFYR + WOLF_{period}$
16 (3)	$SWE_A + HBT + SEASON + SEASON^2 + WOLF_{period}$
17 (4)	$SWE_A + HBT + TIME_{day} + TIME_{day}^2 + WOLF_{period}$
18 (5)	$SWE_A + HBT + GROUP + WOLF_{period}$
19 (6)	$SWE_A + HBT + BFYR + SEASON + SEASON^2 + WOLF_{period}$
20 (7)	$SWE_A + HBT + BFYR + TIME_{day} + TIME_{day}^2 + WOLF_{period}$
21 (8)	$SWE_A + HBT + BFYR + GROUP + WOLF_{period}$
22 (9)	$SWE_A + HBT + BFYR + TIME_{day} + TIME_{day}^2 + SEASON + SEASON^2 + WOLF_{period}$
23 (10)	$SWE_A + HBT + BFYR + TIME_{day} + TIME_{day}^2 + GROUP + WOLF_{period}$
24 (11)	$SWE_A + HBT + BFYR + SEASON + SEASON^2 + GROUP + WOLF_{period}$
25 (12)	$SWE_A + HBT + BFYR + TIME_{day} + TIME_{day}^2 + SEASON + SEASON^2 + GROUP + WOLF_{period}$
26 (13)	$SWE_A + HBT + TIME_{day} + TIME_{day}^2 + SEASON + SEASON^2 + GROUP + WOLF_{period}$

[*] Model number in parentheses represents the associated non-wolf model pair.
SWE_A, local-scale snow water equivalent; HBT, habitat; BFYR, burned forest × year interaction; SEASON, day within the season; $TIME_{day}$, time of day; GROUP, elk group size; and $WOLF_{period}$, pre-reintroduction, colonizing, and established.

APPENDIX 20A.2	Complete Model Suite for the Models Used in the Negative Binomial Regression Analysis to Evaluate the Factors Influencing the Scanning Behavior of Elk in the Madison Headwaters Area of Yellowstone National Park 1998–1999 Through 2005–2006

Model number	Model structure
0	NULL
1	GROUP
2	HBT
3	$TIME_{day} + TIME_{day}^2$
4	$WOLF_{presence}$
5	KILL
6	GROUP + HBT
7	$GROUP + TIME_{day} + TIME_{day}^2$
8	$GROUP + WOLF_{presence}$
9	GROUP + KILL
10	$HBT + TIME_{day} + TIME_{day}^2$
11	$HBT + WOLF_{presence}$
12	HBT + KILL
13	$TIME_{day} + TIME_{day}^2 + WOLF_{presence}$
14	$TIME_{day} + TIME_{day}^2 + KILL$
15	$GROUP + HBT + TIME_{day}^2 + TIME_{day}^2$
16	$GROUP + HBT + WOLF_{presence}$
17	GROUP + HBT + KILL
18	$HBT + TIME_{day} + TIME_{day}^2 + WOLF_{presence}$
19	$HBT + TIME_{day} + TIME_{day}^2 + KILL$
20	$GROUP + TIME_{day} + TIME_{day}^2 + WOLF_{presence}$
21	$GROUP + TIME_{day} + TIME_{day}^2 + KILL$
22	$GROUP + HBT + TIME_{day} + TIME_{day}^2 + WOLF_{presence}$
23	$GROUP + HBT + TIME_{day} + TIME_{day}^2 + KILL$

GROUP, elk group size; HBT, habitat; $TIME_{day}$, time of day; $WOLF_{presence}$, presence or absence of wolves within a drainage on a given day; and KILL, number of wolf-killed ungulates within a drainage on a given day.

CHAPTER 21

Changes in Elk Resource Selection and Distribution with the Reestablishment of Wolf Predation Risk

P. J. White,* Robert A. Garrott,[†] Steve Cherry,[‡] Fred G. R. Watson,[§]
Claire N. Gower,[†] Matthew S. Becker,[†] and Eric Meredith[‡]

*National Park Service, Yellowstone National Park
[†]Fish and Wildlife Management Program, Department of Ecology, Montana State University
[‡]Department of Mathematical Sciences, Montana State University
[§]Division of Science and Environmental Policy, California State University Monterey Bay

Contents

Theme

How animals use resources at local and landscape scales can differ depending on whether starvation is the primary limiting factor or significant predation risk also exists. With predation, individuals face the added and often conflicting demands of obtaining adequate nutrition while reducing predation risk. However, disentangling differences between the respective resource selection strategies in a given system is rarely possible. We evaluated changes in resource selection and prey distribution for a non-migratory elk population formerly limited by winter starvation and now subjected to wolf predation in the Madison headwaters area of Yellowstone National Park. Analyses of elk resource selection prior to wolf establishment (1991–1992 through 1997–1998) demonstrated a dynamic spatial and temporal response to changes in snow pack at the local and landscape level, with elk selecting areas with low snow mass, but forced to occupy areas with higher local snow mass as conditions on the winter range became more severe (Chapter 8 by Messer et al., this volume). We hypothesized that variables affecting elk resource selection prior to wolves would retain their primacy in post-wolf models (1998–1999 through 2005–2006), but that the magnitude of each would change depending on its contribution to predation risk. Broad-scale changes in prey distribution could also occur if the collective attributes of a landscape serve to alternately provide areas with decreased or increased predation risk. Thus, we also evaluated the potential for landscape-scale shifts in the distribution of the elk population among the three drainages of the

The Ecology of Large Mammals in Central Yellowstone
R. Garrott, P. J. White and F. Watson
ISSN 1936-7961, DOI: 10.1016/S1936-7961(08)00221-2

Copyright © 2009, Elsevier Inc.
All rights reserved.

Madison headwaters area if elk selected distinct prey refuges and avoided areas where they were particularly vulnerable to wolf predation. Alternatively, elk distribution shifts among the three drainages could also result from differential predation risk with lower densities of animals in high risk areas due to the removal by wolves of substantial numbers of prey.

I. INTRODUCTION

Predation risk may have an important influence on resource selection by prey because animals often must choose between minimizing the risk of predation and obtaining enough forage to meet nutritional demands and maximize reproductive success (Gilliam and Fraser 1987, Sweitzer 1996, Rachlow and Bowyer 1998, Kie 1999, Grand 2002, Ben-David *et al.* 2004, Gustine *et al.* 2006). The costs of resource selection choices in the form of increased predation will likely vary both spatially and temporally in heterogeneous and diverse landscapes where animals make choices based on biological (*e.g.*, age, reproductive status, nutritional condition), environmental (*e.g.*, vegetation, snow, topography), and social variables (*e.g.*, density, group size; Krebs 1980, Lima and Dill 1990). Animals in these environments should attempt to minimize detrimental effects of the main limiting factors (*e.g.*, food, predation, snow), but may make trade-offs among these if the risks associated with several potential limiting factors are competing and vary with scale (Rettie and Messier 2000, Yasué *et al.* 2003, Dussault *et al.* 2005).

Ungulates in mountainous, temperate, and high latitude environments often use multiple strategies to accrue food supplies during winter (*e.g.*, Johnson *et al.* 2001). These strategies may be a product of the heterogeneous environment or a landscape with dynamic predation risk that can vary at fine and broad temporal and spatial scales (Creel and Winnie 2005, Gude *et al.* 2006). For example, animals may select for areas higher in vegetation quality or with topography and snow characteristics that increase access to forage (Schmidt 1993, Gustine *et al.* 2006). Alternatively, they may minimize predation risk by increasing separation from predators or selecting topography that serves as a form of escape terrain (Bergerud and Page 1987, Barten *et al.* 2001). This plasticity in use of resources can also make animals less predictable in space and time (Gustine *et al.* 2006). Thus, individual and group responses to spatio-temporal variations in food availability and predation risk may affect the distribution of animals, nutrient acquisition, and demography (Sih 1980, 1982; Werner and Hall 1988, Schmitz 1997; Downes 2001; Creel *et al.* 2005, 2007; Dussault *et al.* 2005; Gustine *et al.* 2006).

Elk in the Madison headwaters area of Yellowstone National Park provided an excellent opportunity to evaluate the effects of predation risk on resource selection decisions because they have relatively low reproductive potential, use a heterogeneous landscape to meet stringent seasonal demands, are non-migratory and not subject to human hunting, are demographically sensitive to predation, and in the absence of wolves were strongly limited by winter starvation mortality (Garrott *et al.* 2005; Chapters 11 and 23 by Garrott *et al.*, this volume). We evaluated how elk resource selection changed when wolves were restored and elk were confronted with trade-offs between predation risk and starvation. We compared habitat, snow pack, and topographical attributes at locations used by elk and locations selected at random from an area considered available to the population across several different categories of landscape-scale snow conditions.

Given the constraints of foraging in a severe winter environment, we assumed that the fundamental drivers of elk resource selection prior to wolf recolonization identified by Messer *et al.* (Chapter 8 by Messer *et al.*, this volume) would remain unchanged. Therefore, we predicted the models most supported by the data collected during colonization and after wolves became established in the study system would be similar to the top approximating models for the pre-wolf data. However, we expected the strength of the covariate coefficients would change substantially because the most-effective resource selection strategies for minimizing starvation risk would likely not be optimal when confronted with predation risk (Mech *et al.* 2001, Creel *et al.* 2005, Hebblewhite *et al.* 2005, Gude *et al.* 2006, Kauffman *et al.* 2007). Snow pack has been reported to substantially increase the vulnerability of ungulates to wolf predation by impeding their escape attempts and reducing their body condition due to prolonged nutritional deprivation and periods of high energetic costs (Peterson 1977; Nelson and Mech 1986; Huggard 1993; Post *et al.* 1999; Mech and Peterson 2003; Chapter 9 by White *et al.*, Chapter 11 by Garrott *et al.*, and Chapters 16 and 17 by Becker *et al.*, this volume). Thus, we predicted the decrease in the odds of elk occurrence with increasing snow pack (Chapter 8 by Messer *et al.*, this volume) would become more pronounced after wolves became established in the system. Likewise, we predicted the positive effect of snow heterogeneity on odds of elk occurrence (Chapter 8 by Messer *et al.*, this volume) would strengthen after wolves became established in the system because heterogeneity provides more opportunities for elk to locate areas of low snow pack where they would be less vulnerable to successful attacks. Based on the findings of Messer *et al.* (Chapter 8 by Messer *et al.*, this volume), we anticipated elk would select geothermal and meadow environments and lower elevations, and that this selection would be heightened during and after wolf colonization (Figure 21.1). Though wolves in the Madison headwaters area selected for geothermal areas, meadows, and areas near various types of habitat edges (Bergman *et al.* 2006), these areas still have higher food quantity and quality (Chapter 9 by White *et al.*, this volume). They also have lower snow pack and, in turn, lower energy requirements for foraging and traveling elk. Furthermore, some geothermal habitats support wet meadow areas with high vegetative productivity or plants that photosynthesize year-round (Despain 1990).

In addition to potential changes in resource selection as a consequence of the reestablishment of predation risk for elk in this system, broad distributional shifts were also possible (Lima 1998). Landscapes are complex combinations of many different attributes (Turner and Gardner 1991), some of which may provide areas that serve as effective "refuges" where prey are less detectable or less vulnerable if attacked by predators (Andrewartha and Birch 1984, Berryman and Hawkins 2006). Other combinations of landscape attributes may work in concert to define areas where prey are particularly vulnerable to predation (Lima 1992, Hebblewhite *et al.* 2005, Hopcraft *et al.* 2005). The simple linear combinations of landscape covariates employed in resource selection models may not

FIGURE 21.1 Elk grazing along the Madison River in Yellowstone National Park during late winter. The occurrence of elk was not decoupled from wolves because elk continued to select for areas with high snow heterogeneity even though there was a high risk of predation. This strategy was apparently viable in certain portions of the landscape due to the presence of escape terrain such as rivers adjacent to preferred foraging areas (Photo by Shana Dunkley).

adequately capture the collective attributes of a landscape that define areas where prey are particularly vulnerable and areas prey may be protected from predation. If such areas existed in our study system, then a combination of differential mortality associated with these areas and possible movements of elk out of areas of high perceived predation risk and into areas of lower risk could effectively change the distribution of the population. While we had no strong *a priori* hypotheses with respect to potential refuge areas within our study system, we recognized that snow pack was deeper in upper reaches of the Gibbon and Firehole drainages compared to the Madison drainage (Chapter 6 by Watson *et al.*, this volume), and that small geothermal areas surrounded by deep snow pack could represent environmental traps where elk would be particularly vulnerable to predation. Such areas were scattered throughout the Gibbon and Firehole drainages and were routinely occupied by small groups of elk prior to wolf reestablishment due to their relatively high food availability and quality (Chapter 9 by White *et al.*, this volume). Similarly, we suspected that structural differences in topography and hydrology, and their differing spatial arrangement among the three drainages (Chapter 8 by Messer *et al.*, this volume), offered differing amounts of effective escape habitat for elk when attacked by wolves. Thus, we hypothesized that differences in overall landscape heterogeneity among the three drainages of the study system could result in broad distributional shifts in the population as a consequence of the reestablishment of wolves.

II. METHODS

We used telemetry homing procedures and a stratified random sampling regime to repeatedly visually locate adult female elk fitted with radio collars during daylight hours between November 15 and April 30 of eight consecutive winters (1998–1999 through 2005–2006). For each independent elk location, we selected 20 random locations from within the boundary of the available winter range (Chapter 8 by Messer *et al.*, this volume) and assigned the same date as their corresponding elk location to ensure equal temporal distribution in the response. We used a digital elevation model to estimate elevation (ELEV) and the solar radiation index (SRI; Iqbal 1983, Keating *et al.* 2007) for each actual and random location. We also designated habitat type (HAB) for each location as meadow, burned forest, unburned forest, or geothermal using vegetation cover and geothermal data layers developed with Landsat 7 Enhanced Thematic Mapper satellite imagery (Chapter 2 by Newman *et al.*, this volume). In addition, we used the Langur snow pack model (Chapter 6 by Watson *et al.*, this volume) to estimate the spatial and temporal heterogeneity of the snow pack each day during winter. We calculated mean snow water equivalent (*i.e.*, SWE_A; water content of snow) at the local scale as the average of all 28.5×28.5-m pixels within a 100-m radius of each elk location. We also calculated snow heterogeneity (SNH_A; spatial variability of snow) at the local scale as the standard deviation of estimated SWE of all pixels within a 100-m radius of each elk location. We calculated mean snow water equivalent and snow heterogeneity at the landscape scale (SWE_L, SNH_L) each day using all pixels within the defined winter range we considered available to the elk. In addition, we also calculated an annual index of snow pack severity (SWE_{acc}) by summing the mean daily SWE for the study area (SWE_L) from October 1 to April 30 (Garrott *et al.* 2003), which provided a single metric integrating snow pack depth, density, and duration. Data were partitioned into a colonization period (1998–1999 through 2001–2002) when wolves were initially occupying various river drainages and elk were adjusting to their presence, and an established wolf period (2002–2003 through 2005–2006) when one or more packs consistently used each drainage.

We used log odds ratios (Agresti 1990) to determine the likelihood of an elk occurring at a particular location for the wolf colonizing and established periods, depending on local and landscape-scale snow pack conditions. We sorted all actual and random locations into one of three categories depending on the landscape SWE_L estimate for that date (low 0–0.1 m, medium > 0.1–0.2 m, high > 0.2–0.3 m).

Next, we sorted locations into local SWE_A levels, categorized by 0.05-m increments. We then calculated the log odds and 95% confidence intervals of an elk location occurring in a particular local SWE_A level within each landscape SWE_L category, as described in Messer *et al.* (Chapter 8 by Messer *et al.*, this volume). We also used this approach to calculate log odds ratios of local SNH_A, categorized into 0.02-m increments, within the same landscape SWE_L categories.

Modeling animal distributions based on resource selection or similar types of habitat modeling exercises is a common research activity (Manly *et al.* 2002). The basic purpose of such models is to predict animal distributions. Thus, it is useful to obtain some idea of the generalizability of a model beyond the data that was used in its development (Beutel *et al.* 1999, Vaughan and Ormerod 2005). While resampling of the training data set can be used to evaluate a model's predictive ability (Boyce *et al.* 2002), applying the model to independent data sets provides a more rigorous assessment of a model's generalizability (Vaughan and Ormerod 2005). Messer *et al.* (Chapter 8 by Messer *et al.*, this volume) incorporated landscape attributes into matched case–control logistic regression analyses (also known as conditional logistic regression) to gain insights into elk resource selection patterns prior to wolf reestablishment in the study system. Actual and random locations were matched by date to ensure that snow pack covariates associated with both the used and random locations were drawn from the same daily snow pack predictions from the Langur model. Three *a priori* model suites were evaluated, representing (1) static landscape attributes (ELEV, HAB, SRI; seven models total), (2) temporally dynamic snow pack attributes (SWE_A, SNH_A, SWE_L; seven models total), and (3) combinations of both static landscape and temporally dynamic snow pack attributes (13 models total).

We considered the evaluation of these *a priori* model suites using seven years of elk telemetry data collected before wolf colonization to be a training data set from which there was a single model in each of the three *a priori* model suites that received all the support from the data, as well as an overall top model that included three static landscape covariates and two temporally dynamic snow pack covariates (Chapter 8 by Messer *et al.*, this volume). We used PROC PHREG to fit these same model suites and estimate parameter coefficients (SAS 2000) for two additional, independent data sets including (1) the period when wolves were colonizing the study system, and (2) the period when wolf packs were established in all drainages of the study system. We compared these results with those reported by Messer *et al.* (Chapter 8 by Messer *et al.*, this volume) for the pre-wolf period to evaluate the hypothesis that elk resource selection patterns changed after the establishment of predation risk in the system. We ranked models and compared their relative abilities to explain variation in the data using Akaike's Information Criteria (AIC), with Akaike model weights (w_i) used to address model selection uncertainty (Burnham and Anderson 2002). We also compared the 95% confidence intervals of each variable to evaluate if the coefficients were statistically different among time periods.

An annual estimate of the elk population and the distribution of the population among the three drainages of the study area were obtained each spring from 1997 through 2007 by conducting a series of replicate mark–resight surveys using the radio-collared animals as the marked sample (Chapter 23 by Garrott *et al.*, this volume). Surveys were initiated when elk began aggregating in the meadows adjacent to the rivers to feed on the first green forages of the spring (Figure 21.2) and were conducted by traveling roads that paralleled the rivers and meadows through each drainage. During 1997–2004, 10–11 replicate surveys were conducted each spring. However, 16–33 replicates were needed each spring during 2005–2007 to maintain adequate precision as elk abundance decreased and became more patchily distributed (Chapter 23 by Garrott *et al.*, this volume). Surveys were generally conducted on consecutive days, but low visibility due to spring snowstorms occasionally resulted in 1–3-day breaks between surveys. We used the joint hypergeometric likelihood estimator in Program NOREMARK (White 1996) to calculate a population estimate and confidence limits from these replicate road-based surveys. We estimated the proportion of the elk population in each drainage by summing the total number of animals detected on the replicate surveys in each drainage and dividing by the total number of animals detected throughout the study area during the surveys. The number of elk in each drainage was then

FIGURE 21.2 An aggregation of elk feeding along the Madison River in Yellowstone National Park. Mark–resight surveys to obtain annual population estimates and determine the proportion of the population in each drainage of the study area were conducted each spring when elk were concentrated in the low elevation meadow complexes (Photo by Shana Dunkley).

estimated by multiplying the population estimate by the estimated proportion of the population in each drainage.

We integrated the results of the resource selection analyses and the mark–resight surveys into graphic presentations of changes in relative elk abundance and distribution over the 16 years (1992–2007) we conducted intensive studies of this population. Maps approximating the distribution of elk in the study area for the last winter of the pre-wolf (1997–1998), colonizing (2001–2002), and established (2006–2007) periods were generated by using the most-supported resource selection model for each period to distribute the estimated number of elk that were present in each drainage during the spring mark–resight surveys for each of these years. We fixed snow pack attributes to represent the mean peak landscape SWE_L experienced during the 16 winters of our studies.

III. RESULTS

We collected 3423 locations of 88 adult, female elk (25–30 instrumented animals per year) during eight winters, including 1541 locations during wolf colonization (1998–1999 through 2001–2002) and 1882 locations after wolves became established in the system (2002–2003 through 2005–2006). We compared these data to 4869 locations of 45 adult female elk during seven winter seasons before wolves colonized the system (1991–1992 through 1997–1998; Chapter 8 by Messer *et al.*, this volume).

The annual variation in snow pack severity differed among the three study periods. Winters during the pre-wolf period had an average SWE_{acc} of 3009 and ranged between 1744 and 5690 cm days. Winters during the wolf colonizing period had an average SWE_{acc} of 2265 and ranged between 1042 and 3749 cm days. Winters during the established wolf period had an average SWE_{acc} of 2071 and ranged between 1707 and 2508 cm days. Climate data were available to parameterize the Langur snow pack model (Chapter 6 by Watson *et al.*, this volume) and calculate SWE_{acc} for the most recent 24 winters (1983–1984 through 2006–2007), with an average SWE_{acc} for this time series of 2337 cm days and a range between 1042 and 5690 cm days. Thus, average snow pack severity was higher than the historic average during the pre-wolf period, similar to the historic average during the wolf colonizing period, and lower than the historic average during the established wolf period.

The frequency distribution of observed elk locations compared to the samples of random points drawn from the available winter range before, during, and after wolf colonization suggests elk disproportionately selected areas with lower local snow pack and tended to avoid areas of higher local snow pack. This pattern of selection was consistent for all three study periods despite the differences in winter severity among the periods, as evidenced by the decreasing range of the distributions from the pre-wolf period to the post-wolf period (Figure 21.3). Likewise, log odds ratios demonstrated that elk were less likely to occur in an area as local SWE_A increased with this trend consistent for all three categories of landscape SWE_L (Table 21.1, Figure 21.4). However, the log odds ratios for any given local SWE_A increased with increasing landscape SWE_L, indicating that as snow pack on the winter range became more severe elk were more likely to occupy areas of more severe local SWE_A than they were under less severe snow pack conditions. While the log odds ratio curves were similar among the three study periods at low and moderate landscape SWE_L, there was a clear tendency for elk to more strongly avoid severe snow pack conditions as wolves colonized and became established in the study system (Figure 21.4). Local snow pack heterogeneity also influenced elk occupancy across all three categories of landscape SWE_L, with the likelihood of elk occurrence quickly increasing and then decreasing slightly as local SNH_A levels exceeded 0.06–0.08 m (Table 21.1, Figure 21.5). This selection for increased local heterogeneity increased with increasing landscape SWE_L, but the patterns were similar for all three study periods.

The same landscape and snow pack models selected for the training data set (*i.e.*, pre-wolf; Chapter 8 by Messer *et al.*, this volume) during conditional logistic regression analyses received all of the support from the independent data sets collected during and after wolf colonization. Model comparisons both within and among the static landscape, dynamic snow, and combined landscape-snow model suites resulted in the same top-ranking models with the most-supported model in each suite receiving a model weight of 1.0 (Table 21.2). Coefficient estimates indicated elk selection for areas with lower local snow pack became stronger after wolves became established in the study system, and this selection intensified as landscape SWE_L increased (Table 21.3). Selection of snow heterogeneity was similar among time periods. Elk showed stronger selection for lower elevations after wolves became established in the area, with some evidence of selection for lower exposure to solar radiation (Table 21.3). All habitat coefficients were negative, indicating that elk selection remained strongest for the reference geothermal type during all three study periods. While the coefficient estimates for each of the habitat categories either consistently increased or decreased from the pre-wolf to the post-wolf period, coefficient confidence intervals overlapped among time periods, providing little evidence of fundamental shifts in elk selection of habitat types after changes in selection for snow characteristics were accounted for with other covariates (Table 21.3). In other words, the coefficient estimates derived from each independent data set during and after wolf colonization were similar to the pre-wolf training data set. These results suggest that the resource selection models initially developed in Chapter 8 by Messer *et al.* (this volume) had substantial ability to predict elk occupancy in our study system when applied to other time periods. This assessment is limited to our study system because no independent data sets from other areas were evaluated (Vaughan and Ormerod 2005).

Results of the annual spring replicate mark–resight surveys documented major shifts in the distribution of elk among the three drainages. When surveys were initiated in 1997, just prior to initial recolonization of the study area by wolves, the elk population was approximately equally distributed among the three drainages. Over the next decade, the proportion of the population detected in the Gibbon drainage decreased precipitously, the proportion in the Firehole drainage decreased modestly, and the proportion in the Madison drainage increased substantially (Figure 21.6). The number of elk

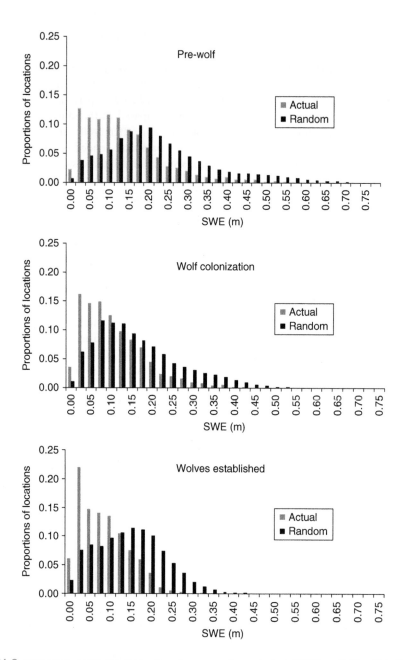

FIGURE 21.3 Histograms comparing the local-scale snow water equivalent (SWE$_A$) available within the study area (random) to the mean SWE$_A$ in the 100-m radius around observed elk locations (actual) before ($n = 4869$), during ($n = 1541$), and after ($n = 1882$) wolf colonization in the Madison headwaters area of Yellowstone National Park during 1991–1992 through 2005–2006.

estimated in both the Gibbon and Firehole drainages also decreased, with estimates suggesting an ephemeral increase in elk abundance in the Madison drainage that was negligible by the end of the study (Figure 21.6). The high values for both the estimated proportion and number of elk in the Madison drainage in 2001 were anomalies in the data due to the appearance of a large group of elk in

TABLE 21.1 Log odds ratios (LOR) and standard errors (SE) determining the likelihood of elk occurrence under changing local- and landscape-scale snow water equivalents (SWE; m) and snow heterogeneity (SNH) in the Madison headwaters area of Yellowstone National Park before, during, and after wolf colonization in the Madison headwaters area of Yellowstone National Park during 1991–1992 through 2005–2006

Landscape SWE categories

Local snow levels	SWE low (0.00–0.10 m)						SWE moderate (0.10–0.20 m)						SWE high (0.20–0.30 m)					
	Pre-wolf		Colonizing		Post-wolf		Pre-wolf		Colonizing		Post-wolf		Pre-wolf		Colonizing		Post-wolf	
	LOR	SE	LOR	SE	LOR	SE	LOR	SE	LOR	SE	LOR	SE	LOR	SE	LOR	SE	LOR	SE
SWE																		
0.00–0.05	0.62	0.09	1.15	0.09	1.79	0.12	1.63	0.05	1.44	0.12	1.52	0.07	1.85	0.08	2.06	0.20	1.71	0.23
0.05–0.10	0.02	0.09	−0.37	0.09	−0.27	0.12	1.21	0.05	0.96	0.10	1.13	0.06	1.46	0.08	1.89	0.21	1.83	0.22
0.10–0.15	−1.35	0.19	−1.90	0.25	−1.59	0.34	0.08	0.05	0.07	0.09	−0.20	0.07	1.08	0.07	1.27	0.19	1.11	0.20
0.15–0.20			−2.24	0.71			−1.20	0.07	−0.89	0.12	−1.15	0.09	0.46	0.07	0.41	0.16	0.83	0.18
0.20–0.25							−1.96	0.14	−1.71	0.28	−2.31	0.22	−0.60	0.08	−1.00	0.20	−1.62	0.35
0.25–0.30							−2.87	0.41	−2.48	0.71	−2.85	0.50	−1.25	0.10	−1.43	0.28	−3.52	1.00
0.30–0.35													−1.83	0.18	−3.29	1.00		
0.35–0.40													−2.84	0.45				
0.40–0.45													−3.53	1.00				
SNH																		
0.00–0.02	−0.58	0.11	−0.56	0.88	−0.17	0.15	−1.08	0.05	−0.92	0.09	−0.75	0.06	−1.17	0.07	−1.52	0.19	−1.01	0.23
0.02–0.04	0.60	0.12	0.55	0.46	0.23	0.16	0.26	0.05	0.56	0.09	0.17	0.06	−0.23	0.06	0.01	0.15	−0.88	0.20
0.04–0.06	0.36	0.30	0.37	0.27	−0.64	0.72	0.94	0.05	0.77	0.12	0.71	0.07	0.67	0.07	1.02	0.17	0.82	0.18
0.06–0.08			−0.09	0.73			1.01	0.08	0.76	0.19	0.65	0.11	0.96	0.08	1.04	0.20	1.03	0.21
0.08–0.10	1.92	1.16					0.70	0.16	0.33	0.46	0.26	0.24	1.04	0.10	1.23	0.23	1.35	0.26
0.10–0.12							0.31	0.46			−0.06	0.59	0.82	0.15	0.53	0.53		
0.12–0.14													0.13	0.33				
0.14–0.16													0.23	0.73				

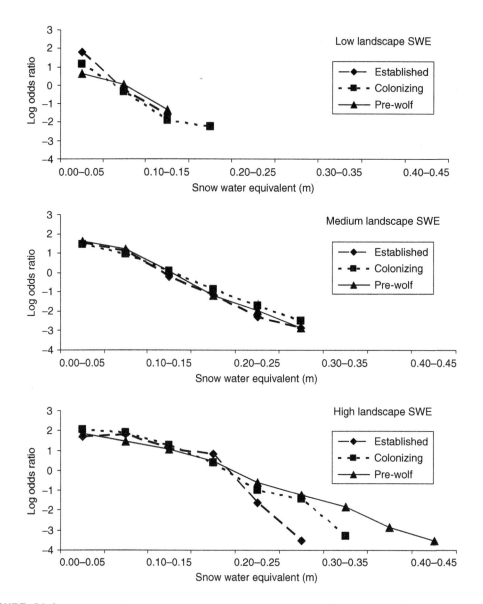

FIGURE 21.4 Log odds of elk occurrence in local snow water equivalent conditions given low (0–0.1), medium (0.1–0.2), and high (0.2–0.3) landscape snow water equivalent (SWE) conditions before, during, and after wolf colonization in the Madison headwaters area of Yellowstone National Park during 1991–1992 through 2005–2006.

the meadows adjacent to the Madison River that remained in the area for several weeks. We believed these animals were early spring migrants that wintered outside of the study area because there were no collared animals associated with this group and it contained a much higher proportion of calves than had been recorded anywhere in the study area that year.

To illustrate the influence of wolf colonization on the distribution of elk in the Madison headwaters, we used our most-supported resource selection models to generate maps depicting the odds of elk occurrence in the Firehole, Gibbon, and Madison drainages before, during, and after wolf colonization. Elk were widely distributed through all three drainages prior to wolves, as evidenced by the relatively

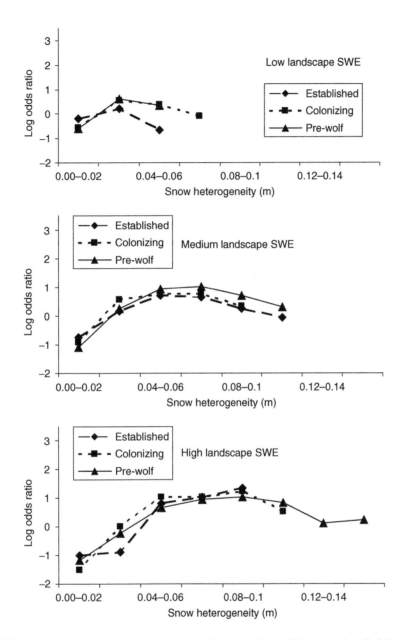

FIGURE 21.5 Log odds of elk occurrence in local snow heterogeneity conditions given low (0–0.1), medium (0.1–0.2), and high (0.2–0.3) landscape snow water equivalent (SWE) conditions before, during, and after wolf colonization in the Madison headwaters area of Yellowstone National Park during 1991–1992 through 2005–2006.

uniform odds of occurrence (Chapter 8 by Messer *et al.*, this volume; Figure 21.7). During wolf colonization, modest dispersal (Chapter 18 by Gower *et al.*, this volume) and substantial elk mortality due to wolves (Chapter 23 by Garrott *et al.*, this volume) resulted in the Gibbon drainage having low odds of elk occurrence, with reduced odds in the Firehole drainage as well (Figure 21.8). Odds of elk occurrence were substantially lower in the Firehole drainage by the end of the study after wolf packs became established in the system (Figure 21.9). The vast majority of elk occurred in the Madison

TABLE 21.2 Selection results of models examining the effects of static landscape and dynamic snow covariates on variation in winter elk distribution before, during, and after wolf colonization in the Madison headwaters area of Yellowstone National Park during 1991–1992 through 2005–2006

Model Structure	K	Pre-wolf Within suite ΔAIC	Pre-wolf Among suite ΔAIC	Wolf colonization Within suite ΔAIC	Wolf colonization Among suite ΔAIC	Post-wolf Within suite ΔAIC	Post-wolf Among suite ΔAIC	Post-wolf Among suite w_i
Snow models								
$SWE_A+SNH_A+(SWE_A\times SWE_L)+$ $(SNH_A\times SWE_L)$	4	0	939	0	525	0	925	0
$SWE_A+SNH_A+(SWE_A\times SNH_A)$	3	449	1388	72	597	91	1016	0
$SWE_A+(SWE_A\times SWE_L)$	2	451	1390	142	667	87	1012	0
SWE_A+SNH_A	2	549	1488	140	665	113	1038	0
SWE_A	1	924	1863	268	793	191	1116	0
$SNH_A+(SNH_A\times SWE_L)$	2	3985	4924	1218	1743	1661	2586	0
SNH_A	1	4094	5033	1216	1741	1670	2595	0
Landscape models								
HBT+ELV+SRI	5	0	1443	0	467	0	419	0
HBT+ELV	4	2	1445	7	474	55	474	0
ELV	1	952	2395	294	761	358	777	0
SRI+ELV	2	953	2396	288	755	298	717	0
HBT+SRI	2	3652	5095	1334	1801	1634	2053	0
HBT	1	3657	5100	1338	1805	2038	2457	0
SRI	1	4855	6298	1648	2115	2242	2661	0
Landscape and snow models								
$SWE_A+SNH_A+HBT+ELV+SRI$	7	0	0	0	0	0	0	1.0
$SWE_A+HBT+ELV+SRI$	6	593	593	194	194	139	139	0
SWE_A+SNH_A+HBT+ $(SWE_A\times SNH_A)+(SWE_L\times HBT)$	9	1284	1284	546	546	917	917	0
SWE_A+SNH_A+HBT+ $(SWE_A\times SNH_A)+(SNH_A\times HBT)$	9	1294	1294	572	572	965	965	0

Model	K							
$SWE_A+SNH_A+HBT+(SWE_A \times SNH_A)+(SWE_A \times HBT)$	9	1360	1360	595	595	976	976	0
$SWE_A+SNH_A+HBT+(SWE_A \times SNH_A)$	6	1360	1360	589	589	978	978	0
$SWE_A+SNH_A+HBT+(SWE_L \times HBT)$	8	1388	1388	606	606	940	940	0
$SWE_A+SNH_A+HBT+(SNH_A \times HBT)$	8	1388	1388	631	631	986	986	0
$SWE_A+SNH_A+HBT+(SWE_A \times HBT)$	8	1460	1460	662	662	999	999	0
SWE_A+SNH_A+HBT	5	1468	1468	660	660	1005	1005	0
$SWE_A+HBT+(SWE_L \times HBT)$	7	1741	1741	726	726	1004	1004	0
$SWE_A+HBT+(SWE_A \times HBT)$	7	1752	1752	768	768	1058	1058	0
SWE_A+HBT	4	1813	1813	776	776	1070	1070	0

SWE_A, local-scale snow water equivalent; SWE_L, landscape-scale snow water equivalent; SNH_A, local-scale snow heterogeneity; SNH_L, landscape-scale snow heterogeneity; SRI, solar radiation index; ELV, elevation; HBT, habitat; K, number of model parameters; $\triangle AIC$, the change in AIC value relative to the best model; AIC model weight.

TABLE 21.3 Coefficient values and confidence intervals for the best-supported models examining the effects of static landscape and dynamic snow covariates examining variation in winter elk distribution before, during, and after wolf colonization in the Madison headwaters area of Yellowstone National Park during 1991–1992 through 2005–2006

Covariate	Pre-wolf			Wolf colonization			Post-wolf		
	β est.	95% Low CI	95% High CI	β est.	95% Low CI	95% High CI	β est.	95% Low CI	95% High CI
Snow model SWE$_A$+SNH$_A$+(SWE$_A$×SWE$_L$)+(SNH$_A$×SWE$_L$)									
SWE$_A$	−20.9	−22.0	−19.9	−26.4	−28.7	−24.0	−39.6	−43.7	−35.5
SNH$_A$	21.3	18.4	24.2	18.4	12.0	24.8	20.4	8.9	32.0
SWE$_A$×SWE$_L$	34.5	31.4	37.7	60.0	50.1	70.0	134.2	109.9	158.4
SNH$_A$×SWE$_L$	−36.4	−45.2	−27.5	−21.7	−48.9	5.6	−43.4	−112.5	25.7
Landscape model HBT+SRI+ELV									
HBT									
Burned forest	−1.70	−1.81	−1.60	−1.67	−1.86	−1.49	−1.75	−1.94	−1.55
Unburned forest	−1.35	−1.44	−1.26	−1.17	−1.32	−1.01	−1.34	−1.30	−0.97
Meadow	−1.27	−1.38	−1.16	−1.29	−1.48	−1.09	−1.53	−1.74	−1.31
SRI	−0.320	−0.623	−0.017	−0.716	−1.184	−0.248	−1.750	−2.186	−1.314
ELV	−0.009	−0.010	−0.009	−0.009	−0.010	−0.009	−0.012	−0.013	−0.011
Combined snow-landscape model SWE$_A$+SNH$_A$+HBT+ELV+SRI									
SWE$_A$	−6.5	−7.0	−6.0	−7.7	−8.8	−6.6	−11.4	−12.8	−10.0
SNH$_A$	14.1	12.9	15.2	17.4	15.0	19.8	18.8	15.7	21.8
HBT									
Burned forest	−0.58	−0.70	−0.46	−0.66	−0.88	−0.45	−0.74	−0.96	−0.52
Unburned forest	−0.38	−0.49	−0.28	−0.33	−0.51	−0.15	−0.28	−0.47	−0.09
Meadow	−0.46	−0.58	−0.33	−0.60	−0.82	−0.38	−0.77	−1.00	−0.54
ELV	−0.007	−0.008	−0.007	−0.008	−0.008	−0.007	−0.010	−0.011	−0.009
SRI	−7.070	−7.450	−6.690	−7.590	−8.204	−6.976	−9.850	−10.531	−9.169

SWE$_A$, local-scale snow water equivalent; SWE$_L$, landscape-scale snow water equivalent; SNH$_A$, local-scale snow heterogeneity; SNH$_L$, landscape-scale snow heterogeneity; SRI, solar radiation index; ELV, elevation; HBT, habitat; K, number of model parameters; ΔAIC, the change in AIC value relative to the best model; AIC model weight.

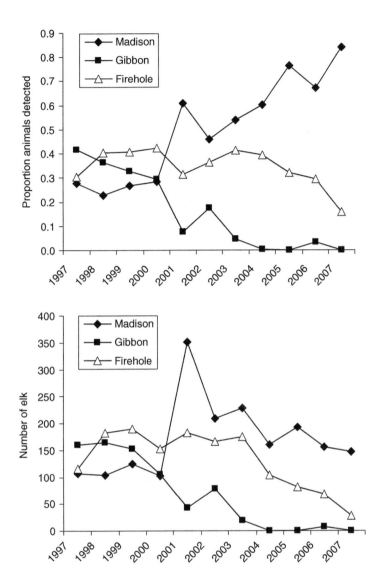

FIGURE 21.6 Estimated changes in the proportion of the elk population and the total number of elk occupying the Madison, Gibbon, and Firehole River drainages from 1997 through 2007. Estimates were based on replicate ($n = 10$–33) mark–resight surveys conducted during the early stages of snow melt-out in late March or April when elk were concentrated in meadows adjacent to the rivers and the road system (Chapter 23 by Garrott *et al.*, this volume).

Canyon of the Madison River drainage, with few elk in the Gibbon drainage and a small, decreasing number of elk in the Firehole drainage.

IV. DISCUSSION

We did not detect profound changes when contrasting habitat selection by elk before, during, and after wolf colonization, though modest and consistent changes in coefficients across the three study periods suggested subtle shifts in selection may have been occurring. Elk continued to select geothermal areas

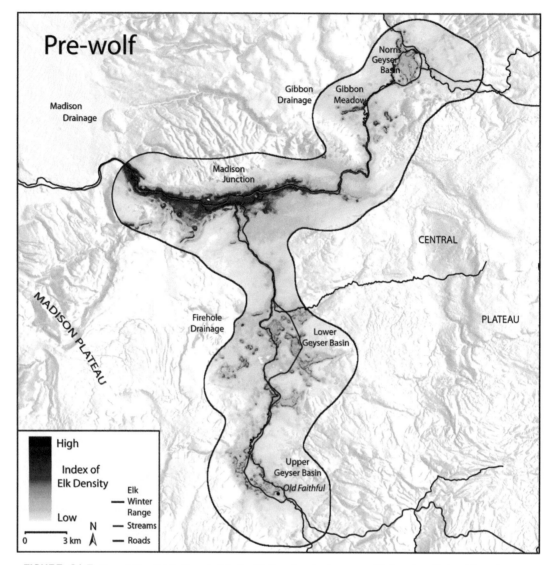

FIGURE 21.7 The relative distribution of elk in the Madison headwaters area of Yellowstone National Park, Montana and Wyoming, USA, before wolves began recolonizing the area in 1998–1999.

with higher food quantity and quality (Chapter 9 by White *et al.*, this volume) even though wolves also selected for these areas (Bergman *et al.* 2006). Similarly, both Mao *et al.* (2005) and Kauffman *et al.* (2007) concluded that data from elk occupying the northern portion of Yellowstone did not yield empirical support for elk substantially altering their habitat selection and movement patterns after wolf restoration. This consistent, strong preference by elk for apparently dangerous habitats appears somewhat paradoxical, but likely reflects the continued need for elk to access as high a quality of forage as available during the severe winter months when they experience lower metabolizable energy intake (Garrott *et al.* 1996; Chapter 9 by White *et al.*, this volume) and diminishing fat reserves (Cook *et al.* 2004). Elk in the Madison headwaters area apparently obtained necessary food resources since there was no indication of any considerable change in foraging time (Chapter 20 by Gower *et al.*, this

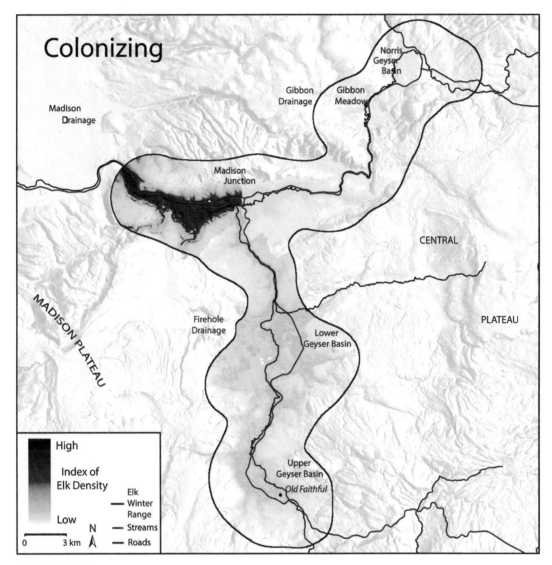

FIGURE 21.8 The relative distribution of elk in the Madison headwaters area of Yellowstone National Park, Montana and Wyoming, USA, during wolf colonization between 1998–1999 through 2001–2002.

volume) and we did not detect any substantial decreases in over-winter nutrition after wolves became established in the system (Chapter 22 by White *et al.*, this volume).

The ability of elk to mediate predation risk through small-scale movements, changes in grouping behavior and activity patterns (Chapters 18–20 by Gower *et al.*, this volume), or the use of landscape features that provide effective escape from predators may explain why they continued similar patterns of habitat selection despite the increased predation risk in geothermal and meadow areas due to high use by wolves (Lima 1992, Hebblewhite *et al.* 2005, Bergman *et al.* 2006; Chapters 18–20 by Gower *et al.*, this volume). Lima (1992) demonstrated that habitat choice can be influenced by the probability of escape as well as the probability of attack itself. Thus, even though elk in the Madison headwaters area were predictable in space and time for wolves to locate, they also occupied areas with landscape

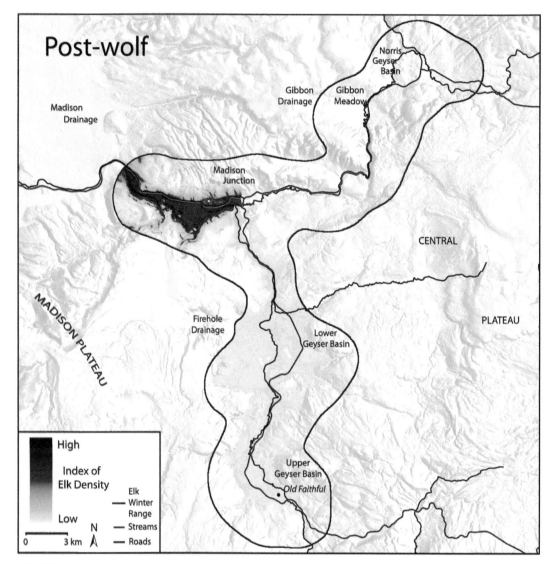

FIGURE 21.9 The relative distribution of elk in the Madison headwaters of Yellowstone National Park, Montana and Wyoming, USA, after wolf packs became established in the area (2002–2003 through 2005–2006).

attributes that reduced vulnerability and served as refuges with lower predation risk. Elk have a relatively high probability of escape when attacked by wolves in open environments because wolves are inefficient predators, being successful in about 20% of their attempts (Mech *et al.* 2001). Also, the large body size and dangerous defensive capabilities of elk can maim or kill wolves when elk have room to maneuver. In addition, elk occurred in larger groups after wolf colonization (Chapter 19 by Gower *et al.*, this volume), which likely enhanced their probability of escape through the dilution effect (Hamilton 1971, Bertram 1978, Treherne and Foster 1982, Dehn 1990) and group detection of predators (Pulliam 1973, Pulliam and Caraco 1984, Elgar 1989). Thus, Mech *et al.* (2001) found that only 1% of elk in groups of up to 150 were killed when attacked by wolves. Similarly, there was a negative correlation ($R^2 = 0.66$) between mean annual group size of Madison headwaters elk and wolf kill rates on elk (Chapter 17 by Becker *et al.*, this volume). Therefore, the collective response of slightly larger group sizes, and more dynamic grouping behavior (Chapter 19 by Gower *et al.*, this volume), may function as an effective strategy when other defensive tactics of elk, such as fleeing, likely do not work as well in deep snow or thick forests, where thick vegetation and downed timber may pose a hindrance to efficient escape.

Nearly all studies of wolf predation in North America report snow pack as an important factor affecting vulnerability of ungulate prey (Mech and Peterson 2003), including ours (Chapter 9 by Gower *et al.*, Chapter 11 by Garrott *et al.*, and Chapter 16 by Becker *et al.*, this volume). Thus, we predicted elk would modify their use of the landscape with respect to snow pack attributes after wolves became established in the system. During our pre-wolf studies, when winter starvation was the predominant source of mortality for elk (Chapter 11 by Garrott *et al.*, this volume), we found that elk selected local areas with lower and more heterogeneous snow pack. We predicted the selection for both of these snow pack attributes would become stronger with the reestablishment of predation risk from wolves. Data collected during and after wolf colonization supported these predictions. We documented an approximate doubling of the SWE_A coefficient from the pre-wolf to the post-wolf period, with a modest increase in selection for more heterogeneous local snow conditions (SNH_A). Similar studies of elk resource selection conducted in the greater Yellowstone ecosystem since the reestablishment of wolves have found no evidence that elk responded to wolf predation risk by altering their selection of snow pack characteristics (Mao *et al.* 2005, Winnie *et al.* 2006). However, these studies incorporated coarse-scale snow pack covariates in their analyses that were not likely to capture variations in snow pack attributes at meaningful spatial and temporal scales for addressing interactions between elk and wolves.

The dramatic shift in the distribution of elk during and after wolf colonization (Figures 21.7–21.9) was not due to a redistribution of elk from the Gibbon and Firehole drainages to the Madison drainage. Intensive tracking of radio-collared elk during the 16-year study demonstrated that individual adult, female elk maintained fidelity to home ranges within a drainage (Chapter 18 by Gower *et al.*, this volume), a behavioral trait commonly documented for elk and many other ungulates (Ortega and Franklin 1995, Bowyer *et al.* 1996, Nicholson *et al.* 1997, Berger 2004). Small numbers of instrumented females were recorded moving temporarily between drainages, but these movements were not common and we documented no animals abandoning a traditional winter home range in the Gibbon or Firehole drainage to establish residency in the Madison drainage (Chapter 18 by Gower *et al.*, this volume). Instead, we attribute the virtual disappearance of elk from the Gibbon drainage and the substantial decrease in numbers of elk occupying the Firehole drainage to a modest level of dispersal by elk from their traditional winter home ranges after wolf recolonization (Chapter 18 by Gower *et al.*, this volume) and, most importantly, to wolf predation removing animals. Of the 312 adult elk killed by wolves during our studies, 62% were found in the Firehole drainage and 19% were found in each of the Gibbon and Madison drainages. However, these numbers are confounded by annual changes in the number of elk that occupied each of the drainages. Using annual autumn estimates of the adult population in each of the drainages (Chapter 16 by Becker *et al.*, this volume) and the total number of wolf-killed adult elk discovered in each drainage, we estimate the average annual percentages of resident adult elk killed by wolves during winters from 1998–1999 through 2006–2007 were 47% (95% CI = 16–77%) in the Gibbon drainage, 17% (95% CI = 11–22%) in the Firehole drainage, and 4% (95% CI = 2–5%) in the Madison drainage. Concurrent demographic studies of this population indicated the amount of predation experienced by elk in the Gibbon and Firehole drainages resulted in substantial decreases in elk numbers in these drainages (Chapters 11 and 23 by Garrott *et al.*, this volume).

These data suggest that the three drainages represented a gradient of predation risk for elk where attributes of the Gibbon drainage resulted in an area of particularly high vulnerability, the Firehole drainage represented an area of moderately high vulnerability, and the Madison drainage potentially serving as a prey refuge where wolves had difficulty successfully attacking elk. Predation risk is a function of where predators hunt and the landscape attributes, individual behaviors, and physiological stressors that render prey more or less susceptible to detection, attack, and predation (Holling 1965, Mech and Peterson 2003, Hebblewhite *et al.* 2005). The changes in strength of resource selection and distribution of the elk population in the study area following colonization by wolves did not result in a disassociation of wolves and elk. Wolves were closely associated with elk wherever they were found in the study system and were constantly coursing through the study system, including areas where prey were at low densities (Bergman *et al.* 2006). Therefore, we contend that the differences in predation risk among the three drainages of our study system were not due to differences in detection or encounter probabilities, but rather differences in the vulnerability of elk once attacked.

When elk are attacked by wolves their primary response is to flee (McNulty *et al.* 2007). Hence, any landscape attributes that differentially reduce the ease of movement and maneuverability by elk to a greater extent than the pursuing wolves would increase their vulnerability. The ubiquitous presence of snow pack on the landscape through winter, and the substantial impediment to movement it can pose, is certainly a key attribute of the landscape that can influence the relatively vulnerability of elk to wolf predation (Mech and Peterson 2003; Chapters 16 and 17 by Becker *et al.*, this volume). Thus, maps of snow pack (Chapter 6 by Watson *et al.*, this volume) capture two important aspects of the landscape with respect to predation risk for elk, snow pack severity as indexed by SWE and the spatial variability of the snow pack. The Gibbon drainage experienced the most severe and uniform snow pack conditions, with a single large geothermal basin (Norris) and numerous small geothermal features scattered through the drainage. The Firehole drainage also experienced deep snow pack, but contained many geothermal basins that created large contiguous areas of reduced snow pack. In contrast, the Madison drainage is at a lower elevation and experienced lower average snow pack. This drainage does not contain any significant geothermal features, but includes a steep south to southwest-facing slope along most of its length where snow pack is significantly reduced. Elk strongly selected for low snow pack areas (Table 21.3) and in the Gibbon drainage this meant primarily small foraging areas surrounded by deep snow and complex forest structure (*e.g.*, dead fall and thick regeneration after the fires of 1988) which made these animals vulnerable to successful attack by wolves. The deep snow pack of the Firehole drainage was somewhat moderated by the relatively large areas of reduced snow pack due to the spatial extent of the geothermal features, thereby providing elk more opportunities to flee for longer distances before confronting deep snow or other obstacles (Bergman *et al.* 2006). The lower overall snow pack of the Madison, combined with the steep southern-oriented slope running most of the length of the drainage provided an environment less constrained by snow.

There are a number of other attributes of the Madison drainage that may have contributed to reduced vulnerability of the elk to wolf predation, the most prominent being the presence of a deep, wide, and ice-free river that provided escape opportunities in certain areas. We frequently observed elk in the Madison headwaters running into rivers when pursued by wolves, which is a common defensive technique of ungulate prey when attacked by wolves (Peterson 1955, Crisler 1956, Carbyn 1974, Nelson and Mech 1981). Large prey with long legs such as elk can stand in water while wolves may need to swim, thereby substantially reducing their maneuverability and conferring a distinct advantage to the prey (Figure 21.10; Mech and Peterson 2003). Thus, elk along the Madison River can likely tolerate a high risk of attack in their preferred habitats because they have a high probability of escape when attacked (Lima 1992). While both the Gibbon and Firehole Rivers are also ice-free and available all winter to elk, these tributaries are shallower and narrower than the main Madison River. The Gibbon River is shallow along its entire course and the Firehole River has only a few reaches where the river channel is deep. Consequently, while elk still used rivers as escape habitat in these drainages, the characteristics of escape terrain (*e.g.*, depth of river), the complexity of the approach terrain, and its distance from preferred elk habitat, likely dictated whether or not it afforded effective escape from wolves, as evidenced by the fact that the majority of wolf kills on elk in the latter years of the study occurred in or next to waterways (Bergman *et al.* 2006).

We cannot discount that the apparent lower predation risk in the Madison drainage may also reflect elk using areas where wolf densities and use were substantially reduced due to high human activity.

FIGURE 21.10 Elk utilizing escape terrain while under attack from wolves. Elk frequently fled into the ice-free rivers, though the success of this strategy was likely dependent on the river characteristics. The narrow and shallow Gibbon river, (top), afforded few deep areas and little protection for elk (Photo by John Felis), while the deep and wide Madison River, (bottom), provided effective escape terrain adjacent to preferred foraging areas throughout its length (Photo by Shana Dunkley).

Large predators such as wolves and grizzly bears may avoid areas of high human activity, thereby reducing predation risk for prey inhabiting such areas. This phenomenon has been reported for a portion of the Bow Valley in Banff National Park, Canada (Gibeau *et al.* 2002, Hebblewhite *et al.* 2005), the headquarters of Yellowstone National Park in Mammoth (Barber-Meyer *et al.* 2008), and urban/suburban areas in the western United States. In addition to intensively developed areas, roads and trails may have similar effects on predator behavior (Thurber *et al.* 1994, Whittington *et al.* 2005). While a road bisects each of the three drainages in our study system, visitor traffic during winter is concentrated on the Madison and Firehole road segments that link the town of West Yellowstone, Montana with the popular Old Faithful area in the upper reaches of the Firehole drainage (National Park Service 2007). The Madison drainage is considerably more restricted than the Firehole drainage, suggesting traffic and human activities might have a stronger affect of inhibiting wolf use in the Madison drainage. However, this disturbance was limited to daylight hours and wolves had a strong propensity to travel and hunt during crepuscular and nocturnal periods (Bergman *et al.* 2006). Also, wolves in Yellowstone National Park tend to be somewhat habituated and demonstrate considerable tolerance of humans (McNay 2002, Smith and Stahler 2003).

The post-wolf redistribution of elk may also reflect their use of areas between wolf territories (Mech 1977). Wolf packs initially established in the Firehole drainage during 1997–1998, with packs expanding into the Gibbon drainage in 2000–2001, and the Cougar Pack establishing along the western boundary of the park and our study area in 2001–2002 (Chapter 15 by Smith *et al.*, this volume). Thus, the Madison drainage was generally located along the periphery of the established wolf pack territories for much of the period of our studies. By the winter of 2004–2005, however, five packs representing approximately 45 wolves were using the entire study area with substantial spatial and temporal overlap between wolf packs (Chapter 15 by Smith *et al.*, this volume). Packs frequently hunted in the Madison drainage, but appeared to have limited success (Chapter 17 by Becker *et al.*, this volume), suggesting that landscape attributes were more responsible for reduced vulnerability of elk to wolf predation than hypothesized mechanisms that may have resulted in wolves tending to avoid the Madison drainage with its abundance of prey.

We suggest that since the extirpation of wolves from Yellowstone National Park approximately 70 years ago, elk in the Madison headwaters area began exploiting areas that previously were too risky to occupy. In the absence of predation risk, geothermal features provided localized niches with reduced snow where animals could minimize energetic expenditures and exploit patches of accessible forage that, in some areas, remained photosynthetically active with high nutritional quality through winter. Thus, such areas adequately met the resource requirements of elk during a period when over-winter starvation was the only substantial limiting factor (Chapter 11 by Garrott *et al.*, this volume). Consequently, these niches were thoroughly exploited as documented by pre-wolf data from this study, as well as previous studies conducted in this system (Craighead *et al.* 1973, Jakubas *et al.* 1994; Chapter 8 by Messer *et al.*, this volume). Over the 8–10 generations that these conditions existed, elk developed strong fidelity to this occupancy pattern and presented what could be considered a naïve and vulnerable prey when wolves recolonized the area (Berger 1999, Berger *et al.* 2001, Sand *et al.* 2006). While we have documented numerous adjustments in elk behavior in this and other chapters (Chapters 18–20 by Gower *et al.*, this volume) that we interpret as responses to the reestablishment of predation risk, we propose that the dramatic changes documented in the distribution of this population were driven primarily by wolves successfully exploiting those vulnerable portions of the elk population occupying the Gibbon and Firehole drainages, with the Madison drainage perhaps serving as an important prey refuge.

Our studies documented the transitional dynamics of an evolving large mammal predator–prey system. Thus, it is unclear if the relatively high density of elk that has persisted in the Madison drainage for at least the past three decades (Chapter 11 by Garrott *et al.*, this volume) will persist or if the animals occupying this apparent prey refuge will be subjected to more intensive predation pressure now that numbers of elk in the Firehole and Gibbon drainages have been substantially reduced (Chapter 23 by Garrott *et al.*, this volume). The existence of prey refuges and their consequences on predator–prey dynamics have been an important area of study since Gause's (1934) classic protozoan experiments. It is well-established from both theoretical and experimental studies that the existence of prey refuges or "enemy-free" space can contribute to the stability of predator–prey interactions (Rosenzweig and MacArthur 1963, Hassell 1978, Strong *et al.* 1984, Price *et al.* 1986, Abrams and Walters 1996), with Berryman and Hawkins (2006) arguing that the refuge concept should be considered one of the integrating concepts in ecology and evolution. The refuge may then serve as the limiting resource in this system, with elk competing to occupy those portions of the landscape where there is significantly reduced risk of being successfully attacked by wolves and wolves competing among themselves to exploit the only high prey density area remaining in our study system. Thus, the stability and persistence of this apparent refuge may dictate the carrying capacity for both predator and prey in this system, as well as influencing the potential for a prey switching response by wolves to exploit the much more formidable, alternate bison prey (Garrott *et al.* 2007; Chapter 16 by Becker *et al.*, this volume).

V. SUMMARY

1. We compared the distributions and resource selection by elk before, during, and after wolves colonized the Madison headwaters area of Yellowstone National Park to evaluate how distribution and selection changed when elk were confronted with trade-offs between predation risk and starvation.
2. We collected 3423 locations from 88 radio-collared elk over eight winter seasons during and after wolf colonization (1998–1999 to 2005–2006), and compared them to 4869 locations from 45 elk during seven winter seasons before wolves colonized the system (1991–1992 to 1997–1998).
3. The data supported our prediction that resource selection responses by elk in the Madison headwaters would be the same before and after wolf colonization, with the strength of the selection being stronger after wolves became established in the system.
4. After wolf recolonization, elk were less likely to occupy deeper snow areas in lower elevations as average snow pack on the landscape increased. This finding likely reflects individual behavioral choices by elk and the selective pressure of wolves differentially removing elk from areas with deeper snow.
5. We did not detect profound changes in habitat selection by elk during and after wolf colonization. Elk continued to select geothermal areas with higher food quantity and quality even though wolves also selected for these areas.
6. The distribution of elk became more constrained during and after wolf colonization. The proportion of elk in the Gibbon drainage decreased from 37% in 1997–1998 to zero by 2003–2004. The proportion of elk in the Firehole drainage was 30–43% during 1997–1998 through 2005–2006, but then decreased to 16% during 2006–2007. As a result, the proportion of elk in the Madison drainage increased from 23% in 1997–1998 to 84% by 2006–2007.
7. Elk apparently minimized predation risk during winter by selecting portions of the landscape that increased their probability of escape if attacked, while still providing relatively high quality vegetation and snow characteristics that allowed access to forage.
8. The change in elk distribution after wolf colonization based on landscape characteristics may account for many of the differences observed in elk responses to wolves among different systems. Fine-scale habitat and topographic features strongly influence elk and wolf habitat selection, creating areas with variable levels of predation risk.

VI. REFERENCES

Abrams, P. A., and C. J. Walters. 1996. Invulnerable prey and the paradox of enrichment. *Ecology* 77:1125–1133.

Agresti, A. 1990. *Categorical Data Analysis*. Wiley, New York, NY.

Andrewartha, H. G., and L. C. Birch. 1984. *The Ecological Web: More on the Distribution and Abundance of Animals*. University of Chicago Press, Chicago, IL.

Barber-Meyer, S. M., L. D. Mech, and P. J. White. 2008. *Survival and cause-specific elk calf mortality following wolf restoration to Yellowstone National Park. Wildlife Monographs*, 169.

Barten, N. L., R. T. Bowyer, and K. J. Jenkins. 2001. Habitat use by female caribou: Tradeoffs associated with parturition. *Journal of Wildlife Management* **65**:77–92.

Ben-David, M., K. Titus, and L. R. Beier. 2004. Consumption of salmon by Alaskan brown bears: A trade-off between nutritional requirements and the risk of infanticide? *Oecologia* **138**:465–474.

Berger, J. 1999. Anthropogenic extinction of top carnivores and interspecific animal behaviour: Implications of the rapid decoupling of a web involving wolves, bears, moose and ravens. *Proceedings Royal Society of London B Biological Sciences* **266**:2261–2267.

Berger, J. 2004. The last mile: How to sustain long-distance migration in mammals. *Conservation Biology* **18**:320–331.

Berger, J., J. E. Swenson, and I.-L. Persson. 2001. Recolonizing carnivores and naïve prey: Conservation lessons from Pleistocene extinctions. *Science* **291**:1036–1039.

Bergerud, A. T., and R. E. Page. 1987. Displacement and dispersion of parturient caribou at calving as antipredator tactics. *Canadian Journal of Zoology* **65**:1597–1606.

Bergman, E. J., R. A. Garrott, S. Creel, J. J. Borkowski, and R. M. Jaffe. 2006. Assessment of prey vulnerability through analysis of wolf movements and kill sites. *Ecological Applications* **16**:273–284.

Berryman, A. A., and B. A. Hawkins. 2006. The refuge as an integrating concept in ecology and evolution. *Oikos* **115**:192–196.

Bertram, B. C. R. 1978. Living in groups: Predator and prey. Pages 64–96 *in* J. R. Krebs and N. B. Davies (Eds.) *Behavioural Ecology: An Evolutionary Approach*. Blackwell, Oxford, United Kingdom.

Beutel, T. S., R. J. S. Beeton, and G. S. Baxter. 1999. Building better wildlife-habitat models. *Ecography* **22**:219–223.

Bowyer, R. T., J. G. Kie, and V. Van Ballenberghe. 1996. Sexual segregation in black-tailed deer: Effects of scale. *Journal of Wildlife Management* **60**:10–17.

Boyce, M. S., P. R. Vernier, S. E. Nielsen, and F. K. A. Schmiegelow. 2002. Evaluating resource selection functions. *Ecological Modeling* **157**:281–300.

Burnham, K. P., and D. R. Anderson. 2002. *Model Selection and Inference*. Springer, New York, NY.

Carbyn, L. N. 1974. *Wolf Predation and Behavioral Interactions with Elk and Other Ungulates in an Area of High Prey Diversity*. Canadian Wildlife Service, Edmonton, Alberta, Canada.

Cook, R. C., J. G. Cook, and L. D. Mech. 2004. Nutritional condition of northern Yellowstone elk. *Journal of Mammalogy* **85**:714–722.

Craighead, J. J., F. C. Craighead, R. L. Ruff, and B. W. O'Gara. 1973. Home ranges and activity patterns of nonmigratory elk of the Madison drainage herd as determined by biotelemetry. *Wildlife Monographs* **33**.

Creel, S., D. Christianson, S. Liley, and J. A. Winnie, Jr. 2007. Predation risk affects reproductive physiology and demography of elk. *Science* **315**:960.

Creel, S., and J. A. Winnie. 2005. Response of elk herd size to fine-scale and temporal variations in the risk of predation by wolves. *Animal Behavior* **69**:1181–1189.

Creel, S., J. A. Winnie, B. Maxwell, K. Hamlin, and M. Creel. 2005. Elk alter habitat selection as an antipredator response to wolves. *Ecology* **86**:3387–3397.

Crisler, L. 1956. Observations of wolves hunting caribou. *Journal of Mammalogy* **37**:337–346.

Dehn, M. M. 1990. Vigilance for predators: Detection and dilution effects. *Behavioral Ecology and Sociobiology* **26**:337–342.

Despain, D. G. 1990. *Yellowstone Vegetation: Consequences of Environment and History in a Natural Setting*. Roberts Rinehart, Boulder, CO.

Downes, S. 2001. Trading heat and food for safety: Costs of predator avoidance in a lizard. *Ecology* **82**:2870–2881.

Dussault, C., J.-P. Ouellet, R. Courtois, J. Huot, L. Breton, and H. Jolicoeur. 2005. Linking moose habitat selection to limiting factors. *Ecography* **28**:619–628.

Elgar, M. A. 1989. Predator vigilance and group size in mammals and birds: A critical review of the empirical evidence. *Biological Reviews of the Cambridge Philosophical Society* **64**:13–33.

Garrott, R. A., J. E. Bruggeman, M. S. Becker, S. T. Kalinowski, and P. J. White. 2007. Evaluating prey switching in wolf-ungulate systems. *Ecological Applications* **17**:1588–1597.

Garrott, R. A., L. L. Eberhardt, P. J. White, and J. Rotella. 2003. Climate-induced variation in vital rates of an unharvested large-herbivore population. *Canadian Journal of Zoology* **81**:33–45.

Garrott, R. A., J. A. Gude, E. J. Bergman, C. Gower, P. J. White, and K. L. Hamlin. 2005. Generalizing wolf effects across the greater Yellowstone area: A cautionary note. *Wildlife Society Bulletin* **33**:1245–1255.

Garrott, R. A., P. J. White, D. B. Vagnoni, and D. M. Heisey. 1996. Purine derivatives in snow-urine as a dietary index for free-ranging elk. *Journal of Wildlife Management* **60**:735–743.

Gause, G. F. 1934. *The Struggle for Existence*. Williams and Wilkins, Baltimore, MD.

Gibeau, M. L., A. P. Clevenger, S. Herrero, and J. Wierzchowski. 2002. Grizzly bear response to human development and activities in the Bow River Watershed, Alberta, Canada. *Biological Conservation* **103**:227–236.

Gilliam, J. F., and D. F. Fraser. 1987. Habitat selection under predation hazard: Test of a model with foraging minnows. *Ecology* **68**:1856–1862.

Grand, T. C. 2002. Alternative forms of competition and predation dramatically affect habitat selection under foraging-predation risk trade-offs. *Behavioural Ecology* **13**:280–290.

Gude, J. A., R. A. Garrott, J. J. Borkowski, and F. King. 2006. Prey risk allocation in a grazing ecosystem. *Ecological Applications* **16**:285–298.

Gustine, D. D., K. L. Parker, R. J. Lay, M. P. Gillingham, and D. C. Heard. 2006. Calf survival of woodland caribou in a multi-predator ecosystem. *Wildlife Monographs* **165**.

Hamilton, W. D. 1971. Geometry for the selfish herd. *Journal of Theoretical Biology* **31**:295–311.

Hassell, M. P. 1978. *The Dynamics of Arthropod Predator–Prey Systems.* Princeton University Press, Princeton, NJ.

Hebblewhite, M., E. H. Merrill, and T. L. McDonald. 2005. Spatial decomposition of predation risk using resource selection functions: An example in a wolf–elk predator–prey system. *Oikos* **111**:101–111.

Holling, C. S. 1965. The functional response of predators to prey density and its role in mimicry and population regulation. *Memoirs of the Entomological Society of Canada* **45**:1–62.

Hopcraft, J. G. C., A. R. E. Sinclair, and C. Packer. 2005. Planning for success: Serengeti lions seek prey accessibility rather than abundance. *Journal of Animal Ecology* **74**:559–566.

Huggard, D. J. 1993. Effect of snow depth on predation and scavenging by gray wolves. *Journal of Wildlife Management* **57**:382–388.

Iqbal, M. 1983. *An Introduction to Solar Radiation.* Academic Press, San Diego, CA.

Jakubas, W. J., R. A. Garrott, P. J. White, and D. R. Mertens. 1994. Fire-induced changes in the nutritional quality of lodgepole pine bark. *Journal of Wildlife Management* **58**:35–46.

Johnson, C. J., K. L. Parker, and D. C. Heard. 2001. Foraging across a variable landscape: Behavioral decisions made by woodland caribou at multiple spatial scales. *Oecologia* **127**:590–602.

Kauffman, M. J., N. Varley, D. W. Smith, D. R. Stahler, D. R. MacNulty, and M. S. Boyce. 2007. Landscape heterogeneity shapes predation in a newly restored predator–prey system. *Ecological Letters* **10**:690–700.

Keating, K. A., P. J. P. Gogan, J. M. Vore, and I. R. Irby. 2007. A simple solar radiation index for wildlife habitat studies. *Journal of Wildlife Management* **71**:1344–1348.

Kie, J. G. 1999. Optimal foraging and risk of predation: Effects on behavior and social structure in ungulates. *Journal of Mammalogy* **80**:1114–1129.

Krebs, J. R. 1980. Optimal foraging, predation risk, and territory defence. *Ardea* **68**:83–90.

Lima, S. L. 1992. Strong preferences for apparently dangerous habitats? A consequence of differential escape from predators. *Oikos* **64**:597–600.

Lima, S. L. 1998. Non-lethal effects in the ecology of predator–prey interactions: What are the ecological effects of anti-predator decision making? *Bioscience* **48**:25.

Lima, S. L., and L. M. Dill. 1990. Behavioral decisions made under the risk of predation: A review and prospectus. *Canadian Journal of Zoology* **68**:619–640.

Manly, B. F. J., L. L. McDonald, D. L. Thomas, T. L. McDonald, and W. P. Erickson. 2002. *Resource Selection by Animals: Statistical Design and Analysis for Field Studies.* Kluwer, Dordrecht, Netherlands.

Mao, J. S., M. S. Boyce, D. W. Smith, F. J. Singer, D. J. Vales, J. M. Vore, and E. H. Merrill. 2005. Habitat selection by elk before and after wolf reintroduction in Yellowstone National Park. *Journal of Wildlife Management* **69**:1691–1707.

McNay, M. E. 2002. Wolf–human interactions in Alaska and Canada: A review of the case history. *Wildlife Society Bulletin* **30**:831–843.

McNulty, D. R., L. D. Mech, and D. W. Smith. 2007. A proposed ethogram of large-carnivore predatory behavior, exemplified by the wolf. *Journal of Mammalogy* **88**:595–605.

Mech, L. D. 1977. Wolf-pack buffer zones as prey reservoirs. *Science* **198**:320–321.

Mech, L. D., and R. O. Peterson. 2003. Wolf–prey relations. Pages 131–160 *in* L. D. Mech and L. Boitani (Eds.) *Wolves: Behavior, Ecology, and Conservation.* University of Chicago Press, Chicago, IL.

Mech, L. D., D. W. Smith, K. M. Murphy, and D. R. MacNulty. 2001. Winter severity and wolf predation on a formerly wolf-free elk herd. *Journal or Wildlife Management* **65**:998–1003.

National Park Service, U.S. Department of the Interior. 2007. *Winter Use Plans Final Environmental Impact Statement for Yellowstone and Grand Teton National Parks and the John D. Rockefeller, Jr. Memorial Parkway.* U.S. Department of the Interior, Denver, CO.

Nelson, M. E., and D. L. Mech. 1981. Deer social organization and wolf predation in northeastern Minnesota. *Wildlife Monographs* 77.

Nelson, M. E., and D. L. Mech. 1986. Mortality of white-tailed deer in northeastern Minnesota. *Journal of Wildlife Management* **50**:691–698.

Nicholson, M. C., R. T. Bowyer, and J. G. Kie. 1997. Habitat selection and survival of mule deer: Tradeoffs associated with migration. *Journal of Mammalogy* **78**:483–504.

Ortega, I. M., and W. L. Franklin. 1995. Social organization, distribution and movements of a migratory guanaco population in the Chilean Patagonia. *Revista Chilena de Historia Natural* **68**:489–500.

Peterson, R. L. 1955. *North American Moose.* University of Toronto Press, Toronto, Ontario, Canada.

Peterson, R. O. 1977. *Wolf Ecology and Prey Relationships on Isle Royale.* National Park Service Scientific Monograph Series 11, Washington, DC.

Post, E., R. O. Peterson, N. C. Stenseth, and B. E. McLaren. 1999. Ecosystem consequences of wolf behavioural response to climate. *Nature* **401**:905–907.

Price, P. W., M. Westoby, B. Rice, P. R. Atsatt, R. S. Fritz, J. N. Thompson, and K. Mobly. 1986. Parasite mediation in ecological interactions. *Annual Review of Ecology and Systematics* **17**:487–505.

Pulliam, H. R. 1973. On the advantage of flocking. *Journal of Theoretical Biology* **38**:419–422.

Pulliam, H. R., and T. Caraco. 1984. Living in groups: Is there an optimal group size? Pages 122–147 *in* J. R. Krebs and N. B. Davies (Eds.) *Behavioural Ecology: An Evolutionary Approach.* Blackwell, Oxford, United Kingdom.

Rachlow, J. L., and R. T. Bowyer. 1998. Habitat selection by Dall's sheep (*Ovis dalli*): Maternal trade-offs. *Journal of Zoology* **245**:457–465.

Rettie, W. J., and F. Messier. 2000. Hierarchical habitat selection by woodland caribou: Its relationship to limiting factors. *Ecography* **23**:466–478.

Rosenzweig, M., and R. H. MacArthur. 1963. Graphical representation and stability conditions of predator–prey interaction. *American Naturalist* **97**:209–223.

Sand, H., C. Wikenros, P. Wabakken, and O. Liberg. 2006. Cross-continental differences in patterns of predation: Will naïve moose in Scandinavia ever learn? *Proceedings Royal Society of London B Biological Sciences* **273**:1421–1427.

SAS. 2000. *SAS/STAT User's Guide.* Version 8.

Schmidt, K. 1993. Winter ecology of non-migratory Alpine red deer. *Oecologia* **95**:226–233.

Schmitz, O. J. 1997. Behaviorally mediated trophic cascades: Effects of predation risk on food web interactions. *Ecology* **78**:1388–1399.

Sih, A. 1980. Optimal behavior: Can foragers balance two conflicting demands? *Science* **210**:1041–1043.

Sih, A. 1982. Foraging strategies and the avoidance of predation by an aquatic insect *Notonecta Hoffmanni. Ecology* **63**:786–796.

Smith, D. W., and D. R. Stahler. 2003. *Management of Habituated Wolves in Yellowstone National Park.* National Park Service, Mammoth, WY.

Strong, D. R., J. H. Lawton, and R. Southwood. 1984. *Insects on Plants: Community Patterns and Mechanisms.* Blackwell, Oxford, United Kingdom.

Sweitzer, R. A. 1996. Predation or starvation: Foraging decisions by porcupines (*Erethizon dorsatum*). *Journal of Mammalogy* **77**:1068–1077.

Thurber, J. M., R. O. Peterson, T. D. Drummer, and S. A. Thomasma. 1994. Grey wolf response to refuge boundaries and roads in Alaska. *Wildlife Society Bulletin* **22**:61–68.

Treherne, J. E., and W. A. Foster. 1982. Group size and anti-predator strategies in a marine insect. *Animal Behaviour* **32**:536–542.

Turner, M. G., and R. H. Gardner. 1991. *Quantitative Methods in Landscape Ecology: The Analysis and Interpretation of Landscape Heterogeneity.* Springer, New York, NY.

Vaughan, I. P., and S. J. Ormerod. 2005. The continuing challenges of testing species distribution models. *Journal of Applied Ecology* **42**:720–730.

Werner, E. E., and D. J. Hall. 1988. Ontogenetic habitat shifts in bluegill: The foraging rate-predation risk trade off. *Ecology* **69**:1352–1366.

White, G. C. 1996. NOREMARK: Population estimation from mark–resighting surveys. *Wildlife Society Bulletin* **24**:50–52.

Whittington, J., C. C. St. Clair, and G. Mercer. 2005. Spatial responses of wolves to roads and trails in mountain valleys. *Ecological Applications* **15**:543–553.

Winnie, J. W., D. Christianson, S. Creel, and B. Maxwell. 2006. Elk decision-making rules simplified in the presence of wolves. *Behavioral Ecology and Sociobiology* **61**:277–289.

Yasué, M., J. L. Quinn, and W. Cresswell. 2003. Multiple effects of weather on the starvation and predation risk trade-off in choice of feeding location in Redshanks. *Functional Ecology* **17**:727–736.

CHAPTER 22

Elk Nutrition after Wolf Recolonization of Central Yellowstone

P. J. White,* Robert A. Garrott,[†] John J. Borkowski,[‡] Kenneth L. Hamlin,[§] and James G. Berardinelli[¶]

*National Park Service, Yellowstone National Park
[†]Fish and Wildlife Management Program, Department of Ecology, Montana State University
[‡]Department of Mathematical Sciences, Montana State University
[§]Montana Fish, Wildlife, and Parks
[¶]Department of Animal and Range Sciences, Montana State University

Contents

Theme

Winter for ungulates in the northern Rocky Mountains is a period of chronic under-nutrition and catabolism of lean muscle and body fat because forage availability and quality are low while energetic demands are high (Moen 1976, Torbit *et al.* 1985, DelGuidice *et al.* 1990). Prior to wolf reintroduction, the primary cause of mortality for elk wintering in Yellowstone was starvation (Houston 1982, Garrott *et al.* 2003, Chapter 10 by Garrott *et al.*, this volume). After wolf reintroduction, predation was the primary cause of death, and survival and recruitment decreased significantly (Garrott *et al.* 2005, White and Garrott 2005, Creel *et al.* 2007, Chapter 23 by Garrott *et al.*, this volume). Anti-predator behaviors such as increased vigilance, shifts in distribution to lower risk areas, and increases in group size could impose additional trade-offs in diet selection due to reduced forage availability, intake, or quality; thereby resulting in decreased nutrition, nutritional condition and, in turn, reproduction and survival (Morgantini and Hudson 1985, Ives and Dobson 1987, Illius and FitzGibbon 1994, Heithaus and Dill 2002, Cowlishaw *et al.* 2004, Creel *et al.* 2007). White *et al.* (Chapter 9, this volume) provided an understanding of the diet composition and nutrition of elk in the Madison headwaters area of Yellowstone National Park when the sole limiting factor was starvation due to chronic under-nutrition and energetic costs, and other limiting factors such as predation and hunting were minimal or absent. We evaluated the nutritional status of these elk after wolves began colonizing the system to determine how elk nutrition changed when they were confronted with tradeoffs between predation risk and starvation.

The Ecology of Large Mammals in Yellowstone
R. Garrott, P. J. White and F. Watson
ISSN 1936-7961, DOI: 10.1016/S1936-7961(08)00222-4

Copyright © 2009, Elsevier Inc.
All rights reserved.

I. INTRODUCTION

The rate of ingestion of assimilable energy and nutrients, and the resulting state of body components, reflect the adequacy of forage quality and quantity (Harder and Kirkpatrick 1994, Parker *et al.* 1999, Cook *et al.* 2004a,b). Furthermore, nutrition and nutritional condition influence strongly the probability of breeding, overwinter survival, lactation yields, and susceptibility to predation (Loudon *et al.* 1983, Hobbs 1989, Kohlmann 1999, Mech *et al.* 2001, Bender *et al.* 2002, Cook 2002). For example, the probability of pregnancy in Yellowstone elk was low if body fat was <6% and death due to starvation greatly increased at <2% body fat (Cook *et al.* 2004b). Moreover, juvenile survival and recruitment were highly sensitive to stochastic environmental perturbations that reduced nutrition and condition (Garrott *et al.* 2003).

Elk populations in Yellowstone National Park were at high densities prior to the restoration of wolves, but the abundance of elk decreased by >50% following restoration (Garrott *et al.* 2005, White and Garrott 2005). This decrease in elk density may have increased *per capita* resources and, possibly, the nutritional quality of diets. However, predation risk is an important variable driving many behavioral decisions and likely exerts strong selection pressures on foraging decisions. Elk in different areas of the Greater Yellowstone Area responded to the risk of predation by wolves with a variety of behaviors, including increased vigilance, reduced foraging time, shifts in distribution to lower risk areas, and increases or decreases in group size (Ripple *et al.* 2001, Creel and Winnie 2005, Creel *et al.* 2005, Fortin *et al.* 2005, Mao *et al.* 2005, Gude *et al.* 2006, Chapters 18 and 19 by Gower *et al.*, this volume). Anti-predator behaviors like these can reduce predation risk, but are also likely to carry costs (Ives and Dobson 1987, Illius and FitzGibbon 1994, Heithaus and Dill 2002, Cowlishaw *et al.* 2004). For example, Creel *et al.* (2007) suggested wolves indirectly affected the reproductive physiology and demography of elk through the costs of anti-predator behavior, likely through effects on nutrition, body condition, and decreased birth rates (Cook *et al.* 2001, 2004b). Such anti-predator costs could have complex and important consequences for elk populations depending on whether they increased the strength of density dependence by concentrating elk in lower risk areas or decreased population growth rates in an independent manner (*e.g.*, decreased pregnancy; Creel *et al.* 2007).

Analysis of urinary metabolite ratios from urine-saturated snow samples is a technique for indexing winter nutritional restriction of ungulates (DelGuidice *et al.* 1989). Urinary metabolite data are reported as ratios of a specific metabolite to creatinine, a compound produced and excreted in relatively constant proportion to muscle mass; thereby allowing comparisons to be made among animals of different body size and hydration, and among samples of different snow dilution (Coles 1980). Allantoin is a metabolite that reliably indexes rumen microbial flow and the amount of metabolizable energy intake by elk during a several-day period prior to the time urine was voided (Vagnoni *et al.* 1996, Garrott *et al.* 1997). Allantoin: creatinine (A:C) ratios were significantly correlated with winter severity and over-winter survival of Yellowstone elk prior to wolf restoration (Chapter 9 by White *et al.*, this volume), and provided a sensitive technique for detecting biologically meaningful differences in nutrition among herds subjected to different management regimes (Garrott *et al.* 1996, 1997; Pils *et al.* 1999).

During winter, elk in Yellowstone face high wolf densities, at least on short time frames and spatial scales (Chapter 17 by Becker *et al.*, this volume). However, they also face severe weather conditions, less nutritious food, and loss of fat reserves and body weight (Craighead *et al.* 1973, Cook 2002). Elk can actively change their foraging strategies to collect more energy or protein because metabolizable energy is maximized by grazing (grasses, rushes, sedges), while dietary protein intake is maximized by browsing (conifers, deciduous trees, shrubs; Robbins 1993, Van Soest 1994). Prior to wolf colonization, elk in the Madison headwaters area were primarily grazers, eating grasses and grass-like plants (Chapter 9 by White *et al.*, this volume), and varied their diet mix in response to seasonal changes in forage quality to maintain consistently high protein content (7%) in their winter diets. Protein was likely needed by rumen microbes to sustain adequate fermentation rates with diets containing 40% digestible carbohydrate (Chapter 9 by White *et al.*, this volume). This demand may increase the importance of access to areas with high-protein plants, thereby making the predation risk-foraging trade-off more obvious than in many other areas or species. Elk altered their distribution and use of the winter landscape after wolves became established in the system, occupying lower-elevation geothermal and meadow areas with less snow along sections of rivers that provided escape terrain (Chapter 21 by White *et al.*, this volume). This altered selection may have limited access to high quality foraging areas and plants with higher protein content. Thus, we expected predation risk by wolves to impose strong selection pressures on foraging decisions and reduce nutrition as indexed by A:C ratios.

II. METHODS

We estimated metabolizable energy intake (nutrition) by elk during and after wolf colonization (1996–1997 through 2005–2006) of the Madison headwaters area of Yellowstone National Park using A:C ratios from fresh urine deposited in snow. We opportunistically collected samples from mixed groups of adult females and calves (*i.e.*, batch samples) as described by Pils *et al.* (1999) during nine 14-day intervals between 4 December and 8 April each winter. Sample storage and assay procedures were described in Chapter 9 by White *et al.*, this volume. We compared samples collected during the winters of 1996–1997 and 1997–1998 using these methods to A:C ratios from samples deposited by adult, female, radio-collared elk during the same winters (Chapter 9 by White *et al.*, this volume) to evaluate if both sampling methods provided similar results. We censored data from 1998–1999 through 2000–2001 because only 1–6 samples per month were collected during mid- to late-winter. We also censored interval one in 2001–2002, and interval nine in 2001–2002 and 2005–2006, because ≤2 samples were collected (Appendix).

Pils *et al.* (1999) reported that mean A:C ratios of samples collected from mixed groups of adult females and calves (*i.e.*, batch samples) were highly variable and consistently overestimated the true mean of A:C ratios for adult female elk. These problems were alleviated by trimming 15% off the right tail of the ordered sample distribution and 20% off the left tail and calculating the mean of the remaining samples. Trimmed estimates of mean A:C ratios were consistently less biased, with lower variability of sample means and smaller confidence intervals than untrimmed means (Pils *et al.* 1999). We then used SAS PROC GLM (SAS 1992) to generate curves of trimmed mean A:C ratios for each winter. We plotted mean A:C ratios by collection date and then integrated, yielding the area under the curves. Additionally, data from early (December 1–31) and mid-winter (15 January to 25 February) time periods were used to compare nutrition among winters: We ran an ANOVA for the following model:

$$A:C = \mu + Year_i + Period_j + (Year \times Period)_{ij} + \varepsilon$$

where μ is the intercept, $Year_i$ is the *i*th year effect ($i = 1–7$), $Period_j$ is the *j*th period ($j =$ early or mid-winter), and ε is the random error. We conducted Bonferroni multiple comparison tests of the 28 pairwise comparisons of means for the seven winters for Period = early, and another 28 pairwise comparisons of means for the seven winters for Period = mid-winter. We used an overall $\alpha = 0.05$ (or $\alpha^* = 0.05/28 = 0.00178$ per pairwise comparison) to evaluate if pairs of means were significantly different from each other.

Trends in A:C ratios during and after colonization by wolves were employed to evaluate the prediction that trade-offs between predation risk and food acquisition would result in decreased nutrition after colonization. Winter maintenance requirements (550–725 kJ metabolizable energy/kg body mass$^{0.75}$) for free-ranging elk, and the A:C ratio (0.34–0.4) that represented this level of metabolizable energy intake, were estimated from experimental feeding trials with groups of captive elk (Garrott *et al.* 1997). The area above the A:C curve for each year and below a threshold A:C level of 0.30 (*i.e.*, submaintenance index) was used to evaluate the relative extent elk nutrition was below maintenance each winter during 18 December through 25 March. We regressed the submaintenance index each winter during and after wolf colonization against accumulated snow water equivalent (SWE_{acc}), which is the sum of mean daily SWE values for the elk winter range during 1 October to 30 April (Garrott *et al.* 2003). We also regressed the submaintenance index against wolf-days, which was calculated as the number of individual wolves located in the elk winter range each day (Chapter 16 by Becker *et al.*, this volume), and adult and calf survival. Adult survival was estimated each year using the proportion of radio-collared females that survived (Evans *et al.* 2006). These estimates may have been confounded somewhat by the age distribution of radio-collared females each year because there was a strong effect of age on survival (*i.e.*, senescence) prior to wolf recolonization (Chapter 11 by Garrott *et al.*, this volume). Calf survival was estimated using the change-in-ratio method (Skalski *et al.* 2005) with autumn and spring calf:cow ratios calculated from groups associated with the first 100 and last 100 radio-collared animals observed during telemetry locations. These estimates were adjusted for adult female survival using a pooled estimate (0.85) for 1996–1997 through 2005–2006 (Chapter 23 by Garrott *et al.*, this volume).

In addition, we used regression techniques in R version 2.5.0 (R Development Core Team 2006) to evaluate the strength of evidence in the data for competing *a priori* regression models describing the nutrition of elk in the Madison headwaters area before and after wolves began colonizing the system (Burnham and Anderson 2002). We fit models for the pre-wolf suite (radio-collar data; 1991–1992 through 1997–1998; Chapter 9 by White *et al.*, this volume), wolf colonization and establishment suite (batch sample data; 1996–1997 through 2005–2006), and the full data set suite (1991–1992 through 2005–2006). We estimated coefficients and used Akaike's Information Criteria corrected for small sample size (AIC_c) as our model selection criterion (Burnham and Anderson 2002).

We chose one warm season and one cold season covariate to capture climate-driven annual variation in density-independent factors that we suspected influenced elk vital rates. Temperature and precipitation regimes during the warm season are fundamental drivers of plant phenology and productivity which, in turn, affect the quality and quantity of forage available to large herbivores through the year (McNaughton 1985, Sala *et al.* 1988, Stephenson 1990). While it has been common to use temperature, precipitation, or some combination of these measures as covariates in ecological studies of large herbivores (Caughley and Gunn 1993, Loison and Langvatn 1998, Clutton-Brock and Coulson 2002, Vucetich *et al.* 2005, Wang *et al.* 2006), the recent availability of AVHRR satellite data (1989–2006) and algorithms for extracting seasonal normalized difference vegetation index (NDVI) parameters provide more direct measures of annual variation in vegetation canopy dynamics (Malingreau 1989, Reed *et al.* 1994, Chapter 7 by Thein *et al.*, this volume) that are biologically relevant to large herbivores. While forests dominate the study area (Chapter 2 by Newman *et al.*, this volume), elk are primarily grazers and, to best index annual variations in forage, we selected the seven largest meadow complexes in the study area, ranging from 1 to 48 km^2 and encompassing a total of approximately 80 km^2. The meadows were distributed across the entire elevational gradient and through the Firehole, Gibbon, and Madison river drainages. AVHRR satellite data at 1-km^2 resolution were processed by the U.S. Geological Survey Center for EROS (Eidenshink 1992, DeFelice *et al.* 2003) with NDVI values composited using the maximum value method over 14-day periods to eliminate spuriously low values caused by cloud contamination, haze, and other atmospheric effects, and further temporally smoothed using the methods of Swets *et al.* (2000). Data included 26 composites per year and program TIMESAT with the Savitsky–Golay filter (Jönsson and Eklundh 2002, 2004) was used to extract NDVI metrics for each polygon and average across all polygons for each year of the study. We evaluated growing season length ($NDVI_{len}$), which indexes the time period when green high-quality vegetation was available (Chapter 7 by Thein *et al.*, this volume).

Snow is a fundamental abiotic ecological force in mid- to high-latitude ecosystems and we used the Langur snow pack model (Watson *et al.* 2006a,b; Chapter 6 by Watson *et al.*, this volume) to generate a cold season covariate that indexed annual variations in snow pack. The snow pack model generated daily estimates of snow water equivalent (SWE; the water content of a column of snow measured in cm) for the study area using temporally dynamic precipitation and temperature data, and spatially explicit slope, aspect, elevation, ground heat, vegetation cover type and mean annual precipitation data. Mean daily SWE was calculated as the average of all pixels in the study area and represented the average water content of the snow pack for that day. An annual index of snow pack severity (SWE_{acc}) was calculated by summing the mean daily SWE values during 1 October to 31 April, thereby providing a single metric integrating snow pack depth, density, and duration. Prior to analyses, SWE_{acc} was scaled by dividing by 1000.

We developed two wolf covariates to assess our predictions about the influence of predation risk on the nutrition of elk. We detected and quantified the presence of wolves from 15 November through 30 April during 1996–1997 to 2006–2007. National Park Service biologists captured wolves to maintain radio-collared animals in each pack (Smith 2005), and we monitored wolf presence and their locations using ground-based telemetry, snow tracking, and visual observations of collared and un-collared individuals. We intensively monitored each drainage for wolf presence daily with crews of 3–4 people using snowmobiles, vehicles, snowshoes, and high points in the landscape to facilitate telemetry and observations. When packs with radio-collared wolves were detected, we estimated telemetry locations using triangulation (White and Garrott 1990) and obtained multiple locations through the day. Snow tracking, visual observations, and aerial monitoring by National Park Service biologists provided estimates of the number of animals per pack and aided our daily assessments of wolf presence or absence (Chapters 16 and 17 by Becker *et al.*, this volume). Detection of un-collared wolves was facilitated by opportunistic observations of tracks and wolves by field personnel that were working close to the roads or in the back country on elk and bison investigations. The covariate WOLF categorized wolves as present or absent in the study area each winter. However, wolves became established in different drainages during different winters (Firehole: 1997–1998, Gibbon: 2000–2001, Madison: 2001–2002). Thus, we used wolf presence data to construct a continuous wolf covariate ($WOLF_{days}$) that summed the number of wolves located in the study area each day during winter.

III. RESULTS

We collected 1323 samples of urine-soaked snow from elk in the Madison headwaters area in seven winters during and after wolf colonization (Appendix I). Overall, mean A:C ratios estimated from trimmed batch samples and samples collected from radio-collared adult females were identical at 0.17 in 1996–1997 and 0.22 in 1997–1998. Likewise, estimates of the relative extent elk nutrition was below maintenance each winter and the lowest A:C ratio each winter were also similar between these collection techniques. Thus, the batch sampling method with trimmed means provided a comparable estimate of A:C ratios as that derived from individually identified female elk.

We detected differences in mean A:C ratios among winters (Tables 22.1 and 22.2), though curves of predicted A:C ratios were generally similar in shape (Figure 22.1). The relative extent elk nutrition was below maintenance each winter, as assessed by A:C ratios from trimmed batch samples during and after wolf colonization (1996–1997 through 2005–2006), was not significantly correlated with SWE_{acc} ($R^2 = 0.06$, $F_{1,5} = 0.31$, $P = 0.60$), $NDVI_{len}$ ($R^2 = 0.02$, $F_{1,5} = 0.10$, $P = 0.77$), $WOLF_{days}$ ($R^2 = 0.01$, $F_{1,5} = 0.02$, $P = 0.90$), adult survival ($R^2 = 0.21$, $F_{1,5} = 1.35$, $P = 0.30$), or calf survival ($R^2 = 0.11$, $F_{1,5} = 0.62$, $P = 0.47$).

The most parsimonious models for elk nutrition from separate analyses of the pre-wolf radio-collared samples and the trimmed batch samples during and after wolf colonization were the same and consisted of the null models (radio-collar: $AIC_c = -24.60$, $w_i = 0.77$; batch: $AIC_c = -21.60$, $w_i = 0.90$; Table 22.3). There was also some support for a model with a significant SWE_{acc} covariate ($\beta_1 = -0.014$, SE = 0.005, $P = 0.05$) from the radio-collared samples prior to wolf recolonization ($\Delta AIC_c = 2.84$, $w_i = 0.18$). However, no other competing models were well supported for the batch samples during

TABLE 22.1 Nutrition, environmental, and survival data for elk in the Madison headwaters area of Yellowstone National Park during the winters of 1991–1992 through 2005–2006

Winter	Mean A:C ratio	Extent A:C below submaintenance	Lowest A:C ratio	SWE_{acc} (m)	Wolf (1 = present)	$Wolf_{days}$ (/1000)	$NDVI_{len}$	Adult survival	Calf survival
1991–1992	0.23	7.84	0.18	2.094	0	0	14.11	1.00	
1992–1993	0.18	11.67	0.15	2.644	0	0	16.22	1.00	0.230
1993–1994	0.23	6.75	0.22	1.744	0	0	14.48	1.00	0.595
1994–1995	0.21	9.04	0.18	3.738	0	0	17.41	0.89	0.235
1995–1996	0.19	11.14	0.15	3.130	0	0	14.56	0.83	0.169
1996–1997	0.17	12.57	0.14	5.693	0	0.060	13.88	0.81	0.000
1997–1998	0.22	7.66	0.18	2.026	1	0.131	17.91	0.97	0.530
2001–2002	0.18	11.08	0.16	2.352	1	1.695	17.93	0.80	0.110
2002–2003	0.21	9.34	0.18	1.707	1	1.526	16.17	0.83	0.300
2003–2004	0.28	3.27	0.24	2.326	1	2.671	16.01	0.61	0.120
2004–2005	0.16	14.30	0.14	1.744	1	3.964	16.80	0.81	0.130
2005–2006	0.17	12.61	0.15	2.509	1	1.346	16.25	0.88	0.170

A:C, allantoin:creatinine; SWE_{acc}, accumulated snow water equivalent; and $NDVI_{len}$, length of the growing season.

TABLE 22.2 Comparison of mean trimmed A:C ratios for mixed calf-cow groups in the Madison headwaters area of Yellowstone National Park, Wyoming, USA, during winters 1996–1997, 1997–1998, and 2001–2002 through 2005–2006 for early and mid-winter periods

Early winter (December 1–31)				Mid-winter (January 15–February 25)					
Mean	Winter			Mean	Winter				
0.2128	2001–2002	A		0.1438	**2004–2005**	A			
0.2162	**2005–2006**	A		0.1604	1996–1997	A	B		
0.2163	**2004–2005**	A		0.1628	**2005–2006**	A	B		
0.2168	1996–1997	A		0.1858	2001–2002		B	C	
0.2726	**2002–2003**		B	0.1865	1997–1998		B	C	
0.3038	**2003–2004**		B	0.1889	**2002–2003**			C	
0.3210	1997–1998		B	0.2361	**2003–2004**				D

We used the Bonferroni multiple comparison test with an overall $\alpha = 0.05$ (or $\alpha^* = 0.05/28 = 0.00178$ per pairwise comparison). Pairs of means connected by the same letter were not considered significantly different from each other. Post-wolf years are identified by bold font.

and after wolves colonized the system (Table 22.3). The most parsimonious model for nutrition from the full data set (1991–1992 through 2005–2006) consisted of the null model ($AIC_c = -44.74$, $w_i = 0.54$; Table 22.3). There was also some support for a model with a SWE_{acc} covariate ($\Delta AIC_c = 1.87$, $w_i = 0.21$; Table 22.3), though the covariate was not significant ($\beta_1 = -0.009$, SE $= 0.009$, $P = 0.34$).

IV. DISCUSSION

Prior to the restoration of wolves, elk wintering in the Madison headwaters area of Yellowstone National Park selected diets dominated by graminoids and browse (59% grass, 33% browse, 8% forbs). Elk selectively varied their diet mix in response to seasonal changes in forage quality. As the difference in protein content of browse and grass increased during October–January, elk ate less grass and

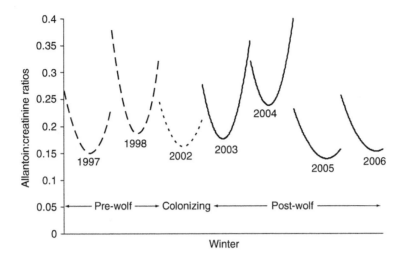

FIGURE 22.1 Predicted allantoin:creatinine (A:C) ratios derived from random collections (*i.e.*, batch samples) of urine-soaked snow samples from elk in the Madison headwaters area of Yellowstone National Park before, during, and after wolf colonization (winters 1996–1997 to 2005–2006). Samples were collected during nine 14-day intervals between 4 December and 8 April each winter. For each sampling period, the largest 20% and the smallest 15% of A:C ratios were trimmed from the dataset prior to analysis.

TABLE 22.3 Model selection results and coefficient values for the best-supported models examining the effects of various covariates on elk nutrition, as estimated by mean allantoin:creatinine ratios, before and after wolf colonization in the Madison headwaters area of Yellowstone National Park during 1991–1992 through 2005–2006

| Model suite | K | Pre-wolf (collar) 1991–1992 to 1997–1998 | | Post-wolf (batch) 1996–1997 to 2005–2006 | | Full data set 1991–1992 to 2005–2006 | |
		ΔAIC_c	w_i	ΔAIC_c	w_i	ΔAIC_c	w_i
Constant	1	0	0.77	0	0.90	0	0.54
SWE_{acc}	2	2.84	0.18	6.61	0.03	1.87	0.21
$NDVI_{len}$	2	6.91	0.02	6.99	0.03	3.65	0.09
WOLF	2	6.86	0.02	6.43	0.04	3.58	0.09
SWE_{acc} + WOLF	3	16.77	0.00	20.20	0.00	6.40	0.02
SWE_{acc} + $NDVI_{len}$	3	16.82	0.00	20.18	0.00	6.17	0.02
SWE_{acc} + $WOLF_{days}$	3	16.82	0.00	20.59	0.00	6.33	0.02
SWE_{acc} + WOLF + $NDVI_{len}$	4	58.76	0.00	61.31	0.00	12.45	0.00
SWE_{acc} + $WOLF_{days}$ + $NDVI_{len}$	4	58.81	0.00	61.97	0.00	12.33	0.00

We fit models for the pre-wolf suite (radio-collar data; 1991–1992 through 1997–1998; Chapter 9 by White *et al.*, this volume), wolf colonization and establishment suite (batch sample data; 1996–1997 through 2005–2006), and the full data set suite (1991–1992 through 2005–2006). Abbreviations: K, number of model parameters; ΔAIC_c, AIC corrected for small sample size; w_i, AIC model weight. AIC_c values for the top models in each suite were pre-wolf (collar) = −24.60, post-wolf (batch) = −27.97, and full data set = −50.68.

more browse with its higher protein content. This trend reversed in spring when the nutritional quality of grasses and grass-like plants increased. Metabolizable energy intake (nutrition), as indexed by A:C ratios, followed a similar trend to grass content in the diet, gradually decreasing during the snow build-up portion of winter, but then increasing rapidly during the snow melt-out phase in spring. Snow pack had a significant influence on elk nutrition among winters and, in turn, nutrition was significantly related to adult female and calf survival (Chapter 9 by White *et al.*, this volume).

There was no indication in our data that the nutrition of elk wintering in the Madison headwaters area was lower after wolves colonized the system. The most parsimonious model for nutrition from the full data set (1991–1992 through 2005–2006) was the null model, indicating our covariates were not adequate in explaining variation in A:C before and after wolf colonization. Also, regression analyses did not detect a relationship between nutrition and wolf use of the study area. However, there was apparently some nutritional effect because snow pack no longer influenced elk nutrition following wolf restoration and nutrition was not significantly related to adult female and calf survival. We suspect there were confounding effects for wolves that we did not cleanly assess.

Many prey species alter their foraging behavior in response to predation risk, trading a reduction in food quality or quantity for increased security (Abramsky *et al.* 2002, Heithaus and Dill 2002). For example, elk in the Gallatin Canyon and lower Madison Valley areas of Montana responded strongly to the presence of wolves in a drainage on a time scale of hours to days and a spatial scale of 1–2 km or less, with strong effects on habitat use (Creel *et al.* 2005, Gude *et al.* 2006). A strong shift in habitat use, movement, activity, or other behavioral patterns in response to predation risk could alter diets or energy budgets, as has been shown for elk that move from grassland to woodland in response to human hunters (Morgantini and Hudson 1985). This is especially true for ungulates in northern temperate latitudes during winter that experience chronic under-nutrition and catabolism of lean muscle and body fat because forage availability and quality are low while energetic demands are high (Moen 1976, Torbit *et al.* 1985, DelGuidice *et al.* 1990).

After wolf recolonization, elk in the Madison headwaters area were less likely to occupy deeper snow areas and used lower elevations as average snow pack on the landscape increased (Chapter 21 by White *et al.*, this volume). However, we did not detect profound changes in habitat selection by elk during and after wolf colonization. Elk continued to select geothermal areas with higher food quantity and quality even though wolves also selected for these areas (Chapter 21 by White *et al.*, this volume). Likewise, there was no indication of a substantial change in foraging time or vigilance after wolf colonization. The effects of snow pack and time of day were more influential on foraging behavior than wolves (Chapter 20 by Gower *et al.*, this volume). Thus, we did not detect strong shifts in habitat use or other behavioral patterns in response to predation risk that likely altered diets or nutrition.

The major effect of wolves on elk in the Madison headwaters area was to constrain their distribution. Prior to wolf colonization, elk were distributed throughout the Madison headwaters area (Chapter 21 by White *et al.*, this volume). As wolf use of the study area increased to high levels by 2004–2005, however, radio-collared elk were detected in larger groups in fairly predictable geothermal areas and meadow complexes (Chapters 18 and 19 by Gower *et al.*, this volume). This redistribution likely reflected individual behavioral choices by elk and the selective pressure of wolves differentially removing elk from areas with deeper snow. The majority of elk remaining in the Madison headwaters area aggregate in flat, valley-bottom geothermal and meadow areas where the open landscape, depth and width of the Madison and Firehole rivers, and close proximity (<200 m) to roads and frequent human activities provide refugia and escape habitat that reduce vulnerability to predation by wolves (Chapter 21 by White *et al.*, this volume). Thus, elk apparently minimized predation risk during winter by selecting portions of the landscape that served as escape terrain, while still providing relatively high quality vegetation and snow characteristics that allowed access to forage.

The diet of elk during winter consists primarily of grasses and flat grasslands generally contain higher food biomass and have more available food than other habitat types (Christianson and Creel 2007). Prior to wolf colonization, however, the protein content of the winter diets of elk in the Madison headwaters area was high (7%) and approximated the crude protein needed by rumen microbes to sustain adequate fermentation rates with their low digestibility (30–40%) diets (Chapter 9 by White *et al.*, this volume). The selection of high protein forages (*e.g.*, browse) during winter may have contributed to increased digestion of protein-deficient grasses and a more-favorable protein balance (Hume *et al.* 1970, Van Gylswyk 1970, Schwartz and Gilchrist 1975, Swick and Benevenga 1977, Mould and Robbins 1981). By constraining elk distribution to primarily grasslands, we thought wolves could potentially limit elk access to foraging areas with high protein plants (*e.g.,* browse); thereby reducing their nutrition somewhat. However, the nutritional indices did not support this prediction and, unfortunately, we do not have quantitative information on elk diets after wolf colonization to assess it further.

Creel *et al.* (2007) suggested that wolves indirectly decreased the birth rates of elk through costs of anti-predator behavior on nutrition and body condition, and that these costs may be large and strongly influence the dynamics of elk populations. Certainly, behavioral responses can carry physiological costs because energy is limiting for elk in winter and anti-predator responses may require an increase in energy expenditure or reduced foraging time and/or efficiency (Caraco 1979, Illius and Fitzgibbon 1994, Brown *et al.* 1999). Thus, it is conceivable that a reduction in the nutrition of elk in the Madison headwaters area after wolf colonization could have had some influence on demography (Cook 2002, Cook *et al.* 2004a,b). However, elk behavioral responses may be quite plastic and allow for minimizing predation risk without evoking major physiological "costs" (Chapter 20 by Gower *et al.*, this volume). Digestible energy intake by elk during the post-wolf winters was not significantly lower than prior to wolf colonization, and several studies have demonstrated that resource limitation (regardless of the cause) does not substantially decrease pregnancy rates of prime-aged females until it becomes rather severe (Fowler 1981, Eberhardt 1977, 2002, Gaillard *et al.* 1998). Also, pregnancy suppression has not been detected in other areas where wolves are established (*e.g.,* White and Garrott 2005, Chapter 25 by Hamlin *et al.*, this volume). Further, the combination of observed pregnancy rates combined with known direct causes of mortality can account for calf to adult female ratios observed at the end of the calves' first year of life in Yellowstone, without invoking pregnancy suppression due to wolves as a mechanism (Chapter 23 by Garrott *et al.*, and Chapter 25 by Hamlin *et al.*, this volume). Thus, we suggest the arguments for substantial demographic consequences of indirect effects of elk behavioral responses to wolf predation risk are not yet substantiated with empirical data and further research will be required before the contribution of indirect predation effects on prey demographic processes will be understood.

V. SUMMARY

1. We used allantoin:creatinine (A:C) ratios to estimate the nutrition of elk during and after wolf colonization of the Madison headwaters area in Yellowstone National Park. We compared these findings to the nutrition of elk before wolves and evaluated the strength of evidence in the data for competing models describing nutrition during 1991–1992 through 2005–2006.

2. There was no indication that the nutrition of elk was lower after wolves colonized the system. The most parsimonious model for nutrition from the full data set (1991–1992 through 2005–2006) was the null model, indicating constant nutrition before and after wolf colonization. Also, regression analyses did not suggest a relationship between nutrition and wolf use of the area.

3. We did not detect strong shifts in habitat use or other behavioral patterns in response to predation risk (Chapter 20 by Gower *et al.*, and Chapter 21 by White *et al.*, this volume) that likely altered diets or nutrition. Also, the survival of adult and calf elk were not significantly correlated with the extent nutrition was below maintenance during and after wolf colonization.

4. Elk apparently minimized predation risk during winter by selecting portions of the landscape that served as escape terrain, while still providing relatively high quality vegetation and snow characteristics that allowed access to forage.

VI. REFERENCES

Abramsky, Z., M. L. Roswnzweig, and A. Subach. 2002. The costs of apprehensive foraging. *Ecology* **83:**1330–1349.

Bender, L. C., E. Carlson, S. M. Schmitt, and J. B. Haufler. 2002. Production and survival of elk (*Cervus elaphus*) calves in Michigan. *American Midland Naturalist* **148:**163–171.

Brown, J. S., J. W. Laundre, and M. Gurung. 1999. The ecology of fear: Optimal foraging, game theory, and trophic interactions. *Journal of Mammalogy* **80:**385–399.

Burnham, K. P., and D. R. Anderson. 2002. *Model Selection and Multi-Model Inference.* Springer, New York, NY.

Caraco, T. 1979. Time budgeting and group size: A test of theory. *Ecology* **60:**618–627.

Caughley, G., and A. Gunn. 1993. Dynamics of large herbivores in deserts: Kangaroos and caribou. *Oikos* **67:**47–55.

Christianson, D. A., and S. Creel. 2007. A review of environmental factors affecting elk winter diets. *Journal of Wildlife Management* **71:**164–176.

Clutton-Brock, T. H., and T. N. Coulson. 2002. Comparative ungulate dynamics: The devil is in the detail. *Philosophical Transactions of the Royal Society B* **357:**1285–1298.

Coles, E. H. 1980. *Veterinary Clinical Pathology.* W. B. Saunders, Philadelphia, PA.

Cook, J. G. 2002. Nutrition and food. Pages 259–349 *in* D. E. Toweill and J. W. Thomas (Eds.) *North American Elk: Ecology and Management.* Smithsonian Institution Press, Washington, DC.

Cook, J. G., B. K. Johnson, R. C. Cook, R. A. Riggs, T. Delcurto, L. D. Bryant, and L. L. Irwin. 2004a. Effects of summer-autumn nutrition and parturition date on reproduction and survival of elk. *Wildlife Monographs* **155.**

Cook, R. C., J. G. Cook, and L. D. Mech. 2004b. Nutritional condition of northern Yellowstone elk. *Journal of Mammalogy* **85:**714–722.

Cook, R. C., D. L. Murray, J. G. Cook, P. Zager, and S. L. Monfort. 2001. Nutritional influences on breeding dynamics in elk. *Canadian Journal of Zoology* **79:**845–853.

Cowlishaw, G. M., M. G. M. Lawes, M. M. Lightbody, A. M. Martin, R. M. Pettifor, and J. M. M. Rowcliffe. 2004. A simple rule for the costs of vigilance: Empirical evidence from a social forager. *Proceedings of Biological Sciences* **271:**27–33.

Craighead, J. J., F. C. Craighead, Jr., R. L. Ruff, and B. W. O'Gara. 1973. Home ranges and activity patterns of nonmigratory elk of the Madison Drainage Herd as determined by biotelemetry. *Wildlife Monographs* **33.**

Creel, S., D. Christianson, S. Liley, and J. A. Winnie, Jr. 2007. Predation risk affects reproductive physiology and demography of elk. *Science* **315:**960.

Creel, S., and J. A. Winnie, Jr. 2005. Responses of elk herd size to fine-scale spatial and temporal variation in the risk of predation by wolves. *Animal Behaviour* **69:**1181–1189.

Creel, S., and J. A. Winnie, Jr, Creel, S., and J. A. Winnie, Jr. 2005.B. Maxwell, K. Hamlin, and M. Creel. 2005. Elk alter habitat selection as an antipredator response to wolves. *Ecology* **86:**3387–3397.

DeFelice, T. P., D. Lloyd, D. J. Meyer, T. T. Baltzer, and P. Pirano. 2003. Water vapour correction of the daily 1 km AVHRR global land dataset: Part I—Validation and use of the Water Vapour input field. *International Journal of Remote Sensing* **24:**2365–2375.

DelGuidice, G. D., L. D. Mech, and U. S. Seal. 1989. Physiological assessment of deer populations by analysis of urine in snow. *Journal of Wildlife Management* **53:**284–291.

DelGuidice, G. D., L. D. Mech, and U. S. Seal. 1990. Effects of winter under-nutrition on body composition and physiological profiles of white-tailed deer. *Journal of Wildlife Management* **54:**539–550.

Eberhardt, L. L. 1977. Optimal policies for the conservation of large mammals, with special reference to marine ecosystems. *Environmental Conservation* **4:**205–212.

Eberhardt, L. L. 2002. A paradigm for population analysis of long-lived vertebrates. *Ecology* **83:**2841–2854.

Eidenshink, J. C. 1992. The 1990 conterminous U.S. AVHRR data set. *Photogrammetric Engineering and Remote Sensing* **58:**809–813.

Evans, S. B., L. D. Mech, P. J. White, and G. A. Sargeant. 2006. Survival of adult female elk in Yellowstone following wolf restoration. *Journal of Wildlife Management* **70:**1372–1378.

Fortin, D., H. L. Beyer, M. S. Boyce, D. W. Smith, T. Duchesne, and J. S. Mao. 2005. Wolves influence elk movements: Behavior shapes a trophic cascade in Yellowstone National Park. *Ecology* **86:**1320–1330.

Fowler, C. W. 1981. Density dependence as related to life history strategies. *Ecology* **62:**602–610.

Gaillard, J. -M., M. Festa-Bianchet, and N. G. Yoccoz. 1998. Population dynamics of large herbivores: Variable recruitment with constant adult survival. *Trends in Ecology and Evolution* **13:**58–64.

Garrott, R. A., J. G. Cook, J. G. Berardinelli, P. J. White, S. Cherry, and D. B. Vagnoni. 1997. Evaluation of the urinary allantoin: creatinine ratio as a nutritional index for elk. *Canadian Journal of Zoology* **75:**1519–1525.

Garrott, R. A., L. L. Eberhardt, P. J. White, and J. Rotella. 2003. Climate-induced variation in vital rates of an unharvested large-herbivore population. *Canadian Journal of Zoology* **81:**33–45.

Garrott, R. A., J. A. Gude, E. J. Bergman, C. Gower, P. J. White, and K. L. Hamlin. 2005. Generalizing wolf effects across the greater Yellowstone area: A cautionary note. *Wildlife Society Bulletin* **33:**1245–1255.

Garrott, R. A., P. J. White, D. B. Vagnoni, and D. M. Heisey. 1996. Purine derivatives in snow-urine as a dietary index for free-ranging elk. *Journal of Wildlife Management* **60:**735–743.

Gude, J. A., R. A. Garrott, J. J. Borkowski, and F. King. 2006. Prey risk allocation in a grazing ecosystem. *Ecological Applications* **16**:285–298.

Harder, J. D., and R. L. Kirkpatrick. 1994. Physiological methods in wildlife research. Pages 275–306 *in* T. A. Bookhout (Ed.) *Research and Management Techniques for Wildlife and Habitats*. The Wildlife Society, Bethesda, MD.

Heithaus, M. R., and L. M. Dill. 2002. Food availability and tiger shark predation risk influence bottlenose dolphin habitat use. *Ecology* **83**:480–491.

Hobbs, N. T. 1989. Linking energy balance to survival in mule deer: Development and test of a simulation model. *Wildlife Monographs* **101**.

Houston, D. B. 1982. *The Northern Yellowstone Elk*. Macmillan, New York, NY.

Hume, I. D., R. J. Moir, and M. Sommers. 1970. Synthesis of microbial protein in the rumen. I. Influence of the level of nitrogen intake. *Australian Journal of Agricultural Research* **21**:283–296.

Illius, A. W., and C. FitzGibbon. 1994. Costs of vigilance in foraging. *Animal Behaviour* **47**:481–484.

Ives, A. R., and A. P. Dobson. 1987. Antipredator behavior and the population dynamics of simple predator–prey systems. *American Naturalist* **130**:431–447.

Jönsson, P., and L. Eklundh. 2002. Seasonality extraction by function fitting to time-series of satellite sensor data. *IEEE Transactions on Geoscience and Remote Sensing* **40**:1824–1832.

Jönsson, P., and L. Eklundh. 2004. TIMESAT A Program for analyzing time-series of satellite sensor data. *Computers and Geosciences* **30**:833–845.

Kohlmann, S. G. 1999. Adaptive fetal sex allocation in elk: Evidence and implications. *Journal of Wildlife Management* **63**:1109–1117.

Loison, A., and R. Langvatn. 1998. Short- and long-term effects of winter and spring weather on growth and survival of red deer in Norway. *Oecologia* **116**:489–500.

Loudon, A. S. I., A. S. McNeilly, and J. A. Milne. 1983. Nutrition and lactation control of fertility in red deer. *Nature* **302**:145–147.

Malingreau, J. P. 1989. The vegetation index and the study of vegetation dynamics. Pages 285–303 *in* F. Toselli (Ed.) *Application of Remote Sensing to Agrometeorology*. ECSC, Brussels and Luxembourg.

Mao, J. S., M. S. Boyce, D. W. Smith, F. J. Singer, D. J. Vales, J. M. Vore, and E. H. Merrill. 2005. Habitat selection by elk before and after wolf reintroduction in Yellowstone National Park. *Journal of Wildlife Management* **69**:1691–1707.

McNaughton, S. J. 1985. Ecology of a grazing ecosystem: The Serengeti. *Ecological Monographs* **55**:259–294.

Mech, L. D., D. W. Smith, K. M. Murphy, and D. R. MacNulty. 2001. Winter severity and wolf predation on a formerly wolf-free elk herd. *Journal of Wildlife Management* **65**:998–1003.

Moen, A. N. 1976. Energy conservation by white-tailed deer in the winter. *Ecology* **57**:192–198.

Morgantini, L. E., and R. J. Hudson. 1985. Changes in diets of wapiti during a hunting season. *Journal of Range Management* **38**:77–79.

Mould, E. D., and C. T. Robbins. 1982. Digestive capabilities in elk compared to white-tailed deer. *Journal of Wildlife Management* **46**:22–29.

Parker, K. L., M. P. Gillingham, T. A. Hanley, and C. T. Robbins. 1999. Energy and protein balance of free-ranging black-tailed deer in a natural forest environment. *Wildlife Monographs* **143**.

Pils, A. C., R. A. Garrott, and J. J. Borkowski. 1999. Sampling and statistical analysis of snow-urine allantoin:creatinine ratios. *Journal of Wildlife Management* **63**:1118–1132.

R Development Core Team, 2006. R: A language and environment for statistical computing. http://www.R-project.org. Accessed. 23 April 2007.

Reed, B. C., J. F. Brown, D. Vander Zee, T. R. Loveland, J. W. Merchant, and D. O. Ohlen. 1994. Measuring phonological variability from satellite imagery. *Journal of Vegetation Science* **5**:703–714.

Ripple, W. J., E. J. Larsen, R. A. Renkin, and D. W. Smith. 2001. Trophic cascades among wolves, elk and aspen on Yellowstone National Park's northern range. *Biological Conservation* **102**:227–234.

Robbins, C. T. 1993. *Wildlife Feeding and Nutrition*. Academic Press, New York, NY.

Sala, O. E., W. J. Parton, L. A. Joyce, and W. K. Lauenroth. 1988. Primary production of the central grassland region of the United States. *Ecology* **69**:40–45.

SAS. 1992. Technical Report P-229. SAS Institute, Cary, NC.

Schwartz, H. M., and F. M. C. Gilchrist. 1975. Microbial interactions with the diet and the host animal. Pages 165–179 *in Digestion and Metabolism in the Ruminant*. Proceedings of the IV International Symposium on Ruminant Physiology. Sydney, Australia.

Skalski, J. R., K. E. Ryding, and J. J. Millspaugh. 2005. *Wildlife Demography: Analysis of Sex, Age, and Count Data*. Elsevier Academic Press, San Diego, CA.

Smith, D. W. 2005. Ten years of Yellowstone wolves. *Yellowstone Science* **13**:7–33.

Stephenson, N. L. 1990. Climatic control of vegetation distribution: The role of the water balance. *American Naturalist* **135**:649–670.

Swets, D. L., B. C. Reed, J. D. Rowland, and S. E. Marko. 2000. *A Weighted Least-Squares Approach to Temporal NDVI Smoothing*. Proceedings of the 1999 ASPRS Annual Conference, From Image to Information. Portland, Oregon, May 17–21, 1999. American Society of Photogrammetry and Remote Sensing, Bethesda, MD.

Swick, R. W., and N. J. Benevenga. 1977. Labile protein reserves and protein turnover. *Journal of Dairy Science* **60**:505–515.

Torbit, S. C., L. H. Carpenter, D. M. Swift, and A. W. Alldredge. 1985. Differential loss of fat and protein by mule deer during winter. *Journal of Wildlife Management* **49**:80–85.

Vagnoni, D. B., R. A. Garrott, J. G. Cook, P. J. White, and M. K. Clayton. 1996. Urinary allantoin:creatinine ratios as a dietary index for elk. *Journal of Wildlife Management* **60**:728–734.

Van Gylswyk, N. O. 1970. The effect of supplementing a low-protein hay on the cellulolytic bacteria in the rumen of sheep and on the digestibility of cellulose and hemicellulose. *Journal of Agricultural Science* **74**:169–180.

Van Soest, P. J. 1994. *Nutritional Ecology of the Ruminant.* Ithaca, New York, NY: Cornell University Press, Ithaca, New York, NY.

Vucetich, J. A., D. W. Smith, and D. R. Stahler. 2005. Influence of harvest, climate, and wolf predation on Yellowstone elk, 1961–2004. *Oikos* **111**:259–270.

Wang, G., N. T. Hobbs, R. B. Boone, A. W. Illius, I. J. Gordon, J. E. Gross, and K. L. Hamlin. 2006. Spatial and temporal variability modify density dependence in populations of large herbivores. *Ecology* **87**:95–102.

Watson, F. G. R., T. N. Anderson, W. B. Newman, S. E. Alexander, and R. A. Garrott. 2006a. Optimal sampling schemes for estimating mean snow water equivalents in stratified heterogeneous landscapes. *Journal of Hydrology* **328**:432–452.

Watson, F. G. R., T. N. Anderson, W. B. Newman, S. E. Alexander, and R. A. Garrott. 2006b. Testing a distributed snowpack simulation model against spatial observations. *Journal of Hydrology* **328**:453–466.

White, G. C., and R. A. Garrott. 1990. *Analysis of Wildlife Radio-Tracking Data.* Academic Press, San Diego, CA.

White, P. J., and R. A. Garrott. 2005. Northern Yellowstone elk after wolf restoration. *Wildlife Society Bulletin* **33**:942–955.

APPENDIX

APPENDIX 22A.1 Frequency table of snow-urine samples collected from mixed groups of adult female and calf elk in the Madison headwaters area of Yellowstone National Park during 2-week intervals in the winters of 1996–1997, 1997–1998, 1999–2000, and 2000–2001 through 2005–2006

Interval	1996–1997	1997–1998	1999–2000	2001–2002	2002–2003	2003–2004	2004–2005	2005–2006	Total
Pre-December 4	0	0	0	0	1	0	11	14	26
December 4–17	14	7	12	2	11	15	14	8	83
December 18–31	4	16	18	18	13	10	18	25	122
January 1–14	25	13	14	23	22	24	31	21	173
January 15–28	33	23	15	28	27	20	18	15	179
January 29–February 11	20	23	9	8	21	24	15	19	139
February 12–25	24	35	6	17	19	17	29	34	181
February 26–March 11	25	25	3	22	32	21	5	29	162
March 12–25	16	25	4	14	11	4	38	25	137
March 26–April 8	6	24	1	7	22	0	13	1	74
Post-April 8	0	0	0	6	6	31	1	3	47
Total	167	191	82	145	185	166	193	194	1323

The Madison Headwaters Elk Herd: Transitioning from Bottom–Up Regulation to Top–Down Limitation

Robert A. Garrott,* P. J. White,[†] and Jay J. Rotella*

*Fish and Wildlife Management Program, Department of Ecology, Montana State University
[†]National Park Service, Yellowstone National Park

Contents

Theme

Predator–prey dynamics encompass an intriguing body of ecological processes that is embedded in all biological communities because virtually every organism is predator, prey, or both. The influence of predators on their prey populations is an active area of research, with the "bottom–up" perspective suggesting predators are controlled by ecological processes flowing from the bottom of the food web up the trophic chain. The "top–down" perspective suggests that predators structure communities by controlling the abundance and distribution of their prey populations with consequences cascading down the trophic chain. Understanding the concurrent, relative influences of bottom–up and top–down processes in wolf–ungulate systems is important for effective management and conservation of these large mammal communities in the diversity of landscapes where they exist or are being restored. However, insights into the dynamics underlying most wolf–ungulate systems have

The Ecology of Large Mammals in Central Yellowstone
R. Garrott, P. J. White and F. Watson
ISSN 1936-7961, DOI: 10.1016/S1936-7961(08)00223-6

Copyright © 2009, Elsevier Inc.
All rights reserved.

typically been confounded by human harvests of ungulates, wolves, or both. The historic debate concerning ungulate overabundance in Yellowstone National Park and the recent reintroduction of wolves into the system ~70 years after their extirpation provided an opportunity to evaluate the potential of large predators to structure ecological communities. The Madison headwaters elk herd remains within the park and is not subjected to human harvest. Demographic studies prior to wolf restoration clearly indicated that this population was food limited and strongly regulated by bottom–up processes (Chapter 11 by Garrott *et al.*, this volume). In this chapter, we describe the changes in vital rates and population dynamics of the elk population as wolves recolonized and eventually completely reoccupied the study system.

I. INTRODUCTION

Predator–prey dynamics is perhaps the most studied and controversial topic in ecology, and there is no consensus as to how such systems behave (Barbosa and Castellanos 2005). Recently, the long-standing controversy that centered over the primacy of "bottom–up" versus "top–down" forces and their impact on herbivores (*e.g.*, Hairston *et al.* 1960, Erlich and Birch 1967) has been succeeded by a more integrated view, in which both forces act in unison to influence herbivore populations (Price *et al.* 1980, Hunter and Price 1992, Stiling and Rossi 1997, Letourneau and Dyer 1998, Denno and Peterson 2000, Forkner and Hunter 2000). Such advances, as well as the foundations of traditional predator–prey theory, have been based primarily on experimental studies of invertebrate and small mammal populations (*e.g.*, Solomon 1949, Holling 1959, Hassell 1978, Abrams 1994, Akcakaya *et al.* 1995, Denno *et al.* 2002). Because of the tractability of such systems, experimentation allows for far less ambiguous and more efficient investigations into community functioning than observational studies or "natural experiments" (Connell 1975, Hairston 1989, Wilbur 1997). However, observational studies are the common research model when working in large mammal predator–prey systems (although there are notable exceptions) because of the inherent difficulties and limitations of conducting experiments in such systems. Thus, ecological understanding is accruing more slowly (Mech and Peterson 2003, Andersen *et al.* 2006). To evaluate the relative strength of top–down and bottom–up forces in wolf–ungulate systems, field studies must focus on addressing two major questions: (1) the proportion of wolf predation that is additive versus compensatory; and (2) the strength of demographic compensation in the prey population if prey densities are reduced due to predation.

In the wildlife ecology literature, the debate about the degree to which predation is additive or compensatory to mortality from other natural causes traces back to the classic work of Errington (1946, 1967), whose studies of mink–muskrat relationships, as well as a number of other terrestrial predator–prey systems, led him to coin the phrase "doomed surplus." Wildlife populations produce a surplus of individuals that cannot be supported with the resources available to them. Hence, a proportion of the individuals in any population are doomed to die from this resource limitation, an observation that traces back to the foundations of evolutionary biology (Darwin 1859). Errington argued that predation tended to remove the doomed individuals and, in the absence of predation, other mortality forces would remove these animals from the population. In this view of predation, the dynamics of the system are bottom–up controlled by available resources, various mortality agents such as starvation and predation simply substitute for one another, and the trajectory of the prey population is similar regardless of how much of the mortality is due to predation. In contrast, if predators are effective at killing substantial numbers of animals that would *not* have died from other causes in the near term, then this proportion of predator-induced mortality would be additive and reduce the prey population below what would have been realized with a lower predation rate or in the absence of predators.

Killing ungulate prey is a formidable challenge to wolves because prey are generally larger and possess a variety of effective defenses (Garrott *et al.* 2007). Thus, wolves tend to seek out and kill the weakest individuals in the population, namely animals that are smaller (young), debilitated by disease, starvation, or other ailments, and the oldest animals in the population compromised by physiological senescence (see review in Mech and Peterson 2003). This segment of the prey population would be considered the doomed surplus, with wolf predation simply replacing other sources of mortality. While there is certainly unequivocal evidence from many wolf–ungulate systems that wolves are effective at detecting and killing weak, debilitated, or otherwise inferior animals (*i.e.*, compensatory; Mech 1970, Mech *et al.* 1998), there is also considerable evidence that at least some proportion of wolf predation is additive to other sources of mortality and limits prey populations (Gasaway *et al.* 1992, Seip 1992, National Research Council 1997). However, what is unresolved is the proportion of wolf predation that is additive, how this proportion changes under varying conditions, and how variable the additive component of wolf predation is among systems.

If wolf predation is additive to other mortality sources, the density of a prey population exposed to wolf predation could be suppressed from what would be realized in the absence of predation. The population will decrease if losses are not compensated for by improved survival or recruitment of new individuals into the population. However, a predator-induced reduction in prey density has the potential to increase per capita resources available to the remaining prey which, in turn, could result in improved physiological condition and increased fecundity and/or survival rates for survivors (Nichols 1991, Boyce *et al.* 1999). Such a demographic response to additive wolf predation could result in an increase in the growth rate of the prey population, at least partially compensating for the losses due to predation.

Eberhardt (1977) proposed that density-dependent changes in vital rates of long-lived large mammal populations are hierarchical, with juvenile survival rate being the most sensitive, followed by age of first reproduction, reproductive rate of prime-age animals, and adult survival rate. A number of extensive literature reviews of large herbivore field studies (Sæther 1997; Gaillard *et al.* 1998, 2000; Eberhardt 2002) found high variability in calf recruitment and relatively high and constant adult survival and fecundity rates in populations not subjected to human harvest, supporting Eberhardt's hypothesis. Given these characteristics of most large herbivore populations, we would expect demographic compensation for additive wolf predation to follow this hierarchy of vital rate responses. However, Nichols (1991) cautioned that species (or age classes) with high survival rates in the absence of a given mortality source would have limited ability to compensate for an additional additive source of mortality. This caution may reasonably be extended to fecundity rates for some large herbivore species such as elk that have a litter size fixed at one, with high pregnancy rates for prime-aged animals reported in most herds (Cook 2002, Raedeke *et al.* 2002; Chapter 25 by Hamlin *et al.*, this volume). As a result, the primary compensatory demographic response to wolf predation that reduces survival rate below that which occurs in a large herbivore population not exposed to predators may be an increase in survival of calves due to factors such as reduced density and increased food resources. Density-dependent demographic compensation for an additive mortality source has a long history of investigation in the field of wildlife management, where a prevalent question has been the impact of human hunting on the population dynamics of game species (*e.g.*, Anderson and Burnham 1976, Nichols 1991, Hilborn *et al.* 1995, Williams *et al.* 1996, Pöysä *et al.* 2004). However, little attention has been focused on demographic compensation for additive predation mortality in long-lived large mammal populations.

Determining what proportion of predation can be attributed to compensatory and additive mortality, as well as the occurrence and means by which prey demographically compensate for additive mortality, has been difficult for ecologists studying wolf–ungulate systems, with field and data limitations often leading to indirect assessment or narrative arguments and interpretations of observations (Peterson 1977, Messier 1994, National Research Council 1997, Hayes and Harestad 2000, Vucetich *et al.* 2005, White and Garrott 2005, Wright *et al.* 2006). The reintroduction of wolves into Yellowstone in the midst of our long-term studies provided a fortuitous ecological experiment resulting in an opportunity to contribute to our knowledge of wolf–ungulate interactions. To the best of our knowledge the elk population in the Madison headwaters existed for at least 7–10 generations in the absence of wolves and with little evidence of significant predation from other large mammal predators (Chapter 11 by Garrott *et al.*, this volume). Data from earlier studies and population monitoring by the National Park Service dating back to the 1960s, along with the seven years of intensive demographic studies conducted during the initial years of our research program, prior to the establishment of a resident wolf pack in summer of 1998, provide strong evidence that this population had been well-regulated by bottom–up processes. Food limitation resulted in varying levels of annual starvation mortality that maintained the population between ∼600 and 800 animals (Craighead *et al.* 1973, Aune 1981; Chapter 11 by Garrrott *et al.*, this volume). This baseline understanding of ecological processes influencing the elk population prior to wolf reestablishment resulted in a rare research opportunity because even the famous long-term studies of the wolf–moose system on Isle Royale did not begin until after wolves were established on the island (Peterson 1977). The goal of this chapter is to compare the demography of the elk population in the Madison headwaters area prior to wolf establishment with population attributes after wolves became part of the ecological community. Specifically, we evaluate the hypotheses that (1) a substantial component of wolf predation would be additive, such that survival rate would be reduced from that observed in the absence of wolves, (2) demographic compensation for this additive mortality would be weak, and (3) the system would transition to an alternate state where elk density would remain considerably lower than realized in the absence of wolves.

II. METHODS

A. Demographic Variables

Intensive field work was conducted from mid-November through early May each year with methodologies established during the initial seven years of field studies (Chapter 11 by Garrott *et al.*, this volume) remaining consistent and uninterrupted for an additional nine years (June 1998 through May 2007) after the first wolf pack became established in the study area (Chapter 15 by Smith *et al.*, this volume). The focus of the field investigations was to collect data for estimating annual adult female survival rates for summer and winter seasons, reproductive rates, recruitment of calves into the adult age class (≥1 year old), over-winter calf survival rates, and population estimates for comparison with similar data collected during the 7-year pre-wolf period (Chapter 11 by Garrott *et al.*, this volume). During the post-wolf studies a total of 95 individual adult female elk were captured, radio collared, and monitored. During each year of the study, 22–39 radio-marked females, ranging from 1 to 17 years of age, were monitored. Twenty-nine collared animals that survived the pre-wolf studies were incorporated into the post-wolf investigations, with three to seven animals added to the instrumented population in each of the subsequent years depending on mortalities during the previous year and the size of the elk population. We monitored 26 instrumented animals for 5–9 years, 32 animals for 3–4 years, and 37 animals for 1–2 years. Small numbers of calf elk (1–15) were also instrumented with mortality-sensing telemetry at the beginning of eight of the nine winter seasons to obtain cause-specific mortality. Animals were arbitrarily selected for instrumentation in an attempt to maintain a collared population reflective of the elk distribution throughout the study area.

Throughout the winter season a sample of the instrumented elk was selected daily using a restricted stratified random sampling scheme (Chapter 11 by Garrott *et al.*, this volume) and relocated via telemetry homing techniques until the collared animals were observed (White and Garrott 1990). This scheme resulted in tracking each instrumented elk ∼2–3 times per week. Instrumented calf elk were not routinely relocated, but radio signals were monitored a minimum of 3–4 times per week through the winter to determine survival status. Mortalities of all instrumented animals were investigated immediately upon detection, with cause of death determined by field necropsy and investigation of the immediate area for signs of predators, blood trails, and struggle (Chapter 16 by Becker *et al.*, this volume). During the summer (mid-May through mid-November) elk were monitored for mortality approximately monthly. Adult female survival data were partitioned into winter (mid-November to mid-May) and summer (mid-May to mid-November) periods and recorded as a dichotomous variable (0-died, 1-lived). A survival record was not recorded unless the instrumented animal was available for monitoring at the start of the period. Likewise, if a collar failed (*e.g.*, premature drop-off) or the animal permanently dispersed from the study area (Chapter 18 by Gower *et al.*, this volume), no survival record was recorded for that period.

During monitoring, instrumented female elk were observed for 5–30 min at a distance (20–300 m) that avoided disturbing the animal's normal activities. Total group counts and compositions (*i.e.*, calf, cow, yearling bull, adult bull), were recorded during each location. If the animal defecated during the observation period, the location of the excreta was noted and a sample collected as soon as animals moved away from the area (White *et al.* 1995). Pregnancy status of each instrumented female was assessed by determining concentration of fecal progestagens (P_4) using fecal samples collected during March or April to maximize pregnancy status discrimination (White *et al.* 1995). Samples from the first 3 years of the study (1999 through 2001) were assayed at the Conservation and Research Center (Smithsonian Institution, Front Royal, Virginia, USA) using fecal hormone extraction and radioimmunoassay (RIA; Brown *et al.* 1994, Wasser *et al.* 1994, Garrott *et al.* 1998). Samples collected during 2002 through 2006 were assayed at Montana State University using hormone extraction and enzyme-linked immunosorbet assay (ELISA) techniques as described by Creel *et al.* (2007). No samples from the final year of study (2007) were assayed. Animals with fecal P_4 concentrations <1.00 μg/g were considered nonpregnant, with concentrations >1.25 μg/g considered indicative of pregnancy (Cook *et al.* 2002). Pregnancy assessments for samples with concentrations between 1.0 and 1.25 μg/g were considered inconclusive and an alternative fecal sample was assayed, if available, to obtain a definitive fecal P_4 concentration.

Independent records of age composition of the randomly sampled elk groups were aggregated (Thompson 1992) to obtain a calf:cow ratio for the periods mid-November through December, January, February, March, April–early May each winter season to evaluate temporal trends in calf:cow ratios, with 45–137 groups used for each ratio. We also calculated a "fall" and "spring" calf:cow ratio for each year using the first and last 100 random elk group records, respectively, to standardize the amount of data used to generate these ratios. The ratios were considered indices of calf "recruitment" to 6 months of age (fall ratio) and "recruitment" to yearling status (spring ratio). We used the two-sample change-in-ratio methods of Hanson (1963) and Paulik and Robson (1969) to estimate over-winter calf survival as described by Skalski *et al.* (2005:212–216), with the calf survival estimates corrected for adult cow survival using the winter survival estimates derived in this chapter.

Historically, winter estimates of the Madison headwaters population were obtained from aerial surveys, which were sometimes augmented by ground counts (Craighead *et al.* 1973, Aune 1981, Cole 1983; National Park Service, unpublished data). These estimates provided information on the minimum number of animals in the population but did not account for undetected animals and provided no confidence limits on the estimates. During the pre-wolf period of this study, we used aerial surveys in conjunction with the instrumented elk to obtain Lincoln–Petersen population estimates, however, this procedure required multiple consecutive days of flying which was difficult to achieve under winter conditions. In addition, the confidence intervals on the population estimates were wide (Eberhardt *et al.* 1998) and the flying required was inherently dangerous. As a consequence, we terminated aerial surveys near the end of the pre-wolf studies and established ground-based mark–resight surveys using two different sampling protocols.

Beginning in spring (April) 1997, we conducted multiple daily mark–resight surveys by traveling the roads along each drainage of the study area when elk were aggregated in the meadows adjacent to the rivers to feed on the first green forages of spring. Surveys were normally conducted on consecutive days each spring until a minimum of 10 mark–resight surveys were completed. Beginning in the 1997–1998 field season, we also incorporated elk mark–resight surveys into our ground-based bison surveys that were performed at 10–14-day intervals through the winter field season (Dec. to early May) (Chapter 12 by Bruggeman *et al.*, this volume), resulting in 8–17 surveys completed each year. These surveys were performed over two consecutive days and involved traversing (foot, snowshoe, truck, snowmobile) 72 delineated survey units that encompassed the entire bison winter range in the Madison headwaters study area (Chapter 11 by Garrott *et al.*, this volume), covering a considerably larger proportion of the study area than could be observed during the road-based spring mark–resight surveys. We used the joint hypergeometric maximum likelihood estimator (MLE) in Program NOREMARK (White 1996) to calculate a population estimate and confidence limits from the replicate road-based surveys, which we termed the "spring" estimate, as well as the independent data collected in conjunction with the bison surveys that we termed the "winter" estimate. We used ln-linear regression (Eberhardt 1987) to obtain an estimate of λ from each time series of the mark–resight population estimates for comparison with the estimates of λ obtained from matrix models constructed and parameterized from the results of our vital rate studies (see below).

B. Vital Rate Covariates, Predictions, Model Development, and Evaluation

We evaluated the potential influence of the same age and climate covariates on vital rates as were assessed in the analyses of the pre-wolf data (Chapter 11 by Garrott *et al.*, this volume). We considered a continuous covariate AGE because survival and reproduction of long-lived large mammals has been demonstrated to be strongly age-dependent (Siler 1979, Eberhardt 1985). AGE was determined by counting cementum structures in the root of a vestigial canine tooth (Hamlin *et al.* 2000) extracted from each cow elk at the time she was collared, with AGE incremented one year if the animal survived to be monitored the succeeding year. The biological year was defined to begin mid-May. Warm season climate covariates were derived from composited and temporally smoothed normalized difference vegetation index (NDVI) parameters obtained from AVHRR satellite data at 1 km^2 resolution (Malingreau 1989, Eidenshink 1992, Swets *et al.* 1999, Jönsson and Eklundh 2002). We generated the warm season covariate $NDVI_{len}$, which indexed length of the growing season and, hence, annual variability in the time period that high-quality green vegetation was available to foraging elk (Chapter 7 by Thein *et al.*, this volume). An alternate metric, $NDVI_{int}$ was also considered in exploratory analyses. $NDVI_{int}$ was the scaled integral of the seasonal NDVI curve and is strongly correlated with net primary productivity (Pettorelli *et al.* 2005). Thus, we considered this metric an index of annual variation in forage quantity. We evaluated the cold season covariate SWE_{acc}, which indexed annual variation in snow pack severity. The covariate was generated by summing the daily mean SWE (snow water equivalent) estimated for the study area from the Langur snow pack model (Chapter 6 by Watson *et al.*, this volume) over the period October 1 to April 30, thus providing a single metric integrating snow pack depth, density, and duration. In addition to these covariates, we also developed a metric that captured the annual variability in the intensity of use of the study area by wolves. Intensive daily monitoring of the presence of wolves in the study area, as described in Chapter 15 by Smith *et al.*, Chapters 16 and 17 by Becker *et al.*, this volume, allowed us to reliably determine the number of wolf packs that used the study area each day of the field season and the number of animals in each pack. We used these data to obtain daily estimates of the total number of wolves occupying the study system each day. The daily estimates were summed over the 167-day winter field season to generate the covariate, $WOLF_{days}$, which was incorporated into exploratory analyses.

Our general analytical approach to assessing variation in vital rates of the study population mimics that employed in the analyses of vital rates for the pre-wolf study period (Chapter 11 by Garrott *et al.*, this volume). We formed hypotheses expressed as *a priori* suites of candidate regression models and evaluated competing models for each vital rate using the data collected in this study and an information-theoretic approach (Burnham and Anderson 2002). The four demographic response variables formally analyzed were adult female survival for the winter and summer periods, yearling recruitment indexed by spring calf-cow ratios, and winter calf survival. The analysis of the pre-wolf vital rate data demonstrated senescence in adult survival, and we expected to detect a similar effect in the analysis of the post-wolf data (Chapter 11 by Garrott *et al.*, this volume). Therefore, we predicted a negative correlation between AGE and adult female survival rate in both the summer and winter periods. The pre-wolf analysis also detected the expected lower reproductive rates in youngest age class animals but found no support for reproductive senescence in the oldest age class animals (Chapter 11 by Garrott *et al.*, this volume). We anticipated evaluating similar age-specific reproductive characteristics using the post-wolf data, but incongruous results between laboratory assays of fecal P_4 concentrations and data collected in the field, led us to conclude that analysis of the pregnancy data would not be a worthwhile exercise (see Section III). In years when summer climate conditions were more beneficial for plant growth, we predicted higher adult female survival rate both in summer and in winter, as well as higher calf recruitment and over-winter survival rates due to better nutrition and body condition. Thus, we predicted a positive correlation between NDVI covariates (NDVI$_{len}$, NDVI$_{int}$) for these vital rates. We predicted lower winter survival of both calves and adults in years with more severe snow pack due to reduced availability of forage buried under the snow and increased energetic costs associated with displacing snow during movement and foraging activities. Thus, we predicted a negative correlation between SWE$_{acc}$ and adult female survival rate over the winter period, calf winter survival rate, and recruitment rate of calves to the yearling age class, as was found in the analysis of the pre-wolf vital rates (Chapter 11 by Garrott *et al.*, this volume). Given that we expected strong AGE effects, we hypothesized that any effects of SWE$_{acc}$ and NDVI$_{len}$ on adult female survival and reproductive rates would be more pronounced for older-aged animals than for prime-aged animals. Thus, we considered interactions in our candidate models. There was no indication of collinearity between SWE$_{acc}$ and NDVI$_{len}$ for the period of this study (1998–2007; $r^2 = 0.11$, $P = 0.21$). Thus, both covariates were considered in the same models.

We used regression techniques in R version 2.5.0 (R Development Core Team 2006) to fit models and estimate coefficients. Annual adult survival and reproduction were dichotomous variables. Thus, we modeled data for these vital rates using a logit link and binomial error structure (logistic regression), whereas standard regression was used to fit calf recruitment and winter calf survival models because these response variables were continuous. Because the spring calf:cow ratio index of calf recruitment and calf survival estimates were proportions the estimates were logit transformed prior to analyses. We used Akaike's Information Criterion corrected for sample size, AIC$_c$, to compare the relative ability of each model to explain variation in the data and Akaike model weights (w_i) to address model-selection uncertainty (Burnham and Anderson 2002). Ninety-five percent confidence intervals were used to assess the degree to which signs of the estimated model parameters were reliably estimated. Because most animals were studied in more than 1 year, we (1) evaluated the possible presence of overdispersion by dividing the residual deviance by the deviance degrees of freedom for the most general model (McCullagh and Nelder 1989) and (2) evaluated goodness of fit for models of binary response variables with the le Cessie–van Houwelingen test (le Cessie and van Houwelingen 1991, Hosmer *et al.* 1997). This test was designed to assess goodness-of-fit for models with continuous covariates and binary responses based on nonparametric kernel methods. Goodness-of-fit of recruitment and calf survival regression models was evaluated using the adjusted coefficient of multiple determination (R^2_{adj}).

C. Constructing Population Models

We used R software (R Development Core Team 2006) to execute modeling of a seasonal prebreeding age-structured population matrix model (Caswell 2001) nearly identical in structure to the model used to integrate vital rates derived from the pre-wolf studies (Chapter 11 by Garrott *et al.*, this volume). Two matrices (**B1** and **B2**) per year were used to project the population through one annual cycle so that we could compare fall and spring calf:cow ratios produced by the matrix model with those observed in the field. **B1** was used to project the population from spring to fall and **B2** was used to project from fall to spring. Each matrix contained vital rate information for calves and females aged 1–17 years old. **B1** contained fecundity information for each age class and the probability that animals in each age class would survive from spring to fall. **B2** contained information on the probability that animals in each age class would survive from fall to spring and transition to the next age class. In the projection from spring to fall with **B1**, calves were created and some adult animals died, but no animals transitioned from one age to the next. In the projection from fall to spring with **B2**, no new calves were created and all animals either died or survived and transitioned to the next age with the constraint that all 17-year-old animals died during the winter. As a consequence, the population vector in spring contained only animals aged 1–17 years old, whereas the fall vector contained calves as well as 1–17 year olds.

   B1 was a deterministic matrix because cow elk survival was consistently high during the summer. Thus, we used the age-specific summer survival estimate obtained from the data pooled across the nine years of monitoring. Fecundity information was obtained using estimates of pregnancy rates, litter size, and birth sex ratio. Birth rates (*br*) in **B1** were estimated using assessments of third trimester pregnancy rates. However, we evaluated two versions of **B1** based on differing birth rate scenarios because of uncertainties experienced in determining post-wolf pregnancy rates. In one version we used the pre-wolf age-specific pregnancy rates, providing birth rate estimates of 0 for yearlings, 0.5 for 2 year olds, 0.6 for 3-year-olds, and 0.9 for animals ≥ 4 years old (Chapter 11 by Garrott *et al.*, this volume). In the alternate version of **B1**, we reduced the pre-wolf age-specific pregnancy rates for each age class by 32% (see Section III). Birth rates were incremented one year to reflect the cow's age at time of parturition. Litter size (*ls*) was assumed to be one and the sex ratio (*sr*) of offspring was assumed to be 0.5 (Raedcke *et al.* 2002). The probability that newborn female calves would survive from spring to fall (*Sjuv_sum*) was not estimated in this study. Therefore, we varied summer calf survival between 0 and 1 until projected fall calf:cow ratios corresponded with the mean observed ratio during the post-wolf studies. The age-specific fecundity values for **B1** were calculated as $br_{age} \times ls \times sr \times Sjuv\_sum$, where br_{age} was an age-specific birth rate. Results for our field studies indicated that survival of adult cows during winter was affected by an animal's age, but neither adult cow nor calf survival after wolf colonization was affected by winter severity. Thus, **B2** was also a deterministic matrix. We obtained both the age-specific winter survival rates for adult cows and the survival rate of calves for **B2** from the regression model that was most supported by the data (see Section III).

   We conducted 10,000 simulations that projected a starting spring population of 375 females (N_0) forward for 10 years (N_{10}). A female population of 375 was chosen to approximate a total elk population of 600 males and females, the approximate size of the herd in the Madison headwaters area when wolves colonized the study area. Rather than choosing a single age structure for the 375 animals in N_0, we generated a set of 10,000 age structures using the stochastic pre-wolf matrix model (Chapter 11 by Garrott *et al.*, this volume). We used the original 10,000 starting age structures from the pre-wolf model that were generated using the median SWE$_{acc}$ experienced during the pre-wolf study period. Each of the 10,000 starting populations was then projected forward using the pre-wolf matrix model and the ordered time series of SWE$_{acc}$ experienced on the study area for the 15 years immediately prior to the start of this study (1984–1998). The age structure at the end of each simulation was then stored and used as the starting age structures for the post-wolf matrix model projections. To conduct each of the 10,000 post-wolf simulations, we first randomly chose a starting age structure for the N_0 of 375 animals from this set of 10,000 random age structures. We then used **B1** and **B2** to project that population forward for 10 years. For each simulation, we recorded population growth rate (spring to spring), and fall and spring calf:cow ratios, and size of the population for each of the 10 projected years. We multiplied the matrix-derived calf:cow ratios by 1.67 to compare the field observations of fall and spring calf:cow ratios with matrix-derived ratios that reflected only female calves, as described in Chapter 11 by Garrott *et al.*, this volume. From the results of the 10,000 simulations, we estimated the arithmetic mean population growth rates for comparison with the direct estimates of population growth rate obtained from mark–resight population estimates. We also projected the population forward using the stochastic pre-wolf model and the ordered time series of SWE_{acc} experienced on the study area for the 9 years of our post-wolf study to provide a prediction of the population trend in the absence of wolves to contrast with the trend in population estimates from the post-wolf studies as well as the projected trend from the post-wolf matrix model. The collective evaluation of our hypotheses that elk survival rates would decrease, demographic compensation would be weak, and elk density would decrease following wolf recolonization and establishment then allowed us to evaluate whether bottom–up effects retained their primacy or whether predation exerted substantial control of community dynamics.

III. RESULTS

A. Snow Pack and Growing Season Length

Although variation in SWE_{acc} experienced during the 9-year study period captured nearly the entire range of annual variation in snow pack described in the available historic data, the study area generally experienced longer growing seasons, as indexed by $NDVI_{len}$, than the average of the historic data. SWE_{acc} during this study ranged from 1042 to 3749 cm days, with a mean of 2075 (CV = 0.38). Langur-generated SWE_{acc} metrics were only available for the winters of 1989–1990 through 2006–2007. However, a SWE_{acc} metric derived from direct SWE measurements from the Madison Plateau automated SNOTEL site located just west of the study area (Garrott *et al.* 2003) was strongly correlated with Langur-derived SWE_{acc} ($R^2 = 0.93$). These data provided a 40-year time series (winters 1967–1968 through 2006–2007) for comparison of historic variation in snow pack severity with that experienced during the nine winters of this study. SNOTEL SWE_{acc} recorded during the study period ranged from 3955 (2000–2001) to 9822 cm days (1998–1999), with a mean of 6182, while the historic data ranged from 2975 (1976–1977) to 12,404 (1996–1997) with a mean of 6888. Thus, while average snow pack severity during the study period was lower than the historic average, the range of snow packs was comparable. Only a single winter experienced lighter snow pack than during 2000–2001, and only five winters in the historic record experienced more severe snow pack than during 1998–1999. $NDVI_{len}$ experienced during this study ranged from 15.37 (1998) to 18.26 (2006), with a mean of 16.54 (CV = 0.06), while the historic data (1989–2006) ranged from 13.88 (1996) to 18.26 (2006), with a mean of 16.07 (CV = 0.08). The two longest growing seasons in the historic recorded occurred during the last two years of this study, while 28% of summers (5 of 18) in the historic record experienced shorter growing seasons than recorded during this study.

B. Adult Female Survival

We accrued 260 animal years of adult female survival monitoring during the summer period and documented 15 mortalities. Five animals were killed by vehicles, two mortalities were attributed to wolf predation, one animal died due to a natural accident, and cause of death could not be determined for seven animals. We evaluated five *a priori* models for adult female survival during the summer period with equivocal results (Table 23.1). The top-ranked model, AGE + $NDVI_{len}$, received 0.46 of the model weight. As predicted, the AGE coefficient was negative and the confidence interval did not overlap zero ($\hat{\beta}_{AGE} = -0.188$, 95% CI = -0.362 to -0.014). The sign of the $NDVI_{len}$ coefficient, in contrast, was negative which was contrary to our prediction, and the 95% confidence interval overlapped zero ($\hat{\beta}_{NDVI} = -0.533$, 95% CI = -1.077 to 0.011). The second-ranked model included only AGE, with the coefficient estimate similar to the top-ranked model ($\hat{\beta}_{AGE} = -0.198$, 95% CI = -0.371 to -0.024),

| TABLE 23.1 | Ranking of *a priori* hypothesized regression models concerning the effects of age (AGE), snow pack severity (SWE_acc), and growing season length (NDVI_len) on demographic vital rates of the elk population in the Madison headwaters area of Yellowstone National Park after wolf colonization (1998–2006) |

Model	K	AIC_c	AIC_c	w_i
Adult Female Summer Survival				
$AGE + NDVI_{len}$	3	111.9	0	0.46
AGE	2	113.4	1.58	0.21
$AGE+NDVI_{len} + (AGE \times NDVI_{len})$	4	113.7	1.84	0.18
$NDVI_{len}$	2	114.6	2.72	0.12
Constant	1	116.7	4.85	0.04
Adult Female Winter Survival				
AGE	2	198.3	0	0.40
$AGE+NDVI_{len}$	3	199.9	1.59	0.18
$AGE+SWE_{acc}$	3	200.3	1.95	0.15
$AGE+SWE_{acc} + (AGE \times SWE_{acc})$	4	200.4	2.05	0.14
$AGE+NDVI_{len} + (AGE \times NDVI_{len})$	4	201.9	3.63	0.06
$AGE+SWE_{acc} + NDVI_{len}$	4	202.0	3.65	0.06
Constant	1	219.6	21.27	0.00
Calf Recruitment (logit)				
Constant	2	33.29	0.00	0.69
$NDVI_{len}$	3	35.31	2.02	0.24
SWE_{acc}	3	37.98	4.69	0.07
Calf Winter Survival (logit)				
Constant	2	34.72	0.00	0.70
$NDVI_{len}$	3	37.26	2.54	0.20
SWE_{acc}	3	38.62	3.90	0.10

while all the coefficient confidence intervals in the third-ranked model overlapped zero. A plot of the predicted age-specific survival and the observed proportions calculated directly from the data for each age class suggested the age model did not fit the data well because the model predicted a decreasing survival probability in the older age classes, while the observed survival proportions for the last three age classes was 1.0. The le Cessie–van Houwelingen goodness-of-fit test also indicated the AGE model did not fit the data well ($P = 0.038$). The constant model provided an estimate of summer survival for data pooled across all post-wolf years of 0.942 (95% CI = 0.907–0.965), which was lower than the 0.987 (95% CI = 0.948–0.997) summer survival estimate for the pre-wolf period (Chapter 11 by Garrott *et al.*, this volume). The confidence interval on the difference between these two survival estimates, however, spanned zero (−0.12 to 0.03, delta method), indicating little support for an additive effect of wolf predation on cow survival during the summer.

Monitoring of instrumented cow elk during the winter period totaled 271 animal years with 40 mortalities documented. Of these mortalities, 27 were attributed to wolf predation, five to starvation, three to mountain lion predation, one animal was killed by a vehicle, and cause of death of four collared cows could not be reliably determined. We evaluated 11 *a priori* models for adult female survival during the winter period using 264 animal years of monitoring with seven animal years of monitoring censored from the analysis because we did not have an estimated age. The AGE-only model was the-top ranked model, receiving a model weight of 0.40 and appeared in all models with any weight (Table 23.1). As predicted, the coefficient was negative ($\hat{\beta}_{AGE} = -0.274$, 95% CI = −0.383 to −0.154) and coefficients estimates and standard errors for all ranked additive models were nearly identical with the model-averaged age-specific survival curve illustrated in Figure 23.1. We did not find support for our predictions of either warm or cold season climate effects, even though $AGE+NDVI_{len}$ and $AGE+SWE_{acc}$ were the second and third ranked models, respectively. Although these models were

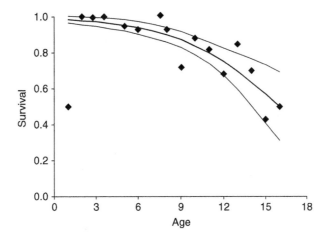

FIGURE 23.1 Observed (diamonds) and predicted (dark line-prediction, light lines-95% confidence interval) age-specific winter survival of cow elk fitted to 264 animal years of monitoring accrued during nine consecutive years after wolves became established in the study system. We used model-averaging techniques (Burnham and Anderson 2002) to generate the predicted age-specific survival curve of the six top-ranked models that received all the model weight (Table 23.1). Sample sizes for age classes 4 through 13 ranged from 14 to 29 animal years while few animals ($n = 1$–3) in each of the youngest and oldest age classes (1, 2, 16, 17) were monitored.

within 2 AIC units of the top-ranked AGE-only model, the confidence intervals of both climate coefficients showed considerable overlap with zero ($\hat{\beta}_{NDVI} = 0.137$, 95% CI $= -0.266$ to 0.539; SWE$_{acc}$ $\hat{\beta}_{SWE} = -0.076$, 95% CI $= -0.542$ to 0.391). Neither adding the wolf covariate, WOLF$_{days}$, to the top-ranked models, nor substituting the alternative warm season climate covariate, NDVI$_{int}$, in *post hoc* exploratory analyses provided improvements over the *a priori* models. There was little evidence of overdispersion in the top-ranked model (deviance/d.f. $= 1.10$) and the goodness-of-fit test supported the null hypothesis that the most general model fit the data ($P = 0.56$).

Over-winter survival of cow elk during the pre-wolf period was dependent on both age and the severity of the snow pack (SWE$_{acc}$), while the data from the post-wolf period support only an age model. Thus, we contrasted the difference between the predicted pre- and post-wolf age-specific survival curves for a mild, moderate, and severe snow pack winter based on the range of snow pack severities experienced during the post-wolf study. The differences in estimates indicated that the survival rate of cow elk over the winter was generally lower after wolves became established in the study area than it was during the seven years immediately prior to wolf establishment. The estimated decrease in cow winter survival was more pronounced during winters with less severe snow pack and for older animals. The confidence intervals of the age-specific survival differences did not include zero except in the oldest age classes for each level of winter severity (Table 23.2). These findings provided support for our hypothesis that a substantial proportion of wolf predation on cow elk in the winter was additive to mortality from other sources, with the magnitude of additive affects most pronounced during mild snow pack winters and diminishing as winter snow pack increased in severity (Figure 23.2).

C. Elk Pregnancy Rates

We assessed the pregnancy status of 23–37 instrumented elk per year based on fecal progestagen concentrations (Table 23.3). Samples from the first 3 years of post-wolf studies (1999–2001) yielded annual pregnancy estimates similar to those documented for pre-wolf years (Chapter 11 by Garrott *et al.*, this volume). However, pregnancy rates estimated for the years 2002 through 2006 were markedly

TABLE 23.2 Differences in the age-specific winter survival estimates for cow elk from the top-ranking model from the 7-year pre-wolf study (Chapter 11 by Garrott *et al.*, this volume) and the survival estimates from the top-ranking model for the 9-year post-wolf period (this study)

		Mild Winter			Moderate Winter			Severe Winter		
Age	Post Surv.	Pre Surv.	Surv. Diff.	Diff. CI	Pre Surv.	Surv. Diff.	Diff. CI	Pre Surv.	Surv. Diff.	Diff. CI
1	0.99	1.00	**0.01**	0.01, 0.02	1.00	**0.01**	0.01, 0.02	1.00	**0.01**	0.01, 0.02
2	0.98	1.00	**0.02**	0.01, 0.03	1.00	**0.02**	0.01, 0.03	1.00	**0.02**	0.01, 0.02
3	0.98	1.00	**0.02**	0.01, 0.03	1.00	**0.02**	0.01, 0.03	1.00	**0.02**	0.01, 0.03
4	0.97	1.00	**0.03**	0.02, 0.04	1.00	**0.03**	0.02, 0.04	1.00	**0.03**	0.01, 0.04
5	0.96	1.00	**0.03**	0.02, 0.05	1.00	**0.03**	0.02, 0.05	0.99	**0.03**	0.02, 0.04
6	0.95	1.00	**0.04**	0.03, 0.06	1.00	**0.04**	0.03, 0.05	0.99	**0.04**	0.02, 0.05
7	0.94	1.00	**0.05**	0.04, 0.07	1.00	**0.05**	0.04, 0.06	0.98	**0.04**	0.02, 0.06
8	0.93	0.99	**0.07**	0.05, 0.08	0.99	**0.06**	0.05, 0.08	0.97	**0.04**	0.02, 0.06
9	0.91	0.99	**0.08**	0.06, 0.09	0.99	**0.07**	0.06, 0.09	0.95	**0.04**	0.01, 0.07
10	0.89	0.98	**0.09**	0.07, 0.11	0.98	**0.08**	0.06, 0.11	0.92	0.03	−0.01, 0.06
11	0.87	0.97	**0.11**	0.08, 0.13	0.96	**0.09**	0.06, 0.12	0.87	0.00	−0.05, 0.05
12	0.84	0.95	**0.12**	0.08, 0.16	0.93	**0.10**	0.05, 0.14	0.79	−0.04	−0.11, 0.20
13	0.80	0.92	**0.12**	0.06, 0.18	0.89	**0.09**	0.02, 0.16	0.69	−0.11	−0.22, −0.01
14	0.76	0.87	**0.11**	0.02, 0.20	0.82	0.06	−0.05, 0.17	0.56	−0.20	−0.34, −0.06
15	0.72	0.80	0.08	−0.05, 0.22	0.73	0.01	−0.14, 0.17	0.43	−0.29	−0.46, −0.12
16	0.67	0.70	0.03	−0.16, 0.23	0.61	−0.06	−0.27, 0.15	0.30	−0.37	−0.55, −0.18
17	0.61	0.57	−0.04	−0.29, 0.21	0.47	−0.14	−0.38, 0.11	0.20	−0.41	−0.58, −0.24

Confidence intervals for the survival differences were calculated using the Delta method (Seber 1982) with differences given in bold interpreted as that proportion of wolf predation mortality that can be considered additive to other mortality sources present prior to wolf reestablishment in the study system.

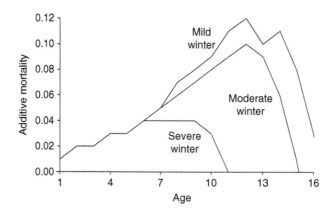

FIGURE 23.2 Estimates of the magnitude of additive mortality due to wolf predation based on a comparison of age-specific survival curves derived from the top-ranking survival models for the pre-wolf and post-wolf studies. Winter severities spanned the range experienced during the post-wolf study (mild: SWE_{acc} = 1612, moderate SWE_{acc} = 2093, severe SWE_{acc} = 3615). Confidence intervals for the point estimates are provided in Table 23.2.

lower, with fecal progestagen concentrations for the last 3 years of collections suggesting that only 2 of 73 cows sampled were pregnant (Table 23.3). We believe these results were erroneous because such low pregnancy rates have never been documented for elk, even in herds that were believed to be severely resource limited (Cook 2002, Raedeke *et al.* 2002). In addition, while fall calf:cow ratios were indeed lower during the latter years of the study compared to pre-wolf years, the ratios were considerably

TABLE 23.3 The age-specific and annual proportion of radio-collared cow elk assessed as pregnant using RIA (1999–2001) or ELISA (2002–2006) techniques to measure fecal progestagens (P_4) from samples collected during the third trimester of gestation. Data from 2002 through 2006 (shaded) were considered unreliable as field observations of calves indicated that the pregnancy rates must have been considerably higher than that indicated by the fecal P_4 assay results

Year	Number	Age 1	2	3	4	5	6	7	8	9	10	11	12	13	14	15	16	Total	Proportion Pregnant[a]
1999	Monitored	0	0	3	3	3	2	2	5	2	2	2	2	1	1	1	0	29	0.86
	Pregnant	0	0	3	2	3	1	1	5	2	2	2	2	1	1	0	0	25	
2000	Monitored	0	0	0	3	5	5	4	2	6	0	2	2	1	1	0	0	31	1.00
	Pregnant	0	0	0	3	5	5	4	2	6	0	2	2	1	1	0	0	31	
2001	Monitored	1	0	1	1	4	4	5	4	3	7	0	1	1	1	0	0	33	0.88
	Pregnant	1	0	1	1	3	4	5	3	2	6	0	1	1	1	0	0	29	
2002	Monitored	2	1	0	1	1	4	5	6	3	3	9	0	1	1	0	0	37	0.68
	Pregnant	1	1	0	1	1	4	3	4	2	1	6	0	1	0	0	0	25	
2003	Monitored	0	2	2	0	1	2	2	3	6	1	1	9	0	1	0	0	29	0.19
	Pregnant	0	0	0	0	1	0	1	0	1	0	0	2	0	0	0	0	5	
2004	Monitored	0	0	2	3	1	2	1	2	1	6	1	0	5	0	0	0	24	0.00
	Pregnant	0	0	0	0	0	0	0	0	0	0	0	0	0	0	0	0	0	
2005	Monitored	0	1	1	2	5	1	2	1	1	0	6	2	0	4	0	0	26	0.08
	Pregnant	0	0	0	0	0	0	0	0	1	0	1	0	0	0	0	0	2	
2006	Monitored	0	0	1	1	2	4	1	2	1	1	0	4	2	0	4	0	23	0.00
	Pregnant	0	0	0	0	0	0	0	0	0	0	0	0	0	0	0	0	0	
Total Monitored[a]		1	0	4	7	12	11	11	11	11	9	4	5	3	3	1	0	232	0.51
Total Pregnant[a]		1	0	4	6	11	10	10	11	10	8	4	5	3	3	0	0	118	
Proportion		1.00	0.00	1.00	0.86	0.92	0.91	0.91	1.00	0.91	0.89	1.00	1.00	1.00	1.00	0.00	0.00		

[a] Proportions for cow ≥3 years old.
[b] Totals exclude 2002–2006 data.

TABLE 23.4	The annual changes in calf:cow ratios from fall to spring for the Madison headwaters area of Yellowstone National Park after wolf colonization (1998–2006). Ratios were calculated from the first and last 100 random elk groups classified at the beginning and end of each winter field season. Survival estimates were based on the two-sample change-in-ratio methods of Hanson (1963) and Paulik and Robson (1969) with the corrected estimate adjusted for annual variation in cow elk over-winter survival

	Fall		Spring		Uncorrected survival estimate[a]	Corrected survival estimate
Year	No. animals classified	Calf:cow ratio	No. animals classified	Calf:cow ratio		
1998–1999	568	0.531	581	0.047	0.09 (0.06–0.12)	0.08
1999–2000	562	0.274	830	0.104	0.38 (0.28–0.48)	0.32
2000–2001	584	0.400	1045	0.293	0.73 (0.59–0.87)	0.63
2001–2002	549	0.248	671	0.031	0.12 (0.07–0.18)	0.11
2002–2003	764	0.142	675	0.050	0.35 (0.23–0.43)	0.30
2003–2004	703	0.147	1209	0.021	0.14 (0.09–0.20)	0.12
2004–2005	1647	0.184	1004	0.028	0.15 (0.10–0.20)	0.13
2005–2006	935	0.330	700	0.064	0.19 (0.14–0.25)	0.17
2006–2007	825	0.269	1051	0.006	0.02 (0.01–0.04)	0.02

[a] 90% confidence interval.

higher than what would have been realized if the fecal progestagen pregnancy assessments were correct (see Table 23.4). Observations of "calves-at-heel," defined as a cow nursing or grooming a calf, collected during routine tracking of collared elk also supported our assessment that fecal P_4 pregnancy assessments in the latter years of our study were unreliable. Approximately 1400 calf-at-heel records were recorded during the 15 years that fecal P_4 assays were conducted. During 1992–2001, when fecal P_4 results indicated 80–100% of the sampled cows were pregnant (excluding the aberrant 1997 results), 98% of the instrumented cows assessed as being with calf-at-heel more than once in a field season were assessed as pregnant the previous spring based on fecal P_4. In contrast, during the 2002–2006 field seasons when fecal P_4 results suggested extremely low pregnancy rates, only 50% of the instrumented cows observed with calf-at-heel more than once were assessed as pregnant the previous spring. This value was reduced to 11% when we included cows observed with a calf-at-heel only once during the field season.

The lower pregnancy rates corresponded to the period when we switched from RIA to ELISA techniques. However, we conducted a blind quantitative comparison of ELISA and RIA results using 40 fecal samples collected in spring 2001. Pregnancy assessments were identical for 39 of the 40 samples. We also reextracted and reassayed the original samples and obtained the same pregnancy assessments for 38 of the 40 samples. A regression of the estimated P_4 concentrations for the first and second analyses yielded a slope estimate of 1.03 and a R^2 of 0.74. Repeating this comparison using a second set of samples from each of the instrumented elk rather than the original samples produced nearly identical results. Thus, the fecal ELISA results were comparable to the RIA results, and the ELISA results were highly repeatable. We have no explanation for why the P_4 results in the last 4–5 years of sampling failed to provide accurate assessments of pregnancy, but note that a similar problem was encountered for one year during our pre-wolf studies (Chapter 11 by Garrott *et al.*, this volume). To obtain an alternate assessment of potential change in pregnancy rates during the last 6 years of our studies, we used serum harvested from blood samples collected when cow elk were immobilized for radio-collaring. We compared serum pregnancy-specific protein B (PSPB, Noyes *et al.* 1997) and serum progesterone concentrations (Weber and Wolfe 1982, Willard *et al.* 1994) for samples pooled over the 1991–2001 period (excluding 1997) and the 2002–2007 period. These data provided pregnancy estimates of 77% for PSPB (47 of 61 samples) and 81% for progesterone (22 of 27 samples) for the

1991–2001 period. Both serum PSPB and progesterone results for the 2002–2007 period provided pregnancy estimates of 54% (PSPB: 21 of 39 samples; progesterone: 26 of 48 samples), which suggests that pregnancy rates may have dropped by approximately 32% during the latter years of the study. Thus, we found no evidence that the elk compensated for lower survival rates through increased pregnancy rates, but on the contrary found some evidence that pregnancy rates may have actually decreased. These results support our hypotheses that demographic compensation for additive wolf predation would be weak.

D. Calf Survival and Yearling Recruitment

Daily telemetry sampling resulted in the observation and classification of 3877 groups of elk during the nine post-wolf winters of field work. The number of groups classified per time period ranged from 39 to 137 with mean group sizes for each period ranging from 4.5 to 15.9. In 17 of the periods, the minimum sample of >80 groups classified was not met. However, similar to the general temporal trend documented in the pre-wolf study (Chapter 11 by Garrott *et al.*, this volume), calf:cow ratios progressively decreased throughout the winter (Figure 23.3), indicating substantial over-winter mortality of calves. A total of 59 calves were instrumented and were monitored for mortality during eight winter field seasons ($n = 15$, 1998–1999; $n = 13$, 1999–2000; $n = 10$, 2000–2001; $n = 5$, 2001–2002; $n = 5$, 2002–2003; $n = 7$, 2003–2004; $n = 1$, 2004–2005; $n = 3$, 2005–2006) with two transmitters prematurely failing. Thirty-five of the remaining 57 calves died (12/15, 1998–1999; 3/12, 1999–2000; 2/10, 2000–2001; 5/5, 2001–2002; 2/4, 2002–2003; 7/7, 2003–2004; 1/1, 2004–2005; 3/3, 2005–2006), with 16 deaths attributed to wolf predation, six deaths due to starvation, and coyote predation was suspected in two deaths. We also recorded 11 mortalities that we suspected were due to wolf predation but for which the carcasses were so completely consumed that cause of death could not be reliable determined. The majority of deaths occurred in February ($n = 11$) and March ($n = 8$), with 5–6 deaths recorded in each of the remaining months (Dec., Jan., Apr.).

A comparison of fall and spring calf:cow ratios from the nine post-wolf years with those recorded during the pre-wolf period and the historic data collected prior to the initiation of our studies clearly demonstrated a tendency for the post-wolf ratios to be lower than those recorded prior to the

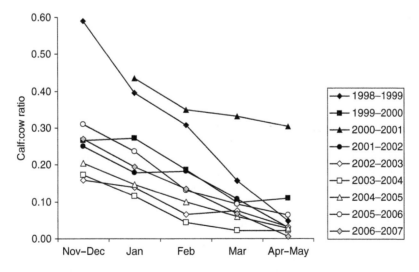

FIGURE 23.3 Over-winter changes in calf:cow ratios in the elk population in the Madison headwaters area of Yellowstone National Park during nine consecutive years after wolves became established in the study system.

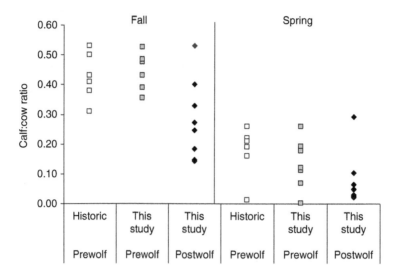

FIGURE 23.4 A comparison of fall and spring calf:cow ratios estimated from historic age-sex composition data and similar data collected during the seven years immediately prior to wolf reestablishment in the Madison headwaters drainages (Chapter 11 by Garrott *et al.*, this volume) and the 9 years post-wolf reestablishment (this chapter).

establishment of wolves in the study system (Figure 23.4). Median fall calf:cow ratios for the historic, pre-wolf, and post-wolf periods were 0.430, 0.453, and 0.269, respectively. Median spring calf:cow ratios for the historic, pre-wolf, and post-wolf periods were 0.190, 0.122, and 0.047, respectively. Similarly, annual estimates of calf winter survival during the post-wolf period, which were derived from the fall and spring calf:cow ratios, were also generally lower than recorded for the pre-wolf period, with the median survival estimates for pre- and post-wolf periods of 0.233 and 0.130, respectively (Tables 11.3 and 23.4). The exception to this pattern was the high recruitment and survival estimates from the winter of 2000–2001. Snow pack during this winter was the lightest recorded in the 16 years that we studied this system, and a large number of calves were consistently observed in the Madison drainage, an area only infrequently visited by the single resident wolf pack (Chapter 15 by Smith *et al.*, this volume) during this relatively early stage of wolf recolonization. In contrast to the pre-wolf studies where snow pack severity (SWE_{acc}) explained a large proportion of the variation in annual recruitment and calf winter-survival estimates, we found little support for either a cold- or warm-season climate effect (Table 23.1, Figure 23.5). For both vital rates, the null model was most supported by the data with a recruitment estimate of 0.044 calf per cow (95% CI = 0.021–0.091) and a calf winter-survival estimate of 0.156 (95% CI = 0.075–0.297). Thus, these results supported our hypothesis that a substantial proportion of wolf predation on calf elk in the winter was additive. However, similar to our results for adult female winter survival, the magnitude of additive affects of winter wolf predation on calves was most pronounced during mild snow pack winters and diminished as winter snow pack increased in severity. Under the harshest winter-severity conditions, our results suggest that wolf predation on elk calves could be considered nearly completely compensatory because nearly all calves perished during the most severe snow pack year experienced during the pre-wolf studies (Figure 23.5).

E. Population Estimates, Growth Rates and Projections

We conducted a total of 259 replicate mark–resight surveys to obtain population estimates, with 8–17 winter surveys and 10–33 spring surveys conducted each field season (Table 23.5). The number of marked animals available during surveys decreased over the years from maximum counts that ranged from 36 to 52 animals during the initial years to maximum counts of 18–24 animals during the latter

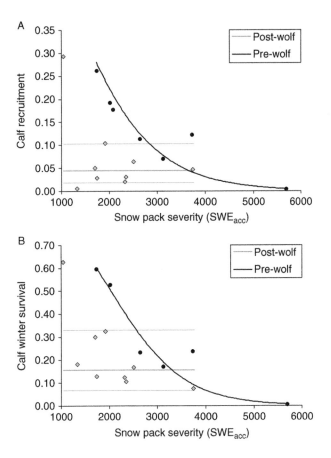

FIGURE 23.5 (A) Pre- and post-wolf relationships between annual variation in an index of snow pack severity (SWE$_{acc}$) for the winter range in the Madison headwaters area of Yellowstone National Park and spring calf:cow ratios, which can be considered an index of annual recruitment of calves to the yearling age class. (B) Pre- and post-wolf relationships between annual variation in an index of snow pack severity (SWE$_{acc}$) and estimated over-winter calf survival based on changes in fall and spring calf:cow ratios (Table 23.4). Points are the observed values, thick lines are the estimated relationships from the top-ranked models (Table 23.1), and thin lines represent the 95% confidence intervals for the post-wolf relationships. Uncertainty about the pre-wolf relationships is presented in Chapter 11 by Garrott *et al.*, this volume (Figure 11.7).

years. However, because of concurrent decreases in elk abundance, the estimated percent of the population marked remained relatively consistent, normally ranging between 7% and 11% with a mean of 9% for both the winter and spring surveys. Elk sightablity was generally lower for winter surveys with a mean of 29% of the marked population sighted, whereas 40% of the marked population was sighted on average during spring surveys. The hypergeometric MLE population estimates from both surveys demonstrated a marked decrease in elk abundance over the study period. At the beginning of wolf restoration, population estimates from the replicate winter surveys generally ranged between 500 and 700 animals with estimates from the replicate spring surveys, after most over-winter mortality had occurred, ranging between 300 and 500 animals. By the final 2 years of this study after the wolf population had been well established in the study system for ~8 years, population estimates from both the winter and spring surveys ranged between 170 and 230 animals (Table 23.5, Figure 23.6). Ln-linear regression of the population estimates yielded λ estimates of 0.85 (95% CI = 0.82–0.89, R^2_{adj} = 0.86) and 0.92 (95% CI = 0.88–0.96, R^2_{adj} = 0.59) for the data from winter and spring surveys, respectively.

TABLE 23.5 Winter and spring estimates of the size of the elk population occupying the Madison headwaters area of Yellowstone National Park

Year	No. Surveys	No. Marked	Mean Est. Prop. Pop. Marked	Mean Est. Prop. Pop. Detected	Min. Known Alive	L-P Pop. Est.	95% Conf. Interval	
							Lower	Upper
Winter								
1997–1998	17	36	0.07	0.29	418	663	598	743
1998–1999	11	40–52	0.06	0.25	302	732	639	851
1999–2000	10	48–52	0.11	0.28	232	490	434	560
2000–2001	9	40–43	0.06	0.13	195	691	546	904
2001–2002	11	31–47	0.07	0.26	255	495	432	576
2002–2003	11	29–33	0.08	0.22	142	389	326	474
2003–2004	9	18–26	0.10	0.48	178	233	209	265
2004–2005	8	23–26	0.09	0.37	179	278	240	330
2005–2006	10	22–26	0.11	0.30	124	223	190	268
2006–2007	9	18–21	0.10	0.33	110	183	154	222
Spring								
1997	10	25–26	0.07	0.25	144	382	315	475
1998	10	36	0.09	0.36	244	449	398	514
1999	11	38	0.09	0.32	299	466	416	530
2000	10	37	0.10	0.35	179	360	318	413
2001	10	41–42	0.08	0.37	353	577	517	652
2002	10	40–41	0.10	0.40	281	454	409	510
2003	10	31–33	0.09	0.35	212	423	371	491
2004	10	25	0.10	0.62	197	264	243	290
2005	16	23–25	0.10	0.41	176	253	230	283
2006	24	22–24	0.10	0.49	143	231	214	251
2007	33	19–20	0.12	0.43	127	174	162	188

Estimates were based on multiple Lincoln–Petersen mark-resight surveys conducted throughout the winter along fixed routes throughout the winter range and independent road-based surveys conducted on consecutive days in spring. Estimates and confidence intervals were calculated using the joint hypergeometric maximum likelihood estimator in program NOREMARK (White 1996).

These results contrast markedly with those from a similar analysis of population estimates obtained over the approximately three decades immediately prior to the wolf reintroduction when the population was estimated to be relatively stable with λ estimated at 1.01 (95% CI = 1.00–1.02, Figure 23.6, Chapter 11 by Garrott *et al.*, this volume).

Integrating our post-wolf vital rate estimates into the deterministic seasonal matrix model provided independent estimates of post-wolf population growth that could be compared with the direct estimates obtained from the mark–resight surveys, as well as with estimates obtained from the pre-wolf matrix model. We developed two versions of **B1** because of uncertainties in determining late-term pregnancy rates in the post-wolf period. For the version that used the pre-wolf pregnancy rates, we found that survival of calves for the first 6 months of life needed to be reduced to 0.40 to mimic the observed mean post-wolf fall calf:cow ratio (0.28). In contrast, the pre-wolf calf survival rate of 0.60 (Chapter 11 by Garrott *et al.*, this volume) adequately projected the observed mean fall calf:cow ratio in the second version of the **B1** matrix that used late-term pregnancy rates reduced by approximately 32% based on differences in serum pregnancy assays from pre-wolf to post-wolf periods. Both versions of the **B1** matrix produced identical calf populations entering the winter and, combined with the deterministic **B2** winter period matrix, yielded an arithmetic mean $\lambda = 0.838$ (95% CI = 0.800–0.876) for the post-wolf period. This λ estimate is similar to the direct estimates obtained from the mark–resight surveys and is strikingly dissimilar to the empirical estimates obtained from the pre-wolf

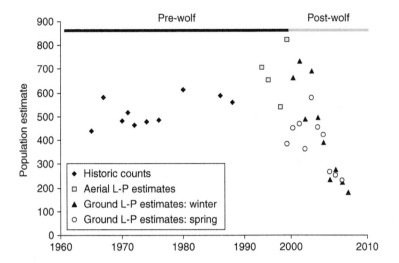

FIGURE 23.6 Population estimates for the elk herd occupying the Madison headwater drainages. Historic estimates are total counts based on aerial flights or a combination of aerial fights and ground surveys conducted during winter to early spring by National Park Service biologists or university researchers (Craighead *et al.* 1973, Aune 1981, Cole 1983, NPS files). The aircraft-based Lincoln–Petersen estimates were obtained during the pre-wolf study period based on aircraft mark-resight surveys of radio-collared elk groups in mid-winter (Eberhardt *et al.* 1998, Chapter 11 by Garrott *et al.*, this volume). The ground-based Lincoln–Petersen estimates were obtained from replicate winter and spring mark-resight surveys initiated near the end of the pre-wolf studies and continued through the post-wolf studies.

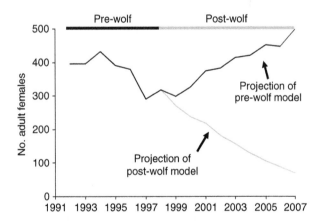

FIGURE 23.7 A comparison of the projected trend of the elk population using the pre-wolf matrix model and the ordered time series of SWE$_{acc}$ experienced on the study area during the 16-year project (*i.e.*, encompassing the pre- and post-wolf study periods) and the projected population trend from the deterministic post-wolf matrix model.

population estimates ($\lambda = 1.01$, 95% CI $= 1.00$–1.02) and the pre-wolf matrix model ($\lambda = 1.03$, 95% CI $= 0.869$–1.195, Chapter 11 by Garrott *et al.*, this volume).

Figure 23.7 contrasts the projected trends in the elk population for the 16 years encompassing the pre-wolf and post-wolf study periods. During the pre-wolf period both calf and cow survival were strongly correlated with snow pack severity (SWE$_{acc}$). Three consecutive winters with above-average to severe snow pack near the end of the pre-wolf studies resulted in a projected decrease in the population. Projecting the population through the post-wolf period using the pre-wolf matrix model predicted a subsequent gradual increase in abundance as a consequence of the mild to average snow pack severities experienced from 2000 to 2007. However, data from the post-wolf studies indicated that both calf and cow survival rates decreased substantially and were no longer correlated with snow pack severity, resulting in the post-wolf matrix model projecting a steady decrease in abundance which closely matched the trend in the actual population estimates during the post-wolf study period (Figure 23.6). Thus, results of population trend from our direct estimates of abundance and those obtained by integrating our vital rate estimates into a matrix model supported our hypotheses that additive wolf predation would limit the elk population and result in a decrease in abundance from pre-wolf levels.

IV. DISCUSSION

Prior to the reintroduction of wolves into the study system the overwhelming cause of mortality for both calf and cow elk was starvation during the winter period, with 93% and 75% of all documented mortalities of radio-collared calves and cows, respectively, attributed to starvation (Chapter 11 by Garrott *et al.*, this volume). This bottom–up, food-limiting process was undoubtedly operating during the post-wolf years of our study (Chapter 22 by White *et al.*, this volume) and annual variability of snow pack experienced during the post-wolf period was substantial with a range of snow pack severities that our pre-wolf study suggested should have resulted in substantial over-winter starvation mortality each year. However, starvation deaths among radio-collared animals during the post-wolf study were infrequent, with only 17% of calf mortalities and 12% of cow mortalities attributed to starvation. In addition, most of these starvation deaths were recorded during the initial years of the post-wolf study when a considerable proportion of the elk population was not yet subjected to significant risk of wolf predation due to the restricted distribution of wolf pack territories (Chapter 15 by Smith *et al.*, this volume). Thus, we documented a nearly complete substitution of wolf predation for starvation mortality in the elk population during the winter period, suggesting that wolves were very effective at identifying and killing those animals that were destined to die each winter (*i.e.*, Errington's "doomed surplus"). These results provided strong evidence that a substantial component of wolf predation on elk was compensatory, that is, wolves were successful in detecting and killing winter-weakened animals before complete depletion of their body reserves and death.

Our conclusion that a proportion of wolf predation was compensatory compliments and corroborates the results of many other wolf predation studies (see review by Mech and Peterson 2003). However, quantifying the proportion of wolf predation that is additive has proven to be a challenge for ecologists (National Research Council 1997, Andersen *et al.* 2006) and wolf predation on calves, in particular, has been identified as a significant knowledge gap (Mech and Peterson 2003). This question is difficult to directly address because, in nearly all wolf–ungulate study systems, wolf predation risk is just one source of mortality competing for the lives of the ungulate prey. As a result, in most systems it has been difficult to impossible to deduce likely prey survival rates in the absence of wolf predation. Perhaps the most productive approach to addressing this question has been the analysis of data from management experiments where wolf densities were temporarily reduced through removal programs and the demographic responses of ungulate prey were monitored (Gasaway *et al.* 1992, Boertje *et al.* 1996, National Research Council 1997, Hayes and Harestad 2000). Though these experiments had limitations (Boutin 1992), they generally provided reasonable support for the idea that a portion of wolf predation is indeed additive to other sources of mortality. Our study contributes additional insights because we estimated vital rates for the elk prey population in the complete absence of wolf predation and identified covariates strongly correlated with annual variation in those vital rates (Chapter 11 by Garrott *et al.*, this volume). By assuming that these non-wolf processes remained unchanged after wolves colonized the study system, we were then able to use those estimated rates and their relationships with environmental conditions to predict survival rates for calves and cows during the nine years of post-wolf studies if these relationships continued to hold. Finally, we were able to compare those predictions with actual estimates for the post-wolf period to provide a direct estimate of the extent to which wolf predation was additive in our study system.

This approach provided strong evidence that a substantial component of wolf predation on both adult cows and calves during the winter period was additive and that the magnitude of the additive predation increased in lighter snow pack years when fewer elk were debilitated by severe nutritional deprivation. Our results indicated that in severe snow pack years additive predation on cows was limited to prime-aged animals and reduced survival probability by a modest 1–3% for each age class. In contrast, during winters with light to moderate snow pack when our pre-wolf study indicated that nearly all adult cows would have survived regardless of age, additive wolf predation was detected for all age classes, with reductions in survival probability ranging from 0.02–0.04 for young cows to a substantial 0.08–0.10 in the oldest-aged cows (Figure 23.2). This pattern was even more pronounced for over-winter calf survival. Pre-wolf data revealed a strong correlation between snow pack severity and calf over-winter survival, with relatively high survival during the lightest snow pack years and the virtual elimination of the entire calf cohort due to starvation during the most severe snow pack year (Chapter 11 by Garrott *et al.*, this volume). With the re-establishment of wolves in the study system, this tight coupling of calf survival and annual variation in snow pack severity was eliminated and calf survival over winter was uniformly low due to additive predation (Figure 23.4). Under mild snow pack conditions, additive wolf predation reduced calf survival to between one-half to one-quarter of the rates realized during pre-wolf years, while wolf predation on calves during severe snow pack conditions was almost entirely compensatory because few calves would have survived in the absence of wolves. These findings are similar to studies across a wide range of systems that indicate environmental effects can substantially temper the respective trophic strengths of food limitation and predation (Menge and Sutherland 1987, Oedekoven and Joern 2000, Sinclair and Krebs 2002, Spiller and Schoener 2008).

The relative impact of wolf predation on cow and calf survival during the summer period was much less clear. While the data pooled across all post-wolf years provided an estimate of cow survival (0.942) that was slightly lower than the pooled survival estimate for pre-wolf years (0.987), confidence limits for the difference between the estimates included zero. In general, cow survival over the summer period was uniformly high throughout our 16-year study with only 17 total deaths documented for 448 animal years of monitoring. Because we relocated instrumented animals on a monthly basis during the summer, we were able to reliably determine cause of death for only nine mortalities, with vehicle collisions responsible for six of these deaths. Therefore, we conclude that wolf predation during the summer period likely had little influence on survival probabilities of cow elk in our study population. These results seem counter-intuitive given that energetic demands of the local wolf population likely increased during the summer months due to requirements for provisioning pups (Mech 1970) and there was little potential for compensatory wolf predation given the high cow survival documented during the pre-wolf period (Chapter 11 by Garrott *et al.*, this volume). While elk in summer are not weakened by sub-maintenance nutrition and were likely more effective in eluding capture in the absence of snow pack impediments, we suspect that numerical and spatial dilution of predation risk was more significant in explaining these results. The central Yellowstone region serves as summer range for a number of migratory elk and mule deer (*Odocoileus hemionus*) populations that winter elsewhere (Singer 1994). Thus, in winter the local wolf population had access to relatively high densities of elk concentrating in the Madison headwaters area (Chapter 8 by Messer *et al.*, this volume), but extremely low elk densities through much of the rest of the central Yellowstone region. As snow pack diminished in spring, the non-migratory Madison headwaters elk were augmented by migratory deer and elk moving into the system from surrounding areas. This migratory influx provided a considerably larger, widely distributed, ungulate prey population throughout central Yellowstone, potentially reducing wolf predation risk for the Madison headwaters elk population.

Additive predation reduces prey densities which, in turn, can increase per capita resources for surviving animals and potentially lead to a compensatory increase in fecundity (Nichols 1991, Boyce *et al.* 1999). However, we found no evidence for such a demographic response. Our primary means of assessing fecundity was determination of late-term pregnancy rates of instrumented cows using fecal P_4 concentrations. This technique was initially validated for elk during the early years of our pre-wolf studies (White *et al.* 1995, Garrott *et al.* 1998), with a subsequent experiment using captive elk providing additional support for the technique's validity (Cook *et al.* 2002). However, application of the technique throughout our studies revealed seven years (1997, 2002–2006) where extremely low pregnancy assessments were obviously erroneous based on ancillary data from sightings of instrumented cows with calves-at-heel, fall calf:cow ratios, and serum PSPB and progesterone assays. Given these unexpected results, we attempted to gain some insight into potential changes in pregnancy rates between pre- and post-wolf study periods using serum pregnancy assays from blood samples collected when we instrumented elk. The data suggested pregnancy rates may have been reduced by approximately 32% during the 2002–2006 period. However, these results must be viewed with caution because they were derived from modest samples each year, including some samples from animals that may have been cycling.

Given this caution, there are two potential explanations for reduced pregnancy rates during the post-wolf years. Creel *et al.* (2007) presented a negative correlation between mean fecal P_4 concentrations and estimated wolf:elk ratios for a number of herds primarily in the vicinity of Yellowstone and suggested that wolves indirectly affected reproductive physiology and reduced pregnancy rates of elk through the costs of anti-predator behavior. An alternative hypothesis is that an extended drought in the region reduced elk nutrition which, in turn, affected demographic performance (Vucetich *et al.* 2005). However, Hamlin *et al.* (Chapter 25 by Hamlin *et al.*, this volume) provided assessments of both these hypotheses using data collected on seven elk populations in the Yellowstone region and found no evidence for either a drought- or wolf-induced reduction in elk pregnancy rates. While we have little confidence in our pregnancy assessments, we are certain that fall calf:cow ratios were lower during the post-wolf period then that recorded prior to wolf reestablishment, thus, we found no evidence for a compensatory increase in fecundity as a response to additive wolf predation. This result was not unexpected because pregnancy rates during the pre-wolf studies approached the biological maxima for the species, thus there was little potential for fecundity to increase.

We did not directly assess calf survival during the summer period and, as a result, insights on potential changes in calf survival due to the reestablishment of wolves were limited to pre-wolf and post-wolf comparisons of fall calf:cow ratios. These comparisons demonstrated a substantial decrease in calf:cow ratios from a mean of 0.456 ($n = 7$, range 0.356–0.531) for the pre-wolf period to a mean of 0.249 ($n = 8$, range 0.142–0.400) for the post-wolf period. This demographic metric is a product of fecundity and survival of calves over the first six months of life. While reduced fecundity may have contributed to lower post-wolf fall calf:cow ratios, the evidence is equivocal, and we suspect that the principal cause of the lower fall ratios was increased predation rates on calves (Chapter 25 by Hamlin *et al.*, this volume). Various field studies conducted in the western United States found that black bears (*Ursus arctos*) and grizzly bears (*Ursus americanus*) were the predominant predators of elk calves up to 45 days old and, where present, wolves were a major predator for older calves (Schlegel 1976, Myers *et al.* 1996, Singer *et al.* 1997, Smith and Anderson 1998, Smith *et al.* 2006, Raithel *et al.* 2007, Barber-Meyer *et al.* 2008). Similar results have been reported for mortality of moose and caribou calves in boreal North America (National Research Council 1997). Over the past decade, both grizzly bear and wolf abundance has increased substantially in the Yellowstone region (Smith *et al.* 2003, Schwartz *et al.* 2006; Chapter 15 by Smith *et al.* and Chapter 25 by Hamlin *et al.*, this volume). Thus, we suggest that both predators likely contributed to the lower fall calf:cow ratios recorded during our post-wolf studies.

We hypothesized that substantial additive predation on elk by wolves and a weak compensatory response in fecundity would result in a reduction in elk abundance. Mark-resight population estimates from the post-wolf period indicated a clear and dramatic change in population trend from relative stability during the three decades prior to wolf restoration to a rapidly decreasing population (Figure 23.6). The change in population trend derived from direct estimates of abundance was corroborated by matrix model projections that integrated our independent estimates of population vital rates. Projecting the pre-wolf matrix model for the post-wolf period provided a prediction of elk population trend for the series of snow pack severities experienced during the nine post-wolf years under the scenario that all wolf predation was compensatory (Figure 23.7). The departure of the actual trend in the population observed provided an assessment of the impact of additive wolf predation. Though our study was not a replicated experiment, we interpret these results as providing evidence that wolf predation strongly limited the elk prey population, exerting a strong top–down influence in a system that was bottom–up regulated prior to wolf reintroduction. The pre-wolf matrix model demonstrated that population growth was most sensitive to changes in calf survival, with the post-wolf data demonstrating the strongest impact of additive wolf predation on this age class (Figure 23.8). Thus, the substantial and consistent predation of calves without concurrent demographic compensation in pregnancy was the primary mechanism responsible for the decrease in elk abundance (Figure 23.8).

Similar conclusions have been reached in numerous other investigations of wolf–ungulate systems (Bergerud and Elliot 1986, Van Ballenberghe 1987, Ballard 1991, Gasaway *et al.* 1992, Dale *et al.* 1994, Boertje *et al.* 1996, Jędrzejewska *et al.* 2002, Hebblewhite *et al.* 2002). However, several authors have argued wolf predation in the Yellowstone region was *primarily* compensatory based on: (1) assessments of bone marrow fat content of wolf-killed ungulates that suggested these animals were in poor nutrition; (2) age distribution of killed ungulates that indicated wolves primarily killed the youngest age class (calves/fawns) and old-age adults with relatively low reproductive value and higher

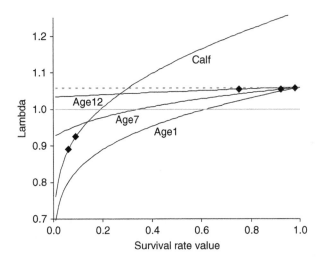

FIGURE 23.8 The sensitivity of population growth rate, λ, to unit changes in annual survival rates for different age classes of female elk derived from a deterministic Leslie matrix model parameterized with vital rates estimated from the pre-wolf study conducted over the seven years immediately prior to wolf recolonization of the study system (Chapter 11 by Garrott *et al.*, this volume). The dashed horizontal line at $\lambda=1.05$ represents the asymptotic population growth rate for the pre-wolf model. The point where each age-specific sensitivity curve intersects and dashed horizontal line represents the estimated pre-wolf survival rate for that age class (calf: 0.30; age 1: 0.98; age 7: 0.97; age 12: 0.90). The solid diamonds represent the estimated annual survival rate for each age class derived from the 9-year study after wolf reestablishment (calf: 0.06/0.09; age 1: 0.98; age 7: 0.92; age 12: 0.75). The differences between the pooled pre- and post-wolf survival rates represent estimates of the additive component of wolf predation (calf: 0.24/0.21; age 1: 0.0; age 7: 0.05; age 12: 0.15) and its relative effect on population growth. Two post-wolf survival estimates are plotted on the calf curve bounding the uncertainty associated with post-wolf calf summer survival.

probabilities of mortality; and (3) differences in annual variation in mortality in ungulate populations exposed to wolf predation compared to populations where large predators were absent (Smith *et al.* 2003, 2004; Vucetich *et al.* 2005; Wright *et al.* 2006). We agree these findings suggest some component of wolf predation is compensatory. Such observations, however, do not necessarily imply that additive wolf predation is subsequently inconsequential, as all of these same indicators were evident in our studies (Chapter 10 by Garrott *et al.* and Chapter 16 by Becker *et al.*, this volume) yet we documented a substantial reduction in survival from pre- to post-wolf study periods that can reasonably be attributed to additive wolf predation. Other ungulate studies addressing the interaction of food limitation, climate, and predators have provided direct evidence that predation may be largely compensatory (Bartmann *et al.* 1992, Tveraa *et al.* 2003), but the primary predators in these studies were mid-sized carnivores that may not be as effective at killing large ungulates as wolves unless the ungulates are debilitated by starvation, disease, or other factors.

During this study, we documented dramatic changes in the demographic characteristics of an ungulate population transitioning from a relatively predator-free to a predator-rich environment. Predictions of the consequences of wolf reintroduction on the high-density ungulate populations of the Greater Yellowstone Ecosystem have varied widely (Boyce 1993, Mack and Singer 1993, Messier *et al.* 1995, Eberhardt 2003, White *et al.* 2003, Varley and Boyce 2006), but nearly all modeling exercises projected some level of reduction in ungulate abundance due to wolf predation. While numerous studies focused on the potential trophic consequences resulting from alterations in elk anti-predator behavior (Berger *et al.* 2001, Beschta 2003, Wilmers *et al.* 2003, Hebblewhite *et al.* 2005), the steady decrease in the abundance of elk in the Madison headwaters area due to additive wolf predation clearly demonstrates that wolves can also exert strong, direct influences by limiting prey. Thus, we suggest direct off-take by wolves can transition systems to alternate low prey density states, with the strength of the effect heavily-mediated by environmental and anthropogenic factors, and with potential consequences throughout the trophic web (Chapter 24 by Garrott *et al.* and Chapter 25 by Hamlin *et al.*, this volume).

While wolves limited elk in the Madison headwaters, the potential for predator regulation of this ungulate prey has not been resolved. Regulation requires some density-dependent mechanism that mediates the impact of predation on the prey population (Messier 1994, Murdoch 1994, Sinclair 2003). We documented a variety of potential regulating mechanisms, including wolf numeric (Chapter 15 by Smith *et al.*, this volume) and functional responses (Chapter 17 by Becker *et al.*, this volume), the use of alternate bison prey (Chapter 16 by Becker *et al.*, this volume), a variety of elk behavioral responses to predation risk (Chapters 18–20 by Gower *et al.*, this volume), and the potential existence of prey refugia (Chapter 21 by White *et al.*, this volume). Eberhardt *et al.* (2003) evaluated the potential consequences of wolf reintroduction on Yellowstone's elk populations and suggested that the transitory dynamics would last approximately one decade. Thus, continued monitoring of the Madison headwaters wolf–ungulate system beyond the studies reported in this book should provide insights into potential top–down regulation of large herbivore populations. We explore these potential regulatory processes in Chapter 24 by Garrott *et al.*, this volume.

Two fundamental aspects of our study enabled us to advance our understanding of wolf–ungulate predator–prey processes, namely the maintenance of an intensive long-term field study and the uniqueness of the Madison headwaters study system. Dynamics can vary within and among years and systems on fine and broad temporal and spatial scales (Sinclair *et al.* 2007), thus maintaining an intensive long-term ecological study allowed for continuous and consistent collection of pertinent data, capturing variation in fundamental aspects of the system that potentially influenced predator–prey processes. These data included detailed descriptions of landscape attributes, climatic variation, a uniquely detailed understanding of spatial and temporal variation in snow pack attributes, and detailed and precise data on the abundance, distribution, and behavior of both predator and prey populations. The study was also unique in that it spanned a substantial period prior to wolf recolonization that allowed a detailed understanding of elk population processes in the absence of a major predator, with convincing evidence that the prey population had been relatively stable near its carrying capacity for nearly three decades. Continuation of these studies for nearly a decade after wolf reestablishment allowed us to document changes in prey population processes that can reasonably be attributed to predation.

While large mammal systems are notoriously difficult to study, the Madison headwaters was tractable, sizeable enough to support significant populations of both predator and prey, but manageable to the extent that we were able to enumerate prey and predator populations precisely and follow all resident wolf packs on a daily basis to obtain more complete and detailed information on kill rate and prey selection than can be realized with typical aircraft-based studies covering broad geographic extents. The unique geothermal environments combined with the extreme severity of winters provided important and strong gradients in spatial and temporal variation and resulted in an abbreviated age structure in the elk population. These attributes magnified and intensified predator–prey interactions, enhancing our ability to characterize and understand variability in these processes. While the outcome of this fortuitous ecological experiment is yet to be determined, and almost certainly will not be directly applicable to other wolf–ungulate systems, we believe insights gained from the ecological processes and themes underlying this predator–prey system can be generalized to other systems and represent a contribution to our knowledge of wolf–ungulate interactions and community ecology.

V. SUMMARY

1. We documented the transition of the Madison headwaters elk population in Yellowstone National Park from a population that was clearly "bottom–up" regulated by food limitation in the absence of wolves to a population that is now strongly "top–down" limited by predation.
2. Wolf predation during winters experiencing mild snow pack conditions was strongly additive because few elk died under these conditions in the absence of wolves. Conversely, wolf predation during winters with severe snow pack was primarily compensatory because starvation mortality was pervasive under these conditions prior to wolf reestablishment.
3. We found no evidence for a compensatory increase in pregnancy rates as a result of additive wolf predation. The magnitude of additive predation overwhelmed any capacity for demographic compensation in vital rates and, as a consequence, the abundance of elk decreased substantially.
4. This study documented the transitional dynamics of an elk population due to the reestablishment of wolves, but whether wolf predation will be regulatory is uncertain.
5. The ultimate state of this predator–prey system will depend on the strength and interaction of a number of ecological processes, including wolf functional and numeric responses to the decreasing abundance of elk, the influence of the alternate bison prey on wolf predation, and the effectiveness of apparent prey refuges and anti-predatory responses of elk.

VI. REFERENCES

Abrams, P. A. 1994. The fallacies of ratio-dependent predation. *Ecology* **75**:1842–1850.

Akcakaya, H. R., R. Arditi, and L. R. Ginzburg. 1995. Ratio-dependent predation: An abstraction that works. *Ecology* **76**:995–1004.

Andersen, R., J. D. Linnell, and E. J. Solberg. 2006. The future role of large carnivores in terrestrial trophic interactions: The northern temperate view. Pages 413–448 *in* K. Danell, R. Bergström, P. Duncan, and J. Pastor (Eds.) *Large Herbivore Ecology, Ecosystem Dynamics and Conservation.* Cambridge University Press, Cambridge, United Kingdom.

Anderson, D. R., and K. P. Burnham. 1976. Population ecology of the mallard. VI. The effect of exploitation on survival. *U.S. Fish and Wildlife Service Resource Publication* **128**:1–110.

Aune, K. E. 1981. Impacts of winter recreationists on wildlife in a portion of Yellowstone National Park, Wyoming. Thesis. Montana State University, Bozeman, Montana.

Ballard, W. B. 1991. Management of predators and their prey: The Alaskan experience. *Transactions of the North American Wildlife and Natural Resources Conference* **56**:527–538.

Barber-Meyers, S. M., L. D. Mech, and P. J. White. 2008. Survival and cause-specific elk calf survival following wolf restoration to Yellowstone National Park. *Wildlife Monographs*.

Barbosa, P., and I. Castellanos. 2005. *Ecology of Predator–Prey Interactions*. Oxford University Press, New York, NY.

Bartmann, R. M., G. C. White, and L. H. Carpenter. 1992. Compensatory mortality in a Colorado mule deer population. *Wildlife Monographs* **121**.

Berger, J., P. B. Stacey, L. Bellis, and M. P. Johnson. 2001. A mammalian predator–prey balance: Grizzly bear and wolf extinction affect neotropical migrants. *Ecological Applications* **11**:947–960.

Bergerud, A. T., and J. P. Elliot. 1986. Dynamics of caribou and wolves in northern British Columbia. *Canadian Journal of Zoology* **64**:1515–1529.

Beschta, R. L. 2003. Cottonwoods, elk, and wolves in the Lamar Valley of Yellowstone National Park. *Ecological Applications* **13**:1295–1309.

Boertje, R. D., P. Valkenburg, and M. E. McNay. 1996. Increases in moose, caribou, and wolves following wolf control in Alaska. *Journal of Wildlife Management* **60**:474–489.

Boutin, S. 1992. Predation and moose population dynamics: A critique. *Journal of Wildlife Management* **56**:116–127.

Boyce, M. S. 1993. Predicting the consequences of wolf recovery to ungulates in Yellowstone National Park. Pages 234–269 *in* R. S. Cook (Ed.) *Ecological Issues on Reintroducing Wolves into Yellowstone National Park*. U.S. Department of Interior, National Park Service Science Monograph NPS/NRYELL/NRSM-93/22.

Boyce, M. S., A. R. E. Sinclair, and G. C. White. 1999. Seasonal compensation of predation and harvesting. *Oikos* **87**:419–426.

Brown, J. L., S. K. Wasser, D. E. Wildt, and L. H. Graham. 1994. Comparative aspects of steroid hormone metabolism and ovarian activity in felids, measured noninvasively in feces. *Biology of Reproduction* **51**:776–786.

Burnham, K. P., and D. R. Anderson. 2002. *Model Selection and Multimodel Inference: A Practical Information Theoretic approach*. Springer-Verlag, New York, NY.

Caswell, H. 2001. *Matrix Population Models: Construction, Analysis, and Interpretation*. Sinauer Associates, Sunderland, MA.

Cole, G. F. 1983. A naturally regulated elk population. Pages 62–81 *in* F. L. Bunnell, D. S. Eastman, and J. M Peek (Eds.) *Symposium on Natural Regulation of Wildlife Populations*. Proceedings No. 14, Forest Wildlife and Range Experiment Station, University of Idaho, Moscow, ID.

Connell, J. H. 1975. Some mechanisms producing structure in natural communities: A model and evidence from field experiments. Pages 460–490 *in* M. L. Cody and J. Diamond (Eds.) *Ecology and Evolution of Communities*. Harvard University Press, Cambridge, MA.

Cook, J. G. 2002. Nutrition and food. Pages 259–349 *in* D. E. Toweill and J. W. Thomas (Eds.) *North American Elk: Ecology and Management*. Smithsonian Institution Press, Washington, DC.

Cook, R. C., J. G. Cook, R. A. Garrott, L. L. Irwin, and S. L. Monfort. 2002. Effects of diet and body condition on fecal progestagen excretion in elk. *Journal of Wildlife Diseases* **38**:558–565.

Craighead, J. J., F. C. CraigheadJr., R. L. Ruff, and B. W. O'Gara. 1973. Home ranges and activity patterns of nonmigratory elk of the Madison drainage herd as determined by biotelemetry. *Wildlife Monographs* **33**.

Creel, S., D. Christianson, S. Lily, and J. A. Winnie, Jr. 2007. Predation risk affects reproductive physiology and demography of elk. *Science* **315**:960.

Dale, B. W., L. G. Adams, and R. T. Bowyer. 1994. Functional response of wolves preying on barren-ground caribou in a multiple-prey ecosystem. *Journal of Animal Ecology* **63**:644–652.

Darwin, C. 1859. *On the Origin of Species by Means of Natural Selection*. Down, Bromley, and Kent, London, United Kingdom.

Denno, R. F., and M. A. Peterson. 2000. Caught between the devil and the deep blue sea: Mobile planthoppers elude natural enemies and deteriorating host plants. *American Entomologist* **46**:95–109.

Denno, R. F., C. Gratton, M. A. Peterson, G. A. Langellotto, D. L. Finke, and A. F. Huberty. 2002. Bottom–up forces mediate natural-enemy impact in a phytophagous insect community. *Ecology* **83**:1443–1458.

Eberhardt, L. L. 1977. Optimal policies for conservation of large mammals with special reference to marine ecosystems. *Environmental Conservation* **4**:205–212.

Eberhardt, L. L. 1985. Assessing the dynamics of wild populations. *Journal of Wildlife Management* **49**:997–1012.

Eberhardt, L. L. 1987. Population projections from simple models. *Journal of Applied Ecology* **24**:103–118.

Eberhardt, L. L. 2002. A paradigm for population analysis of long-lived vertebrates. *Ecology* **83**:2841–2854.

Eberhardt, L. L., R. A. Garrott, P. J. White, and P. J. Gogan. 1998. Alternative approaches to aerial censusing of elk. *Journal of Wildlife Management* **62**:1046–1055.

Eberhardt, L. L., R. A. Garrott, D. W. Smith, P. J. White, and R. O. Peterson. 2003. Assessing the impacts of wolves on ungulate prey. *Ecological Applications* **13**:776–783.

Eidenshink, J. C. 1992. The 1990 conterminous U.S. AVHRR data set. *Photogrammetric Engineering and Remote Sensing* **58**:809–813.

Erlich, P. R., and L. C. Birch. 1967. The balance of nature and population control. *American Naturalist* **101**:97–107.

Errington, P. L. 1946. Predation and vertebrate populations. *Quarterly Review of Biology* **21**:147–177.

Errington, P. L. 1967. *Of Predation and Life.* Iowa State University Press, Ames, IA.

Forkner, R. E., and M. D. Hunter. 2000. What goes up must come down? Nutrient addition and predation pressure on oak herbivores. *Ecology* **81**:1588–1600.

Gaillard, J. -M., M. Festa-Bianchet, and N. G. Yoccoz. 1998. Population dynamics of large herbivores: Variable recruitment with constant adult survival. *Trends in Ecology and Evolution* **13**:58–63.

Gaillard, J. -M., M. Festa-Bianchet, N. G. Yoccoz, A. Loison, and C. Toïgo. 2000. Temporal variation in fitness components of population dynamics in large herbivores. *Annual Review of Ecology and Systematics* **31**:367–393.

Garrott, R. A., S. L. Monfort, P. J. White, K. L. Mashburn, and J. G. Cook. 1998. One-sample pregnancy diagnosis in elk using fecal steroid metabolites. *Journal of Wildlife Diseases* **34**:126–131.

Garrott, R. A., L. L. Eberhardt, P. J. White, and J. Rotella. 2003. Climate-induced limitation of a large herbivore population. *Canadian Journal of Zoology* **81**:33–45.

Garrott, R. A., J. E. Bruggeman, M. S. Becker, S. T. Kalinowski, and P. J. White. 2007. Evaluating prey switching in wolf–ungulate systems. *Ecological Applications* **17**:1588–1597.

Gasaway, W. C., R. D. Boertje, D. V. Grangaard, D. G. Kelleyhouse, R. O. Stephenson, and D. G. Larsen. 1992. The role of predation in limiting moose at low densities in Alaska and Yukon and implications for conservation. *Wildlife Monographs* **120**.

Hairston, N. G. 1989. *Ecological Experiments: Purpose, Design, and Execution.* Cambridge University Press, New York, NY.

Hairston, N. G., F. E. Smith, and L. B. Slobodkin. 1960. Community structure, population control, and competition. *American Naturalist* **44**:421–425.

Hamlin, K. L., D. F. Pac, C. A. Sime, R. M. DeSimone, and G. L. Dusek. 2000. Evaluating the accuracy of ages obtained by two methods for Montana ungulates. *Journal of Wildlife Management* **64**:441–449.

Hanson, W. R. 1963. Calculation of productivity, survival, and abundance of selected vertebrates from sex and age ratios. *Wildlife Monographs* **9**.

Hassell, M. P. 1978. *The Dynamics of Arthropod Predator–Prey Systems.* Princeton University Press, Princeton, NJ.

Hayes, R. D., and A. S. Harestad. 2000. Wolf functional response and regulation of moose in the Yukon. *Canadian Journal of Zoology* **78**:60–66.

Hebblewhite, M., D. H. Pletscher, and P. C. Paquet. 2002. Elk population dynamics in areas with and without predation by recolonizing wolves in Banff National Park, Alberta. *Canadian Journal of Zoology* **80**:789–799.

Hebblewhite, M., C. A. White, C. G. Nietvelt, J. A. McKenzie, T. E. Hurd, J. M. Fryxell, S. E. Bayley, and P. C. Paquet. 2005. Human activity mediates a trophic cascade caused by wolves. *Ecology* **86**:2135–2144.

Hilborn, R., C. J. Walters, and D. Ludwig. 1995. Sustainable exploitation of renewable resources. *Annual Review of Ecology and Systematics* **26**:45–67.

Holling, C. S. 1959. The components of predation as revealed by a study of small mammal predation of the European pine sawfly. *Canadian Entomologist* **91**:293–320.

Hosmer, D. W., T. Hosmer, S. le Cessie, and S. Lemeshow. 1997. A comparison of goodness-of-fit tests for the logistic regression model. *Statistics in Medicine* **16**:965–980.

Hunter, M. D., and P. W. Price. 1992. Playing chutes and ladders: Heterogeneity and the relative roles of bottom–up and top–down forces in natural communities. *Ecology* **73**:724–732.

Jędrzejewski, W., K. Schmidt, J. Theuerkauf, B. Jędrzejewska, N. Selva, K. Zub, and L. S. Szymura. 2002. Kill rates and predation by wolves on ungulate populations in Białowieza Primeval Forest (Poland). *Ecology* **83**:1341–1356.

Jönsson, P., and L. Eklundh. 2002. Seasonality extraction by function fitting to time-series of satellite sensor data. *IEEE Transactions on Geoscience and Remote Sensing* **40**:1824–1832.

le Cessie, S., and J. C. van Houwelingen. 1991. A goodness-of-fit test for binary regression models, based on smoothing methods. *Biometrics* **47**:1267–1282.

Letourneau, D. K., and L. A. Dyer. 1998. Experimental test in lowland tropical forest shows top–down effects through four trophic levels. *Ecology* **79**:1678–1687.

Mack, J. A., and F. J. Singer. 1993. Using Pop II models to predict effects of wolf predation and hunter harvest on elk, mule deer, and moose on the Northern Range. Pages 45–74 *in* R. S. Cook (Ed.) *Ecological Issues on Reintroducing Wolves into Yellowstone National Park.* U.S. Department of Interior, National Park Service Science Monograph NPS/NRYELL/NRSM-93/22.

Malingreau, J. P. 1989. The vegetation index and the study of vegetation dynamics. Pages 285–303 *in* F. Toselli (Ed.) *Application of Remote Sensing to Agrometeorology.* ECSC, Brussels, Belgium and Luxembourg.

McCullagh, P., and J. A. Nelder. 1989. *Generalized Linear Models.* Chapman and Hall, New York, NY.

Mech, L. D. 1970. *The Wolf: The Ecology and Behavior of an Endangered Species.* Natural History Press, Garden City, New York, NY.

Mech, L. D., L. G. Adams, T. J. Meier, J. W. Burch, and B. W. Dale. 1998. *The Wolves of Denali.* University of Minnesota Press, Minneapolis, MN.

Mech, L. D., and R. O. Peterson. 2003. Wolf-prey relations. Pages 131–160 *in* L. D. Mech and L. Boitani (Eds.) *Wolves: Behavior, Ecology, and Conservation*. University of Chicago Press, Chicago, IL.

Menge, B. A., and J. P. Sutherland. 1987. Community regulation: Variation in disturbance, competition, and predation in relation to environmental stress and recruitment. *American Naturalist* **130**:730–757.

Messier, F. 1994. Ungulate population models with predation: A case study with the North American moose. *Ecology* **75**:478–488.

Messier, F., W. C. Gasaway, and R. O. Peterson. 1995. *Wolf–Ungulate Interactions in the Northern Range of Yellowstone: Hypotheses, Research Priorities, and Methodologies*. Midcontinent Ecological Science Center, National Biological Service, Fort Collins, CO.

Murdoch, W. W. 1994. Population regulation in theory and practice. *Ecology* **75**:271–287.

Myers, W. L., B. Lyndaker, P. E. Fowler, and W. Moore. 1996. Investigations of calf elk mortalities in southeast Washington: A progress report 1992–1996. Pittman-Roberson Program Report. Washington Department of Wildlife, Olympia, WA.

National Research Council. 1997. *Wolves, Bears, and Their Prey in Alaska: Biological and Social Challenges in Wildlife Management*. National Academy Press, Washington, DC.

Nichols, J. D. 1991. Responses of North American duck populations to exploitation. Pages 498–525 *in* C. M. Perrins, J.-D. Lebreton, and G. J. M. Hirons (Eds.) *Bird Population Studies: Relevance to Conservation and Management*. Oxford University Press, Oxford, United Kingdom.

Noyes, J. H., R. G. Sasser, B. K. Johnson, L. D. Bryant, and B. Alexander. 1997. Accuracy of pregnancy detection by serum protein (PSPB) in elk. *Wildlife Society Bulletin* **25**:695–698.

Oedekoven, M. A., and A. Joern. 2000. Plant quality and spider predation affects grasshoppers (Acrididae): Food-quality-dependent compensatory mortality. *Ecology* **81**:66–77.

Paulik, G. J., and D. S. Robson. 1969. Statistical calculations for change-in-ratio estimators of population parameters. *Journal of Wildlife Management* **33**:1–27.

Peterson, R. O. 1977. National Park Service Scientific Monograph Series 11. Wolf Ecology and Prey Relationships on Isle Royale. Washington, DC.

Pettorelli, N., A. Mysterud, N. G. Yoccoz, R. Langvatn, and N. C. Stenseth. 2005. Importance of climatological downscaling and plant phenology for red deer in heterogeneous landscapes. *Proceedings of the Royal Society B: Biological Sciences* **272**:2357–2364.

Pöysä, H., J. Elmberg, G. Gunnarsson, P. Nummi, and K. Sjöberg. 2004. Ecological basis of sustainable harvesting: Is the prevailing paradigm of compensatory mortality still valid? *Oikos* **104**:612–615.

Price, P. W., C. E. Bouton, P. Gross, B. A. McPheron, J. N. Thompson, and A. E. Weis. 1980. Interactions among three trophic levels: Influence of plants on interactions between insect herbivores and natural enemies. *Annual Review of Ecology Systematics* **11**:41–65.

R Development Core Team. 2006. R: A language and environment for statistical computing. R Foundation for Statistical Computing, Vienna, Austria, http://www.R-project.org.

Raedeke, K. J., J. J. Millspaugh, and P. E. Clark. 2002. Population characteristics. Pages 449–491 *in* D. E. Toweill and J. W. Thomas (Eds.) *North American Elk: Ecology and Management*. Smithsonian Institution Press, Washington, DC.

Raithel, J. D., M. J. Kauffman, and D. H. Pletscher. 2007. Impacts of spatial and temporal variation in calf survival on the growth of elk populations. *Journal of Wildlife Management* **71**:795–803.

Sæther, B. E. 1997. Environmental stochasticity and population dynamics of large herbivores: A search for mechanisms. *Trends in Ecology and Evolution* **12**:143–149.

Schlegel, M. 1976. Factors affecting calf elk survival in north central Idaho. A progress report. *Proceedings of the Western Association of State Game Fish Commission* **56**:342–355.

Schwartz, C. C., M. A. Haroldson, G. C. White, R. B. Harris, S. Cherry, K. A. Keating, D. Moody, and C. Servheen. 2006. Temporal, spatial and environmental influences on the demographics of grizzly bears in the Greater Yellowstone Ecosystem. *Wildlife Monographs* **161**.

Seber, G. A. F. 1982. *The Estimation of Animal Abundance and Related Parameters*. 2nd edn. Wiley, New York, NY.

Seip, D. R. 1992. Factors limiting woodland caribou populations and their interrelationships with wolves and moose in southeastern British Columbia. *Canadian Journal of Zoology* **70**:1494–1503.

Siler, W. 1979. A competing-risk model for animal mortality. *Ecology* **60**:750–757.

Sinclair, A. R. E. 2003. Mammal population regulation, keystone processes and ecosystem dynamics. *Philosophical Transactions of the Royal Society of London, B* **358**:1729–1740.

Sinclair, A. R. E., and C. J. Krebs. 2002. Complex numerical responses to top–down and bottom–up processes in vertebrate populations. *Philosophical Transactions of the Royal Society of London* **357**:1221–1231.

Sinclair, A. R. E., S. A. R. Mduma, J. G. C. Hopcraft, J. M. Fryxell, R. Hilborn, and S. Thirgood. 2007. Long-term ecosystem dynamics in the Serengeti: Lessons for conservation. *Conservation Biology* **21**:580–590.

Singer, F. J. 1994. The ungulate prey base for wolves in Yellowstone National Park. Pages 323–348 *in* R. B. Keiter and M. S. Boyce (Eds.) *The Greater Yellowstone Ecosystem: Redefining America's Wilderness Heritage*. Yale University Press, New Haven, CT.

Singer, F. J., A. Harting, K. K. Symonds, and M. B. Coughenour. 1997. Density dependence, compensation, and environmental effects on elk calf mortality in Yellowstone National Park. *Journal of Wildlife Management* **61**:12–25.

Skalski, J. R., K. E. Ryding, and J. J. Millspaugh. 2005. *Wildlife Demography: Analysis of Sex, Age, and Count Data.* Elsevier Academic Press, San Diego, CA.

Smith, B. L., and S. H. Anderson. 1998. Juvenile survival and population regulation of the Jackson Elk Herd. *Journal of Wildlife Management* **62:**1036–1045.

Smith, B. L., E. S. Williams, K. C. McFarland, T. L. McDonald, G. Wang, and T. D. Moore. 2006. *Neonatal Mortality of Elk in Wyoming: Environmental, Population, and Predator Effects.* U.S. Department of Interior, U.S. Fish and Wildlife Service, Biological Technical Publication BTP-R6007–2006, Washington, DC.

Smith, D. W., R. O. Peterson, and D. B. Houston. 2003. Yellowstone after wolves. *BioScience* **53:**330–340.

Smith, D. W., T. D. Drummer, K. M. Murphy, D. S. Guernsey, and S. B. Evans. 2004. Winter prey selection and estimation of wolf kill rates in Yellowstone National Park, 1995–2000. *Journal of Wildlife Management* **68:**153–166.

Solomon, M. E. 1949. The natural control of animal populations. *Journal of Animal Ecology* **18:**1–35.

Spiller, D. A., and T. W. Schoener. 2008. Climatic control of trophic interaction strength: The effect of lizards on spiders. *Oecologia* **154:**763–771.

Stiling, P., and A. M. Rossi. 1997. Experimental manipulations of top–down and bottom–up factors in a tri-trophic system. *Ecology* **78:**1602–1606.

Swets, D. L., B. C. Reed, J. D. Rowland, and S. E. Marko. 1999. *A Weighted Least-Squares Approach to Temporal NDVI Smoothing* Proceedings of the 1999 ASPRS Annual Conference, From Image to Information. Portland, Oregon, May 17–21, 1999, American Society of Photogrammetry and Remote Sensing, Bethesda, MD.

Thompson, S. K. 1992. *Sampling.* Wiley, New York, NY.

Tveraa, R., P. Fauchald, C. Henaug, and N. G. Yoccoz. 2003. An examination of a compensatory relationship between food limitation and predation in semi-domestic reindeer. *Oecologia* **137:**370–376.

VanBallenberghe, B. 1987. Effects of predation on moose numbers: A review of recent North American studies. *Swedish Wildlife Research* Supplement **1:**431–460.

Varley, N., and M. S. Boyce. 2006. Adaptive management for reintroductions: Updating a wolf recovery models for Yellowstone National Park. *Ecological Modelling* **193:**315–339.

Vucetich, J. A., D. W. Smith, and D. R. Stahler. 2005. Influence of harvest, climate and wolf predation on Yellowstone elk, 1961–2004. *Oikos* **111:**259–270.

Wasser, S. K., S. L. Monfort, J. Souther, and D. E. Wildt. 1994. Excretion rates and metabolite of oestradiol and progesterone in baboons (*Papio cynocephalus cynocephalus*) faeces. *Journal of Reproduction and Fertility* **101:**213–220.

Weber, B. J., and M. L. Wolfe. 1982. Use of serum progesterone levels to detect pregnancy in elk. *Journal of Wildlife Management* **46:**835–837.

White, G. C. 1996. NOREMARK: Population estimation from mark-resighting surveys. *Wildlife Society Bulletin* **24:**50–52.

White, G. C., and R. A. Garrott. 1990. *Analysis of Wildlife Radio-Tracking Data.* Academic Press, San Diego, CA.

White, P. J., and R. A. Garrott. 2005. Yellowstone's ungulates after wolves-expectations, realization, and predictions. *Biological Conservation* **125:**141–152.

White, P. J., R. A. Garrott, and L. L. Eberhardt. 2003. *Evaluating the Consequences of Wolf Recovery on Northern Yellowstone Elk.* YCR-NR-2004-02, U.S. Department of Interior, National Park Service, Yellowstone Center for Resources, Yellowstone National Park, Wyoming.

White, P. J., R. A. Garrott, J. F. Kirkpatrick, and E. V. Berkeley. 1995. Diagnosing pregnancy in free-ranging elk using fecal steroid metabolites. *Journal of Wildlife Diseases* **31:**514–522.

Wilbur, H. M. 1997. Experimental ecology of food webs: Complex systems in temporary ponds. *Ecology* **78:**2279–2302.

Willard, S. T., R. G. Sasser, J. C. Gillespie, J. T. Jaques, T. H. WelshJr., and R. D. Randel. 1994. Methods for pregnancy determinations and the effects of body condition on pregnancy status in Rocky Mountain elk. *Theriogenology* **42:**1095–1102.

Williams, B. K., F. A. Johnson, and K. Wilkins. 1996. Uncertainty and adaptive management of waterfowl harvests. *Journal of Wildlife Management* **60:**223–232.

Wilmers, C. C., R. L. Crabtree, D. Smith, K. M. Murphy, and W. M. Getz. 2003. Trophic facilitation by introduced top-predators: Gray wolf subsidies to scavengers in Yellowstone National Park. *Journal of Animal Ecology* **72:**909–916.

Wright, G. J., R. O. Peterson, D. W. Smith, and T. O. Lemke. 2006. Selection of northern Yellowstone elk by grey wolves and hunters. *Journal of Wildlife Management* **70:**1070–1078.

CHAPTER 24

Apparent Competition and Regulation in a Wolf-Ungulate System: Interactions of Life History Characteristics, Climate, and Landscape Attributes

Robert A. Garrott,* P. J. White,[†] Matthew S. Becker,* and Claire N. Gower*

*Fish and Wildlife Management Program, Department of Ecology, Montana State University
[†]National Park Service, Yellowstone National Park

Contents

Theme

The nonmigratory elk population in the Madison headwaters area of Yellowstone National Park appeared to be regulated near ecological carrying capacity by food limitation for at least three decades prior to the reestablishment of wolves. Eight years of post-wolf data indicated a substantial proportion of wolf predation was additive and overwhelmed any potential for the elk population to demographically compensate. Thus, wolf predation resulted in a 60–70% decrease in elk abundance and the system transitioned from being bottom-up regulated in the absence of a significant predator to strong top-down limitation due to wolf predation. However, it is uncertain if predation will ultimately regulate the elk population at a lower, alternate state or if predation and other factors influencing elk vulnerability will interact to result in further decreases in elk abundance. Fundamental to this question is the role of bison as an alternative prey for wolves. We discuss regulatory processes in predator–prey systems and present conceptual models that provide contrasting predictions for wolf-ungulate dynamics. We also characterize various aspects of prey vulnerability that may influence the effects of alternative prey on predation of the preferred prey and, in turn, the stability or

The Ecology of Large Mammals in Central Yellowstone
R. Garrott, P. J. White and F. Watson
ISSN 1936-7961, DOI: 10.1016/S1936-7961(08)00224-8

Copyright © 2009, Elsevier Inc.
All rights reserved.

instability of wolf multiple-prey systems. We conclude by evaluating support for these processes in data we have collected on Madison headwaters elk and predicting the future trajectory of the herd.

I. INTRODUCTION

There are two fundamental questions to address when evaluating the effects of a predator on its prey. First, does predation limit the prey population below the level of abundance expected in the absence of predators? This question is answered by determining the proportion of predation that is additive to other forms of prey mortality and whether the population compensates for the additional mortality by increased reproductive output. If the limiting effect of additive predation is present then the subsequent question is whether predation is regulatory? That is, are there density-dependent feedback mechanisms operating to reduce predation as prey density decreases? Regulatory mechanisms are not likely to be manifested over the entire range of prey densities because large herbivore populations in the absence of predators are food limited only at high densities near carrying capacity (Eberhardt 1977, Fowler 1981, Caughley and Sinclair 1994). Predation, in contrast, is generally regulatory only at low prey densities (Messier 1994, 1995). Therefore, predator-prey dynamics must be evaluated over a wide range of prey densities to assess the relative roles and strengths of bottom-up and top-down regulatory processes.

The influence of prey density on predation can be decomposed into two responses. The functional response describes how the per capita kill rate of the predator changes with prey density, while the numerical response describes how predator numbers change with prey density. Holling (1959) defined three basic types of functional responses: linear (type I); asymptotic (type II); and sigmoidal (type III), though it is extremely difficult to distinguish between type II and type III responses using field data (Dale *et al.* 1994, Van Ballenberghe and Ballard 1994, Marshal and Boutin 1999, Sinclair and Krebs 2002, Mech and Peterson 2003). Numerical responses are also likely to assume one of these functional forms. The multiplicative effect of the functional and numerical responses dictates the total response that describes the relationship between the proportion of the prey population killed by predators and prey density. The total response can assume a variety of forms depending on the shapes of the functional and numerical responses (Messier 1995).

Theoretical and empirical findings suggest the functional and numerical responses in wolf-ungulate systems must be type II or type III due to factors such as satiation at higher prey densities and strong wolf territoriality (Messier 1994), as well as the potential of prey-switching (Murdoch 1969), selective foraging and learning by the predator (Tinbergen 1960, Real 1979), or use of refuges by prey (Taylor 1984). In a one-prey system, the numerical response is assumed to converge at the origin because wolves cannot exist in the absence of prey. All combinations of type II and type III functional and numerical responses that pass through the origin result in a total response curve that is density-dependent at low prey densities (regulatory) and inversely density dependent, termed "depensatory," at intermediate to high prey densities (Figure 24.1, Caughley and Sinclair 1994). However, alternative prey in multiple-prey systems can support a positive numerical response when the primary prey is at low densities or even absent. Thus, a numerical response with a y-intercept can result in a total response that is depensatory over the entire range of densities for the primary prey (Figure 24.1).

We documented that a substantial proportion of wolf predation on elk in the Madison headwaters study system was additive, with predation decreasing the survival rate of adults and calves (Chapter 23 by Garrott *et al.*, this volume). Also, we did not find any evidence for a compensatory increase in fecundity due to additive wolf predation. As a consequence, the trajectory of the elk population transitioned from relative stability prior to the reestablishment of wolves in the study system to a consistently and rapidly decreasing population (Chapter 23 by Garrott *et al.*, this volume, Figure 23.6), providing strong evidence that wolf predation in our study system was limiting the elk population. We detected density-dependent relationships in the functional and numerical responses of wolves, and the shape of the functional response was best described by a type II asymptotic model (Chapter 17 by Becker *et al.*, this volume, Table 17.5, Figure 17.6), We also documented a strong numerical response, with wolf numbers rapidly increasing after reestablishment, peaking during the winter of 2004–05, and decreasing in subsequent years as elk densities decreased (Figure 24.2), apparently due to the effects of additive predation.

Direct assessment of the proportion of the elk population killed (total response) from our post-wolf studies reveals a strong depensatory total response over the range of elk densities experienced in the Madison headwaters (Figure 24.3). Thus, we have not detected any signal of regulatory predation, perhaps because we have not reached the critical low elk densities where a density-dependent

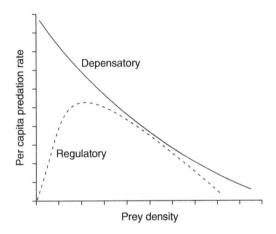

FIGURE 24.1 The impact of wolf predation on their ungulate prey depends on the form of the total response curve. A density-dependent total response at low prey densities is required for predation to be stabilizing and regulate the prey population. Alternatively, predation could be inversely density-dependent (depensatory) across the entire range of prey densities and, possibly, result in extirpation of a prey population (adapted from Sinclair and Krebs 2002).

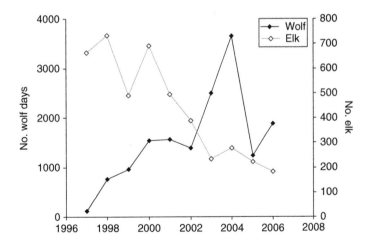

FIGURE 24.2 The relationship between elk numbers determined through replicate mark-resight surveys conducted at regular intervals throughout each field season (Chapter 23 by Garrott *et al.*, this volume) and wolf presence within the Madison headwater study area, as indexed by total wolf days for each 167 day winter field season (Chapter 16 by Becker *et al.*, this volume). A wolf day is one wolf detected within the study area for one day.

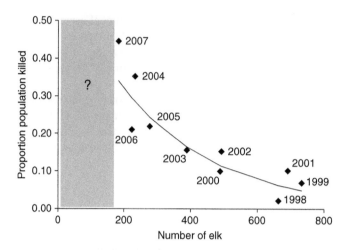

FIGURE 24.3 Total predation rate of wolves on elk in the Madison headwaters study area. Annual estimates of elk abundance were obtained through replicate mark-resight surveys conducted at regular intervals throughout each field season (Chapter 23 by Garrott *et al.*, this volume) with the estimates of the proportion of the elk population killed by wolves calculated based on the number of wolf-killed elk discovered each field season (Chapter 16 by Becker *et al.*, this volume).

regulatory response might be expected. We have documented density-related changes in the functional and numerical responses of wolves for elk (Chapter 17 by Becker *et al.*, this volume, Figure 24.2), suggesting there is some potential for the total response to transition to a density-dependent relationship at elk densities lower than those experienced thus far. Alternatively, we have also documented that wolves were effective predators of bison, with bison representing a substantial proportion of the total prey killed in some winters (Chapter 16 by Becker *et al.*, this volume). Thus, we cannot dismiss the possibility that wolf predation will remain depensatory over the entire range of elk densities. Given these two very different scenarios for the total response, the future dynamics of this predator–prey system is uncertain. However, alternate outcomes can be explored using a series of conceptual models.

II. CONCEPTUAL WOLF-UNGULATE MODELS

Messier (1994) presented four conceptual models for ungulate regulation that have been the centerpiece of subsequent debates on the role of wolf predation limiting and/or regulating ungulate populations (Figure 24.4). The Food Model predicts that in the absence of an effective predator an ungulate population is regulated at carrying capacity (K) by density-dependent food limitation. Evidence from our elk demographic studies prior to wolf restoration strongly supports this model (Chapter 11 by Garrott *et al.*, this volume). Theoretically, this model would also be applicable to a predator–prey system where predation is completely compensatory, with predation limited to Errington's (1946) "doomed surplus." The three models that incorporate additive predation provide alternative predictions of ungulate regulation and equilibrium points. The shape of the total predation response curve is identical in all three models, with the only difference being the magnitude of the predation response relative to the potential growth rate of the ungulate population. The Predation-Food/1-State Model predicts that additive predation decreases the ungulate population growth rate, but ungulate population growth remains positive across the entire range of prey densities (Figure 24.4). Thus, the ungulate population remains regulated by density-dependent food limitation and stabilizes at a high density

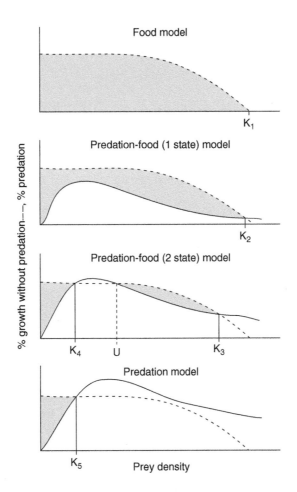

FIGURE 24.4 Messier's (1994) conceptual models of ungulate population regulation when wolf predation is assumed to be regulatory. The points on the *x* axis denoted by $\mathbf{K}_x$ represent various stable equilibrium prey densities predicted by the various models, with the point label **U** representing an unstable equilibrium point.

near carrying capacity. The Predation-Food/2-State Model predicts that additive predation exceeds the potential growth rate of the prey population at an intermediate range of ungulate densities and the prey population is maintained in a "predator pit" at low densities. However, if a perturbation (*e.g.,* from the weather or reduction in predators) allows the ungulate density to exceed this intermediate range, then an alternate, relatively high, prey density equilibrium can be attained (Figure 24.4). The Predation Model predicts that additive predation exceeds the potential growth rate of the prey population across nearly the entire range of prey densities and, as a result, the ungulate population is maintained at a low-density equilibrium (Figure 24.4).

There is an implicit assumption in all three of these predation models that predation is density dependent at low prey densities and depensatory at intermediate to high prey densities (Messier 1994). While these four models may be adequate to describe the potential dynamics in systems with only a single prey species, the presence of multiple ungulate prey species could contribute to a total predation response that is depensatory across the entire range of prey densities (Messier 1995, Sinclair and Krebs 2002). Most wolf-ungulate predator–prey systems are composed of multiple ungulate prey species (Dale *et al.* 1994, Andersen *et al.* 2006, Garrott *et al.* 2007). Thus, we propose broadening these conceptual models to include three additional depensatory predation models that compliment the

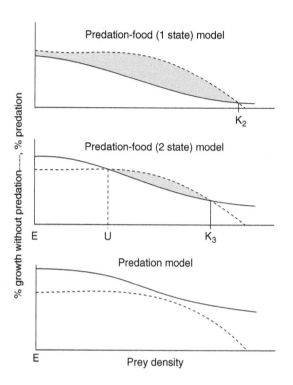

FIGURE 24.5 An adaptation of Messier's (1994) conceptual models of ungulate population regulation when wolf predation is assumed to be depensatory across the entire range of prey densities. The points on the *x* axis denoted by K_x represent various stable equilibrium prey densities predicted by the various models, with the point label **U** representing an unstable equilibrium point, and **E** indicating extirpation of the prey.

density-dependent predation models (Figure 24.5). Similar to Messier's Predation-Food/1-State Model, the Depensatory Predation-Food/1-State Model predicts the ungulate population remains regulated by density-dependent food limitation and stabilizes at a high density near carrying capacity (Figure 24.5). The Depensatory Predation-Food/2-State Model is unstable because it predicts that at intermediate to high ungulate densities predation will have modest impacts and the ungulate population is maintained at a relatively high density equilibrium. However, if a perturbation such as severe weather, disease, or an aggressive human harvest reduces ungulate densities substantially, predation could exceed the demographic capacity of the ungulate population, driving it to extinction (Figure 24.5). The Depensatory Predation Model predicts that predation will exceed the demographic potential of the ungulate prey across the entire range of prey densities, thereby resulting in the extirpation of the prey population (Figure 24.5). This dynamic could be observed in systems where a new predator or prey species is introduced, such as the restoration of an extirpated native species (wolves) in the Madison headwaters, or the invasion of an exotic species (Courchamp *et al.* 2000, Norbury 2001, Roemer *et al.* 2002, Matter and Mannan 2005). This dynamic may also be applicable in existing native predator–prey systems where anthropogenic activities fundamentally modify the ecosystem in ways that alter predator–prey relationships, as has been suggested for wolf-ungulate systems in western Canada where woodland caribou (*Rangifer tarandus caribou*) populations are imperiled (McLoughlin *et al.* 2003, Wittmer *et al.* 2005, Hebblewhite *et al.* 2007), as well as throughout the circumpolar range of caribou due to global climate change (Bergerud *et al.* 2008).

Evidence from the Madison headwaters system indicates that additive wolf predation at intermediate to high prey densities was sufficient to result in consistently decreasing elk abundance (Chapter 23 by Garrott *et al.*, this volume). While the naivety of elk to predation risk (Berger *et al.* 2001) may have initially resulted in an exaggerated wolf predation effect, we documented a number of behavioral and distributional changes in the elk population over the subsequent years of study (Chapters 18 and 20 by Gower *et al.*, Chapter 21 by White *et al.*, this volume), suggesting that any exaggerated affect of wolf predation due to naïve prey was ephemeral. Thus, we suggest the conceptual density-dependent and depensatory predation models represent the two alternatives likely for our study system, with the density-dependent model predicting wolf predation on elk will be regulatory and maintain the elk population at some low equilibrium density, and the depensatory model predicting extirpation of the elk population. In large part, the realized outcome will be dependent on the role of bison as an alternative prey. While wolf-ungulate systems containing multiple prey species are common, the indirect effects of shared predation among prey species, termed apparent competition, has received little attention. However, there is a rich literature on apparent competition in predator–prey systems with smaller taxa that we can use to develop predictions of the ultimate state of our wolf-ungulate system.

III. APPARENT COMPETITION

The rate of predation experienced by a focal prey species can be profoundly influenced by the indirect effects of alternative prey sustaining predator populations at densities higher than allowed by the focal prey alone (Holt and Lawton 1994). Apparent competition occurs when two prey species share a common predator and an increasing abundance of one prey species results in a decreasing abundance of the second prey species due to a numerical response by the predator (Holt 1977). Apparent competition is arguably one of the most influential indirect ecological phenomena because it can substantially affect community structuring and higher-level processes such as food web persistence (Holt 1977, Sih *et al.* 1985, Abrams *et al.* 1995, Menge 1995). Theory predicts that such interactions are inherently unstable and can eventually lead to the extinction of the prey species less able to sustain the predation pressure. While these predictions of extinction are identical to those from direct competition theory, the avenues by which they occur differ dramatically (Holt 1977). Thus, empirically demonstrating the occurrence of apparent competition is difficult due to its complexity (Chaneton and Bonsall 2000). Nevertheless, considerable advances have been made primarily through experimental studies of tractable invertebrate food webs, wherein the importance of apparent competition has been established in a variety of systems (Schmitt 1987; Bonsall and Hassell 1997, Morris *et al.* 2001, 2004).

Manipulation of large mammal systems to evaluate the indirect effects of shared predators is difficult due to the spatial scale of such systems, less tractable organisms, and societal concerns for animal welfare. Thus, the thorough experimental designs described by Chaneton and Bonsall (2000) are not typically possible. However, this limitation should not preclude studies of large herbivore-predator food webs because similar observational studies capitalizing on natural experiments have already provided indications that apparent competition is a strong force in terrestrial vertebrate trophic interactions. For example, Wittmer *et al.* (2005) presented convincing evidence that recent range-wide decreases in the abundance of woodland caribou throughout the Canadian Rocky Mountain region are a result of apparent competition with moose (*Alces alces*), supporting hypotheses suggested by various investigators (Bergerud and Elliot 1986, Seip 1991,1992). Wolves prey primarily on moose but also use caribou as a secondary prey due to an apparent weakening of habitat segregation between the two prey species caused by anthropogenic landscape alterations (James *et al.* 2004) and moose range expansion due to global climate change (Bergerud *et al.* 2008). Apparent competition has also been implicated in the decreased abundance of roan antelope (*Hippotragus equinus*) in South Africa due to shared lion (*Panthera leo*) predation with wildebeest (*Connochaetes taurinus*) and zebra (*Equus burchelli*) attracted to artificial waterholes (Harrington *et al.* 1999, McLoughlin and Owen-Smith 2003). Furthermore, observational studies of smaller taxa in a variety of island systems affected by exotic invasions indicate that apparent competition has been a significant ecological force (Courchamp *et al.* 2000). For instance, Norbury (2001) demonstrated the presence of exotic rabbit (*Oryctolagus cuniculus*) prey in New Zealand allowed for introduced cat (*Felis catus*) and ferret (*Mustela furo*) predators to reach very high densities and subsequently depress or eliminate native skink (*Oligosoma spp.*) species. Similarly, the introduction of feral pigs (*Sus scrofa*) on California's Channel Islands led to the colonization of the island by golden eagles (*Aquila chrysaetos*), which in turn preyed upon endemic island foxes (*Urocyon littoralis*), nearly driving them to extinction while facilitating a competitive release for the islands' skunk (*Spilogale gracilis amphiala*) populations (Roemer *et al.* 2002). Apparent competition studies are also difficult in established systems because the most affected prey species are typically already absent (Holt and Lawton 1994). Thus, newly-established large-mammal predator–prey systems present unique research opportunities.

There are six pairwise interactions in a predator–prey food web in which two prey species share a common predator (Figure 24.6). These food web interactions can be broken down into two direct interactions between the predator and each of the prey species and two indirect interactions where one prey species may affect the other prey species through the influence of the shared predator. Theory predicts reciprocal negative indirect effects (-, -) between two prey species that share a common predator (*i.e.*, apparent competition; Holt 1977). These effects can be symmetric if the relative strengths of the reciprocal effects are similar. However, empirical studies have generally found asymmetrical indirect effects (-, 0; Chaneton and Bonsall 2000) where the presence of an alternative prey strengthens the negative effect of the shared predator on the focal prey, but the focal prey has no influence on the direct interaction between the predator and the alternative prey (Holt and Lawton 1993). These asymmetrical cases of apparent competition are known as indirect amensalism (Lawton and Hassell 1981) and have been well documented in a variety of small taxa systems (Chaneton and Bonsall 2000).

There are statistical and biological mechanisms that may result in indirect amensalism. The statistical mechanism occurs when there is a pronounced difference in the size of the two prey populations that leads to undetectable indirect effects in one direction. The biological mechanism occurs when there are large differences in the nutritional value of two prey species due to differential body size or differences in physical and/or behavioral attributes of the two prey types that may result in strong predator preference for a focal prey type (Chaneton and Bonsall 2000). Thus, vulnerability of species and individuals, and the characteristics of the environment that can lessen or increase vulnerability of one prey type relative to the other prey type, can exert considerable indirect effects on food webs and ultimately determine whether shared predation regulates or destabilizes a focal prey population.

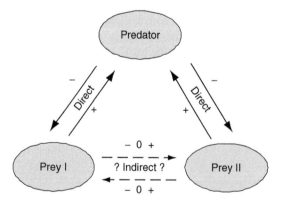

FIGURE 24.6 Conceptual schematic of direct (solid arrows) and indirect (dashed arrows) interactions for a one predator-two prey system (adapted from Chaneton and Bonsall 2000).

IV. PHYSICAL, BEHAVIORAL, AND ENVIRONMENTAL VULNERABILITY

Because apparent competition involves shared predation, the respective characteristics of each prey species that collectively determine their vulnerability and, thus, the extent of their use by a predator is of fundamental importance. The magnitude of differences in vulnerability between species can determine predator functional responses, the existence and type of apparent competition, and subsequent prey population trajectories. Vulnerability is typically considered in physical and behavioral contexts and can be affected by certain physical attributes characteristic of a prey species such as size, morphology, coloration, speed, agility, and physical and chemical defenses (Edmunds 1974, Paine 1976, Peckarsky 1984, Sih 1987, Sinclair *et al.* 2003). Species-specific behavioral attributes include elusive behavior to reduce encounter rates or differing responses to attacks such as flight, grouping, aggression and migration (Schaller 1972, Lima and Dill 1990, Caro 2005) and can have equal or even greater influence on vulnerability. All these attributes can also vary fundamentally within a species to such a degree that certain age classes could be considered different prey types (Chapter 16 by Becker *et al.*, this volume). For example, vertebrates in younger age classes typically necessitate parental protection because they are smaller, possess fewer defenses, and are less experienced and more prone to risky behaviors when attacked (Curio 1976, Vitale 1989, FitzGibbon and Lazarus 1995). Furthermore, while decreasing body condition universally increases vulnerability, other individual attributes such as disease, injuries, deformities, behavior, and phenotypic expressions of what collectively determines the fitness of an individual also contribute to vulnerability (Errington 1946, Temple 1987, Cresswell 1994, Wilson *et al.* 1994, Sinclair and Arcese 1995, Sih *et al.* 2004).

Dramatic differences in morphology, defenses, and behavior exist among ungulate species killed by wolves in temperate systems (Mech and Peterson 2003, Garrott *et al.* 2007). While some species such as white-tailed deer (*Odocoileus virginianus*) and mule deer (*Odocoileus hemionus*) are very similar in physical characteristics, others such as moose and caribou can differ substantially in size, strength, and morphology. Likewise, the behavioral responses of particular ungulate species to predators can differ, with some species typically employing flight while others exhibit aggression (Murie 1944, Carbyn 1974, Carbyn *et al.* 1993, Lingle 2002). Collectively, these physical and behavioral life history characteristics influence relative vulnerability among species, and the type of apparent competition that could occur between species with a shared predator. Symmetric apparent competition could occur between prey species that are similar in size and defenses (*e.g.*, mule and white-tailed deer), while the likelihood of asymmetric apparent competition should increase with increasing disparities in size and defenses among prey species (*e.g.*, deer and bison; Figure 24.7A).

There are pronounced differences in physical and behavioral vulnerability between elk and bison in the Madison headwaters (Chapter 16 by Becker *et al.*, this volume). Adult elk are considerably smaller than adult bison and elk typically use flight as an anti-predator response (Carbyn 1974, MacNulty *et al.* 2007). In contrast, bison adults are large and have substantially more body reserves that may mediate the effects of chronic nutritional deprivation compared to elk. Also, bison are one of the most formidable prey for wolves, frequently employing aggressive group defense and likely benefiting from thick, protective hides (Carbyn *et al.* 1993, Garrott *et al.* 2007, MacNulty *et al.* 2007). Adults of both species are more capable than calves of inflicting injuries to wolves due to their larger size, strength, experience, and the presence of horns or antlers. Although calves of both species do not differ substantially in size, bison calves are considerably less vulnerable to wolves than elk calves due to the benefits of adult group defense (Chapter 16 by Becker *et al.*, this volume, Figure 24.7B). Collectively, these disparities in physical and behavioral vulnerability between species are likely to result in indirect amensalism.

Apparent competition relationships between prey species with a shared predator may also be influenced by the degree to which vulnerability is differentially affected by attributes of the environment. The magnitude of differences in "environmental vulnerability" between prey species should be positively related to the probability of asymmetric interactions. Factors that alter the effectiveness of an

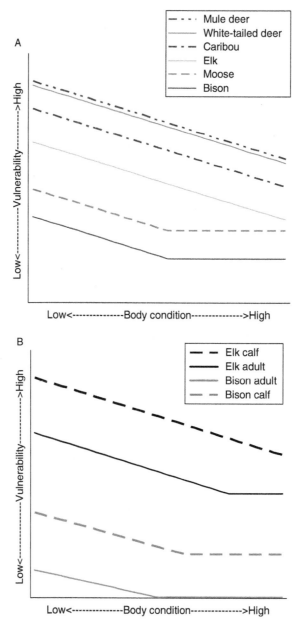

FIGURE 24.7 A conceptual model of the relationship between physical and behavioral vulnerability and body condition for six species of North American ungulates (A). While body condition and vulnerability are negatively related for all species, the similarities between species are also likely to be positively related to the probability of symmetric apparent competition relationships. Thus symmetric relationships are likely between deer species, whereas asymmetric relationships are likely between deer and bison. Similar relationships can be considered between species and age classes (B), with adults of both species, particularly bison, relatively invulnerable in good condition, and calves of both species similar in size and having fewer reserves than adults but differing in the significant amount of parental protection provided by bison. Asymmetric apparent competition relationships would be expected between these two species due to the substantial differences in vulnerability.

animal's defenses against detection such as moonlight, temperature, turbidity, and habitat complexity (Christian and Tracy 1981, Crowder and Cooper 1982, Clarke 1983, Gregory and Levings 1998, Hopcraft *et al.* 2005) can be considered agents of environmental vulnerability. Other factors can reduce prey defenses upon encounter and attack such as snow pack, distance to cover, and habitat structure inhibiting flight or aggression (Savino and Stein 1989, Lima 1992). Conversely, environmental attributes may act to mediate predation effects by reducing the hunting efficiency of predators. Patches of vegetative cover facilitate hiding and reduce the probability of detection and capture (Kotler and Brown 1988, Lima and Dill 1990). Similarly, attributes of the environment such as habitat complexity and topography may hinder the ability of predators to pursue and catch escaping prey (Crowder and Cooper 1982, Lewis and Eby 2002, Warfe and Barmuta 2004).

Environmental vulnerability appears to significantly influence the interactions between wolves, elk, and bison in the Madison headwaters (Chapter 16 by Becker *et al.*, Chapter 18 by Gower *et al.*, Chapter 21 by White *et al.*, this volume). The heterogeneous landscape of this system includes complex mosaics of coniferous forest in various stages of regeneration, narrow lower elevation meadow complexes and riparian corridors, and extensive geothermal areas (Chapter 2 by Newman *et al.*, Chapter 4 by Watson *et al.*, this volume). The diversity of habitats and topographic variations in slope and elevation, combined with the relatively severe winters characteristic of the area, results in a highly heterogeneous snow pack across the landscape (Chapter 2 by Newman *et al.*, Chapter 3 by Watson *et al.*, Chapters 4 and 6 by Watson *et al.*, this volume) that creates heterogeneity in predation risk (Chapter 21 by White *et al.*, this volume). Deep snow pack hinders movement and escape of ungulates in the event of an attack, and complex structured forests with downed timber and dense regeneration have similar and compounding effects on fleeing ungulates as they encounter hard edges between habitats where prey mobility is substantially curtailed (Bergman *et al.* 2006). While these environmental conditions certainly increase predation risk for individuals that are already physically vulnerable, perhaps the most significant effect of environmental vulnerability is to facilitate a predator's ability to kill prey that would otherwise be relatively invulnerable, effectively killing "good animals in bad places." Large predators often select for areas of high environmental vulnerability, where a prey's ability to effectively escape is compromised, over areas of high prey density (Hopcraft *et al.* 2005, Bergman *et al.* 2006). Thus, the vulnerability of elk to wolf predation is substantially influenced by landscape characteristics because fleeing elk are often caught in environmental "traps" created from interactions between habitat complexity and snow pack (Figure 24.8). Conversely, bison are much less vulnerable in areas of the landscape with environmental traps because they typically exhibit aggressive defense and do not employ flight to the same degree as elk (Figure 24.9). Thus, the effect of environmental vulnerability is greater on elk than bison, and the interactions of this vulnerability with elk physical and behavioral vulnerability (Figure 24.10) serve to magnify differences between prey species and make indirect amensalistic relationships even more likely.

In contrast, other areas of the landscape such as rocky promontories and wide, deep rivers can serve as escape terrain for ungulates and hinder wolf attacks (Murie 1944, Peterson 1955, Crisler 1956, Nelson and Mech 1981, Stephens and Peterson 1984; Chapter 21 by White *et al.*, this volume). Although high risk areas for elk have been heavily exploited by wolves throughout the Madison headwaters (Bergman *et al.* 2006), elk continue to select for high risk habitats such as meadows and geothermal areas in some areas due to their proximity to escape terrain (Chapter 21 by White *et al.*, this volume). Lima (1992) demonstrated that prey can tolerate high risk of encounter by predators in a particular habitat provided that there is also a high probability of escape. For example, elk in the Madison canyon continued to use preferred habitats in close proximity to rivers that served as effective escape terrain (Figure 24.11A), thereby potentially making this area a refuge from wolf predation (Chapter 21 by White *et al.*, this volume). This dynamic suggests that relatively fine-scale landscape and snow characteristics could strongly influence whether wolf predation will be regulatory or depensatory. We detected a gradient of environmental vulnerability for elk among the various river drainages of the Madison headwaters (Chapter 8 by Messer *et al.*, Chapter 21 by White *et al.*, this volume). The Gibbon River drainage is characterized by high environmental vulnerability for elk to wolf predation owing to

FIGURE 24.8 Landscape characteristics in concert with snow pack in the Madison headwaters frequently create terrain traps where predation risk is substantially increased for prey such as elk that generally try to flee when attacked by wolves. Bull elk feeding in environmental trap (A) (Photo by Jeff Henry) and, wolf-killed bull elk in environmental trap (B) (Photo by Chris Kenyon).

FIGURE 24.9 When bison are attacked by wolves they generally respond by running only short distances, aggregating into a tight group defense formation, standing their ground, and confronting the threat, hence, landscape characteristics do not influence bison vulnerability to wolf predation to the extent realized by prey that depend on flight to escape wolves (Photo by Jon Felis).

relatively uniform deep snow pack, dense forest regeneration, patchy geothermal areas, and a narrow, shallow river that does not deter wolves (Figures 24.10 and 24.12A). While the Firehole River drainage shares many of the attributes of the Gibbon, some segments of the river are wide and deep enough to provide protection from attacking wolves and the presence of a number of large geothermal basins provide low snow areas of sufficient size to allow effective escape from wolf attacks by fleeing elk (Figure 24.12B). In contrast, the Madison River drainage has relatively low environmental vulnerability for elk owing to the juxtaposition of extensive steep south facing slopes with reduced snow pack, and high forage quality meadows directly adjacent to a uniformly deep, wide river that provides an effective refuge from attacking wolves (Figure 24.12C).

Environmental vulnerability is also likely to differ on a broader spatial scale (e.g., ecosystems, winter ranges) due to differences in landscape characteristics and local climate. For example, the Lower Madison Valley, which is located only 40 km downstream from the Madison headwaters, has a strikingly different landscape dominated by continuous and extensive wind-swept grasslands (Figure 24.11B, Garrott *et al.* 2005, Chapter 25 by Hamlin *et al.*, this volume). Thus, snow pack does not typically impede elk movements and there are relatively few environmental traps and refuges across the homogeneous landscape (Figure 24.12D). Consequently, the severity of snow pack and the spatial distribution and abundance of areas on the landscape that can serve as refuges, decreasing predation risk, or traps increasing predation risk, can determine a prey species' distribution (Chapter 21 by White *et al.*, this volume), abundance and, ultimately, their persistence in a given system.

V. PREDICTING THE FUTURE OF THE MADISON HEADWATERS PREDATOR–PREY SYSTEM

Interactions between landscape, climate, and life history characteristics can determine relative vulnerability among prey species and, ultimately, the type of apparent competition that is likely to occur in a given system. If a multiple-prey system contains an abundant prey species with low vulnerability relative to a less abundant, but more vulnerable prey species, then indirect amensalism would be likely, possibly leading to extirpation of the more vulnerable prey. Alternatively, if disparities in vulnerability between prey species decrease via changes in the effectiveness of prey antipredator responses or changes in the predator's prey preference, then stabilizing effects such as refuge use and/or prey-switching can occur and exert stabilizing effects on the system. The extreme differences shown between the vulnerability of elk and bison in the Madison headwaters provide insight into the type of apparent competition we expect to occur. While bison are considerably more abundant within the system, they are also far more formidable prey for wolves and less influenced by landscape characteristics that create traps and refuges. Conversely, elk are much more vulnerable to wolf predation than bison, both physically and behaviorally, and the success of this species' flight strategy is far more dependent upon the local landscape characteristics. Thus, in this two-prey system with prey differing substantially in their relative physical, behavioral, and environmental vulnerability, we predict indirect amensalism

FIGURE 24.10 (Continued)

FIGURE 24.10 Two wolves attacking a cow and calf elk in the Gibbon river (A–D). Elk typically resort to flight when attacked but frequently utilize escape terrain in rivers and bluffs when long-distance flight is not possible. Upon being encountered by wolves these elk fled into the shallow riverbed, backing against one of the few small, but very deep, pools (A). Without aggressive maternal protection the calf was separated from the cow (B) and fled into the deep pool (C). Undeterred, and able to worry the calf from both banks, the wolves pursued it downstream where the river quickly shallowed, and eventually killed it (D). Interactions between deep snow pack, shallow and narrow riverbeds, and complex forest structure substantially increased environmental vulnerability for elk in the Gibbon drainage, likely driving their virtual disappearance in this area (Photos by Jon Felis).

competition with bison having a strong negative effect on elk, but elk have negligible indirect effects on bison (Figure 24.13).

If we are correct in our prediction of strong indirect amensalism, then bison will continue to subsidize the wolf population, thereby dampening the wolf numerical response to lower elk abundance. In turn, this will result in the total predation rate continuing to be depensatory as elk abundance

FIGURE 24.11 The collective attributes of the landscape and climate serve to create heterogeneity in environmental vulnerability with the availability and need for elk refuge habitat depending on the system. Elk bedded along the Madison river, possibly for quick access to this aquatic escape terrain after multiple days of wolf attacks on the herd in this area (A) (Photo by Shana Dunkley). A large herd dispersed across the relatively homogenous wind-swept benches of the lower Madison valley, Montana (B). The relative lack of environmental traps created by landscape and snow pack heterogeneity in this system likely make the effect of environmental vulnerability considerably less for elk in this system (Photo by Anne and Ron Carlson).

decreases further, with the prospect that wolf predation may eventually extirpate the Madison headwaters elk population. The countervailing phenomenon of an effective prey refuge in the Madison drainage (Chapter 21 by White *et al.*, this volume) could mediate the effects of indirect amensalism and allow the elk population to persist at low densities and with a more constrained distribution in the

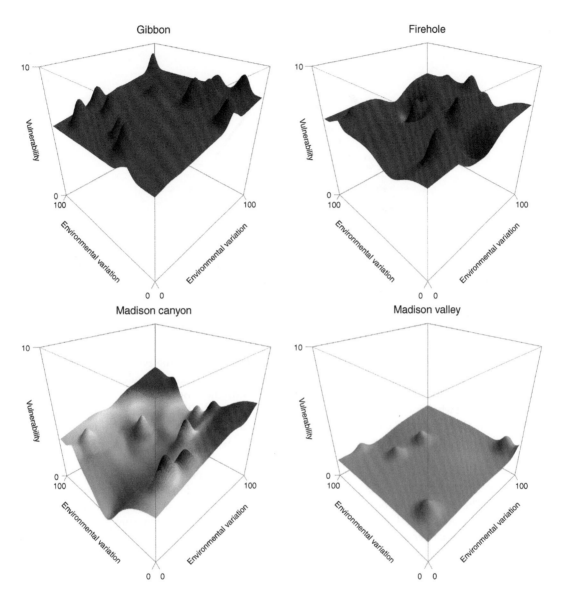

FIGURE 24.12 Conceptual models of environmental vulnerability for two wolf-ungulate systems of the Madison headwaters (A–C) and the Lower Madison Valley (D). These three-dimensional relationships demonstrate the interactions between various landscape and climate variables (generically termed environmental variables) that create traps and refuges across a system (denoted by elevated and depressed areas respectively). The Gibbon drainage (A), with deep snow pack, complex forest structure, and little escape terrain had an abundance of environmental traps for elk. Although the Firehole (B) was similar in landscape and climate characteristics creating an abundance of traps, the large, interconnected thermal and meadow complexes with moderate escape terrain provided some modest refuge effects for elk. The Madison drainage (C), with lower and more heterogeneous snow pack as well as abundant escape terrain in the deep, wide river provided considerable refuge for elk compared to the other drainages. In contrast, the relatively homogenous Lower Madison Valley system (D) has comparatively few areas where climate and landscape create traps, and its ungulate species are less likely to have strong environmental vulnerability effects.

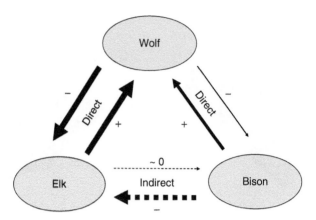

FIGURE 24.13 Predicted direct (solid arrows) and indirect (dashed arrows) interaction between wolves, elk, and bison in the Madison headwaters predator–prey system. There are pronounced differences in vulnerability between bison and elk resulting from disparities in body size, behavioral defenses and interactions between their respective life history characteristics and the characteristics of the system's landscape and climate. These disparities, combined with substantial disparity in the size of the two prey populations, lead us to suspect strong asymmetric indirect effects and continued depensatory predation on elk in this system (adapted from Chaneton and Bonsall 2000).

study system than that documented prior to wolf reintroduction. However, we suspect that the protection afforded to elk in the Madison drainage may not be sufficient to allow persistence of the elk population because annual calf recruitment remained extremely low in the latter years of our post-wolf studies even when most of the elk population was found in the Madison drainage (Chapter 23 by Garrott *et al.*, this volume). Thus, we expect that the Depensatory Predation Model (Figure 24.5) may best describe predator–prey interactions for the wolf-elk-bison system in the Madison headwaters study area.

If the elk population is ultimately extirpated it raises the prospect that the unique nonmigratory Madison headwaters elk herd likely did not exist prior to colonization of the Yellowstone region by Europeans. The extreme depletion of the region's ungulate populations during the era of unregulated commercial harvest during the mid to late 1800s, followed by government sponsored predator control, both within the park to enhance ungulate restoration and outside the park to protect livestock, resulted in the regional extirpation of wolves by the early 1900s. With the removal of predation risk elk may have begun exploiting the Madison headwaters area in winter due to the existence of areas of low snow pack and photosynthesizing plant communities associated with the concentration of geothermal features in these drainages. In the absence of predation risk, the Madison headwaters area provided an ecological niche for wintering elk that resulted in the development of a population unit that was strongly bottom-up regulated by the interacting constraints of the severe snow pack environment and the finite areas of reduced snow pack caused by the area's unique geothermal geology that served as relief from the most severe winter conditions (Chapter 11 by Garrott *et al.*, this volume). The reintroduction of wolves into Yellowstone (Chapter 15 by Smith *et al.*, this volume), combined with the successful restoration of bison and consequent expansion of the central bison herd's winter range to include the Madison headwaters area (Chapter 11 by Garrott *et al.*, Chapter 12 by Bruggeman *et al.*, this volume), set the stage for the reestablishment of strong top-down processes (Chapter 23 by Garrott *et al.*, this volume) in the system, which we suspect may be rapidly reverting the system to a state that may have existed in the early 1800s prior to the major human manipulations of the region's large mammal communities. While bison are considerably less vulnerable than elk, wolves subsist almost entirely on bison in areas such as Wood Buffalo National Park (Carbyn *et al.* 1993) and historical accounts from the Lewis and Clark expedition described wolves as being very abundant and strongly associated with bison to the extent that they were described as "shepherds" of the herds (Burroughs 1995). In contrast elk, while abundant, were largely seasonally absent from many of the mountainous regions (Burroughs 1995), perhaps due to the high predation risk associated with these areas in the winter. Thus it is possible that central Yellowstone will develop largely into a wolf-bison system or that we could expect a substantial wolf population decline with continued elk declines.

Regardless, we suspect that additional changes in prey distribution and abundance are likely in many newly-established wolf-ungulate systems within and outside protected areas, with the magnitude of these changes and their subsequent ecosystem effects dependent on the respective system's land management, prey species assemblages, and landscape and climate variables (Chapter 25 by Hamlin *et al.*, this volume). Our studies further corroborate the need for historical perspectives in studies of large mammal dynamics in these systems in order to distinguish and evaluate the influences of direct and indirect anthropogenic disturbances on ecosystem dynamics. Thus we concur with Sinclair *et al.* (2007) that differentiating between human-induced and natural ecosystem change is essential and long-term studies of protected areas are critical in providing ecological baselines for effective conservation and management of ecosystems (Chapter 30 by White *et al.*, this volume).

VI. SUMMARY

1. Following wolf reestablishment in the Madison headwaters we documented a substantial proportion of predation on elk was additive, overwhelming any potential for demographic compensation, and resulting in a 60–70% decrease in elk abundance. Thus, wolf predation limited the elk population below the level of abundance expected in the absence of predators.

2. Direct assessment of the proportion of the elk population killed annually by wolves, known as the total predation response, revealed a depensatory relationship, that is, as the elk population declined wolves killed a larger proportion of the population. Such an inverse density-dependent response is destabilizing, however, predatory-prey theory suggests that at low prey densities the total response may become density dependent and thus be regulatory, resulting in persistence of the predator–prey system at a lower prey density than realized in the absence of wolves.

3. While predator regulation of prey may be expected in systems containing only a single prey species, in multiple prey systems the rate of predation experienced by a focal prey species can be profoundly influenced by the indirect effects of alternative prey sustaining predator populations at densities higher than allowed by the focal prey alone, an ecological phenomenon known as apparent competition.

4. Bison, with their formidable size, body reserves, and group defense are considerably less vulnerable physically and behaviorally than elk, which are smaller and prone to flight when attacked. Flight also makes environmental vulnerability in the form of landscape attributes considerably more important for elk, as pursued animals are often caught in environmental "traps" created by habitat complexity and snow pack.

5. We predict these disparities in vulnerability of bison and elk to wolf predation create asymmetric apparent competition, where bison will continue to subsidize the wolf population, thereby dampening the wolf numerical response to lower elk abundance. In turn, this will result in the total predation response continuing to be depensatory as elk abundance decreases further, with the prospect that wolf predation may eventually extirpate the Madison headwaters elk population.

6. If the elk population is ultimately extirpated we suggest that the unique nonmigratory Madison headwaters elk herd likely did not exist prior to European colonization of the Yellowstone region, suggesting some changes in ungulate distribution and abundance following wolf restoration may be the result of direct and indirect anthropogenic alterations of ecosystems.

VII. REFERENCES

Abrams, P. A., B. A. Menge, G. G. Mittelbach, D. Spiller, and P. Yodzis. 1995. The role of indirect effects in food webs. Pages 371–395 *in* G. Polis and K. O. Winemiller (Eds.) *Food Webs: Integration of Patterns and Dynamics.* Chapman & Hall, New York, NY.

Andersen, R., J. D. Linnell, and E. J. Solberg. 2006. The future role of large carnivores in terrestrial trophic interactions: The northern temperate view. Pages 413–448 *in* K. Danell, R. Bergström, P. Duncan, and J. Pastor (Eds.) *Large Herbivore Ecology, Ecosystem Dynamics and Conservation.* Cambridge University Press, Cambridge, United Kingdom.

Berger, J., J. E. Swenson, and I.-L. Persson. 2001. Recolonizing carnivores and naive prey: Conservation lessons from Pleistocene extinctions. *Science* **291**:1036–1039.

Bergerud, A. T., and J. P. Elliot. 1986. Dynamics of caribou and wolves in northern British Columbia. *Canadian Journal of Zoology* **64**:1515–1519.

Bergerud, A. T., S. N. Luttich, and L. Camps. 2008. *The Return of Caribou to Ungava.* McGill-Queens University Press, Montreal, Quebec, Canada.

Bergman, E. J., R. A. Garrott, S. Creel, J. J. Borkowski, and R. M. Jaffe. 2006. Assessment of prey vulnerability through analysis of wolf movements and kill sites. *Ecological Applications* **16**:273–284.

Bonsall, M. B., and M. P. Hassell. 1997. Apparent competition structures ecological assemblages. *Nature* **338**:371–373.

Burroughs, R. D. (Ed.). 1995. *The Natural History of the Lewis and Clark Expedition.* Michigan State University Press, East Lansing, MI.

Carbyn, L. N. 1974. *Wolf Predation and Behavioral Interactions with Elk and Other Ungulates in an Area of High Prey Diversity.* Canadian Wildlife Service, Edmonton, Alberta, Canada.

Carbyn, L. N., S. M. Oosenbrug, and D. W. Anions. 1993. *Wolves, Bison and the Dynamics Related to the Peace-Athabasca Delta in Canada's Wood Buffalo National Park.* Canadian Circumpolar Institute, Edmonton, Alberta, Canada.

Caro, T. 2005. *Antipredator Defenses in Birds and Mammals.* University of Chicago Press, Chicago, IL.

Caughley, G., and A. R. E. Sinclair. 1994. *Wildlife Ecology and Management.* Blackwell, Boston, MA.

Chaneton, E. J., and M. B. Bonsall. 2000. Enemy-mediated apparent competition: Empirical patterns and the evidence. *Oikos* **88**:380–394.

Christian, K. A., and C. R. Tracy. 1981. The effect of the thermal environment on the ability of hatchling Galapagos land iguanas to avoid predation during dispersal. *Oecologia* **49**:218–223.

Clarke, J. A. 1983. Moonlight's influence on predator/prey interactions between short-eared owls (*Asio flammeus*) and deermice (*Peromyscus maniculatus*). *Behavioral Ecology and Sociobiology* **13**:205–209.

Courchamp, F., M. Langlais, and G. Sugihara. 2000. Rabbits killing birds: Modeling the hyperpredation process. *Journal of Animal Ecology* **69**:154–164.

Cresswell, W. 1994. Age-dependent choice of Redshank (*Tringa totanus*) feeding location: Profitability or risk? *Journal of Animal Ecology* **63**:589–600.

Crisler, L. 1956. Observations of wolves hunting caribou. *Journal of Mammalogy* **37**:337–346.

Crowder, L. B., and W. E. Cooper. 1982. Habitat structural complexity and the interaction between bluegills and their prey. *Ecology* **63**:1802–1813.

Curio, E. 1976. *Ethology of Predation.* Springer, Berlin, Germany.

Dale, B. W., L. C. Adams, and R. T. Bowyer. 1994. Functional response of wolves preying on barren ground caribou in a multiple-prey ecosystem. *Journal of Animal Ecology* **63**:644–652.

Eberhardt, L. L. 1977. Optimal policies for conservation of large mammals with special reference to marine ecosystems. *Environmental Conservation* **4**:205–212.

Edmunds, M. 1974. *Defence in Animals.* , New York, NY: Longman, New York, NY.

Errington, P. L. 1946. Predation and vertebrate populations. *Quarterly Review of Biology* **21**:145–177, 221–245.

FitzGibbon, C. D., and J. Lazarus. 1995. Anti-predator behavior of Serengeti ungulates: Individual differences and population consequences. Pages 274–296 *in* A. R. E. Sinclair and P. Arcese (Eds.) *Serengeti II: Dynamics, Management, and Conservation of an Ecosystem.* University of Chicago Press, Chicago, IL.

Fowler, C. W. 1981. Density dependence as related to life history strategy. *Ecology* **62**:602–610.

Garrott, R. A., J. E. Bruggeman, M. S. Becker, S. T. Kalinowski, and P. J. White. 2007. Evaluating prey switching in wolf-ungulate systems. *Ecological Applications* **17**:1588–1597.

Garrott, R. A., J. A. Gude, E. J. Bergman, C. Gower, P. J. White, and K. L. Hamlin. 2005. Generalizing wolf effects across the greater Yellowstone area: A cautionary note. *Wildlife Society Bulletin* **33**:1245–1255.

Gregory, R. S., and C. D. Levings. 1998. Turbidity reduces predation on migrating juvenile Pacific salmon. *Transactions of the American Fisheries Society* **127**:275–285.

Harrington, R., N. Owen-Smith, P. C. Viljoen, H. C. Biggs, D. R. Mason, and P. Funston. 1999. Establishing the causes of the roan antelope decline in the Kruger National Park, South Africa. *Biological Conservation* **90**:69–78.

Hebblewhite, M., J. Whittington, M. Bradley, G. Skinner, A. Dibb, and C. A. White. 2007. Conditions for caribou persistence in the wolf-elk-caribou systems of the Canadian Rockies. *Rangifer Special Issue No.* **17**:79–91.

Holling, C. S. 1959. The components of predation as revealed by a study of small mammal predation of the European pine sawfly. *Canadian Entomology* **91**:293–320.

Holt, R. D. 1977. Predation, apparent competition, and the structure of prey communities. *Theoretical Population Biology* **12**:197–229.

Holt, R. D., and J. H. Lawton. 1993. Apparent competition and enemy-free space in insect host-parasitoid communities. *American Naturalist* **142**:623–645.

Holt, R. D., and J. H. Lawton. 1994. The ecological consequences of shared natural enemies. *Annual Review of Ecology and Systematics* **25**:495–520.

Hopcraft, J. G. C., A. R. E. Sinclair, and C. Packer. 2005. Planning for success: Serengeti lions seek prey accessibility rather than abundance. *Journal of Animal Ecology* **74**:559–566.

James, A. R. C., S. Boutin, D. M. Hebert, and A. B. Rippin. 2004. Spatial separation of caribou from moose and its relation to predation by wolves. *Journal of Wildlife Management* **68**:799–809.

Kotler, B. P., and J. S. Brown. 1988. Environmental heterogeneity and the coexistence of desert rodents. *Annual Review of Ecology and Systematics* **19**:281–307.

Lawton, J. H., and M. P. Hassell. 1981. Asymmetrical competition in insects. *Nature* **289**:793–795.

Lewis, D. B., and L. A. Eby. 2002. Spatially heterogeneous refugia and predation risk in intertidal salt marshes. *Oikos* **96**:119–129.

Lima, S. L. 1992. Strong preferences for apparently dangerous habitats? A consequence of differential escape from predators. *Oikos* **64**:597–600.

Lima, S. L., and L. M. Dill. 1990. Behavioral decisions made under the risk of predation: A review and prospectus. *Canadian Journal of Zoology* **68**:619–640.

Lingle, S. 2002. Coyote predation and habitat segregation of white-tailed deer and mule deer. *Ecology* **83**:2037–2048.

MacNulty, D. R., L. D. Mech, and D. W. Smith. 2007. A proposed ethogram of large-carnivore predatory behavior, exemplified by the wolf. *Journal of Mammalogy* **88**:595–605.

Marshal, J. P., and S. Boutin. 1999. Power analysis of wolf-moose functional response. *Journal of Wildlife Management* **63**:396–402.

Matter, W. J., and R. W. Mannan. 2005. How do prey persist? *Journal of Wildlife Management* **69**:1315–1320.

McLoughlin, C. A., and N. Owen-Smith. 2003. Viability of a diminishing roan antelope population: Predation is the threat. *Animal Conservation* **6**:231–236.

McLoughlin, P. D., E. Dzus, B. Wynes, and S. Boutin. 2003. Declines in populations of woodland caribou. *Journal of Wildlife Management* **67**:755–761.

Mech, L. D., and R. O. Peterson. 2003. Wolf-Prey relations. Pages 131–160 *in* L. D. Mech and L. Boitani (Eds.) *Wolves: Behavior, Ecology, and Conservation.* University of Chicago Press, Chicago, IL.

Menge, B. A. 1995. Indirect effects in marine rocky intertidal interaction webs: Patterns and importance. *Ecological Monographs* **65**:21–74.

Messier, F. 1994. Ungulate population models with predation: A case study with the North American moose. *Ecology* **75**:478–488.

Messier, F. 1995. On the functional and numerical responses of wolves to changing prey density. Pages 187–197 *in* L. N. Carbyn, S. H. Fritts, and D. R. Seip (Eds.) *Ecology and Conservation of Wolves in a Changing World.* Canadian Circumpolar Institute, University of Alberta, Edmonton, Alberta, Canada.

Morris, R. J., O. T. Lewis, and C. J. Godfray. 2004. Experimental evidence for apparent competition in a tropical forest food web. *Nature* **428**:310–312.

Morris, R. J., C. B. Muller, and C. J. Godfray. 2001. Field experiments testing for apparent competition between primary parasitoids mediated by secondary parasitoids. *Journal of Animal Ecology* **70**:301–309.

Murdoch, W. W. 1969. Switching in general predators: experiments on predator specificity and stability of prey populations. *Ecological Monographs* **39**:335–354.

Murie, A. 1944. The wolves of Mount McKinley. *Fauna of the National Parks of the United States Fauna Series* **5**:1–238.

Nelson, M. E., and D. L. Mech. 1981. Deer social organization and wolf predation in northeastern Minnesota. *Wildlife Monographs* **77**.

Norbury, G. 2001. Conserving dryland lizards by reducing predator-mediated apparent competition and direct competition with introduced rabbits. *Journal of Applied Ecology* **38**:1350–1361.

Paine, R. T. 1976. Size-limited predation: An observational and experimental approach with the *Mytilus-pisaster* interaction. *Ecology* **57**:858–873.

Peckarsky, B. L. 1984. Predator–prey interactions among aquatic insects. Pages 196–254 *in* V. H. Resh and D. M. Rosenberg (Eds.) *The Ecology of Aquatic Insects.* Praeger, New York, NY.

Peterson, R. L. 1955. *North American Moose.* University of Toronto Press, Toronto, Ontario, Canada.

Real, L. A. 1979. Ecological determinants of functional response. *Ecology* **60**:481–485.

Roemer, G. W., C. J. Donlan, and F. Courchamp. 2002. Golden eagles, feral pigs, and insular carnivores: How exotic species turn native predators into prey. *Proceedings National Academy of Sciences* **99**:791–796.

Savino, J. F., and R. A. Stein. 1989. Behavioral interactions between fish predators and their prey: Effects of plant density. *Animal Behaviour* **37**:311–321.

Schaller, G. B. 1972. *The Serengeti Lion: A Study of Predator–prey Relations.* University of Chicago Press, Chicago, IL.

Schmitt, R. J. 1987. Indirect interactions between prey: Apparent competition, predator aggregation, and habitat segregation. *Ecology* **68**:1887–1897.

Seip, D. R. 1991. Predation and caribou populations. *Rangifer Special Issue* **7**:46–52.

Seip, D. R. 1992. Factors limiting woodland caribou populations and their interrelationship with wolves and moose in southeastern British Columbia. *Canadian Journal of Zoology* **70**:1494–1503.

Sih, A. 1987. Predators and prey lifestyles: An evolutionary and ecological overview. Pages 203–224 *in* W. C. Kerfoot and A. Sih (Eds.) *Predation: Direct and Indirect Impacts on Aquatic Communities.* University Press of New England, Hanover, NH.

Sih, A., A. M. Bell, and J. C. Johnson. 2004. Behavioral syndromes: An ecological and evolutionary overview. *Trends in Ecology and Evolution* **19**:372–378.

Sih, A., P. Crowley, M. McPeek, J. Petranka, and K. Strohmeier. 1985. Predation, competition and prey communities: A review of field experiments. *Annual Review of Ecology and Systematics* **16**:269–311.

Sinclair, A. R. E., and P. Arcese. 1995. Population consequences of predation-sensitive foraging: The Serengeti wildebeest. *Ecology* **76**:882–891.

Sinclair, A. R. E., and C. J. Krebs. 2002. Complex numerical response to top-down and bottom-up processes in vertebrate populations. *Philosophical Transactions of the Royal Society of London B* **357**:1221–1231.

Sinclair, A. R. E., S. Mduma, and J. S. Brashares. 2003. Patterns of predation in a diverse predator–prey system. *Nature* **425**:288–290.

Sinclair, A. R. E., S. Mduma, G. Hopcraft, J. M. Fryxell, R. Hilborn, and S. Thirgood. 2007. Long-term ecosystem dynamics in the Serengeti: Lessons for conservation. *Conservation Biology* **21**:580–590.

Stephens, P. W., and R. O. Peterson. 1984. Wolf-avoidance strategies of moose. *Holarctic Ecology* **7**:239–244.

Taylor, R. J. 1984. *Predation.* , New York, NY: Chapman & Hall, New York, NY.

Temple, S. A. 1987. Do predators always capture substandard individuals disproportionately from prey populations? *Ecology* **68**:669–674.

Tinbergen, L. 1960. The natural control of insects in pine woods. Factors influencing intensity of predation by songbirds. *Archives Neerlandaises de Zoologie* **13**:265–343.

Van Ballenberghe, V., and W. B. Ballard. 1994. Limitation and regulation of moose populations: The role of predation. *Canadian Journal of Zoology* **72**:2071–2077.

Vitale, A. F. 1989. Changes in the anti-predator responses of wild rabbits, *Oryctolugus cuniculus* (L.) with age and experience. *Behaviour* **110**:47–61.

Warfe, D. M., and L. A. Barmuta. 2004. Habitat structural complexity mediates the foraging success of multiple predator species. *Oecologia* **141**:171–178.

Wilson, D. S., A. B. Clark, K. Coleman, and T. Dearstyne. 1994. Shyness and boldness in humans and other animals. *Trends in Ecology and Evolution* **9**:442–446.

Wittmer, H. U., A. R. E. Sinclair, and B. N. Mclellan. 2005. The role of predation in the decline and extripation of woodland caribou. *Oecologia* **144**:257–267.

CHAPTER 25

Contrasting Wolf–Ungulate Interactions in the Greater Yellowstone Ecosystem

Kenneth L. Hamlin,* Robert A. Garrott,† P. J. White,‡ and Julie A. Cunningham*

*Montana Fish, Wildlife, and Parks
†Fish and Wildlife Management Program, Department of Ecology, Montana State University
‡National Park Service, Yellowstone National Park

Contents

Theme

Considerable research attention has focused on large herbivore population dynamics in protected natural areas to better understand the roles of resource limitation (a "bottom–up" process) and predation (a "top–down" process). This emphasis on landscapes with minimal human impact is particularly true for investigations of wolf–elk dynamics in the Rocky mountain states, where the vast majority of research occurs within the national park system. Although such studies largely avoid the confounding effects of humans, Andersen *et al.* (2006) argued that evaluation of large herbivore dynamics on multiple-use landscapes, where human influence is typically pervasive, is also needed. Initial investigations of this kind (Garrott *et al.* 2005) indicated that the effects

The Ecology of Large Mammals in Central Yellowstone
R. Garrott, P. J. White and F. Watson
ISSN 1936-7961, DOI: 10.1016/S1936-7961(08)00225-X

Copyright © 2009, Elsevier Inc.
All rights reserved.

of wolves (*Canis lupus*) on elk (*Cervus elaphus*) dynamics could vary considerably within the same river drainage system and distances less than 40 km apart, making generalizations strictly from protected areas equivocal. Here, we expand this landscape approach to compare population dynamics among seven elk populations in the Greater Yellowstone Ecosystem, all with winter ranges within 115 km of each other and whose summer ranges partially overlap. Land ownership, land use, vegetation communities, large predator densities and management, and local landscape and environmental conditions vary considerably across this area. Similarly, elk harvest management strategies also vary and reflect different migratory patterns, harvest availabilities, and habitats of the elk populations. Our use of data collected extensively for management purposes, rather than intensively for research, was necessarily more descriptive compared to presentation of data collected intensively on one area. However, we believe information collected across multiple-use landscapes allows insights and conclusions not possible from one or a few intensively studied protected areas.

I. INTRODUCTION

Population dynamics of large herbivores can be influenced by an array of biotic and abiotic factors (Chapter 8 by Messer *et al.*, and Chapter 23 by Garrott *et al.*, this volume) and, particularly when predation is limiting, the collective attributes of the local landscape and climate can exert considerable influence on these dynamics (Chapter 21 by White *et al.*, and Chapter 24 by Garrott *et al.*, this volume). Consequently, the dynamics of a particular predator and their focal prey can be expected to vary across systems. Discerning the respective influences of resources and predation on large ungulates in temperate systems is often heavily confounded by human harvests (Figure 25.1) and habitat alteration (Eberhardt 1997). Therefore, most investigations occur in systems that are managed for resource protection and experience only minimal human interference. Nevertheless, most ungulate species are widely distributed throughout a mosaic of landscapes with different uses, ownerships, and policies, where human impacts are a pervasive and significant component. Insights gained from protected areas and generalized to these multiple-use areas may be misleading due to the significant differences in human influence. Thus, developing effective management plans necessitates an evaluation of ungulate dynamics in multiple-use landscapes.

Even in protected areas the spatial and temporal scales at which large herbivore population dynamics occur make evaluations particularly difficult. While experimentation is the strongest science, it is problematic at a landscape scale because true controls are rare, replicates are difficult to obtain, and experiments take years to complete (Hobbs and Hilborn 2006). Although there are notable exceptions, such as partially controlled field manipulations or treatments (Platt 1964, Romesburg 1981) that have been conducted at the landscape scale (*e.g.*, Boutin 1992, Krebs *et al.* 1995), their feasibility in most systems is limited. These limitations are obvious for national parks, which are managed to minimize human intervention, but also affect research on many other multiple-use public and private lands due to multiple land ownerships and the differing goals and expectations of different segments of the public. Furthermore, true experimentation is rarely possible when major influential species of study such as the wolf and grizzly bear (*Ursus arctos*) are protected from adverse effects under the Endangered Species Act.

Because true experiments are typically not feasible, biologists are often faced with the problem of how to obtain results useful for the planning and management decisions that will be made regardless of study method, whether the information was arrived at by the strong inference (deduction) of experimentation, or whether there is information available at all. Planned studies with random sampling and reasonable sample sizes can provide sound inferences (Cochran 1983, Eberhardt and Thomas 1991) and the method of induction can use the "general-purpose data" collected by "game agencies" to obtain reliable information (Romesburg 1981). Thus, we can gain reliable knowledge

FIGURE 25.1 A successful elk hunter on the Lower Madison winter range where late-season management hunts are conducted to reduce elk numbers (Photo by Julie Cunningham).

(Romesburg 1981) by repeating observational studies many times, in many places, to sequentially gain understanding of basic ecological processes (Eberhardt 2003).

Wolf restoration to the greater Yellowstone, central Idaho, and northwest Montana recovery areas has been spectacularly successful. Goals for numbers of breeding pairs and total wolves, along with distribution goals for the States of Idaho, Montana, and Wyoming were met by the end of 2002 (U.S. Fish and Wildlife Service *et al.* 2006). As a result, the U.S. Fish and Wildlife Service proposed delisting wolves in the northern Rocky Mountains and transferring management to state wildlife agencies pursuant to approved wolf management plans (72 Federal Register 36939). The increasing numbers and distribution of wolves make it imperative that state agencies managing wolves and game animals, and responding to livestock depredations, have the best available information on factors influencing wolf predation. Ideally, this information should be as broadly applicable as possible, gathered and synthesized across the range of wolves and ungulates densities as well as across a diversity of ecological and management situations.

Results presented elsewhere in this book for the Madison headwaters area represent substantial new knowledge of elk and wolf dynamics in Yellowstone National Park (Yellowstone). Previously, most research and attention was focused on elk and wolves that wintered in the northern portion of Yellowstone (*e.g.*, Houston 1982, White and Garrott 2005, Vucetich *et al.* 2005, Smith *et al.* 2006a, Varley and Boyce 2006). However, areas within Yellowstone are managed pursuant to preservation policies and under conditions that are quite different than those affecting elk and predators outside the park (though portions of the northern Yellowstone elk herd are subject to hunting when animals migrate outside the park during winter). Thus, management goals, population densities, hunting harvest, local climate, and other conditions vary substantially for both elk and wolves in Montana within short distances of central and northern Yellowstone. The interactions of wolves and elk could vary considerably depending on the ecological and human-influenced characteristics of the area they occupy (Garrott *et al.* 2005) and results determined for the central (Madison headwaters) and northern Yellowstone case studies might not be widely applicable.

In 2001, the Montana Department of Fish, Wildlife, and Parks, Montana State University, National Park Service, and U.S. Fish and Wildlife Service began a cooperative project to provide mutual assistance among some intensive studies of wolves and ungulates and initiate systematic collection and examination of past, present, and future "general-purpose data" (Romesburg 1981) for elk populations in southwestern Montana (Hamlin 2005). Our goal was to use population monitoring data routinely collected for many years before and after wolf restoration across a multiple-use landscape to provide insights that could contribute to management decisions regarding wolves and elk. We compared and contrasted vital rates such as yearling and adult pregnancy rates, calf survival and recruitment, and the survival of adult female elk among areas relative to the strength of influencing factors such as hunter harvest, weather, and levels of wolf and grizzly bear populations. Eberhardt (1977), Hanks (1981), and Gaillard *et al.* (1998) suggested that a decrease in physical condition through increased ungulate population densities (or other stressors) leads to the following sequential demographic changes: (1) increased juvenile mortality, (2) increased age at sexual maturity, (3) decreased fecundity of mature females, and (4) increased adult mortality. However, this sequence and its interpretation, especially for increased juvenile mortality, can be complicated by the presence of effective predators (Bergerud 1971, Beasom 1974, Keith 1974, Bergerud 1978, Stout 1982, Gasaway *et al.* 1983, 1992, Hamlin and Mackie 1989, Chapter 23 by Garrott *et al.*, this volume).

For many presentations, we will use numbers of wolves and grizzly bears per 1000 elk to represent the potential influence of this variable because predator to prey ratios are likely most ecologically relevant (Eberhardt 1997). We also compared population trends among the seven elk populations relative to these changes in vital rates and influencing factors. We believe conclusions based on this information will be useful to guide planning and management decisions throughout the current and potential future distribution of wolves in the northern Rocky Mountain region of the United States.

II. METHODS

A. Elk Populations and Study Site Descriptions

We assessed the status of elk on seven areas with different predator guilds and management paradigms in the greater Yellowstone area, including the Madison headwaters and northern winter range of Yellowstone National Park (northern Yellowstone), the Wall Creek Wildlife Management Area (Wall Creek) and Blacktail–Robb Ledford Wildlife Management Areas (Blacktail–Robb Ledford) of the Gravelly-Snowcrest mountains, Gallatin Canyon (Gallatin), Lower Madison drainage (Hunting District 362), and the Yellowstone Valley (Hunting District 314) in Montana (Figure 25.2). The relative proximity of the seven areas (Figure 25.2, most winter ranges within 80 km of each other, the furthest 115 km) resulted in similar broad-scale climate patterns. Thus, all areas experienced sustained drought since 1995 and relatively mild winters since 1998.

Preservation was the primary land use for areas used by the Madison headwaters and northern Yellowstone elk populations (Table 25.1). The Madison headwaters population remained in Yellowstone, but a portion of the northern Yellowstone elk population was subject to multiple-use management, including hunting, when elk seasonally migrated into Montana. The Gallatin Canyon elk population was subject to public multiple-use management, with preservation on some summer range but no significant agricultural influence. The Wall Creek and Blacktail–Robb Ledford elk populations were managed under public multiple-use, with surrounding areas of agricultural ownership. The Lower Madison and Yellowstone Valley elk populations wintered primarily on private agricultural lands (Figure 25.3). Thus, different levels of tolerance for wolves occurred among the seven areas. The Montana Department of Fish, Wildlife, and Parks collected elk demographic data in these areas for at least several decades prior to wolf reintroduction.

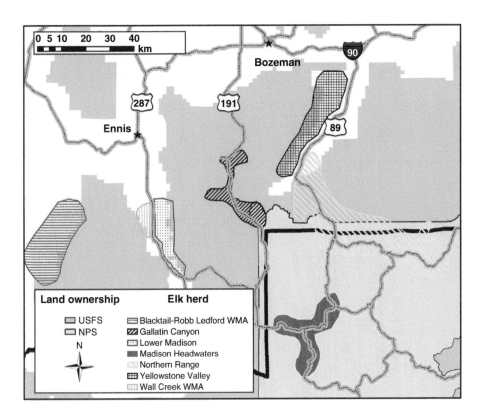

FIGURE 25.2 Proximity and landownership patterns of the winter ranges for seven elk herds in the greater Yellowstone area. White areas depict private lands.

Primary winter habitat for five of the seven elk populations was predominantly grasslands (Table 25.1, Figure 25.4), with inclusions of sagebrush (*Artemisia spp.*) steppe. Conifer forest occurred on the upper fringes of winter range. The Madison headwaters and Gallatin Canyon elk winter ranges (Figure 25.5) occurred within a predominantly conifer forest area, with limited inclusions of grasslands and sagebrush along some stream courses and meadows near thermal features in the Madison headwaters. Thus, the winter distribution of elk after wolf restoration was limited to smaller, predictable areas in the Madison headwaters and Gallatin Canyon areas compared to the other five areas (Chapter 18 by Gower *et al.*, and Chapter 21 by White *et al.*, this volume). These broad habitat differences among areas were also related to differences in winter severity faced by elk, with the most-severe winter conditions occurring in the Madison headwaters and Gallatin Canyon areas. Winter severity ranged from moderate to mild for the northern Yellowstone elk population, with severity decreasing to the north of the park. Winter severity was comparatively mild for the other four elk populations (Table 25.1).

Total range in estimated elk population size among the seven areas since wolf restoration was approximately 200–17,000 elk (Table 25.1). The smallest elk populations were in the Madison headwaters (200–600) and Gallatin Canyon (700–1500), while the largest population was in northern Yellowstone (6700–16,700). Elk numbers in the other four populations (1500–4800) ranged from 10–20 times greater than the low estimate for the Madison headwaters population and one-half to one-third of the lowest count for the northern Yellowstone population. Annual levels of hunter harvest also varied considerably among populations, from no harvest of the nonmigratory Madison headwaters population to an average harvest of 65% of the bulls and 15% of the cows from the Blacktail–Robb

TABLE 25.1 Description and comparison of landscape and ecological characteristics of 7 study areas in southwestern Montana and northwestern Wyoming (Yellowstone National Park)

Characteristics	Madison Headwaters	Northern Range	Wall Creek WMA	Blacktail–Robb Ledford WMA	Gallatin Canyon	Lower Madison	Yellowstone Valley
Primary Winter Habitat	Conifer Forest	Grassland & Sage Steppe	Grasslands & Sage Steppe	Grasslands & Sage Steppe	Conifer Forest	Grasslands & Sage Steppe	Grasslands & Sage Steppe
Winter Severity	Severe	Moderate to Mild	Mild	Mild	Moderate	Mild	Mild
Land Ownership	Public—Preservation	Public—Preservation & Multiple Use	Public—Multiple Use	Public—Multiple Use	Public—Multiple Use	Private—Agricultural	Private—Agricultural
Winter Elk Counts	200–600	6700–16,700	1500–2700	1700–2700	700–1500	1500–3500	3000–4800
Hunting Pressure	None	Bulls—low (<10%), Cows—(2–13%) low recently	Bulls—High (50%), Cows moderate (10–12%)	Bulls—Highest (65%), Cows moderate to high (15%)	Bulls—High (45%), Cows (1–25%) low recently	Bulls—Moderate (30%), Cows (3–20%) lower recently	Bulls—High (40%), Cows moderate (10%)
Alternate Prey	High densities of bison	Moderate densities of bison, low of mule deer, pronghorn	Low densities of mule and white tail deer	Low densities of mule and white tail deer, pronghorn and moose	Low densities of mule deer and moose	Low densities of pronghorn and mule deer	High densities of white-tail deer and mule deer
Grizzly Bear Presence[a]	Highest: 105.7	Moderate: 9.25	Transients: rare	Transients: rare	High: 56.9	Low: 7.1	Low to moderate
Winter Wolf Presence[b]	Highest: 85.3	Moderate: 9.5	Transients: rare	Low: 2.6	High: 23.4	Low: 1.9	Low: 2.1

Areas

[a] Highest estimated grizzly numbers per 1000 elk.
[b] Highest estimated wolf numbers per 1000 elk.

FIGURE 25.3 Cattle being moved across private land elk winter range on the Lower Madison area toward elk summer range. Wolf depredations of livestock reduce tolerance for wolves on and near private lands (Photo by Julie Cunningham).

FIGURE 25.4 Elk winter range in the Lower Madison Valley, looking from the east side (typical of Lower Madison) to the west side (typical of the Wall Creek Wildlife Management Area). This habitat is also typical of Blacktail–Robb Ledford Wildlife Management Area, Yellowstone Valley, and the north portion of Yellowstone National Park (Montana Fish, Wildlife, and Parks photo by Craig Jourdonnais).

Ledford population. Hunter harvest for the other elk populations (Table 25.1) varied between these extremes, with harvests of cows decreasing to an average of 3% of estimated preseason numbers in the Gallatin Canyon and northern Yellowstone populations since 2004. More detailed descriptions of the characteristics of the seven study areas are provided in Appendix 25A.1, and more detailed descriptions for estimating elk numbers and population parameters are in Appendix 25A.2.

FIGURE 25.5 Elk winter range in the Gallatin Canyon (Montana Fish, Wildlife, and Parks photo by Craig Jourdonnais).

B. Estimates of Predator Numbers

The annual number of wolves estimated for each area and numbers killed in control actions or other known mortalities due to humans were determined from December numbers reported in the Cooperative Rocky Mountain Wolf Recovery Annual Reports (*e.g.*, U.S. Fish and Wildlife Service *et al.* 2007), Yellowstone National Park research reports (Smith *et al.* 2001, 2003, 2004a,b, 2005, 2006a, 2007, Smith and Guernsey 2002), and consultations with field biologists from Montana Fish, Wildlife, and Parks. After consultation with a Montana wolf management specialist (Mike Ross), we estimated wolf numbers by elk range each winter to fractions of 0.25 in some areas because some wolves and wolf packs used multiple elk ranges. For the Blacktail–Robb Ledford, we used Global Positioning System telemetry locations to estimate that the wolf pack spent 34% of its time with this elk herd and fractionalized annual wolf numbers for that pack accordingly. Similarly, intensive fieldwork in the Madison headwaters allowed estimation of wolf-days on the area and fractional wolf numbers equivalent to estimates for the other areas (Chapter 15 by Smith *et al.*, Chapter 16 and 17 by Becker *et al.*, this volume).

Grizzly bears have been implicated in substantial mortality of neonatal elk calves in this region (Singer *et al.* 1997, Barber-Meyer 2006, Creel *et al.* 2007). Thus, we attempted to quantify relative grizzly bear numbers by area and year. Our index of grizzly bear numbers represents relative differences among areas and years rather than accurate, absolute numbers. We used information presented in the annual reports of the Interagency Grizzly Bear Study Team (*e.g.*, Knight *et al.* 1990, Knight and Blanchard 1996, Haroldson 2006, Haroldson and Frey 2006) to construct our estimates. More detailed descriptions of estimating wolf and grizzly bear numbers and population parameters are in Appendix 25A.2.

C. Estimates of Elk Vital Rates

1. Pregnancy Rate

Hunters at the Gardiner (northern Yellowstone) and Gallatin check stations were asked to report the pregnancy, sex of fetus, and lactation status of their harvests during 1985–2007. The reported pregnancy rates are underestimates because hunters vary in their anatomical skills and, also, miss some small fetuses. However, this underestimate should be consistent among years and yield a good relative annual index of pregnancy. Check station personnel at the Gardiner and Gallatin check stations also weighed hunter-harvested calves to the nearest pound and we converted these weights to kg for analysis. Male calves consistently weighed more (70.53 kg, 95% CI = 69.81, 71.25) than female calves (67.04 kg, 95% CI = 66.45, 67.63) and we adjusted female weights upward based on this average difference to construct an annual average sex-adjusted weight for all calves.

When adult female elk were captured to attach radio-transmitter collars on our study areas, we collected blood samples and used pregnancy-specific protein B (PSPB) assays to assess pregnancy status (Sasser *et al.* 1986). Earlier tests, where both blood samples and reproductive tracts were obtained (Palmisciano 1986) or blood samples and rectal palpation occurred (Hamlin and Ross 2002), indicated 100% accuracy ($n = 78$) for PSPB assays to estimate pregnancy.

The hunter-reported pregnancy data were binomial and required an analytical method accounting for nonnormal error structure. We used the logit-transform (on proportions) rather than logistic regression because, although sampling was random within years, it was nonrandom among years. For example, sample sizes for adult females from the northern Yellowstone population varied from 37 to 1456 among years and were related to factors that could affect pregnancy rates such as increased samples during severe winters and decreased sample sizes with increasing numbers of wolves and reduced hunting. For the January 2003 hunter-reported sample, there were no pregnancies among 19 yearlings. We used 0.01 (lower than any other year) as the pregnancy rate for logit-transformation. We used logistic regression to analyze pregnancy rates estimated using the PSPB assay because samples were random and similar in size among years. We used the R statistical package, version 2.5.0 (R Development Core Team 2006) for all analyses.

2. Elk Calf Recruitment

Montana Fish, Wildlife, and Parks biologists recorded total numbers and bulls during all aerial surveys of elk. Numbers of calves and cows were also recorded during many surveys. However, separate classifications of cows and calves were commonly conducted from the ground for many of the areas. When a mixture of techniques occurred, we used aerial counts to determine proportions of bulls and ground classifications to determine ratios of calves per 100 cows. These classifications were not available for a few years in the northern Yellowstone, Gallatin Canyon, and Lower Madison. However, age structure data were available from check stations. We did not directly use numbers of harvested calves and cows to estimate the ratio of calves per 100 cows in the populations because hunters may select against calves (Hamlin and Ross 2002). Rather, we used proportion of yearling cows in the next year's harvest (YC_{t+1}) to estimate the recruitment of female calves per 100 adult cows during the previous year. We added estimated recruited male calves (MC_{t+1}) to YC_{t+1}, where $MC_{t+1} = (YC_{t+1})$ (MC_t/FC_t). Thus, ($YC_{t+1} + MC_{t+1}$) per 100 cows $\geq$2-years old = recruitment of total calves per 100 adult females (t). This estimate could be slightly lower than comparable classifications of calves per 100 cows because it included any mortality occurring between the end of March and the next hunting season. We assumed this mortality was low except during severe winters.

We used the average Palmer Drought Severity Index (PDSI) during May and June as an index of soil moisture conditions during the early growing season (Palmer 1968). The PDSI incorporates precipitation, temperature, and evapotranspiration, all correlated with production and nutritive quality of plants (Sala *et al.* 1988). Thus, we expected PDSI to be positively related to elk nutrition and elk calf survival because dry years (*i.e.*, low PDSI) would decrease plant production, thereby decreasing fat reserves for elk entering winter and resulting in lower calf survival.

All of our graphics display relationships relative to observed calves per 100 cows for ease of understanding. However, we used the proportional calf per cow ratio and logit-transformed this quantity for statistical analysis.

3. Adult Female Survival

Ages of hunter-harvested elk were estimated by eruption-wear (Quimby and Gaab 1957) and recorded at Montana Fish, Wildlife, and Parks check stations. Since 1989, incisors also were collected from elk >2 years of age for aging via examination of annuli in the cementum of incisor root tips (Hamlin *et al.* 2000). For years when only eruption-wear ages were available, and for portions of annual samples that included some elk aged by eruption-wear only among samples mostly consisting of incisor annuli aged elk, we constructed area and time period specific matrices (*e.g.*, Table 3 in Hamlin *et al.* 2000) to proportionally correct eruption-wear ages to their expected distribution as incisor determined ages. We then added incisor-aged and corrected eruption-wear ages for the time period samples. We natural-log transformed the age–frequency distributions of adult females recorded at check stations and fit linear regressions against age (Hamlin and Ross 2002, White and Garrott 2005). We used the slope of these regressions to estimate survival ($S = e^{\beta}$, where e = inverse of natural log and β = slope of the regression line). For years and areas when population trend did not indicate relative stability (Caughley 1977), we estimated rate of change in population size by regressing natural log of raw population counts against year to determine the instantaneous rate of change (r). We then adjusted age–frequency distributions by the equation-adjusted number of age $x = f_x e^{rx}$ (Caughley 1977). We then proceeded as above to use slopes of the regression of natural log of age distribution against age to determine the estimate of survival adjusted for population growth rate. We used the different time periods and areas as factors in multiple linear regression models (Program R) to test for differences in survival among periods by area and period, and allow interactions between ln(count) and age. We compared periods, and areas within periods, using all-subsets regression. The *P*-value of the regression coefficients on the interaction terms represented the significance level of the difference between the survival rate for the periods and areas. The age–frequency technique for determining survival closely matched survival data determined by telemetry for elk in the Gravelly-Snowcrest mountains during 1989–1996 (Hamlin and Ross 2002). Also, the survival rate we determined by age structure for the northern Yellowstone during 2001–2006 (0.81) was similar to the rate estimated by telemetry (0.80) during March 2000 to February 2004 (Evans *et al.* 2006). Thus, the age structure technique for estimating survival appeared to be relatively accurate.

We used annual reports of the Montana Statewide Harvest Questionnaire (Cada 1985; Montana Fish, Wildlife, and Parks annual reports) to estimate hunter harvest by area as calves, adult females, and adult males. Additionally, annual reports were available for the mandatory check of northern Yellowstone elk harvested during late season hunts (T. Lemke, Montana Fish, Wildlife, and Parks, unpublished reports). More detailed descriptions of methods used to determine harvest by area, adjust for a year of missing data, and adjust for estimated wounding loss are available in Appendix 25A.2.

To estimate preseason adult female numbers for comparison to hunter harvest and wolf-kill rates, we multiplied raw post-season counts by correction factors for sightability (Appendix 25A.2) and used classification data to estimate the proportion or number that were adult females, calves, and males. To obtain the estimated preseason number of adult female elk, we added the estimated harvest of adult females to the estimated number of adult females and adjusted for estimated wounding loss (Appendix 25A.2).

We used estimates provided in Smith *et al.* (2004a) and Becker *et al.* (Chapter 17 by Becker *et al.*, this volume) to estimate total wolf-killed elk as 0.060 elk killed per wolf day during the 181 day (November 1–April 30) winter period and reduced summer kill rates (184 days, May 1–October 31) to 70% (0.042 elk killed per wolf day) of winter rates (Messier 1994). We used 25% (0.015 adult female elk killed per wolf day in winter and 0.0105 in summer) as the estimate for percent of wolf-kill that was adult females based on averages in reports for the northern Yellowstone (Smith *et al.* 2004a, 2005, 2006a, 2007), Gallatin Canyon (Creel *et al.* 2007), and Lower Madison (Hamlin 2005; R. Garrott, unpublished data). Thus, we estimated the number of adult female elk killed by wolves each year in each area as the [(number of wolves)×(0.015×181)]+[(number of wolves)×(0.0105×184)].

III. RESULTS

A. Predator Numbers

Wolves first established in the northern portion of Yellowstone in 1995 and colonization proceeded to the Gallatin Canyon, Madison headwaters, Lower Madison, and Yellowstone Valley (Table 25.2). Wolves formally colonized the last of our seven areas (Blacktail–Robb Ledford) in 2000. Though a breeding pack did not establish at the Wall Creek Wildlife Management Area, transient wolves have been noted since 1999. Pack tenure was longest, and the maximum number of packs and wolves was highest, in northern Yellowstone, followed by the Madison headwaters and Gallatin Canyon (Table 25.2). As would be expected, wolf deaths due to agency control actions related to depredations were highest in private agricultural valleys and closely associated public lands (Table 25.2, average = 21–49% of wolves). Wolf deaths due to agency control actions ranged from 0 to 4% in the central and northern portions of Yellowstone to 11% on public land in the Gallatin Canyon where there were also no significant agricultural conflicts. Smith *et al.* (2006a) reported that during the first nine years after wolf restoration to Yellowstone, including the Madison headwaters and northern Yellowstone areas, wolf mortality in the park from all causes averaged 18% per year and dispersal averaged 28% per year, totaling 45% annual loss from the population. We do not have this type of information for the other areas, but based on numbers and growth rates, wolf mortality was much higher overall (at least 21–49% annual loss) in the agricultural valleys surrounding Yellowstone than inside the park or on adjacent public lands without agricultural conflicts.

The broad range of 0–85 wolves per 1000 elk among areas, and among years within areas (Table 25.2), is of major ecological significance in interpreting our results. Public lands managed primarily for preservation (*e.g.*, Yellowstone National Park) and public lands without agricultural conflicts (*e.g.*, Gallatin Canyon) reached the highest levels of 10–85 wolves per 1000 elk. In contrast, fewer than three wolves per 1000 elk were sustained in valleys with a significant agricultural component. In addition, the absolute and relative number of predators on elk varied substantially among the seven areas (Table 25.1). The ratio of wolves per 1000 elk reached the highest level in the Madison headwaters (85) and was four times greater than the second highest ratio (23) in the Gallatin Canyon. Wolves per 1000 elk in northern Yellowstone (10) were one-half the maximum ratio recorded for the Gallatin Canyon, with the other four areas reaching only minor levels (0–3) of wolves relative to the number of their main prey, the elk. Grizzly bears, which were determined to be a major predator of neonatal elk calves in the northern Yellowstone (Singer *et al.* 1997, Barber-Meyer 2006) and Gallatin Canyon (Creel *et al.* 2007) populations, increased in the Yellowstone ecosystem (Schwartz *et al.* 2007) coincident with wolf restoration. As with wolves, the index of grizzly bear numbers per 1000 elk reached the highest levels in the Madison headwaters (106), followed by the Gallatin Canyon (57; Table 25.1). The index indicated relatively moderate levels of grizzly bears per 1000 elk occurred in northern Yellowstone (9), with lower numbers in the Lower Madison (7) and low to rare numbers elsewhere (Table 25.1). Alternate prey for these predators is relatively low on most areas (Table 25.1), with fairly high densities of harder to kill bison (*Bison bison*) in the Madison headwaters and northern Yellowstone areas and relatively high densities of mule deer (*Odocoileus hemionus*) and white-tailed deer (*O. virginianus*) in the Yellowstone Valley.

B. Pregnancy Rate

Hunter-reported pregnancy rates were determined from records for 1714 yearling female elk in northern Yellowstone during 1985–2005. Annual yearling pregnancy rates were not related to numbers of wolves per 1000 elk during the prior winter ($R^2 = 0.08$, $F_{1,19} = 1.74$, $P = 0.20$, Figure 25.6A) or prior

TABLE 25.2 Description and characteristics of the status of wolves on 7 study areas in southwestern Montana and northwestern Wyoming (Yellowstone National Park), 1995–2007

				Areas			
		Madison Headwaters	Northern Range	Wall Creek WMA	Blacktail–Robb Ledford WMA	Gallatin Canyon	Lower Madison	Yellowstone Valley
Wolf pack characteristics	Year of Wolf Colonization	1998[a]	1995	Transients	2000	1996	1999	1999[a]
	# Packs[b]	1 to 3	5 to 10	None	1 to 2	1 to 3	1	1 to 2
	# Wolves[c]	30 to 40	21 to 106	Transients	3 to 14	2 to 19	5	2 to 11
	Average wolf loss to control actions	None (0%)	None to Low (4.3%)	Transients	Highest (48.5%)	Low (10.9%)	High (33.5%)	Moderate (21.2%)
	Maximum pack tenure	8 years	11+ years[d]	None	5 years	10 years[e]	2 years	2.5 years
	Range in Wolves:1000 elk (time adjusted)	0.5–85.3	1.0–9.5	Transients	0.3–2.6	1.0–23.4	0.5–1.9	0.1–2.1

[a] Some use by transient wolves earlier.

[b] Maximum number of packs in a year.

[c] The total number of wolves with a home range that crossed the specific wintering elk populations.

[d] There are two packs in the northern range that have existed since reintroduction: the Leopold and Druid packs.

[e] The Chief Joseph pack began in 1996 with little territory outside YNP, but gradually increased presence outside YNP in the Gallatin National Forest with time. The Chief Joseph pack was counted as a southwestern Montana pack in 2005 and 2006.

summer ($R^2 = 0.02$, $F_{1,19} = 0.44$, $P = 0.51$). Also, annual hunter-reported pregnancy rates for 13,750 female elk $\geq$2-years old in northern Yellowstone during 1985–2007 and 812 female elk $\geq$2-years old in the Gallatin Canyon during 1989–2004 were not correlated with the number of wolves per 1000 elk the previous winter ($R^2 < 0.01$, $F_{1,36} = 0.01$, $P = 0.91$, Figure 25.6B) or previous summer ($R^2 < 0.01$, $F_{1,36} = 0.001$, $P = 0.97$). Pregnancy rates determined using PSPB assays (northern Yellowstone, $n = 167$; Gallatin Canyon, $n = 62$; and Lower Madison, $n = 59$) for all ages of elk across three of our elk populations were not significantly related to numbers of wolves per 1000 elk in the winter prior to conception ($\beta_w = 0.0997$, SE $= 0.063$, $P = 0.11$, Figure 25.6C) or during the prior summer ($\beta_s = 0.0854$, SE $= 0.057$, $P = 0.14$). A possible 32% decrease in pregnancy rate occurred on the Madison headwaters area during the post-wolf period (Chapter 23 by Garrott *et al.*, this volume), but results were equivocal.

C. Elk Calf Recruitment in Relation to Wolves and Grizzly Bear

Recruitment of elk calves across all seven areas, as indexed by ratios of calves per 100 cows observed in late winter or spring, decreased exponentially as the numbers of wolves per 1000 elk, grizzly bears per 1000 elk, and combined wolves and grizzly bears per 1000 elk increased (Figure 25.7A–C). Analysis with logit-transformed ratios of calves per 100 cows indicated a significant correlation with wolves per 1000 elk ($R^2 = 0.47$, $F_{1,110} = 97.3$, $P < 0.001$), grizzly bears per 1000 elk ($R^2 = 0.56$, $F_{1,90} = 114$, $P < 0.001$), and wolves and grizzly bears per 1000 elk ($R^2 = 0.60$, $F_{3,88} = 46.7$, $P < 0.001$). These statistics excluded the data from 1989 and 1997 when the effects of extensive fires and a severe winter, respectively, resulted in high malnutrition and starvation. However, the relationships remained significant when 1989 and 1997 were included (wolves per 1000 elk: $R^2 = 0.38$, $F_{1,122} = 74.8$, $P < 0.001$; grizzly bears per 1000 elk: $R^2 = 0.45$, $F_{1,100} = 82.1$, $P < 0.001$; and wolves and grizzly bears per 1000 elk: $R^2 = 0.49$, $F_{3,98} = 31.4$, $P < 0.001$). The increase in numbers of wolves and grizzly bears was coincident, so that they covaried strongly ($R^2 = 0.71$, $F_{1,107} = 261.4$, $P < 0.001$) and interpretation of relative effects is problematic. However, regional studies determining elk calf mortality based on telemetry indicate that mortality of calves due to bears is almost entirely within the first 45 days of life (Singer *et al.* 1997, Barber-Meyer 2006, Creel *et al.* 2007, Zager *et al.* 2007b) and mortality due to wolves continues through the year.

D. Elk Calf Recruitment in Relation to Condition, Drought, and Forage

Annual mean weights recorded for 1906 northern Yellowstone elk calves harvested during 1985–2005 late hunts were generally at or above the long-term mean after wolf restoration (Figure 25.8), indicating no wolf-induced decreases in nutrition. Further, variation in nutritional level did not affect calf survival because numbers of calves per 100 cows (logit-transformed) in spring were not positively related to weights of harvested calves ($\beta_{cw} = 0.012$, SE $= 0.024$, $R^2 = 0.0135$, $P = 0.62$, Figure 25.9).

There was a significant positive relationship between the regional PDSI and recruitment of elk calves (*i.e.*, low PDSI = dry years and lower calf recruitment) as indexed by logit-transformed ratios of calves per 100 cow in spring across our seven elk populations during 1968–2007 ($\beta_{all} = 0.098$, SE $= 0.019$, $R^2 = 0.14$, $P < 0.001$), but residuals were large and predictability low (Figure 25.10). The relationship remained positive after data were partitioned into areas and years when wolves and/or grizzly bears were greater than four per 1000 elk ($\beta_{great} = 0.069$, SE $= 0.039$, $R^2 = 0.07$, $P = 0.08$) or less than four per 1000 elk ($\beta_{less} = 0.025$, SE $= 0.014$, $R^2 = 0.02$, $P = 0.08$). Other factors affecting calf survival apparently overwhelmed regional drought effects or these effects did not affect nutrition sufficiently to

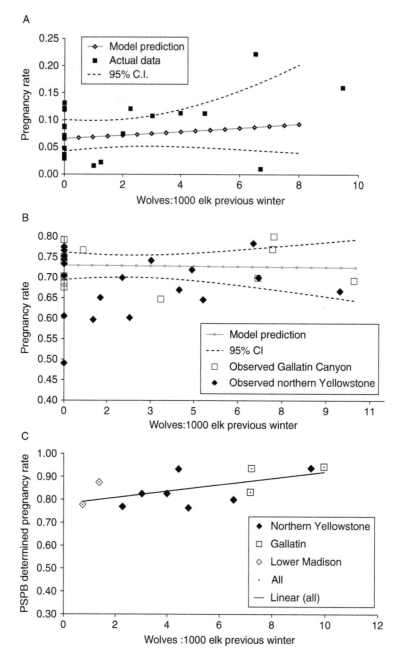

FIGURE 25.6 Pregnancy rates of elk in relation to number of wolves:1000 elk on winter range. Panel A—hunter-reported pregnancy rate for yearling females migrating outside the northern portion of Yellowstone National Park during winter ($n = 14$–250). Panel B—hunter-reported pregnancy rate for females $\geq$2-years-old outside the northern portion of Yellowstone National Park ($n = 37$–1456) and in the Gallatin Canyon ($n = 11$–114). Panel C—pregnancy rate determined by PSPB for all female elk outside the northern portion of Yellowstone National Park ($n = 15$–40), in the Gallatin Canyon ($n = 12$–32), and in the Lower Madison areas ($n = 27$–32).

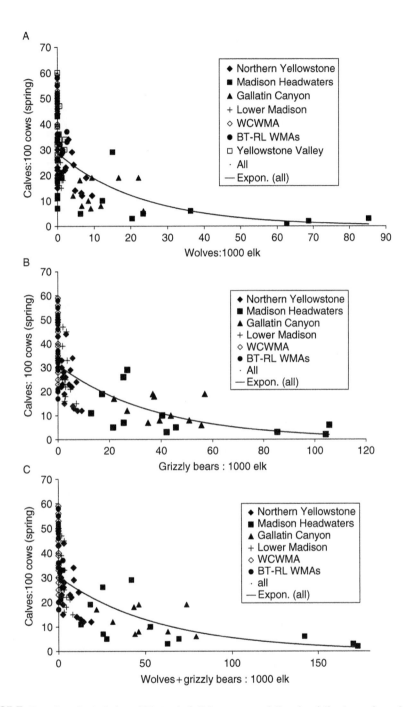

FIGURE 25.7 Recruitment rate (calves:100 cows) of elk in seven populations in relation to numbers of wolves and/or grizzly bears:1000 elk (excluding 1989 and 1997). Panel A—calves:100 cows recruited relative to numbers of wolves:1000 elk on winter range for 7 elk populations, 1986–2007. Panel B—calves:100 cows recruited relative to numbers of grizzly bear:1000 elk on winter range for 6 elk populations (Yellowstone Valley not included), 1986–2006. Panel C—calves:100 cows recruited relative to numbers of wolves and grizzly bear:1000 elk on winter range for 6 elk populations (Yellowstone Valley not included), 1986–2006.

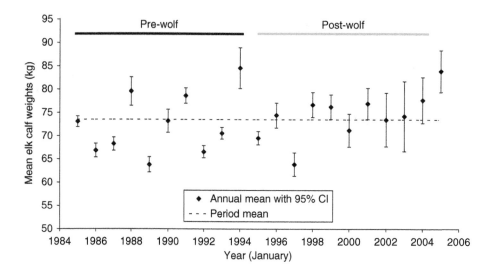

FIGURE 25.8 Mean weight (kg) of calves harvested during the late hunt for northern Yellowstone elk during 1985–2005 ($n = 12$–253).

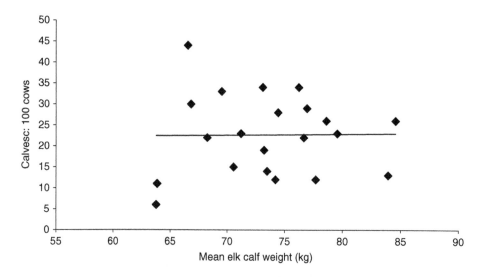

FIGURE 25.9 Recruitment rate (calves:100 cows) of elk calves relative to mean weight (kg) of elk calves recorded in early winter during the late hunt for northern Yellowstone elk, 1985–2005.

result in a strong calf survival response, even in areas with few or no wolves or grizzly bears. This conclusion also was indicated for the Madison headwaters, where the data supported constant nutrition for elk during the pre-wolf and post-wolf drought period (Chapter 22 by White *et al.*, this volume). The lack of a strong drought effect was not unexpected because the recent period of drought was similar across our entire area of study. Also, our findings suggest the effects of predation may have exceeded the effects of moisture and nutrition because areas and years with fewer than four wolves and/or grizzly bears per 1000 elk had significantly higher ratios of calves per 100 cows than areas and

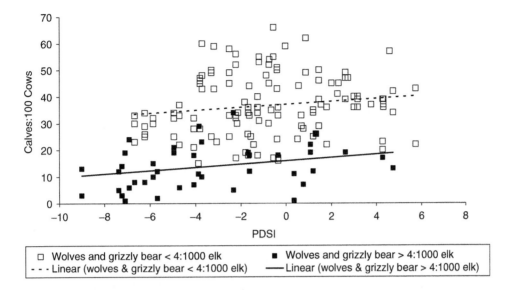

FIGURE 25.10 Recruitment rate of elk calves (calves:100 cows) across seven elk populations relative to the average PDSI during May and June of their birth year and relative to numbers of wolves and grizzly bear per 1000 elk in their herd range, 1968–2007. Yellowstone Valley not included for grizzly bear indexes.

years with greater than four wolves and/or grizzly bears per 1000 elk, regardless of the drought severity level (Welch 2-sample *t*-test, $t = 9.9905$, $P < 0.001$, Figure 25.10).

E. Adult Female Elk Survival

Survival estimates for adult females ranged from 0.76 to 0.88 among areas and periods for the four areas with data to estimate survival based on age structure. Survival of adult females was generally lower during the post-wolf period (Figure 25.11), but only consistently for elk in the Gallatin Canyon. Survival of adult females in the northern Yellowstone and Gravelly-Snowcrest mountains was lower during the 1997–2000 period, which included effects of the relatively severe winter of 1996–1997. For the Lower Madison area, adult female survival was significantly higher ($P < 0.001$) during the 2001–2006 post-wolf period compared to the 1989–1996 pre-wolf period, but this coincided with a substantial reduction in hunting pressure (Table 25.3), indicating hunting effects on adult female survival.

Survival estimates for adult female elk in northern Yellowstone during the 2001–2006 post-wolf period were not significantly different ($P \geq 0.19$) than during the 1983–1988 and 1989–1996 pre-wolf periods (Figure 25.11), despite a reduction in hunting mortality after 2003 (Table 25.3). Survival for adult female elk in the Gallatin Canyon was not significantly different among pre-wolf periods (1973–1975, 1983–1988, and 1989–1996, $P \geq 0.19$), but was significantly lower ($P \leq 0.004$) during the 2001–2006 post-wolf period (Figure 25.11), which included a reduction in hunting mortality after 2003 (Table 25.3). For the Gravelly-Snowcrest mountains, including the Wall Creek and Blacktail–Robb Ledford Wildlife Management Areas, survival of adult female elk was not different between the pre-wolf (1989–1996) and post-wolf (2001–2006) periods ($P = 0.41$), but survival during 1997–2000 was significantly lower ($P = 0.004$) than during both other periods (Figure 25.11). Survival of adult females decreased during the post-wolf period in the Madison headwaters compared to the pre-wolf period (Chapters 11 and 23 by Garrott *et al.*, this volume).

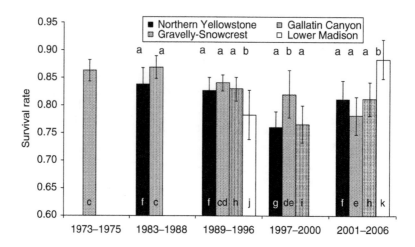

FIGURE 25.11 Estimated survival rate of adult female elk by time period and area as determined by age structure, 1973–2006. Different letters at top (a, b) represent significant differences ($P \leq 0.05$) among areas within time periods. Different letters at bottom (c—k, within bars) represent significant differences ($P \leq 0.05$) among time periods within areas.

F. Hunter Harvest and Estimated Wolf-Kill of Adult Female Elk

There was no hunter harvest of adult females or other elk for the Madison headwaters population except for those that dispersed to other populations (Chapter 18 by Gower *et al.*, this volume). Hunter harvest of adult female elk in northern Yellowstone varied considerably among years (Table 25.3, Figure 25.12A) with snow conditions and permit levels. We did not have age structure data to calculate survival for the adjacent Yellowstone Valley population, but hunter harvest rates of adult females were generally always higher than for the northern Yellowstone population. Estimated hunter harvests of adult females averaged 14% (range 8–21%) of estimated preseason populations in the Yellowstone Valley during 1989–1996, 11% (range 8–14%) during 1997–2000, and 10% (range 9–12%) during 2001–2003. Estimated hunter harvest of adult females in the Gallatin Canyon was generally high prior to 2001 (Table 25.3, Figure 25.12B). Except for an increase during 2003–2004 (Figure 25.12B), hunter harvest as a percent of the estimated preseason Gallatin Canyon adult female elk population decreased after 2000 (Table 25.3). We believe that the harvest estimates we portray for the Gallatin Canyon are overestimates to some degree, and estimates for the Lower Madison are slight underestimates, because some of the adult females harvested in the Gallatin Canyon were migratory to Madison Valley wintering herds (including the Lower Madison). Though hunter harvests of adult female elk in the Gravelly-Snowcrest Mountains (Wall Creek and Blacktail–Robb Ledford) decreased during 2001–2006 compared to historic levels (Table 25.3), they remained higher than for other areas. Based on historical trends in hunter harvest (Hamlin and Ross 2002), the harvest rate for the Blacktail–Robb Ledford population would have been higher than the Gravelly-Snowcrest average (Table 25.3) and the harvest rate for the Wall Creek population lower than the average.

Estimates of adult female elk killed by wolves were less than 2% of pre-hunting season numbers for all areas except the Gallatin Canyon, northern Yellowstone, and Madison headwaters areas (Table 25.3). The recent reduction in harvest mortality has not reduced overall mortality of adult females in the northern Yellowstone and Gallatin Canyon populations (Table 25.3, Figure 25.12A and B). Kills by wolves and deaths due to other factors (*e.g.*, other predators, malnutrition/winter-kill, accidents, disease) appear to have replaced the reduced deaths due to hunting (Table 25.3). Hunter harvests of adult females in northern Yellowstone and the Gallatin Canyon are currently at insignificant levels, but further years of data collection will be necessary to determine if survival will increase in the face of significant wolf presence.

TABLE 25.3 Rates and causes of mortality for adult female elk and comparison of average recruitment of calves with replacement level necessary for stable populations on four areas in southwestern Montana and northwestern Wyoming, 1983–2006

| Area | Period | Mortality Rate (M) from Age Structure | Mortality Rate | | | | | | Average Recruitment | |
			Hunting	Wolf-kill	Other[a]	% Hunting	% Wolf-Kill	% Other[a]	Calves:100 Cows to replace Mortality[b]	Calves:100 Cows Observed
Gallatin	1989–1996	0.16	0.18[c]	0.00	0	100[c]	0	0	32	22
	1997–2000	0.18	0.18[c]	0.04	0	100[c]	22	0	37	16
	2001–2006	0.22	0.07[c]	0.09	0.06	32[c]	41	27	47	10
Lower Madison	1989–1996	0.22	0.20	0	0.02	91	0	9	47	38
	1997–2000		0.07	0.01						
	2001–2006	0.12	0.07	0.01	0.04	58	8	34	23	20
Gravelly-Snowcrest[d]	1989–1996	0.17	0.15	0	0.02	88	0	12	34	46
	1997–2000	0.23	0.16	0.01	0.06	70	4	26	50	36
	2001–2006	0.19	0.10	0.01	0.08	53	5	42	39	30
Northern Yellowstone	1983–1988	0.16	0.05	0	0.11	31	0	69	32	31
	1989–1996	0.17	0.07	0.01	0.09	41	6	53	34	25
	1997–2000	0.24	0.11	0.02	0.11	46	8	46	53	23
	2001–2006	0.19	0.05[e]	0.04[e]	0.10	26	21	53	39	17

[a] Other = winter-kill/malnutrition, other predators, accidents, disease, etc.
[b] Replacement = $((M/0.6)/(1-M)) \times 100$. M = proportion of cows to replace, M/0.6 = total calves of both sexes, and 1-M = surviving cows.
[c] Because an unknown proportion of harvested elk were migrating through to Madison Valley winter ranges, estimated mortality rate to the Gallatin population from hunting is likely high.
[d] Estimated mortality rate from age structure only available for entire Elk Management Unit. Estimates for hunting and wolf-kill an average of only WCWMA and BT-RL WMAs.
[e] Hunter harvest decreased throughout the period and wolf-kill increased to exceed hunter-kill.

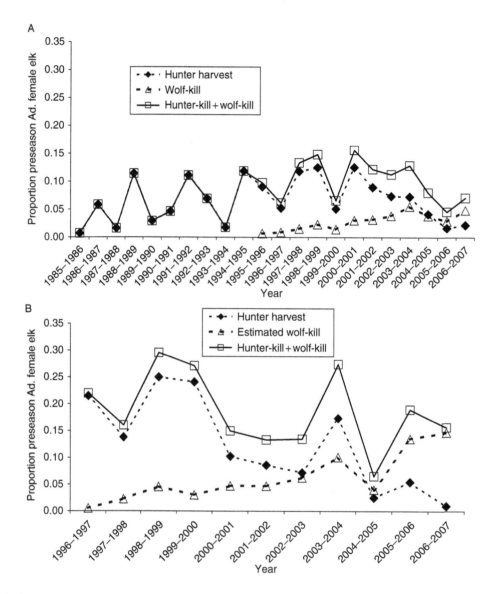

FIGURE 25.12 Estimated proportion of preseason adult female elk population killed by hunters and wolves. Panel A—estimated proportion of preseason adult female elk in the northern Yellowstone population killed by hunters (1985–2007), by wolves (1995–2007), and total off-take to both causes (1985–2007). Panel B—estimated proportion of preseason adult female elk in the Gallatin Canyon population killed by hunters, by wolves, and total off-take due to both causes (1996–2007).

G. Elk Population Trend

Elk populations in most of Montana and the greater Yellowstone area have increased substantially since the 1940s, with some exceptions. Long-term trends were increasing for five of the seven populations we examined, with a downward trend for the Gallatin Canyon and Madison headwaters elk populations (Figure 25.13). Thus, our seven elk populations have mirrored regional long-term trends.

The long-term trend for the Madison headwaters elk population (Figure 25.13A, Chapters 11 and 23 by Garrott *et al.*, this volume) indicated relative long-term stability with a recent decrease following

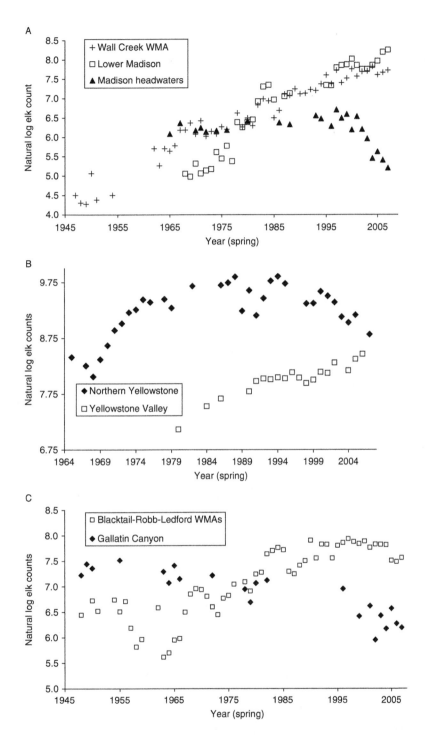

FIGURE 25.13 Long-term trend of elk population counts (natural log-transformed) for our seven study populations, 1947–2007. Panel A—trend of natural log-transformed counts of elk in Madison Valley populations: Madison headwaters, Lower Madison, and the Wall Creek Wildlife Management Area. All populations are within approximately 40 km. Panel B—trend of natural log-transformed counts of elk in Yellowstone Valley populations. The northern end of the northern Yellowstone elk winter range and southern end of the Yellowstone Valley winter range are immediately adjacent. Panel C—trend of natural log-transformed counts of elk in the Gallatin Canyon and the Blacktail–Robb Ledford Wildlife Management Area.

wolf recolonization, whereas there is a continuing increasing trend for elk wintering in the Lower Madison and an increasing trend with recent stability for the Wall Creek population. Both of these populations are only 40 km downstream from the Madison headwaters (Figure 25.13A, Garrott *et al.* 2005). The northern Yellowstone elk population (Figure 25.13B) increased from lows after intense culling during 1961–1968 to a peak in the mid-1990s and has decreased since that time. Elk counted in the immediately adjacent Yellowstone Valley population have continued to increase since 1980 (Figure 25.13B), exceeding Montana Fish, Wildlife, and Parks objective levels. Similar to the close-proximity Madison River elk populations, the two Yellowstone River populations in close proximity displayed differing population trends. The Gallatin Canyon elk population showed relative long-term stability, perhaps with a slight decrease through the 1980s. A reduction in long-term numbers from about 2500 to 1500 elk counted was an objective of Montana Fish, Wildlife, and Parks and late hunts were held, after elk migrated from Yellowstone. Recently, the Gallatin population has decreased substantially below objective (Figure 25.13C), and late hunts ceased in 2004 with only youth allowed to hunt antlerless elk during the general season. The elk population wintering on the Blacktail–Robb Ledford increased substantially since the 1960s. However, population growth has slowed or stabilized recently (Figure 25.13C), in line with Montana Fish, Wildlife, and Parks management objectives of a reduction in numbers through harvests.

Population growth rate estimates during 1994–2007 support these differing population trends among the seven areas (Figure 25.13), with substantial decreases in the Madison headwaters ($r = -0.085$, $\lambda = 0.919$), Gallatin Canyon ($r = -0.077$, $\lambda = 0.926$), and northern Yellowstone ($r = -0.070$, $\lambda = 0.933$) populations. Elk populations wintering in the Lower Madison ($r = 0.074$, $\lambda = 1.077$) and Yellowstone Valley ($r = 0.032$, $\lambda = 1.033$) increased during the post-wolf period. Elk populations on the Wall Creek ($r = 0.018$, $\lambda = 1.019$) and Blacktail–Robb Ledford ($r = -0.019$, $\lambda = 0.982$) were relatively stable (Figure 25.13).

IV. DISCUSSION

We detected strong demographic differences among the elk populations from the seven different areas examined during this study that we interpret were due to varying site-specific effects of predation, land use and management regimes, and landscape characteristics. The close geographic proximity between sites allowed for similar influential effects of broad-scale climatic patterns such as drought. Conversely, predator densities, moderated by land management practices and local climatic and landscape conditions, varied considerably among elk populations and the effect of wolves was best described by predator:prey ratios. Decreases in elk calf recruitment with increasing predator:elk ratios were the most pronounced demographic effect and contributed most significantly to population trends within sites. Other demographic parameters such as adult female survival were negatively affected by predator densities. However, pregnancy rates, nutrition, and calf weights were not related to predator:elk ratios. Consequently, these comparative results illustrate that management actions cannot be applied universally over a large geographical scale.

Based on average progesterone concentrations in fecal samples, Creel *et al.* (2007) argued that pregnancy rates decreased with increasing ratios of wolves to elk as a consequence of elk behavioral responses to wolf predation risk. However, the data reported here from directly observed pregnancy rates reported by hunters (Figure 25.6A and B) and PSPB assays of blood samples (Figure 25.6C) that were collected from many of the same elk populations during the same time periods sampled by Creel *et al.* (2007), did not find evidence for such a decrease. Post-wolf pregnancy rates, as estimated using PSPB assays on samples from northern Yellowstone, Gallatin Canyon, and Lower Madison elk (Figure 25.6C) were equal to, or higher, than the average 80.7% pre-wolf pregnancy rate for all elk (including yearlings) reported for seven elk populations across Montana (Hamlin and Ross 2002).

These results are not surprising because reductions in pregnancy are unlikely to occur unless nutritional limitation, including by stress or disruption of foraging, becomes severe (Fowler 1981, Gaillard *et al.* 1998, Eberhardt 2002). Zager *et al.* (2007a) also found that elk pregnancy rates in Idaho were within the normal species range regardless of predator abundance. Fecal hormone assays suggested a post-wolf decrease in pregnancy rates in the Madison headwaters. However, ancillary data from sightings of radio-collared cows with calves-at-heel, fall calf:cow ratios, and serum PSPB and progesterone assays provided convincing evidence that the extremely low pregnancy assessments for these years were erroneous (Chapter 23 by Garrott *et al.*, this volume). Many of the Madison headwaters fecal samples were assayed in the same lab and frequently intermixed with the fecal samples reported in Creel *et al.* (2007). We are not certain of the reasons for the discrepancy between pregnancy assessments presented by Creel *et al.* (2007) and this study, but suggest some unexplained failure of the fecal hormone assays for progesterone (*e.g.*, physiological binding of metabolic by-products of progesterone in the gut or interference by them such that progesterone was not recognized in the enzyme immunoassays; Chapter 23 by Garrott *et al.*, this volume). The results reported in Creel *et al.* (2007) may also have been confounded with the use of fecal samples that likely included calves, immature females, and males not adequately identified and censored with the procedures used in that study.

Survival of elk calves appeared to be the vital rate most affected by increasing numbers of wolves and the coincident increase in grizzly bears relative to numbers of elk (Figure 25.7). Also, relative numbers of predators to prey appeared to be more important than absolute numbers of predators in affecting elk calf survival. Thus, the lowest survival of elk calves occurred in the smaller elk populations (Madison headwaters and Gallatin Canyon) with relatively high numbers of both wolves and grizzly bears. These two elk populations also occurred in areas with the greatest mixture of conifer cover and scattered small openings, the overall greatest level of landscape heterogeneity, and most severe snow conditions. This increased edge and restricted winter distribution of elk possibly made searches for, and killing of, prey easier for wolves (Bergman *et al.* 2006, Chapter 24 by Garrott *et al.*, this volume). Regional studies of neonatal elk calf survival indicated that predation by bears accounted for most mortality during the first 30–45 days (Singer *et al.* 1997, Barber-Meyer 2006, Creel *et al.* 2007, Zager *et al.* 2007b). Predation by wolves occurred through the year, perhaps increasing after calves became more active and visible.

Zager *et al.* (2007b) concluded from a study employing experimental manipulation of predator numbers and control areas that predation of elk calves, primarily by black bears (*U. americanus*) and mountain lions (*Puma concolor*), was largely additive. Our information indicated that nutritional factors played little or no role in elk calf mortality (Figures 25.8–25.10), which also indicated that most or much of elk calf mortality resulting from predation in the post-wolf period was likely additive. Likewise, nutritional effects were not detected during intensive studies of the Madison headwaters elk population (Chapter 22 by White *et al.*, and Chapter 23 by Garrott *et al.*, this volume). A correlative analysis indicated that ratios of calves per 100 cows in northwestern Wyoming were depressed to the greatest extent within herds with the greatest wolf presence (Wyoming Game and Fish Department 2007). Smith *et al.* (2006b), however, suggested that increasing predation of elk calves in the Jackson elk herd during 1990–1992 and 1997–1999 was partially compensatory because of nutritional factors, though mortality rates were lower there than reported for northern Yellowstone (Singer *et al.* 1997, Barber-Meyer 2006) or Gallatin Canyon (Creel *et al.* 2007). Thus, conclusions are generally similar regarding the effects of predation on elk calf survival across a variety of study approaches and within the broader area of wolf restoration.

Other information also indicated that it was unnecessary to invoke reduced pregnancy as a cause for low observed elk calf survival because observed normal pregnancy rates combined with telemetry-determined mortality rates explained observed survival. Average pregnancy rates for northern Yellowstone elk during 2000–2006 were estimated at 82% using the PSPB assay. Approximate average annual mortality of neonatal elk calves during 2003 through 2005 (Barber-Meyer 2006) was 76%, with approximately 58% of mortality attributed to bears, 16% to wolves, 19% to other unidentified predators (including unidentified wolves and bears), 5% to natural causes, and 2% to hunting.

Thus, for every 100 cows, 82 calves were born, 62 (82 × 0.76) died during the year, and 20 were recruited into the population. Average observed recruitment during 2003–2004 through 2005–2006 for northern Yellowstone elk was 16 calves per 100 cows. Thus, the observed low recruitment level was substantially explained by observed normal pregnancy rates and telemetry-determined calf mortality rates when also considering normal sampling error inherent in determinations of pregnancy, mortality rates of calves and cows, and spring classifications of calves per 100 cows.

Survival of adult females in most large mammal populations is generally high without human harvest (Eberhardt 2002). For example, annual survival of adult female elk averaged 0.83 during 1989–1996 in the Gravelly-Snowcrest mountains (Wall Creek and Blacktail–Robb Ledford), and 91% of this mortality was hunting related (Hamlin and Ross 2002). High survival rates in the absence of hunting (0.015 natural mortality in Gravelly-Snowcrest mountains) allows for limited operation of compensation in adult female mortality at densities below resource limitation (Nichols 1991). Thus, both human harvest and increased predation would likely decrease adult female survival, especially given the mild winter conditions in the region since 1997.

Survival of adult female elk was significantly lower in the Gallatin Canyon after wolf colonization compared to the pre-wolf period (Figure 25.11, Table 25.3), despite a reduction in hunting mortality to less than 3% of preseason numbers after 2003. Survival of adult female elk was also significantly lower in the Madison headwaters area following wolf colonization (Chapters 11 and 23 by Garrott *et al.*, this volume). For the other areas, adult female survival was not significantly lower during the post-wolf period (Figure 25.11, Table 25.3). High rates of hunter harvest, mortality during the relatively severe winter of 1996–1997, and low recruitment contributed to decreasing elk populations in the Gallatin Canyon and northern Yellowstone during the early post-wolf years (Figures 25.12 and 25.13, Table 25.3). However, adult female survival did not increase in northern Yellowstone after a reduction in hunting mortality to less than 3% of preseason numbers following 2003–2004. An increase in the estimated kill by wolves (Figure 25.12, Table 25.3) suggested that, similar to the Gallatin Canyon and Madison headwaters populations (Chapter 23 by Garrott *et al.*, this volume), some wolf-kills of adult female elk were additive to other sources of mortality, at least after 2003.

Observed vital rates and their relative changes (Table 25.3) generally explained elk population trends (Figure 25.13). Recruitment of calves was consistently low in historical terms during the post-wolf (and increased grizzly bear) period, and substantially below replacement levels for concurrent adult female mortality in the Madison headwaters, Gallatin Canyon, and northern Yellowstone populations. This consistently low recruitment of calves during the post-wolf period appeared to be the major factor influencing population trends. Recruitment of calves decreased somewhat during the post-wolf period on the other areas, but not to the extent observed in the high wolf and bear areas (Table 25.3). Elk populations were stable to increasing in the Wall Creek, Blacktail–Robb Ledford (Gravelly-Snowcrest mountains), Lower Madison, and Yellowstone Valley areas where ratios of wolves and/or grizzly bears per 1000 elk did not consistently exceeded about 4–5 per 1000 elk. Thus, the presence of wolves and/or grizzly bears does not automatically signify significant consequences for elk populations, which can sustain at least moderate hunter harvest with low levels of wolf and bear predation.

Our conclusions are similar to those of many other studies of wolf-bear-ungulate systems in the north temperate region (Van Ballenberghe 1987, Ballard 1991, Gasaway *et al.* 1992, Crête 1998, 1999, Kunkel and Pletscher 1999, Hebblewhite *et al.* 2002, 2005). However, our findings do not support the conclusion of Vucetich *et al.* (2005) that hunting and weather alone explain the decrease in the northern Yellowstone elk. Their analysis through spring 2004 did not have the benefit of insight provided by the more recent reductions in hunter harvest. Also, their use of raw elk counts, uncorrected by sightability estimates, in combination with estimates of actual harvest overestimated the effects of hunting by 25–60%. We do not imply that hunting was not an important source of mortality, only that it was overestimated and other mortality sources, including the unexamined portion contributed by wolves and bears, are necessary to explain the observed population trend. High hunter harvest during 1997–2004 was implemented by Montana Fish, Wildlife, and Parks to reduce elk numbers wintering in

Montana to population objectives and, as intended, contributed to a significant decrease in the northern Yellowstone elk population. Population modeling that included effects of wolves along with winter severity/forage production and hunting (Varley and Boyce 2006) estimated about equal contributions by wolves and hunting (~38% each) and a contribution of approximately 23% by winter severity/forage production to the decrease in elk numbers in northern Yellowstone. The impact of bears on neonatal mortality of elk calves has generally been unaddressed in all models to date and should be considered in the future (Barber-Meyer 2006).

The effects of wolf restoration on elk populations in Montana and Yellowstone National Park have varied considerably by area, habitat, landownership, and management objectives. In Yellowstone and other public lands with few agricultural conflicts (*e.g.*, Gallatin Canyon), wolves increased to high densities relative to elk numbers. Also, grizzly bear numbers simultaneously increased where conflicts with humans were minimal. In these situations, elk numbers have decreased. In contrast, on public multiple-use lands surrounded by private agricultural lands, and in valleys that contain largely private agricultural ownership, wolves have consistently depredated domestic livestock and been intensively controlled. Though wolves have persisted in these areas, regular control actions have maintained them at low absolute numbers and, also, low numbers relative to elk. We did not detect significant effects of wolves on numbers of elk in these areas. In fact, elk populations have remained above Montana Fish, Wildlife, and Parks objective levels in the agricultural valleys of Montana.

The differences we observed in the effects of wolves on elk populations and agricultural operations across our study region adds to the caution by Garrott *et al.* (2005) that one management prescription will not apply in all areas. Montana Fish, Wildlife, and Parks and the public will need to consider ecological factors such as predator to prey ratios, calf survival, and adult female survival in determining appropriate hunting regulations for elk. Similar factors should be considered in any management of wolves in which Montana, Idaho, and Wyoming becomes involved. Landscape, social, and economic factors will also be important in determining viable and realistic management options for both wolves and elk. Where elk numbers are above population objectives in agricultural environments, wolves are unlikely to contribute significantly to elk population reduction and varying strategies to increase hunter harvest will be necessary. However, where elk populations are small and wolf and other predator numbers are high, hunter harvest regulations must carefully consider potential "surplus" levels. These types of areas may support some regulated hunter harvest of wolves. However, land ownership type, management objectives, and social considerations will be important in determining management options for wolves.

Our investigational process was observational and correlative and, therefore, there are limitations to the strength of our conclusions (Boutin 1992). However, preliminary results from a more experimental, manipulative, multiple-area approach in the region (Zager *et al.* 2007b) indicate similar conclusions to our correlative approach. We believe that we have contributed to increasing reliable knowledge (Romesburg 1981) by repeating observational studies on a broader landscape scale (*i.e.*, many places; Andersen *et al.* 2006) to aid in the sequential gain in understanding of ecological processes (Eberhardt 2003). Also, we believe that our results and conclusions can aid in the management of both wolves and elk as the process of gaining reliable knowledge continues.

V. SUMMARY

1. We documented the effects of wolf restoration on elk populations in the greater Yellowstone area, which varied considerably with variations in ecological, landscape, and land use factors.

2. We found no correlation between wolf:elk ratios and the proportion of adult cows pregnant. Pregnancy rates were uniformly high for all herds, approaching the maximal levels that could be expected for this species. Thus, reduced pregnancy was unlikely to have substantially contributed to low indices of recruitment (*i.e.*, ratios of calves per 100 adult females) observed in some herds after wolf establishment.

3. We found a strong negative correlation between the ratio of predators to prey and indices of calf recruitment and attribute this relationship to additive predation effects that reduced calf survival below levels that would have been experienced in the absence of predators.

4. There was some evidence the survival of adult female elk decreased at high numbers of wolves relative to elk, and that a portion of this increased mortality was likely additive to other causes.

5. Elk populations decreased in areas where combined high numbers of wolves and grizzly bears occurred in relation to numbers of elk. However, elk populations remained stable or increased where consistently low numbers of wolves and/or grizzly bears coexisted with elk and moderate levels of hunter harvest occurred.

6. The effects of wolves on elk populations varied depending on the predominant land use. Wolves reached high numbers relative to elk populations where preservation was the main land use (*e.g.*, Yellowstone) and/or there were few conflicts with agricultural activities (*e.g.*, Gallatin Canyon). However, in areas where agriculture was the predominant land use, consistent depredations by wolves resulted in control actions that maintained low wolf to elk ratios.

VI. REFERENCES

Andersen, R., J. D. Linnell, and E. J. Solberg. 2006. The future role of large carnivores in terrestrial trophic interactions: The northern temperate view. Pages 413–448 *in* K. Danell, R. Bergström, P. Duncan, and J. Pastor (Eds.) *Large Herbivore Ecology, Ecosystem Dynamics and Conservation.* Cambridge University Press, Cambridge, United Kingdom.

Ballard, W. B. 1991. Management of predators and their prey: The Alaskan experience. *Transactions of the North American Wildlife and Natural Resources Conference* **56:**527–538.

Barber-Meyer, S. M. 2006. *Elk Calf Mortality Following Wolf Restoration to Yellowstone National Park.* Thesis. University of Minnesota, St. Paul, MN.

Beasom, S. L. 1974. Relationships between predator removal and white-tailed deer net productivity. *Journal of Wildlife Management* **38:**854–859.

Bergerud, A. T. 1971. The population dynamics of Newfoundland caribou. *Wildlife Monographs* **25:**1–55.

Bergerud, A. T. 1978. Caribou. Pages 83–101 *in* J. L. Schmidt and D. L. Gilbert (Eds.) *Big Game of North America.* Stackpole Books, Harrisburg, PA.

Bergman, E. J., R. A. Garrott, S. Creel, J. J. Borkowski, R. Jaffe, and F. G. R. Watson. 2006. Assessment of prey vulnerability through analysis of wolf movements and kill sites. *Ecological Applications* **16:**273–284.

Boutin, S. 1992. Predation and moose population dynamics: A critique. *Journal of Wildlife Management* **56:**116–127.

Cada, J. D. 1985. Evaluations of the telephone and mail survey methods of obtaining harvest data from licensed sportsmen in Montana. Pages 117–128 *in* S. L. Beasom and S. F. Roberson (Eds.) *Game Harvest Management.* Caesar Kleberg Wildlife Research Institute, Kingsville, TX.

Caughley, G. 1977. *Analysis of Vertebrate Populations.* Wiley, New York, NY.

Cochran, W. G. 1983. *Planning and Analysis of Observational Studies.* Wiley, New York, NY.

Cole, G. F. 1971. An ecological rationale for the natural or artificial regulation of native ungulates in parks. *Transactions of the North American Wildlife Conference* **36:**417–425.

Coughenour, M. B., and F. J. Singer. 1996. Elk population processes in Yellowstone National Park under the policy of natural regulation. *Ecological Applications* **6:**573–593.

Craighead, J. J., J. R. Varney, and F. C. Craighead. 1974. A population analysis of the Yellowstone grizzly bears. Montana Forest and Conservation Experiment Station, School of Forestry, Bulletin No. 40. University of Montana, Missoula, MT.

Creel, S., D. Christianson, S. Lily, and J. A. Winnie,Jr. 2007. Predation risk affects reproductive physiology and demography of elk. *Science* **315:**960.

Creel, S., and J. A. Winnie, Jr. 2005. Responses of elk herd size to fine-scale spatial and temporal variation in the risk of predation by wolves. *Animal Behaviour* **69:**1181–1189.

Crête, M. 1998. Ecological correlates of regional variation in life history of the moose *Alces alces*. Comment. *Ecology* **79**:1836–1838.

Crête, M. 1999. The distribution of deer biomass in North America supports the hypothesis of exploitation ecosystems. *Ecology Letters* **2**:223–227.

Eberhardt, L. L. 1977. Optimal policies for conservation of large mammals with special reference to marine ecosystems. *Environmental Conservation* **4**:205–212.

Eberhardt, L. L. 1997. Is wolf predation ratio-dependent? *Canadian Journal of Zoology* **75**:1940–1944.

Eberhardt, L. L. 2002. A paradigm for population analysis of long-lived vertebrates. *Ecology* **83**:2841–2854.

Eberhardt, L. L. 2003. What should we do about hypothesis testing? *Journal of Wildlife Management* **67**:241–247.

Eberhardt, L. L., and J. M. Thomas. 1991. Designing environmental field studies. *Ecological Monographs* **61**:53–73.

Evans, S. B., L. D. Mech, P. J. White, and G. A. Sargeant. 2006. Survival of adult female elk in Yellowstone following wolf restoration. *Journal of Wildlife Management* **70**:1372–1378.

Fowler, C. W. 1981. Density dependence as related to life history strategies. *Ecology* **62**:602–610.

Gaillard, J. M., M. Festa-Bianchet, and N. G. Yoccoz. 1998. Population dynamics of large herbivores: Variable recruitment with constant adult survival. *Trends in Ecology and Evolution* **13**:58–63.

Garrott, R. A., J. A. Gude, E. J. Bergman, C. Gower, P. J. White, and K. L. Hamlin. 2005. Generalizing wolf effects across the greater Yellowstone area: A cautionary note. *Wildlife Society Bulletin* **33**:1245–1255.

Garrott, R. A., L. L. Eberhardt, P. J. White, and J. Rotella. 2003. Climate-induced limitation of a large herbivore population. *Canadian Journal of Zoology* **81**:33–45.

Gasaway, W. C., R. D. Boertje, D. V. Grangaard, D. G. Kellyhouse, R. O. Stephenson, and D. G. Larsen. 1992. The role of predation in limiting moose at low densities in Alaska and Yukon and implications for conservation. *Wildlife Monographs* **120**:1–59.

Gasaway, W. C., R. O. Stephenson, J. L. Davis, P. E. Sheppard, and O. E. Burris. 1983. Interrelationships of wolves, prey, and man in interior Alaska. *Wildlife Monographs* **84**:1–50.

Grigg, J. L. 2007. Gradients of predation risk affect distribution and migration of a large herbivore. Montana State University, Bozeman, MT.

Gude, J. A., R. A. Garrott, J. J. Borkowski, and F. King. 2006. Prey risk allocation in a grazing ecosystem. *Ecological Applications* **16**:285–298.

Hamlin, K. L. 2005. Monitoring and assessment of wolf–ungulate interactions and trends within the greater Yellowstone ecosystem and associated areas of southwestern Montana. Montana Fish, Wildlife, and Parks, Federal Aid in Wildlife Restoration Project, Helena, MT.

Hamlin, K. L., D. F. Pac, C. A. Sime, R. M. DeSimone, and G. L. Dusek. 2000. Evaluating the accuracy of ages obtained by two methods for Montana ungulates. *Journal of Wildlife Management* **64**:441–449.

Hamlin, K. L., and R. J. Mackie. 1989. Mule deer in the Missouri River Breaks, Montana: a study of population dynamics in a fluctuating environment. Montana Fish, Wildlife, and Parks, Federal Aid in Wildlife Restoration ProjectW-120-R-. Helena, MT.

Hamlin, K. L., and M. S. Ross. 2002. Effects of hunting regulation changes on elk and hunters in the Gravelly-Snowcrest Mountains, Montana. Montana Department of Fish, Wildlife, and Parks, Federal Aid in Wildlife Restoration ProjectW-120-R-. Helena, MT.

Hanks, J. 1981. Characterization of population condition. Pages 47–73 *in* C. W. Fowler and T. D. Smith (Eds.) *Dynamics of Large Mammal Populations*. Wiley, New York, NY.

Haroldson, M. A. 2006. Unduplicated females. Pages 11–16 *in* C. C. Schwartz, M. A. Haroldson, and K. West (Eds.) *Yellowstone Grizzly Bear Investigations: Annual Report of the Interagency Grizzly Bear Study Team, 2005*. U. S. Geological Survey, Bozeman, MT.

Haroldson, M. A., and K. Frey. 2006. Grizzly bear mortalities. Pages 25–30 *in* C. C. Schwartz and M. A. Haroldson (Eds.) *Yellowstone Grizzly Bear Investigations: Annual Report of the Interagency Grizzly Bear Study Team, 2005*. U.S. Geological Survey, Bozeman, MT.

Hebblewhite, M., D. H. Pletcher, and P. C. Paquet. 2002. Elk population dynamics in areas with and without predation by recolonizing wolves in Banff National Park, Alberta. *Canadian Journal of Zoology* **80**:789–799.

Hebblewhite, M., C. A. White, C. G. Nietvelt, J. A. McKenzie, T. E. Hurd, J. M. Fryxell, S. E. Bayley, and P. C. Paquet. 2005. Human activity mediates a trophic cascade caused by wolves. *Ecology* **86**:2135–2144.

Hobbs, N. T., and R. Hilborn. 2006. Alternative to statistical hypothesis testing in ecology: A guide to self teaching. *Ecological Applications* **16**:5–19.

Houston, D. B. 1982. *The Northern Yellowstone Elk*. Macmillan, New York, NY.

Keith, L. B. 1974. Some features of population dynamics in mammals. *Proceedings of International Congress of Game Biologists* **11**:17–58.

Knight, R. R., and B. M. Blanchard. 1996. *Yellowstone Grizzly Bear Investigations: Annual Report of the Interagency Study Team, 1995*. National Biological Service, Bozeman, MT.

Knight, R. R., B. M. Blanchard, and D. J. Mattson. 1990. *Yellowstone Grizzly Bear Investigations: Annual Report of the Interagency Study Team, 1989*. National Park Service, Bozeman, MT.

Krebs, C. J., S. Boutin, R. Boonstra, A. R. E. Sinclair, J. N. M. Smith, M. R. T. Dale, K. Martin, and R. Turkington. 1995. Impact of food and predation on the snowshoe hare cycle. *Science* **269**:1112–1115.

Kunkel, K. E., and D. H. Pletscher. 1999. Species-specific population dynamics of cervids in a multipredator system. *Journal of Wildlife Management* **63**:1082–1093.

Lemke, T. O., J. A. Mack, and D. B. Houston. 1998. Winter range expansion by the northern Yellowstone elk herd. *Intermountain Journal of Science* **4**:1–9.

Messier, F. 1994. Ungulate population models with predation: A case study with the North American moose. *Ecology* **75**:478–488.

Nichols, J. D. 1991. Responses of North American duck populations to exploitation. Pages 498–525 *in* C. M. Perrins, J. D. Lebreton, and G. J. M. Hirons (Eds.) *Bird Population Studies: Relevance to Conservation and Management*. Oxford University Press, Oxford, United Kingdom.

Palmer, W. C. 1968. Keeping track of crop moisture conditions, nationwide: The new crop moisture index. *Weatherwise* **21**:156–161.

Palmisciano, D. P. 1986. Wildlife investigations laboratory annual report. Montana Department Fish, Wildlife, and Parks, Federal Aid in Wildlife Restoration ProjectW-120-R-. Helena, MT.

Platt, J. R. 1964. Strong inference—Certain systematic methods of scientific thinking may produce much more rapid progress than others. *Science* **146**:347–353.

Quimby, D. C., and J. E. Gaab. 1957. Mandibular dentition as an age indicator in Rocky Mountain elk. *Journal of Wildlife Management* **21**:435–451.

R Development Core Team.2006. *R: A language and environment for statistical computing*. R Foundation for Statistical Computing, Vienna, Austria<http:www.R-project.org>.

Romesburg, H. C. 1981. Wildlife science: Gaining reliable knowledge. *Journal of Wildlife Management* **45**:293–313.

Sala, O. E., W. J. Parton, L. A. Joyce, and W. K. Lauenroth. 1988. Primary production of the central grassland region of the United States. *Ecology* **69**:40–45.

Sasser, R. G., C. A. Ruder, K. A. Ivani, J. E. Butler, and W. C. Hamilton. 1986. Detection of pregnancy by radioimmunoassay of a novel pregnancy-specific protein in serum of cows and a profile of serum concentration during gestation. *Biology of Reproduction* **35**:936–942.

Schwartz, C. C., M. A. Haroldson, and K. West (Eds.). 2007. *Yellowstone Grizzly Bear Investigations: Annual Report of the Interagency Grizzly Bear Study Team, 2006*. U.S. Geological Survey, Bozeman, MT.

Singer, F. J., A. Harting, K. K. Symonds, and M. B. Coughenour. 1997. Density dependence, compensation, and environmental effects on elk calf mortality in Yellowstone National Park. *Journal of Wildlife Management* **61**:12–25.

Smith, D. W., and D. S. Guernsey. 2002. Yellowstone wolf project: Annual report, 2001. YCR-NR-2002-03, Yellowstone Center for Resources. Yellowstone National Park, WY.

Smith, D. W., K. M. Murphy, and D. S. Guernsey. 2001. Yellowstone wolf project: Annual report, 2000. YCR-NR-2001-02, Yellowstone Center for Resources. Yellowstone National Park, WY.

Smith, D. W., D. R. Stahler, and D. S. Guernsey. 2003. Yellowstone wolf project: Annual report, 2002. YCR-NR-2003-04, USDI National Park Service, Yellowstone Center for Resources. Yellowstone National Park, WY.

Smith, D. W., T. D. Drummer, K. M. Murphy, D. S. Guernsey, and S. B. Evans. 2004a. Winter prey selection and estimation of wolf kill rates in Yellowstone National Park, 1995–2000. *Journal of Wildlife Management* **68**:153–166.

Smith, D. W., D. R. Stahler, and D. S. Guernsey. 2004b. Yellowstone wolf project: Annual Report, 2003. YCR-NR-2004-04, USDI National Park Service, Yellowstone Center for Resources. Yellowstone National Park, WY.

Smith, D. W., D. R. Stahler, and D. S. Guernsey. 2005. Yellowstone wolf project: Annual report, 2004. YCR-NR-2005-02, USDI National Park Service, Yellowstone Center for Resources. Yellowstone National Park, WY.

Smith, D. W., D. R. Stahler, and D. S. Guernsey. 2006a. Yellowstone wolf project: Annual report, 2005. YCR-NR-2006-04, USDI National Park Service, Yellowstone Center for Resources. Yellowstone National Park, WY.

Smith, D. W., D. R. Stahler, D. S. Guernsey, M. Metz, A. Nelson, E. Albers, and R. McIntyre. 2007. Yellowstone wolf project: Annual report, 2006. YCR-NR-2007-01, USDI National Park Service, Yellowstone Center for Resources. Yellowstone National Park, WY.

Smith, B. L., E. S. Williams, K. C. McFarland, T. L. McDonald, G. Wang, and T. D. Moore. 2006b. *Neonatal Mortality of Elk in Wyoming: Environmental, Population, and Predator Effects*. U.S. Department of Interior. U.S. Fish and Wildlife Service. Biological Technical Publication, BTP-R6007-2006, Washington, DC.

Stout, G. G. 1982. Effects of coyote reduction on white-tailed deer productivity on Fort Sill, Oklahoma. *Wildlife Society Bulletin* **10**:329–332.

U. S. Fish and Wildlife Service, Nez Perce Tribe, National Park Service, and USDA Wildlife Services. 2000. Rocky mountain wolf recovery 1999 annual report. , T. Meier (Ed.). 2000. United States Fish and Wildlife Service, Ecological Services, Helena, MT.

U. S. Fish and Wildlife Service, Nez Perce Tribe, National Park Service, and USDA Wildlife Services. 2001. Rocky mountain wolf recovery 2000 annual report. , T. Meier (Ed.). 2001. United States Fish and Wildlife Service, Ecological Services, Helena, MT.

U. S. Fish and Wildlife Service, Nez Perce Tribe, National Park Service, and USDA Wildlife Services. 2002. Rocky mountain wolf recovery 2001 annual report. , T. Meier (Ed.). 2002. United States Fish and Wildlife Service, Ecological Services, Helena, MT.

U. S. Fish and Wildlife Service, Nez Perce Tribe, National Park Service, and USDA Wildlife Services. 2003. Rocky mountain wolf recovery 2002 annual report. , T. Meier (Ed.). 2003. United States Fish and Wildlife Service, Ecological Services, Helena, MT.

U. S. Fish and Wildlife Service, Nez Perce Tribe, National Park Service, and USDA Wildlife Services. 2004. Rocky mountain wolf recovery 2003 annual report. , T. Meier (Ed.). 2004. United States Fish and Wildlife Service, Ecological Services, Helena, MT.

U. S. Fish and Wildlife Service, Nez Perce Tribe, National Park Service, Montana Fish,Wildlife and Parks, Idaho Fish and Game, and USDA Wildlife Services. 2005. Rocky mountain wolf recovery 2004 annual report., D. Boyd (Ed.). 2005. United States Fish and Wildlife Service, Ecological Services, Helena, MT.

U. S. Fish and Wildlife Service, Nez Perce Tribe, National Park Service, Montana Fish,Wildlife and Parks, Idaho Fish and Game, and USDA Wildlife Services. 2006. Rocky mountain wolf recovery 2005 annual report., C. A. Sime, and E. E. Bangs (Eds.) 2006. United States Fish and Wildlife Service, Ecological Services, Helena, MT.

U. S. Fish and Wildlife Service, Nez Perce Tribe, National Park Service, Montana Fish,Wildlife and Parks, Idaho Fish and Game, and USDA Wildlife Services. 2007. Rocky mountain wolf recovery 2006 annual report., C. A. Sime, and E. E. Bangs (Eds.) 2007. United States Fish and Wildlife Service, Ecological Services, Helena, MT.

VanBallenberghe, V. 1987. Effects of predation on moose numbers: A review of recent North American studies. *Swedish Wildlife Research* (Suppl 1):431–460.

Varley, N., and M. S. Boyce. 2006. Adaptive management for reintroductions: Updating a wolf recovery model for Yellowstone National Park. *Ecological Modeling* 193:315–339.

Vucetich, J. A., D. W. Smith, and D. R. Stahler. 2005. Influence of harvest, climate, and wolf predation on Yellowstone elk, 1961–2004. *Oikos* 111:259–270.

Wang, G., N. T. Hobbs, R. B. Boone, A. W. Illius, I. J. Gordon, J. E. Gross, and K. L. Hamlin. 2006. Spatial and temporal variability modify density dependence in populations of large herbivores. *Ecology* 87:95–102.

White, P. J., and R. A. Garrott. 2005. Yellowstone's ungulates after wolves—Expectations, realizations, and predictions. *Biological Conservation* 125:141–152.

Wyoming Game and Fish Department, Wildlife Division. 2007. *An Assessment of Changes in Elk Calf Recruitment Relative to Wolf Reestablishment in Northwest Wyoming*. Wyoming Game and Fish Department, Cheyenne, WY.

Zager, P., G. Pauley, M. Hurley, and C. White. 2007a. Study I. Population performance of mule deer and elk populations. Job 1. Survival, cause-specific mortality, and pregnancy rates of elk and mule deer in Idaho. Federal Aid in Wildlife Restoration, Job Progress Report, W-160-R-34. Idaho Department of Fish and Game, Boise, ID.

Zager, P., C. White, and G. Pauley. 2007b. Study IV. Factors influencing elk calf recruitment. Federal Aid in Wildlife Restoration, Job Progress Report, W-160-R-33, Subproject 31. Idaho Department of Fish and Game, Boise, ID.

APPENDIX

Appendix 25A.1

Description of Study Areas

We assessed the status of elk on seven areas with different predator guilds and management paradigms in the greater Yellowstone area of the western United States, including: the Madison headwaters (Madison-Firehole, MF) and Northern Range (NR) areas of Yellowstone National Park (YNP), the Wall Creek Wildlife Management Area (WCWMA) and Blacktail–Robb Ledford Wildlife Management Areas (BT-RL WMAs) of the Gravelly-Snowcrest Mountains, Gallatin Canyon (Gallatin), Lower Madison drainage (HD 362), and the Yellowstone Valley (HD 314) in Montana (Figure 25.2). The relative proximity of the 7 areas (most winter ranges within 80 km of each other, the furthest 115 km) results in similarities in broad climate patterns. Thus, all areas have experienced the recent period of drought and mild winters.

The Madison headwaters range is well described elsewhere in this book (Chapters 11 and 23 by Garrott *et al.*, this volume). It supported 200–600 elk along the confluence of the Firehole and Gibbon rivers and the uppermost stretches of the Madison River. This nonmigratory herd was managed under the National Park Service's policy of natural regulation, whereby elk were not culled and numbers fluctuated in response to weather, predators, and resource limitations (Cole 1971, Chapters 11 and 23 by Garrott *et al.*, this volume). Snow depths and winter conditions are the most severe of the 7 areas discussed here. Coniferous forest covered 82% of this range, half of which burned during summer 1988. The remainder consisted of meadows (14%) and geothermal areas (4%) that reduced snow depths for wintering ungulates (Garrott *et al.* 2003). Approximately 1500–2000 ungulates wintered in the area, 66% of which were bison (*Bison bison*) and 34% of which were elk. One wolf pack became resident in 1997 and by 2002 there were multiple packs totaling 30–40 animals.

The NR (Figure 25.2) has recently supported 6000–14,000 elk along a decreasing elevation gradient extending approximately 60 km along the Yellowstone River (Houston 1982, Lemke *et al.* 1998). These migratory elk were managed under the National Park Service's policy of natural regulation (Cole 1971). However, portions of the population that winter outside YNP boundaries have been subjected to varying levels of late season hunter harvest in Montana. Lower-elevation vegetation was primarily grassland and sagebrush steppe, with coniferous forests interspersed at higher elevations (Houston 1982). Other ungulates on this range included<150 bighorn sheep (*Ovis canadensis*), 700–1000 bison, <50 moose (*Alces alces*), <300 mule deer (*Odocoileus hemionus*), and 200–300 pronghorn (*Antilocapra americana*). The number of wolves residing in this area increased from 21 in 1996 to a high of 106 in 2003 (Smith *et al.* 2005). Hunter harvest of bulls is generally low, with the exception of 1991, when heavy harvest of bulls occurred as they migrated out of YNP due to early season snow conditions. Harvest of antlerless elk has varied considerably, from 1200–2400 during 1988–1989, 1994–1995, 1997–1998, 1998–1999, and 2000–2001 (10–13% of preseason estimates) to less than 140 antlerless elk during 2005–2006 and 2006–2007 (2%). An overall average antlerless harvest since 1986 is 6–7% of the preseason antlerless population.

The Wall Creek Wildlife Management Area (Figure 25.2) supported 1500–2700 migratory elk during winter on the west side of the Madison River Valley directly across the river from the Lower Madison drainage winter range (HD 362). This area was primarily managed to provide winter range for elk, though a rest-rotation cattle grazing system was also implemented in 1984. Vegetation on wintering areas was primarily open grassland benches and footslopes, which were kept relatively snow-free during most winters by strong winds. Summer and fall ranges throughout the Gravelly-Snowcrest mountains are a relatively open mixture of coniferous forests and grass-forb meadows (Hamlin and Ross 2002). Elk were the predominant ungulate, with less than 50 each of mule deer, white-tailed deer (*O. virginianus*), and pronghorn antelope also using the wintering area. Compared to elk, numbers of these species also are relatively low on summer and fall range, but additionally, a few moose occur on summer and fall range. Wolves were only transitory on the wintering area. Their attempts to establish a den site have been unsuccessful thus far for various reasons. However, wolves are established on summer and fall range of this population. The elk herd is relatively intensively hunted (10–12% of antlerless and about 50% of adult males, Hamlin and Ross 2002) to prevent long-term damage to vegetation and minimize complaints from private landowners.

The Blacktail–Robb Ledford Wildlife Management Areas (Figure 25.2) generally supported 1700–2700 elk during winter. Snow depths on winter range were generally low because periodic high winds blew much of the ridge top areas clear of snow during much of winter. Most of the valley and foothills were grassland with some inclusions of sagebrush. Similar to elk wintering on the WCWMA, elk on this area are scattered throughout a mixture of relatively open forests in the Gravelly-Snowcrest and Centennial mountains during summer and fall (Hamlin and Ross 2002). Elk comprised >85% of the wintering ungulate community. The remainder consisted of primarily mule deer, with a few white-tailed deer, pronghorn antelope, and moose. Up to 2 wolf packs totaling 3–14 wolves have used the area in a transitory manner since 2000. Locations from a pack member with a Global Positioning System radio-transmitter collar in 2005–2006 (5 wolves) indicated that about 34% of its time during winter was spent on and near the Blacktail–Robb Ledford WMAs. Some control of wolves because of conflicts with livestock has occurred in most years on and near this area. Hunter harvest of elk may average higher here than all the other areas, averaging about 15% of antlerless elk and 60% of antlered elk (Hamlin and Ross 2002).

The Gallatin Canyon range (Figures 25.2 and 25.5) historically supported 1000–3000 elk in the Daly/ Tepee Creek, Porcupine Creek (including the Porcupine WMA), and Taylor Fork drainages on a combination of national forest, national park, private, and state lands. Recently, less than 1000 elk were counted during aerial trend flights. Snow depths on winter ranges here are lower than in the Madison headwaters, but higher than all other areas described here. Valley bottoms were primarily sagebrush and grassland, with small riparian zones, coniferous forest, and small meadows on the slopes above (Creel and Winnie 2005). Elk comprised >90% of the wintering ungulate community, with the remainder consisting of about 50 moose, <100 mule deer, and a few white-tailed deer. At least one wolf pack totaling 8–13 wolves frequented this range since 1996–1997, though their presence was often transitory (Smith *et al.* 2005). In 2005 and 2006, 2 other wolf packs (6–10 wolves) also used the area on a consistent, but transitory basis. Some wolves in these packs were controlled when they moved into the Madison or Yellowstone Valleys and depredated livestock. Hunters harvested an average of 150 bull elk annually (about 45%) during 1994–2006. Harvest of antlerless elk averaged about 200 annually (about 15%) during 1994–1999, about 75 annually (about 8%) during 2000–2003, and was less than 30 annually (about 3%) during 2004–2006.

The Lower Madison drainage (HD 362, Figure 25.2 and 25.4), approximately 40 km downstream from the Madison headwaters range (Garrott *et al.* 2005) and directly east of and across the Madison River from the WCMWA, supported 1500–3500 migratory elk. This wintering area is a matrix of private (86%) and public (14%) lands managed primarily for livestock production, with 95% of the area supporting livestock at some time during the year. Average snow depth was low due to high winds that reduced snow accumulation (Gude *et al.* 2006). Grasslands comprised 58% of the area, with the remainder consisting of montane and riparian forests (32%) and sagebrush steppe (10%). Ungulates wintering in the area included 90% elk, 6% pronghorn, and 4% mule deer. Similarly, only low-moderate numbers of these other ungulates, including a few moose, occur on summer and fall range. Summer and fall range is a mixture of coniferous forest and meadows at higher elevations in the Madison-Hilgaard Range, western YNP, and portions of Idaho (Grigg 2007). One wolf pack became resident to the winter range in 1999, but the area never supported more than a single pack totaling 5 animals due to poor recruitment, low adult survival, and control due to livestock depredation (Garrott *et al.* 2005). Some other wolf packs, including some associated with the Gallatin elk herd, have also used the area on a transitory basis. Hunter harvest of bull elk varied from 100–200 annually (about 30%) during 1996–2006. Harvest of antlerless elk varied from 350–550 during the early 1990s (about 20%), to 50–100 (about 3%) during 1994–1996 and 160–320 annually (about 7%) during 1997–2006.

The western portion of the Yellowstone Valley is outside YNP, just northwest of the northern end of the NR, extending north to near Livingston, Montana (Figure 25.2). Winter range is mostly private lands and consists of open grasslands and foothills that are relatively snow-free during most winters due to strong winds. Summer and fall range includes much National Forest lands in the Gallatin range that are a mixture of coniferous forest and meadows. Little is known of extent of summer range for this population, but elk wintering in the southern portion of this area may overlap with portions of the Gallatin and NR herds in YNP during summer. During the period since wolf restoration, 3000–4800 elk were counted during aerial trend counts of this population. We do not have comparable population size information for mule deer and white-tailed deer, but annual harvests of each species are about 85% and 67%, respectively, of annual elk harvest. This indicated that the Yellowstone Valley has a greater compliment of other small ungulates in proportion to elk than the other 6 areas studied. Hunter harvest of bull elk ranged from about 170–450 since 1996 (about 40%) and harvest of antlerless elk has ranged from about 250–425 annually (about 9%).

Appendix 25A.2

Methods

Predator Numbers

The annual number of wolves estimated for each area and numbers killed in control actions or other known mortalities due to humans were determined from December numbers reported in the Cooperative Rocky Mountain Wolf Recovery Annual Reports (*e.g.*, U.S. Fish and Wildlife Service *et al.* 2007),

YNP research reports (Smith and Guernsey 2002; Smith *et al.* 2001, 2003, 2004a,b, 2005, 2006a, 2007), and consultations with field biologists from Montana Fish, Wildlife, and Parks. After consultation with Montana wolf management specialist Mike Ross, we estimated winter wolf numbers by elk range to fractions of 0.25 in some areas because some wolves and wolf packs used multiple elk ranges. For the Blacktail–Robb Ledford, we used Global Positioning System telemetry locations to estimate that the areas wolf pack spent 34% of the time with this elk herd and fractionalized annual wolf numbers for that pack accordingly. Similarly, intensive fieldwork in the Madison headwaters allowed estimation of wolf-days on the area and fractional wolf numbers equivalent to estimates for the other areas (Chapters 15 by Smith *et al.*, Chapter 16 and 17 by Becker *et al.*, this volume).

Since restoration, estimated numbers of wolves (time adjusted) associated with our 7 elk popula-tions (at least during winter) have ranged from: NR (21–106); Madison headwaters (0.4–23.7); Gallatin Canyon (1.5–17.75); Yellowstone Valley (0.5–10.5); BT-RL WMAs (1.01–10.03); Lower Madison (2–7); and WCWMA (minor transient use).

Wolf population growth rate relative to elk numbers followed an increasing linear trend for wolves associated with Madison headwaters (r = 0.45, λ = 1.57), Gallatin Canyon (r = 0.23, λ = 1.26), and Northern Range elk (r = 0.18, λ = 1.20). A minor increasing linear trend occurred (r = 0.07, λ = 1.07) for wolves associated with the Lower Madison elk, but total numbers were small (2–7). Numbers of wolves:1000 elk associated with the BT-RLWMAs and Yellowstone Valley increased to 2003–2004 and then declined. No wolves directly associated with the WCWMA elk wintering area have successfully denned and established a pack. Minor transient use of the WCWMA has occurred by wolves associated with the immediately adjacent Lower Madison elk population and other unknown wolves, and minor numbers of wolf-killed elk were observed. Similarly, relatively small numbers of wolves occur on summer/fall range of elk wintering on WCWMA (mostly wolves associated with BT-RL WMAs during winter).

Although absolute numbers of wolves on the Northern Yellowstone elk range were far above all other areas (peak = 106, next highest area = 23.7), numbers of wolves:1000 elk were highest on the Madison headwaters (peak = 85.3:1000), Gallatin Canyon (peak = 23.4:1000), followed by the NR (peak = 9.5:1000). Wolves:1000 elk on the other areas ranged from "none" to 2.6 wolves:1000 elk.

Numbers of grizzly bears also have made a remarkable recovery in the GYA from an estimated 136 (82–233) in 1974 (Craighead *et al.* 1974) to a minimum estimate of 405 in 2006 (Schwartz *et al.* 2007). Because bears, especially grizzly bears, have been implicated in substantial mortality of neonatal elk calves in this region (Singer *et al.* 1997, Barber-Meyer 2006, Creel *et al.* 2007), we thought it important to attempt to quantify relative grizzly bear numbers by area and year. Our index of grizzly bear numbers represents relative differences among areas and years rather than accurate, absolute numbers. We used information presented in the annual reports of the Interagency Grizzly Bear Study Team (*e.g.*, Knight *et al.* 1990, Knight and Blanchard 1996, Haroldson 2006, Haroldson and Frey 2006) to construct our index. Annual sightings of unduplicated females with cubs-of-the-year (COY) were a good predictor of the minimum population estimate for the entire Greater Yellowstone ecosystem (y = 7.5861x+56.363, R^2 = 0.85). Therefore, we used map locations of initial sightings of females with COY in the IGBST annual reports to partition sightings within elk calving/summer range among four of our study areas: the Gallatin Canyon, Northern Yellowstone, Lower Madison, and Madison head-waters. Bear Management Units (BMUs) 1 and 2 represented the Gallatin elk range, BMUs 3, 4, 5, 8, and 9 the Northern Yellowstone, BMUs 11 and 12 the Lower Madsion, and BMUs 10 and 13 the Madison headwaters. Some sightings occurred on or near BMU boundaries and we divided those fractionally among the areas. A high proportion of the sightings in BMU 10 (Madison headwaters) occurred near the eastern boundary. Thus, it is likely that grizzly bear numbers affecting the Madison headwaters elk calves may be somewhat overestimated in our index. However, to maintain consistency of application of technique to all areas we included (or fractionalized) all BMU 10 sightings in the Madison headwaters estimate. To help smooth estimates, we summed the proportions that females with COY in our four areas comprised of the entire GYA each year and multiplied the average

(42.4%) times the GYA annual minimum estimate to obtain a minimum annual total bear estimate for our 4 areas (4AAME). Further, because of the relatively long reproductive cycle of bears, we used a 3-year-running average (RA) of females with COY (3YRACOY) to smooth the calculated estimated bear numbers. To construct the 3-year-RA for 1987 and 1988, we assumed that system-wide bear numbers in 1985 and 1986 were equal to those in 1987. Estimated annual numbers of grizzly bears by area was: (3YRACOY area/sum 3YRACOY all 4 areas) × 4AAME. Similar to wolves, from this estimate, we calculated estimated number of grizzly bears:1000 elk by area and year.

Our estimated numbers for 1985–2006 ranged from 6–24 grizzly bears for the Madison headwaters, 32–104 for the Northern Yellowstone, 18–45 for the Gallatin Canyon, and 3–21 for the Lower Madison. Relative to elk numbers, however, peak numbers were 105.7 grizzly bears:1000 elk in the Madison headwaters, 56.9:1000 in the Gallatin Canyon, 9.25:1000 on the NR, and 7.1:1000 in the Lower Madison. Estimated rates of population increase from 1985–2006 were: Madison headwaters ($r = 0.099$, $\lambda = 1.10$), Gallatin Canyon ($r = 0.087$, $\lambda = 1.09$), Northern Yellowstone ($r = 0.087$, $\lambda = 1.09$), and Lower Madison ($r = 0.029$, $\lambda = 1.03$). We had no method to estimate grizzly bear numbers for the Yellowstone Valley because females with COY were not recorded there. We believe that the southern portion would be similar to the Gallatin Canyon, the middle portion similar to the Lower Madison, and the northern portion would contain no or few bears. Single grizzly bear are sometimes observed in the Gravelly-Snowcrest Mountains (WCWMA and BT-RL WMAs) but females with COY have not been observed there in historic times.

We had no data to estimate levels of other predators (black bear, mountain lions, coyotes) and assumed that they were at relatively the same level for all areas.

Pregnancy, early calf survival, and condition

Hunters at the Gardiner (Northern Yellowstone) and Gallatin check stations were asked to report the pregnancy, sex of fetus, and lactation status of their harvest. Although the pregnancy rates reported may be underestimates because hunters vary in their anatomical skills and also miss some small fetuses, this underestimate should be consistent among years and yield a good relative annual index of pregnancy. Check station personnel at the Gardiner and Gallatin check stations also weighed hunter-harvested calves to the nearest pound and we converted these weights to kg for analysis. Male calves consistently weighed more than female calves and we adjusted female weights upward based on this average difference to construct an annual average sex-adjusted weight for all calves.

When adult female elk were captured to attach radio-transmitter collars on our study areas, we collected blood samples and used pregnancy-specific protein B (PSPB) assays to assess pregnancy status (Sasser *et al.* 1986). Earlier tests, where both blood samples and reproductive tracts were obtained (Palmisciano 1986) or blood samples and rectal palpation occurred (Hamlin and Ross 2002), indicated 100% accuracy for PSPB assays to estimate pregnancy ($n = 78$).

The hunter-reported pregnancy data were binomial and required an analytical method accounting for nonnormal error structure. We used the logit-transform (as proportions) rather than logistic regression because, although sampling was random within years, it was nonrandom among years. For example, sample sizes for adult females from the northern Yellowstone population varied from 37–1456 among years and were related to factors that could affect pregnancy rates such as increased samples during severe winters and decreased sample sizes with increasing numbers of wolves. For the January 2003 hunter-reported sample, there were no pregnancies among 19 yearlings. We used 0.01 (lower than any other year) as the pregnancy rate for logit-transformation. We used logistic regression to analyze pregnancy rates estimated using the PSPB assay because samples were random and similar in size among years. We used R version 2.5.0 (R Development Core Team 2006) for all analyses.

We used the average PDSI during May and June as an index of soil moisture conditions during the early growing season (Palmer 1968). The PDSI incorporates precipitation, temperature, and evapotranspiration, all correlated with production and nutritive quality of plants (Sala *et al.* 1988). Thus, we expected PDSI to be positively related to elk nutrition and elk calf survival (*i.e.*, wet years and high PDSI = higher nutrition and elk calf survival).

All of our graphics display relationships relative to observed calves per 100 cows for ease of understanding. However, for statistical analysis, we logit-transformed calf:100 cow ratios as proportions.

Elk Population Size and Composition

Montana Fish, Wildlife, and Parks biologists conducted population trend counts for elk annually by using fixed-wing Piper Super Cub aircraft and Bell Jet Ranger helicopters. Counts were conducted as near as possible to the same mid-late winter time period each year. However, counts were not accomplished in some years due to weather conditions and scheduling conflicts. We recognize that single annual counts are particularly subject to substantial sampling variation among years, but variances of measurement errors of counts on the WCWMA, BT-RL WMAs, and NR were relatively small (Wang *et al.* 2006). For some analyses we present differences in natural log transformed raw counts, without corrections for missing years (*r*, Caughley 1977, $\lambda = e^r$), as measures of rates of change directly comparable among time periods. Where only a few years of missing counts occurred, for some other analyses we estimated the data for the missing year as the midpoint between the preceding and succeeding year. Where some indication of "true population size" was necessary (such as determining harvest rates where harvest estimates are of "true" numbers) and calculating predator:prey ratios, we used established correction factors to roughly estimate "true" population size.

For the Madison headwaters elk population, we used census figures reported elsewhere in this book (Chapters 11 and 23 by Garrott *et al.*, this volume). For the northern Yellowstone population, we used correction factors and estimates provided by Coughenour and Singer (1996) and Singer *et al.* (1997) through 1991–1992. After that period, we used the mean corrections factors established by sightability surveys (Coughenour and Singer 1996) to adjust counts upward by a factor of 1.322 (76% observed) during good survey conditions and 1.863 (54% observed) during poor survey conditions. For the WCWMA and BT-RL WMA populations, we used correction factors of 1.25 (80% observed) during good survey conditions, 1.67 (60% observed) during poor survey conditions, and 1.43 (70% observed) for average conditions (Hamlin and Ross 2002). These ranges were very similar to corrections determined for the Northern Yellowstone population (Coughenour and Singer 1996). For 1990 and 2005, when aerial counts were not conducted on the WCWMA, we used the high ground counts conducted by MFWP to represent the expected aerial count. Historically, aerial counts on this area averaged 110% of high ground counts conducted near the same time (Hamlin and Ross 2002). For 2003, when no aerial count was conducted on the BT-RL WMAs, we estimated the aerial count as the average of the prior and following year. For the Lower Madison and Yellowstone Valley, where terrain and cover conditions are similar to WCWMA and BT-RL WMAs, we used 1.25 (80% observed) as the correction factor for all years. For the Gallatin, aerial counts were not available for the period 1982–1995 and we used only counts from 1996 to 2007 for comparisons. For 1997, 2000 and 2002, when counts were not conducted, we estimated the aerial count as the average of the prior and following year. Because the Gallatin winter range has a greater portion of timber cover than the other Montana winter ranges, we used 1.43 (70% observed) as the correction factor for all years when estimating the "true" population.

Total numbers and bulls were recorded for all aerial counts of elk; for many, numbers of calves and cows were also recorded. However, for many areas and years, separate classifications of cows and calves were conducted from the ground. Where a mixture of techniques occurred, we used aerial counts to determine proportions of bulls and ground classifications to determine calf:100 cow ratios. For a few years in the Northern Yellowstone, Gallatin, and Lower Madison, calf:100 cow classifications were not available, but age structure data were available from check stations. We did not directly use calves:100 cows harvested to estimated calves:100 cows because hunters may select against calves

(Hamlin and Ross 2002). For those years, we used proportion of yearling cows in the next years harvest (YC_{t+1}) to estimate the previous years female calf:100 adult cow recruitment. To YC_{t+1} we added estimated recruited male calves (MC_{t+1}), where $MC_{t+1} = (YC_{t+1})(MC_t/FC_t)$. Thus, ($YC_{t+1}$+ MC_{t+1}):100 cows ≥2-years = recruitment of total calves per 100 adult females (t). This estimate could be slightly lower than comparable calf:100 cow classifications because it included any mortality occurring between the end of March and the next hunting season. We expect this mortality to be low in most years excepting those with severe winters.

To estimate preseason adult female numbers for comparison of hunter harvest and wolf-kill rates, we multiplied raw post-season counts by correction factors for sightability and used classification data to estimate the proportion/number that were adult females, calves, and males. To this estimate for adult females, we added estimated harvest of adult females, adjusted for estimated wounding loss (see later), which became the estimated preseason numbers of adult female elk.

Adult Female Elk Survival

Ages of hunter-harvested elk were estimated by eruption-wear (Quimby and Gaab 1957) and recorded at MFWP check stations. Increasingly since 1989, incisors also were collected for elk over 2 years of age and ages determined by examination of annuli in the cementum of incisor root tips (Hamlin *et al.* 2000). For years when only eruption-wear ages were available and for portions of annual samples that included some elk aged by eruption-wear only among samples mostly consisting of incisor annuli aged elk, we constructed area and time period specific matrices (*e.g.*, see Table 3, Hamlin *et al.* 2000) to proportionally correct eruption-wear ages to their expected distribution as incisor determined ages. We then added incisor-aged and corrected eruption-wear ages for the time period samples. We natural log transformed the age–frequency distributions of adult females recorded at check stations and fit linear regressions against age (Hamlin and Ross 2002, White and Garrott 2005). We used the slope of these regressions to estimate survival ($S = e^\beta$, where e = inverse of natural log and β = slope of the regression line). For years and areas when population trend did not indicate relative stability (Caughley 1977), we estimated rate of change in population size by regressing natural log of raw population counts against year to determine instantaneous rate of change (*r*). We then adjusted age–frequency distributions by the equation, adjusted number of age x = $f_x e^{rx}$ (Caughley 1977). We then proceeded as above to use slopes of the regression of natural log of age distribution against age to determine the estimate of survival adjusted for population growth or decline. To test for differences in survival among periods by area and period we used the different time periods and areas as factors in multiple linear regression models (Program R) allowing interactions between ln(count) and age. We compared periods to each other and areas within periods to each other using all-subsets regression. The *P*-value of the regression coefficients on the interaction terms represented the significance level of the difference between the survival rate for the periods and areas. The age structure technique for estimating survival appears to be relatively accurate. During 1989–1996, for the Gravelly-Snowcrest Mountains, the age–frequency technique for determining survival closely matched survival data determined by telemetry (Hamlin and Ross 2002). Also, survival we determined by age structure for the Northern Yellowstone during 2001–2006 (0.81, this study) was the same as telemetry-determined survival (0.80) during March 2000 to February 2004 (Evans *et al.* 2006).

Hunter Harvest

We used annual reports of the Montana Statewide Harvest Questionnaire (SWHQ, Cada 1985, MFWP annual reports) to estimate hunter harvest as calves, adult females, and adult males. Additionally, for the Northern Yellowstone, annual reports were available for the mandatory check of late season elk harvest (T. Lemke, MFWP unpublished reports). Only raw results from the hunter questionnaire sample, without multiplication by expansion factors, were issued for 1998. No elk harvest report was issued for 1997. For 1998, we averaged expansion factors reported for each hunting district in the 1996 and 1999 reports and multiplied those by the reported raw figures to estimate 1998 harvest. To estimate harvest for 1997 we used linear regression of harvest recorded at check stations against harvest estimated by the SWHQ for prior and subsequent years to estimate expected 1997 SWHQ harvest. For HD 314, where no check station was available, we used the mean estimate from all other areas to estimate SWHQ harvest (1997 SWHQ harvest = 0.471×1996 SWHQ harvest).

Hamlin and Ross (2002) reported that to account for wounding loss in the Gravelly-Snowcrest Mountains, it was necessary to multiply kill reported by the SWHQ by 1.22 for antlerless elk and by 1.1 for bulls. This figure likely varies among areas, years and conditions. Here, to estimate total loss due to hunting, we multiplied SWHQ reported kill by 1.15 for antlerless and 1.075 for bulls. For the more controlled Northern Yellowstone late hunt, we multiplied antlerless harvest by 1.1 and bull harvest by 1.05.

Most harvest for the WCWMA elk population occurred in HD 323 and most for the BT-RL WMAs population occurred in HD 324. However, because harvest of these populations also occurred in other Gravelly-Snowcrest hunting districts (Hamlin and Ross 2002) we partitioned harvest within all Gravelly-Snowcrest hunting districts by the percent distribution of WCWMA and BT-RL WMAs female elk during the hunting season (Hamlin and Ross 2002, p. 142, Table 9.2). We then summed harvests attributable to the WCWMA and BT-RL WMAs populations across hunting districts 323, 324, 325, 326, 327, and 330 for the total harvest estimate for each population.

Estimates of Wolf-killed Adult Female Elk

We used estimates provided in Smith *et al.* (2004a) and Becker *et al.* (Chapter 17 by Becker *et al.*, this volume) to estimate total wolf-killed elk as 0.060 elk killed per wolf day during the 181 day (1 November–30 April) winter period and reduced summer kill rates (184 days, 1 May–31 October) to 70% (0.042 elk killed per wolf day) of winter rates (Messier 1994). We used 25% (0.015 adult female elk killed per wolf day in winter and 0.0105 in summer) as the estimate for percent of wolf-kill that was adult females based on averages in reports for the Northern Yellowstone (Smith *et al.* 2004b, 2005, 2006), Gallatin Canyon (Creel *et al.* 2007), and Lower Madison (Hamlin 2005, Garrott unpublished data). Thus, estimated wolf-killed adult female elk annually by area was: [(Number of wolves)× (0.015×181)+(Number of wolves) × (0.0105×184)].

Wolf Survival/Mortality

Smith *et al.* (2006a) indicated that during the first 9 years of wolf restoration to YNP (including both the NR and Madison headwaters) wolf mortality averaged 18% per year and dispersal averaged 28% per year, totaling 45% annual loss to the population. This total was more than exceeded by production and survival of pups in most years, resulting in increasing populations. Except for the Sheep Mountain pack, wolves associated with YNP were not subject to agency control actions.

We did not have telemetry-determined survival information for wolves on our other areas, but information on human-caused mortality for wolves during 1999–2006 (U.S. Fish and Wildlife Service *et al.* 2000, 2001, 2002, 2003, 2004, 2005, 2006, 2007) in these areas indicated the relative degree of human-caused (agency lethal control, private citizen 10j regulation removal, and other unknown/suspected illegal kill) mortality varied substantially among the populations. Human removal accounted for 34 (48.5%) of 70.17 cumulative wolves associated with the BT-RL WMA during years they were present. Percentages for other areas were: 33.5% (18.5/55.25, Lower Madison), 21.2% (11/52, Yellowstone Valley), 10.9% (10.5/96.5, Gallatin), and 4.3% (27/626, Northern Yellowstone).

Human removal of wolves did not occur within the Madison headwaters elk range. The high human removal rates in addition to unknown natural mortality and dispersal for the small populations associated with BT-RL WMAs, Lower Madison, and Yellowstone Valley likely accounts for their low/negative growth rates compared to the NR, Madison headwaters, and Gallatin where human removal was none to very low.

Section 6

Human Wildlife Interactions

CHAPTER 26

Wildlife Responses to Park Visitors in Winter

P. J. White,* John J. Borkowski,[†] Troy Davis,* Robert A. Garrott,[‡]
Daniel P. Reinhart,* and D. Craig McClure*

*National Park Service, Yellowstone National Park
[†]Department of Mathematical Sciences, Montana State University
[‡]Fish and Wildlife Management Program, Department of Ecology, Montana State University

Contents

Theme

Executive Orders and regulations of the National Park Service stipulate that snowmobile use must cease if it is determined to be causing considerable adverse effects to resources in national park units, including the disturbance of wildlife or their habitats (Exec. Order No. 11644 § 3(2); Exec. Order No. 11989 § 2; 36 C.F.R. § 2.18(c)). In 2000, the National Park Service concluded that "snowmobile use at current levels adversely affects wildlife" in Yellowstone National Park and proposed the elimination of this activity (65 Federal Register 80915). However, less than 3 years later the Service proposed to continue snowmobile recreation based on new requirements for cleaner, quieter snowmobiles and guided group tours (68 Federal Register 51533, 69268). The U.S. District Court for the District of Columbia chastised the Service for this reversal and mandated the elimination of snowmobiles which, in turn, led to additional litigation, conflicting legal decisions (District of Columbia 2003, District of Wyoming 2004), and development of another final Environmental Impact Statement and Record of Decision to implement revised winter use plans (72 Federal Register 70781). We sampled the behavioral responses of bald eagles (*Haliaeetus leucocephalus*), bison (*Bison bison*), elk (*Cervus elaphus*), and trumpeter swans (*Olor buccinator*) to snowmobiles and coaches in Yellowstone to: (1) document human activities associated with snowmobiles and coaches, (2) quantify responses of these wildlife to human activities,

The Ecology of Large Mammals in Central Yellowstone
R. Garrott, P. J. White and F. Watson
ISSN 1936-7961, DOI: 10.1016/S1936-7961(08)00226-1

Copyright © 2009, Elsevier Inc.
All rights reserved.

(3) identify conditions that increased the likelihood of wildlife responses, and (4) assess the effects of human disturbance on their distribution and demography.

I. INTRODUCTION

Shifts in the distributions of birds of prey, ungulates, and waterbirds away from areas of human disturbance have been widely documented (Klein *et al.* 1995, Stalmaster and Kaiser 1998, Rowland *et al.* 2000). The magnitude of potential avoidance by wildlife, especially along roads or waterways used by motorized vehicles, appears to be influenced by the frequency of human use or rate of traffic (Stalmaster and Kaiser 1998, Wisdom *et al.* 2005b). If human disturbance significantly alters the activities or distributions of animals, then there may be increased vigilance and movement levels, reduced foraging rates, and reduced levels of parental care; each of which could reduce reproduction, survival, and population size (Stalmaster and Kaiser 1998, Duchesne *et al.* 2000, Steidl and Anthony 2000, González *et al.* 2006). As a result, extensive closures of areas or roads have been implemented to mitigate the presumed reduction in wildlife use of habitat near areas of human disturbance (Gill *et al.* 2001b, Wisdom *et al.* 2005a).

Similar to predation risk, an animal's decision whether to move away from a disturbed area is influenced by factors such as the quality of the occupied site, the distance to and quality of other suitable sites, the relative risk of predation or competition at alternate sites, the investment the individual has made in a site (*e.g.*, energy expenditure), and the individual's nutritional condition (Gill *et al.* 2001b). Animals should move away from disturbed sites if fitness costs of disturbance are high or, alternatively, fitness costs of disturbance are low but there is a high availability of alternate habitat elsewhere. However, there may be little change in numbers with increasing disturbance if there is little alternate habitat where animals can move (Gill *et al.* 2001b). Thus, studies of disturbance need to identify major factors influencing the behavior and distribution of affected species and assess the role of disturbance in altering these relationships (Gill *et al.* 2001a,b; Wisdom *et al.* 2005b).

Human disturbance of wildlife is widely considered a serious conservation problem when animals demonstrate negative responses to human presence (Boyle and Sampson 1985, Klein *et al.* 1995, Knight and Gutzwiller 1995). However, behavioral responses and apparent avoidance may not reduce the number of animals supported in an area because animals displaced from disturbed sites in the short term may return later, resulting in the overall use of these sites being unchanged (Gill *et al.* 2001a). Also, animals may habituate to human disturbances by not responding to repeated stimuli that are not biologically meaningful. Animals are more likely to become conditioned to human activity when it is controlled, predictable, and not harmful to the animals (Schultz and Bailey 1978, Thompson and Henderson 1998). Predictability permits animals to live at the lowest possible energy costs, while conserving energy and nutrients for reproduction and survival (Geist 2007).

However, even if wildlife demonstrate few movement responses to recreational activities, disturbances could increase energetic costs in relation to their total daily energy expenditures or cause physiological responses such as elevated heart rate, blood pressure, breathing rate, and release of adrenaline. The adrenal cortex secretes glucocorticoids that alter metabolic pathways for the production of adenosine 5′-triphosphate (ATP) and divert energy from physiological processes not required for immediate survival. The secretion of glucocorticoids is beneficial to an animal in the short term, but chronic elevation can inhibit digestion and growth, result in decreased resistance to disease, and produce an array of pathologies, including reproductive suppression, ulcers, and muscle wasting (Munck *et al.* 1984). Thus, frequent, severe, or prolonged responses to human disturbance could have fitness costs and studies of disturbance should evaluate if disturbances adversely affect the demography or population dynamics of wildlife (Gill *et al.* 2001b).

We sampled the behavioral responses of bald eagles, bison, elk, and trumpeter swans in Yellowstone National Park to snowmobiles and coaches during 2002–2003 through 2005–2006 to test if the odds of eliciting a response by these wildlife increased as: (1) wildlife group size and distance from road decreased, (2) human–wildlife interaction time and numbers of snowmobiles and coaches increased, (3) human activity became more directed and pronounced towards wildlife, (4) wildlife were in open habitats compared with forested habitats, and (5) cumulative traffic increased. We also assessed the effects of recreation on the distribution and demography of these wildlife.

II. METHODS

Human activity in the Madison headwaters area occurred in a predictable pattern during winter. A network of paved, two-lane roads paralleled the Firehole, Gibbon, and Madison Rivers through the study area. Vehicular travel was restricted to paved, two-lane roads that were closed in early November each year, and wheeled traffic was limited to park staff. Once sufficient snow accumulated, traffic on the roads transitioned from wheeled vehicles to snowmobiles and coaches. Roads were groomed (*i.e.*, snow packing) at least every other night and the park was open to public snowmobiles and coaches during mid-December through mid-March. Over-snow vehicle (OSV) traffic each winter consisted of commercially guided groups of snowmobiles or coaches, unguided groups of snowmobiles, and administrative snowmobiles and coaches operated by park staff and concessionaires. Most groups of snowmobiles were unguided during 2003, but in winter 2004 the park initiated regulations requiring all groups of recreational snowmobiles to be guided by a trained operator. Roads were plowed from early March through early April, after which they were only open to wheeled vehicles operated by park staff or concessionaires. Roads were opened to public wheeled vehicles on the third Friday of April for the summer season.

Surveys were conducted at least twice a week during daylight by a pair of observers snowmobiling ≤50 km/h along the following nine road segments: (1) Madison to West Yellowstone (22 km); (2) Madison to Old Faithful (26 km); (3) Madison to Norris (23 km); (4) Norris to Mammoth (26 km); (5) Canyon Village to Lake Butte (26 km); (6) Fishing Bridge to West Thumb (34 km); (7) Canyon Village to Norris (19 km); (8) Old Faithful to West Thumb (27 km); and (9) South Entrance to West Thumb (43 km). Segments 1–4 were used frequently by snowmobiles and coaches (collectively referred to as OSVs) and surveyed during all winters. Lower-use segments 5–6 were surveyed during 2002–2003 through 2004–2005. Segment 7 was only surveyed in 2002–2003, while segments 8 and 9 were only surveyed in 2003–2004.

Our sampling unit was an interaction between a group of OSVs and associated humans and a group of wildlife within 500 m of the road. Observers traveled a given road segment until a group (*i.e.*, ≥1 animal) of wildlife was detected. The observers stopped at a location where approaching OSVs could be observed without disturbing the animals. For each interaction, observers recorded the most extreme human group activity as: (1) no visible reaction; (2) stopped to observe animals; (3) dismounted or exited the OSV; (4) approached animals on foot; or (5) impeded or hastened movement by chasing animals or by forcing animals ahead of vehicles. Observers recorded the response behaviors of wildlife as: (1) no visible reaction to vehicles or humans; (2) looked at OSVs or humans and then resumed their behavior; (3) alert posture, including rising from bed or agitation; (4) traveled away from OSVs or humans; (5) flight (*i.e.*, quick movement away); or (6) defense (*i.e.*, attacked or charged). They also recorded the number of OSVs and if the group consisted of commercially guided snowmobiles or coaches, unguided snowmobiles, or administrative snowmobiles and coaches operated by park staff and concessionaires. Once an interaction was complete, the observers continued the survey along the road segment to locate the next group of wildlife.

We modeled the most common wildlife group response observed during a human–wildlife interaction. We combined the travel, flight, and defense responses ($n = 486$) into a single "movement" category that represented activities requiring the greatest amount of energy expenditure. We combined the alert and look-and-resume responses ($n = 1556$) into a single "vigilance" category because they required less energy than a movement response. If the wildlife group did not respond to the human activity ($n = 3646$), then it was categorized as "none." We obtained daily measurements of snow water equivalent from the Natural Resources Conservation Service's automated SNOTEL site at West Yellowstone (2042 m) and summed measurements from October 1 through April 31 for each winter to obtain a daily cumulative value (Garrott *et al.* 2003). We also summed the daily number of OSVs entering the west, south, or east gates for each day of the winter recreation season. Data from 25 surveys in winter 2002–2003 were inadvertently omitted from analyses. The dates and results of these surveys are summarized in Appendix 26A.1.

A. Model Formulation

Candidate sets of multinomial logistic regression models were formulated from a base model with terms that were included in every *a priori* model based on their importance in previous analyses (Borkowski *et al.* 2006). Base model terms included: (1) year; (2) animal group size; (3) distance from road; (4) human group activity; (5) human–wildlife interaction time; (6) number of snowmobiles; (7) number of coaches; (8) group size by distance interaction (bison, elk, and swans only due to sample size constraints); and (9) an on-road indicator for the wildlife group. We then added combinations of the following covariate effects to this base model: (1) habitat; (2) cumulative OSV visitation; and (3) the interactions group size by habitat, year by cumulative visitation, distance by human activity, group size by human activity, interaction time by distance, and/or interaction time by group size.

Multinomial logits regression (Hosmer and Lemeshow 2000) was used to fit each model for each species because there were three response categories (none, vigilance, and movement). Two logits $L_a(\mathbf{x}) = \log\left[\pi_a(\mathbf{x})/\pi_2(\mathbf{x})\right]$ ($a = 0, 1$) were modeled where $\pi_0(\mathbf{x})$, $\pi_1(\mathbf{x})$, and $\pi_2(\mathbf{x})$ were the probabilities of a movement response, vigilance response, and no response, respectively, given $\mathbf{x} = (x_1, x_2, \ldots, x_p)$ is a vector of model covariates. We treated no response as the baseline response by selecting $\pi_2(\mathbf{x})$ to be in the denominator of each odds ratio. We fit the logit parameters using the SAS LOGISTIC procedure and used Akaike's Information Criteria corrected for small sample size (AIC_c) as model-selection criterion (Burnham and Anderson 2002, SAS 2002).

We postulated three forms of logit model effects ($\beta_i x_i^*$) for quantitative covariates x_i, including $x_i^* = x_i$ (linear), $x_i^* = x_i$ (moderated), and $x_i^* = x_i$ for $x_i \leq T_i$ and $x_i^* = T_i$ for $x_i > T_i$ (threshold, where T_i is the maximum value). We created a candidate set of 86 multinomial logit models for bison and elk, 36 models for swans, and 12 models for bald eagles. The general form for each fitted logit model was $\hat{L}_a(x) = b_{0a} + \sum_i b_{ia} x_i^* + \sum_i \sum_j b_{ija} x_i^* x_j^*$, where $a = 0, 1$. For each model there were $3^6 = 729$ possible combinations of quantitative covariate effect forms. The x_i^* (or x_j^*) in each multinomial logits model for the categorical variables year, on-road, habitat, and human activity, were indicator variables corresponding to categorical levels of that variable.

We used a sequential model selection process to address the problem of high dimensionality (Borkowski *et al.* 2006). We initially fitted all *a priori* models with linear forms for the quantitative covariates and selected the "best" models for each species having the smallest AIC_c values. Next, we replaced the linear form of one covariate with its moderated form and calculated AIC_c values for these models. Similarly, the same linear effect was replaced with a threshold form and AIC_c values for the resulting models were calculated. We estimated the threshold value by checking a set of potential threshold values and retaining the one that yielded the lowest AIC_c value. Once AIC_c values corresponding to the best models for linear, moderated, and threshold forms of the first covariate were determined, variable forms yielding models with the best AIC_c values were selected and the sequential assessment process was repeated for the second quantitative covariate, yielding another set of AIC_c values. This process was repeated until all quantitative covariates were examined with linear, moderated, and threshold forms. The final models were assessed to determine relationships between covariates, interactions, and the wildlife group responses.

We conducted *post hoc* analyses to see if removing model main effects or interactions with large *P*-values improved the current best AIC_c value. Also, we combined two levels of a categorical covariate having similar parameter estimates. We replaced cumulative visitation with the quantitative covariate cumulative snow water equivalent because of a strong correlation between these covariates. In addition, we included the categorical covariate human group type indicating whether OSVs were driven by park staff and concessionaires (*i.e.*, administrative traffic) or commercially guided or unguided visitors.

B. Odds Ratios

We used parsimonious *post hoc* models to generate odds ratios from the logit estimates. For covariate vectors $\mathbf{x}_1$ and $\mathbf{x}_2$, the odds of a wildlife group response requiring energy expenditure for movement and a wildlife group response requiring negligible or no energy expenditure under condition $\mathbf{x}_1$ to that under condition $\mathbf{x}_2$ is denoted $OR_1(x_1, x_2) = \left((\hat{\pi}_0(x_1)/\hat{\pi}_2(x_1))/(\hat{\pi}_0(x_2)/\hat{\pi}_2(x_2)) \right)$. Likewise, the odds of a wildlife group response requiring energy expenditure for vigilance and a wildlife group response requiring negligible or no energy expenditure under condition $\mathbf{x}_1$ to that under condition $\mathbf{x}_2$ is denoted $OR_2(x_1, x_2) = \left((\hat{\pi}_1(x_1)/\hat{\pi}_2(x_1))/(\hat{\pi}_1(x_2)/\hat{\pi}_2(x_2)) \right)$. For quantitative covariate estimates and their interactions, the odds ratios (OR) associated with one-unit of measurement increases were found by exponentiation of the parameter estimate (*i.e.*, $OR = e^{estimate}$). We took reciprocals (1/OR) to get the odds ratios associated with one-unit decreases (*i.e.*, $1/OR = 1/e^{estimate} = e^{-estimate}$). For categorical variable estimates and their interactions, the odds ratios were found by exponentiation of the differences in estimates between the effects of interest and the baselines (*i.e.*, $OR = e^{effect-baseline}$ or $1/OR = e^{baseline-effect}$, where the baseline was on the road for the on-road indicator, no response for human activity, and meadow for habitat). The OR value compares an effect to the baseline, while the 1/OR value compares a baseline to the effect. No response was the baseline group response in all analyses.

C. Abundance, Demography, and Displacement of Wildlife

Bald eagles, elk, and trumpeter swans in the west-central portion of Yellowstone were not subject to human harvests or management removals. Thus, we regressed estimates of abundance for these species (Garrott *et al.* 2005, McEneaney 2006) directly on cumulative visitation each winter during 1965–1966 through 2005–2006. Conversely, there were large, irregularly spaced management removals of central Yellowstone bison that significantly influenced their population trend during this period (Fuller *et al.* 2007). We removed the effects of these culls from the population trend by calculating annual growth rates (r_t) as $r_t = \ln(\text{count}_t) - \ln(\text{count}_{t-1} - \text{removals}_{t-1})$. We regressed these growth rates on cumulative visitation each winter during 1965–1966 through 2005–2006.

We also regressed counts of fledgling eagles and cygnets (McEneaney 2006) on cumulative visitation. However, for elk in central Yellowstone there was a significant negative relationship between spring calf ratios and accumulated snow water equivalent (SWE_{acc}; April calf:cow ratio = $350 \times \exp(-SWE_{acc} \times 0.00046)$, $R^2 = 0.91$, Garrott *et al.* 2003). Likewise, spring calf ratios for central Yellowstone bison were significantly and negatively related to SWE_{acc} (spring calf:cow ratio = $0.27 + (-0.000022 \times SWE_{acc})$, $R^2 = 0.20$, Fuller 2006). We used these relationships to calculate differences between predicted and observed calf ratios across the range of observed SWE_{acc} and regressed these residuals, for which the effects of SWE were removed, on cumulative visitation.

III. RESULTS

The public OSV season lasted 72 days in 2002–2003, 89 days in 2003–2004, 72 days in 2004–2005, and 82 days in 2005–2006. The mean and standard deviation (SD) of daily OSVs entering the West Entrance Station were 320 ± 114 in 2002–2003, 178 ± 59 in 2003–2004, 156 ± 70 in 2004–2005, and 181 ± 56 in 2005–2006. Maximum daily numbers were 573 on February 20, 2003; 330 on February 15, 2004; 324 on February 23, 2005; and 338 on December 30, 2005 (winter 2005–2006). Peak visitation typically occurred on weekends and holidays, while fewer vehicles entered the park on weekdays. Cumulative OSVs entering the West Entrance Station totaled 23,073 in 2002–2003, 15,846 in 2003–2004, 11,199 in 2004–2005, and 14,856 in 2005–2006. Peak SWE (cm) at the West Yellowstone SNOTEL site were 20.8 in 2002–2003, 30.7 in 2003–2004, 25.7 in 2004–2005, and 36.1 in 2005–2006, compared to a 37-year average of 26.6 cm. Ambient temperatures were moderate in winters 2002–2003 through 2005–2006, with minimum daily temperatures during December through April ranging from −41 to 3 °C at the West Yellowstone SNOTEL site.

We observed 5688 interactions between groups of wildlife and OSVs (Figure 26.1). Sixty-six percent of humans on OSVs showed no visible reaction to wildlife groups and did not stop, 19% stopped to observe but remained on or inside their OSV, 8% stopped and dismounted their OSV, and 7% approached, impeded, or hastened the movement of bison or elk. For bison, 80% of responses to OSVs and associated human activities were characterized as no apparent response, 9% look-resume, 3% alert, 5% travel, 2% flight, and <1% defensive. For elk, 48% of responses were characterized as no response, 27% look-resume, 17% alert, 5% travel, 2% flight, and <1% defensive. For swans, 57% of responses were characterized as no response, 21% look-resume, 12% alert, 9% travel, and 1% flight. For bald eagles, 17% of responses were characterized as no response, 64% look-resume, 9% alert, 4% travel, and 6% flight.

For bison, the two best models included year, group size, distance from road, on-road indicator, human activity, interaction time, number of snowmobiles and coaches, habitat, and cumulative visitation, with threshold values of eight bison for group size, 85 m for distance, 3.0 min for interaction time, eight snowmobiles, three coaches, and 28,500 OSVs for cumulative visitation (Table 26.1). They

FIGURE 26.1 Snowmobiles encountering a group of bison on the groomed roadway near the Madison River in Yellowstone National Park, Wyoming, during 2007 (Photo by Bob Weselmann).

TABLE 26.1	Model selection results for the top four *a priori* models for bison, elk, trumpeter swans, and bald eagles

Bison	Model = *year* + *sppnum* (T = 8) + *dist* (T = 85) + *onroad* + *hact* + *intxn* (T = 3.0) + *sb* (T = 8) + *coach* (T = 3) + *sppnum*dist*	K	ΔAIC$_c$	w$_k$
1	*hab* + *cumvis* (T = 28500) + *dist*hact* + *intxn*sppnum* + *year*cumvis*	56	0.00	0.45
2	*hab* + *cumvis* (T = 28500) + *intxn*sppnum* + *year*cumvis*	48	1.98	0.1
3	*hab* + *cumvis* (T = 28500) + *dist*hact* + *intxn*sppnum* + *sppnum*hab* + *year*cumvis*	64	2.71	0.1
4	*hab* + *cumvis* (T = 28500) + *dist*hact* + *intxn*dist* + *intxn*sppnum* + *year*cumvis*	58	3.31	0.0
Elk	Model = *year* + *sppnum* (T = 8) + *dist* + *onroad* + *hact* + *intxn* (T = 5.7) + *sb* + *coach* (T = 1) + *sppnum*dist*	K	ΔAIC$_c$	w$_k$
1	*hab* + *cumvis* + *sppnum*hact* + *intxn*dist* + *intxn*sppnum*	48	0.00	0.27
2	*hab* + *cumvis* + *sppnum*hact* + *intxn*dist* + *intxn*sppnum* + *sppnum*hab* + *year*cumvis*	56	0.39	0.22
3	*hab* + *cumvis* + *sppnum*hact* + *intxn*dist* + *intxn*sppnum* + *year*cumvis*	54	1.08	0.16
4	*hab* + *cumvis* + *sppnum*hact* + *intxn*dist* + *intxn*sppnum* + *sppnum*hab* + *year*cumvis*	62	2.08	0.10
Swan	Model = *year* + *sppnum* (T = 7) + *dist* + *hact* + *intxn* (T = 8.0) + *sb* + *coach* (T = 1) + *sppnum*dist*	K	ΔAIC$_c$	w$_k$
1	*cumvis* + *intxn*dist* + *year*cumvis*	36	0.00	0.30
2	*cumvis* + *intxn*sppnum* + *year*cumvis*	36	1.14	0.17
3	*cumvis* + *dist*hact* + *intxn*sppnum* + *year*cumvis*	42	2.14	0.10
4	*cumvis* + *intxn*dist*	30	2.22	0.10
Eagle	Model = *year* + *sppnum* + *dist* (T = 250) + *hact* + *intxn* (T = 1.4) + *sb* (T = 18) + *coach* (T = 1)	K	ΔAIC$_c$	w$_k$
1	*hab* + *intxn*dist* + *year*cumvis*	30	0.00	0.61
2	*hab* + *cumvis* + *intxn*dist*	32	1.89	0.24
3	*hab*	28	3.52	0.10
4	*hab* + *cumvis*	30	5.27	0.04

Each model for a given species contains the base model and the terms indicated. Abbreviations are: *year* (year of study); *dist* (distance from road in meters); *sppnum* (group size); *onroad* (on or off road indicator); *hact* (human activity); *intxn* (interaction time in minutes); *sb* (number of snowmobiles); *coach* (number of coaches); *cumvis* (cumulative daily number of OSVs); *hab* (habitat); T (variable threshold); K (number of model parameters); ΔAIC$_c$ (AIC corrected for small sample size); and w$_k$ (AIC model weight).

also included the following interactions with Akaike predictor weights (w_p): group size by distance (1), distance by human activity (0.69), interaction time by group size (1), and year by cumulative visitation (0.97); though the distance by human activity interaction was not statistically significant ($P = 0.15$). The estimated odds of no response relative to a movement response were 17 times greater for each additional 10 animals in the group and 912 times greater for each 100-m increase in distance from the road (Table 26.2). However, the effect of increasing distance from roads at reducing the odds of a movement response decreased as bison groups got larger. The odds of a movement response were 1.1 times greater for each additional snowmobile, 1.5 times greater for each additional coach, two times greater when humans approached, seven times greater when OSVs impeded or hastened bison, 17 times greater when bison were on the road, and 17 and 1.6 times greater in aquatic or thermal habitat, respectively (Figure 26.2).

For elk, the three best models included year, group size, distance from road, on-road indicator, human activity, interaction time, number of snowmobiles and coaches, habitat, and cumulative

TABLE 26.2	Estimates, standard errors, and *P*-values for the movement and vigilance logits from the *a priori* models for bison best supported by the data

Effect	Movement logit			Vigilance logit		
	Estimate	SE	P	Estimate	SE	P
intercept	−0.23	0.63	0.776	0.82	0.42	0.055
year = 2003	2.20	0.42	<0.0001	0.74	0.25	0.003
= 2004	−0.08	0.40	0.839	−0.03	0.27	0.912
= 2005	−0.08	0.48	0.545	0.21	0.30	0.478
= 2006	−1.83	0.63	0.004	−0.92	0.31	0.003
sppnum	−2.90	0.72	<0.0001	1.53	0.54	0.004
dist	−6.82	2.61	0.009	−2.32	0.81	0.004
hact = IH	1.41	0.33	<.0001	1.12	0.31	<0.001
= AP	0.33	0.65	0.608	−0.53	0.49	0.282
= D	−1.47	0.50	0.003	0.08	0.30	0.782
= S	0.20	0.27	0.456	−0.07	0.22	0.738
= N	−0.47	0.30	0.112	−0.60	0.23	0.009
intxn	0.27	0.21	0.206	0.49	0.14	<0.001
onroad = OFF	−1.43	0.20	<0.0001	−0.50	0.12	<0.0001
= ON	1.43	0.20	<0.0001	0.50	0.12	<0.0001
sb	0.11	0.04	0.002	0.08	0.03	0.002
coach	0.41	0.15	0.009	0.32	0.12	0.009
hab = A	1.90	0.53	<0.001	1.28	0.34	<0.001
= BF	−0.37	0.27	0.161	−0.27	0.19	0.156
= F	−0.11	0.21	0.588	0.08	0.14	0.581
= TH	−0.47	0.24	0.047	−0.66	0.17	<0.0001
= M	−0.95	0.22	<0.0001	−0.43	0.14	0.002
cumvis	0.00	0.02	0.825	−0.04	0.01	0.001
*sppnum*dist*	4.85	1.25	<0.001	−0.74	0.72	0.305
*dist*hact* = IH	1.06	3.77	0.779	−2.10	2.47	0.395
= AP	−11.62	8.87	0.190	−0.11	1.29	0.935
= D	5.38	2.47	0.029	0.70	0.84	0.406
= S	2.59	2.39	0.279	0.85	0.75	0.256
= N	2.59	2.38	0.277	0.66	0.72	0.364
*intxn*sppnum*	0.01	0.32	0.966	−0.66	0.22	0.003
*cumvis*year* = 2003	−0.07	0.02	0.005	0.03	0.02	0.095
= 2004	0.07	0.03	0.786	−0.03	0.02	0.110
= 2005	0.03	0.04	0.496	−0.01	0.03	0.656
= 2006	0.03	0.04	0.399	0.02	0.02	0.392

For each categorical variable, results are presented relative to the baseline level (*i.e.*, 2006 for *year*, N for *hact*, ON for *onroad*, and M for *hab*) and each estimate represents a departure from zero (*i.e.*, no effect). Abbreviations are: *year* (year of study); *dist* (distance of group from road); *sppnum* (group size); *onroad* (on or off road); *hact* (human activity); *intxn* (interaction time); *sb* (number of snowmobiles); *coach* (number of coaches); *hab* (habitat), *cumvis* (cumulative daily number of OSVs); IH (impeded/hastened animal movement); AP (approached animals); D (dismounted or exited OSVs); S (stopped to observe animals); N (no visible reaction); ON (animals on the road); OFF (animals off the road); A (aquatic); BF (burned forest); F (unburned forest); TH (thermal); and M (meadow).

visitation, with a moderated form for distance and threshold values of eight elk for group size, 5.7 min for interaction time, and one snow coach (Table 26.1). They also included interactions (w_p) for group size by distance (1), group size by human activity (1), interaction time by distance (0.80), interaction time by group size (1), group size by habitat (0.49), and year by cumulative visitation (0.30). The estimated odds of a vigilance or movement response relative to no response were 3–4 times greater for each 1-min increase in interaction time (Table 26.3). However, the effect of increasing interaction time at increasing the odds of a movement response decreased as elk groups got larger or the elk groups were farther from the road. The odds of a movement response were 31 times greater when elk were on the

FIGURE 26.2 Snow coach approaching a group of bison on the groomed roadway near Indian Creek in Yellowstone National Park, Wyoming, during 2007 (Photo by Bob Weselmann).

road. The effects of human activities and habitat on elk were dependent on elk group size, with larger groups magnifying the odds of a vigilance or movement response when humans approached or the habitat was aquatic or unburned forest. Conversely, the odds of a movement response decreased with larger groups in thermal habitat.

For trumpeter swans, the two best models included year, group size, distance from road, human activity, interaction time, number of snowmobiles and coaches, and cumulative visitation, with a moderated form for distance and threshold values of seven swans for group size, 8.0 min for interaction time, and one snow coach (Table 26.1). They also included interactions (w_p) for group size by distance (1), interaction time by distance (0.59), interaction time by group size (0.52), and year by cumulative visitation (0.71). The estimated odds of no response relative to movement responses were six times greater for each additional 10 birds in the group and eight times greater for each 100-m increase in distance from the road (Table 26.4). The odds of a movement response were 1.2 times greater for each 1-min increase in interaction time, 1.1 times greater for each additional snowmobile, and three times greater when humans approached.

For bald eagles, the two best models included year, group size, distance from road, human activity, interaction time, number of snowmobiles and coaches, habitat, and cumulative visitation, with a threshold values of 250 m for distance, 1.4 min for interaction time, 18 snowmobiles, and one snow coach (Table 26.1). They also included interactions (w_p) for distance by interaction time (0.85) and year by cumulative visitation (0.01). The estimated odds of no response relative to a movement response were four times greater for each 100-m increase in distance from the road (Table 26.5). The odds of a vigilance response were 60 times greater for each 1-min increase in interaction time and 54 times greater when they were in burned forest than meadow habitat. The odds of a movement response were 1.3 times greater for each additional snowmobile and five times greater when humans approached.

There were no significant changes in the odds of movement responses for any wildlife species as cumulative OSV traffic increased through winter. However, we observed significant decreases in the odds of vigilance responses by bison (1.04) and swans (1.08) for each additional 1000 OSVs (Tables 26.2 and 26.4). In contrast, there was a significant increase in the odds (1.03) of vigilance responses by elk for each additional 1000 OSVs (Table 26.3). The year by cumulative visitation interaction indicated the effects of cumulative OSV traffic varied across winters for bison and trumpeter swans. The odds of a movement response by bison as cumulative visitation increased were greater during winters 2002–2003 through 2004–2005 compared to winter 2005–2006, whereas the opposite pattern occurred for swans.

| TABLE 26.3 | Estimates, standard errors, and *P*-values for the movement and vigilance logits from the *a priori* models for elk best supported by the data |

	Movement logit			Vigilance logit		
Effect	**Estimate**	**SE**	***P***	**Estimate**	**SE**	***P***
intercept	−2.59	0.84	0.002	−0.73	0.55	0.185
year = 2003	0.41	0.31	0.193	0.37	0.16	0.019
= 2004	0.23	0.22	0.300	−0.03	0.11	0.741
= 2005	1.12	0.30	<0.001	0.63	0.15	<0.0001
= 2006	−1.76	0.36	<0.0001	−0.96	0.15	<0.0001
sppnum	2.03	1.53	0.184	0.94	0.93	0.310
dist	−1.18	0.73	0.108	0.19	0.32	0.559
hact = AP	0.69	0.64	0.283	0.42	0.53	0.422
= D	−0.83	0.77	0.281	0.31	0.51	0.548
= S	−0.64	0.50	0.201	−0.55	0.31	0.077
= N	0.78	0.49	0.111	−0.18	0.31	0.574
intxn	1.45	0.24	<0.0001	0.90	0.17	<0.0001
onroad = OFF	−1.29	0.36	<0.001	0.00	0.30	0.987
= ON	1.29	0.36	<0.001	−0.00	0.30	0.987
sb	0.00	0.02	0.862	−0.01	0.01	0.526
coach	0.63	0.33	0.054	−0.06	0.17	0.714
hab = A	0.62	0.31	0.043	0.44	0.15	0.003
= BF	−0.36	0.28	0.199	−0.09	0.12	0.452
= F	0.91	0.28	0.001	0.41	0.15	0.007
= TH	−1.25	0.51	0.015	−0.54	0.19	0.006
= M	0.08	0.33	0.818	−0.22	0.14	0.124
cumvis	0.03	0.02	0.224	0.03	0.01	0.007
sppnum * *dist*	0.87	1.20	0.467	−1.21	0.64	0.057
sppnum * *hact* = AP	0.65	1.11	0.556	−0.45	0.87	0.608
= D	1.90	1.32	0.150	−0.69	0.89	0.438
= S	1.23	0.94	0.192	1.43	0.57	0.012
= N	−3.78	1.08	<0.001	−0.29	0.59	0.622
intxn * *dist*	−0.28	0.18	0.127	−0.31	0.12	0.009
intxn * *sppnum*	−1.62	0.33	<0.0001	−0.74	0.21	<0.001

For each categorical variable, results are presented relative to the baseline level and each estimate represents a departure from zero. Abbreviations are as in Table 26.2.

The *post hoc* exploratory analyses did not result in any improvement when cumulative visitation was replaced with SWE_{acc} or human group type was added to the models. Also, the exploratory model for bison was the same as the *a priori* model. The best *a priori* model for elk ($AIC_c = 1824.7$) was improved by removing distance to road ($P = 0.14$), human activity ($P = 0.15$), number of snowmobiles ($P = 0.70$), and the group size by distance interaction ($P = 0.06$), and replacing group size ($P = 0.36$) and habitat ($P = 0.0004$) with the group size by habitat interaction ($AIC_c = 1808.0$). The best *a priori* model for trumpeter swans ($AIC_c = 1872.7$) was improved by removing interactions for group size by distance ($P = 0.70$) and distance by interaction time ($P = 0.26$; $AIC_c = 1866.6$). The best *a priori* model for bald eagles ($AIC_c = 412.4$) was improved by removing group size ($P = 0.31$; $AIC_c = 403.7$).

A. Abundance, Demography, and Displacement of Wildlife

Bison counts increased exponentially with cumulative visitation during 1965–1966 through 1993–1994 (Appendix 26A.2, Figure 1), but population growth rates from which the effects of management culls were removed were not significantly related to cumulative visitation during 1965–1966 through 2005–2006 ($R^2 = 0.01$, $F_{1,26} = 0.1$, $P = 0.78$). Bison calf ratios were not significantly correlated with

TABLE 26.4 Estimates, standard errors, and *P*-values for the movement and vigilance logits from the *a priori* models for trumpeter swans best supported by the data

Effect	Movement logit			Vigilance logit		
	Estimate	SE	P	Estimate	SE	P
intercept	0.64	1.03	0.534	2.40	0.65	<0.001
year = 2003	1.42	0.56	0.011	0.29	0.39	0.467
= 2004	0.73	0.77	0.343	1.46	0.48	0.002
= 2005	−2.01	0.97	0.038	0.05	0.45	0.917
= 2006	−0.14	0.64	0.824	−1.79	0.51	<0.001
sppnum	−3.10	1.73	0.073	−2.78	1.15	0.016
dist	−2.12	1.09	0.052	−2.50	0.65	<0.001
hact = AP	0.31	0.35	0.376	0.32	0.32	0.307
= D	0.38	0.28	0.180	−0.31	0.27	0.252
= S	0.10	0.22	0.668	0.23	0.18	0.210
= N	−0.78	0.23	<0.001	−0.24	0.18	0.201
intxn	0.38	0.15	0.013	−0.04	0.13	0.776
sb	0.08	0.03	0.012	0.07	0.02	0.004
coach	0.53	0.18	0.004	0.46	0.14	0.001
cumvis	−0.10	0.07	0.141	−0.08	0.04	0.062
*sppnum*dist*	1.58	2.12	0.455	0.78	1.42	0.580
intxn dist*	−0.22	0.18	0.211	0.08	0.14	0.578
*cumvis*year* = 2003	−0.10	0.09	0.252	0.01	0.06	0.928
= 2004	−0.08	0.12	0.497	−0.18	0.07	0.016
= 2005	0.39	0.15	0.010	0.08	0.07	0.268
= 2006	−0.20	0.11	0.062	0.09	0.08	0.240

For each categorical variable, results are presented relative to the baseline level and each estimate represents a departure from zero. Abbreviations are as in Table 26.2.

TABLE 26.5 Estimates, standard errors, and *P*-values for the movement and vigilance logits from the *a priori* models for bald eagles best supported by the data

Effect	Movement logit			Vigilance logit		
	Estimate	SE	P	Estimate	SE	P
intercept	0.37	1.40	0.791	−3.11	1.31	0.017
year = 2003	1.04	0.45	0.022	0.05	0.38	0.895
= 2004	−0.24	0.51	0.633	0.77	0.34	0.023
= 2005	0.66	0.45	0.141	0.38	0.37	0.297
= 2006	−1.46	0.56	0.009	−1.20	0.35	<0.001
sppnum	−0.18	0.17	0.296	−0.43	0.33	0.190
dist	−1.21	0.72	0.094	0.67	0.55	0.229
hact = AP	0.80	0.85	0.346	−0.27	0.75	0.715
= D	0.13	0.55	0.812	−0.72	0.51	0.158
= S	−0.14	0.49	0.774	−0.32	0.42	0.444
= N	−0.79	0.66	0.232	1.32	0.54	0.014
intxn	0.19	1.27	0.883	4.40	1.17	<0.001
sb	0.23	0.08	0.003	0.19	0.07	0.005
coach	1.41	0.67	0.036	0.56	0.51	0.267
hab = BF	0.34	0.37	0.363	2.06	0.34	<0.0001
= F	−0.81	0.40	0.045	−0.39	.33	0.232
= M	0.47	0.49	0.338	−1.67	0.50	<0.001
*intxn*dist*	0.15	0.67	0.822	−1.37	0.59	0.019

For each categorical variable, results are presented relative to the baseline level and each estimate represents a departure from zero. Abbreviations are as in Table 26.2.

cumulative visitation ($R^2 = 0.01$, $F_{1,16} = 0.2$, $P = 0.68$). Elk abundance was positively correlated with cumulative winter visitation during 1966–1967 through 2005–2006 (Appendix 26A.2, Figure 2; $R^2 = 0.21$, $F_{1,17} = 4.5$, $P = 0.05$), but calf ratios were not correlated with cumulative visitation during 1991–1992 through 2005–2006 after the effects of snow water equivalent on calf recruitment were removed ($R^2 = 0.02$, $F_{1,13} = 0.3$, $P = 0.60$). Numbers of nesting and fledgling bald eagles in Yellowstone increased incrementally during 1987–2005 (Appendix 26A.2, Figure 3), but were not significantly correlated with cumulative visitation ($R^2 < 0.05$, $F_{1,17} < 0.8$, $P > 0.40$). Numbers of resident adult/subadult and cygnet trumpeter swans decreased during 1966–2005 (Appendix 26A.2, Figure 4) and were negatively correlated with cumulative visitation (adults: $R^2 = 0.37$, $F_{1,22} = 13.0$, $P = 0.002$; cygnets: $R^2 = 0.13$, $F_{1,22} = 3.1$, $P = 0.09$).

IV. DISCUSSION

Bald eagles, bison, elk, and trumpeter swans in Yellowstone behaviorally responded to OSVs and associated human activities with increased vigilance, travel and, occasionally, flight or defense. The likelihood and intensity of these responses differed by species, with bison responding less frequently (20%) than swans (43%), elk (52%), or bald eagles (83%). This difference was due to increased vigilance responses by swans (33%), elk (44%), and bald eagles (73%) compared to bison (12%). The frequency of higher-intensity movement responses was similar (8–10%) among species. Similar to other studies, we found the odds of a movement response significantly increased when humans stopped, dismounted, and approached wildlife (Stalmaster and Kaiser 1998, Steidl and Anthony 2000, González *et al.* 2006).

The likelihood and intensity of responses by bald eagles, bison, elk, and trumpeter swans increased significantly if animals were on or near roads, animal groups were smaller, interaction times increased, or numbers of snowmobiles and coaches increased. However, there were thresholds on the odds of eliciting a response by wildlife for several of these covariates. For example, the effects of group size at reducing the odds of a response reached a maximum at 7–8 animals for bison, elk, and swans. Conversely, the effects of interaction time at increasing the odds of a response were reached at quite different thresholds, ranging from 1 min for eagles to 8 min for swans. The odds of eliciting a movement response by wildlife were higher for coaches than snowmobiles, but the maximum effect was reached at a threshold of ≤3 coaches. In contrast, there was no threshold for elk and swans with each additional snowmobile and the threshold was 7–18 snowmobiles for bison and eagles.

Borkowski *et al.* (2006) sampled interactions between groups of bison and elk and OSVs in Yellowstone during 1998–1999 through 2002–2003 when visitation was 2–3 times higher than during 2003–2004 through 2005–2006. Similar to our study, they found elk responded three times more (52%) than bison (19%) during interactions with OSVs due to increased vigilance responses (elk: 44%; bison: 10%). In addition, the frequency of higher-intensity movement responses by bison and elk were similar (8–9%) and relatively low compared to other studies reporting substantially higher degrees of avoidance and responses to snowmobiles by bison (Fortin and Andruskiw 2003), caribou (*Rangifer tarandus*; Seip *et al.* 2007), moose (*Alces alces*; Colescott and Gillingham 1998), mule deer (*O. hemionus*; Freddy *et al.* 1986), reindeer (*Rangifer tarandus*; Tyler 1991, Reimers *et al.* 2003), and white-tailed deer (*Odocoileus virginianus*; Dorrance *et al.* 1975, Richens and Lavigne 1978, Eckstein *et al.* 1979). For example, Fortin and Andruskiw (2003) reported 51% of bison in Prince Albert National Park, Saskatchewan, Canada, reacted to human presence by fleeing the area.

Likewise, trumpeter swans at Harriman State Park in Idaho had more pronounced reactions to human disturbance than swans in Yellowstone. Swans in Idaho took flight when approached by a person on skis or snowmobile and often moved several kilometers to another stretch of river (Shea 1979). Conversely, swans in Yellowstone generally reacted to humans by swimming farther away while continuing to feed. The responses of bald eagles in Yellowstone (9% alert, 6% flight) were similar to those (3–17% alert/flight) of nesting eagles in Voyageurs National Park, Minnesota to watercraft passing within 800 m of their nests (Grubb *et al.* 2002). Several studies reported eagle response rates were affected by distance, number of vehicles per event, interaction duration and rates, and time of day (Grubb *et al.* 2002, González *et al.* 2006). Thus, buffer zones 400–800 m wide, where watercraft or vehicles are not permitted to stop, have been recommended for sensitive foraging areas and nesting sites of eagles elsewhere (Stalmaster and Kaiser 1998, Grubb *et al.* 2002, González *et al.* 2006). During our study, there were few responses by eagles when the distance to OSVs and associated humans was >250 m. Thus, the buffer zone that prohibits stopping within 400 m of the eagle nest between West Yellowstone and Madison Junction has likely reduced disturbances to this pair. However, the majority of OSVs entering the park still drove by the nest at a distance of ~55 m and it was impossible for visitors to keep >250 m from foraging or perched eagles in the Firehole and Madison drainages because, even at their widest points, rivers were often <250 m from the road.

The comparatively less frequent and lower intensity responses by the wildlife species we studied in Yellowstone suggest there is a certain level of habituation to OSVs and associated human activities (Figure 26.3). Aune (1981) and Hardy (2001) concluded bison, elk, and swans habituated to the presence and patterns of snowmobile activity in Yellowstone. Also, Borkowski *et al.* (2006) reported the likelihood of an active response by bison during 1999–2003 in Yellowstone decreased within winters having the highest visitation. Similarly, Shea (1979) reported swans wintering within 55 m of the road along the Madison River, which had high snowmobile traffic, showed more tolerance to winter visitors than swans on the Yellowstone River where levels of traffic were lower. We did not detect significant increases or decreases in the odds of movement responses as cumulative OSV traffic increased through winter. However, the effect of increased cumulative visitation at increasing the odds of movement by bison was less in the high visitation winter of 2002–2003 (72,560 visitors) compared to relatively low visitation years of 2003–2004 through 2005–2006 (<49,000 visitors). Also, we observed a decrease in the odds of vigilance responses by bison and swans as cumulative OSV traffic increased through the winter. Thus, there appeared to be some habituation by bison and swans with increasing exposure to vehicles during a season.

There are several characteristics of winter recreation in Yellowstone that likely facilitate behavioral habituation by wintering animals to OSV traffic (Aune 1981, Hardy 2001). All OSVs traveled through

FIGURE 26.3 A wolf in the groomed roadway howls as winter visitors watch near Alum Creek in Yellowstone National Park, Wyoming, during 2007 (Photo by Bob Weselmann).

our study area in predictable ways, remaining confined to roads and typically without humans threatening or harassing wildlife. Few people ventured far from roads, established trails, or areas of concentrated human activities (*e.g.*, warming huts, geyser basin trails). Also, the minimal risk of mortality associated with human presence likely reduces avoidance responses by these species (Gill *et al.* 2001a). Hunting is illegal in the park (National Park Service Protective Act of 1894; 16 USC 1, 5§26) and few animals were hit and killed by OSVs each winter. Gunther *et al.* (1999) reported 10 bison, three elk, two coyotes (*Canis latrans*), one red fox (*Vulpes vulpes*), and one pine marten (*Martes martes*) were killed by snowmobiles in Yellowstone during 1989–1998. For comparison, 98 bison, 427 elk, 75 coyotes, 84 moose, and 406 other large mammals (*e.g.*, bears, bighorn sheep *Ovis canadensis*, deer *Odocoileus sp.*, pronghorn *Antilocapra americana*, wolves *Canis lupus*) were killed by wheeled vehicles in Yellowstone during these years.

There are complex species-specific differences and interactions among covariates that affect the degree of habituation by wildlife in Yellowstone and which complicate interpretation of recreation effects. For example, elk exhibited a significant increase in the odds of vigilance responses as OSV traffic increased through winter. Also, the effect of increased cumulative visitation at increasing the odds of movement by trumpeter swans was less in 2005–2006 than 2002–2003 through 2004–2005, perhaps due to the reduced availability of ice-free areas during the relatively harder winter of 2005–2006. Animals can probably only habituate to particular types and levels of human disturbance up to some threshold (Steidl and Anthony 2000), above which a change in the amount or type of disturbance may drastically alter the likelihood and intensity of their responses (Stalmaster and Newman 1978, White and Thurow 1985). Hence, winter recreational activities in Yellowstone should continue to be conducted in a predictable manner.

There is some evidence animals were displaced away from roads with OSV traffic in Yellowstone. Aune (1981) and Hardy (2001) indicated elk were temporarily displaced about 60 m from busy road segments (*e.g.*, Madison to Old Faithful) as cumulative traffic increased, while Shea (1979) reported swans in Yellowstone often retreated from visitors that stopped by swimming away. However, human disturbance did not appear to be a primary factor influencing the distribution and movements of the wildlife species we studied, suggesting behavioral responses and apparent avoidance of humans in the vicinity of the road were apparently short-term changes that were later reversed. Bison, elk, and swans in Yellowstone used the same core winter ranges during the past three decades despite large winter-to-winter variability in cumulative exposure to OSVs (Craighead *et al.* 1973, Shea 1979, Aune 1981, Hardy 2001). Also, Bruggeman (2006) found that factors influencing resource availability, including snow pack, population density, and drought, provided the primary impetus for variability in the distribution, movements, and foraging behavior of central Yellowstone bison during winter. Similarly, Messer (2003) reported the distribution of elk in central Yellowstone during winter was primarily influenced by snow mass and heterogeneity. During 2002–2006, a pair of bald eagles took up residence in a large tree nest located 55 m from the road and ~11 km east of West Yellowstone (McEneaney 2006). Despite high OSV and wheeled-vehicle traffic, this pair maintained the territory throughout the year and fledged 1–2 eaglets in 2002, 2003, and 2005 (McEneaney 2006). Furthermore, counts of swans on Firehole and Madison Rivers increased during January when open water sections of the Yellowstone River diminished, despite increasing OSV use during this period (T. McEneaney, National Park Service (retired), unpublished data).

Any adverse behavioral, energetic, or stress effects of OSV recreation to bald eagles, bison, and elk were apparently compensated for at the population level. Hardy (2001) found no consistent or obvious trends between daily traffic of OSVs and fecal glucocorticoid levels of bison and elk in the Madison headwaters area of Yellowstone during winters of 1998–1999 and 1999–2000. Counts of central Yellowstone bison increased exponentially during 1980–1994 despite a 20-fold increase in cumulative visitation and annual survival of adult females was high (96%) and constant during 1995–2001 (Fuller 2006). Meagher (1998) hypothesized this enhanced population growth was due to energy saved by bison traveling on roads groomed (*i.e.*, snow packed) for snowmobile recreation and better access to foraging habitat. Simulations by Coughenour (2005) suggested minor energetic savings by bison traveling on groomed roads could cumulatively result in increased population growth over three decades. Thus, the possibility that the energetic savings could have accelerated population growth cannot be dismissed. However, several independent analyses concluded road grooming did not change the population growth rates of bison relative to what may have been realized in the absence of road grooming (Gates *et al.* 2005, Bruggeman *et al.* 2006, Fuller 2006, Wagner 2006). In fact, the population growth rates estimated by Fuller *et al.* (2007) using primarily Dr. Meagher's data were well below the biological potential of the species. Also, population estimates for central Yellowstone elk fluctuated around a dynamic equilibrium of ~550 elk during 1968–2004 (λ = 0.99–1.01; Garrott *et al.* 2005, Chapter 11 by Garrott *et al.*, this volume). The annual survival of adult female elk in this population exceeded 90% and early winter calf ratios indicated healthy reproductive rates prior to colonization by wolves in 1998 (Garrott *et al.* 2003, Chapter 11 by Garrott *et al.*, this volume). In addition, numbers of nesting bald eagles and fledglings in Yellowstone increased incrementally during 1987–2005 (McEneaney 2006). Fortin and Andruskiw (2003) reached a similar conclusion for bison in Prince Albert National Park, Saskatchewan, Canada. They found no evidence that the frequency of disturbance imposed on bison by snowmobiles, trucks, or foot traffic had an important effect on resource use or bison density among meadows.

The significant, negative correlation between swan counts and OSV traffic in Yellowstone was likely spurious because numbers of swans decreased regionally throughout the Greater Yellowstone Area during the past several decades, including the productive Centennial Valley of Montana (McEneaney 2006). It is unlikely that poor regional production resulted from winter recreation in Yellowstone, especially when swans generally incubate eggs in May and hatch young in late June, which is well after the OSV recreation period. Also, the area located <100 m from the road along the Madison River approximately 11 km east of West Yellowstone has been a traditional nesting area for decades and at least 23 cygnets have fledged from this site since 1983, making it one of the more productive nesting areas in Yellowstone (McEneaney 2006). Moreover, as winter progressed and open water areas in the park diminished, the proportion of the swan population within Yellowstone decreased compared to areas outside the park (T. McEneaney, unpublished data). Thus, relatively few swans were exposed to motorized winter use in the park.

Monitoring during winters 1998–1999 through 2005–2006 in Yellowstone documented the vast majority of winter visitors traveling on OSVs remained on groomed roads, behaved appropriately when viewing wildlife, and rarely approached wildlife except when animals were on or immediately adjacent to the road. These attributes allowed wildlife in Yellowstone to habituate somewhat to OSV recreation, commonly demonstrating no observable response and rarely displaying "fight or flight" responses when animals were off road. Further, available data provides no evidence that levels and patterns of OSV traffic during the past 35 years significantly affected the demography or population dynamics of the wildlife we studied, either by substantially decreasing or increasing growth rates. Thus, we suggest regulations restricting the levels and travel routes of OSVs during recent years were effective at reducing disturbances to wildlife below a level that would cause measurable fitness effects.

V. SUMMARY

1. A conflict between protecting park resources and the desires of many visitors to experience the park via snowmobile arose in Yellowstone National Park during the late 1990s. Rigorous studies evaluating the effects of snowmobiling on the behavior, distribution, and demography of bison and other wildlife were needed to evaluate the merits of conflicting claims.

2. This study sampled 5688 interactions between groups of bald eagles, bison, elk, and trumpeter swans and groups of snowmobiles and coaches during winters 2002–2003 through 2005–2006 to identify conditions leading to behavioral responses.

3. Bison responded less frequently (20%) to snowmobiles and coaches than swans (43%), elk (52%), or bald eagles (83%) due to fewer vigilance responses. The frequency of higher-intensity movement responses was similar among species (8–10%). Responses increased significantly if animals were on or near roads, animal groups were smaller, humans approached animals, interaction times increased, or numbers of vehicles in a group increased.

4. We did not detect significant changes in the odds of movement responses for any species as cumulative OSV traffic increased through winter. Vigilance responses by bison decreased within the winter having the largest visitation, suggesting some habituation. Conversely, vigilance responses by elk increased as cumulative visitation increased.

5. Human disturbance was not a primary factor influencing the distribution and movements of bison, elk, and swans. The risk of vehicle-related mortality was quite low and these species used the same core winter ranges despite large winter-to-winter variability in cumulative exposure to snowmobiles and coaches.

6. There was no evidence that snowmobile use during the past 35 years significantly affected the demography of bald eagles, bison, elk, or trumpeter swans. Regulations were effective at reducing disturbances below a level that would cause measurable fitness effects.

VI. REFERENCES

Aune, K. E. 1981. Impacts of winter recreationists on wildlife in a portion of Yellowstone National Park, Wyoming. Thesis. Montana State University, Bozeman, MT.

Borkowski, J. J., P. J. White, R. A. Garrott, T. Davis, A. R. Hardy, and D. J. Reinhart. 2006. Behavioral responses of bison and elk in Yellowstone to snowmobiles and snow coaches. *Ecological Applications* **16**:1911–1925.

Boyle, S. A., and F. B. Sampson. 1985. Effects of nonconsumptive recreation on wildlife: A review. *Wildlife Society Bulletin* **13**:110–116.

Bruggeman, J. E. 2006. Spatio-temporal dynamics of the central bison herd in Yellowstone National Park. Dissertation. Montana State University, Bozeman, MT.

Bruggeman, J. E., R. A. Garrott, D. D. Bjornlie, P. J. White, F. G. R. Watson, and J. J. Borkowski. 2006. Temporal variability in winter travel patterns of Yellowstone bison: The effects of road grooming. *Ecological Applications* **16**:1539–1554.

Burnham, K. P., and D. R. Anderson. 2002. *Model Selection and Multimodel Inference: A Practical Information-Theoretic Approach.* New York, NY: Springer-Verlag, New York, NY.

Colescott, J. H., and M. P. Gillingham. 1998. Reaction of moose (*Alces alces*) to snowmobile traffic in the Greys River Valley, Wyoming. *Alces* **34**:120–125.

Coughenour, M. B. 2005. *Spatial-Dynamic Modeling of Bison Carrying Capacity in the Greater Yellowstone Ecosystem: A Synthesis of Bison Movements, Population Dynamics, and Interactions with Vegetation.* Final report to U.S. Geological Survey Biological Resources Division, Bozeman, MT.

Craighead, J. J., F. C. Craighead, Jr., R. L. Ruff, and B. W. O'Gara. 1973. Home ranges and activity patterns of nonmigratory elk of the Madison drainage herd as determined by biotelemetry. *Wildlife Monographs* **33**.

District of Columbia. 2003. The Fund for Animals v. Norton, 294 F. Supp. 2d 92, 115. December 16, 2003. U.S. District Court, Washington, DC.

District of Wyoming 2004. International Snowmobile Manufacturers Association v. Norton, 304 F. Supp. 2d 1278, 1285. February 10, 2004. U.S. District Court, Cheyenne, WY.

Dorrance, M. J., P. J. Savage, and D. E. Huff. 1975. Effects of snowmobiles on white-tailed deer. *Journal of Wildlife Management* **39**:563–569.

Duchesne, M., S. D. Côté, and C. Barrette. 2000. Responses of woodland caribou to winter ecotourism in the Charlevoix Biosphere Reserve, Canada. *Biological Conservation* **96**:311–317.

Eckstein, R. G., T. F. O'Brien, O. J. Rongstad, and J. G. Bollinger. 1979. Snowmobile effects on movements of white-tailed deer: A case study. *Environmental Conservation* **6**:45–51.

Fortin, D., and M. Andruskiw. 2003. Behavioral response of free-ranging bison to human disturbance. *Wildlife Society Bulletin* **31**:804–813.

Freddy, D. J., M. B. Whitcomb, and M. C. Fowler. 1986. Responses of mule deer to disturbances by persons afoot and snowmobiles. *Wildlife Society Bulletin* **14**:63–68.

Fuller, J. A. 2006. Population demography of the Yellowstone National Park bison herds. Thesis. Montana State University, Bozeman, MT.

Fuller, J. A., R. A. Garrott, and P. J. White. 2007. Emigration and density dependence in Yellowstone bison. *Journal of Wildlife Management* **71**:1924–1933.

Garrott, R. A., L. L. Eberhardt, P. J. White, and J. Rotella. 2003. Climate-induced variation in vital rates of an unharvested large-herbivore population. *Canadian Journal of Zoology* **81**:33–45.

Garrott, R. A., J. A. Gude, E. J. Bergman, C. Gower, P. J. White, and K. L. Hamlin. 2005. Generalizing wolf effects across the greater Yellowstone area: A cautionary note. *Wildlife Society Bulletin* **33**:1245–1255.

Gates, C. C., B. Stelfox, T. Muhly, T. Chowns, and R. J. Hudson. 2005. *The Ecology of Bison Movements and Distribution in and Beyond Yellowstone National Park*. University of Calgary, Alberta, Canada.

Geist, V. 2007. How close is too close? Wildlife professionals grapple with habituating wildlife. *Wildlife Professional* , Spring 2007.

Gill, J. A., K. Norris, and W. J. Sutherland. 2001a. The effects of disturbance on habitat use by black-tailed godwits *Limosa limosa*. *Journal of Applied Ecology* **38**:846–856.

Gill, J. A., K. Norris, and W. J. Sutherland. 2001b. Why behavioural responses may not reflect the population consequences of human disturbance. *Biological Conservation* **97**:265–268.

González, L. M., B. E. Arroyo, A. Margalida, R. Sánchez, and J. Oria. 2006. Effect of human activities on the behaviour of breeding Spanish imperial eagles (*Aquila adalberti*): Management implications for the conservation of a threatened species. *Animal Conservation* **9**:85–93.

Grubb, T. G., W. L. Robinson, and W. W. Bowerman. 2002. Effects of watercraft on bald eagles nesting in Voyageurs National Park, Minnesota. *Wildlife Society Bulletin* **30**:156–161.

Gunther, K. A., M. J. Biel, R. A. Renkin, and H. N. Zachary. 1999. *Influence of Season, Park Visitation, and Mode of Transportation on the Frequency of Road-Killed Wildlife in Yellowstone National Park*. Yellowstone National Park, Mammoth, WY.

Hardy, A. R. 2001. Bison and elk responses to winter recreation in Yellowstone National Park. Thesis. Montana State University, Bozeman, MT.

Hosmer, D. W., and S. Lemeshow. 2000. *Applied Logistic Regression*. Wiley, New York, NY.

Klein, M. L., S. R. Humphrey, and H. F. Percival. 1995. Effects of ecotourism on distribution of waterbirds in a wildlife refuge. *Conservation Biology* **9**:1454–1465.

Knight, R. L., and K. J. Gutzwiller (Eds.). 1995. *Wildlife and Recreationists: Coexistence Through Management and Research*. Island Press, Washington, DC.

McEneaney, T. 2006. Yellowstone bird report, 2005. National Park Service, Yellowstone Center for Resources. Yellowstone National Park, WY.

Meagher, M. 1998. Recent changes in Yellowstone bison numbers and distribution. Pages 107–112 *in* L. Irby and J. Knight (Eds.) *The Pelican Bison and the Domino Effect*. Proceedings of the International Symposium on Bison Ecology and Management in North America, June 4–7, 1997, Bozeman, MT.

Messer, M. A. 2003. Identifying large herbivore distribution mechanisms through application of fine-scale snow modeling. Thesis. Montana State University, Bozeman, MT.

Munck, A., P. Guyre, and N. Holbrook. 1984. Physiological functions of glucocorticoids in stress and their relation to pharmacological actions. *Endocrine Reviews* **5**:25–48.

Reimers, E., S. Eftestol, and J. E. Colman. 2003. Behavior responses of wild reindeer to direct provocation by a snowmobiler or skier. *Journal of Wildlife Management* **67**:747–754.

Richens, V. B., and G. R. Lavigne. 1978. Response of white-tailed deer to snowmobiles and snowmobile trails in Maine. *Canadian Field Naturalist* **92**:334–344.

Rowland, M. M., M. J. Wisdom, B. K. Johnson, and J. G. Kie. 2000. Elk distribution and modeling in relation to roads. *Journal of Wildlife Management* **64**:672–684.

SAS. 2002. *SAS Help and Documentation*. version 9.00. Cary, NC.

Schultz, R. D., and J. A. Bailey. 1978. Responses of national park elk to human activity. *Journal of Wildlife Management* **42**:91–100.

Seip, D. R., C. J. Johnson, and G. S. Watts. 2007. Displacement of mountain caribou from winter habitat by snowmobiles. *Journal of Wildlife Management* **71**:1539–1544.

Shea, R. 1979. The ecology of the trumpeter swan in Yellowstone National Park and vicinity. Thesis. University of Montana, Missoula, MT.

Stalmaster, M. V., and J. L. Kaiser. 1998. Effects of recreational activity on wintering bald eagles. *Wildlife Monographs* **137**.

Stalmaster, M. V., and J. R. Newman. 1978. Behavioral responses of wintering bald eagles to human activity. *Journal of Wildlife Management* **42**:506–513.

Steidl, R. J., and R. G. Anthony. 2000. Experimental effects of human activity on breeding bald eagles. *Ecological Applications* **10**:258–268.

Thompson, M. J., and R. E. Henderson. 1998. Elk habituation as a credibility challenge for wildlife professionals. *Wildlife Society Bulletin* **26**:477–483.

Tyler, N. C. 1991. Short-term behavioural responses of Svalbard reindeer *Rangifer tarandus* to direct provocation by a snowmobile. *Biological Conservation* **56**:179–194.

Wagner, F. H. 2006. *Yellowstone's Destabilized Ecosystem: Elk Effects, Science, and Policy Conflict.* Oxford University Press, New York, NY.

White, C. M., and T. L. Thurow. 1985. Reproduction of ferruginous hawks exposed to controlled disturbance. *Condor* **87**:14–22.

Wisdom, M. J., A. A. Ager, H. K. Preisler, N. J. Cimon, and B. K. Johnson. 2005a. Effects of off-road recreation on mule deer and elk. Pages 67–80 *in* M. J. Wisdom (Ed.) *The Starkey Project: A Synthesis of Long-Term Studies of Elk and Mule Deer.* Reprinted from the 2004 Transactions of the North American Wildlife and Natural Resource Conference, Alliance Communications Group, Lawrence, KS.

Wisdom, M. J., N. J. Cimon, B. K. Johnson, E. O. Garton, and J. W. Thomas. 2005b. Spatial partitioning by mule deer and elk in relation to traffic. Pages 53–66 *in* M. J. Wisdom (Ed.) *The Starkey Project: A Synthesis of Long-Term Studies of Elk and Mule Deer.* Reprinted from the 2004 Transactions of the North American Wildlife and Natural Resource Conference, Alliance Communications Group, Lawrence, KS.

APPENDIX

Appendix 26A.1 Summary of Data from 25 Surveys (265 interactions) During Winter 2002–2003 That Were Inadvertently Omitted from Analyses

Dates of Omitted Surveys

December 31, 2002	February 1, 2003
January 10, 2003	February 17, 2003
January 13, 2003	February 19, 2003
January 15, 2003	February 20, 2003
January 28, 2003	February 21, 2003
January 29, 2003	February 25, 2003
January 30, 2003	February 26, 2003
January 31, 2003	February 27, 2003

Human Responses

Activity	% Response
None	51
Stop/Watch	20
Dismount	17
Approach/Impede	13

Wildlife Responses

Activity	% Response			
	Bison	Bald Eagle	Elk	Trumpeter Swan
None	67	6	21	44
Look/resume	20	71	60	33
Travel	8	12	8	19
Alert/attention	3	0	12	3
Flight	2	12	0	0

Appendix 26A.2 Estimates of Abundance and Recruitment for Bison, Elk, Trumpeter Swans, and Bald Eagles Plotted Against Numbers of Visitors on Snowmobiles and Coaches During 1965–1966 Through 2005–2006

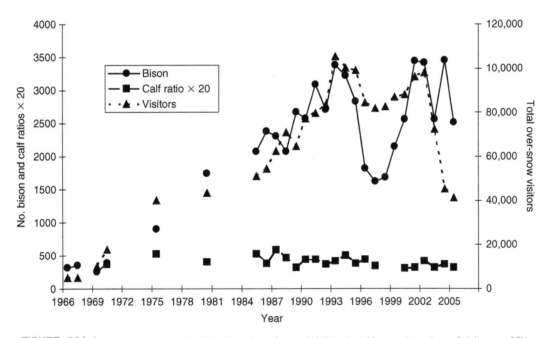

FIGURE 26A.1 Summer counts and calf ratios (20×) of central Yellowstone bison and numbers of visitors on OSVs during 1966–2006 in Yellowstone National Park, Montana and Wyoming, USA.

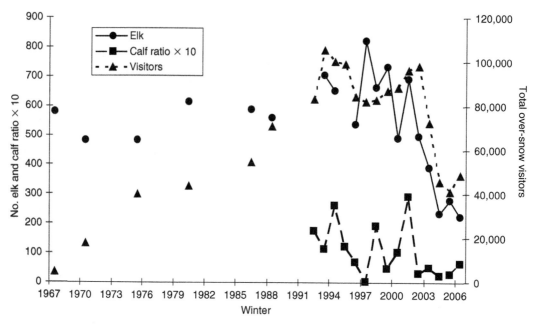

FIGURE 26A.2 Winter counts and calf ratios (10×) of central Yellowstone elk and numbers of visitors on OSVs during 1966–1967 through 2005–2006 in Yellowstone National Park, Montana and Wyoming, USA.

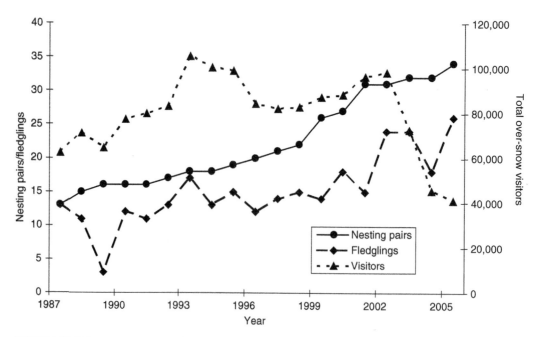

FIGURE 26A.3 Nesting pairs and fledglings of bald eagles and numbers of visitors on OSVs during 1986–1987 through 2004–2005 in Yellowstone National Park, Montana and Wyoming, USA (McEneaney 2006).

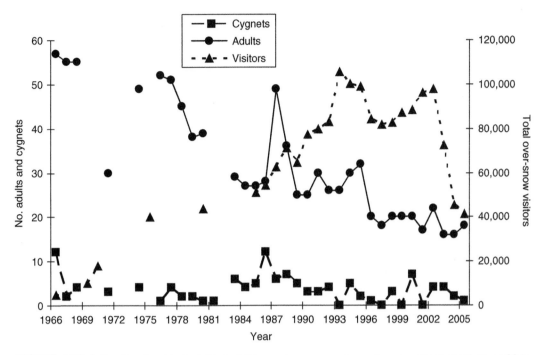

FIGURE 26A.4 Number of resident adult/subadult trumpeter swans and cygnets and numbers of visitors on OSVs during 1965–1966 through 2004–2005 in Yellowstone National Park, Montana and Wyoming, USA (McEneaney 2006).

CHAPTER 27

Bison Winter Road Travel: Facilitated by Road Grooming or a Manifestation of Natural Trends?

Jason E. Bruggeman,* Robert A. Garrott,* P. J. White,[†] Daniel D. Bjornlie,*
Fred G. R. Watson,[‡] and John J. Borkowski[§]

*Fish and Wildlife Management Program, Department of Ecology, Montana State University
[†]National Park Service, Yellowstone National Park
[‡]Division of Science and Environmental Policy, California State University Monterey Bay
[§]Department of Mathematical Sciences, Montana State University

Contents

Theme

The effects of road grooming on bison (*Bison bison*) distribution and movements in Yellowstone National Park have been debated since the early 1990s. Meagher (1993) expressed concern that energy saved by bison traveling on packed snow, in combination with better access to foraging habitat, resulted in enhanced population growth and increased movements to boundary areas. Thus, she recommended prohibiting road grooming to reduce the number and rate of bison leaving the park and induce them to revert to their traditional (*i.e.*, pre-road grooming) distributions (Meagher 2003). Conversely, Bjornlie and Garrott (2001) suggested that grooming of roads during winter did not have a major influence on bison ecology because of minimal use of roads for travel (19%) compared to off-road areas, decreased use of roads during the grooming season, and short distances traveled on roads. In 2003, the U.S. District Court for the District of Columbia rebuked the National Park Service

The Ecology of Large Mammals in Central Yellowstone
R. Garrott, P. J. White and F. Watson
ISSN 1936-7961, DOI: 10.1016/S1936-7961(08)00227-3

Copyright © 2009, Elsevier Inc.
All rights reserved.

for not identifying which argument it found persuasive and instructed the agency to gather necessary data to make a reasoned choice (294 F. Supp. 2d 92, 115). To aid in this decision, we documented (1) spatial patterns in bison road travel, (2) landscape attributes affecting these trends, and (3) abiotic and biotic factors influencing temporal variability in travel on and off roads during November-May from 1996–1997 through 2005–2006 to evaluate if trends in road travel were facilitated by road grooming or a manifestation of general bison travel patterns throughout the landscape.

I. INTRODUCTION

The effects of anthropogenic alterations to wildlife habitat are controversial (Cole and Landres 1996, Peterson *et al.* 2004). While habitat destruction is an extreme deleterious scenario, the coexistence of humans and wildlife is an important conservation issue, particularly owing to recent increased popularity of outdoor recreation (Cordell and Super 2000). Impacts of recreation on wildlife are often documented as detrimental to animal behavior, survival, and population level processes (Boyle and Samson 1985, Knight and Cole 1995, Taylor and Knight 2003). Further, roads may enable recreation and lead to avoidance of roads or inhibition of movement by some wildlife species (DeMaynadier and Hunter 2000, Trombulak and Frissell 2000, Laurance *et al.* 2004, Whittington *et al.* 2005) or, alternatively, the facilitation of travel by other species (St. Clair 2003).

Managers of Yellowstone National Park must conserve resources, while providing for their use and/ or enjoyment by people (National Park Service Organic Act of 1916; 16 USC 1, 2–4). However, the desires of people to view awe-inspiring geothermal features and wildlife at close range in Yellowstone may disrupt ecological processes, alter the behavior and distributions of wildlife, increase their energetic costs, or change their demography (Knight and Gutzwiller 1995). Thus, managers must address the effects of recreation on wildlife to ensure the integrity of these resources, and the ecosystem processes on which they depend, are not harmed (National Park Service 2006). The debate regarding motorized, over-snow, vehicle recreation in Yellowstone exemplifies the dilemma posed to managers by this dual mandate. Snow coaches and snowmobiles were first used in the park during 1955 and 1963, respectively, and park staff began grooming (*i.e.*, packing) snow-covered roads in 1971 to facilitate their safe passage. Snowmobile use increased dramatically in the following decades to more than 100,000 riders per year during the early 1990s (Yochim 1998). During this same period, the bison population increased from 700 to >4000 animals and expanded its range outside the park (National Park Service 2000). Many Yellowstone bison carry the pathogenic bacterium *Brucella abortus*, which produces abortions in bison, cattle, and elk (*Cervus elaphus*) and can be transmitted among these species (Thorne *et al.* 1978, Rhyan *et al.* 1994). This disease (brucellosis) has been the subject of a national eradication program for more than 70 years and has cost approximately $3.5 billion in public and private funds. Thus, starting in the mid-1980s, federal and state agencies negotiated a series of management agreements to manage bison moving outside the park that included hazing bison back into the park, capture and slaughter of bison that repeatedly left the park, culling of bison by agency personnel, and hunting of bison outside the park (National Park Service 2000).

The severe winter of 1996–1997 induced a large migration of bison out of the park, where >1000 were captured and removed from the population. This removal compelled several plaintiffs to file suit against the Department of Interior to end road grooming and snowmobiling, alleging the Department failed to adequately consider the effects of these activities on the behavior, distribution, and demography of bison and other wildlife (District of Columbia 2003). The plaintiffs contended the increased abundance, distribution, and culling of bison were direct consequences of energy savings provided by bison traveling on the packed road system (Figure 27.1) that led to better access to foraging habitat, increased survival, and enhanced movements outside the park (Meagher 1993). Thus, they sought an injunction prohibiting road grooming and snowmobiling to reduce the number and rate of bison leaving

FIGURE 27.1 Bison traveling on a groomed road in the Fountain Flats area of the Firehole drainage during winter in Yellowstone National Park (Photo by Jeff Henry).

the park and to induce bison to revert to their traditional, pre-road grooming distributions (District of Columbia 2003, Meagher 2003). In turn, the International Snowmobile Manufacturer's Association contested this petition as an unsupported ban on snowmobiling (District of Wyoming 2004).

The lack of rigorous studies to evaluate the merits of these opposing claims led to conflicting legal decisions and corresponding reactive changes in winter recreation regulations (National Park Service 2004). It is impossible to determine through retrospective analyses if groomed roads facilitated increased abundance and range expansion of bison because no detailed data on bison travel patterns existed prior to road grooming and bison are now familiar with destination ranges in the expanded range. Instead, biologists explored several alternate approaches to understand how road grooming and over-snow vehicle recreation currently influence the behavior, demography, distribution, and/or travel of bison and other wildlife in central Yellowstone. Bjornlie and Garrott (2001) and Bruggeman *et al.* (2006) concluded changes in bison spatial dynamics in Yellowstone National Park during the past three decades were the result of density-dependent range expansion, rather than road grooming. In addition, an extensive review of available information regarding bison movements and distribution found no evidence that groomed roads changed population growth rates of bison relative to what may have happened in the absence of road grooming (Gates *et al.* 2005). Bison were generally not constrained by winter snow pack because they maintained trails (*i.e.*, trenches in snow) in most corridors in the absence of road grooming. However, bison use of groomed roads aligned with natural movement corridors likely facilitated travel within traditional foraging areas and between ranges (Gates *et al.* 2005). Also, the Madison-Mammoth corridor appeared to be a barrier to movements during non-road grooming periods. Thus, road grooming of this corridor may have facilitated the increasing numbers of bison migrating from the central range to the northern range since the early 1980s (Meagher 1998, Gates *et al.* 2005).

Bison travel along roads appears more frequent in late winter and early spring, possibly in relation to snow depth and spring green-up (Bjornlie and Garrott 2001, Gates *et al.* 2005). However, previous assessments of the influence of road grooming on bison travel patterns were limited by the lack of a spatially explicit snow pack model to fully evaluate the effects of snow pack and the springtime melt on bison travel patterns. We addressed this limitation using a snow pack simulation model (Watson *et al.* 2006a,b, Chapter 6) and an information-theoretic approach to (1) determine the importance of snow pack, snow heterogeneity, bison density, forage accessibility, road grooming, and over-snow vehicle (snowmobile, coach) traffic on temporal variability in bison travel on and off roads, (2) evaluate competing models to determine topography and habitat attributes influencing spatial variability in the amount of road travel, and (3) examine bison activity patterns to evaluate the importance of travel behavior in bison ecology. The views and opinions in this chapter are those of the authors and should not be construed to represent any views, determinations, or policies of the National Park Service.

II. METHODS

We used the ground-based survey methods described by Bjornlie and Garrott (2001) and Bruggeman *et al.* (2006) to record the number, distribution, and activities of bison wintering in the Madison headwaters area of Yellowstone National Park every 10–14 days from late autumn until late spring during the winters of 1997–1998 through 2005–2006. We recorded the location, composition, and activity (*i.e.*, proportion foraging, resting, or traveling) of each group. Also, we recorded if traveling animals were on or off roads or trails and quantified the proportion of bison displacing snow during foraging and traveling activities.

In addition, bison travel on roads was recorded daily by four observers traveling independently on snowmobiles or trucks. Roads were traveled in early morning, before the daily influx of visitors, to record bison tracks on the freshly groomed roads and evaluate nocturnal travel. Observers also recorded the composition of all bison groups encountered on the road during their travels along road segments during the day (Bjornlie and Garrott 2001, Bruggeman *et al.* 2006). We mapped the locations of bison groups encountered traveling on the road during winters 2002–2003 through 2005–2006 by dividing the 72.6-km road network into 52 segments based on topographical similarities.

A. Analysis of Temporal Travel Patterns

We defined a road travel response variable (ρ_{ij}) with units of bison groups observed per 100 km of road surveyed for each 2-week time interval (i; $1 \leq i \leq 14$) during November through May of each winter (j; $1 \leq j \leq 9$). We used data from our road use surveys to calculate the response as $\rho_{ij} = \beta_{ij} \sigma_{ij}$, with the β_{ij} calculated by summing the number of bison groups observed traveling on roads for the *ij*th period and dividing by the total distance of road surveyed for that period. The unitless σ_{ij}, defined as the road use weighting factor for each period, was calculated as the total number of individual bison in road traveling groups for the *ij*th period divided by the total number of individual bison documented in road traveling groups for the entire season. This weighting factor was necessary because using β_{ij} by itself would treat all bison groups equivalently, whether the group consisted of two or 100 bison, and not provide an accurate quantification of road travel (Bruggeman *et al.* 2006).

Similarly, we defined an off-road travel response variable (τ_{ij}) using data from our ground distribution surveys. We defined the off-road response as $\tau_{ij} = \alpha_{ij} \gamma_{ij}$, where α_{ij} is the total number of bison groups observed traveling off-road per ground distribution survey for the *ij*th period. We defined γ_{ij} as a unitless off-road travel weighting factor calculated as the number of bison observed traveling off-road during the *ij*th period divided by the total number of bison observed traveling off-road for the entire season during ground distribution surveys.

We used a validated snow pack simulation model (Watson *et al.* 2006a,b, Chapter 6) to compute daily estimates of snow pack and forage availability metrics, with covariates calculated by averaging daily values across each period. We defined a covariate (SWE) to provide the mean snow water equivalent for all pixels within the bison range in the Madison headwaters study area and the Cougar Meadows complex. We defined another covariate (SNH) as the standard deviation of snow water equivalent estimates for all pixels within the bison range at elevations <2070 m to provide a measure of snow pack heterogeneity and "patchiness" at lower elevations. We defined a third covariate (COHE-SION) as the connectivity of snow-free patches at elevations <2070 m, with COHESION = 100 denoting a completely snow-free landscape (McGarigal and Marks 1995). The covariates SNH and COHESION provided indices of forage availability as patches of vegetation on lower-elevation, non-geothermal areas began to emerge during the springtime melt, which led to the subsequent green-up of nutrient-rich vegetation (Chapter 7).

We used the total number of bison counted during ground surveys for each period to define another covariate (BISON) that measured the influence of bison density on travel. For the road travel analysis, we defined a covariate (GROOM) to indicate if roads were groomed (0 = ungroomed and 1 = groomed) and another covariate (TRAFFIC) to provide a measure of the mean daily number of over-snow vehicles entering the park's West Yellowstone entrance during each period.

We developed *a priori* hypotheses to compare the relative contributions of snow pack, bison numbers, forage availability and, for the road travel analysis, road grooming and over-snow vehicle traffic to temporal variations in bison travel. Hypotheses for the on- and off-road analyses were expressed as 24 and eight candidate models, respectively, in the form of regression equations consisting of covariate main effects and interactions (Appendix). Owing to the complexity of the Yellowstone ecosystem, relationships between the amount of travel and each covariate may not be linear (Bruggeman *et al.* 2006). Therefore, we hypothesized linear and nonlinear forms for each continuous covariate and predicted the direction of correlation for each coefficient.

We predicted travel would be positively correlated with bison density and a quadratic form of this covariate (β_{1i}*BISON+β_{2i}*BISON^2) since increased competition for forage would result in a higher frequency of redistribution. Also, we predicted travel would be positively correlated with a moderated form of mean snow water equivalent (β_i*SWE$^{1/2}$), based on results from Bruggeman *et al.* (2006) depicting high amounts of bison travel when snow pack was near its peak. We anticipated snow effects would interact with bison density (SWE*BISON) such that travel would be accentuated during periods of high snow pack or bison density. We hypothesized that travel would be positively correlated with a moderated form of the standard deviation of snow water equivalent (β_i*SNH$^{1/2}$) because a more-variable snow pack would contain patches of snow-free vegetation during the springtime melt that bison could access for forage. Similarly, we predicted travel would be positively correlated with a quadratic form of the covariate for connectivity of snow-free patches at lower elevations (β_{1i}* COHESION+β_{2i}* COHESION^2) because bison would increase travel as patches of snow-free forage became larger and more connected. Also, we hypothesized that travel would be positively correlated with a BISON*COHESION interaction effect. Further, we predicted that road travel would be negatively correlated with a pseudo-threshold form for the mean daily number of over-snow vehicles (*ln* TRAFFIC) because bison would attempt to minimize negative interactions with over-snow vehicles. We expected that road travel would decrease during road grooming periods (GROOM = 1) because previous studies indicated bison traveled less during winter (Bruggeman *et al.* 2006). We also predicted attenuated effects of bison density and snow pack on travel during road grooming periods, with road travel negatively correlated with BISON*GROOM and SWE*GROOM interaction effects.

We used regression techniques in R version 2.3.1 (R Development Core Team 2006) to fit models and estimate coefficients. We censored one outlying, influential data point from both the on- and off-road data sets, leaving 91 and 96 observations, respectively. Residual and normal probability plots for both on- and off-road models demonstrated non-constant error variance and departures from normality in the error terms. Therefore, we applied a square-root transform on both response variables to stabilize variance and normalize errors. To allow comparisons of parameter coefficients on a similar scale, each continuous covariate was centered and scaled prior to analyses by subtracting the midpoint and dividing by half of the range resulting in values between −1 and 1. We also calculated variance inflation factors (VIFs) to assess potential multicollinearity between model predictors, including interactions (Neter *et al.* 1996). Any model containing a predictor that had a VIF > 5, given the other covariates in a model, was removed from the model list. Some biologically plausible interactions were not evaluated owing to multicollinearity. We used a sequential model fitting technique that incorporated our *a priori* candidate model list with the linear and nonlinear covariate functional forms (Borkowski *et al.* 2006). After finding the appropriate functional form for each covariate in each model (Bruggeman *et al.* 2006), we ranked and selected the best approximating models using ΔAIC_c values and calculated Akaike weights (w_i; Burnham and Anderson 2002).

B. Analysis of Spatial Travel Patterns

We defined a response variable (η_{kn}) to quantify the amount of bison travel in each of the 52 defined road segments (k) for each of the four years (n). We calculated η_{kn} with units of bison groups observed per segment per 100 km surveyed as the total number of bison groups observed traveling in the kth segment divided by the survey effort for that segment (*i.e.*, total km traveled by observers in the segment).

We used a Geographic Information System road layer to define nodes at 400 m intervals along the 72.6-km road network and calculated six covariates at each node to characterize the topography and habitat. A U.S. Geological Survey digital elevation model was used to calculate topography covariates, while habitat type covariates were determined using vegetation cover type and geothermal data layers (Watson *et al.* 2006a, 2007, Chapters 2 and 4). Topography covariates were calculated based on mean pixel values within a 200-m radius of the node location. Mean slope (SLOPE) and slope heterogeneity (SLHG), the standard deviation of the slope estimates within the circle, provided measures of landscape roughness (Bruggeman 2006). Also, we calculated the distances to nearest stream (DST), burned forest (DBF), unburned forest (DUF), and foraging area (DFA) from the node location with foraging areas defined as meadows ≥ 25 ha. Streams were defined using National Hydrographic Dataset Levels 3, 4, or 5, while excluding streams of Level 6 or smaller. We averaged covariate node values across each of the 52 segments to obtain one value for each covariate per road segment. Finally, we used indicator variables to define a covariate for year (*e.g.*, for 2002–2003, YEAR = 1 if data collected in 2002–03; YEAR = 0 otherwise).

We expressed our *a priori* hypotheses as 21 multiple regression models consisting of covariate main effects and interactions (Appendix) with VIFs < 5 to estimate the effects of topography and habitat type attributes on η. For each continuous covariate, we hypothesized one linear and nonlinear form and predicted the direction of correlation for each coefficient. We hypothesized η would be negatively correlated with exponential forms of distances to nearest stream, burned forest, and unburned forest (*e.g.*, β_i*exp DST) because bison would travel more in road segments that parallel streams or are closer to edges of forested habitats. We expected the effect of burned forest habitat on the amount of travel would be accentuated closer to streams (*i.e.*, negative DBF*DST interaction). Also, we predicted η would be positively correlated with a moderated form of distances to nearest foraging area because road segments farther from meadows would serve as travel corridors connecting important foraging areas. Next, we expected η to be negatively correlated with a moderated form of slope because bison generally avoid traveling across steep slopes. Further, we hypothesized a moderated form for slope heterogeneity, which we predicted to be positively correlated with η because road segments that pass through rugged topography (*i.e.*, canyons) would be more likely to constrain bison travel to the road. Likewise, we expected the impact of burned forest habitat to be greater in areas of rough topography, such that η would be positively correlated with a DBF*SLHG interaction. Finally, we expected YEAR to be positively correlated with η because bison abundance increased during 2002–2006.

We used regression techniques in R version 2.3.1 (R Development Core Team 2006) to fit models and estimate coefficients using centered and scaled continuous covariates. We censored two outlying data points, leaving 206 observations for the analysis, and applied a square-root transform on η because residual and normal probability plots demonstrated non-constant error variance and departures from normality in the error terms. We used a sequential model fitting technique to evaluate linear and nonlinear covariate functional forms (Borkowski *et al.* 2006), and ranked and selected the top models using ΔAIC_c values (Burnham and Anderson 2002).

III. RESULTS

We conducted 104 ground distribution surveys during the winters of 1997–1998 through 2005–2006, enumerating a low of 205 bison in January 2001 and a high of 1538 bison in April 2005. The number of bison migrating from the summer range to winter in the Madison headwaters study area generally increased through each winter before peaking in late March–April, with an average increase throughout each winter of 835 ± 80 (SE) bison (Figure 27.2A).

Snow pack began accumulating in late October, built throughout the winter, and generally peaked in late March, with a mean annual peak in snow water equivalent of 0.129 ± 0.016 m (Figure 27.2B). After peaking, the melt-out occurred rapidly, beginning in the lower elevation meadows in the Madison drainage and Cougar Meadows area, and gradually extended to higher elevations in the upper Firehole and Gibbon drainages. Heterogeneity in the snow pack increased through the winter before peaking around mid-March (mean annual peak $= 0.0014 \pm 0.0002$) when the snow pack began to melt and patches of snow-free vegetation appeared amidst remnant snow (Figure 27.3A). The covariate indexing connectivity of snow-free patches (COHESION) ranged from 32.4–99.7 (mean $= 73.3 \pm 1.6$) and generally reached a minimum in January, when there were few snow-free patches, before increasing as snow pack melted (Figure 27.3B).

A. Temporal Variation in Bison Travel Behaviors

Seasonal patterns of bison traveling on roads were similar among years with the amount of travel decreasing through the middle of winter during the over-snow vehicle season and peaking in the spring, typically after road grooming ceased (Figure 27.4A). We observed 3602 bison groups traveling on the road during 1997–1998 through 2005–2006 (range $= 221$–656, mean $= 400 \pm 50$). The annual total number of individual bison in these groups ranged from 3479–8538 (mean $= 5005 \pm 618$). Road survey effort ranged from 15,067–34,464 km (mean $= 23,495 \pm 1965$). Early morning road surveys before visitor traffic detected 349 sets of bison tracks, including 41% from nocturnal travel bouts, 45% from diurnal travel, and 14% from travel at an unknown time.

We found three top approximating models with $\Delta AIC_c < 2$ for the road travel analysis. The top model ($AIC_c = -86.56$, $K = 7$, $w_i = 0.27$) included GROOM, moderated SNH, quadratic BISON, and quadratic COHESION effects, and had a relative likelihood of 1.5 compared to the second model ($\Delta AIC_c = 0.77$, $K = 5$, $w_i = 0.18$) that included SWE, BISON, GROOM, and a SWE*BISON interaction (Table 27.1). The third model ($\Delta AIC_c = 1.68$, $K = 6$, $w_i = 0.12$) included significant SWE, GROOM, and SWE*BISON effects.

Seasonal patterns of bison observed traveling off roads during ground distribution surveys (Figure 27.4B), were similar to those observed for bison traveling on roads during road surveys. Ground distribution surveys ($n = 104$) during winters 1997–1998 through 2005–2006 detected 808 bison groups traveling off roads, ranging from 68–125 groups (mean $= 90 \pm 6$) per winter. The number of individual bison in groups traveling off roads varied from 426–2260 (mean $= 845 \pm 185$) per winter. Survey effort varied between 8–17 surveys (mean $= 12 \pm 1$) per winter.

There were two top approximating models for the off-road analysis. The top model ($AIC_c = 89.47$, $K = 3$, $w_i = 0.52$) included BISON and moderated SNH, and had a relative likelihood of 2.3 compared to the second model (Table 27.1). The second model ($\Delta AIC_c = 1.70$, $K = 4$, $w_i = 0.22$) also included a COHESION effect, though confidence intervals spanned zero.

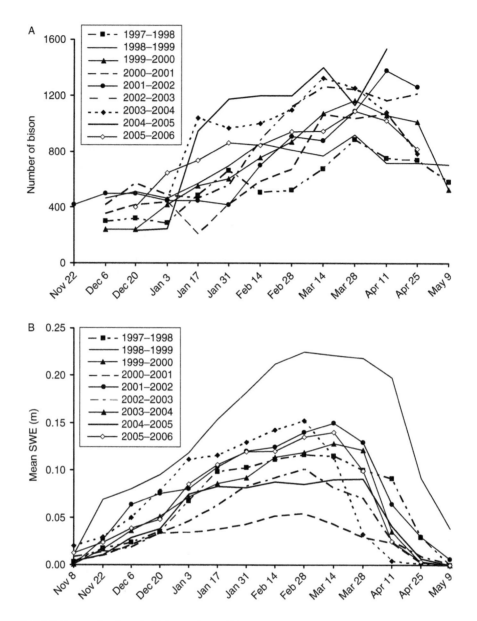

A

1600

Number of bison

1200

800

400

0

— ■-- 1997–1998
— 1998–1999
— ▲ 1999–2000
— - 2000–2001
— ● 2001–2002
— - 2002–2003
— ◆ 2003–2004
— 2004–2005
— ◇ 2005–2006

Nov 22 Dec 6 Dec 20 Jan 3 Jan 17 Jan 31 Feb 14 Feb 28 Mar 14 Mar 28 Apr 11 Apr 25 May 9

FIGURE 27.2 Temporal patterns in (A) the numbers of bison counted during ground distribution surveys and (B) mean snow water equivalent (SWE) in the Madison headwaters study area of Yellowstone National Park during the winters of 1998–2006.

B. Spatial Variation in Bison Road Travel

Road surveys during winters of 2002–2003 through 2005–2006 detected 1890 bison groups traveling on roads, ranging between 285–656 groups (mean = 473 ± 77) per winter. Road survey effort varied between 25,031–34,464 km (mean = 28,653 ± 2057) per winter. Bison travel on roads varied greatly among segments, with relatively high amounts of travel occurring in canyons and the lower Firehole drainage (Figure 27.5). The amount of bison travel per segment ranged between 0.0–24.2 bison groups observed per segment per 100 km traveled in that segment (mean = 3.6 ± 0.3).

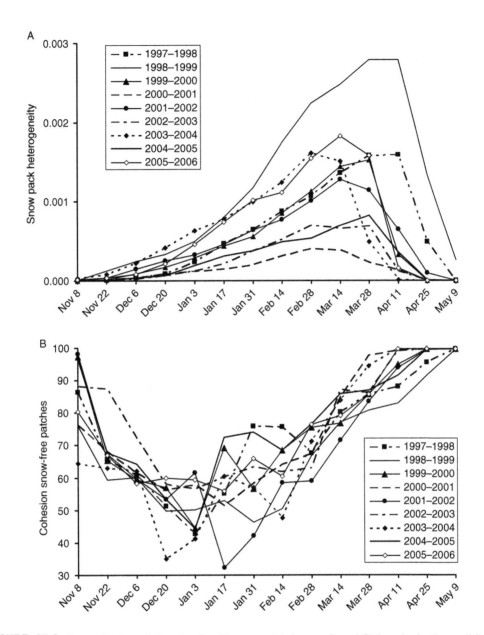

FIGURE 27.3 Temporal patterns in low elevation (A) snow pack heterogeneity and (B) the cohesion (connectivity) of snow-free patches in the Madison headwaters study area of Yellowstone National Park during the winters of 1998–2006.

We found three top approximating models for spatial variability in bison road travel. The best model (AIC$_c$ = 427.88, K = 9, w_i = 0.29) had a relative likelihood of 1.3 compared to the second (ΔAIC$_c$ = 0.46, K = 9, w_i = 0.23) and third (ΔAIC$_c$ = 0.52, K = 9, w_i = 0.22) models. All three models contained negative DST and DUF covariates and a positive YEAR effect for winters 2003–2004 and 2005–2006 (Table 27.2). Additionally, the best model included a positive DBF effect, moderated SLHG effect, and a negative DBF*SLHG interaction. The second model also included a positive DFA effect, while the third model included a positive DBF*DST interaction.

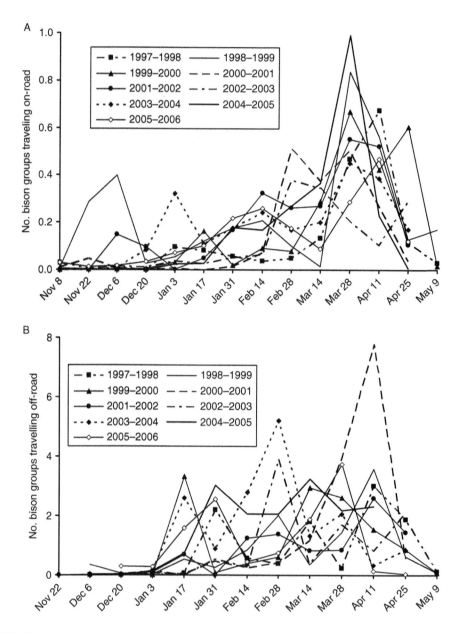

FIGURE 27.4 Temporal variability in bison travel (A) on roads and (B) off roads in the Madison headwaters study area of Yellowstone National Park during the winters of 1998–2006. Number of bison groups per 100 km of road surveyed (ρ_{ij}), calculated from our road use surveys, and number of bison groups traveling off-road per survey (τ_{ij}), calculated from our ground distribution surveys, are presented for each time interval (i) and year (j).

C. Activity Patterns

Sixty-seven percent of the 76,901 bison observed during ground distribution surveys were foraging, compared to 22% resting and 11% traveling. Thirty-one percent of foraging bison were displacing snow, compared to 7% of traveling bison. The majority of bison travel occurred off roads and trails (mean groups per survey = 6.4, $n = 695$), compared with on roads (mean = 1.8, $n = 225$) or on trails (mean = 1.3, $n = 173$, $P < 0.001$).

TABLE 27.1 Coefficient values (βi) and 95% confidence limits for covariates from the best approximating models of temporal variability in bison travel on and off roads

Covariate	Model 1 (β_i)	Model 2 (β_i)	Model 3 (β_i)
Travel On Roads			
SNH	**0.204 (0.079, 0.329)**		
SWE		**0.154 (0.077, 0.231)**	**0.161 (0.083, 0.240)**
BISON	**0.211 (0.121, 0.302)**	0.105 (−0.006, 0.216)	0.076 (−0.045, 0.198)
BISON²	−0.180 (−0.328, 0.032)		
GROOM	**−0.194 (−0.293, −0.094)**	**−0.186 (−0.261, −0.112)**	**−0.145 (−0.248, −0.042)**
COHESION	**−0.143 (−0.267, −0.019)**		
COHESION²	**0.132 (0.010, 0.255)**		
TRAFFIC			−0.052 (−0.143, 0.039)
SWE*BISON		**−0.265 (−0.465, −0.064)**	**−0.289 (−0.494, −0.084)**
Travel Off Roads			
SNH	**0.545 (0.305, 0.784)**	**0.579 (0.318, 0.839)**	
BISON	**0.782 (0.611, 0.953)**	**0.726 (0.487, 0.964)**	
COHESION		0.070 (−0.135, 0.274)	

Values in bold denotes significant coefficients at $\alpha = 0.05$. *Abbreviations:* SNH (standard deviation of snow water equivalent); SWE (snow water equivalent); BISON (bison density); GROOM (road grooming indicator); COHESION (connectivity of snow-free patches); and TRAFFIC (daily number of over-snow vehicles from the West Yellowstone entrance).

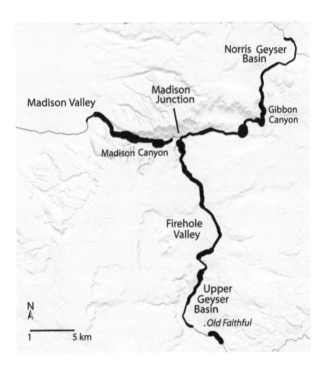

FIGURE 27.5 Spatial variability in bison travel on roads in the Madison headwaters study area of Yellowstone National Park during the winters of 2002–2003 through 2005–2006. Roads are depicted in dark gray and the width of the black lines depicts mean bison travel on each road segment.

TABLE 27.2 Coefficient values (β_i) and 95% confidence limits for covariates from the three best approximating models of spatial variability in bison travel on roads

Covariate	Model 1 (β_i)	Model 2 (β_i)	Model 3 (β_i)
DST	**−0.691 (−0.968, −0.413)**	**−0.500 (−0.742, −0.259)**	**−0.502 (−0.744, −0.261)**
DUF	**−0.745 (−0.968, −0.522)**	**−0.646 (−0.873, −0.418)**	**−0.724 (−0.944, −0.504)**
SLHG	**0.757 (0.161, 1.352)**	0.094 (−0.142, 0.331)	0.122 (−0.109, 0.354)
DBF	**0.581 (0.212, 0.950)**	**0.266 (0.042, 0.490)**	−0.096 (−0.434, 0.243)
DFA		**0.251 (0.031, 0.472)**	
YEAR 2004	**0.599 (0.342, 0.856)**	**0.599 (0.342, 0.856)**	**0.599 (0.342, 0.856)**
YEAR 2005	0.030 (−0.226, 0.287)	0.030 (−0.227, 0.287)	0.030 (−0.227, 0.287)
YEAR 2006	**0.560 (0.301, 0.819)**	**0.560 (0.301, 0.820)**	**0.559 (0.300, 0.819)**
DBF*SLHG	**−0.974 (−1.671, −0.277)**		
DBF*DST			**0.297 (0.035, 0.559)**

Values in bold denotes significant coefficients at $\alpha = 0.05$. *Abbreviations*: DST (distance to nearest stream); DUF (distance to unburned forest); SLHG (slope heterogeneity); DBF (distance to burned forest); DFA (distance to foraging area); and YEAR 2004 (data collected in winter 2003–2004).

IV. DISCUSSION

Road travel by bison was negatively correlated with the road grooming period, suggesting grooming did not facilitate bison travel during winter. The decrease in road travel did not appear to be an avoidance response to over-snow vehicle traffic because the covariate measuring the mean daily number of over-snow vehicles entering the park's West Yellowstone entrance (TRAFFIC) was not significant in the top models. Other work also indicates bison were somewhat habituated to over-snow vehicles and generally only responded if they were on or near roads (Borkowski *et al.* 2006, Chapter 26). Further, bison in Prince Albert National Park, Canada, responded to human disturbance, but did not alter their distribution and resource use following negative interactions with people (Fortin and Andruskiw 2003). We suspect decreased on- and off-road travel by Yellowstone bison during winter reflects that foraging areas (except geothermal basins) were covered with deep snow and contained senescent vegetation. As a result, bison had little impetus to expend energy to travel from one area to another in search of energy efficient foraging opportunities.

Temporal trends in road travel were likely a manifestation of general travel patterns driven by factors affecting resource availability. Bison density, snow pack heterogeneity, and cohesion of snow-free patches affected temporal variation in bison travel both on and off roads. Bison traveled less, both on and off roads, during winter compared to spring, when snow pack begins to melt and patches of snow-free vegetation provide for efficient foraging (Figure 27.6). In other regions, bison movements and habitat use are affected by seasonality, snow pack, and forage availability and quality (Campbell and Hinkes 1983, Larter and Gates 1991, Fortin *et al.* 2003, Rouys 2003). Resource availability influences ungulate movements and distributions in all seasons (*e.g.*, Wilmshurst *et al.* 1999, Friar *et al.* 2005, Fryxell *et al.* 2005), but limitations imposed by snow pack and climate during winter are especially important to animal survival.

Additional evidence that trends in road travel are an extension of natural bison travel patterns comes from spatial data documenting large variation in bison use of the road network depending on the habitat attributes and topography surrounding road segments, with some segments acting as travel corridors (Figure 27.5). The top spatial travel models indicated topography and distances to streams, forested habitats, and foraging areas significantly influenced the amount of bison road travel, with

FIGURE 27.6 Bison foraging during the spring melt period along the Firehole River near Midway Geyser Basin. Most animals at this time foraged in patches of vegetation amid remnant snow pack (Photo by Shana Dunkley).

more travel occurring on road segments closer to streams and unburned forest, farther from foraging areas, and passing through canyons. Additionally, a negative interaction between distance to burned forest and slope heterogeneity, combined with a positive distance to burned forest main effect, predicted increased road travel closer to burned forests. Analyses of bison GPS data revealed similar trends, with streams acting as natural travel pathways, travel more likely in canyons and closer to forested habitats, and certain road segments used as high use travel corridors (Bruggeman *et al.* 2007, Chapter 28).

The road network through the winter range for the central bison herd traverses a heterogeneous landscape consisting of canyons, meadow complexes, geothermal basins, and burned and unburned forest. Therefore, bison travel on roads may reflect both proximal and distant features. Amounts of road travel were highest through the Firehole, Gibbon, and Madison Canyons, which offer no alternate, energy-efficient travel pathways. Road segments through these canyons were surrounded by burned forest that contains numerous downed trees which make travel difficult. The road segments through canyons served as bison travel corridors connecting key foraging areas. Similar high use travel corridors existed distant from roads along the Gneiss Creek and Nez Perce Creek trails (Bruggeman *et al.* 2006, Figure 27.7). Road travel was lowest in large meadow complexes, through which many travel paths are possible and where road segments led away from foraging areas (Figure 27.5).

Bison travel on and off roads was positively correlated with snow pack heterogeneity as a moderated effect, suggesting bison responded to the springtime melt by increasing their travel to lower elevation meadows where snow pack was partially melted and forage exposed. Additionally, both on- and off-road travel were positively correlated with snow-free patch cohesion in top models. The snow pack is most heterogeneous at the beginning of the melt, when small snow-free patches appear amidst deep, remnant snow. As the spring progresses and snow-free patches become more connected, vegetation begins to green-up, which attracts bison owing to nutritious, higher quality forage (Van Soest 1994).

FIGURE 27.7 A large herd of bison traveling from Gneiss creek trail into Cougar meadows on a self-maintained trail to forage on newly-exposed vegetation. With the onset of the melt period this heavily-used trail from the Madison canyon to these extensive meadow complexes was one of the major travel routes used by bison during late winter and spring (Photo by Shana Dunkley).

Similar to other herbivores (McNaughton 1990, Albon and Langvatn 1992, Pettorelli *et al.* 2005), bison appear to follow vegetation green-up across the landscape, starting with lower elevation meadows in the Madison drainage and Cougar Meadows complex, and ending in the upper Firehole drainage.

Some top models for bison traveling on roads included a positive effect of snow water equivalent rather than a snow pack heterogeneity effect. The springtime peak in snow water equivalent coincided with the beginning of melt-out in lower elevations. Thus, the positive correlation between bison road travel and increasing snow pack probably resulted from high snow water equivalent levels at the time when bison redistributed themselves to partially melted-out foraging areas. This spurious correlation illustrates the difficulty in using average landscape scale metrics, such a mean snow water equivalent across the winter range, to describe ecologically complex processes that are partially influenced by small-scale factors (Hobbs 2003).

Bison density was positively correlated with on- and off-road travel, suggesting the frequency of redistribution increases as more bison migrate to the winter range. Travel bouts resulting from this redistribution owe to increased competition for resources because as more bison enter a foraging area, "optimal" foraging conditions may no longer exist at some occupancy threshold. Density-related movement responses have been documented for bison in other regions (Gates and Larter 1990) and other ungulates (Heard and Calef 1986, Reynolds 1998, Mahoney and Schaefer 2002, Ramp and Coulson 2002). However, the effect of bison density on travel cannot be evaluated without considering interactions with other abiotic and biotic factors because of the importance of multiple scale dependent effects on bison spatial dynamics (Bruggeman 2006). Top road travel models included a negative SWE*BISON interaction that attenuated the positive main effects of snow and bison density at high levels of either covariate. It is likely that significant interactions between bison density and snow heterogeneity also exist, however this interaction was not evaluated owing to multicollinearity. The coupling of these ecological processes has important effects on bison travel patterns because peak in travel occurs when the number of bison on the winter range is near its maximum, which coincides with the beginning of springtime melt when snow pack is most heterogeneous.

Activity scans indicated travel comprised a small amount (11%) of bison activity and few (7%) traveling animals actually were displacing snow. This finding is not the result of bison traveling on groomed roads to avoid energy expenditures from displacing snow (Parker *et al.* 1984) because activity scans documented off-road travel to comprise the majority (79%) of bison travel. Rather, Yellowstone bison generally used "self-groomed," off-road, trail networks (Figure 27.7; Bjornlie and Garrott 2001, Bruggeman *et al.* 2006, 2007) similar to bison in other regions (Gates *et al.* 2001). As a result, foraging is the most time-intensive, energetically costly activity for bison during winter.

The ecological effects of road grooming on current bison travel patterns appear minimal, with movements driven by abiotic and biotic factors affecting resource availability. Bison travel on and off roads, and their spatial use of the road network depends on landscape attributes surrounding the road. While roads may facilitate bison travel in certain areas (Bruggeman *et al.* 2007, Chapter 28), winter road grooming does not currently appear to promote bison movements beyond park boundaries because, in part, the majority of these movements occur during spring after grooming has ceased.

V. SUMMARY

1. The effects of road grooming on bison distribution and movements in Yellowstone National Park have been debated since the early 1990s. Opponents claim energy saved by bison traveling on packed snow, in combination with better access to foraging habitat, results in enhanced population growth and increased movements to boundary areas.

2. We collected spatial and temporal data on bison travel on and off roads during the winters of 1997–1998 through 2005–2006 to evaluate if road travel was facilitated by road grooming or a manifestation of general bison travel patterns throughout the landscape.

3. Road travel was negatively correlated with road grooming, suggesting grooming did not facilitate bison travel during winter. Temporal trends in road travel were likely a manifestation of general travel patterns because travel on and off roads was driven by factors affecting resource availability, including bison density, snow pack heterogeneity, and cohesion of snow-free patches.

4. Bison use of roads varied depending on habitat attributes and topography surrounding road segments, with certain segments acting as travel corridors. Topography and distances to streams, forested habitats, and foraging areas were significant influences on the amount of bison road travel, with more travel occurring on road segments closer to streams and unburned forest, farther from foraging areas, and passing through canyons.

5. Foraging was the most time-intensive activity (67%) by bison during winter. Traveling comprised a small amount (11%) of bison activity, with the majority (79%) of travel occurring off road. Thirty-one percent of foraging bison displaced snow, compared to only 7% of traveling animals.

6. The ecological effects of road grooming on current bison travel patterns appear minimal, with no evidence that grooming facilitates bison movements beyond park boundaries.

VI. REFERENCES

Albon, S. D., and R. Langvatn. 1992. Plant phenology and the benefits of migration in a temperate ungulate. *Oikos* **65**:502–513.

Bjornlie, D. D., and R. A. Garrott. 2001. Effects of winter road grooming on bison in Yellowstone National Park. *Journal of Wildlife Management* **65**:560–572.

Borkowski, J. J., P. J. White, R. A. Garrott, T. Davis, A. Hardy, and D. J. Reinhart. 2006. Wildlife responses to motorized winter recreation in Yellowstone National Park. *Ecological Applications* **16**:1911–1925.

Bruggeman, J. E. 2006. Spatio-temporal dynamics of the central bison herd in Yellowstone National Park. Dissertation, Montana State University, Bozeman, MT.

Bruggeman, J. E., R. A. Garrott, D. D. Bjornlie, P. J. White, F. G. R. Watson, and J. J. Borkowski. 2006. Temporal variability in winter travel patterns of Yellowstone bison: The effects of road grooming. *Ecological Applications* **16**:1539–1554.

Bruggeman, J. E., R. A. Garrott, P. J. White, F. G. R. Watson, and R. W. Wallen. 2007. Covariates affecting spatial variability in bison travel behavior in Yellowstone National Park. *Ecological Applications* **17**:1411–1423.

Burnham, K. P., and D. R. Anderson. 2002. *Model Selection and Multi-Model Inference.* Springer, New York, NY.

Boyle, S. A., and F. B. Samson. 1985. Effects of nonconsumptive recreation on wildlife: A review. *Wildlife Society Bulletin* **13**:110–116.

Campbell, B. H., and M. Hinkes. 1983. Winter diets and habitat use of Alaska bison after wildfire. *Wildlife Society Bulletin* **11**:16–21.

Cole, D. N., and P. B. Landres. 1996. Threats to wilderness ecosystems: Impacts and research needs. *Ecological Applications* **6**:168–184.

Cordell, K. H., and G. R. Super. 2000. Trends in Americans' outdoor recreation. Pages 133–144 *in* W. C. Gartner and D. W. Lime (Eds.) *Trends in Outdoor Recreation, Leisure, and Tourism.* CABI Publishing, New York, NY.

DeMaynadier, P. G., and M. L. Hunter. Jr. 2000. Road effects on amphibian movements in a forested landscape. *Natural Areas Journal* **20**:56–65.

District of Columbia. 2003. The Fund for Animals v. Norton, 294 F. Supp. 2d 92, 115. December 16, 2003. U.S. District Court, Washington, DC.

District of Wyoming. 2004. International Snowmobile Manufacturers Association v. Norton, 304 F. Supp. 2d 1278, 1285. February 10, 2004. U.S. District Court, Cheyenne, WY.

Fortin, D., and M. Andruskiw. 2003. Behavioral response of free-ranging bison to human disturbance. *Wildlife Society Bulletin* **31**:804–813.

Fortin, D., J. M. Fryxell, L. O'Brodovich, and D. Frandsen. 2003. Foraging ecology of bison at the landscape and plant community levels: The applicability of energy maximization principles. *Oecologia* **134**:219–227.

Friar, J. L., E. H. Merrill, D. R. Visccher, D. Fortin, H. L. Beyer, and J. M. Morales. 2005. Scales of movement by elk (*Cervus elaphus*) in response to heterogeneity in forage resources and predation risk. *Landscape Ecology* **20**:273–287.

Fryxell, J. M., J. F. Wilmshurst, A. R. E. Sinclair, D. T. Haydon, R. D. Holt, and P. A. Abrams. 2005. Landscape scale, heterogeneity, and the viability of Serengeti grazers. *Ecology Letters* **8**:328–335.

Gates, C. C., J. Mitchell, J. Wierzchowski, and L. Giles. 2001. *A Landscape Evaluation of Bison Movements and Distribution in Northern Canada.* AXYS Environmental Consulting Ltd., Calgary, Alberta, Canada.

Gates, C. C., B. Stelfox, T. Muhly, T. Chowns, and R. J. Hudson. 2005. *The Ecology of Bison Movements and Distribution in and Beyond Yellowstone National Park.* University of Calgary, Alberta, Canada.

Gates, C. C., and N. C. Larter. 1990. Growth and dispersal of an erupting large herbivore population in northern Canada: The Mackenzie wood bison (*Bison bison athabascae*). *Arctic* **43**:231–238.

Heard, D. C., and G. W. Calef. 1986. Population dynamics of the Kaminuriak caribou herd, 1968–1985. *Rangifer Special Issue* **1**:159–166.

Hobbs, N. T. 2003. Challenges and opportunities in integrating ecological knowledge across scales. *Forest Ecology and Management* **181**:223–238.

Knight, R. L., and D. N. Cole. 1995. Wildlife Responses to Recreationists. Pages 51–69 *in* R. L. Knight and K. J. Gutzwiller (Eds.) *Wildlife and Recreationists: Coexistence Through Management and Research.* Island Press, Washington, DC.

Knight, R. L., and K. J. Gutzwiller (Eds.). 1995. *Wildlife and recreationists: Coexistence Through Management and Research.* Island Press, Washington, DC.

Larter, N. C., and C. C. Gates. 1991. Diet and habitat selection of wood bison in relation to seasonal changes in forage quantity and quality. *Canadian Journal of Zoology* **69**:2677–2685.

Laurance, S. G. W., P. C. Stouffer, and W. F. Laurance. 2004. Effects of road clearings on movement patterns of understory rainforest birds and central Amazonia. *Conservation Biology* **18**:1099–1109.

Mahoney, S. P., and J. A. Schaefer. 2002. Long-term changes in demography and migration of Newfoundland caribou. *Journal of Mammalogy* **83**:957–963.

McGarigal, K., and B. J. Marks. 1995. *FRAGSTATS: Spatial pattern analysis program for quantifying landscape structure.* General Technical Report PNW-GTR-351. U.S. Department of Agriculture, Forest Service, Pacific Northwest Research Station, Portland, OR.

McNaughton, S. J. 1990. Mineral nutrition and seasonal movements of African migratory ungulates. *Nature* **345**:613–615.

Meagher, M. 1973. The bison of Yellowstone National Park. National Park Service Scientific Monograph Series No. 1.

Meagher, M. 1993. Winter recreation-induced changes in bison numbers and distribution in Yellowstone National Park. Unpublished report, Yellowstone National Park, WY.

Meagher, M. 1998. Recent changes in Yellowstone bison numbers and distribution. Pages 107–112 *in* L. Irby and J. Knight (Eds.) *International Symposium on Bison Ecology and Management in North America.* Montana State University, Bozeman, MT.

Meagher, M. M. 2003. Declaration to the United States District Court for the District of Columbia, CA 02-2367(EGS), Executed September 30, 2003, in Gardiner, MT .

National Park Service.2000. Bison Management Plan for the State of Montana and Yellowstone National Park—Final Environmental Impact Statement. U.S. Department of the Interior, Denver, CO.

National Park Service. 2004. Special regulations, areas of the national park system, final rule. *Federal Register* **69**:65348–65366.

National Park Service. 2006. Management policies. U.S. Department of the Interior, Washington, DC.

Neter, J., M. H. Kutner, C. J. Nachtsheim, and W. Wasserman. 1996. *Applied Linear Statistical Models.* , New York, NY: McGraw-Hill, New York, NY.

Parker, K. L., C. T. Robbins, and T. A. Hanley. 1984. Energy expenditures for locomotion by mule deer and elk. *Journal of Wildlife Management* **48**:474–488.

Peterson, M. N., S. A. Allison, M. J. Peterson, T. R. Peterson, and R. R. Lopez. 2004. A tale of two species: Habitat conservation plans as bounded conflict. *Journal of Wildlife Management* **68**:743–761.

Pettorelli, N., A. Mysterud, N. G. Yoccoz, R. Langvatn, and N. C. Stenseth. 2005. Importance of climatological downscaling and plant phenology for red deer in heterogeneous landscapes. *Proceedings of the Royal Society B* **272**:2357–2364.

R Development Core Team. 2006. R: A language and environment for statistical computing. R Foundation for Statistical Computing, Vienna, Austria. ISBN 3-900051-00-3. Available at http://www.R-project.org.

Ramp, D., and G. Coulson. 2002. Density dependence in foraging habitat preference of eastern grey kangaroos. *Oikos* **98**:393–402.

Reynolds, P. E. 1998. Dynamics and range expansion of a reestablished muskox population. *Journal of Wildlife Management* **62**:734–744.

Rhyan, J. C., W. J. Quinn, L. S. Stackhouse, J. J. Henderson, D. R. Ewalt, J. B. Payeur, M. Johnson, and M. Meagher. 1994. Abortion caused by *Brucella abortus* biovar 1 in a free-ranging bison (*Bison bison*) from Yellowstone National Park. *Journal of Wildlife Diseases* **30**:445–446.

Rouys, S. 2003. Winter movements of Europoean bison in the Białowieza Forest, Poland. *Mammalian Biology* **68**:122–125.

St. Clair, C. C. 2003. Comparative permeability of roads, rivers, and meadows to songbirds in Banff National Park. *Conservation Biology* **17**:1151–1160.

Taylor, A. R., and R. L. Knight. 2003. Wildlife responses to recreation and associated visitor perceptions. *Ecological Applications* **13**:951–963.

Thorne, E. T., J. K. Morton, F. M. Blunt, and H. A. Dawson. 1978. Brucellosis in elk. II. Clinical effects and means of transmission as determined through artificial infections. *Journal of Wildlife Diseases* **14**:280–291.

Trombulak, S. C., and C. A. Frissell. 2000. Review of ecological effects of roads on terrestrial and aquatic communities. *Conservation Biology* **14**:18–30.

Van Soest, P. J. 1994. *Nutritional Ecology of the Ruminant.* , Ithaca, New York, NY: Cornell University Press, Ithaca, New York, NY.

Watson, F. G. R., W. B. Newman, J. C. Coughlan, and R. A. Garrott. 2006a. Testing a distributed snow pack simulation model against spatial observations. *Journal of Hydrology* **328**:453–466.

Watson, F. G. R., T. N. Anderson, W. B. Newman, S. E. Alexander, and R. A. Garrott. 2006b. Optimal sampling schemes for estimating mean snow water equivalents in stratified heterogeneous landscapes. *Journal of Hydrology* **328**:432–452.

Watson, F. G. R., R. E. Lockwood, W. B. Newman, T. N. Anderson, and R. A. Garrott. 2007. Development and comparison of Landsat radiometric and snowpack model inversion techniques for estimating geothermal heat flux. *Remote Sensing of Environment* doi:10.1016/j.rse.2007.05.010.

Whittington, J., C. C. St. Clair, and G. Mercer. 2005. Spatial responses of wolves to roads and trails in mountain valleys. *Ecological Applications* **15**:543–553.

Wilmshurst, J. F., J. M. Fryxell, B. P. Farm, A. R. E. Sinclair, and C. P. Henschel. 1999. Spatial distribution of Serengeti wildebeest in relation to resources. *Canadian Journal of Zoology* **77**:1223–1232.

Yochim, M. J. 1998. The development of snowmobile policy in Yellowstone National Park. Thesis. University of Montana, Missoula, MT.

APPENDIX

APPENDIX 27A.1	Candidate *a priori* models for analysis of bison on-road travel in Yellowstone National Park during winters 1997–1998 through 2005–2006

Model	Structure
1	SWEMGF + BISON + GROOM
2	SWEMGF + BISON + GROOM + SWEMGF*BISON
3	SWEMGF + BISON + GROOM + BISON*GROOM
4	SWEMGF + BISON + GROOM + SWEMGF*GROOM
5	SWEMGF + BISON + GROOM + TRAFFIC
6	SWEMGF + BISON + GROOM + TRAFFIC + SWEMGF*BISON
7	SWEMGF + BISON + GROOM + TRAFFIC + BISON*GROOM
8	SWEMGF + BISON + GROOM + TRAFFIC + SWEMGF*GROOM
9	SWEMGF + BISON + GROOM + COHESION
10	SWEMGF + BISON + GROOM + COHESION + SWEMGF*BISON
11	SWEMGF + BISON + GROOM + COHESION + BISON*GROOM
12	SWEMGF + BISON + GROOM + COHESION + SWEMGF*GROOM
13	SWEMGF + BISON + GROOM + TRAFFIC + COHESION
14	SWEMGF + BISON + GROOM + TRAFFIC + COHESION + SWEMGF*BISON
15	SWEMGF + BISON + GROOM + TRAFFIC + COHESION + BISON*GROOM
16	SWEMGF + BISON + GROOM + TRAFFIC + COHESION + SWEMGF*GROOM
17	SHGLOW + BISON + GROOM
18	SHGLOW + BISON + GROOM + BISON*GROOM
19	SHGLOW + BISON + GROOM + TRAFFIC
20	SHGLOW + BISON + GROOM + TRAFFIC + BISON*GROOM
21	SHGLOW + BISON + GROOM + COHESION
22	SHGLOW + BISON + GROOM + COHESION + BISON*GROOM
23	SWEMGF + BISON + GROOM + TRAFFIC + COHESION
24	SHGLOW + BISON + GROOM + TRAFFIC + COHESION + BISON*GROOM

The response variable is the amount of bison road travel (bison groups observed per 100 km of road surveyed). Covariates are described in the text and linear forms are denoted (see text for nonlinear forms used in substitutions).

APPENDIX 27A.2	Candidate *a priori* models for analysis of bison off-road travel in Yellowstone National Park during winters 1997–1998 through 2005–2006

Model	Structure
1	SWEMGF + BISON
2	SWEMGF + BISON + SWEMGF*BISON
3	SWEMGF + BISON + COHESION
4	SWEMGF + BISON + COHESION + BISON*COHESION
5	SWEMGF + BISON + COHESION + SWEMGF*BISON
6	SHGLOW + BISON
7	SHGLOW + BISON + COHESION
8	SHGLOW + BISON + COHESION + BISON*COHESION

The response variable is the amount of bison off-road travel (bison groups observed travelling off-road per ground distribution survey). Covariates are described in the text and only linear forms are denoted (see text for nonlinear forms used in substitutions).

| APPENDIX 27A.3 | Candidate *a priori* models for analysis of bison on-road spatial travel patterns in Yellowstone National Park during winters 2002–2003 through 2005–2006 |

Model	Structure
1	DST + DUF + SLHG
2	DST + DUF + SLHG + DBF
3	DST + DUF + SLHG + SLOPE
4	DST + DUF + SLHG + DFA
5	DST + DUF + SLHG + YEAR
6	DST + DUF + SLHG + DBF + SLOPE
7	DST + DUF + SLHG + DBF + DFA
8	DST + DUF + SLHG + DBF + YEAR
9	DST + DUF + SLHG + SLOPE + DFA
10	DST + DUF + SLHG + SLOPE + YEAR
11	DST + DUF + SLHG + DFA + YEAR
12	DST + DUF + SLHG + DBF + SLOPE + DFA
13	DST + DUF + SLHG + DBF + SLOPE + YEAR
14	DST + DUF + SLHG + DBF + DFA + YEAR
15	DST + DUF + SLHG + SLOPE + DFA + YEAR
16	DST + DUF + SLHG + DBF + DBF*DST
17	DST + DUF + SLHG + DBF + DBF*SLHG
18	DST + DUF + SLHG + DBF + DFA + DBF*DST
19	DST + DUF + SLHG + DBF + DFA + DBF*SLHG
20	DST + DUF + SLHG + DBF + YEAR + DBF*DST
21	DST + DUF + SLHG + DBF + YEAR + DBF*SLHG

The response variable is the amount of bison road travel per road segment (bison groups observed per segment per 100 km surveyed in that segment). Covariates are described in the text and only linear forms are denoted (see text for nonlinear forms used in substitutions).

CHAPTER 28

Effects of Snow and Landscape Attributes on Bison Winter Travel Patterns and Habitat Use

Jason E. Bruggeman,* Robert A. Garrott,* P. J. White,† Fred G. R. Watson,‡ and Rick W. Wallen†

*Fish and Wildlife Management Program, Department of Ecology, Montana State University
†National Park Service, Yellowstone National Park
‡Division of Science and Environmental Policy, California State University Monterey Bay

Contents

Theme

Movement patterns evolved among animal species in response to diverse ecological pressures, including to find new resources, escape predation pressure, find new mates, and improve reproductive potential (Dobson 1982). As the bison (*Bison bison*) population in Yellowstone National Park increased following the cessation of intensive management in the late 1960s, bison began moving from isolated herds in the interior of the park to contiguous habitat along the western and northern boundaries. Climate (*e.g.*, winter severity) may have exacerbated such movements by reducing the availability of forage (Clutton-Brock *et al.* 1985, Gaillard *et al.* 2000, Sæther 1997). Also, human actions to the landscape (*e.g.*, roads, fences) can modify habitats in ways that result in profound changes in ungulate distribution. However, the underlying ecology of bison movements and the influence of natural and anthropogenic features in the Yellowstone landscape are not well documented in

The Ecology of Large Mammals in Central Yellowstone
R. Garrott, P. J. White and F. Watson
ISSN 1936-7961, DOI: 10.1016/S1936-7961(08)00228-5

Copyright © 2009, Elsevier Inc.
All rights reserved.

the peer-reviewed literature (Gates *et al.* 2005). We instrumented 30 adult female bison from the central herd with Global Positioning System collars during three winters (2003–2004 through 2005–2006) to quantify how snow, topography, habitat attributes, and roads influenced bison travel patterns and non-traveling activities (*i.e.*, foraging, resting). We used a behaviorally-based resource selection analysis to document: (1) spatial trends in bison travel and non-traveling activities, (2) dynamic and static landscape attributes affecting these patterns, and (3) the influence of snow pack on the odds of bison travel and non-traveling activities.

I. INTRODUCTION

Patterns in animal travel are a crucial aspect of ecology affecting population-level processes. Migration, dispersal, and smaller-scale movements all influence population dynamics through direct or indirect causes (Taylor and Taylor 1977, Dobson and Jones 1985, Dingle 1996). Thus, it is important to understand how an animal arrived at a given location to relate resource selection to population processes in a spatially heterogeneous environment. Climate, predators, and anthropogenic influences affect survival and can influence an animal's choice of habitat use and travel routes (Fraser *et al.* 1995, Ferguson and Elkie 2004). In addition, topography and habitat characteristics have been shown to affect the movements of fish (Meyer and Holland 2005), insects (Turchin 1991), mammals (Johnson *et al.* 2002), and birds (Williams *et al.* 2001). Gradients in elevation, topographic constraints, and habitat heterogeneity may guide animals to travel along paths that form natural travel corridors (Crist *et al.* 1992, Zollner and Lima 1999, Selonen and Hanski 2003, Dickson *et al.* 2005). For example, mountain ranges and steep slopes may impede travel (Wall *et al.* 2006) and form barriers to dispersal, thereby affecting population-level processes (Gebeyehu and Samways 2006, Pe'er *et al.* 2006). Also, rugged topography may channel movements into corridors that exist through canyon drainages or along ridges (Shkedy and Saltz 2000, Kie *et al.* 2005, Meyer and Holland 2005). Repeated use of these routes can form travel networks for both migratory and small-scale movements (Sinclair 1983, Haddad 1999, Cronin 2003, Flamm *et al.* 2005). Thus, understanding the mechanisms that influence the movement paths of animals is essential for comprehending behavior and predicting use of travel corridors.

Human effects on wildlife travel paths range from facilitation of movement for some species through the development of recreational trails, to hindrance by habitat degradation and fragmentation (Bruns 1977, Hilty and Merenlender 2004). Roads are controversial, because some species use them as major pathways, whereas others avoid them owing to traffic and/or human presence (Brody and Pelton 1989, Cole *et al.* 1997, Dyer *et al.* 2002, Ito *et al.* 2005, Whittington *et al.* 2005). Interactions between wildlife, roads, and outdoor recreation are high-profile issues, as the negative aspects on animals, such as disturbance, stress, lowered survival, and habitat degradation become the focus (Trombulak and Frissell 2000, Taylor and Knight 2003). The effects of winter recreation on large mammals (Freddy *et al.* 1986, Borkowski *et al.* 2006) are particularly controversial owing to the added physiological stresses of deep snow, cold temperatures, restricted forage, and predators (Moen 1976, Gabrielsen and Smith 1995, Garrott *et al.* 2005).

Ungulate movements are potentially affected not only by landscape attributes that are relatively unchanging over time, but also by dynamic processes (Fryxell *et al.* 2004, Fortin *et al.* 2005). The influence of snow on the population dynamics of ungulates is well documented, with increasing snow pack and winter severity resulting in reduced adult survival and recruitment (Garrott *et al.* 2003, Jacobson *et al.* 2004). Snow may also influence ungulate distribution and foraging behavior (Larter and Gates 1991, Fortin *et al.* 2003, Doerr *et al.* 2005). For example, muskoxen (*Ovibos moschatus*) preferred foraging patches with low snow cover and adjusted the amount of time spent in areas depending on energetic costs of travel through snow to reach other patches (Schaefer and Messier 1995a,b). This influence of snow pack on large herbivore foraging relates directly to resource limitations and increased energetic costs needed to displace snow (Parker *et al.* 1984), both of which may ultimately affect large-scale population processes. However, little research has examined the effects of snow on smaller-scale (*i.e.*, nonmigratory) movements. In part, this owes to the difficulty in obtaining small-scale estimates of snow attributes, which change daily owing to accumulation and ablation. Also, ungulates have home ranges across large spatial scales, thereby making the sampling of snow at fine spatial and temporal scales across entire ranges logistically unfeasible.

The influence of winter recreation on bison in Yellowstone has been a subject of debate since park staff began grooming (*i.e.*, packing) snow on interior park roads in 1971 to facilitate the passage of visitors on over-snow vehicles (*e.g.*, snowmobiles, coaches) from December to March. Motorized winter recreation increased substantially during the decades that followed, from 2000 to 100,000 riders per winter during the mid-1990s (Gates *et al.* 2005). Concurrently, counts of central Yellowstone bison increased from 500 to 3000 animals (National Park Service 2000). As the population grew, bison expanded their range through the Madison headwaters area and beyond Yellowstone's boundaries. Meagher (1993, 1998) attributed this expansion to purported energy savings from groomed roads, which allowed bison to better survive winters and produce healthy calves in the spring, resulting in an unnatural population increase and alteration of bison spatial dynamics.

Litigation has resulted in conflicting legal decisions from different courts, primarily owing to a lack of rigorous empirical studies to evaluate the merits of opposing claims. Previous attempts to address the effects of road grooming on travel by bison have been criticized for making strong inferences in the absence of rigorous experimental designs (*e.g.*, controls, replicates). Such studies are problematic in Yellowstone because shutting down sections of roads in winter reduces public access to the park and affects economic concerns by gateway communities and contracts with concessionaires. Also, potential annual variability in abiotic and biotic factors may confound any grooming effect. Detailed data on bison distribution and travel patterns were not collected before road grooming began and, thus, no true experimental control case of bison road travel exists before bison gained knowledge of foraging areas in the Madison headwaters area and elsewhere. As a result, it is impossible to conclusively determine through retrospective analyses why bison use of groomed roads began or if groomed roads enabled range expansion.

Given these constraints, we explored an alternate approach to quantify the influence of landscape and habitat type attributes on bison spatial use of travel routes in central Yellowstone and to gain insight into how roads may currently affect bison travel during winter. We used detailed movement data collected from bison instrumented with GPS radio collars in conjunction with a snow pack estimation model (Chapter 6 by Watson *et al.*, this volume) and other landscape data (Chapter 2 by Newman and Watson, this volume) to examine winter travel patterns and non-traveling activities (*i.e.*, foraging, resting) of bison in the central winter range (Figure 28.1). We used an information-theoretic approach to evaluate competing hypotheses and quantify the relative contributions of snow, topography, habitat, and roads in influencing the odds of bison travel and non-traveling activities. We used the top approximating models to predict the odds of bison travel and non-traveling activities given various snow conditions during a winter and developed maps to display predicted patterns. This chapter represents an extension of studies published by Bruggeman *et al.* (2007) that used data from the first year of deployments of GPS collars on bison and did not include consideration of temporally-dynamic snow pack attributes because the snow pack model described in Chapter 6 by Watson *et al.*, this volume, was not yet refined. This second generation of data and analyses enhances our understanding of how snow, topographical variation, habitat attributes, and roads affect bison movements and spatial ecology during winter. The views and opinions in this chapter are those of the authors and should not be construed to represent any views, determinations, or policies of the National Park Service.

II. METHODS

Fifteen telemetry collars (GPS/VHF Model TGW 3700, Telonics, Mesa, Arizona) were deployed on adult female bison beginning in November 2003 and then redeployed on different adult female bison beginning in November 2004. Some collars deployed in 2004 remained on animals until spring 2006.

FIGURE 28.1 Four bull bison travel along a self maintained trail on the West side of the Madison River after a fresh snowfall (Photo by Claire Gower).

Collars were distributed on bison in the Hayden and Pelican Valley summer range and on early migrants to the winter range using ground darting with Carfentanil. From November 2003 to mid-March 2004, locations were recorded every 30 min from 0700 to 1900 with fixes also taken at 2300 and 0300. From mid-March onward, locations were recorded every 30 min from 0600 to 2300 with fixes also recorded at 0100 and 0300. During November 2004–2006, locations were recorded every 48 min during both day and night. For our analysis we used only locations from winter (November–April) of each year.

We categorized the location data through a series of steps to separate traveling locations from non-traveling locations associated with foraging and resting activities. First, we removed all locations obtained >50 min apart to obtain the most accurate movement paths (*i.e.*, segments) between consecutive locations. Second, we calculated Euclidean distances (d) for each segment and turning angles (α ; $0° \leq \alpha \leq 180°$) between segments, enabling us to define threshold values of ≥ 800 m and $\leq 90°$, respectively, to indicate a significant movement (Bruggeman 2006, Bruggeman *et al.* 2007). The travel data set was comprised of vectors meeting both criteria, while segments with $d < 800$ m and/or α $> 90°$ comprised the non-traveling data set. After censoring, our travel and non-travel data sets consisted of the actual point locations associated with traveling and non-traveling vectors, respectively. We mapped each location into a Geographic Information System layer and sampled for covariates. Additionally, we created random data by taking each "used" travel or non-travel location and assigning 20 random locations that were each matched with the date of the used location. We obtained the random locations from the available sampling universe encompassing the central herd's winter range, including the central plateau. Each random location was then sampled for covariates. Locations were assigned coded binary response variables to be analyzed as use (1) versus availability (0) data using logistic regression techniques (Manly *et al.* 2002).

A. Snow Pack, Topography, and Habitat Type Covariates

We used a snow pack simulation model (Chapter 6 by Watson *et al.*, this volume) to calculate three covariates that estimated large- and small-scale snow water equivalents (SWEs) throughout the sampling universe. We calculated mean snow water equivalent at the landscape scale (SWE$_L$; water content of snow) as the daily average of all 28.5×28.5-m^2 pixels within the sampling universe and a mean snow water equivalent at the local scale, SWE_A, using all pixels within a 200-m radius around the location for that date. We also calculated snow heterogeneity at the landscape scale (SNH_L; spatial variability of snow) as the standard deviation of estimated SWE of all pixels within the sampling universe.

We calculated eight covariates to characterize topography and habitat type attributes for each location using Geographic Information System data layers. A U.S. Geological Survey digital elevation model was used to calculate topography covariates, while habitat type covariates were determined using vegetation cover type and geothermal data layers (Chapter 2 by Newman and Watson, and Chapter 4 by Watson *et al.*, this volume). Topography covariates (average slope, SLOPE; slope heterogeneity, $SLOPE_{het}$) were calculated based on averages of pixel values within a circle of 200-m radius from the location and provided measures of landscape roughness. We classified each location into one of five habitat (HAB) categories as meadow, burned forest, unburned forest, geothermal, or other (*i.e.*, talus or aquatic). Additionally, we calculated the nearest distances to stream ($DIST_{stream}$), burned forest ($DIST_{burn}$), unburned forest ($DIST_{unburn}$), foraging area ($DIST_{forage}$), and road ($DIST_{road}$) from the point location. Foraging areas were defined as meadows ≥ 25 ha and streams were defined from the National Hydrographic Dataset and included streams of levels 3, 4, or 5, while excluding streams of level 6 or smaller. A network of paved, two-lane roads paralleled the Madison, Gibbon, Firehole, and Yellowstone Rivers through the study area as described in Bruggeman *et al.* (2006). Roads were groomed daily for snowmobile and snow coach travel by visitors during mid-December until early March, at which point the roads were plowed and then opened in late April for the summer visitation season. Roads were open to visitor travel in wheeled vehicles from mid-April until early November, at which point they were closed to visitors to allow snow accumulation for the motorized winter recreation season.

B. Log Odds Ratios

We used log odds ratios (Agresti 2002) to determine the likelihood of a bison occurring at a particular location depending on local- and landscape-scale snow pack conditions. We sorted all actual and random locations into one of four categories depending on the landscape SWE_L (0.000–0.059, 0.060–0.119, 0.120–0.179, and ≥ 0.180 m) estimated for that date. Next, we sorted locations into local SWE_A levels, categorized by 0.05 m increments between 0.000 and 0.849 m. We then calculated the log odds and 95% confidence intervals of a bison location occurring in a particular local SWE_A level within each landscape SWE_L category, as described in Chapter 8 by Messer *et al.*, this volume. We also used this approach to calculate log odds ratios of local SWE_A within landscape SNH_L categories, which were divided into four 0.04 m increments (0.000–0.039, 0.040–0.079, 0.080–0.119, and ≥ 0.120 m).

C. *A Priori* Hypotheses

We formulated *a priori* hypotheses for each covariate regarding the direction of their effects on the log odds responses by bison for traveling and non-traveling activities (*e.g.*, feeding, resting). First, we hypothesized the odds of bison traveling and non-traveling activities would be negatively correlated with SLOPE because bison generally avoid steep slopes. Second, we predicted $SLOPE_{het}$ would positively affect the odds of travel since areas of variable topography would be more likely to influence bison choice of travel routes and restrict travel to corridors. For similar reasons, we predicted $SLOPE_{het}$ would be negatively correlated with the odds of non-traveling activities. Third, we hypothesized the odds of travel and non-traveling activities would be positively correlated with meadow and thermal habitats because bison often forage, rest, and establish travel networks through and connecting these habitats. Conversely, we expected the odds of travel and non-traveling activities would be negatively correlated with burned forest, unburned forest, and talus since bison often avoid traveling or feeding in burned areas containing downed trees, heavily forested regions lacking suitable forage, and rocky talus areas. Fourth, we predicted $DIST_{stream}$, $DIST_{burn}$, and $DIST_{unburn}$ would have negative effects on the odds of travel since bison would develop travel networks along stream corridors and near the edges of burned and unburned forest habitats. Conversely, we predicted these covariates would have positive effects on the odds of non-traveling activities since bison may move away from travel networks to feed and rest. Fifth, we hypothesized $DIST_{forage}$ would be negatively correlated with the odds of travel and non-traveling activities because bison travel routes would be less likely farther from foraging areas and bison predominately use foraging areas for eating and resting. We predicted $DIST_{road}$ would be negatively correlated with the odds of travel since bison are known to use some road corridors as travel routes. Conversely, we predicted $DIST_{road}$ would be positively correlated with the odds of non-traveling activities since bison would be more likely to forage or rest farther away from travel routes along some roads.

We hypothesized SWE_A would be negatively correlated with the odds of travel and non-traveling activities because bison would be less likely to travel or feed through deep snow at local scales. We hypothesized three interactions involving landscape scale SWE_L and distances to habitat types. First, we predicted the odds of travel and non-traveling activities would be negatively correlated with a $SWE_L \times DIST_{stream}$ interaction effect because bison would be less likely to travel or feed farther away from geothermally-influenced streams that reduce snow accumulation as SWE_L increased on the landscape. Second, we expected the odds of travel and non-traveling activities to be negatively correlated with a $SWE_L \times DIST_{forage}$ interaction effect because bison would be less likely to travel or feed farther away from foraging areas as landscape SWE_L increased. Finally, we hypothesized a $SWE_L \times DIST_{road}$ interaction would be negatively correlated with the odds of travel and non-travel activities because deeper snows at higher elevations confine bison to lower elevation meadows closer to roads during winter and because bison use some road corridors for travel. We also hypothesized three interactions involving landscape scale snow heterogeneity (SNH_L) and distances to habitat types. We predicted the odds of travel and non-traveling activities would be positively correlated with a $SNH_L \times DIST_{stream}$ interaction effect because bison would be more likely to travel and feed farther away from travel corridors along streams as variability in the snow pack increased across the landscape. Second, we expected the odds of travel to be positively correlated with a $SNH_L \times DIST_{forage}$ interaction effect because bison would be more likely to travel farther away from the main foraging areas as snow pack heterogeneity increased, thereby affording better foraging opportunities in certain meadows with lower SWE. For similar reasons, we predicted this interaction would be negatively correlated with non-traveling activities. Finally, we hypothesized a $SNH_L \times DIST_{road}$ interaction would be positively correlated with the odds of travel and non-traveling activities because bison would travel more and use more meadows and foraging areas away from roads as variability in the snow pack increased.

D. Model Development and Statistical Analyses

We compared *a priori* hypotheses, expressed as multiple logistic regression models, to estimate the relative influences of snow, topography, and habitat type attributes on the probabilities of bison travel and non-traveling activities in two separate modeling efforts. While forming our model list, we calculated variance inflation factors to quantify multicollinearity between model predictors, including interactions (Neter *et al.* 1996). We removed models containing predictors having a variance inflation factor >5 from our *a priori* list and, as a result, we could only evaluate six interactions for both the travel and non-travel analysis. Hypotheses for each modeling exercise were expressed as 48 candidate models consisting of biologically plausible combinations of additive main effects and interactions of covariates (Table 28A).

We used a conditional logistic regression analytical approach (Hosmer and Lemeshow 2000, Chapter 8 by Messer *et al.*, this volume) and fit models and estimated parameter coefficients using PROC PHREG in SAS version 9.1.3 (Allison 1999, Hosmer and Lemeshow 2000, SAS 2004). We stratified (*i.e.*, matched) the used and available locations by the used location date to account for changing landscape snow pack conditions. We also used a robust variance estimator (Lin and Wei 1989, Allison 2005) to account for any dependence among consecutive locations for each bison. By defining locations for each bison as an independent "cluster" (*i.e.*, travel/non-travel for one collared bison was not influenced by that of another), the robust variance estimator corrected for dependence within clusters by adjusting the standard errors of parameter estimates, but not the estimates themselves. All continuous covariates were centered and scaled prior to analysis by subtracting the midpoint and dividing by half of the range resulting in values between -1 and 1. We calculated the Akaike Information Criterion (AIC) for each model in each modeling effort, ranked and selected the best approximating models using ΔAIC values, and calculated Akaike weights (w_i) to obtain a measure of model selection uncertainty (Burnham and Anderson 2002). We used the top models to develop probability maps of bison travel and non-traveling activities given various snow pack conditions through a winter.

III. RESULTS

A total of 145,181 locations were obtained from 30 collared bison during November–April of winters 2003–2004 through 2005–2006. We censored 11,955 locations obtained >50 min apart and 19,165 locations that were outside of the central winter range. Applying distance and angle criteria to the remaining 114,061 locations provided data sets of 2841 travel and 111,220 non-travel locations. Because of the large amount of non-travel data, we selected every tenth location to be evaluated, which resulted in 11,122 non-travel locations. We then generated random data sets consisting of 56,820 and 222,440 available travel and non-travel locations, respectively. Travel and non-travel locations were distributed throughout the bison range (Figure 28.2), but there were increased concentrations of travel locations in canyons and corridors along the Mary Mountain trail, lower Firehole Canyon, Gibbon Canyon, and Madison Canyon. Non-travel locations were concentrated in meadow complexes in the Hayden Valley, Fountain Flats, and Cougar Meadows.

Snow pack in the central winter range during 2003–2004 through 2005–2006 increased during the winter, peaked in mid-March to early April, and then decreased rapidly as the spring time melt began (Figure27.2B). The peak average SWE_L for the central Yellowstone landscape was 0.24 m during 2003–2004, 0.19 m during 2004–2005, and 0.27 m during 2005–2006. For "used" locations, the maximum local SWE_A in which a bison was traveling was 0.34 m compared to 0.31 m for non-traveling activities. The maximum SWE_A for "available" traveling and non-traveling locations was 0.74 m and 0.83 m, respectively, due to the inclusion of the high-elevation Central Plateau and Mary Mountain in the available area.

As landscape-scale SWE increased, the odds of bison traveling or not traveling through differing snow pack on a local scale were altered because, in general, bison had to occupy areas of higher local scale SWE (Tables 28.1 and 28.2). Within a level of landscape SWE (*e.g*, 0.120–0.179 m) the odds of bison traveling or performing non-traveling activities (*e.g.*, feeding) decreased as local scale SWE_A increased (Figure 28.3). For example, at low levels of landscape SWE_L (0.000–0.059 m) the log odds ratio of bison travel decreased from 1.69 (95% CI: 1.17, 2.20) to -1.58 (-2.09, -1.06) as local SWE_A increased from 0.00–0.049 to 0.050–0.099 m. Thus, the likelihood of a point on the landscape being used by bison for traveling decreased as local SWE_A increased. Similarly, at a landscape SWE_L of 0.000–0.059 m the log odds ratio of bison non-traveling activities decreased from 1.05 (0.90, 1.19) to -0.94 (-1.08, -0.79) as local SWE_A increased from 0.000–0.049 to 0.050–0.099 m. Owing to small sample sizes of used locations at $SWE_A \geq 0.25$ m, we removed them from consideration for Figure 28.3. At lower levels of landscape SWE_L (<0.119 m), and at local SWE_A between 0.000 and 0.049 m, the odds of bison traveling were higher than those for non-traveling activities. However, at moderate and high landscape SWE_L (≥ 0.120 m) the odds of bison non-traveling activities were higher than those for travel at low local SWE_A amounts (*e.g.*, <0.10 m).

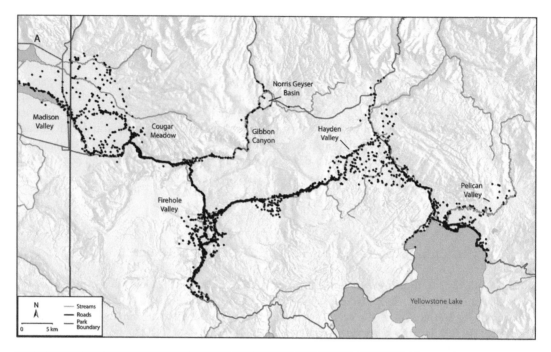

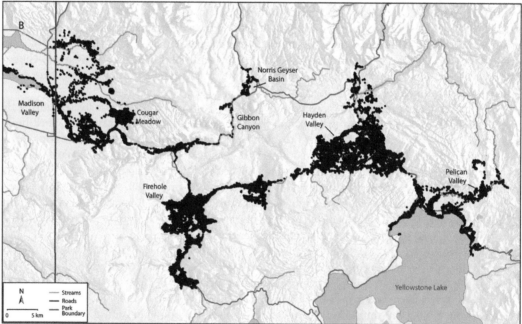

FIGURE 28.2 The distribution of Global Positioning System locations for bison in the central portion of Yellowstone National Park that were (A) traveling or (B) engaged in non-traveling (*i.e.*, foraging, resting) activities during winters (November-April) of 2003–2004 through 2005–2006.

TABLE 28.1	Log odds ratios with 95% confidence levels (in parentheses) and odds ratios with 95% confidence levels (in parentheses) for actual bison traveling locations versus random locations with changing local-scale snow water equivalent (SWE$_A$) for four different categories of landscape-scale snow pack (SWE$_L$)

| Local | Landscape-scale SWE$_L$ (m) | | | |
SWE$_A$ (m)	0.000–0.059	0.060–0.119	0.120–0.179	≥ 0.180
Log odds ratios				
0.000–0.049	1.69 (1.17, 2.20)	−1.06 (0.73, 1.39)	1.86 (1.74, 1.98)	2.07 (1.86, 2.28)
0.050–0.099	−1.58 (−2.09, −1.06)	−0.02 (−0.34, 0.30)	1.11 (0.98, 1.25)	1.89 (1.69, 2.10)
0.100–0.149		−1.55 (−2.19, −0.90)	−0.28 (−0.42, −0.13)	0.85 (0.64, 1.06)
0.150–0.199			−1.95 (−2.23, −1.68)	−0.81 (−1.08, −0.55)
0.200–0.249			−3.04 (−3.70, −2.39)	−2.04 (−2.47, −1.61)
Odds ratios				
0.000–0.049	5.40 (3.23, 9.04)	2.89 (2.08, 4.00)	6.42 (5.69, 7.24)	7.94 (6.43, 9.81)
0.050–0.099	0.21 (0.12, 0.35)	0.98 (0.71, 1.35)	3.04 (2.66, 3.48)	6.62 (5.39, 8.13)
0.100–0.149		0.21 (0.11, 0.41)	0.76 (0.65, 0.88)	2.34 (1.90, 2.88)
0.150–0.199			0.14 (0.11, 0.19)	0.44 (0.34, 0.58)
0.200–0.249			0.05 (0.02, 0.09)	0.13 (0.08, 0.20)

TABLE 28.2	Log odds ratios with 95% confidence levels (in parentheses) and odds ratios with 95% confidence levels (in parentheses) for actual bison non-traveling locations (*e.g.*, feeding, resting) versus random locations with changing local-scale snow water equivalent (SWE$_A$) for four different categories of landscape-scale snow pack (SWE$_L$).

| Local | Landscape-scale SWEL (m) | | | |
SWE$_A$(m)	0.000–0.059	0.060–0.119	0.120–0.179	≥ 0.180
Log odds ratios				
0.000–0.049	1.05 (0.90, 1.19)	0.33 (0.22, 0.45)	2.25 (2.19, 2.31)	2.40 (2.30, 2.51)
0.050–0.099	−0.94 (−1.08, −0.79)	0.93 (0.82, 1.03)	1.16 (1.09, 1.22)	2.16 (2.07, 2.25)
0.100–0.149		−2.03 (−2.25, −1.80)	−0.27 (−0.34, −0.20)	0.87 (0.78, 0.96)
0.150–0.199		−3.27 (−4.40, −2.13)	−2.69 (−2.86, −2.53)	−0.46 (−0.56, −0.36)
0.200–0.249				−1.69 (−1.84, −1.54)
Odds ratios				
0.000–0.049	2.85 (2.46, 3.29)	1.40 (1.25, 1.56)	9.47 (8.91, 10.06)	11.05 (9.97, 12.26)
0.050–0.099	0.39 (0.34, 0.45)	2.53 (2.27, 2.81)	3.18 (2.99, 3.39)	8.68 (7.94, 9.48)
0.100–0.149		0.13 (0.11, 0.16)	0.76 (0.72, 0.82)	2.39 (2.19, 2.61)
0.150–0.199		0.04 (0.01, 0.12)	0.07 (0.06, 0.08)	0.63 (0.57, 0.70)
0.200–0.249				0.19 (0.16, 0.22)

Snow pack heterogeneity in the central herd's winter range during 2003–2004 through 2004–2005 increased steadily during the winter, peaked in late April when the spring melt was underway, and then decreased rapidly. The peak average snow heterogeneity (SNH$_L$) for the landscape was 0.13 m in

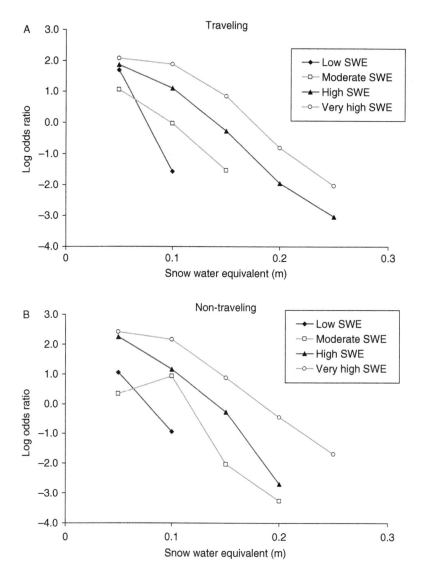

FIGURE 28.3 Changes in the log odds ratios for bison (A) traveling locations versus random locations, and (B) non-traveling locations versus random locations with changing local-scale snow water equivalent (SWE$_A$) for four different categories of landscape-scale SWE$_L$ (0.000–0.059, 0.060–0.119, 0.120–0.179, >0.180). Local-scale SWE$_A$ is divided into five categories: 0.000–0.049 m, 0.050–0.099 m, 0.100–0.149 m, 0.150–0.199 m, and 0.200–0.249 m.

2003–2004, 0.11 m in 2004–2005, and 0.16 in 2005–2006. Changes in landscape SNH$_L$ also influenced the odds of bison travel and non-travel at various local SWE$_A$ levels (Tables 28.3 and 28.4). The odds of bison traveling and not traveling decreased as local-scale SWE$_A$ increased within a level of landscape SNH$_L$ (Figure 28.4). Again, because of small sample sizes of used locations at SWE$_A$ ≥ 0.3 m, we removed them from evaluation for Figure 28.4. At low SNH$_L$ (0.000–0.039 m), the log odds ratio of bison travel decreased from 1.02 (95% CI: 0.80, 1.29) to −0.68 (−0.93, −0.43) to −1.48 (−2.15, −0.82). Likewise, the log odds ratio of non-traveling activities decreased from 0.49 (0.42, 0.57) to 0.09

(0.01, 0.16) to −2.38 (−2.68, −2.07) as local SWE_A increased from 0.000–0.049 m to 0.050–0.099 m to 0.100–0.149 m. However, as landscape SNH_L increased, bison did not occupy areas of increasing local SWE_A. Rather, bison adjusted their use of the landscape to occupy areas of lower local SWE_A as landscape SNH_L increased beyond a moderately low level (0.040–0.079 m), which would often be present the majority of the winter.

A. Modeling and Predicting Spatial Variation in Bison Travel and Non-Travel Activities

We found one top-approximating model (AIC = 13,749.4, $K = 13$, $w_i = 0.944$) for the bison travel analysis that had a relative likelihood of 18 compared to the second-best model (ΔAIC = 5.8, $K = 13$, $w_i = 0.053$). The top model contained significant, positive habitat type (HAB), distance to burned forest ($DIST_{burn}$), and slope heterogeneity ($SLOPE_{het}$) covariates, and an interaction between daily landscape SWE heterogeneity (SNH_L) and distance to stream ($DIST_{stream}$; Table 28.5). Negative, significant covariates in the top model included local-scale SWE_A, average slope (SLOPE), and distances to stream ($DIST_{stream}$), foraging area ($DIST_{forage}$), unburned forest ($DIST_{unburn}$), and road ($DIST_{road}$). Local-scale SWE_A and slope heterogeneity ($SLOPE_{het}$) were the most influential negative and positive covariates, respectively, on the odds of bison travel.

The distribution of areas predicted to have the highest odds of travel decreased in breadth as snow pack across the landscape increased (Figure 28.5). In November (Figure 28.5A), predicted odds of travel was both higher and more widely distributed than in March (Figure 28.5B). During both months the odds of travel was highest along streams and in canyons, most notably in the Madison and Firehole valleys, Madison and Gibbon canyons, and along Nez Perce Creek and the Yellowstone River (Figure 28.5).

There were two top-approximating models for the bison non-travel analysis with the first (AIC = 52,752.9, $K = 13$, $w_i = 0.513$) and second (ΔAIC = 0.10, $K = 14$, $w_i = 0.487$) best models

TABLE 28.3 Log odds ratios with 95% confidence levels (in parentheses) and odds ratios with 95% confidence levels (in parentheses) for actual bison traveling locations versus random locations with changing local-scale snow water equivalent (SWE_A) for four different categories of landscape-scale snow pack heterogeneity (SNH_L).

Local SWE_A(m)	Landscape-scale snow pack heterogeneity (m)			
	0.000–0.039	0.040–0.079	0.080–0.119	≥ 0.120
Log odds ratios				
0.000–0.049	1.04 (0.80, 1.29)	2.49 (2.28, 2.71)	2.25 (2.08, 2.42)	3.04 (2.63, 3.46)
0.050–0.099	−0.68 (−0.93, −0.43)	1.72 (1.58, 1.87)	1.00 (0.80, 1.20)	−0.20 (−0.70, 0.30)
0.100–0.149	−1.48 (−2.15, −0.82)	0.33 (0.19, 0.47)	−0.80 (−1.13, −0.47)	−1.31 (−2.13, −0.50)
0.150–0.199		−1.43 (−1.66, −1.21)	−1.56 (−1.94, −1.19)	−2.92 (−4.31, −1.53)
0.200–0.249		−2.37 (−2.83, −1.91)	−2.32 (−2.89, −1.74)	
Odds ratios				
0.000–0.049	2.84 (2.22, 3.62)	12.12 (9.74, 15.07)	9.49 (7.97, 11.28)	21.00 (13.84, 31.88)
0.050–0.099	0.51 (0.39, 0.65)	5.61 (4.85, 6.48)	2.72 (2.24, 3.31)	0.82 (0.50, 1.35)
0.100–0.149	0.23 (0.12, 0.44)	1.40 (1.21, 1.61)	0.45 (0.32, 0.63)	0.27 (0.12, 0.61)
0.150–0.199		0.24 (0.19, 0.30)	0.21 (0.14, 0.30)	0.05 (0.01, 0.22)
0.200–0.249		0.09 (0.06, 0.15)	0.10 (0.06, 0.18)	

TABLE 28.4 Log odds ratios with 95% confidence levels (in parentheses) and odds ratios with 95% confidence levels (in parentheses) for actual bison non-traveling locations versus random locations with changing local-scale snow water equivalent (SWE_A) for four different categories of landscape-scale snow pack heterogeneity (SNH_L)

Local SWE_A(m)	Landscape-scale snow pack heterogeneity (m)			
	0.000–0.039	0.040–0.079	0.080–0.119	≥ 0.120
Log odds ratios				
0.000–0.049	0.49 (0.42, 0.57)	3.12 (3.04, 3.20)	1.81 (1.72, 1.91)	3.43 (3.14, 3.72)
0.050–0.099	0.09 (0.01, 0.16)	1.70 (1.64, 1.76)	1.41 (1.30, 1.51)	−0.55 (−0.90, −0.20)
0.100–0.149	−2.38 (−2.69, −2.08)	0.16 (0.10, 0.22)	0.11 (−0.02, 0.23)	−1.77 (−2.34, −1.20)
0.150–0.199		−1.70 (−1.80, −1.60)	−0.82 (−0.96, −0.67)	−3.92 (−5.31, −2.53)
0.200–0.249		−2.88 (−3.10, −2.65)	−1.43 (−1.63, −1.24)	−4.54 (−6.50, −2.58)
Odds ratios				
0.000–0.049	1.63 (1.52, 1.76)	22.66 (20.83, 24.65)	6.13 (5.57, 6.75)	30.92 (23.06, 41.47)
0.050–0.099	1.09 (1.01, 1.18)	5.47 (5.15, 5.81)	4.08 (3.69, 4.52)	0.58 (0.41, 0.82)
0.100–0.149	0.09 (0.07, 0.13)	1.18 (1.11, 1.25)	1.11 (0.98, 1.26)	0.17 (0.10, 0.30)
0.150–0.199		0.18 (0.17, 0.20)	0.44 (0.38, 0.51)	0.02 (0.00, 0.08)
0.200–0.249		0.06 (0.04, 0.07)	0.24 (0.20, 0.29)	0.01 (0.00, 0.08)

receiving comparable support in the data. The third best model received little support ($\Delta AIC = 309.3$, $K = 14$, $w_i \ll 0.001$). The top two models contained positive meadow habitat type (HAB = meadow), distance to burned forest ($DIST_{burn}$), and slope heterogeneity ($SLOPE_{het}$) covariates, and an interaction between daily landscape SWE_L and distance to road ($DIST_{road}$), with confidence intervals not spanning zero (Table 28.5). These models also included significant, negative local-scale SWE_A, average slope (SLOPE), and distances to stream ($DIST_{stream}$), foraging area ($DIST_{forage}$), and road ($DIST_{road}$). Local-scale SWE_A and the landscape SWE_L by distance to road ($DIST_{road}$) interaction were the most influential negative and positive covariates, respectively, on the odds of bison non-travel.

Similar to predicted patterns in bison travel, the extent of areas predicted to have the highest odds of non-travel also changed seasonally as snow pack increased (Figure 28.6). Meadows located both away from and near streams had the highest predicted odds of non-travel in both November (Figure 28.6a) and March (Figure 28.6b). However, the extent of these areas decreased as winter progressed and bison were more constrained to lower elevation valley bottoms owing to deep snow at higher elevations and near the periphery of the bison range (Figure 28.6). Meadow complexes that were predicted to have the highest odds of non-travel included those in the Hayden Valley, Fountain Flats area in the Firehole Valley, Cougar Meadows, Madison Valley, and Gibbon Meadows.

IV. DISCUSSION

Previous studies have reported that snow affects the foraging behavior, distribution, and temporal travel patterns of bison (*e.g.*, Fortin *et al.* 2003, Bruggeman 2006, Bruggeman *et al.* 2006) and other large herbivores (Schaefer and Messier 1995a,b, Doerr *et al.* 2005). However, this is the first work to document how changing snow pack at small and large scales simultaneously affects spatial patterns in

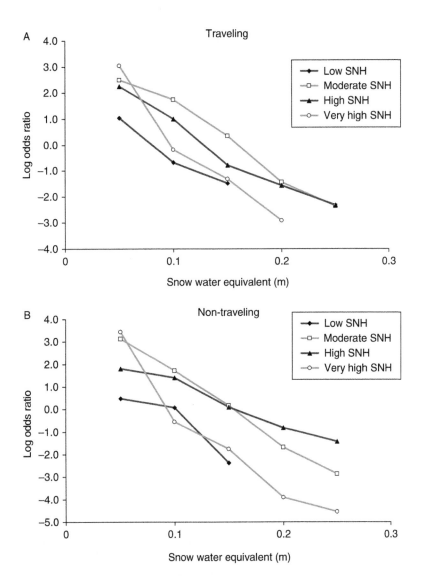

FIGURE 28.4 Changes in the log odds ratios for bison (A) traveling locations versus random locations, and (B) non-traveling locations versus random locations with changing local-scale snow water equivalent (SWE_A) for four different categories of landscape-scale snow pack heterogeneity (SNH_L; 0.000–0.039, 0.040–0.079, 0.080–0.119, >0.120). Local-scale SWE_A is divided into five categories: 0.000–0.049 m, 0.050–0.099 m, 0.100–0.149 m, 0.150–0.199 m, and 0.200–0.249 m.

bison travel and feeding behavior. As snow pack increased at a given location, bison were less likely to use that area for either traveling or foraging, likely because the displacement of snow for movement or foraging becomes more energetically costly as snow water equivalent increases (Parker *et al.* 1984). Bison in central Yellowstone spend the majority (67%) of their time foraging during winter (Bjornlie and Garrott 2001, Bruggeman *et al.* 2006, Chapter 27 by Bruggeman *et al.*, this volume), with the displacement of snow to access forage occupying nearly one-third of that time (Figure 28.7). Thus, it is not surprising that the negative influence of snow was slightly greater on the odds of bison

TABLE 28.5 Coefficient values (β_i) and 95% confidence limits for covariates from the top approximating model of spatial variability in bison travel and top two models for non-travel

Covariate	Bison Travel (βi) Model #1	Bison Non-Travel (βi) Model #1	Bison Non-Travel (βi) Model #2
SWE_A	**−6.82 (−8.45, −5.18)**	**−9.66 (−12.60, −6.73)**	**−9.65 (−12.57, −6.74)**
HAB = meadow	**2.89 (2.40, 3.39)**	**0.83 (0.02, 1.63)**	**0.82 (0.02, 1.62)**
HAB = thermal	**2.16 (1.37, 2.95)**	0.12 (−0.81, 1.05)	0.12 (−0.81, 1.04)
HAB = burned forest	**2.09 (1.67, 2.52)**	−0.51 (−1.24, 0.23)	−0.51 (−1.25, 0.22)
HAB = unburned forest	**2.17 (1.76, 2.57)**	−0.10 (−0.84, 0.64)	−0.12 (−0.85, 0.62)
$DIST_{stream}$	**−3.87 (−4.72, −3.01)**	**−2.18 (−2.97, −1.39)**	**−2.17 (−2.96, −1.37)**
$DIST_{forage}$	**−0.68 (−1.08, −0.29)**	**−4.06 (−5.64, −2.48)**	**−4.08 (−5.65, −2.50)**
$DIST_{unburn}$	**−0.58 (−1.15, −0.01)**		−0.07 (−0.62, 0.49)
$DIST_{burn}$	**0.84 (0.58, 1.09)**	**1.33 (0.85, 1.82)**	**1.33 (0.84, 1.82)**
$DIST_{road}$	**−0.41 (−0.79, −0.04)**	**−0.86 (−1.34, −0.38)**	**−0.86 (−1.34, −0.38)**
SLOPE	**−3.67 (−4.80, −2.54)**	**−1.47 (−2.25, −0.68)**	**−1.48 (−2.28, −0.69)**
$SLOPE_{het}$	**3.37 (2.17, 4.57)**	**0.97 (0.48, 1.46)**	**0.96 (0.44, 1.48)**
SWE_L		N/A	N/A
SNH_L	N/A		
SWE_L*DIST_{road}		**1.67 (0.49, 2.86)**	**1.67 (0.48, 2.86)**
SNH_L*DIST_{stream}	**3.11 (1.29, 4.94)**		

Bold notation denotes significant coefficients at $\alpha = 0.05$. Abbreviations are: $DIST_{burn}$ (distance to burned forest); $DIST_{forage}$ (distance to foraging area); $DIST_{road}$ (distance to road); $DIST_{stream}$ (distance to stream); $DIST_{unburn}$ (distance to unburned forest); HAB (habitat type); SLOPE (average slope); $SLOPE_{het}$ (slope heterogeneity); SNH_L (daily snow water equivalent heterogeneity); SWE_A (daily local snow water equivalent); and SWE_L (daily landscape snow water equivalent).

non-traveling activities (unscaled coefficient $= -23.2$; 95% CI: -30.3, -16.2) compared to travel (unscaled coefficient $= -18.5$; 95% CI: -22.9, -14.1). Overall, bison traveling and non-traveling locations were concentrated in lower-elevation meadows, geothermal areas, and canyons where snow pack would be less, and away from higher elevation plateaus where substantial snow accumulation would inhibit foraging and travel.

Snow pack characteristics at the landscape scale also affected bison travel and non-traveling activities. Bison were more likely to travel farther away from travel corridors along streams as variability in the snow pack increased, which occurred primarily during April at the onset of the springtime melt. Contrary to our predictions, there was an increased probability of non-traveling behaviors (*e.g.*, feeding, resting) farther from roads as winter progressed and snow accumulated. There was also an increase in bison occupancy of local areas with higher snow pack as overall snow pack on the landscape increased. At a log odds ratio of zero there is an equal likelihood of a point on the landscape being a used versus a random traveling or non-traveling location. As landscape snow pack increased, the local snow pack at which this equal likelihood occurred increased from ~0.07 to 0.18 m for both traveling and non-traveling locations. Given the negative correlation between local snow pack and the odds of travel and non-travel, it is likely that bison are forced to occupy these deeper or wetter snow areas as opposed to choosing these locations.

Changes in landscape-scale snow heterogeneity also altered the likelihood of bison traveling or foraging in locations with differing local snow pack. At low levels of snow heterogeneity (0.000–0.039 m), which are present in November–December at the beginning of winter, the odds of a location being used for travel compared to a random location were equally likely at a local SWE of 0.08 m. At moderate snow heterogeneity levels (0.040–0.079 m), present for the majority of winter, the local SWE for equal likelihood shifted to 0.16 m. However, as snow heterogeneity increased to

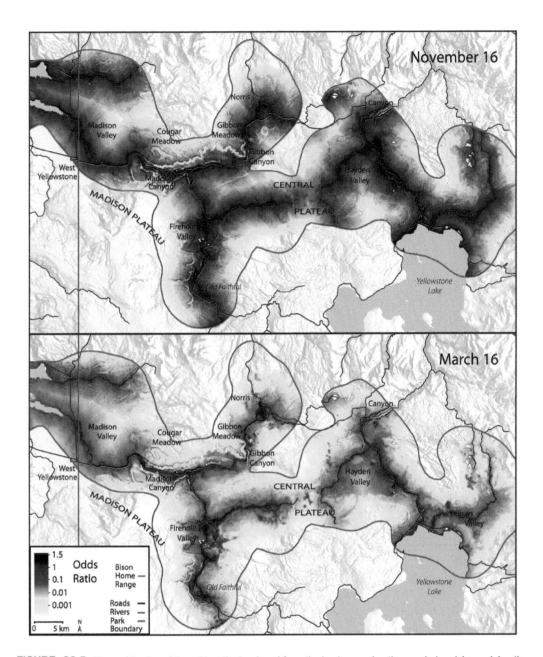

FIGURE 28.5 Maps of the log of the odds ratio developed from the best approximating *a priori* model examining the effects of snow pack, topography, and habitat type attributes on bison travel patterns in central Yellowstone National Park during the 2004–2005 winter with snow pack attributes estimated on (A) November 16 and (B) March 16.

high levels, which occur only during the spring melt in late March–April when the landscape is comprised of a number of snow-free patches situated amidst remnant snow, the local SWE for equal likelihood decreased to ~0.13 and 0.10 m. This finding is in agreement with previous work documenting that bison increase their travel during the spring (Bruggeman *et al.* 2006, Chapter 27 by Bruggeman *et al.*, this volume) and preferentially seek out partially melted-out, lower-elevation areas

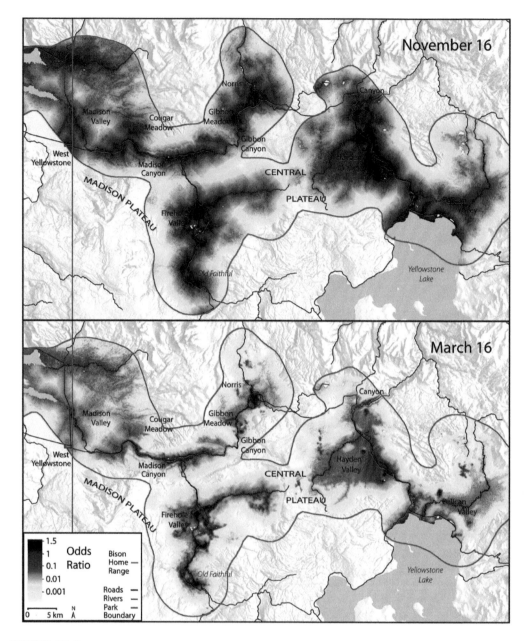

FIGURE 28.6 Maps of the log of the odds ratio developed from the best approximating *a priori* model examining the effects of snow pack, topography, and habitat type attributes on bison non-travel (*i.e.*, foraging, resting) habitat use in central Yellowstone National Park for the 2004–2005 winter with snow pack attributes estimated on (A) November 16 and (B) March 16.

to take advantage of energy efficient foraging opportunities with exposed vegetation (Bruggeman 2006). This shift in landscape use was even more pronounced for non-traveling (*e.g.*, feeding) compared to random locations, which was expected because of the importance of partially melted areas for bison foraging. At high snow heterogeneity levels (>0.12 m) the local SWE for equally likely

FIGURE 28.7 A bull bison displaces snow while feeding in the Upper Geyser Basin of the Firehole drainage. While traveling often occurred on a network of trails, constant foraging through deep snow represented one the major energy costs for wintering bison (Photo by Matt Becker).

used versus random foraging locations was approximately 0.09 m, which was slightly below that for low landscape snow heterogeneity. Maps of predicted odds of travel and non-travel document this spatial redistribution of bison to lower elevations as winter progresses (Figures 28.5 and 28.6). In March, this shift may be affected by both increased heterogeneity of snow pack from melted patches at lower elevations and increased SWE at higher elevations along the edges of the bison range that constrain bison to valley bottoms.

Streams were the most influential natural landscape features affecting bison travel. Streams guide animal movements (Noss 1991) and bison regularly establish and use travel routes along them. Our results suggest the bison travel network throughout central Yellowstone is spatially defined largely by the presence of streams connecting foraging areas. Areas receiving the most bison travel often paralleled major streams, and while segments of some of these streams parallel roads, other portions are located away from any roads, such as streams along the west park boundary and the Mary Mountain trail. In addition to the natural pathways provided by streams, many streams in Yellowstone remain unfrozen during winter, a result of runoff from geothermal features entering waterways. These "heated" open streams reduce snow pack along the riverbanks and afford bison easier travel routes and foraging than nearby areas of deep snow. Streams with predicted high odds of travel included portions of the Madison, Gibbon, Firehole, and Yellowstone rivers, Nez Perce and Duck creeks on the west-side, and Alum and Pelican creeks on the east-side (Figure 28.5).

Habitat attributes such as distance to foraging area and distance to stream were also influential and negatively correlated with the odds of bison foraging behavior. While the former result is intuitive because bison are obviously less likely to forage farther away from foraging areas, the latter is likely the result of the high number of streams that pass through meadow complexes that serve as foraging areas. Some of the meadows predicted to have high odds of foraging and resting behavior include Fountain Flats, portions of the Hayden Valley, and Cougar Meadows, through which streams such as the Firehole and Yellowstone rivers and Duck Creek pass (Figure 28.6). The probability of bison travel increased in meadow and thermal habitats, indicating that bison develop some travel routes through inter-connected foraging areas, which is how bison use the western part of the Mary Mountain trail that passes through a series of meadows. Thermal areas may also have increased importance as travel corridors because the magnitude of effect of thermal habitats was greater for long-distance than general travel. In winter, thermal habitats, which also provide some forage, facilitate bison travel because of minimal snow pack accumulation in these areas. Bruggeman *et al.* (2007) documented that movements in large meadow complexes are generally of short distance and likely related to travel between patches of suitable forage. In contrast, bison travel behavior is different in habitats lacking adequate foraging areas (*i.e.*, corridors), with travel consisting of long-distance sustained movements (Bruggeman 2006, Bruggeman *et al.* 2007). Our findings agree qualitatively with other studies that document associations between animal movement behavior and habitat, many of which found travel speed or habitat preferences to be influenced by the availability of forage or prey (Ferguson and Elkie 2004, Dickson *et al.* 2005). For example, Frair *et al.* (2005) documented that elk (*Cervus elaphus*) were more likely to be traveling in areas close to linear clearings and across steeper slopes, foraging when in areas that offered adequate biomass, and resting when in areas with cool microclimatic conditions. Also, woodland caribou (*Rangifer tarandus*) selected different cover types when moving within foraging patches as opposed to between patches (Johnson *et al.* 2002).

Topographical attributes significantly affected winter travel patterns of bison, with the odds of travel positively and negatively correlated with slope heterogeneity and slope, respectively. The probability of travel was higher in areas of variable topography that constrained movements, such as canyons in portions of the Madison, Gibbon, and Firehole drainages, which were also areas predicted to have a high odds of travel (Figure 28.5). The probability of travel was lower in areas with steep slopes, indicating bison generally avoided traversing steep slopes and preferred to travel along gentle gradients in elevation. Again, these results concur with the effect of topography on year-round bison travel patterns (Bruggeman *et al.* 2007). The probability of non-traveling activities by bison was also influenced by topography in a similar manner to travel, though the magnitude of influence of these covariates on non-traveling activities was less than on travel patterns. The influence of topography on travel patterns and habitat use of a variety of species has been documented (Klimley *et al.* 2002, Whittington *et al.* 2004, 2005, Dickson *et al.* 2005, Meyer and Holland 2005). For instance, Kie *et al.* (2005) discovered elk had a higher probability of moving parallel to drainages, particularly when close to streams and in valley bottoms, which is a result that generally concurs with our findings.

Though distance to road was a significant, negative effect in the top bison travel model, it was not as influential as found in previous work on year-round travel patterns, particularly when compared with the effect of streams on travel routes (Bruggeman 2006, Bruggeman *et al.* 2007). Distance to stream was greater than nine times more influential than distance to road on the odds of bison travel locations, while earlier work found this ratio to be only 1.7 (Bruggeman 2006, Bruggeman *et al.* 2007). This difference likely reflects the inclusion of snow pack covariates in our models that help account for bison use of geothermal areas and lower-elevation meadows, many of which are situated near roads, in which snow accumulation is reduced. This effect was also present in the top models of non-traveling activities that contained a significant, negative correlation between distance to road and the odds of bison foraging behavior. Again, this is likely the result of many meadows throughout bison winter foraging habitat being situated close to, or along, roads.

Nonetheless, roads in portions of the landscape may facilitate bison travel, either alone or in conjunction with streams. Road segments predicted to have a high probability of bison travel included in the Madison Canyon, assorted portions of the Firehole and Gibbon drainages, and along the Yellowstone River in the Hayden Valley. An analysis of spatial patterns in bison road travel found variability in the amount of travel on roads depending on topographic and habitat attributes surrounding road segments (Bruggeman *et al.* 2007, Chapter 27 by Bruggeman *et al.*, this volume). Portions of the road network acted as travel corridors and more travel occurred on road segments closer to streams and unburned forest, farther from foraging areas, and passing through canyons

(Chapter 27 by Bruggeman *et al.*, this volume). Road segments that passed through meadow complexes without topographic restrictions, or circumvented important foraging areas, received below average bison travel (Chapter 27 by Bruggeman *et al.*, this volume). Of those road segments documented to have a high amount of use (Chapter 27 by Bruggeman *et al.*, this volume), only the lower Firehole drainage road segment was not predicted to have odds of use as high as expected, suggesting some facilitation of bison travel through this portion of the landscape.

Bison appear to travel on roads that align with natural travel corridors or when convenient, suggesting that road grooming has minimal influence on bison travel compared to snow, landscape, and habitat attributes (Bjornlie and Garrott 2001, Gates *et al.* 2005, Bruggeman *et al.* 2007, Chapter 27 by Bruggeman *et al.*, this volume). Bison travel corridors exist both along portions of groomed roads as well as distant from roads, such as along the Mary Mountain trail migratory route that traverses the central plateau (Figure 28.8), and along the Madison River near the west park boundary. Corridors along roads in the Madison and Gibbon canyons offer the most direct travel routes along rivers to reach large meadows, specifically Cougar Meadows and Gibbon Meadows, respectively. Our findings suggest natural dynamic and static landscape characteristics, including snow pack, topography, and habitat attributes, are the primary factors affecting bison choice of travel routes and habitat use for foraging and resting activities.

Our results suggest bison use of certain road segments as travel corridors would persist whether or not roads were groomed during winter, owing to repeated use of learned travel routes and the necessity of movements to access foraging areas. Use of spatial memory to revisit foraging areas has been documented for several ungulate species (Bailey *et al.* 1989, Hewitson *et al.* 2005) and is likely used by bison given their well-defined travel corridors. Bison use the Mary Mountain trail the entire winter to migrate and facilitate movements between foraging areas despite deep snow. Repeated use

FIGURE 28.8 Bison traveling single-file into the Firehole drainage from the Mary Mountain trail. This migration route from the Hayden valley was facilitated by over 22 miles of self-maintained routes, allowing efficient travel with comparatively little energy expenditure (Photo by Matt Becker).

of the trail by bison traveling in single file lines maintains it in a "self-groomed" state, an adaptation for saving energy while traveling in snow (Telfer and Kelsall 1984). There are two alternate routes along groomed roads that would allow bison to migrate into the Madison headwaters area, and neither received bison travel, possibly owing to a combination of factors including lack of foraging areas and geothermal habitats, route length, and deep snow. Corridors along roads through canyons in the Madison headwaters area offer the most direct travel routes along rivers to reach large meadows. Alternate paths are not likely because of topography and habitat constraints (*i.e.*, plateaus or burned forest).

Though our study cannot fully resolve the debate over effects of roads and winter road grooming on bison travel in central Yellowstone, it offers novel insights into bison spatial dynamics. Completely separating the effect of roads on bison travel is impossible because bison use travel corridors along portions of roads, and there are areas where roads may have initially facilitated movements. Previous analyses by Bruggeman *et al.* (2006) documented that temporal patterns in the amount of bison road travel were negatively correlated with the road-grooming period, and found no evidence that bison preferentially used groomed roads during winter. Temporal trends in bison road travel were influenced by similar abiotic and biotic factors as trends in off-road travel (Bruggeman *et al.* 2006). We suggest that, like temporal trends in bison road travel, most spatial patterns in road use are likely a manifestation of general spatial travel trends through the landscape, as topography and habitat type attributes alone predicted the majority of bison travel corridors throughout central Yellowstone. Beyond investigating bison travel in Yellowstone, our study provides conceptual and analytical frameworks for examining animal movement patterns using a behaviorally based resource selection analysis. Given the increasing use of Global Positioning System and Geographic Information System technology in wildlife research, we anticipate the methodology presented here will be applicable to studies of behaviorally influenced resource selection for a variety of species.

V. SUMMARY

1. The influence of winter road grooming on bison travel patterns in Yellowstone National Park has been debated for more than two decades. We radio collared 30 adult, female bison from the central herd during three winters to quantify how snow, topography, habitat attributes, and roads influenced bison travel patterns and non-traveling activities (*i.e.*, foraging, resting).
2. Bison were less likely to use a point on the landscape for traveling or feeding as snow pack increased. However, bison used local areas with deeper snow as the overall snow pack increased on the landscape.
3. Distance to stream was the most influential habitat covariate, with the spatial travel network of bison being largely defined by streams connecting foraging areas. Distances to foraging areas and streams also significantly influenced non-traveling activities, being negatively correlated with the odds of bison foraging or resting.
4. Topography significantly affected bison travel patterns, with the probability of travel being higher in areas of variable topography that constrained movements (*e.g.*, canyons). Distance to road had a significant, negative effect on bison travel, but was nine times less influential compared to the impact of streams.
5. Road grooming has a minimal influence on bison travel and habitat use given the importance of natural dynamic and static landscape characteristics such as snow pack, topography, and habitat attributes on bison choice of travel routes and habitat use for foraging and resting.

VI. REFERENCES

Agresti, A. 2002. *Categorical Data Analysis.* Wiley, Hoboken, NJ.

Allison, P. D. 1999. *Logistic Regression Using the SAS System: Theory and Application.* Cary, NC.

Allison, P. D. 2005. *Fixed Effects Regression Methods for Longitudinal Data.* Cary, NC.

Bailey, D. W., L. R. Rittenhouse, R. H. Hart, and R. W. Richards. 1989. Characteristics of spatial memory in cattle. *Applied Animal Behaviour Science* **23**:331–340.

Bjornlie, D. D., and R. A. Garrott. 2001. Effects of winter road grooming on bison in Yellowstone National Park. *Journal of Wildlife Management* **65**:560–572.

Borkowski, J., P. J. White, R. A. Garrott, T. Davis, A. Hardy, and D. J. Reinhart. 2006. Wildlife responses to motorized winter recreation in Yellowstone National Park. *Ecological Applications* **16**:1911–1925.

Brody, A. J., and M. R. Pelton. 1989. Effects of roads on black bear movements in western North Carolina. *Wildlife Society Bulletin* **17**:5–10.

Bruggeman, J. E. 2006. Spatio-temporal Dynamics of the Central Bison Herd in Yellowstone National Park. Dissertation. Montana State University, Bozeman, MT.

Bruggeman, J. E., R. A. Garrott, D. D. Bjornlie, P. J. White, F. G. R. Watson, and J. J. Borkowski. 2006. Temporal variability in winter travel patterns of Yellowstone bison: the effects of road grooming. *Ecological Applications* **16**:1539–1554.

Bruggeman, J. E., R. A. Garrott, P. J. White, F. G. R. Watson, and R. W. Wallen. 2007. Covariates affecting spatial variability in bison travel behavior in Yellowstone National Park. *Ecological Applications* **17**:1411–1423.

Bruns, E. H. 1977. Winter behavior of pronghorns in relation to habitat. *Journal of Wildlife Management* **41**:560–571.

Burnham, K. P., and D. R. Anderson. 2002. *Model Selection and Multi-Model Inference.* Springer-Verlag, New York, NY.

Clutton-Brock, T. H., M. Major, and F. E. Guinness. 1985. Population regulation in male and female red deer. *Journal of Animal Ecology* **54**:831–846.

Cole, E. K., M. D. Pope, and R. G. Anthony. 1997. Effects of road management on movement and survival of Roosevelt elk. *Journal of Wildlife Management* **61**:1115–1126.

Crist, T. O., D. S. Guertin, J. A. Wiens, and B. T. Milne. 1992. Animal movement in heterogeneous landscapes: an experiment with *Eleodes* beetles in short grass prairie. *Functional Ecology* **6**:536–544.

Cronin, J. T. 2003. Movement and spatial population structure of a prairie planthopper. *Ecology* **84**:1179–1188.

Dickson, B. G., J. S. Jenness, and P. Beier. 2005. Influence of vegetation, topography, and roads on cougar movement in southern California. *Journal of Wildlife Management* **69**:264–276.

Dingle, H. 1996. *Migration.* Oxford University Press, New York, NY.

Dobson, F. S. 1982. Competition for mates and predominant juvenile male dispersal in mammals. *Animal Behavior* **30**:1183–1192.

Dobson, F. S., and W. T. Jones. 1985. Multiple causes of dispersal. *American Naturalist* **126**:855–858.

Doerr, J. G., E. J. DeGayner, and G. Ith. 2005. Winter habitat selection by Sitka black-tailed deer. *Journal of Wildlife Management* **69**:322–331.

Dyer, S. J., J. P. O'Neill, S. M. Wasel, and S. Boutin. 2002. Quantifying barrier effects of roads and seismic lines on movements of female woodland caribou in northeastern Alberta. *Canadian Journal of Zoology* **80**:839–845.

Ferguson, S. H., and P. C. Elkie. 2004. Habitat requirements of boreal forest caribou during the travel season. *Basic and Applied Ecology* **5**:465–474.

Flamm, R. O., B. L. Weigle, I. E. Wright, M. Ross, and S. Aglietti. 2005. Estimation of manatee (*Trichechus manatus latirostris*) places and movement corridors using telemetry data. *Ecological Applications* **15**:1415–1426.

Fortin, D., J. M. Fryxell, L. O'Brodovich, and D. Frandsen. 2003. Foraging ecology of bison at the landscape and plant community levels: the applicability of energy maximization principles. *Oecologia* **134**:219–227.

Fortin, D., H. L. Beyer, M. S. Boyce, D. W. Smith, T. Duchesne, and J. S. Mao. 2005. Wolves influence elk movements: behavior shapes a trophic cascade in Yellowstone National Park. *Ecology* **86**:1320–1330.

Frair, J. L., E. H. Merrill, D. R. Visscher, D. Fortin, H. L. Beyer, and J. M. Morales. 2005. Scales of movement by elk (*Cervus elaphus*) in response to heterogeneity in forage resources and predation risk. *Landscape Ecology* **20**:273–287.

Fraser, D. F., J. F. Gilliam, and T. Yiphoi. 1995. Predation as an agent of population fragmentation in a tropical watershed. *Ecology* **76**:1461–1472.

Freddy, D. J., W. M. Bronaugh, and M. C. Fowler. 1986. Responses of mule deer to disturbance by persons afoot and snowmobiles. *Wildlife Society Bulletin* **14**:63–68.

Fryxell, J. M., J. F. Wilmshurst, and A. R. E. Sinclair. 2004. Predictive models of movement by Serengeti grazers. *Ecology* **85**:2429–2435.

Gabrielsen, G. W., and E. N. Smith. 1995. Physiological responses of wildlife to disturbance. Pages 95–107 *in* R. L. Knight and K. J. Gutzwiller (Eds.) *Wildlife and Recreationists: Coexistence Through Management and Research.* Island Press, Washington, DC.

Gaillard, J. M., M. Festa-Bianchet, N. G. Yoccoz, A. Loison, and C. Toïgo. 2000. Temporal variation in fitness components and population dynamics of large herbivores. *Annual Review of Ecology and Systematics* **31**:367–393.

Garrott, R. A., L. L. Eberhardt, P. J. White, and J. J. Rotella. 2003. Climate-induced variation in vital rates of an unharvested large-herbivore population. *Canadian Journal of Zoology* **81**:33–45.

Garrott, R. A., J. A. Gude, E. J. Bergman, C. Gower, P. J. White, and K. L. Hamlin. 2005. Generalizing wolf effects across the greater Yellowstone area: a cautionary note. *Wildlife Society Bulletin* **33**:1245–1255.

Gates, C. C., B. Stelfox, T. Muhly, T. Chowns, and R. J. Hudson. 2005. *The Ecology of Bison Movements and Distribution in and beyond Yellowstone National Park*. University of Calgary, Alberta, Canada.

Gebeyehu, S., and M. J. Samways. 2006. Topographic heterogeneity plays a crucial role for grasshopper diversity in a southern African megabiodiversity hotspot. *Biodiversity and Conservation* **15**:231–244.

Haddad, N. M. 1999. Corridor and distance effects on interpatch movements: a landscape experiment with butterflies. *Ecological Applications* **9**:612–622.

Hewitson, L., B. Dumot, and I. J. Gordon. 2005. Response of foraging sheep of variability in the spatial distribution of resources. *Animal Behaviour* **69**:1069–1076.

Hilty, J. A., and A. M. Merenlender. 2004. Use of riparian corridors and vineyards by mammalian predators in northern California. *Conservation Biology* **18**:126–135.

Hosmer, D. W., and S. Lemeshow. 2000. *Applied Logistic Regression*. Wiley, New York, NY.

Ito, T. Y., N. Miura, B. Lhagvasuren, D. Enkhbileg, S. Takatsuki, A. Tsunekawa, and Z. Jiang. 2005. Preliminary evidence of a barrier effect of a railroad on the migration of Mongolian gazelles. *Conservation Biology* **19**:945–948.

Jacobson, A. R., A. Provenzale, A. von Hardenverg, B. Bassano, and M. Festa-Bianchet. 2004. Climate forcing and density dependence in a mountain ungulate population. *Ecology* **85**:1598–1610.

Johnson, C. J., K. L. Parker, D. C. Heard, and M. P. Gillingham. 2002. A multiscale behavioral approach to understanding the movements of woodland caribou. *Ecological Applications* **12**:1840–1860.

Kie, J. G., A. A. Ager, and R. T. Bowyer. 2005. Landscape-level movements of North American elk (*Cervus elaphus*): Effects of habitat patch structure and topography. *Landscape Ecology* **20**:289–300.

Klimley, A. P., S. C. Beavers, T. H. Curtis, and S. J. Jorgensen. 2002. Movements and swimming behavior of three species of sharks in La Jolla Canyon, California. *Environmental Biology of Fishes* **63**:117–135.

Larter, N. C., and C. C. Gates. 1991. Diet and habitat selection of wood bison in relation to seasonal changes in forage quantity and quality. *Canadian Journal of Zoology* **69**:2677–2685.

Lin, D. Y., and L. J. Wei. 1989. The robust inference for the Cox proportional hazards model. *Journal of the American Statistical Association* **84**:1074–1078.

Manly, B. F., L. L. McDonald, D. L. Thomas, T. L. McDonald, and W. P. Erickson. 2002. *Resource Selection by Animals: Statistical Design and Analysis for Field Studies*. Chapman and Hall, New York, NY.

Meagher, M. 1993. *Winter Recreation-Induced Changes in Bison Numbers and Distribution in Yellowstone National Park*. Yellowstone National Park, WY.

Meagher, M. 1998. Recent changes in Yellowstone bison numbers and distribution. Pages 107–112 *in* L. Irby and J. Knight (Eds.) *International Symposium on Bison Ecology and Management in North America*, Montana State University, Bozeman, MT.

Meyer, C. G., and K. N. Holland. 2005. Movement patterns, home range size and habitat utilization of the bluespine unicornfish, *Naso unicornis* (Acanthuridae) in a Hawaiian marine reserve. *Environmental Biology of Fishes* **73**:201–210.

Moen, A. N. 1976. Energy conservation by white-tailed deer in the winter. *Ecology* **57**:192–198.

National Park Service. 2000. *Winter use plans for the Yellowstone and Grand Teton national parks and John D. Rockefeller, Jr., memorial parkway—final environmental impact statement*. U.S. Department of the Interior, Denver, CO.

Neter, J., M. H. Kutner, C. J. Nachtsheim, and W. Wasserman. 1996. *Applied Linear Statistical Models*. McGraw-Hill, New York, NY.

Noss, R. F. 1991. Landscape connectivity: Different functions at different scales. Pages 27–39 *in* W. E. Hudson (ed.) Landscape Linkages and Biodiversity, Island Press, Washington, DC.

Parker, K. L., C. T. Robbins, and T. A. Hanley. 1984. Energy expenditure for locomotion by mule deer and elk. *Journal of Wildlife Management* **48**:474–488.

Pe'er, G., S. K. Heinz, and K. Frank. 2006. Connectivity in heterogeneous landscapes: analyzing the effect of topography. *Landscape Ecology* **21**:47–61.

Sæther, B. E. 1997. Environmental stochasticity and population dynamics of large herbivores: a search for mechanisms. *Trends in Ecology and Evolution* **12**:143–149.

SAS. 2004. *SAS/STAT 9.1 User's Guide*. Cary, NC.

Schaefer, J. A., and F. Messier. 1995a. Winter foraging by muskoxen: A hierarchical approach to patch residence time and cratering behavior. *Oecologia* **104**:39–44.

Schaefer, J. A., and F. Messier. 1995b. Habitat selection as a hierarchy: The spatial scales of winter foraging by muskoxen. *Ecography* **18**:333–344.

Selonen, V., and I. K. Hanski. 2003. Movements of the flying squirrel *Pteromys volans* in corridors and in matrix habitat. *Ecography* **26**:641–651.

Shkedy, Y., and D. Saltz. 2000. Characterizing core and corridor use by Nubian ibex in the Negev Desert, Israel. *Conservation Biology* **14**:200–206.

Sinclair, A. R. E. 1983. The function of distance movements in vertebrates. Pages 240–258 *in* I. R. Swingland and P. J. Greenwood (Eds.) *The Ecology of Animal Movement.* Clarendon Press, Oxford, United Kingdom.

Taylor, A. R., and R. L. Knight. 2003. Wildlife responses to recreation and associated visitor perceptions. *Ecological Applications* **13**:951–963.

Taylor, L. R., and R. A. Taylor. 1977. Aggregation, migration and population mechanics. *Nature* **265**:415–421.

Telfer, E. S., and J. P. Kelsall. 1984. Adaptation of some large North American mammals for survival in snow. *Ecology* **65**:1828–1834.

Trombulak, S. C., and C. A. Frissell. 2000. Review of ecological effects of roads on terrestrial and aquatic communities. *Conservation Biology* **14**:18–30.

Turchin, P. 1991. Translating foraging movements in heterogeneous environments into the spatial distribution of foragers. *Ecology* **72**:1253–1266.

Wall, J., I. Douglas-Hamilton, and F. Vollrath. 2006. Elephants avoid costly mountaineering. *Current Biology* **16**:R527–R529.

Whittington, J., C. C. St Clair, and G. Mercer. 2004. Path tortuosity and the permeability of roads and trails to wolf movement. *Ecology and Society* **9**:4.

Whittington, J., C. C. St Clair, and G. Mercer. 2005. Spatial responses of wolves to roads and trails in mountain valleys. *Ecological Applications* **15**:543–553.

Williams, T. C., J. M. Williams, P. G. Williams, and P. Stokstad. 2001. Bird migration through a mountain pass studied with high resolution radar, ceilometers, and census. *Auk* **118**:389–403.

Zollner, P. A., and S. L. Lima. 1999. Search strategies for landscape-level interpatch movements. *Ecology* **80**:1019–1030.

APPENDIX

APPENDIX 28A.1 Candidate *a priori* models for analyses of travel and non-travel (*e.g.*, feeding, resting) locations for bison in Yellowstone National Park during winters 2003–2004 through 2005–2006

Model	Structure
1	$SWE_A + HAB + DIST_{stream} + DIST_{forage} + SLOPE + SLOPE_{het}$
2	$SWE_A + SWE_L + HAB + DIST_{stream} + DIST_{forage} + SLOPE + SLOPE_{het} + SWE_L*DIST_{stream}$
3	$SWE_A + SWE_L + HAB + DIST_{stream} + DIST_{forage} + SLOPE + SLOPE_{het} + SWE_L*DIST_{forage}$
4	$SWE_A + SNH_L + HAB + DIST_{stream} + DIST_{forage} + SLOPE + SLOPE_{het} + SNH_L*DIST_{stream}$
5	$SWE_A + SNH_L + HAB + DIST_{stream} + DIST_{forage} + SLOPE + SLOPE_{het} + SNH_L*DIST_{forage}$
6	$SWE_A + HAB + DIST_{stream} + DIST_{forage} + DIST_{unburn} + SLOPE + SLOPE_{het}$
7	$SWE_A + SWE_L + HAB + DIST_{stream} + DIST_{forage} + DIST_{unburn} + SLOPE + SLOPE_{het} + SWE_L*DIST_{stream}$
8	$SWE_A + SWE_L + HAB + DIST_{stream} + DIST_{forage} + DIST_{unburn} + SLOPE + SLOPE_{het} + SWE_L*DIST_{forage}$
9	$SWE_A + SNH_L + HAB + DIST_{stream} + DIST_{forage} + DIST_{unburn} + SLOPE + SLOPE_{het} + SNH_L*DIST_{stream}$
10	$SWE_A + SNH_L + HAB + DIST_{stream} + DIST_{forage} + DIST_{unburn} + SLOPE + SLOPE_{het} + SNH_L*DIST_{forage}$
11	$SWE_A + HAB + DIST_{stream} + DIST_{forage} + DIST_{burn} + SLOPE + SLOPE_{het}$
12	$SWE_A + SWE_L + HAB + DIST_{stream} + DIST_{forage} + DIST_{burn} + SLOPE + SLOPE_{het} + SWE_L*DIST_{stream}$
13	$SWE_A + SWE_L + HAB + DIST_{stream} + DIST_{forage} + DIST_{burn} + SLOPE + SLOPE_{het} + SWE_L*DIST_{forage}$
14	$SWE_A + SNH_L + HAB + DIST_{stream} + DIST_{forage} + DIST_{burn} + SLOPE + SLOPE_{het} + SNH_L*DIST_{stream}$
15	$SWE_A + SNH_L + HAB + DIST_{stream} + DIST_{forage} + DIST_{burn} + SLOPE + SLOPE_{het} + SNH_L*DIST_{forage}$
16	$SWE_A + HAB + DIST_{stream} + DIST_{forage} + DIST_{road} + SLOPE + SLOPE_{het}$

(*continued*)

APPENDIX 28A.1   (*continued*)

Model	Structure
17	$SWE_A + SWE_L + HAB + DIST_{stream} + DIST_{forage} + DIST_{road} + SLOPE + SLOPE_{het} + SWE_L*DIST_{stream}$
18	$SWE_A + SWE_L + HAB + DIST_{stream} + DIST_{forage} + DIST_{road} + SLOPE + SLOPE_{het} + SWE_L*DIST_{forage}$
19	$SWE_A + SWE_L + HAB + DIST_{stream} + DIST_{forage} + DIST_{road} + SLOPE + SLOPE_{het} + SWE_L*DIST_{road}$
20	$SWE_A + SNH_L + HAB + DIST_{stream} + DIST_{forage} + DIST_{road} + SLOPE + SLOPE_{het} + SNH_L*DIST_{stream}$
21	$SWE_A + SNH_L + HAB + DIST_{stream} + DIST_{forage} + DIST_{road} + SLOPE + SLOPE_{het} + SNH_L*DIST_{forage}$
22	$SWE_A + SNH_L + HAB + DIST_{stream} + DIST_{forage} + DIST_{road} + SLOPE + SLOPE_{het} + SNH_L*DIST_{road}$
23	$SWE_A + HAB + DIST_{stream} + DIST_{forage} + DIST_{unburn} + DIST_{burn} + SLOPE + SLOPE_{het}$
24	$SWE_A + SWE_L + HAB + DIST_{stream} + DIST_{forage} + DIST_{unburn} + DIST_{burn} + SLOPE + SLOPE_{het} + SWE_L*DIST_{stream}$
25	$SWE_A + SWE_L + HAB + DIST_{stream} + DIST_{forage} + DIST_{unburn} + DIST_{burn} + SLOPE + SLOPE_{het} + SWE_L*DIST_{forage}$
26	$SWE_A + SNH_L + HAB + DIST_{stream} + DIST_{forage} + DIST_{unburn} + DIST_{burn} + SLOPE + SLOPE_{het} + SNH_L*DIST_{stream}$
27	$SWE_A + SNH_L + HAB + DIST_{stream} + DIST_{forage} + DIST_{unburn} + DIST_{burn} + SLOPE + SLOPE_{het} + SNH_L*DIST_{forage}$
28	$SWE_A + HAB + DIST_{stream} + DIST_{forage} + DIST_{unburn} + DIST_{road} + SLOPE + SLOPE_{het}$
29	$SWE_A + SWE_L + HAB + DIST_{stream} + DIST_{forage} + DIST_{unburn} + DIST_{road} + SLOPE + SLOPE_{het} + SWE_L*DIST_{stream}$
30	$SWE_A + SWE_L + HAB + DIST_{stream} + DIST_{forage} + DIST_{unburn} + DIST_{road} + SLOPE + SLOPE_{het} + SWE_L*DIST_{forage}$
31	$SWE_A + SWE_L + HAB + DIST_{stream} + DIST_{forage} + DIST_{unburn} + DIST_{road} + SLOPE + SLOPE_{het} + SWE_L*DRD$
32	$SWE_A + SNH_L + HAB + DIST_{stream} + DIST_{forage} + DIST_{unburn} + DIST_{road} + SLOPE + SLOPE_{het} + SNH_L*DST$
33	$SWE_A + SNH_L + HAB + DIST_{stream} + DIST_{forage} + DIST_{unburn} + DIST_{road} + SLOPE + SLOPE_{het} + SNH_L*DIST_{forage}$
34	$SWE_A + SNH_L + HAB + DIST_{stream} + DIST_{forage} + DIST_{unburn} + DIST_{road} + SLOPE + SLOPE_{het} + SNH_L*DIST_{road}$
35	$SWE_A + HAB + DIST_{stream} + DIST_{forage} + DIST_{burn} + DIST_{road} + SLOPE + SLOPE_{het}$
36	$SWE_A + SWE_L + HAB + DIST_{stream} + DIST_{forage} + DIST_{burn} + DIST_{road} + SLOPE + SLOPE_{het} + SWE_L*DIST_{stream}$
37	$SWE_A + SWE_L + HAB + DIST_{stream} + DIST_{forage} + DIST_{burn} + DIST_{road} + SLOPE + SLOPE_{het} + SWE_L*DIST_{forage}$
38	$SWE_A + SWE_L + HAB + DIST_{stream} + DIST_{forage} + DIST_{burn} + DIST_{road} + SLOPE + SLOPE_{het} + SWE_L*DIST_{road}$
39	$SWE_A + SNH_L + HAB + DIST_{stream} + DIST_{forage} + DIST_{burn} + DIST_{road} + SLOPE + SLOPE_{het} + SNH_L*DIST_{stream}$
40	$SWE_A + SNH_L + HAB + DIST_{stream} + DIST_{forage} + DIST_{burn} + DIST_{road} + SLOPE + SLOPE_{het} + SNH_L*DIST_{forage}$
41	$SWE_A + SNH_L + HAB + DIST_{stream} + DIST_{forage} + DIST_{burn} + DIST_{road} + SLOPE + SLOPE_{het} + SNH_L*DIST_{road}$
42	$SWE_A + HAB + DIST_{stream} + DIST_{forage} + DIST_{unburn} + DIST_{burn} + DIST_{road} + SLOPE + SLOPE_{het}$
43	$SWE_A + SWE_L + HAB + DIST_{stream} + DIST_{forage} + DIST_{unburn} + DIST_{burn} + DIST_{road} + SLOPE + SLOPE_{het} + SWE_L*DIST_{stream}$
44	$SWE_A + SWE_L + HAB + DIST_{stream} + DIST_{forage} + DIST_{unburn} + DIST_{burn} + DIST_{road} + SLOPE + SLOPE_{het} + SWE_L*DIST_{forage}$
45	$SWE_A + SWEL + HAB + DIST_{stream} + DIST_{forage} + DIST_{unburn} + DIST_{burn} + DIST_{road} + SLOPE + SLOPE_{het} + SWE_L*DIST_{road}$

(*continued*)

APPENDIX 28A.1   (*continued*)

Model	Structure
46	$SWE_A + SNH_L + HAB + DIST_{stream} + DIST_{forage} + DIST_{unburn} + DIST_{burn} + DIST_{road} + SLOPE + SLOPE_{het} + SNH_L*DIST_{stream}$
47	$SWE_A + SNH_L + HAB + DIST_{stream} + DIST_{forage} + DIST_{unburn} + DIST_{burn} + DIST_{road} + SLOPE + SLOPE_{het} + SNH_L*DIST_{forage}$
48	$SWE_A + SNH_L + HAB + DIST_{stream} + DIST_{forage} + DIST_{unburn} + DIST_{burn} + DIST_{road} + SLOPE + SLOPE_{het} + SNH_L*DIST_{road}$

The response variable is either the log odds of bison travel or log odds of bison non-travel. Covariates are described in the text.

Section 7

Communicating Ecological Knowledge and Contributing to Natural Resource Management

CHAPTER 29

Communicating Ecological Knowledge to Students and the Public

Susan E. Alexander,* Thor N. Anderson,* Fred G. R. Watson,* Sally Plumb,† Wendi B. Newman,* Simon S. Cornish,* Jon Detka,* and Robert A. Garrott‡

*Division of Science and Environmental Policy, California State University Monterey Bay
†National Park Service, Yellowstone National Park
‡Fish and Wildlife Management Program, Department of Ecology, Montana State University

Contents

Theme

The effective conservation and management of our natural resources depends upon education and communication. The theme of the National Park Service Interpretation and Education Division is "connecting people to parks." This theme centers on creating unique and powerful learning experiences that increase understanding and inspire personal values. Recent technological and internet advances have allowed scientists to communicate complex and dynamic ecological processes more easily and to broader audiences than ever before. Our applied, interdisciplinary approach for communicating ecological knowledge links science, management, and policy through a visualization-based integration of climate, landscape, animal data, and humanity in the central Yellowstone ecosystem. We developed visualization products that enable students and the public to explore and understand ecological patterns and relationships in Yellowstone National Park. By integrating these visualization products with innovative pedagogy and web technology, we hope to engage and excite a broader audience in the ecological research and natural resource management issues of the central Yellowstone

The Ecology of Large Mammals in Central Yellowstone
R. Garrott, P. J. White and F. Watson
ISSN 1936-7961, DOI: 10.1016/S1936-7961(08)00229-7

Copyright © 2009, Elsevier Inc.
All rights reserved.

ecosystem. We illustrate this approach by presenting a diverse set of educational and outreach products, including: (1) a simple, web-based "Data Mapper" visualization tool; (2) an undergraduate education module; (3) a series of informal education "Virtual Interpretive Trail" visualizations; (4) a variety of data mapping and analysis services; and (5) a decision-support tool for resource managers.

I. INTRODUCTION

The importance of communicating ecological knowledge and discoveries extends beyond the traditional scientific community. Bridging the communication gap between scientists, policy- and decision-makers, natural resource managers, students, and the public has emerged as an important environmental conservation issue of the twenty-first century (Bazzaz *et al.* 1998, Lubchenco 1998, Parson 2001, Palmer *et al.* 2005). By empowering people to understand scientific processes and the connection of those processes to their everyday lives, individuals may collectively make more informed decisions regarding environmental issues in today's world (Worcester 2002, Millennium Ecosystem Assessment 2005). Scientific interpretation and education are integral to the National Park Service mission of conserving park resources and values, while providing for their enjoyment by the people of the United States (National Park Service Organic Act of 1916 [16 USC 1, 2–4]; National Park Service General Authorities Act of 1970 [16 USC 1a–1 through 1a–8]). Environmental issues surrounding these resources typically involve diverse stakeholders, who may learn or obtain information in different ways. Innovative approaches that combine traditional and nontraditional modes of education can be used to facilitate a better understanding of complex ecological processes which, in turn, may enhance enjoyment of national park lands and provide motivation for democratic engagement in natural resource conservation and management. Indeed, the resilience and sustainability of wildlife populations and their habitats in jewels such as Yellowstone National Park may depend upon such transfer of information.

Finding effective ways of translating ecological knowledge into accessible and understandable components is challenging. With advancements in technology, many datasets are large, multidimensional, and multilayered, making it difficult for the untrained eye to see and explore ecological patterns and processes. In the past decade, interactive computer visualizations and 3-dimensional images have emerged as powerful outreach and educational tools to help people understand complex and dynamic ecological processes. At the same time, the advance in internet communications has provided an unparalleled opportunity to reach a large and diverse audience.

We present examples of our applied, interdisciplinary approach for communicating ecological knowledge that links science, management, and policy through a visualization-based integration of climate, landscape, animal data, and humanity in the central Yellowstone ecosystem. Using this technological approach, we highlight a diverse set of outreach and educational products specifically targeted to students and Yellowstone National Park visitors (both on-site and virtual visitors). By integrating these visualization products with innovative pedagogy and web technology, we hope to engage and excite a broader audience in the ecological research and management issues of the central Yellowstone ecosystem. We intend that these products may serve as examples or templates for others seeking to link science, education, and management through visualization technologies.

II. APPROACH AND TECHNOLOGICAL FRAMEWORK

We applied the earth systems science approach to learning (NASA 1986, Ireton *et al.* 1996, National Science Foundation 2003, Johnson *et al.* 1997), which emphasizes the integration of data and information to link the natural sciences with the physical sciences, economics, policy, and other disciplines.

One of the most powerful aspects of our educational and outreach products is their use of visualization tools and technology to showcase complex and dynamic ecological processes that are often difficult to convey to the general public. These tools are developed within the Tarsier modeling framework (Watson and Rahman 2003), which allows the user to combine and visualize scientific data sets from multiple sources across multiple scales of space and time. For instance, within the Tarsier software framework a user can create visualizations that show dynamic spatial patterns of wildlife movement within the context of a similarly dynamic landscape of changing snow pack and plant phenology as influenced by climate, terrain, fire-affected vegetation, and geothermal heat flow. The computer animations, visualizations, maps, and graphics produced within the Tarsier framework are incorporated in different ways into our outreach products.

Depending on the target audience, visualization products can be developed into a variety of media to enhance understanding of the Yellowstone ecosystem. Visualizations targeted for teachers, students, and park visitors are often combined with real video footage, music, and descriptive narration and are interactive or coupled with learning activities. Visualizations that are created for researchers and managers often use 3-dimensional renderings of data sets and model outputs of interest, supported by spreadsheets of the underlying information. Such visualizations rely as little as possible on artificial proxy. Whenever possible, they are created with real data and imagery, and are augmented with models where necessary. Each educational product is designed for a particular use (*e.g.*, formal education, information education, research, management, public outreach) and audience (*e.g.*, K-12, undergraduate, interested public, park visitors, scientists), with a specific technology format (*e.g.*, web-based, interactive DVD, film) and location (*e.g.*, visitor center kiosk, virtual visitor center, auditorium, classroom). Additionally, all products are archived and described on the *EcoViz: Yellowstone* website (http://ynp.csumb.edu/) with the goal of engaging individuals with different interests and levels of scientific knowledge in the ecological processes of the central Yellowstone ecosystem. In the remainder of this chapter, we describe the following specific products and services: (1) a simple web-based "Data Mapper" tool for exploration of ecological and landscape data; (2) an interactive, ecological teaching module designed for formal education use in undergraduate science courses; (3) a series of "Virtual Interpretive Trail" visualizations for informal education and outreach via the world wide web and in Visitor Education Centers; (4) a variety of data mapping and analysis services such as the *Windows Into Wonderland* electronic field trip site; and (5) a decision-support tool for resource managers, in the form of web-based disseminations of targeted snow pack nowcasts.

III. SIMPLE WEB-BASED "DATA MAPPER" TOOL

We developed a simple web-based tool that allows rapid, interactive access to basic spatial data on landscape pattern and wildlife distribution (http://ynp.csumb.edu/mapper). The tool is intended for use in formal educational settings, and is designed to allow users to compare different characteristics of the landscape to form and examine questions about landscape ecological processes. It is effectively a very simple Geographic Information System (GIS), but it places far less technical demand on the user than either a conventional personal computer-based GIS such as ArcMap or a web-based GIS server such as ArcIMS (Environmental Systems Research Institute, Redlands, CA). As a result, nontechnical students can interactively obtain information from the tool within seconds in hands-on classroom settings. The trade-off for this simplicity is relative inflexibility to the addition of new data, and lack of features such as zoom functionality. Thus, the tool is best suited for predesigned teaching scenarios with specific ecological learning outcomes.

There are three types of data on the Data Mapper tool: raster, vector, and point. In particular, the tool provides access to the maps of landscape characteristics described in Chapter 2 by Newman and

Watson, Chapter 3 by Watson and Newman, and Chapters 4–6 Watson *et al.*, including vegetation (land use cover), digital elevation, terrain slope, snow pack (snow water equivalent) in different years, burn intensity from the 1988 fires, and geothermal heat intensity, as well as water, roads, and bison locations in different seasons and years (Figure 29.1). A key feature of the tool is the ability to facilitate rapid side-by-side comparison of spatial data sets such as climate in different years or different aspects of wildlife distribution. In this side-by-side viewing mode, two maps are juxtaposed and the user can independently populate each map with different data using single clicks on the browser. The user can also switch to a "single-map" viewing mode to get a larger and more detailed image. The Data Mapper tool is intended for simple, visual data exploration, but it also provides links for downloading the underlying data sets and metadata. Users wishing to perform more rigorous, quantitative analysis can obtain the data in standard GIS formats for subsequent analysis.

The Data Mapper tool provides a unique link between researchers, educators, and the public. Researchers often produce prodigious amounts of data that are of interest to students and the public. Typically specialized training and large investments of time would be required to access and learn directly from these data. With the Data Mapper tool, we have made a simple, web-based map interface that opens the door to the existence and general appearance of the data, with the ability to drill-down further if desired. This communication concept and synthesis effort could easily be expanded to a wide range of data sets, issues, and scales.

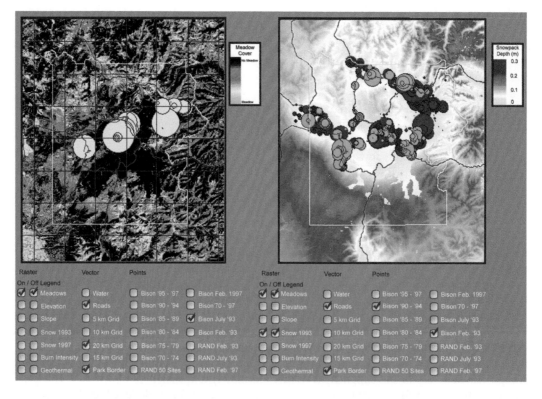

FIGURE 29.1 Screen-grab from the web-based "Data Mapper" tool. New information is interactively and instantaneously displayed using single clicks of data layers and associated legends. Multiple layers can be viewed at the same time.

IV. "BISON ECOLOGY AND MANAGEMENT" UNDERGRADUATE EDUCATION MODULE

We developed the "Bison Ecology and Management" education module to provide undergraduate students with scientific information and sequential sets of data in a case-study framework designed to explore ecological patterns, relationships, and management issues in Yellowstone. We chose bison as the focus of the module because they embody an important issue involving wildlife response to landscapes and their responses to the landscapes are often more clearly expressed than the dynamics of other species. This makes them well suited as a case study for introductory ecology students. In addition, the bison dataset covers a long timescale and is publicly available. We use the module in our own undergraduate classrooms for environmental science and ecology majors at the sophomore through senior level.

The web-based module is student-centered, makes use of the Data Mapper tool, and uses scientific inquiry, data exploration, geospatial technology, and group collaboration to create an active learning environment. Students generate and test simple hypotheses using visualizations and analyses of real field data, remote sensing data, and computer modeling data. The case-study module is composed of multiple parts or submodules that can be tailored to meet a variety of requirements and constraints in an undergraduate classroom. The common goal is to engage the students by having them simulate the role of a research scientist and place their findings in the context of important management dilemmas. The highlighted module, "Bison Ecology & Management," includes an introduction and overview of the ecosystem, a web-based student research project, and a synthesis and communication of project results (Table 29.1). Each section of "Bison Ecology & Management" can be expanded or contracted to fit the instructor's timeline. The key features of this module are built upon the web-based spatial data viewer accessible through *EcoViz:Yellowstone*.

Section I seeks to motivate the student to explore specific ecological relationships, including herbivore movement, predator–prey interactions, animal population dynamics and management, animal and climate interactions, plant phenology patterns, human–wildlife conflicts, and their relevance to management decisions in Yellowstone National Park. This case-study module begins with a "point-of-view" to help the student think about the intersection of science, management, and policy in Yellowstone (Plumb and Aune 2002, Gates *et al.* 2005, Chapter 30 by White *et al.*, this volume):

> Bison are a strong symbol of the American West. They once numbered in the millions, roaming the North American plains. Eventually, millions dwindled to thousands, and thousands dwindled to

TABLE 29.1 Overview of "Bison Ecology and Management" education module showing main sections, associated activities, and approximate classroom time required to complete the module

Section	Activities	Classroom time
Introduction	Background Research—Related Websites	Outside of classroom
	Background Research—Films	1–2 h
	Background Research—Scientific Paper Discussion	1–2 h (plus preparation time)
	Development of Climate Diagrams	1–2 h (plus preparation time)
Research project	Population Dynamics—Excel	2 h (plus preparation time)
	Spatial Dynamics – Web GIS	2–4 h (plus preparation time)
	Habitat Dynamics—Web GIS	2–4 h (plus preparation time)
Conclusions	Student Research Papers	Outside of classroom
	Student Oral Presentations	2 h (plus preparation time)
	Classroom Group Discussion	2–4 h (plus preparation time)

hundreds as the pressures of European westward expansion, hunting, and land use change converged. By 1894 only 23 wild bison remained in the United States. Those few animals found refuge in the remote Pelican Valley of Yellowstone National Park. Over the next century, extensive restoration efforts in the park brought bison back from the brink of extinction. Wild bison in Yellowstone National Park now number in the multiple thousands. They are a true conservation success story. Yet, they are also a management dilemma.

Bison, like all wild animals, do not recognize invisible boundaries, such as those of Yellowstone National Park. In certain winters, they roam beyond boundaries, crossing roads and entering populated areas. Of their own free will, they most likely would return to the resource-rich park in a few weeks or months. But, weeks or months is not soon enough in the eyes of most people. In addition to the physical risk they impart on humans, bison pose a risk to livestock through the potential transmission of the disease brucellosis. This disease can wreak havoc on domesticated livestock, causing cattle to abort their young and diseased animals to weaken the herd. The current management policy that addresses these concerns involves efforts to contain, and then cull animals as they reach the park boundaries.

Are current policies the "best" policies? Can we increase our understanding of bison ecology and behavior to better inform management decisions to decrease culling? Specifically, what factors cause bison to leave the boundaries of Yellowstone National Park? By using spatial technology to visualize scientific field data, computer modeling data, and remote sensing data, we can begin to explore patterns of bison distribution and movement.

There are numerous factors that influence bison. In addition to natural environmental factors and the risk of disease, there is an ongoing debate surrounding winter recreation in the park, and the effects of winter road use by snowmobiles on bison movements and range expansion, including trans-boundary movements, bison condition, and population dynamics. What impact do groomed winter roads have on bison? Do they facilitate bison leaving the park? Or maybe they don't make any difference to bison.

This case study provides an introduction to the integration of science, technology, and policy as a framework to improve understanding of environmental management issues in Yellowstone National Park. We analyze scientific data and information using web-based spatial technologies to better understand current bison management dilemmas and controversies. Specifically, we address bison distribution and movement in relation to: (1) the risk to livestock of transmission of the disease brucellosis from bison moving beyond park boundaries; and (2) criticism of the effects of winter road use by snowmobiles on bison movements and range expansion, including trans-boundary movements, bison condition, and population dynamics.

The student is then directed to explore related Yellowstone National Park website links, films, and peer-reviewed scientific papers to better understand the ecological, management, and political systems. Class time can be used for some or all of these activities. A useful exercise at this stage of the module is to have students conduct a scientific critique of a related, peer-reviewed research paper.

The next activity involves understanding Yellowstone's abiotic factors through the construction of Climate Diagrams (Walter 1985, Molles 2007). Understanding climate patterns in Yellowstone National Park helps students to develop a baseline for understanding more complex ecological processes and explore the relationship between the type and distribution of terrestrial ecosystems and climate. Climate patterns also influence the seasonal spatial dynamics of animal populations in the park, which, in turn, have major influences on their population dynamics. Climate Diagrams summarize annual variation in temperature and precipitation, the length and intensity of wet and dry seasons, and the portions of the year during which average minimum temperature is above and below 0 °C (Figure 29.2). Students construct Climate Diagrams and compare the climate (temperature, precipitation, snow depth) at locations with variable elevation or other factors within Yellowstone National Park (*e.g.*, West Yellowstone, Old Faithful, Lake, Mammoth). Climate data are available through the Western Regional Climate Center's Western U.S. Climate Historical Summaries Database (http://www.wrcc.dri.edu/CLIMATEDATA.html). All Climate Diagrams are plotted on set scales, allowing students to compare climate patterns throughout Yellowstone and the surrounding areas.

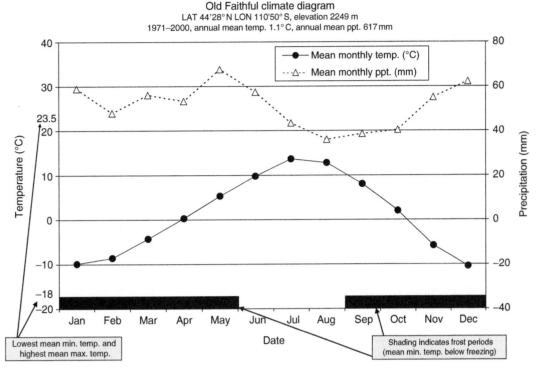

FIGURE 29.2 Example "Climate Diagram" shows seasonal variability in average monthly temperature and average monthly precipitation for Old Faithful in Yellowstone National Park. This diagram indicates moist conditions year-round because the precipitation line is located above the temperature line.

The main "Research Project" builds upon this background information and centers around two bison management issues: (1) bison leaving the park in certain winters and potentially transmitting the disease brucellosis to cattle in the surrounding area; and (2) bison leaving the park due, potentially, to the grooming of roads and associated use of snowmobiles. To better understand bison ecology and management, students analyze historical and recent data to answer the following sets of questions:

Population Dynamics

- How has bison population size changed over the past century?
- What factors have influenced changes in population size over the past century?

Spatial Dynamics

- What changes in bison distribution have occurred during the past century?
- What factors have influenced changes in bison distribution during the past century?

Habitat Dynamics

- What type of habitat do bison select as their preferred place to live?
- What type of habitat facilitates bison movement?

In the Population Dynamics section, the students explore historical datasets of bison counts and culling in the park between 1900 and 2000 (Meagher 1998, Taper *et al.* 2000, Fuller *et al.* 2007) and use this information to graph fluctuations in the park's bison populations. Students then identify periods with relative levels of bison culling (*e.g.*, none, low, high) from the data and research the specific park management policies during the same time period. The combined information is used by the students to explore the population growth patterns of both the central and northern herds over the past century with respect to park management policies.

In the Spatial Dynamics section, students explore the spatial patterns of bison relative to abiotic, biotic, and human factors. This web-based, GIS activity integrates field data, computer modeling, and remote sensing data. Using the Data Mapper visualization tool, different maps can be layered (*e.g.*, view elevation, roads, and water together) or viewed side-by-side for comparison. Students explore the maps and propose hypotheses relating bison data to landscape data to investigate factors that may have influenced bison distribution during the past century. Example hypotheses include "the range of the bison population has increased over the past 30 years" and "most bison spend the summer in meadow areas." Each hypothesis is then evaluated by viewing, comparing, layering, and studying the maps and overlaying gridlines. Students can count grid cells to quantitatively answer as many hypotheses as possible.

The final, and most complex, part of the project involves exploring bison Habitat Dynamics through another activity using the same web-based, GIS, Data Mapper tool. Students pose multiple hypotheses that address the question "What type of habitat do bison select as their preferred place to live?" Ultimately, this question is explored for three seasons: average winters, harsh winters, and summers. Rather than using statistical software, students develop simple "use-availability" models based on each hypothesis that compare habitat conditions in areas that have been used by bison against all areas that are available for use by bison.

The covariates that students can choose to examine in their habitat use availability models are vegetation (land use cover), digital elevation, slope, snow pack (snow water equivalent) in different years, burn intensity from the 1988 fire, geothermal heat intensity, water, and roads. Example hypotheses can include single covariates or combinations of covariates:

> **H1**: In an average winter, bison prefer to occupy low-elevation areas over high-elevation areas;
> **H2**: In an average winter, bison prefer to occupy areas that are influenced by geothermals over areas that are not influenced by geothermals;
> **H3**: In an average winter, bison prefer to occupy areas that have light snow cover over areas that have heavy snow cover;
> **H4**: In an average winter, bison prefer to occupy areas that have both light snow cover and are near waterways over areas that have heavy snow cover and are not near waterways.

The quantitative analysis is conducted by comparing side-by-side maps of the covariate (*i.e.*, elevation in H1) in the Data Mapper. On one map, a specific number of random points are overlaid on the raster image. On the other map, the same number of known bison location points (generated randomly from all known bison location data points in the selected season and/or year) are overlaid on the raster image. Students obtain a quantitative value for the covariate at each point (both random points and bison location points) and use these data to choose "cut-off" values (*e.g.*, the value above which we will consider elevation to be "high" for bison). For each cut-off value, the student counts the number of times the "available" data (*i.e.*, random points) fall above the specified value and the number of times the "use" data (*i.e.*, bison location points) fall below the specified value (or vice versa if the particular covariate dictates). The combination of these counts provides a quantitative estimate of the "likelihood" of the model. Students perform quantitative analyses of each model and display results in a graphical format. Finally, the use-availability model results are compared to determine which model is most likely to explain habitat selection by bison.

The "Conclusions" section of the project involves student papers, presentations, and group discussions to synthesize the students' results and thoughts on questions posed in the three main sections (Population Dynamics, Spatial Dynamics, Habitat Dynamics). Graphs, tables, and charts that summarize their data and results are helpful. The group discussion may center on what the students have learned and how their ideas and results may inform the current bison management plan or other related issues in the central Yellowstone ecosystem. Ultimately, it is hoped that the student comes away from the project with a greater understanding of: (1) how science, management, and policy are interlinked; (2) how useful and fun it is to explore scientific data in a visual and dynamic manner; and (3) some specific issues affecting wildlife and management in the central Yellowstone ecosystem. While bison were chosen as the focus for this educational module, the framework could be easily modified for other species of interest such as bears, elk, or wolves.

V. "VIRTUAL INTERPRETIVE TRAIL" VISUALIZATIONS

In support of our goal to communicate knowledge and discoveries of the Yellowstone ecosystem to the visiting public, we have developed a diverse set of informal education and outreach products based on the previously described technological framework. Our visualization products for Yellowstone National Park visitors employ what we refer to as a "Virtual Interpretive Trail" (VIT) metaphor. A VIT takes viewers on a virtual journey through environments that may be otherwise inaccessible to them. The virtual reality is made more appealing and intuitive because it is designed to feel like something that is familiar to most visitors: a walk along a physical interpretive trail, with interpretive signage along the way. In a VIT, visitors interactively fly through a 3-dimensional representation of the environment, stopping at predefined places to interactively learn about ecosystem processes relevant to that place. VIT products are not intended as a substitute for visiting the park and immersing oneself in the Yellowstone ecosystem. Rather, they enhance the physical visitor experience by providing more information about the park, including access to spatial and temporal scales that may not be available to most visitors. Many of Yellowstone's dynamic ecological processes occur over large spatial and temporal scales that visitors typically do not see in a single visit. Additionally, some areas of interest are very remote and difficult to access. For example, few visitors are capable of walking the entire length of the bison migration route, or of being in the park for the entire year that it takes to complete a cycle of migration. Via a VIT, visitors can virtually witness this phenomenon, and relate it to their own personal experiences with bison in the park. Therefore, a VIT can become a powerful educational tool for teachers, students, and visitors eager to learn more about Yellowstone's unique ecosystems, and consequently to strengthen and stimulate their physical experience.

The Yellowstone VIT explores two themes: (1) bison migration; and (2) the upper geyser basin. These products are deployed in a physical kiosk at the Canyon Visitor Education Center in Yellowstone National Park, on the park's educational website, *Windows Into Wonderland* (www. windowsintowonderland.org), and in the forthcoming Old Faithful Virtual Visitor Education Center (to be available via the park's official website: http://www.nps.gov/yell).

A. Bison Migration Virtual Interpretive Trail

Unlike museums, national park visitor education centers are not the primary destination of interest to the visitor. Rather, the physical world of the park itself is the primary interest. Therefore, interpretive exhibits in parks must be brief and highly engaging. This creates a challenge for science communication – to balance depth of content against typical attention spans and, in the case of Yellowstone, the physical constraints of facilitating access for over three million visitors annually. As scientists, we strive to infuse as much ecological insight as possible into the visitor experience. We want to enable visitors to interactively "drill down" past the charisma of the wildlife and access the back stories – the researchers, management, and history.

The Bison Migration VIT kiosk deployment at the Canyon Visitor Education Center occurred within a larger interpretive design context for the Center as a whole, which placed specific stipulations on visitor sensory overload. To address this, we used relatively short video segments and substituted audio sound effects and narration for on-screen text-based narration. The web deployments have full audio, music, and the unmistakable sounds of bison.

The Bison Migration VIT, "Bison on a Volcano," allows the user to "fly" through a virtual Yellowstone landscape (based on real data), following the central bison herd's annual migration (Figure 29.3). There are four predetermined stops along the migration route that highlight important bison life cycle processes at different locations and times of year. At each stop, the user may choose to view a short interpretive story or to continue on to the next predetermined stop in the dynamic landscape (Figure 29.4). The short interpretive stories include footage for a particular location and season. During summertime in the Hayden Valley, the footage showcases bison congregating in great numbers for the annual rut, bulls courting cows, bulls fighting, and bison foraging on green grass. This use of documentary-style wildlife footage is an informative, familiar medium to those visitors who are more used to seeing nature documentaries than virtual reality computer games.

FIGURE 29.3 View of the Canyon Visitor Education Center and the Bison Migration "Virtual Interpretive Trail" ("Bison on a Volcano") interactive touch-screen kiosk that allows viewers to interactively "fly" through a 3-dimensional representation of the central bison herds' annual migration.

FIGURE 29.4 Menu and button choices found at each of the four predetermined stops on the Bison Migration "Virtual Interpretive Trail." At each stop, the left menu button lets the user view a short video of typical bison activity for that season and location. The right menu button takes the user on one of four virtual flights that link together to reveal a known bison migration route. During each flight, the changing landscape is indicated by dynamic variations in snow pack distribution.

If the user chooses to skip to the next location, the program continues a virtual flyover along the Mary Mountain bison migration corridor to the next stop (Figure 29.5). This 20-mile journey, compressed into a one minute flight on the VIT, allows the user to experience bison migration and associated landscape processes from the animals' point of view. The flight is accompanied by narration

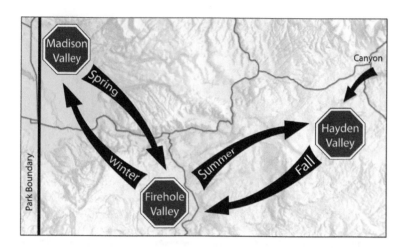

FIGURE 29.5 Image shows the virtual flight paths between the four predetermined stops in the Bison Migration "Virtual Interpretive Trail." The virtual flight paths cover the central bison herds' annual migration route from their high elevation summer range to their lower elevation winter range.

or narrative text that interprets the landscape and highlights ecological processes affecting bison migration. After the visitor has traveled to all four stops, the VIT continues to loop or, if left untouched, returns to the start menu.

A more in-depth version of the Bison Migration VIT incorporates three additional stories at each stop to supplement the existing bison life cycle story. The following choices are available at each stop: (1) Bison Life Cycle; (2) Bison History and Management; (3) Bison Research; and (4) Bison Habitat. The Bison History and Management stories focus on a range of issues surrounding Yellowstone bison including human safety, bison restoration, and the effects of winter recreation on bison. The stories contain interviews with National Park Service interpretive rangers and wildlife biologists, and have descriptive narration. For example, a wildlife biologist at Yellowstone National Park, states that "... to provide managers within all the jurisdictions better information to make decisions, we need to know both why and how bison move on the landscape."

Both the bison research and bison habitat stories are part of a series of in-the-field interviews with wildlife and landscape researchers. The researchers describe their fieldwork and how it pertains to that season and location. For example, a graduate student explains that "for the past decade, we've conducted bi-weekly bison population estimates in the Madison-Firehole and we quite simply ski, snowshoe, and snowmobile specific routes where we record bison distribution, their location, habitat, group counts and composition. Because of the geothermal influence in the Madison-Firehole, bison come over from Hayden Valley in winter. In early winter, we have 200–300 bison in the Madison-Firehole, but by early spring, we can expect to see 1200–1500 bison in the area."

B. Upper Geyser Basin Virtual Interpretive Trails

The Upper Geyser Basin is home to the world's largest concentration of geysers, including the famous Old Faithful geyser, and is the most-visited area of Yellowstone National Park. Though the physical landscape of the basin is quite accessible seasonally in a spatial sense, a first-hand appreciation of the temporal dynamics is limited by the short amount of time that most visitors spend there. The National Park Service has developed an on-line Old Faithful Virtual Visitor Education Center, soon to be available through the official website of Yellowstone National Park (http://www.nps.gov/yell) to provide extended educational and interpretive opportunities for park enthusiasts and students, and complement the Old Faithful Visitor Education Center.

Three VIT products have been developed for inclusion in the Old Faithful Virtual Visitor Education Center. The first serves as an introduction to the virtual center, communicating the vastness of the Yellowstone landscape and providing a geographic context for the Upper Geyser Basin. It is designed to hook a viewer's interest and entice further exploration of the site. This flyover sequence begins in a remote part of western Yellowstone and progresses over the park's Madison Plateau towards the Firehole Valley and the Upper Geyser Basin. The flight continues through the front doors of a virtual visitor center, through the lobby, and out into the Upper Geyser Basin, culminating in an eruption of Old Faithful. The entire sequence is accompanied by the evocative sounds of wolves, geese, elk, bison, and the geyser's eruption.

Most of the predictable geysers of the Upper Geyser Basin erupt at least once each day, but many have multiple-hour intervals between eruptions. The second VIT feature is a "24-hour Loop" that provides a 360° overhead perspective of the Upper Geyser Basin and compresses 24 hour into a 90-second, narrated visualization. The virtual eruptions of the major geysers are animated with correct timing, height, and location (Figure 29.6). Narration and geyser sounds accompany the VIT to enhance the user experience. This VIT feature enables viewers to see the longer-term picture through a virtual experience and provide a context to better inform their (usually) shorter-term physical experiences.

FIGURE 29.6 Screen grabs of the 24-hour Geyser Loop "Virtual Interpretive Trail" which compresses 24 hours of real time into 90 seconds of virtual time to illustrate the timing and frequency of geysers in the Upper Geyser Basin. The main image shows Old Faithful Geyser erupting in the foreground.

A closer look at individual hot water features is provided by the third VIT, "Upper Geyser Basin Tour," which features photographs of eight hydrothermal features set within a landscape visualization context. The features include Old Faithful, Beehive, Castle, Grand, Giant, Grotto, and Riverside geysers, as well as Morning Glory Pool. The VIT allows the viewer to "fly" low to the ground over the boardwalk and trail network in the Upper Geyser Basin, stopping at the various hydrothermal features. At each stop, a photograph of the feature appears, along with the sound effect of a camera shutter. The user may then choose to learn more about the highlighted feature or continue on to the next stop. If the viewer chooses to learn more, he or she may access a slide show accompanied by informative narration of the feature.

VI. ELECTRONIC FIELD TRIPS—*WINDOWS INTO WONDERLAND*

The *Windows Into Wonderland* web site was established by Yellowstone National Park to provide detailed, interactive, web-based, educational experiences for viewers around the world (Figure 29.7; www.windowsintowonderland.org). The website hosts a variety of electronic field trips (eTrips), especially designed for middle school students, which explore the park's unique resources. The programs are reaching far more than the target audience because individual adults, university students, elementary schools, and families are also logging onto *Windows Into Wonderland* programs. Currently the eTrips' audience includes viewers from all 50 states and over 170 countries, geographically as diverse as South Africa, Thailand, Brazil, Germany, and Australia. The eTrips provide teachers with an attractive alternative to traditional methods of instruction and address educational inequities by eliminating geographical and economic barriers. They also appeal to different learning styles and allow students to learn at a rate compatible with individual background knowledge or aptitude. The eTrips conform to the Americans with Disabilities Act, are free of charge, and available at any time to anyone who has access to the Internet.

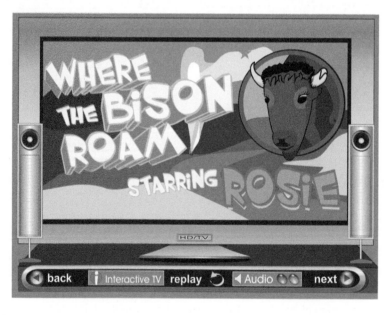

FIGURE 29.7 Screen grab from the National Park Service *Windows Into Wonderland* website, introducing the interactive, electronic field trip "Where the Bison Roam."

Over the years, the format of the eTrips has changed from simple text with scanned slides to productions with plots, characters, animations, video clips, graphics, audible dialogue, and interactive components. The presentations also feature an "Ask an Expert" forum which, for a limited time during the premiere of each new trip, allows students to pose questions to subject matter experts. A special "For Teachers" section in each trip includes web links, references, glossaries, suggested discussion points, and pre- and post-visit lesson plans. The *Windows Into Wonderland* eTrips are guided by National Education Standards for grades 5–8, and correlations are provided to National Standards for Science, Mathematics, Technology, and other subjects.

Beginning in 2003, the Informal Science Education Program of the National Science Foundation funded six new eTrips, focusing on the volcanic forces that have shaped the Yellowstone region, the area's hydrothermal resources, and the attendant complex ecosystems resulting from this unique geology. The eTrips were developed with extensive input and review from scientific researchers. As part of our data mapping and analysis service, we collaborated with the *Windows Into Wonderland* team to convert research-level spatial data directly into simple maps (*e.g.*, a shaded relief map, thermal heat flux map, and snow depth map) designed to support an activity within an eTrip entitled "Where the Bison Roam." Students are encouraged to hypothesize and explore the bison migration route between Hayden Valley and the Firehole River.

> Meet Rosie, a young bison alone in a Yellowstone winter, and learn how she is affected by the park's thermal areas. How will these sites influence Rosie's behavior or sway her migration choices? What advantages do they possess and what threats do they pose? Will these areas determine whether Rosie lives or dies? It's a question of survival, Yellowstone style!

Students select and study the various maps during this interactive exercise. Based on the information in the images, they are then invited to "draw" an expected migration route using their computer mouse and compare their drawing with a map of the actual bison migration route. These maps, plus an additional digital elevation model of the Mary Mountain migration corridor, are also provided as downloadable files for an extended classroom curriculum. The curriculum objectives seek to enable students to correlate information between different models or maps, reference geographical features, identify the Mary Mountain bison migration corridor, and express an inference on the influence of hydrothermal activity on bison migration with specific references to information obtained from the maps and models. The Bison Migration VIT (described in the previous section) is included for reference on the "For Teachers" page of the "Where the Bison Roam" eTrip.

VII. DECISION SUPPORT TOOL FOR MANAGERS

The ability to quickly visualize and integrate datasets and model outputs is important not only to educate students and the public, but also to help resource managers as they tackle difficult and sometimes contentious resource issues. We developed and deployed a web-based tool for comparison of current snow pack conditions (simulated "nowcasts") at the park boundary with previous conditions. These nowcasts have been used in recent years to assist natural resource managers in the decision making process surrounding trans-boundary bison management.

To reduce the transmission of brucellosis from bison to cattle, the Interagency Bison Management Plan includes hazing, "capture and release," and culling of bison when the animals exit Yellowstone National Park (USDOI *et al.* 2000a,b). This requires operating bison capture facilities when the hazing of the animals at park boundaries is no longer safe or effective. The decision on when (or whether) to release the bison is not straightforward. If bison are released too early (*e.g.*, before vegetation green-up), it is possible that they will not return to higher elevations in the park and may instead need to be recaptured. Thus, it is important that bison managers use tools that help them make an objective decision on the timing of release for bison from capturing facilities.

Snow pack dynamics have a key influence on bison movement (Meagher 1973, Chapter 12, 27, and 28 by Bruggeman *et al.*, this volume). It is generally understood that during periods when the herd size is large, deeper snow packs will force more bison to migrate to lower elevations near the boundary of the park. Similarly, those bison will tend to move back into the park once the snow pack conditions have ameliorated. Managing this situation in real time requires knowledge of snow pack conditions, which the snow pack nowcast product addresses.

We used the Langur snow pack model (Chapter 6 by Watson *et al.*, this volume) to implement snow pack nowcasts which estimate maps of snow pack depth throughout the entire northern boundary region daily during late winter. Simulation-based nowcasts are appropriate in this situation because real-time observations are sparse and do not provide a spatially integrated estimate, and remotely sensed observations may be obfuscated by cloud cover during critical periods. The Langur model takes real-time point observations of precipitation and temperature and uses these to estimate spatial snow pack distribution over large areas.

FIGURE 29.8 Snow pack "nowcast" decision-support tool for comparison of current snow pack conditions near the park boundary with previous conditions. Web-based users can interactively select any two days to compare side-by-side, ranging from current conditions back through the past few years.

We disseminated nowcasts directly to park staff in tabular form, averaged over core bison use areas, and in a publicly accessible interactive web-visualization form (Figure 29.8). The latter provides an intuitive visual picture of the model estimates that can be rapidly compared with visualizations from any day in recent years. The visualizations help users understand what the model does, and how model estimates compare to what can be seen in the real landscape from the roadside. This comparative functionality is more important than the absolute accuracy of the model. Park staff can compare any two days, answering questions such as "Is there at least as much snow in core use areas now as there was on days in past years when bison were documented to have still been moving in a generally out-of-park direction?"

The nowcasts are produced automatically each spring. Each day, a dedicated personal computer automatically downloads the latest climate input data, runs the model, visualizes the results, and posts the resulting images to a web site. When a user visits the website, he or she may access the latest image of the nowcast area. The user can navigate back and forth in time with the help of an embedded calendar, compare two dates side-by-side, and zoom into a more detailed image for a single day.

VIII. CONCLUSIONS

The visualization products presented in this chapter are intended to facilitate a better understanding of complex ecological issues in the central Yellowstone ecosystem. We used an applied, interdisciplinary approach for communicating ecological knowledge that links science, management, and policy through a visualization-based integration of climate, landscape, animal data, and humanity in the central Yellowstone ecosystem. With the development of each product, we faced the challenge of identifying and understanding the needs, constraints, and abilities of the target audience. For all audiences, we embraced the importance of visual data exploration as a means of translating ecological knowledge into accessible and understandable components. For visitors to Yellowstone National Park and park websites, these products provide large-scale, spatial perspectives for understanding and enjoying the park's natural resources. Within the undergraduate-level education module, the visualization products provide opportunities for students to explore real-world ecological questions and natural resource management issues using the same landscape and animal data and model outputs that are being used by natural resource managers. For park biologists, access to visualizations of current conditions help them tease out patterns in the data and recommend more-informed courses of action. We hope that each of these products will stimulate similar ideas in others seeking to integrate ecological research with education and outreach.

IX. SUMMARY

1. We present a diverse set of visualization products for education and outreach that enable students and the public to explore and understand ecological patterns and relationships in Yellowstone National Park.
2. Through the use of innovative pedagogy and web technology, each visualization product was designed to address the needs, constraints, and abilities of specific target audiences.
3. The web-based "Data Mapper" tool, for use in formal education settings, allows rapid, interactive access to basic spatial data on landscape pattern and wildlife distribution and allows users to compare different characteristics of the landscape to form and examine questions about landscape ecological processes.

4. The "Bison Ecology and Management" undergraduate education module uses scientific inquiry, data exploration, geospatial technology, and group collaboration within a case-study framework to create an active learning environment where students generate and test simple hypotheses using visualizations and analyses of real field data, remote sensing data, and computer modeling data.

5. The visualization products based upon the "Virtual Interpretive Trail" metaphor take viewers on a virtual journey through environments in Yellowstone that may otherwise be inaccessible to them, including the annual "Bison Migration" route, and a 24-hour time-lapse of the "Upper Geyser Basin" geothermal features.

6. Through the data mapping and analysis service, research-level spatial data was converted directly into simple maps designed to support an activity within the *Windows Into Wonderland* electronic field trip entitled "Where the Bison Roam."

7. The "nowcast" decision-support tool compares current snow pack conditions near the park boundary with previous conditions to assist natural resource managers in the decision-making process surrounding trans-boundary bison management.

X. REFERENCES

Bazzaz, F., G. Ceballos, M. Davis, R. Dirzo, P. R. Ehrlich, T. Eisner, S. Levin, J. H. Lawton, J. Lubchenco, P. A. Matson, H. A. Mooney, P. H. Raven, *et al.* 1998. Ecological science and the human predicament. *Science* **282:**879.

Fuller, J. A., R. A. Garrott, and P. J. White. 2007. Emigration and density dependence in Yellowstone bison. *Journal of Wildlife Management* **71:**1924–1933.

Gates, C. C., B. Stelfox, T. Muhly, T. Chowns, and R. J. Hudson. 2005. *The Ecology of Bison Movements and Distribution in and Beyond Yellowstone National Park.* University of Calgary, Alberta, Canada.

Ireton, M. F., C. A. Manduca, and D. W. Mogk. 1996. Shaping the future of undergraduate Earth science education: Innovation and change using an earth system approach. American Geophysical Union Report. http://www.agu.org/sci_soc/spheres/. Accessed September 16, 2005.

Johnson, D. R., M. Ruzek, and M. Kalb. 1997. What is Earth system science? *Proceedings of the 1997 International Geoscience and Remote Sensing Symposium,* Singapore, p. 688–691.

Lubchenco, J. 1998. Entering the century of the environment: A new social contract for science. *Science* **279:**491–497.

Meagher, M. 1973. The bison of Yellowstone National Park. National Park Service Scientific Monograph Series No.1.

Meagher, M. 1998. Recent changes in Yellowstone bison numbers and distribution. Pages 107–112 *in* L. Irby and J. Knight (Eds.) International symposium on bison ecology and management in North America. Montana State University, Bozeman, MT.

Millennium Ecosystem Assessment. 2005. *Ecosystems and Human Well-being: Synthesis.* Island Press, Washington, DC.

Molles, M. 2007. *Ecology: Concepts and Applications.* McGraw-Hill, New York, NY.

NASA [National Acronatics and Space Administration]. 1986. *Earth System Science—A Closer View. Report of the Earth System Science Committee.* NASA Advisory Council, Washington, DC.

National Science Foundation. 2003. Complex environmental systems: Synthesis for earth, life and society in the 21st century. National Science Foundation, Washington, DC.

Palmer, M. A., E. S. Bernhardt, E. A. Chornesky, S. L. Collins, A. P. Dobson, C. S. Duke, B. D. Gold, R. B. Jacobson, S. E. Kingsland, R. H. Kranz, M. J. Mappin, M. L. Martinez, *et al.* 2005. Ecological science and sustainability for the 21st century. *Frontiers in Ecology and the Environment* **3:**4–11.

Parson, W. 2001. Practical perspective: Scientists and politicians: the need to communicate. *Public Understanding of Science* **10:**303–314.

Plumb, G., and K. Aune. 2002. The interagency bison management plan for Yellowstone National Park and Montana. Pages 136–145 *in* T. J. Kreeger (Ed.) *Brucellosis in Elk and Bison in the Greater Yellowstone Area.* Wyoming Game and Fish Department, Cheyenne, WY.

Taper, M. L., M. Meagher, and C. L. Jerde. 2000. *The Phenology of Space: Spatial Aspects of Bison Density Dependence in Yellowstone National Park.* U.S. Geological Service, Biological Resources Division, Bozeman, MT.

U.S. Department of the Interior (USDOI), National Park Service, and U.S. Department of Agriculture (USDA), Forest Service. 2000a. Bison management for the state of Montana and Yellowstone National Park, Final Environmental Impact Statement for the Interagency Bison Management Plan for the State of Montana and Yellowstone National Park. Denver, CO.

U.S. Department of the Interior (USDOI), National Park Service, and U.S. Department of Agriculture (USDA), Forest Service. 2000b. Record of decision for the final environmental impact statement and bison management plan for the state of Montana and Yellowstone National Park. Denver, CO.

Walter, H. 1985. Vegetation of the Earth. Springer-Verlag, New York, NY.

Watson, F. G. R., W. Newman, J. C. Coughlan, and R. A. Garrott. 2006. Testing a distributed snowpack simulation model against diverse observations. *Journal of Hydrology* **328**:453–466.

Watson, F. G. R., and J. M. Rahman. 2003. Tarsier: A practical software framework for model development, testing and deployment. *Environmental Modeling Software* **19**:245–260.

Worcester, R. 2002. Public understanding of science. *Biologist* **49**:143.

CHAPTER 30

Science in Yellowstone: Contributions, Limitations, and Recommendations

P. J. White,* Robert A. Garrott,† and S. Thomas Olliff*

*National Park Service, Yellowstone National Park
†Fish and Wildlife Management Program, Department of Ecology, Montana State University

Contents

Theme

The National Park Service has long been criticized for emphasizing scenery and tourism management, while being indifferent, if not hostile, to the support of basic research and science-informed decision making (Sellers 1997, Parsons 2004). Congress provided the Service with a clear mandate to support and conduct research that benefits park management and society as a whole when it passed the 1998 National Parks Omnibus Management Act (Public Law 105-391, Title II, Sections 201–207). At about the same time, the Service obtained a major budget initiative to support scientific management, promulgated guidance for managing biological resources, established a standardized system for research and collecting permits, initiated a network of university-based cooperative units and a Sabbatical-in-the-Parks program, and developed several in-park research centers in larger, flagship parks (Parsons 2004). These changes provided the impetus for Yellowstone National Park to implement concurrent changes that supported research and incorporated scientific information into the decision-making process. Resource management in Yellowstone has benefited substantially from the description, interpretation, prediction, and monitoring of resources and ecological processes. Also, scientific information developed in the park has provided necessary information for partners outside the park to use in managing wildlife and their habitats. However, linking science and management continues to be a challenge due to the complexity of landscape-scale ecological issues and disagreements with stakeholders regarding relevant knowledge and values (Hall 2004, Hobbs and Hilborn 2006). We conducted an introspective examination of the contributions and limitations of science to resource management in Yellowstone and make recommendations for strengthening this link based on our experiences with wildlife issues. The views and opinions in this chapter are those of the authors and should not be construed to represent any views, determinations, or policies of the National Park Service.

The Ecology of Large Mammals in Central Yellowstone
R. Garrott, P. J. White and F. Watson
ISSN 1936-7961, DOI: 10.1016/S1936-7961(08)00230-3

Copyright © 2009, Elsevier Inc.
All rights reserved.

I. EXPECTATIONS

Yellowstone National Park is a national treasure that attracts 3 million people each year to experience its unique geothermal features, wildlife populations, natural vistas, and cultural resources. Managers have been quite successful at providing enjoyment for these visitors, while protecting the resources for future generations. There is a full and thriving complement of species that existed prior to European settlement and large-scale natural processes such as fire (Figure 30.1) and winterkill are allowed to exert their forces on the landscape. Also, nearly 90% of the land is managed as wilderness. Despite this success, management of the park is continually embroiled in controversy, including recent and highly-publicized debates on wild fire management, appropriate population levels of wildlife, disease management (*e.g.*, brucellosis), conservation of threatened and endangered species, and the effects of winter recreation on wildlife populations and the environment (Figure 30.2).

To help address these issues, there was a substantial initiative in Yellowstone over the last two decades to encourage research in the park and incorporate scientific information into the decision-making process. In 1993, the National Park Service established a multidisciplinary, in-park, resource

FIGURE 30.1 The large spatial extent of Yellowstone National Park and the public mandate to preserve this largest remaining relatively intact temperate ecosystem in as natural a state as possible provides the pubic and scientists the opportunity to witness large-scale ecological processes that can be dramatic. The major wildfires that swept across Yellowstone and the surrounding area in 1988 received attention worldwide and generated a tremendous amount of controversy and dialogue about public policies on forest fire management and the history and role of large-scale forest fires in the ecosystem (Photo by Jeff Henry).

FIGURE 30.2 Road sign illustrating the controversy surrounding bison management in Yellowstone National Park (National Park Service Photo by Steve Kelley).

management center in Yellowstone dedicated to producing objective science with the goal of advancing our knowledge of the Yellowstone ecosystem, supporting sound resource management, and communicating knowledge and discoveries to the visiting public to enhance their experience and enjoyment of the park. This Center for Resources is comprised of a team of interdisciplinary staff with the expertise and experience to provide park leadership with scientific information regarding increasingly difficult issues related to the management of natural systems.

In addition, park leadership recognized that other expertise and diversity were needed to develop efficient and timely research that would contribute to management decisions and address contentious public debates about policy. Thus, the Center for Resources solicited nongovernment scientists for research on both basic scientific questions and applied research in support of management. The high-profile, controversial issues mentioned previously stimulated interest by nongovernmental scientists and led to new research initiatives. These issues also led to recognition by federal funding agencies such as the National Science Foundation and National Aeronautics and Space Administration that national parks are excellent research laboratories, which resulted in additional resources for developing new research initiatives. The Center for Resources provides facilities, funding, and logistical support for many scientists working in the park, and has established extensive collaborations and partnerships with academic, federal, private, and state organizations, as well as university-based cooperative units. This emphasis on science in the park resulted in the number of permitted research projects increasing from approximately 60 in 1975 to over 200 by 2000.

There is general consensus that science is critical to making successful management decisions for natural resources and should be a part of the organizational culture and valued during the decision-making process (Walters 1986, Likens 1989, Williams *et al.* 2002, Wright 2004, National Park Service 2006). Science is an endeavor to discover knowledge using rational logic, objectivity, and quantification, while research is the activity implemented to discover this knowledge (Ozawa 1996, Behan 1997). Thus, the recent increased research presence in the park was intended to improve the quality of science and decision making by contributing to: (1) objective interpretations of data; (2) reliable knowledge of how the system works; (3) dissemination of accurate information to the public; and (4) effective conservation and management.

II. CONTRIBUTIONS

The emphasis on science in Yellowstone has led to increased recognition that understanding resources is essential for their preservation and that science can contribute to this understanding. Science is a powerful, authoritative institution in the modern world (Ozawa 1996). Scientists are trained to maintain a skeptical, questioning perspective and to employ logical analysis to scrutinize propositions. Thus, they can point out the limitations of existing knowledge, identify areas of uncertainty, and provide objective interpretations of generated data (Hall 2004). The increased emphasis on rigorous data collection and evaluation in Yellowstone has enhanced the credibility of park positions during many proceedings and deliberations with other federal and state agencies and stakeholders. More importantly, scientific information developed in the park has benefited other agencies responsible for resource management outside the park by providing necessary information for use in devising harvest regulations, managing wildlife and their habitat, and supporting the delisting of bald eagles (*Haliaeetus leucocephalus*), grizzly bears (*Ursus arctos*), peregrine falcons (*Falco peregrinus*), and wolves (*Canis lupus*) pursuant to the Endangered Species Act. Effective management of migratory or dispersing species that cross jurisdictional boundaries cannot be accomplished in isolation and requires strong partnerships that include the cooperative collection and sharing of information, discussion of alternate management approaches, and the implementation of proactive measures to ensure that unacceptable changes to the ecosystem do not occur.

Resource management in Yellowstone has also benefited from the description, interpretation, prediction, and monitoring of resources and ecological processes (Hall 2004). Descriptive knowledge from observational studies has resulted in new perspectives and innovative ideas that were later brought to bear on specific management issues. For example, extensive reviews of existing information by scientists led to the development of several new approaches to understand the causes, timing, and routes of migratory movements by bison (*Bison bison*) that, in turn, have been applied to management decisions regarding boundary control measures and the potential for brucellosis transmission to cattle (Bjornlie and Garrott 2001, Ferrari and Garrott 2002, Coughenour 2005, Gates *et al.* 2005, Gogan *et al.* 2005, Borkowski *et al.* 2006, Bruggeman *et al.* 2006, 2007, Fuller *et al.* 2007a, Olexa and Gogan 2007, Chapter 26 by White *et al.*, Chapters 27 and 28 by Bruggeman *et al.*, this volume). These studies also provide the best available scientific information from which adjudicators, park managers, and stakeholders can infer the potential effects of winter recreation and road grooming on bison and other wildlife. Also, research on pronghorn (*Antilocapra americana*), which are a native species of special concern (Figure 30.3) and face a serious risk of local extirpation, generated new insights into recent changes in migration patterns, diet, and population dynamics that led to increased scrutiny of key migration corridors and summering areas, habitat restoration efforts, and collaborations to improve connectivity between the park and historic winter ranges to the north (Boccadori 2002, Byers 2002, Keating 2002, Northern Yellowstone Cooperative Wildlife Working Group 2006, White *et al.* 2007a,b). In addition, research on the environmental, spatial, and temporal factors influencing the demography and distribution of grizzly bears (Figure 30.4) identified the potential for a source–sink dynamic in the Greater Yellowstone Ecosystem which, coupled with concerns for managing sustainable mortality, led to new ideas about how management agencies might approach long-term conservation of this threatened species (Schwartz *et al.* 2006c).

Science has been particularly good at describing baseline conditions and estimating natural variability in ecological processes through reliable, clearly articulated methods; thereby permitting the detection of changes in resources over time and comparisons of changes among areas. For example, the implementation of standardized observations of wildlife responses to snowmobiles and coaches has provided extensive baseline information for assessing the effects and effectiveness of future changes in recreation policy (Borkowski *et al.* 2006, Chapter 26 by White *et al.*, this volume). Also, the use of standardized collection protocols or metrics enabled comparisons of elk (*Cervus elaphus*) nutrition and demography among areas with different management regimes and predator complexes; thereby

FIGURE 30.3 Pronghorn buck northwest of Gardiner, Montana, during January 2006 (National Park Service Photo by Jim Peaco).

FIGURE 30.4 Grizzly bear near Trout Creek in Yellowstone National Park around the 1960s (National Park Service Photo by William Keller).

demonstrating that the effects of wolf predation on elk populations differ substantially over relatively small spatial scales depending on a complex suite of interacting factors (Pils *et al.* 1999, Garrott *et al.* 2005, Hamlin 2006, USFWS *et al.* 2007, Chapter 25 by Hamlin *et al.*, this volume). In addition, extensive studies of biomass and nutrient dynamics on large grasslands have demonstrated the promotion of aboveground grassland production and stimulation of soil carbon and nitrogen processes by abundant migratory grazers (Frank and McNaughton 1992, 1993; Frank *et al.* 1994; Merrill *et al.* 1994, Frank and Groffman 1998). Further, science has been used to successfully quantify the strength, function, and interaction of relationships among variables to predict how systems will respond if variables are manipulated or changed under alternate environmental conditions or management scenarios. For example, the collection of intensive data on wolf functional (*e.g.*, kill rates) and numerical (*e.g.*, abundance) responses and elk population demographics and behavioral adjustments (Mech *et al.* 2001, Creel and Winnie 2005, Smith 2005b, Vucetich *et al.* 2005, White and Garrott 2005) has enabled biologists to provide managers with revised predictions of the effects of wolf restoration on elk (Eberhardt *et al.* 2003, White *et al.* 2003, Varley and Boyce 2006, Kauffman *et al.* 2007). Science has also been highly effective at describing changes in food sources (army cutworm moths *Euxoa auxiliaries*, cutthroat trout *Oncorhynchus clarki*, elk calves, ungulate carcasses) and use by grizzly bears (Mattson 1997, Haroldson *et al.* 2005, Schwartz *et al.* 2006b).

Science has been used to design critical tests of competing explanations and interpret mechanisms (*i.e.*, causal relationships) by which effects to resources occur. True experimentation, with the use of replication and randomized controls and treatments, is difficult in the park because it is managed to minimize human intervention (National Park Service 2006) and is generally not suited for random assignments of treatments (*i.e.*, manipulations) due to disruptions of park operations, visitor expectations regarding access, contracts with concessionaires, and economic concerns by gateway communities. However, partially controlled field manipulations, natural events, and observational studies with random sampling and respectable sample sizes have provided sound inferences about the degree of spatial or temporal differences. For example, the massive fires during summer 1988 created an unusual opportunity to observe the responses of large mammals to large-scale fire by digitizing the spatial pattern of burn severity (Figure 30.5) into geographic information systems and comparing it to grazing intensity, habitat use, and distribution (Singer *et al.* 1989, Pearson *et al.* 1995, Norland *et al.* 1996, Scott and Geisser 1996). Researchers also examined the relative roles of elk browsing and fire suppression on vegetation regeneration and landscape heterogeneity by mapping the effects of fire size and pattern and sampling attributes from paired burned and unburned plots (Turner *et al.* 1994a,b, 1997, Romme *et al.* 1995, Singer and Harter 1996). In addition, scientists have used manipulations of hydrological regimes (*e.g.*, water impoundment, wells), fenced exclosures, and simulated browsing treatments to evaluate the influences of climate, herbivory, and water on aspen (*Populus tremuloides*) and willow (*Salix* spp.) communities (Renkin and Despain 1996, Johnston *et al.* 2007, Wolf *et al.* 2007). Further, scientists compared attributes of elk demography and distribution before and after wolf restoration to

FIGURE 30.5 Mosaic burn pattern in the Madison Canyon of Yellowstone National Park from the 1988 fires (National Park Service Photo by Jim Peaco).

examine the relative roles of predators, climate, and human harvest (Barber *et al.* 2005, Garrott *et al.* 2005, Mao *et al.* 2005).

Science has vastly improved our predictive capability for identifying and preparing alternate management actions for foreseeable or potential problems. For example, fisheries biologists conducted exposure studies to determine the prevalence and spatial extent of the exotic parasite *Myxobolus cerebralis* and predict the susceptibility of Yellowstone cutthroat trout (*O. clarki bouvieri*) to whirling disease (Figure 30.6; Koel *et al.* 2006b). Biologists also conducted experiments using slate tiles and whole stream measures of carbon and nitrogen fluxes to predict colonization by the exotic New Zealand mudsnail (*Potamopyrgus antipodarum*) and its effects on primary productivity and native macroinvertebrates, respectively (Hall *et al.* 2003, 2006, Kerans *et al.* 2005). Also, science has been highly effective at reducing the number of human–bear conflicts and human-caused bear mortalities in and near the park by predicting the causes, types, and locations of conflicts (Gunther 1994, Gunther *et al.* 2004). In addition, science has contributed to identifying and reducing factors that increase the frequency of road-killed wildlife, including increased traffic, vehicle speeds, and certain road designs (Gunther *et al.* 1998, 2000). Further, scientists have conducted extensive studies of the potential effects of herbivory on vegetation and nitrogen processes using comparisons between fenced exclosures and paired grazed sites (Coughenour 1991, Singer and Renkin 1995, Singer *et al.* 1998, Wambolt and Sherwood 1999). Moreover, scientists have used simulation models to predict possible climate change effects on predator, prey, and scavenger dynamics (Wilmers and Getz 2005, Wilmers and Post 2006), and identify the implications of changes in land cover and use on ecological function and biodiversity in the Greater Yellowstone Ecosystem (Hansen *et al.* 1999, 2002, Parmenter *et al.* 2003, Gude *et al.* 2007, Hansen and DeFries 2007).

Simulations using validated models have enhanced our ability to describe fine-scale spatial and temporal variations in environmental covariates that influence the behavior, demography, and movements of wildlife (Turner *et al.* 1994b, Coughenour 2005, Watson *et al.* 2006a,b; Chapter 2 by Newman *et al.*, Chapters 3–6 by Watson *et al.*, Chapter 7 by Thein *et al.*, this volume). Recent innovations have also enabled biologists to fit multiple sources of observed data and provide estimates of unobserved parameters to identify gaps in knowledge, spurious trends, and key vital rates for long-term monitoring of ungulates (Coughenour and Singer 1996, Taper and Gogan 2002, Fuller *et al.* 2007a,b, Kauffman *et al.* 2007). Further, modeling of available data has allowed biologists to make

FIGURE 30.6 Cutthroat trout from Yellowstone National Park (National Park Service Photo).

testable predictions regarding Yellowstone's diverse predator–prey system based on varying assumptions about the strength of density-related rates, prey selection in multiple prey systems, trade-offs between the risk of predation and food acquisition, and climate change (Wilmers and Getz 2005, Varley and Boyce 2006, Garrott *et al.* 2007, Kauffman *et al.* 2007).

Science has also been effective at monitoring the effects and effectiveness of implemented management actions to ensure resources are not impaired (*i.e.*, harmed). Through repeated assessments of key attributes or quantities within a defined area over a specified time period (Thompson *et al.* 1998), scientists have been able to measure progress towards meeting management objectives regarding the (1) restoration of wolves and cutthroat trout, (2) recovery of bighorn sheep (*Ovis canadensis*), grizzly bear, pronghorn, and trumpeter swan (*Olor buccinator*) populations, (3) status or spread of invasive lake trout (*Salvelinus namaycush*) and rainbow trout (*Oncorhynchus mykiss*), mountain goat (*Oreamnos americanus*), and various weedy plant populations, (4) implementation of the interagency bison and grizzly bear management plans, (5) status of rare forest carnivores such as Canada lynx (*Lynx canadensis*) and wolverines (*Gulo gulo*), and (6) recovery of federally-listed populations (National Park Service 2000, Olliff *et al.* 2001, Lemke 2004, Koel *et al.* 2005, 2006a, Smith 2005b, Copeland *et al.* 2006, McEneaney 2006, Murphy *et al.* 2006, Schwartz *et al.* 2006a, White *et al.* 2007a,c). General and targeted surveillance has also been implemented to detect or assess the prevalence and distribution of various diseases (Plumb and Aune 2002, Koel *et al.* 2006b, Anderson and Southers 2005, Plumb *et al.* 2007, White and Davis 2007). Thus, managers have more information on which to base decisions, including the condition of resources, how resources have and are changing over time, what actions are needed to preserve resources (including foreseeable crises), and the effectiveness of management decisions and natural resource policy.

III. LIMITATIONS

Despite these contributions, the new science-friendly atmosphere and science-informed management envisioned by Congress and the National Park Service has not become completely ingrained in the culture of Yellowstone, in part because such a change requires a culture shift in the attitudes, receptivity, and training of park managers (Pringle and Collins 2004). Also, research in the park is not without its pitfalls and limitations, which at times tend to discourage the use of scientific information by park leadership (Lewis 2007). Most importantly, science cannot resolve issues where policy is driven by values judgments and perceptions about what is appropriate in the park (Hall 2004). People have strong values and beliefs about the purpose of the park (*e.g.*, recreation versus conservation), what is natural (*e.g.*, pre- versus post-European settlement), and what is an acceptable level of impact to resources (*e.g.*, individual versus population). Science can quantify or predict effects to resources and help managers understand the implications of alternate actions, but it cannot specify the appropriate or acceptable level of these effects for various stakeholders with different values. With complex ecological issues, stakeholders rarely agree on the criteria to be used for assessing effects, design of tests, nature of data to be collected, and interpretation of data. Even if stakeholders can agree on an objective and what data are necessary to achieve that end, however, deciding what levels of impact are acceptable remains a value judgment outside the sole purview of science (Hall 2004). Social scientists, who are best equipped to give meaningful input into these values-based issues, are largely absent from Yellowstone. Only three of 211 (1.4%) research permits issued park-wide in 2006 were for social science studies (Lewis 2007). These values-based dilemmas are exemplified by the debates over the appropriate forms and levels of motorized winter recreation and the management of bison in Yellowstone.

Though science has contributed extensive information regarding the causes, timing, and routes of movements and responses by wildlife to motorized winter recreation (*e.g.*, Chapter 26 by White *et al.*, Chapters 27 and 28 by Bruggeman *et al.*, this volume), there is still no resolution of this issue because various constituencies have strong values and beliefs about acceptable levels of impact to wildlife (Borrie *et al.* 2002, Yochim 2004, Davenport and Borrie 2005, Layzer 2006). At one extreme, it is argued that ungulate responses to activities associated with snowmobiles are minor and of little consequence given the absence of a measurable decrease in demographic rates and abundance (Creel *et al.* 2002, Borkowski *et al.* 2006). At the other extreme, it is argued that human activities which induce any behavioral and stress responses should be curtailed (Coalition of National Park Service Retirees 2007). Policies and regulations provide little guidance for managers to resolve this debate because the management policies for biological resources in park units clearly focus on management at the species and community level (National Park Service 2006), while the Code of Federal Regulations regarding snowmobiles prohibits their use in park units if they "disturb wildlife" (36 CFR §2.18), which could be construed to refer to individual animals.

Likewise, the management of bison in Yellowstone is stymied by values disagreements among stakeholders over the ultimate goal of brucellosis management. The successful conservation of bison in Yellowstone National Park from 44 animals in 1901 to a high of almost 5000 in 2005 has led to bison migration outside the park during winter and the potential for transmission of brucellosis to cattle with widespread economic consequences. Livestock producers and cattle regulatory agencies contend that any risk of brucellosis transmission is unacceptable for public health and economic reasons ($3.5 billion spent since 1934; Cheville *et al.* 1998). Thus, they advocate the eradication of brucellosis from wild populations of bison and elk. Conversely, the National Park Service and environmental groups contend the elimination of brucellosis from wildlife in the Greater Yellowstone Ecosystem is not an achievable goal given current information, technical capabilities (*e.g.*, lack of reliable tools to determine infectious versus exposed animals), the amount of time and resources required to capture and handle >40,000 elk that are not readily accessible to health professionals, and the continuation of winter feed grounds for elk in Wyoming (Cheville *et al.* 1998, Cross *et al.* 2007). Thus, the National Park Service has advocated managing the risk of brucellosis transmission from bison to livestock on winter ranges outside the park through hazing animals back into the park (Figure 30.7), capture and culling of bison that repeatedly try to leave the park and test positive for brucellosis and, possibly, the implementation of a remote vaccination program (Cheville *et al.* 1998, National Park Service 2000). The irony in this debate is that brucellosis management remains focused

FIGURE 30.7 Rangers hazing bison near Undine Falls in Yellowstone National Park during February 2003 (National Park Service Photo by Jim Peaco).

on bison even though every known brucellosis transmission to cattle in the Greater Yellowstone Ecosystem where the species could be reasonably assigned has been attributed to elk (Cheville *et al.* 1998, Galey *et al.* 2005; T. Roffe, U.S. Fish and Wildlife Service, Bozeman, Montana, personal communication). This disparity occurs, in part, because bison have no direct economic value to states surrounding the park, while elk have significant economic value to many stakeholder groups, including hunters, guides, associated businesses that provide services to hunters, state wildlife agencies that obtain revenue through hunting license sales, and hunter advocacy, tourism, and conservation groups. Thus, state livestock and wildlife agencies have declined to adopt brucellosis test-and-slaughter programs for elk in most areas.

Another problem limiting the integration of science and management in Yellowstone is the lack of an established science strategy that specifically addresses the role of science for informing decision makers, providing a process to identify and prioritize science needs and addressing the need to communicate those needs to science providers, stakeholders, and visitors (Wright 2004, Lewis 2007). Thus, there are no mechanisms in place to ensure long-term data collection on foundation species and key ecological processes, even though the benefits of long-term monitoring and research are well known (Likens 1989). Scientists generally come and go over time, with scientific studies driven by their particular interests and success at obtaining funding. This contributes to a lack of continuity in data collection and changes in sampling methods that make temporal comparisons difficult. Furthermore, staff and funding for long-term data collection are often considered for reduction or elimination during tight budget periods because administrators are skeptical of ongoing work and believe we already have sufficient knowledge for wise management decisions (Smith 2005a, Lewis 2007). Some of these issues have been partially resolved by the creation of the Greater Yellowstone Inventory and Monitoring Network, one of 32 networks of parks organized to establish indicators of ecosystem health, called vital signs, under the auspices of a National Park Service initiative called the Natural Resource Challenge (National Park Service 1999). The vital signs chosen include a suite of four related to air and climate, seven related to geology and soils, 11 related to water, 19 related to biological integrity, three related to human use, and three related to ecosystem pattern and processes. These vital signs serve as a template for guiding science monitoring efforts and form the basis for funding long-term data collection (Jean *et al.* 2005). However, funding limitations have allowed only 12 of the 47 vital signs to be fully developed. The remaining vital signs are either monitored through existing park- or partner-based programs or must await development until further funding is available.

A third problem is that managers are often frustrated by the lack of unambiguous results from science. Ecological research often produces partial support for competing views, rather than the unambiguous rejection of one over another, because interactions are complex at the landscape scale (Hobbs and Hilborn 2006). The lack of unequivocal scientific conclusions is difficult for managers to interpret, especially given the level of rigor necessary for study design and statistics that make reports difficult to read and understand. This problem is exacerbated by the fact that obtaining the best available scientific information rarely equates to consensus among scientists on what is the "ecologically correct" policy or management action to take. Scientists frequently disagree on interpretations of data and often provide diametrically opposing recommendations that are of little help to park managers in making decisions or resolving issues. Thus, managers inevitably need to act without the luxury of complete knowledge or scientific consensus, using the available information and contrasting interpretations to weigh the potential benefits and costs of alternate management actions. Decision-makers want to take courses of action with high likelihood of success rather than implementing actions based on data or analyses with a lot of uncertainty or models with low predictive power. Thus, when scientists cannot agree on scientific findings, politics and values will impart more influence on the choice of action.

Scientists can also create distrust in managers when they are not sensitive to the impacts of their activities on resources or adopt advocacy positions regarding park policies. Many park staff do not support activities such as capturing and radio-collaring animals, over-flights, and destructive sampling of vegetation in the park, and injuries to animals, poorly fitting radio collars, disturbances to wildlife or visitors, or man-made structures such as instrument towers, fences, and exclosures all contribute to increased calls for limiting research. Scientists can also undermine managers when they practice advocacy science by "marketing" preliminary results, aggressively proclaiming their position or interpretation of data is "correct," and publicly prodding managers to adopt favored positions even though the evidence may be under-whelming and alternative positions may have equal or better support. While this type of advocacy is not new and relatively rare, the passionate and highly-publicized debates in Yellowstone occasionally attract and favor advocates who claim to know the answer and look for evidence to support it, while posing as unbiased scientists testing competing hypotheses. Thus, scientists of all disciplines working in Yellowstone should heed Hilborn's (2006) challenge to question whether faith-based, agenda-driven support for ideas has replaced critical and skeptical analyses of evidence.

IV. RECOMMENDATIONS

The integrated science program for central Yellowstone certainly did not avoid all of these pitfalls, but it did achieve some modest success at minimizing these limitations by:

- maintaining long-term research and monitoring of foundation species that can be ecologically dominant;
- developing collaborations with scientists and professionals with diverse expertise, and establishing strong partnerships with park staff at all levels;
- contributing to basic understanding of ecological processes and the development of important Geographic Information System (GIS) databases and validated predictive models that can be used by the entire professional community;
- being responsive to the needs and concerns of park managers, accumulating key information, and working through the peer-review publication process so that managers could feel comfortable using this information in decision-making processes;
- providing objective interpretations, new perspectives, and innovative approaches for collecting useful information (data) and accruing new knowledge; and
- contributing to the park's educational mission by developing specific products for communicating the results of research to stakeholders and the public.

Despite our successes, we feel fortunate to have sustained this research over the past 16 years given that we often struggled through institutional hurdles, funding shortages, and quarrels with park staff and other scientists. At times, our research seemed to persist merely due to good fortune, stubbornness, and timely assistance from sympathetic park staff and other benefactors. Thus, we think there are a number of institutional changes that could enhance the prospect of more long-term and integrated science programs becoming established to support the mission of the park.

The first step we recommend to improve the contribution of science in decision-making is for park leadership to give careful consideration to the question of how science can successfully be used to influence decisions and resolve disputes. The types of resource management problems that science has proven effective in contributing to successful decision-making generally have two key characteristics in common, namely agreement about the desired objectives and the types of relevant knowledge needed to attain those end states (Hisschemoeller and Hoppe 1996, Hall 2004). Science can help resolve issues when problems are structured such that all stakeholders agree on the desirable goal to be attained, the nature of the data that is needed to achieve this goal, and the levels of impacts to resources that are acceptable (Hall 2004). However, science is ineffective at resolving issues when problems are characterized by strife between stakeholders over objectives or the relevant knowledge needed to attain the objectives (*i.e.*, unstructured problems; Hall 2004). Unfortunately, many high-profile resource management issues in Yellowstone fall into this latter category and placing faith in scientists to completely resolve such conflicts will not lead to an accepted resolution (Hall 2004, Lewis 2007).

This is not to say that science should not be used to obtain better understanding of all complex resource issues and processes. Science is always needed once goals are established, to lay out the process to get to that goal. Also, even if science can only contribute modestly to defining an appropriate natural resource goal or objective, it should be integral to defining the way that goal is pursued. The plan of action to reach a goal should be based on our best ecological and sociological knowledge, and we should evaluate how well the plan is working to achieve that goal. If the plan is ineffective, then it should be revised to make progress towards the goal and accumulate new knowledge in the process (*i.e.*, adaptive management; Walters 1986). However, managers should still develop a process that evaluates the nature of each problem to determine if science should be the dominant force guiding decisions or if disputes regarding the means and end will limit its contribution. In the latter cases, negotiation and mediation may be necessary or desirable if resolution through such means will not conflict with the preservation mission of the National Park Service.

Second, park leadership should develop an explicit science strategy that provides a process for identifying foreseeable issues and crises, prioritizing science needs, and facilitating research well in advance through funding, permits, and other measures. The scientific process is often slow because it takes a lot of time to design studies, gather and analyze data, prepare reports, conduct peer-review, and disseminate information. Spending money on scientific projects once there is already a management or policy crisis is generally not an effective strategy because the decision will likely have to be made before the scientific process is complete. Some progress has been made on this issue, with recent workshops on how to define priorities for research regarding winter recreation and wildlife disease issues. To further facilitate effective long-term research, we recommend adopting an approach similar to the model used by the Colorado Division of Wildlife during the 1980's, whereby long-term funding was provided for 10-year studies in three parts: (1) synthesis of the current state of knowledge and development of a peer-reviewed study plan; (2) data collection; and (3) data analysis, write-up, peer-review, and dissemination. Scientists, including park staff, should be required to provide deliverables pursuant to clearly defined time tables at each step in the process or their permit for research in the park would be rescinded. Too often, research is conducted in the park by scientists that leave without ever providing park managers with a detailed record or data from their activities. Thus, park biologists, managers, and policy makers must have stronger assurances that they will have access to data collected on natural resources in the park for future use. In turn, park staff must develop a centralized data management system that integrates diverse data and facilitates collaborative analyses, while still protecting the privilege of independent researchers that collected the data.

Third, park managers need to take the time and responsibility for reading and understanding key findings that scientists summarize in peer-reviewed reports. There is only so much distilling and simplification scientists can do without losing the meaning and appropriate application of research and it is discouraging for them to work hard at producing pertinent results, only to see them ignored or misinterpreted during planning and decision-making processes (Wright 2004). While we acknowledge that science alone is not sufficient for making effective decisions on complex resource issues that invariably involve values judgments about what is desirable or appropriate, science seems critical for making decisions that have some chance of succeeding in achieving the preservation of resources, which is the primary mission of the National Park Service (Hisschemoeller and Hoppe 1996, Adams and Hairston 1996). Thus, it seems reasonable for park managers to put significant effort into understanding and considering the best available science in their decision-making process, and interacting with their staff and independent scientists to learn and understand what we know and do not know about various ecological processes.

Likewise, scientists can improve the acceptance of science in the decision-making process by implementing several actions. First, researchers should coordinate with park staff regarding information needs and management issues before they develop funding proposals and implementation plans. Likewise, after the research is completed scientists need to translate the findings into meaningful information for park managers, visitors, funding sources, and other interested parties. Second, scientists should compile tight-knit research teams, including park staff, with complimentary skills and expertise from diverse disciplines (*e.g.*, spatial ecology, predator–prey dynamics, statistics, hydrology, remote sensing, geographic information systems, and modeling). This approach facilitates equitable involvement of resources and frequent, close communications that develop trust and efficiency. It also reduces duplication of effects to resources and synergizes our understanding of ecological processes in the system. Further, this approach facilitates the acquisition of funding from diverse sources, which is necessary to maintain long-term efforts.

Third, we recommend scientists and park leadership consider an approach for integrating research with management that transitions from baseline monitoring to intensive research and then to expanded monitoring of key attributes of natural resources and ecological processes once the intensive studies are completed. The initial phase is baseline monitoring of basic ecological information and long-term trends on foundation or sensitive species and ecological processes that have particularly strong, ramifying effects in the system. This monitoring catalyzes work on issues and accumulates relevant data that the park can use to quickly expand studies and respond to unforeseen problems. Baseline monitoring also enables scientists to develop and demonstrate the success and relatively low impacts of their activities to skeptical park staff. The second phase is intensive, inferential research to address specific knowledge gaps of ongoing or foreseeable issues. This research is typically expensive, intrusive, and relatively risky for park managers when it involves the handling or manipulation of resources and higher levels of risk for scientists during implementation. Thus, these studies must have well-defined time periods and specific, realistic objectives. The third phase is expanded monitoring, whereby key vital rates and limiting factors identified during intensive research are integrated into long-term monitoring of vital signs and ecosystem change. Examples include indices of predator–prey dynamics, climate drivers in the system, and human impacts on ecological processes.

While our integrated science program in central Yellowstone has been relatively successful, it was a small effort focused on a narrow range of ecological issues in one area of the park. There are many more issues confronting effective conservation and management of Yellowstone's natural resources, including:

- diseases such as brucellosis and chronic wasting disease, which is getting closer and closer to the park;
- the effects of nonnative species such as mountain goats, lake trout, and invasive weeds;
- the integration of management practices across administrative boundaries owing to the seasonal migration of wildlife outside the park; and
- evaluating the cumulative effects of climate change, development, and recreation inside and outside park.

Many more integrated science programs will be needed to effectively meet these complex challenges. We hope that some of you will bring your expertise, enthusiasm, and passion for science and conservation to Yellowstone.

V. SUMMARY

1. We conducted an introspective examination of the contributions and limitations of science to resource management in Yellowstone National Park and made recommendations for strengthening this link based on our experiences with wildlife issues.

2.   The increased emphasis on science in Yellowstone since 1993 has enhanced the credibility of park positions during many proceedings and benefited other agencies responsible for resource management outside the park.

3.   Resource management in Yellowstone has also benefited from the description, interpretation, prediction, and monitoring of resources and ecological processes that resulted in new perspectives and innovative ideas that were later brought to bear on management issues.

4.   Science-based management has not become ingrained in the culture of Yellowstone because (a) such a change requires a culture shift in the attitudes of park managers, (b) limitations of science tend to discourage its use by park leadership, and (c) science cannot resolve issues where policy is driven by values judgments about what is appropriate in the park.

5.   We recommend park leadership establish a science strategy that specifies the role of science for informing decision makers, provides a process for prioritizing science needs, facilitates necessary research well in advance, and addresses communication needs to stakeholders.

6.   Park leadership should consider an approach for integrating research with management that transitions from baseline monitoring to intensive research of specific knowledge gaps and then to expanded monitoring of key attributes of natural resources and ecological processes.

VI. REFERENCES

Adams, P. W., and A. B. Hairston. 1996. Calling all experts: Using science to direct policy. *Journal of Forestry* **94**:27–30.

Anderson, N., and J. Southers. 2005. *CWD Surveillance 1998–2004.* Montana Department of Fish, Wildlife, and Parks, Bozeman, MT.

Barber, S. M., L. D. Mech, and P. J. White. 2005. Yellowstone elk calf mortality following wolf restoration – Bears remain top summer predators. *Yellowstone Science* **13**:37–44.

Behan, R. W. 1997. Scarcity, simplicity, separatism, science and systems. Pages 411–417 *in* K. A. Kohm and J. F. Franklin (Eds.) *Creating a Forestry for the 21st Century: The Science of Ecosystem Management.* Island Press, Washington, DC.

Bjornlie, D. D., and R. A. Garrott. 2001. Effects of winter road grooming on bison in Yellowstone National Park. *Journal of Wildlife Management* **65**:560–572.

Boccadori, S. J. 2002. *Effects of Winter Range on a Pronghorn Population in Yellowstone National Park, Wyoming.* Thesis. Montana State University, Bozeman MT.

Borkowski, J. J., P. J. White, R. A. Garrott, T. Davis, A. Hardy, and D. J. Reinhart. 2006. Wildlife responses to motorized winter recreation in Yellowstone National Park. *Ecological Applications* **16**:1911–1925.

Borrie, W., W. Freimund, and M. Davenport. 2002. Winter visitors to Yellowstone National Park: Their value orientations and support for management actions. *Human Ecology Review* **9**:41–48.

Bruggeman, J. E., R. A. Garrott, D. D. Bjornlie, P. J. White, F. G. R. Watson, and J. J. Borkowski. 2006. Temporal variability in winter travel patterns of Yellowstone bison: the effects of road grooming. *Ecological Applications* **16**:1539–1554.

Bruggeman, J. E., R. A. Garrott, P. J. White, F. G. R. Watson, and R. W. Wallen. 2007. Covariates affecting spatial variability in bison travel behavior in Yellowstone National Park. *Ecological Applications* **17**:1411–1423.

Byers, J. A. 2002. Fecundity and fawn mortality of northern Yellowstone pronghorn. Final report CA1443CA157099002 dated December 4, 2002. University of Idaho, Moscow, ID.

Cheville, N. F., D. R. McCullough, L. R. Paulson, N. Grossblatt, K. Iverson, and S. Parker. 1998. *Brucellosis in the Greater Yellowstone Area.* National Academy Press, Washington, DC.

Coalition of National Park Service Retirees. 2007. NPS retirees: Wildlife in country's first national park threatened by rewritten standards and "burial" of scientists' recommendations. February 28, 2007, press release. Tucson, AZ.

Copeland, J., K. Murphy, and J. Wilmot. 2006. *Absaroka-Beartooth Wolverine Project.* Unpublished newsletter, USDA Forest Service, Rocky Mountain Research Station, Missoula, MT.

Coughenour, M. B. 1991. Biomass and nitrogen responses to grazing of upland steppe on Yellowstone's northern winter range. *Journal of Applied Ecology* **28**:71–82.

Coughenour, M. B. 2005. Spatial-dynamic modeling of bison carrying capacity in the greater Yellowstone ecosystem: a synthesis of bison movements, population dynamics, and interactions with vegetation. Natural Resource Ecology Laboratory, Colorado State University, Fort Collins, CO.

Coughenour, M. B., and F. J. Singer. 1996. Elk population processes in Yellowstone National Park under the policy of natural regulation. *Ecological Applications* **6**:573–593.

Creel, S., J. E. Fox, A. Hardy, J. Sands, B. Garrott, and R. O. Peterson. 2002. Snowmobile activity and glucocorticoid stress responses in wolves and elk. *Conservation Biology* **16**:809–814.

Creel, S., and J. A. Winnie. 2005. Responses of elk herd size to fine-scale spatial and temporal variation in the risk of predation by wolves. *Animal Behaviour* **69**:1181–1189.

Cross, P. C., W. H. Edwards, B. M. Scurlock, E. J. Maichak, and J. D. Rogerson. 2007. Effects of management and climate on elk brucellosis in the Greater Yellowstone Ecosystem. *Ecological Applications* **17**:957–964.

Davenport, M., and W. Borrie. 2005. The appropriateness of snowmobiling in national parks: An investigation of the meanings of snowmobiling experiences in Yellowstone National Park. *Environmental Management* **35**:151–160.

Eberhardt, L. L., R. A. Garrott, D. W. Smith, P. J. White, and R. O. Peterson. 2003. Assessing the impact of wolves on ungulate prey. *Ecological Applications* **13**:776–783.

Ferrari, M. J., and R. A. Garrott. 2002. Bison and elk: brucellosis seroprevalence on a shared winter range. *Journal of Wildlife Management* **66**:1246–1254.

Frank, D. A., and P. M. Groffman. 1998. Ungulate vs. landscape control of soil C and N processes in grasslands of Yellowstone National Park. *Ecology* **79**:2229–2241.

Frank, D. A., R. S. Inouye, N. Huntly, G. W. Minshall, and J. E. Anderson. 1994. The biogeochemistry of a north-temperate grassland with native ungulates: Nitrogen dynamics in Yellowstone National Park. *Biogeochemistry* **26**:163–188.

Frank, D. A., and S. J. McNaughton. 1992. The ecology of plants, large mammalian herbivores, and drought in Yellowstone National Park. *Ecology* **73**:2043–2058.

Frank, D. A., and S. J. McNaughton. 1993. Evidence for the promotion of aboveground grassland production by native large herbivores in Yellowstone National Park. *Oecologia* **96**:157–161.

Fuller, J. A., R. A. Garrott, and P. J. White. 2007a. Emigration and density dependence in Yellowstone bison. *Journal of Wildlife Management* **71**:1924–1933.

Fuller, J. A., R. A. Garrott, P. J. White, K. E. Aune, T. J. Roffe, and J. C. Rhyan. 2007b. Reproduction and survival of Yellowstone bison. *Journal of Wildlife Management* **71**:2365–2372.

Galey, F., *et al* 2005. Wyoming Brucellosis Coordination Team report and recommendations. Report presented to Governor Dave Freudenthal on January 11, 2005. Cheyenne, WY. http://wyagric.state.wy.us/relatedinfo/govbrucecoordinati.htm.

Garrott, R. A., J. E. Bruggeman, M. S. Becker, S. T. Kalinowski, and P. J. White. 2007. Evaluating prey switching in wolf-ungulate systems. *Ecological Applications* **17**:1588–1597.

Garrott, R. A., J. A. Gude, E. J. Bergman, C. Gower, P. J. White, and K. L. Hamlin. 2005. Generalizing wolf effects across the greater Yellowstone area: a cautionary note. *Wildlife Society Bulletin* **33**:1245–1255.

Gates, C. C., B. Stelfox, T. Muhly, T. Chowns, and R. J. Hudson. 2005. *The Ecology of Bison Movements and Distribution in and Beyond Yellowstone National Park.* University of Calgary, Alberta, Canada.

Gogan, P. J. P., K. M. Podruzny, E. M. Olexa, H. I. Pac, and K. L. Frey. 2005. Yellowstone bison fetal development and phenology of parturition. *Journal of Wildlife Management* **69**:1716–1730.

Gude, P. H., A. J. Hansen, and D. A. Jones. 2007. Biodiversity consequences of alternative future land use scenarios in Greater Yellowstone. *Ecological Applications* **17**:1004–1018.

Gunther, K. A. 1994. Bear management in Yellowstone National Park, 1960–93. *International Conference of Bear Research and Management* **9**:549–560.

Gunther, K. A., M. J. Biel, and H. L. Robison. 1998. Factors influencing the frequency of road-killed wildlife in Yellowstone National Park. International Conference on Wildlife Ecology and Transportation, Missoula, MT. February 9–12, 1998.

Gunther, K. A., M. J. Biel, and H. L. Robison. 2000. Influence of vehicle speed and vegetation cover-type on road-killed wildlife in Yellowstone National Park. Pages 42–51 *in* T. A. Messmer and B. West (Eds.) *Wildlife and Highways: Seeking Solutions to an Ecological and Socio-economic Dilemma.* Nashville, TN. September 12–16, 2000.

Gunther, K. A., M. A. Haroldson, K. Frey, S. L. Cain, J. Copeland, and C. C. Schwartz. 2004. Grizzly bear-human conflicts in the greater Yellowstone ecosystem, 1992–2000. *Ursus* **15**:10–22.

Hall, R. O., M. F. Dybdahl, and M. C. VanderLoop. 2006. Extremely high secondary production of introduced snails in rivers. *Ecological Applications* **16**:1121–1131.

Hall, R. O., J. L. Tank, and M. F. Dybdahl. 2003. Exotic snails dominate carbon and nitrogen cycling in a highly productive stream. *Frontiers in Ecology and the Environment* **1**:408–411.

Hall, T. E. 2004. Recreation management decisions: what does science have to offer? Pages 10–15 *in* D. Harmon, B. M. Kilgore, and G. E. Vietzke (Eds.) *Protecting Our Diverse Heritage: The Role of Parks, Protected Areas, and Cultural Sites.* Proceedings of the 2003 George Wright Society/National Park Service Joint Conference, Hancock, MI.

Hamlin, K. L. 2006. *Monitoring and Assessment of Wolf-Ungulate Interactions and Population Trends Within the Greater Yellowstone Area, Southwestern Montana, and Montana Statewide.* Montana Department of Fish, Wildlife, and Parks, Bozeman, MT.

Hansen, A. J., and R. DeFries. 2007. Ecological mechanisms linking protected areas to surrounding lands. *Ecological Applications* **17**:974–988.

Hansen, A. J., R. Rasker, B. Maxwell, J. Rotella, A. W. Parmenter, U. Langner, W. Cohen, R. Lawrence, and J. Johnson. 2002. Ecological causes and consequences of demographic change in the New West: A case study from greater Yellowstone. *BioScience* **52**:151–162.

Hansen, A. J., J. J. Rotella, M. P. V. Kraska, and D. Brown. 1999. Dynamic habitat and population analysis: an approach to resolve the biodiversity manager's dilemma. *Ecological Applications* **9**:1459–1476.

Haroldson, M. A., K. A. Gunther, D. P. Reinhart, S. R. Podruzny, C. Cegelski, L. Waits, T. Wyman, and J. Smith. 2005. Changing numbers of spawning cutthroat trout in tributary streams of Yellowstone Lark and estimates of grizzly bears visiting streams from DNA. *Ursus* **16**:167–180.

Hilborn, R. 2006. Faith-based fisheries. *Fisheries* **31**:554–555.

Hisschemoeller, M., and R. Hoppe. 1996. Coping with intractable controversies: the case for problem structuring in policy design and analysis. *Knowledge and Policy* **8**:50–60.

Hobbs, N. T., and R. Hilborn. 2006. Alternative to statistical hypothesis testing in ecology: a guide to self teaching. *Ecological Applications* **16**:5–19.

Jean, C., A. M. Schrag, R. E. Bennetts, R. Daley, E. A. Crowe, and S. O'Ney. 2005. *Vital Signs Monitoring Plan for the Greater Yellowstone Network.* National Park Service, Greater Yellowstone Network, Bozeman, MT.

Johnston, D. B., D. J. Cooper, and N. T. Hobbs. 2007. Elk browsing increases aboveground growth of water-stressed willows by modifying plant architecture. *Oecologia* **154**:467–478.

Kauffman, M. J., N. Varley, D. W. Smith, D. R. Stahler, D. R. MacNulty, and M. S. Boyce. 2007. Landscape heterogeneity shapes predation in a newly restored predator–prey system. *Ecological Letters* **10**:1–11.

Keating, K. 2002. *History of Pronghorn Population Monitoring, Research, and Management in Yellowstone National Park.* U.S. Geological Survey, Northern Rocky Mountain Science Center, Bozeman, MT.

Kerans, B. L., M. F. Dybdahl, M. M. Gangloff, and J. E. Jannot. 2005. *Potamopyrgus antipodarum*: Distribution, density, and effects on native macroinvertebrate assemblages in the greater Yellowstone ecosystem. *Journal of North American Benthological Society* **24**:123–138.

Koel, T. M., J. L. Arnold, P. E. Bigelow, P. D. Doepke, B. D. Ertel, D. L. Mahony, and M. E. Ruhl. 2006a. Yellowstone fisheries & aquatic sciences: Annual report, 2005. YCR-2006–09, Yellowstone National Park, WY.

Koel, T. M., P. E. Bigelow, P. D. Doepke, B. D. Ertel, and D. L. Mahony. 2005. Non-native lake trout result in Yellowstone cutthroat trout decline and impacts to bears and anglers. *Fisheries* **30**:10–19.

Koel, T. M., D. L. Mahoney, K. L. Kinnan, C. J. Rasmussen, C. J. Hudson, S. Murcia, and B. L. Kerans. 2006b. *Myxobolus cerebralis* in native cutthroat trout of the Yellowstone Lake ecosystem. *Journal of Aquatic Animal Health* **18**:157–175.

Layzer, J. 2006. *The Environmental Case: Translating Values into Policy.* CQ Press, Washington, DC.

Lemke, T. O. 2004. Origin, expansion, and status of mountain goats in Yellowstone National Park. *Wildlife Society Bulletin* **32**:532–541.

Lewis, S. 2007. The role of science in National Park Service decision-making. *George Wright Forum* **24**:36–40.

Likens, G. E. (Ed.). 1989. *Long-Term Studies in Ecology: Approaches and Alternatives.* Springer-Verlag, New York, NY.

Mao, J. S., M. S. Boyce, D. W. Smith, F. J. Singer, D. J. Vales, J. M. Vore, and E. H. Merrill. 2005. Habitat selection by elk before and after wolf reintroduction in Yellowstone National Park. *Journal of Wildlife Management* **69**:1691–1707.

Mattson, D. J. 1997. Use of ungulates by Yellowstone grizzly bears *Ursus arctos. Biological Conservation* **81**:161–177.

McEneaney, T. 2006. Yellowstone bird report, 2005. YCR-2006–2. National Park Service, Yellowstone Center for Resources, Yellowstone National Park, WY.

Mech, L. D., D. W. Smith, K. M. Murphy, and D. R. MacNulty. 2001. Winter severity and wolf predation on a formerly wolf-free elk herd. *Journal of Wildlife Management* **65**:998–1003.

Merrill, E. H., N. L. Stanton, and J. C. Hak. 1994. Responses of bluebunch wheatgrass, Idaho fescue, and nematodes to ungulate grazing in Yellowstone National Park. *Oikos* **69**:231–240.

Murphy, K. M., T. Potter, J. C. Halfpenny, K. A. Gunther, M. T. Jones, P. A. Lundberg, and N. D. Berg. 2006. Distribution of Canada lynx in Yellowstone National Park. *Northwest Science* **80**:199–206.

National Park Service. 1999. *Natural Resource Challenge: The National Park Service's Action Plan for Preserving Natural Resources.* National Park Service, Washington, DC.

National Park Service. 2000. *Bison Management Plan for the State of Montana and Yellowstone National Park, Final Environmental Impact Statement.* U.S. Department of Interior, National Park Service, Denver, CO.

National Park Service. 2006. *Management Policies.* U.S. Department of the Interior, National Park Service, Washington, DC.

Norland, J. E., F. J. Singer, and L. Mack. 1996. Effects of the Yellowstone fires of 1988 on elk habitats. Pages 223–232 *in* J. Greenlee (Ed.) *Ecological Implications of Fire in Greater Yellowstone.* International Association of Wildland Fire, Fairfield, WA.

Northern Yellowstone Cooperative Wildlife Working Group. 2006. Annual report. Unpublished report issued October 13, 2006, by the National Park Service (Yellowstone National Park), Montana Fish, Wildlife, and Parks, U.S. Forest Service (Gallatin National Forest), and U.S. Geological Service (Northern Rocky Mountain Science Center). Copy on file at the Yellowstone Center for Resources. Yellowstone National Park, WY.

Olexa, E. M., and P. J. P. Gogan. 2007. Spatial population structure of Yellowstone bison. *Journal of Wildlife Management* **71**:1531–1538.

Olliff, S. T., R. Renkin, C. McClure, P. Miller, D. Price, D. Reinhart, and J. Whipple. 2001. Managing a complex exotic vegetation management program in Yellowstone National Park. *Western North American Naturalist* **61**:347–358.

Ozawa, C. P. 1996. Science in environmental conflicts. *Sociological Perspectives* **39**:219–230.

Parmenter, A. W., A. Hansen, R. E. Kennedy, W. Cohen, U. Langner, R. Lawrence, B. Maxwell, A. Gallant, and R. Aspinall. 2003. Land use and land cover change in the greater Yellowstone ecosystem: 1975–1995. *Ecological Applications* **13**:687–703.

Parsons, D. J. 2004. Supporting basic ecological research in U.S. national parks: challenges and opportunities. *Ecological Applications* **14**:5–13.

Pearson, S. M., M. G. Turner, L. L. Wallace, and W. H. Romme. 1995. Winter habitat use by large ungulates following fire in northern Yellowstone National Park. *Ecological Applications* **5**:744–755.

Pils, A. C., R. A. Garrott, and J. J. Borkowski. 1999. Sampling and statistical analysis of snow-urine allantoin:creatinine ratios. *Journal of Wildlife Management* **63**:1118–1132.

Plumb, G., and K. Aune. 2002. The interagency bison management plan for Yellowstone National Park and Montana. Pages 136–145 *in* T. J. Kreeger (Ed.) *Brucellosis in Elk and Bison in the Greater Yellowstone Area*. Wyoming Game and Fish Department, Cheyenne, WY.

Plumb, G., L. Babiuk, J. Mazet, S. Olsen, P.-P. Pastoret, C. Rupprecht, and D. Slate. 2007. Vaccination in conservation medicine. *Revue Scientifique et Technique Office International des Epizooties* **26**:229–241.

Pringle, C. M., and S. L. Collins. 2004. Needed: A unified infrastructure to support long-term scientific research on public lands. *Ecological Applications* **14**:18–21.

Renkin, R., and D. Despain. 1996. Preburn root biomass/basal area influences on the response of aspen to fire and herbivory. Pages 95–103 *in* J. Greenlee (Ed.) *Ecological Implications of Fire in Greater Yellowstone*. International Association of Wildland Fire, Fairfield, WA.

Romme, W. H., M. G. Turner, L. L. Wallace, and J. S. Walker. 1995. Aspen, elk, and fire in northern Yellowstone National Park. *Ecology* **76**:2097–2106.

Schwartz, C. C., M. A. Haroldson, K. A. Gunther, and D. Moody. 2006a. Distribution of grizzly bears in the greater Yellowstone ecosystem in 2004. *Ursus* **17**:63–66.

Schwartz, C. C., M. A. Haroldson, and K. West (Eds.). 2006. *Yellowstone Grizzly Bear Investigations: Annual Report of the Interagency Grizzly Bear Study Team, 2005*. U.S. Geological Survey, Bozeman, MT.

Schwartz, C. C., M. A. Haroldson, G. C. White, R. B. Harris, S. Cherry, K. A. Keating, D. Moody, and C. Servheen. 2006c. Temporal, spatial, and environmental influences on the demographics of grizzly bears in the greater Yellowstone ecosystem. *Wildlife Monographs* **161**.

Scott, M. D., and H. Geisser. 1996. Pronghorn migration and habitat use following the 1988 Yellowstone fires. Pages 123–132 *in* J. M. Greenlee (Ed.) *Ecological Implications of Fire in the Greater Yellowstone*. Proceedings Second Biennial Conference on the Greater Yellowstone Ecosystem. National Park Service, Yellowstone National Park, Mammoth, WY.

Sellers, R. W. 1997. *Preserving Nature in the National Parks: A History*. Yale University Press, New Haven, CT.

Singer, F. J., and M. K. Harter. 1996. Comparative effects of elk herbivory and 1988 fires on northern Yellowstone National Park grasslands. *Ecological Applications* **6**:185–199.

Singer, F. J., and R. A. Renkin. 1995. Effects of browsing by native ungulates on the shrubs in big sagebrush communities in Yellowstone National Park. *Great Basin Naturalist* **55**:201–212.

Singer, F. J., W. Schreier, J. Oppenheim, and E. O. Garton. 1989. Drought, fires, and large mammals: evaluating the 1988 severe drought and large-scale fires. *BioScience* **39**:716–722.

Singer, F. J., L. C. Zeigenfuss, R. G. Cates, and D. T. Barnett. 1998. Elk, multiple factors, and persistence of willows in national parks. *Wildlife Society Bulletin* **26**:419–428.

Smith, D. W. 2005a. Mixed messages about opportunistic carnivores. *Conservation Biology* **19**:1676–1678.

Smith, D. W. 2005b. Ten years of Yellowstone wolves. *Yellowstone Science* **13**:7–33.

Taper, M. L., and P. J. P. Gogan. 2002. The northern Yellowstone elk: density dependence and climatic conditions. *Journal of Wildlife Management* **66**:106–122.

Thompson, W. L., G. C. White, and C. Gowan. 1998. *Monitoring Vertebrate Populations*. Academic Press. New York, NY.

Turner, M. G., W. W. Hargrove, R. H. Gardner, and W. H. Romme. 1994a. Effects of fire on landscape heterogeneity in Yellowstone National Park, Wyoming. *Journal of Vegetation Science* **5**:731–742.

Turner, M. G., W. H. Romme, R. H. Gardner, and W. W. Hargrove. 1997. Effects of fire size and pattern on early succession in Yellowstone National Park. *Ecological Monographs* **67**:411–433.

Turner, M. G., Y. Wu, L. L. Wallace, W. H. Romme, and A. Brenkert. 1994b. Simulating winter interactions among ungulates, vegetation and fire in northern Yellowstone Park. *Ecological Applications* **4**:472–496.

U.S. Fish and Wildlife Service, Nez Perce Tribe, National Park Service, Montana Fish, Wildlife and Parks, Idaho Fish and Game, and USDA Wildlife Services [USFWS *et al.*]. 2007. Rocky Mountain Wolf Recovery 2006 Annual Report, C. A. Sime, and E. E. Bangs (Eds.). USFWS, Ecological Services, Helena, MT.

Varley, N., and M. S. Boyce. 2006. Adaptive management for reintroductions. Updating a wolf recovery model for Yellowstone National Park. *Ecological Modeling* **193**:315–339.

Vucetich, J. A., D. W. Smith, and D. R. Stahler. 2005. Influence of harvest, climate, and wolf predation on Yellowstone elk, 1961–2004. *Oikos* **111**:259–270.

Walters, C. J. 1986. *Adaptive Management of Renewable Resources*. MacMillian Press, New York, NY.

Wambolt, C. L., and H. W. Sherwood. 1999. Sagebrush response to ungulate browsing in Yellowstone. *Journal of Range Management* **52**:363–369.

Watson, F. G. R., T. N. Anderson, W. B. Newman, S. E. Alexander, and R. A. Garrott. 2006a. Optimal sampling schemes for estimating mean snow water equivalents in stratified heterogeneous landscapes. *Journal of Hydrology* **328**:432–452.

Watson, F. G. R., W. B. Newman, J. C. Coughlan, and R. A. Garrott. 2006b. Testing a distributed snow pack simulation model against spatial observations. *Journal of Hydrology* **328**:453–466.

White, P. J., J. E. Bruggeman, and R. A. Garrott. 2007a. Irruptive population dynamics in Yellowstone pronghorn. *Ecological Applications* **17**:1598–1606.

White, P. J., and T. L. Davis. 2007. Chronic wasting disease – planning for an inevitable dilemma. *Yellowstone Science* **15**:8–10.

White, P. J., T. L. Davis, K. K. Barnowe-Meyer, R. L. Crabtree, and R. A. Garrott. 2007b. Partial migration and philopatry of Yellowstone pronghorn. *Biological Conservation* **135**:518–526.

White, P. J., and R. A. Garrott. 2005. Northern Yellowstone elk after wolf restoration. *Wildlife Society Bulletin* **33**:942–955.

White, P. J., R. A. Garrott, and L. L. Eberhardt. 2003. Evaluating the consequences of wolf recovery on northern Yellowstone elk. Unpublished report dated October 2003 on file at the Yellowstone Center for Resources, Yellowstone National Park, Mammoth, WY.

White, P. J., T. O. Lemke, D. B. Tyers, and J. A. Fuller. 2007c. Bighorn sheep demography following wolf restoration. *Wildlife Biology* **14**:138–146.

Williams, B. K., J. D. Nichols, and M. J. Conroy. 2002. *Analysis and Management of Animal Populations*. Academic Press, San Diego, CA.

Wilmers, C. C., and W. M. Getz. 2005. Gray wolves as climate change buffers in Yellowstone. *PLoS Biology* **3**:571–576.

Wilmers, C. C., and E. Post. 2006. Predicting the influence of wolf-provided carrion on scavenger community dynamics under climate change scenarios. *Global Change Biology* **12**:403–409.

Wolf, E. C., D. J. Cooper, and N. T. Hobbs. 2007. Hydrologic regime and herbivory stabilize an alternative state in Yellowstone National Park. *Ecological Applications* **17**:1572–1587.

Wright, V. 2004. Barriers to science-based management: What are they and what can we do about them (session summary). Pages 34–37 *in* D. Harmon, B. M. Kilgore, and G. E. Vietzke (Eds.) *Protecting Our Diverse Heritage: The Role of Parks, Protected Areas, and Cultural Sites.* Proceedings of the 2003 George Wright Society/National Park Service Joint Conference, Hancock, MI.

Yochim, M. J. 2004. *Compromising Yellowstone: The Interest Group-National Park Service Relationship in Modern Policy-Making.* Dissertation. University of Wisconsin, Madison, WI.

Index

Printed and bound by CPI Group (UK) Ltd, Croydon, CR0 4YY

08/05/2025

01864920-0003